THE CARDIOVASCULAR SYSTEM: DEVELOPMENT, PLASTICITY AND PHYSIOLOGICAL RESPONSES

This is Volume 36B in the

FISH PHYSIOLOGY series
Edited by Anthony P. Farrell and Colin J. Brauner
Honorary Editors: William S. Hoar and David J. Randall

A complete list of books in this series appears at the end of the volume

THE CARDIOVASCULAR SYSTEM: DEVELOPMENT, PLASTICITY AND PHYSIOLOGICAL RESPONSES

Fish Physiology

A. KURT GAMPERL
Departments of Ocean Sciences and Biology
Memorial University of Newfoundland
St. John's, Newfoundland and Labrador
Canada

TODD E. GILLIS
Department of Integrative Biology
University of Guelph
Guelph, Ontario
Canada

ANTHONY P. FARRELL
Department of Zoology, and Faculty of Land and Food Systems
The University of British Columbia
Vancouver, British Columbia
Canada

COLIN J. BRAUNER
Department of Zoology
The University of British Columbia
Vancouver, British Columbia
Canada

Academic Press is an imprint of Elsevier
50 Hampshire Street, 5th Floor, Cambridge, MA 02139, United States
525 B Street, Suite 1800, San Diego, CA 92101–4495, United States
The Boulevard, Langford Lane, Kidlington, Oxford OX5 1GB, United Kingdom
125 London Wall, London, EC2Y 5AS, United Kingdom

First edition 2017

ISBN: 978-0-12-804164-2
ISSN: 1546-5098

Publisher: Zoe Kruze
Acquisition Editor: Sam Mahfoudh
Editorial Project Manager: Ana Claudia Garcia
Production Project Manager: Stalin Viswanathan
Cover Designer: Mark Rogers

Typeset by SPi Global, India

Cover Images: Starting at top and moving clockwise. Top, Cardiovascular morphology of a 3 day post-fertilization zebrafish embryo (Burggren, Dubansky and Bautista, 2017; Chapter 2); Bottom right, Influence of cold acclimation on collagen content of the zebrafish heart (Gillis and Johnston, 2017; Chapter 3); Bottom middle, Visible xenomas on the surface of an Atlantic cod heart (Powell and Yousaf, 2017; Chapter 7); Bottom left, Effects of acute changes in temperature on the resting heart rate (f_H) of cold- (blue lines) and warm- (red lines) acclimated rainbow trout (Eliason and Anttila, 2017; Chapter 4).

CONTENTS

CONTRIBUTORS

KATJA ANTTILA *(235), University of Turku, Turku, Finland*

NAIM M. BAUTISTA *(107), University of North Texas, Denton, TX, United States*

COLIN J. BRAUNER *(1), University of British Columbia, Vancouver, BC, Canada*

WARREN W. BURGGREN *(107), University of North Texas, Denton, TX, United States*

BENJAMIN DUBANSKY *(107), University of North Texas, Denton, TX, United States*

ERIKA J. ELIASON *(235), University of California, Santa Barbara, Santa Barbara, CA, United States*

TODD E. GILLIS *(185), University of Guelph, Guelph, ON, Canada*

TILL S. HARTER *(1), University of British Columbia, Vancouver, BC, Canada*

JOHN P. INCARDONA *(373), Northwest Fisheries Science Center, National Marine Fisheries Service, National Oceanic and Atmospheric Administration, Seattle, WA, United States*

ELIZABETH F. JOHNSTON *(185), University of Guelph, Guelph, ON, Canada*

MARK D. POWELL *(435), Institute of Marine Research; Institute for Biology, University of Bergen, Bergen, Norway*

NATHANIEL L. SCHOLZ *(373), Northwest Fisheries Science Center, National Marine Fisheries Service, National Oceanic and Atmospheric Administration, Seattle, WA, United States*

JONATHAN A.W. STECYK *(299), Department of Biological Sciences, University of Alaska Anchorage, Anchorage, AK, United States*

MUHAMMAD N. YOUSAF *(435), Norwegian Veterinary Institute, Harstad, Norway*

ABBREVIATIONS

2,3-DPG 2,3-diphosphoglycerate
4-HT 4-hydroxytamoxifen
5-HD sodium 5-hydroxydecanoic acid
5-HT 5-hydroxytryptamine (serotonin)
A atrium
AA arachidonic acid
A_o ventral aorta
ABA afferent branchial arteries
ACE angiotensin-converting enzyme
ACH acetylcholine
ACV anterior cardinal vein
AD adrenaline
ADO adenosine
ADP adenosine diphosphate
ADP_{50} ratio of atrial to ventricular action potential duration
AE anion exchange or Cl^-/HCO_3^- exchanger
AFA afferent filamental arteries
AGD amoebic gill disease
AIP AHR-interacting protein
AKAP A-kinase anchoring protein
AHR aryl hydrocarbon receptor
ALAs afferent lamellar arterioles
AMP adenosine monophosphate
AMPA alpha-amino-3-hydroxy-5-methylisoxazole-4-propionic acid
AMPK AMP-activated protein kinase
AMs adrenomedullins
ANG-1 angiopoietin 1
AngII angiotensin II
ANP atrial natriuretic peptide
ANRT AHR nuclear receptor translocator

AOP adverse outcome pathway
APC antigen-presenting cells
AP action potential
AR adrenoreceptor
ARC activity-regulated cytoskeleton-associated protein
ASCV Atlantic salmon calcivirus
ASN anterior spinal nerve
ASR aquatic surface respiration
AT_1 angiotensin 1 receptor
AT_2 angiotensin 2 receptor
ATP adenosine triphosphate
A.U. arbitrary units
$A\text{-}VO_2$ difference between the oxygen content of arterial and venous blood, also known as tissue oxygen extraction
AV atrioventricular
AVP atrioventricular plexus
AVR atrioventricular region
AVT arginine vasotocin
Ax axillary body
AZ acetazolamide
α-AR alpha adrenoreceptor
β_2-AR β_2-type adrenoreceptor
BA bulbus arteriosus
β_b capacitance of blood for O_2
BCR branchial cardiac ramus
BCT branchiocardiac nerve trunk
BKs bradykinins
$BL\,s^{-1}$ swimming speed expressed as body lengths per second
B_{max} receptor density
BMP bone morphogenic protein
BN branchial nerve
β-NHE β-adrenergic Na^+–H^+ exchanger
BNP brain natriuretic peptide
bpm beats per minute
BV branchial vein
BZ benzolamide
C conus or compliance
CA carbonic anhydrase
Ca^{2+} calcium
$[Ca^{2+}]_i$ intracellular free Ca^{2+} concentration
CAM chorioallantoic membranes
cAMP cyclic adenosine monophosphate
CaN calcineurin

C_aO_2 arterial O_2 content
CASQ2 calsequestrin 2
CAT catecholamines
CaTF Ca^{2+} sensitive transcription factors
CBF coronary blood flow
CBS cystathionine beta-synthase
CCO cytochrome *c* oxidase
CCK cholecystokinin
CD73 ecto-5′-nucleotidase
CdA caudal artery
CDβ β COMMA-D cell line engineered to express β-galactosidase
CdV caudal vein
CgA chromogranin A
cGMP cyclic guanosine monophosphate
CGRP calcitonin gene-related peptide
ChAT choline acetyltransferase
CICR Ca^{2+}-induced Ca^{2+}-release
CK creatine kinase
cKit receptor tyrosine kinase
cmlc myosin light chain polypeptide
CMA coeliacomesenteric artery
CMS cardiomyopathy syndrome
cNOS constitutive nitric oxide synthase
CNP C-type natriuretic peptide
CNS central nervous system
CO carbon monoxide
CO_2 carbon dioxide
C_{O_2} content of O_2 in blood
CoA coenzyme A
COL1A1 collagen Type I alpha 1 chain gene
COX cyclooxygenase
CP creatine phosphate
CPCs cardiac progenitor cells
CPO cardiac power output
CPO_{max} maximum cardiac power output
CPO_{sys} systemic cardiac power output
CPT carnitine palmitoyltransferase
Cr creatine
CR cagal cardiac ramus
CrA carotid arteries
CS citrate synthase
C_s the slope of the capacitance curve
CSE cystathionine gamma-lyase

CSPN cardiac spinal pre-ganglionic neuron
CSQ calsequestrin
CSQ2 cardiac isoform of calsequestrin
CST catestatin
CTGF connective tissue growth factor
CT_{max} critical thermal maximum
cTnC cardiac troponin C
cTnI cardiac troponin I
cTnT cardiac troponin T
CV conal valves
C_vO_2 venous O_2 content
CVS central venous sinus
CVPN cardiac vagal pre-ganglionic neuron
CYP1A cytochrome P450 1A
CYS cysteine
d day
DA dorsal aorta
DC ductus of Cuvier
DDT dichlorodiphenyltrichloroethane
DIDS 4,4′-diisothiocyano-2,2′-stilbenedisulfonic acid
DLCs dioxin-like compounds
DMO 5,5-dimethyl-2,4-oxazolidinedione
DNA deoxyribonucleic acid
DPI days post-injury
DPF days post-fertilization
DPT days post-treatment
dsRNA double-stranded RNA
ΔG_{ATP} Gibbs free energy per mole of hydrolyzed ATP
ΔP change in pressure (pressure difference)
ΔpH_{a-v} arterial–venous pH difference
ΔP_V pressure gradient that drives venous return
$\Delta S_{a-v}O_2$ change in arterial–venous Hb–O_2 saturation (see below)
ΔV change in volume
$E_{a-v}O_2$ extraction of O_2 at the tissues
EBA efferent branchial arteries
E–C excitation–contraction
ECG electrocardiogram
ECM extracellular matrix
ECs endocardial cushions
EDCF endothelium-derived contracting factors
EDRF endothelium-derived relaxing factors
EE endocardial endothelium
EFAs efferent filamental arteries

EGCs eosinophilic granular cells
E_{ion} equilibrium potential
EIPA ethyl isopropyl amiloride
E_{K+} equilibrium potential for K^+
ELAs efferent lamellar arterioles
EM electron microscopy
E_m membrane potential
EMT epithelial-to-mesenchymal transition
E_{Na+} equilibrium potential for Na^+
eNOS endothelial nitric oxide synthase
EPA eicosapentanoic acid
EPDCs epicardial-derived cells
EPO erythropoietin
EPOR EPO receptor
ERG erythroblast transformation-specific -related gene
ERG channel ether-à-go-go-related gene K^+ channel
ERK extracellular signal-regulated kinase
ET endothelins
E–T excitation–transcription
ET-1 endothelin-1
ETA eicosatetranoic acid
F Faraday's constant
F_C flow rate through chamber
FGF fibroblast growth factor
f_H frequency of the heartbeat, or heart rate
f_{Hmax} maximum cardiac frequency or maximum heart rate
FP flow probe
FKBP12 12-kDa FK506-binding protein
f_R respiratory frequency
Fs Furans
G6P glucose-6-phosphate
Gα13 G protein subunit Gα13
GA gill arch
GBF gonad blood flow
GC guanylate cyclase
GF gill filaments
G_i inhibitory G protein
G_o gonad
GLUTs glucose transporter proteins
GPCR G protein-coupled receptor
GPI glycophosphatidylinositol
G_s stimulatory G protein
GTP guanosine triphosphate

G_u gut
H heart
h hour
H^+ hydrogen ion
Hb hemoglobin
[Hb] hemoglobin concentration
Hct hematocrit
HCN channel family of hyperpolarization-activated, cyclic nucleotide-gated, ion channels involved in controlling the pacemaker
HEK cells human embryonic kidney cells
HEP high energy phosphate
HH hedgehog
HIF-1 hypoxia inducible factor 1
HIF 1-α hypoxia inducible factor 1-α
HK hexokinase
HO heme oxygenase.
HOAD β-hydroxyacyl CoA dehydrogenase
hpf hours post-fertilization
HPV hepatic portal vein
HSMI heart and skeletal muscle inflammation
HSP heat shock proteins
H_2S hydrogen sulfide
HV hepatic vein
I_{Ca} Ca^{2+} current
I_{CaL} L-type Ca^{2+} current
I_{CaT} T-type Ca^{2+} current
ICN intracardiac neuron
ICNS intracardiac nervous system
I_f pacemaker current, "funny" current
IFNα interferon alpha
IFNγ interferon gamma
IGF-1 insulin-like growth factor 1
IgM immumoglobulin
I_{KACH} acetylcholine-activated inward rectifier current
I_{KATP} ATP-sensitive inward rectifier current
I_{K1} inward-rectifier K^+ current
I_{Kr} delayed-rectifier K^+ current
I_{Na} Na^+ current
I_{NCX} Na^+–Ca^{2+} exchange current
iNOS inducible NO synthase
ION_i intracellular "free" ion concentration
ION_e extracellular "free" ion concentration
IP_2 inositol diphosphate

IP_5 inositol pentaphosphate
IRK2 inwardly rectifying K^+ channel 2
I/R ischemia–reperfusion
ISA infectious salmon anemia
ISG15 interferon-stimulated gene 15
ISL1 transcription factor islet 1
ISO isoproterenol
JAK1 janus kinase 1
JNK C-jun NH_2 terminal kinase
JV jugular vein
K_{ATP} ATP-sensitive K^+ channels
K_{Ca} Ca^{2+}-activated K^+ channels
K_{cat} turnover number
K_d receptor binding affinity
kDa kilo Dalton
K_{eq} equilibrium constant
K^+ potassium ion
K^+/Cl^- potassium chloride co-transport
$[K^+]_e$ extracellular K^+ concentration
$[K^+]_i$ intracellular K^+ concentration
K_H hydration rate constant
KH7 blocker of soluble guanylate cyclase
k_i inhibition constant
Kir inward rectifier K^+ channels
L vessel length
La lamella
L-AA L-ascorbic acid
L-ARG L-arginine
LCV lateral cutaneous vein
LDA length-dependent activation
LDH lactate dehydrogenase
L-NMMA L-N^G-monomethyl-L-arginine
LTCC L-type Ca^{2+} channels
M_2R muscarinic type-2 receptor
MAPK mitogen-activated protein kinase
max dP/dt_{sys} maximal rate of pressure change during systole
max dP/dt_{dia} minimum rate of pressure change during diastole
M_b body mass
MCFP mean circulatory filling pressure
MCT monocarboxylate transporters
MD medulla
MHC major histocompatibility complex
min minute

miRNAs microRNAs
miR-133 microRNA 133
MLP muscle LIM protein
MMPs matrix metalloproteinases
$\dot{M}O_2$ rate of oxygen consumption
$\dot{M}O_{2max}$ maximum rate of oxygen consumption
MPP5 membrane palmitoylated protein 5
mRNA messenger RNA
ms milliseconds
M_V mass of ventricle
MY million years
Na^+ sodium
NAD noradrenaline
NANC non-adrenergic non-cholinergic
NCX Na^+/Ca^{2+}-exchanger
NECs neuroepithelial cells
NeKA neurokinin A
NF-κB nuclear factor kappa light chain enhancer of activated B cells
η_H hill coefficient
NHE Na^+–H^+ exchanger
NKA Na^+–K^+–ATPase
NMDA *N*-methyl-D-aspartate
NPs natriuretic peptides (ANP, BNP, CNP, CNP, VNP)
NPR natriuretic peptide receptor (A, B, C, D, V)
NO nitric oxide
NO_2^- nitrite
NOAA National Oceanic and Atmospheric Association
NOK novel oncogene with kinase domain
NOS nitric oxide synthase
NPY neuropeptide Y
NTP nucleotide triphosphate
O ostium of sinoatrial valve
O_2 oxygen
OEC O_2 equilibrium curve
OFT outflow tract
P atrioventricular plug
P_{50} partial pressure of O_2 at which 50% of hemoglobin is bound to oxygen
P_A arterial blood pressure
Pa_{50} arterial P_{50}
PACs polycyclic aromatic compounds
PACA plasma-accessible carbonic anhydrase
P_aCO_2 arterial partial pressure of carbon dioxide

PAF1 polymerase-associated factor 1
PAHs polycyclic aromatic hydrocarbons
PAK p21-activated kinase
P_aO_2 arterial partial pressure of O_2
pCa calcium concentration, expressed as −log
pCa$_{50}$ P_{Ca} for half maximal activation
PCBs polychlorinated biphenyls
PCDD polychlorinated dibenzo-*p*-dioxins
PCNA proliferating cell nuclear antigen
PCO_2 partial pressure of CO_2
PCR polymerase chain reaction
PCS posterior cardinal sinus
PCV posterior cardinal veins
P_{CV} central venous blood pressure
PD pancreas disease
P_{DA} dorsal aortic pressure
PDH pyruvate dehydrogenase
PE phenylephrine
PFK 6-phosphofructokinase
PGE2 prostaglandin E2
pH_a arterial pH
pH_e extracellular pH
pH_i intracellular pH
PHZ phenylhydrazine
P_i inorganic phosphate
PICA plasma inhibitors of carbonic anhydrase
PI–PLC phosphatidylinositol specific phospholipase C
P_{in} input pressure
PK pyruvate kinase
PKC protein kinase C
PKG protein kinase G
PLB phospholamban
PMCA plasma membrane Ca^{2+}–ATPase
PMCV piscine myocarditis virus
PO_2 partial pressure of O_2
P_{out} output pressure
proANF atrial natriuretic factor prohormone
PRV piscine reovirus
PTx pertussis toxin
PUFA polyunsaturated fatty acids
PV pulmonary vein
P_V plasma volume

P_{ven} venous pressure
Pv_{50} venous P_{50}
P_VO_2 venous partial pressure of oxygen
P_{VA} blood pressure in the ventral aorta
$\dot{Q}$ cardiac output
$\dot{Q}_{max}$ maximum cardiac output
Q_{10} temperature quotient; the ratio of a rate function over a 10°C temperature difference
qPCR quantitative polymerase chain reaction
r vessel radius
R vascular resistance
RA retinoic acid
RAS renin–angiotensin system
RBC red blood cell
R_{cor} vascular resistance of the coronary circulation
RD2 retinaldehyde dehydrogenase 2
ReA renal artery
R_{gill} vascular resistance of the gill circulation
RMP resting membrane potential
RNA-Seq high-throughput RNA sequencing
ROS reactive oxygen species
RQ respiratory quotient
R_{sys} vascular resistance of the systemic circulation
rTNF-α recombinant tumor necrosis factor-α
R_{tot} total peripheral vascular resistance
RyR ryanodine receptor
R_v resistance to venous return
RVM relative ventricular mass (mass of the ventricle relative to body mass, expressed as a percentage)
RVR resistance to venous return
S1P lysosphingolipid siphogosin-1-phosphate
S1Pr2 lysosphingolipid siphogosin-1-phosphate receptor 2
SA sinoatrial
sAC soluble adenylyl cyclase
SA node sinoatrial node
SAP sinoatrial plexus
SAR sinoatrial region
SAV salmonid alphavirus
$S_{a-v}O_2$ difference in arterial–venous Hb–O_2 saturation
SBV stressed blood volume
SC spinal cord
SCA subclavian artery

SCP salmon cardiac peptide
SCV subclavian vein
SD sleeping disease
SDS sodium dodecyl sulfate
SEM scanning electron microscope
SERCA sarco–endoplasmic reticulum Ca^{2+}-ATPase
SeV subepithelial vein
SG sympathetic ganglia
SGC soluble guanylate cyclase
SIV supraintestinal vein
SL sarcolemma or sarcolemmal
SL_n sarcomere length
SMLC2 slow myosin light chain 2
SN spinal nerve
SNP sodium nitroprusside
SNS sympathetic nervous system
SPN spinal pre-ganglionic neurons
SO_2 Hb-O_2 saturation (arterial S_aO_2 and venous S_vO_2)
SP substance P
SR sarcoplasmic reticulum
SUR sulfonylurea receptor
SV sinus venosus
$t_{1/2}$ half-time
TBX T-box transcription factor
TCA cycle tricarboxylic acid cycle
TCDD 2,3,7,8-tetrachlorodibenzo-*p*-dioxin
TCR T-cell receptor
TDEE temperature-dependent deterioration of electrical excitation
TdT dUTP nick-end labeling
TEFs toxicity equivalence factors
TGF-β_1 transforming growth factor-beta 1
TIMP tissue inhibitor of metalloproteinase
TM tropomyosin
TMAC transmembrane adenylyl cyclase
TnC troponin C
TnI troponin I
TNF-α tumor necrosis factor-alpha
TO_{2max} maximum arterial O_2 transport
T_{optAS} optimal temperature for aerobic scope
TUNEL terminal deoxynucleotidyl transferase
UI urotensin I
UII urotensin II

U_{crit} critical swimming speed of a fish
UDP uridine diphosphate
UK United Kingdom
Us urotensins
USBV unstressed blood volume
UTP uridine triphosphate
UTR 3′-untranslated region
V ventricle
V_b blood volume
$\dot{V}_b$ blood flow
VC vena cava
VCR visceral cardiac ramus of vegas
VEGF vascular endothelial growth factor
VIP vasoactive intestinal peptide
V_m resting membrane potential
VMHC ventricular myosin heavy chain
VNP ventricular natriuretic peptide
VO vascular occluder
$\dot{V}O_2$ O_2 uptake per unit time
V_R total ventilation volume
VS vasostatin
V_s stroke volume, volume of blood pumped with each heartbeat
V_{smax} maximum stroke volume
VS-1 vasostatin 1
VS-2 vasostatin 2
VST vagosympathetic trunk
WAF water accommodated fraction
W_S stroke work
WT1 Wilms' tumor protein 1
X cardiac vagus rami
$X1$ lateral vagal motor neuron group
X_{br} vagal ramus interconnecting gill arches
X_m medial vagal motor neuron group
X_{mr} rostral components of the medial vagal motor neuron group
X_{mc} caudal components of the medial vagal motor neuron group
X_r vagal root
Φ Bohr-coefficient
η fluid viscosity

PREFACE

The cardiovascular system plays a critical role in fishes, as it does in all vertebrates. It transports oxygen, carbon dioxide, metabolic substrates and wastes, hormones and various immune factors throughout the body. As such, it is one of the most studied organ systems in fishes, and significant insights into the evolution of this taxonomic group have emerged through this research. The cardiovascular system also determines, in part, the ecological niche that can be occupied by many fish species and how they will potentially be impacted by future abiotic and biotic challenges. For example, over the past decade considerable evidence has accumulated that metabolic scope (the difference between resting and maximum metabolic rate) is a key determinant of fish thermal tolerance, and that this parameter is largely determined by the capacity of the fish to deliver enough oxygen to the tissues to meet the rising metabolic demand as temperature increases. Likewise, the heart needs its own oxygen supply to function properly, and so, the increasing examples of aquatic hypoxia are of great concern for fish distributions.

Twenty-five years ago, two volumes (Volumes 12A and 12B) titled *The Cardiovascular System* were published in the *Fish Physiology* series. These volumes provided a synthesis of the state of the field and played a significant role in driving this research area forward; indeed they are required reading for any scientist interested in comparative cardiovascular physiology. However, molecular and cellular approaches have tremendously advanced the field, and consequently, we now have a much more complete understanding of the mechanistic underpinnings of cardiovascular function. It is, therefore, critical that we take stock of the current state of knowledge in this very active and growing field of research. The two current volumes (Volumes 36A and 36B) contain 14 contributions from 25 world experts in various aspects of fish cardiovascular physiology. As such, they provide an up-to-date and comprehensive coverage and synthesis of the field. The 14 chapters, collectively, highlight the tremendous diversity in cardiovascular morphology and function among the various fish taxa, and the anatomical and physiological plasticity shown by this system when faced with different abiotic and biotic challenges. These chapters also integrate molecular and cellular data with the growing body of knowledge on heart (i.e., whole organ) and *in vivo* cardiovascular function and, as a result, provide insights into some of the most interesting, and important, questions that remain to be answered in this field.

The two volumes are organized to complement each other. The first volume (36A) titled *The Cardiovascular System: Morphology, Control and Function* summarizes our current understanding of the fish heart and vasculature, and how they work. The first chapter of Volume 36A provides an extensive review of the morphology and structure of the fish heart and highlights the tremendous diversity in form and function displayed by various taxa. The second chapter focuses on the morphology and function of cardiomyocytes, and relates form and function to the mechanistic basis of cardiac contraction. Chapter 3 describes, in significant detail, the electrical excitability of the fish heart. This chapter's major focus is on recent advances in our understanding of cardiomyocyte ion currents, their molecular basis and their regulation by the autonomic nervous system, and on how this enables the heart to respond to changes in physiological requirements. The fourth chapter provides a comprehensive account of the form, function and physiology of the fish heart, including information on cardiac innervation and the intracardiac nervous system, and provides a general overview of how this organ responds to various physiological and environmental challenges. Chapter 5 reviews what is known of how hormones and locally released (autacoid) substances regulate cardiac function under both normal and stress-induced conditions, and the biochemical pathways that underpin this control. The sixth chapter examines our current understanding of the heart's energy requirements, and the biochemical pathways utilized by various fish species to meet cardiac energy demand and to maintain cardiac energy state. Finally, the seventh chapter of this volume provides a review of what is currently known about the vasculature in different fish species and the mechanisms that regulate its function.

The second volume (36B) titled *The Cardiovascular System: Development, Plasticity and Physiological Responses* is focused on how cardiac phenotype and the cardiovascular system change during development and in response to variations in physiological and environmental conditions. It covers the impact of a variety of challenges from development, to environmental stressors, to disease on cardiovascular function, and the mechanisms that provide this system with the plasticity and capacity to respond to such challenges. The first chapter is focused on the role of the erythrocytes in maintaining O_2 delivery to the tissues under a range of physiological conditions. Further, it describes the heterogeneous distribution of plasma-accessible carbonic anhydrase, and the role that this enzyme may play in enhancing tissue oxygen extraction in various tissues and species. The second chapter summarizes our current understanding of cardiovascular development and function in embryonic and larval fishes, as well as the plasticity shown during this process and how it is influenced by environmental change. In Chapter 3, the authors discuss the hypoxic preconditioning response of the fish heart, as well as the remodeling and/or regenerative capacity of the hearts of some fish species following exposure to various physiological stressors. Chapter 4 summarizes the influence of acute and chronic changes in temperature on cardiovascular function and the capacity of some fish species to compensate for changes in their thermal environment. Chapter 5 details cardiovascular responses to oxygen limitation and their regulation in water-breathing and air-breathing fishes that are intolerant of hypoxia, as well as the strategies that hypoxia/anoxia-tolerant species use to survive prolonged periods of oxygen deprivation. The sixth chapter summarizes the consequences of toxicant exposure for the development and function of the heart, with an emphasis on the effects of dioxins, polychlorinated biphenyls, polycyclic aromatic hydrocarbons and crude oil exposure. Finally, Chapter 7 describes the effects of morphological abnormalities, pathological conditions, and various parasites and pathogens (i.e., diseases) on the function of the fish heart.

These volumes represent the collective effort of a large number of individuals. We wish to thank all the authors for the significant work that was needed to write their respective chapters, and all the anonymous reviewers who ensured that each chapter was of the highest quality. We also acknowledge the staff at Elsevier for their assistance in constructing these volumes and efforts to keep us on schedule. Finally, we recognize the tremendous contributions that numerous individuals have made to this very interesting and important discipline of fish physiology, and encourage them to keep asking key questions and challenging dogma in the field, and to keep inspiring those who will follow in their footsteps.

Todd E. Gillis and A. Kurt Gamperl

1

THE O_2 AND CO_2 TRANSPORT SYSTEM IN TELEOSTS AND THE SPECIALIZED MECHANISMS THAT ENHANCE HB–O_2 UNLOADING TO TISSUES

TILL S. HARTER[1]
COLIN J. BRAUNER

University of British Columbia, Vancouver, BC, Canada
[1]Corresponding author: harter@zoology.ubc.ca

Teleost fishes comprise nearly half of all vertebrate species, which perhaps represents the most extensive adaptive radiation in vertebrate evolution. This success is in part through adaptations in the modes that they use to take up, transport, and deliver oxygen (O_2) to the tissues. The present chapter reviews the basic concepts of O_2 transport in fishes with a focus on hemoglobin (Hb) and red blood cell (RBC) function, and associated features that enhance the

The Cardiovascular System: Development, Plasticity and Physiological Responses, Volume 36B
FISH PHYSIOLOGY

DOI: https://doi.org/10.1016/bs.fp.2017.09.001

capacitance of blood for O_2 (β_b): the Bohr and Root effects, RBC β-adrenergic sodium proton exchangers (RBC β-NHE), and the *retia mirabilia* (teleost vascular counter-current exchangers). The RBC microenvironment plays an important role in modulating Hb characteristics in vivo during challenges to O_2 transport such as hypoxia, exercise, and increases in temperature. In addition, a novel mechanism of enhanced Hb–O_2 unloading in teleosts is described and discussed, one that relies on the heterogeneous distribution of plasma-accessible carbonic anhydrase (PACA). This is an intrinsic characteristic of the cardiovascular system, and a vastly understudied area not only in fishes. Thus, the available data on PACA distribution in fishes are presented in a phylogenetic context, concluding with the putative sequence of events that may have led to this remarkable mode of Hb–O_2 unloading in modern teleosts. In combination with other well-studied teleost Hb and RBC characteristics, a heterogeneous distribution of PACA in the cardiovascular system may greatly enhance β_b, and thus tissue O_2 delivery, and support a higher rate of tissue O_2 consumption without increasing [Hb] or blood flow. This is a research area worthy of further investigation that may improve our understanding of O_2 transport in this most specious group of vertebrates.

1. INTRODUCTION

Teleosts are the most speciose aquatic vertebrates living today and, at over 30,000 named species (Nelson et al., 2006), they comprise more than 95% of all extant fishes. What made them so successful? The answer to this question is likely found in an era more commonly associated with extinction rather than the diversification of life. Caused by extensive volcanic activity and global fires, the Permian crisis (also known as the Permian–Triassic extinction) sets the stage for the teleost story of success. During the Permian crisis atmospheric oxygen (O_2) levels decreased dramatically and remained low for another 100 million years (Clack, 2007). This led to the loss of 96% of all marine fish species. Teleosts, however, experienced an explosive adaptive radiation, perhaps the most extensive one in vertebrate evolution (Helfman et al., 2009). Among the most pivotal adaptations in teleosts was a system of enhanced Hb–O_2 unloading that allowed the swim bladder to be filled at depth and for a supply of O_2 to the avascular retina of the eye (Wittenberg and Wittenberg, 1974); innovations that lead to the lifestyle, habitats, and morphologies of modern teleost species. In addition, it has been argued that a greater hypoxia tolerance was key to not only surviving this mass extinction but also to thrive thereafter (Randall et al., 2014). All of these adaptations were only possible in association with the unique characteristics of teleost red blood cells (RBC) and hemoglobin (Hb).

The following chapter will review the characteristics of cardiovascular O_2 transport in fishes, and discuss those properties of the RBC and Hb that allow teleosts to fine-tune and enhance O_2 uptake, transport, and delivery to the tissues. In addition, a mechanism for enhanced Hb–O_2 unloading that hinges on the heterogeneous distribution of plasma-accessible carbonic anhydrase (PACA) in the circulation is described and discussed, because it is likely unique to teleosts. The distribution of PACA is an intrinsic property of the cardiovascular system, and its presence or absence in various tissues may change the interactions between the blood and the cardiovascular system with great benefit to O_2 transport. Finally, the distribution of PACA is discussed in a phylogenetic context, and within the sequence of events proposed by Berenbrink et al. (2005) that led to the unique mode of O_2 and CO_2 transport in modern teleosts.

2. BLOOD O_2 TRANSPORT

Numerous excellent reviews have been written on blood O_2 transport in vertebrates, including those in several volumes of the Fish Physiology series. These include Volume 17 (Respiration) and those dedicated to specific fish groups: tropical fishes (Brauner and Val, 2005; Nilsson and Östlund-Nilsson, 2005), polar fishes (Steffensen, 2005; Wells, 2005), primitive fishes (Brauner and Berenbrink, 2007); zebrafish (Pelster and Bagatto, 2010), and elasmobranchs (Morrison et al., 2015). Thus, we only briefly review the basic concepts of O_2 transport (Section 2) in preparation for a more in-depth discussion of: the in vivo response of animals to environmental or metabolic challenges to O_2 transport (Section 3); the interaction between O_2 and carbon dioxide (CO_2) transport (Section 4); and a specialized system of Hb–O_2 unloading in teleosts that hinges on the vascular distribution of PACA (Section 5).

To meet the O_2 demand of metabolically active tissues, O_2 is taken up from the environment and transported via the circulatory system to the mitochondria in tissues by a series of diffusive and convective steps (see, e.g., Dejours, 1981). This chapter will focus on the convective transport of O_2 in the cardiovascular system, which is described by the Fick equation (Fick, 1870):

$$\dot{M}O_2 = \dot{V}_b \times (C_aO_2 - C_vO_2)$$

where $\dot{M}O_2$ is the rate of O_2 consumption (used to infer metabolic O_2 demand), $\dot{V}_b$ is blood flow, and C_aO_2 and C_vO_2 are the content of O_2 in the arterial and venous blood, respectively. The theoretical maximal O_2 transport capacity can then be calculated as the product of blood flow and arterial blood O_2 content, assuming that C_vO_2 is zero:

$$\dot{M}O_2 = \dot{V}_b \times C_aO_2$$

In fish without significant cutaneous respiration the maximal capacity for arterial O_2 transport is a function of cardiac output and arterial O_2 content of the blood ($\dot{Q} \times C_aO_2$), and this generally represents a good predictor of maximal $\dot{M}O_2$ in fish (but see Farrell et al., 2014). Thus, it has been suggested that $\dot{Q}$ and C_aO_2 are primary loci for evolutionary adaptations that determine a species' maximum $\dot{M}O_2$ (Gallaugher et al., 2001). Clearly, cardiovascular adjustments to $\dot{Q}$ (and $\dot{V}_b$) are central mechanisms by which animals regulate O_2 transport and these are discussed elsewhere (Butler and Metcalfe, 1989; Farrell, 1992, 2007; Farrell and Jones, 1992; Fritsche and Nilsson, 1993; Gamperl and Driedzic, 2009; Randall and Daxboeck, 1984; also see Chapter 4, Volume 36A: Farrell and Smith, 2017). According to Henry's law, C_aO_2 is a function of the capacitance of blood for O_2 (β_b) and the arterial partial pressure of O_2 (P_aO_2):

$$C_aO_2 = \beta_b \times P_aO_2$$

where the capacitance of blood is defined as the increment in C_aO_2 per increment in P_aO_2, and includes both the physically dissolved O_2 and O_2 bound to respiratory pigments such as Hb:

$$\beta_b = \frac{\Delta C_aO_2}{\Delta P_aO_2}$$

Therefore, convective O_2 transport in the cardiovascular system can be expressed as:

$$\dot{M}O_2 = \dot{Q} \times \beta_b(P_aO_2 - P_vO_2)$$

Hemoglobin in the blood, the principal O_2 carrier, greatly increases β_b and reduces the requirements on $\dot{Q}$ proportionally. Hb also allows for O_2 to be taken up from the water against its concentration gradient. By providing a large sink for O_2 within the animal, the PO_2 gradient between the venous blood and the water remains large, which allows for an efficient uptake of O_2 within the short residence time that blood has in the gill lamellae (about 1–3 s; Cameron and Polhemus, 1974; Booth, 1978; Hughes et al., 1981; Randall, 1982b). The content of physically dissolved O_2 in the plasma is typically of minor importance; however, its contribution increases at lower [Hb], lower temperatures and higher PO_2. In anemic starry flounder (hematocrit, Hct = 13%) at 10°C about 12.5% of the total O_2 in air-equilibrated blood was physically dissolved in the plasma, but contributed less than 8% to tissue O_2 delivery, due to the linear relationship between plasma O_2 content and PO_2 (Wood et al., 1979a,b). However, in Hb-less icefishes that live at sub-zero temperatures, O_2 transport in the cardiovascular system relies entirely on physically dissolved O_2 in the plasma (Ruud, 1954), which is somewhat facilitated by the higher solubility of O_2 in plasma at low temperatures. The effect of temperature and Hct on the solubility of O_2 in human plasma has been described by Christoforides and Hedley-Whyte (1969), and Boutilier et al. (1984) provide an excellent summary of O_2 solubility in the plasma of humans as compared to fish.

2.1. Hematocrit

Hemoglobin is contained within the erythrocytes or red blood cells (RBC), the functional unit of O_2 transport in vertebrates. Hematocrit is the percentage of blood volume that is occupied by the RBCs. In mature RBCs [Hb] is close to its solubility limit (Riggs, 1976), and thus, there is typically a good correlation between [Hb] and Hct (Fänge, 1992; Farrell, 1991, 1992); however, exceptions have been described (see Brauner and Berenbrink, 2007; Wells, 2005; Wells and Baldwin, 1990). Recent data indicate that high [Hb] may severely limit the diffusivity of gases through the RBC cytoplasm and set limits on the maximal [Hb] that can be contained within RBCs of a given size and shape (Richardson and Swietach, 2016). Hematocrit in fish varies greatly (from 0% to >50%) between, but also within, fish species and factors that influence Hct have been reviewed in a previous volume of this series (Gallaugher and Farrell, 1998).

In adult vertebrates, RBCs are produced continuously (a process termed erythropoiesis) to balance the loss due to lysis and senescence. Anemia in fish stimulates erythropoiesis, and as in mammals, this mechanism is likely mediated by the hormone erythropoietin (Lai et al., 2006; Weinberg et al., 1976). An increase in Hct will increase β_b, and is a common response where increases in C_aO_2 are needed to offset challenges to O_2 transport. However, the benefits of changes in Hct with respect to O_2 transport are limited to a narrow range in this parameter. Broadly speaking, while anemia can limit the metabolic scope of fish, which has been shown in salmonids (Gallaugher et al., 1995; Jones, 1971; Thorarensen et al., 1993), polycythemia may set limits to cardiac function as blood viscosity, and thus cardiac work, increase with Hct (Egginton, 1996; Gallaugher et al., 1995; Graham and Fletcher, 1983; Wells and Weber, 1991). Consequently, and analogous to the situation in mammals, it has been proposed that an optimal Hct must exist that balances the requirements for O_2 carrying capacity of the blood and its impact on cardiac function (Richardson and Guyton, 1959; Wells and Weber, 1991). However, outside of the extreme scenarios of pathological anemia and extreme polycythemia, this concept fails to explain the observed variability in Hct within and between fish species, and the idea of an optimal Hct in fishes has been criticized (Gallaugher and Farrell, 1998).

2.2. Hemoglobin O_2-Binding

Hemoglobin is perhaps one of the best-studied proteins, and excellent reviews describe in detail the structure of fish Hbs in relation to their O_2 binding properties (Brittain, 2005; Jensen et al., 1998; Riggs, 1970; Weber and Fago, 2004). Briefly, most fishes possess tetrameric Hb (agnathans have monomeric Hb and some Antarctic icefishes do not express Hb), a globular protein consisting of two α and two β chains that form two α–β dimers. Each subunit has a heme group with a central iron atom that is capable of reversibly binding one molecule of O_2. For most physiological applications it suffices to describe Hb–O_2 binding by a two-state allosteric model (Monod et al., 1965), in which Hb is in equilibrium between two alternate conformations: the low affinity (T)ense-state, and the high affinity (R)elaxed-state. The affinity of Hb–O_2 binding is described by the Hb P_{50} value (the PO_2 at which Hb is 50% saturated), where higher P_{50} values represent a lower binding affinity. In the absence of ligand binding, the Hb molecule will be present in the thermodynamically more stable T-state and a transition to the R-state occurs with the binding of O_2; this necessarily requires a shift in the position of several amino acids, and thus, structural changes that propagate into the protein's tertiary conformation. As a consequence, interchain contacts and the interaction between the subunits (quaternary structure) are altered, and this provides the molecular mechanism for cooperativity and the linkage between Hb (de)oxygenation and the binding of allosteric effectors (Nagai et al., 1985; Perutz, 1970; Perutz et al., 1987). The relationship between PO_2 and Hb–O_2 saturation (SO_2) is expressed by the O_2 equilibrium curve (OEC). The cooperative nature of Hb–O_2 binding in vertebrates arises from a decrease in P_{50} during the T–R transition, which is described by the Hill coefficient (n_H), and this results in the characteristic sigmoidal shape of the OEC.

2.2.1. The Oxygen Equilibrium Curve

In addition to an increase in [Hb], a decrease in Hb P_{50} will increase β_b as more O_2 is bound to Hb at any given PO_2 (up to full saturation), and thus, transported in the blood (Jensen, 1991; Jensen et al., 1993; Nikinmaa, 1990; Weber and Jensen, 1988). However, Hb P_{50} represents a trade-off between O_2 loading at the gas exchange surface and unloading at the tissues. Therefore, in terms of P_{50}, β_b should be viewed as the slope of the OEC between the points of O_2 loading and unloading. This slope is dependent not only on P_{50} but also on n_H and the position on the OEC. Thus, changes in P_{50} may increase β_b by ensuring that O_2 transport occurs over an appropriate range of the OEC that depends on the prevailing environmental and metabolic conditions.

Based on the insightful review by Tenney (1995), the following analysis illustrates the influence of P_{50} on O_2 delivery under different environmental conditions (see also the analyses of Malte and Weber, 1985, 1987, 1989). Three species were chosen for this analysis that provide a large contrast in P_{50} values: the hypoxia-tolerant carp (*Cyprinus carpio*, $P_{50}=3.8$ mm Hg, $n_H=1.3$; Brauner et al., 2001); the active rainbow trout (*Oncorhynchus mykiss*, $P_{50}=24.1$ mm Hg, $n_H=2.1$; Tetens and Lykkeboe, 1981); and the hypoxia-intolerant air-breathing mouse (*Mus musculus*, $P_{50}=50$ mm Hg, $n_H=2.6$; Newton and Peters, 1983). All calculations are based on the equations described in Willford et al. (1982), which are derived from the original Hill equation (Hill, 1910):

$$SO_2 = \frac{P_aO_2^{n_H}}{(P_aO_2^{n_H} + P_{50}^{n_H})}$$

where SO_2 is Hb–O_2 saturation (in %), P_aO_2 is the arterial partial pressure of O_2, P_{50} is the PO_2 at which Hb is 50% saturated, and n_H is the Hill coefficient. Further, the following analysis is performed at equivalent P_aO_2 values for all species (100 and 40 mm Hg during normoxia and hypoxia, respectively), and assuming a constant Hb–O_2 unloading at the tissues. Hb–O_2 unloading is expressed as the difference between arterial and venous SO_2 ($S_{a-v}O_2$), and was set to 50% of maximal Hb–O_2 carrying capacity. Thus, half of the theoretically maximal amount of O_2 transported in the blood is unloaded during capillary transit.

In normoxia, at a P_aO_2 of 100 mm Hg, the Hb of the two fish species is close to fully saturated, but the high P_{50} of the mouse does not allow for full Hb saturation at this PO_2 (Fig. 1). However, assuming a $S_{a-v}O_2$ of 50% at the tissues, the high P_{50} in the mouse yields a venous PO_2 ($P_vO_2=40$ mm Hg) that is higher than that of the trout (22 mm Hg) and the carp (4 mm Hg). Likewise, if P_vO_2 was maintained at the same value in all species, $S_{a-v}O_2$ in the mouse would greatly exceed that of the fishes, and thus, more O_2 would be unloaded at the tissues for the same $\dot{V}_b$ and P_aO_2. Finally, the low P_{50} and large n_H of the carp result in a high β_b and a steep slope of the OEC, and unloading of most O_2 occurs over a relatively narrow range of P_vO_2 values.

Of course, the higher P_vO_2 in the mouse would enhance diffusion of O_2 to the mitochondria. This is because P_vO_2 is the primary determinant of the diffusion gradient for O_2 from the blood to the mitochondria in the tissues, and assuming that diffusion distance and conductance of the tissue remain unchanged, there must be some minimum P_vO_2 that is necessary to sustain aerobic respiration. In fishes this is especially pertinent for the heart as it is downstream of all other tissues. While many fish, especially more active species, have a coronary supply of arterial blood to the compact myocardium, the heart of all fishes, to varying degrees, relies on the O_2 supplied by venous

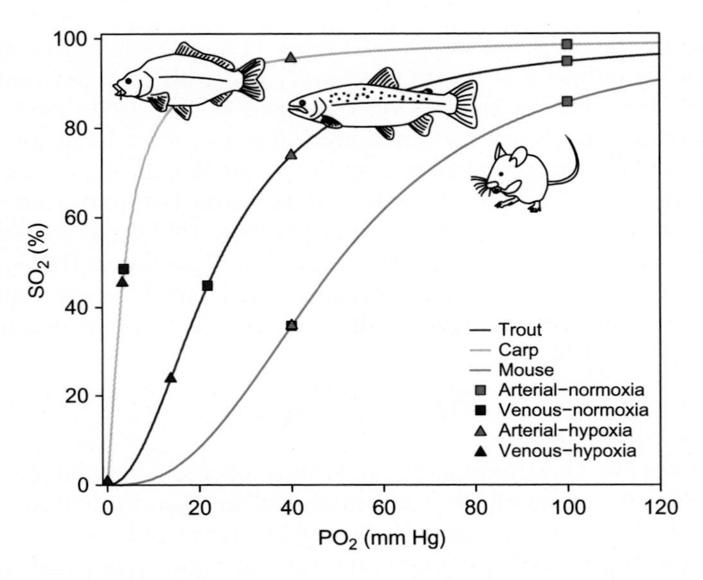

Fig. 1. Oxygen equilibrium curves (OEC) for carp (*Cyprinus carpio*), rainbow trout (*Oncorhynchus mykiss*), and mouse (*Mus musculus*) blood calculated using the Hill equation (Hill, 1910) and previously reported blood parameters for these species (Brauner et al., 2001; Newton and Peters, 1983; Tetens and Lykkeboe, 1981). The arterial hemoglobin (Hb) oxygen saturation (S_aO_2) during normoxia (at an arterial partial pressure of O_2, $P_aO_2 = 100$ mm Hg) and hypoxia ($P_aO_2 = 40$ mm Hg) are indicated by *red squares* and *triangles*, respectively. Corresponding venous values (S_vO_2) are shown in *blue*, and were calculated assuming a constant Hb–O_2 unloading among species. The latter is expressed as the difference between S_aO_2 and S_vO_2 ($S_{a-v}O_2$), which was set to 50% of maximal Hb–O_2 carrying capacity. Thus half of the theoretically maximal amount of O_2 transported in the blood is unloaded during capillary transit.

return (Farrell and Jones, 1992). Therefore, the teleost heart may set the lower limits for P_vO_2 (Davie and Farrell, 1991; Farrell and Clutterham, 2003), and thus, the lower limits for P_{50}.

A high P_{50} value is beneficial for O_2 unloading at the tissues, as long as a high S_aO_2 at the gas exchange surface can be sustained. Therefore, the situation changes during exposure to hypoxia (e.g., a P_aO_2 of 40 mm Hg; Fig. 1). Now S_aO_2 is severely compromised in the mouse (36%), while the trout maintains an intermediate S_aO_2 (73%) and Hb in the carp remains close to full saturation (96%). Clearly, a low P_{50} aids in O_2 uptake from the environment, and this benefit becomes increasingly important as P_aO_2 decreases. The implications for P_vO_2 are evident; when maintaining Hb–O_2 unloading at the tissues constant ($S_{a-v}O_2 = 50\%$), P_vO_2 in the mouse falls to zero. In contrast, P_vO_2 in the carp is maintained within narrow, albeit low, limits over nearly the entire range of P_aO_2. This situation is beneficial for an animal that frequently experiences hypoxia in its habitat, and is part of a suite of adaptations that contribute to the hypoxia tolerance of carp (Vornanen et al., 2009). Nevertheless, the downside of a low P_{50} is the resulting low P_vO_2, which greatly reduces the PO_2 gradient from the blood to the mitochondria, and requires compensation by other physiological or morphological adaptations. If hypoxia is not an immediate threat, as for the mouse, which would rarely encounter this condition in its natural environment, a low P_{50} is of little benefit. Instead, a higher P_vO_2 can be sustained by having a high P_{50}, and this facilitates O_2 diffusion into the tissues.

In fishes, functional hypoxia may occur when tissue O_2 consumption increases during exercise and exceeds circulatory O_2 supply, at which point anaerobic energy production is required to sustain swimming (Farrell and Richards, 2009). During normoxic exercise, where O_2 loading at the gill is typically not compromised, a high P_{50} will enhance O_2 supply to the tissues by ensuring a high P_vO_2 and a large diffusion gradient. Under routine conditions rainbow trout utilize less than half of their total O_2 carrying capacity; Hb is nearly fully saturated at the gill (S_aO_2 ~100%), but venous S_vO_2 is still ~68% (Kiceniuk and Jones, 1977). Therefore, a large amount of O_2 is circulated without being unloaded and this O_2 store is termed the venous reserve. During exercise this venous reserve is increasingly utilized, i.e., tissue O_2 extraction increases (or the arterial–venous O_2 difference increases), and at maximal swimming speed (U_{crit}) S_vO_2 is $<15\%$ while S_aO_2 remains largely unchanged (Steinhausen et al., 2008). Therefore, at rest, ~30% of the total blood O_2 carrying capacity is utilized compared to $>80\%$ at U_{crit}. Animals with a high P_{50} can exploit the venous reserve to a larger extent as a critical P_vO_2 is reached at a lower S_vO_2. Therefore, during exercise a high P_{50} can sustain a high β_b, a situation that is beneficial for active species that are not likely to encounter environmental hypoxia.

Clearly, Hb P_{50} represents a trade-off between O_2 loading at the gill and unloading at the tissues, within the constraints set by the available environmental PO_2, the metabolic condition of the animal, and the minimum critical P_vO_2. Based on these conflicting requirements it follows that, for a suite of conditions, there must be a theoretical P_{50} value that maximizes the O_2 flux from the water to the tissues, an optimal P_{50}.

2.2.2. The Optimal P_{50}

As a word of caution, optimality models have limited relevance for predicting the *in vivo* conditions in animals. Morphological structures and molecules often perform multiple roles, and in the following analysis the importance of Hb for CO_2 transport, proton (H^+) buffering, and other functions will not be considered. In addition, selective pressures are not constant, and genetic and ontogenetic constraints may preclude a biological system from reaching optimality under the prevailing conditions. However, the concept of optimality is a powerful tool for developing hypotheses and fits well within the comparative approach, as long as the adaptive value of the results is not overinterpreted or indiscriminately translated *in vivo* (for reviews see Dudley and Gans, 1991; Wells, 1990). A number of excellent studies have addressed the idea of an optimal P_{50} in mammals (Tenney, 1995; Turek et al., 1973; West and Wagner, 1980; Willford et al., 1982) and in fishes (Brauner and Wang, 1997; Wang and Malte, 2011), and the analysis provided here is a synthesis of previous findings.

At the tissues O_2 is extracted from the blood and CO_2 is added, and therefore, the resulting changes in PO_2 and PCO_2 at the capillaries must be considered dependent variables in mathematical models. Two factors are critically important for maximizing the flux of O_2 from the blood to the tissues: (i) Hb–O_2 unloading, expressed as the difference between arterial and venous Hb–O_2 saturation, $S_{a-v}O_2$; and (ii) the P_vO_2 at which O_2 is unloaded, which approximates the PO_2 gradient between the blood and the mitochondria. Willford et al. (1982) quantified the problem mathematically by rearranging the Hill equation and applying a series of derivative tests. The condition of optimality was defined in terms of (i) maximal $S_{a-v}O_2$, as a function of P_aO_2 and P_vO_2:

$$\text{optimum}\,P_{50} = (P_aO_2 \times P_vO_2)^{\frac{1}{2}} \tag{1}$$

or in terms of (ii) maximal P_vO_2, as a function P_aO_2 and $S_{a-v}O_2$:

$$\text{optimum}\,P_{50} = P_aO_2\left(\frac{1 - S_{a-v}O_2}{1 + S_{a-v}O_2}\right)^{\frac{1}{n_H}} \tag{2}$$

Both equations will return the same optimal P_{50} value, whether P_{50} is defined in terms of maximal $S_{a-v}O_2$ or P_vO_2. Using Eq. (1) and the previously calculated P_vO_2 for trout, carp, and mouse, one can calculate theoretical optimal P_{50} values for the previous normoxic and hypoxic scenarios (Table 1).

It is remarkable that in normoxia, the calculated optimum P_{50} in all three species is higher than that measured in the animal, which may indicate that the selective pressures that determine P_{50} in these species do not include normoxic O_2 transport at rest. The difference between optimal and measured P_{50} is largest in the hypoxia-tolerant carp and smallest in the mouse. During hypoxia, when P_aO_2 falls to 40 mm Hg, the optimal P_{50} decreases in carp and trout, and we cannot calculate a value for the mouse since P_vO_2 is already zero. In carp, the theoretical optimal P_{50} is still higher than the measured P_{50}, indicating that there is some capacity to tolerate a further decrease in PO_2. In the hypoxic rainbow trout on the other hand, the theoretical optimal P_{50} closely matches the measured P_{50}. Therefore, $S_{a-v}O_2$ is close to maximal and O_2 loading and unloading occurs over the steep part of the OEC where β_b is large, a

Table 1

Measured blood P_{50} values (in mm Hg) for carp (*Cyprinus carpio*; Brauner et al., 2001), rainbow trout (*Oncorhynchus mykiss*; Tetens and Lykkeboe, 1981), and mouse (*Mus musculus*; Newton and Peters, 1983) during normoxia (arterial partial pressure of O_2, $P_aO_2=100$ mm Hg) and hypoxia ($P_aO_2=40$ mm Hg); and theoretical optimal P_{50} values calculated according to Willford et al. (1982) and using the parameters provided in the latter studies (see text for further details)

		Theoretical optimal P_{50}	
	Measured P_{50}[1]	Normoxia	Hypoxia
Carp	3.8	19	12
Trout	24.1	47	23
Mouse	50.0	63	—

[1]P_{50} is the PO_2 at which hemoglobin is 50% saturated.

situation that alleviates respiratory and cardiovascular adjustments to hypoxia (Gamperl and Driedzic, 2009; Holeton and Randall, 1967; Perry et al., 2009; Randall, 1982a, 1990; Smith and Jones, 1982).

The way in which P_{50} alters the balance between O_2 loading and unloading can be visualized by plotting S_aO_2 and S_vO_2 as a function of P_{50}, in this case using the OEC for rainbow trout (Fig. 2A). The difference between S_aO_2 and S_vO_2 ($S_{a-v}O_2$) represents the Hb–O_2 unloading at the tissue, which by definition is maximal at the optimal P_{50}. The optimal P_{50} is strongly influenced by the shape of the OEC and the range over which O_2 loading and unloading occur. A more pronounced sigmoidal shape (larger n_H) will lead to a more narrow window over which P_{50} is optimal, compared to an OEC with a more parabolic shape (see Kobayashi et al., 1994). During hypoxia, the optimal P_{50} shifts to a lower PO_2 value, and the maximal $S_{a-v}O_2$ is reduced due to the lower S_aO_2 (Fig. 2B). Therefore, the fraction of the total O_2 carrying capacity that can be realized during hypoxia is lower (Wang and Malte, 2011).

Using Eq. (2), P_vO_2 can be expressed as a function of P_aO_2 by keeping $S_{a-v}O_2=50\%$ (Fig. 3), and it becomes evident how during normoxia high P_{50} values will sustain a higher P_vO_2 and favor O_2 diffusion to the tissues. However, the benefit of a high P_{50} decreases with P_aO_2, and during hypoxia a lower P_{50} can sustain a higher P_vO_2. Such is the case in the carp, where P_vO_2 is largely independent of P_aO_2; however, the maximal P_vO_2 that can be realized is low as well. The influence of P_{50} on P_vO_2 becomes even more prominent as Hb–O_2 unloading at the tissues is increased to $S_{a-v}O_2=75\%$ (see dashed lines in Fig. 3). Many tissues, such as the heart in fishes, indeed have exceptionally high extraction efficiencies, which are associated with a high myoglobin content (Bailey et al., 1990; Farrell and Jones, 1992). Extraction efficiency will also increase when $\dot{M}O_2$ is elevated during exercise and animals use the venous reserve.

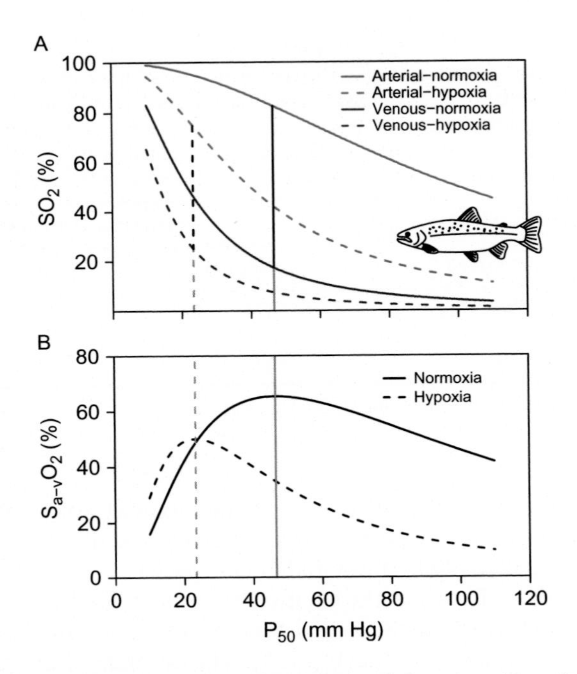

Fig. 2. (A) Hemoglobin (Hb) oxygen saturation (SO_2) in rainbow trout (*Oncorhynchus mykiss*) as a function of Hb P_{50} during normoxia (arterial partial pressure of O_2, $P_aO_2=100$ mm Hg) and hypoxia ($P_aO_2=40$ mm Hg). P_{50} is the PO_2 at which Hb is 50% saturated. Venous PO_2 (P_vO_2) values were as calculated in Fig. 1 (22 mm Hg in normoxia and 14 mm Hg in hypoxia). Hb–O_2 unloading at the tissues is calculated as the difference between S_aO_2 and S_vO_2 ($S_{a-v}O_2$) and represents the vertical distance between the two curves. (B) $S_{a-v}O_2$ as a function of P_{50} during normoxia and hypoxia, at constant P_aO_2 and P_vO_2. $S_{a-v}O_2$ is maximal at the theoretical optimal P_{50}, which is 47 mm Hg in normoxia and 23 mm Hg in hypoxia.

Experimental evidence is generally in line with the previous modeling results. Studies that experimentally altered blood P_{50} in mammals found a beneficial effect of higher Hb P_{50} on tissue O_2 supply during normoxia (Bakker et al., 1976; Harken and Woods, 1976; Moores et al., 1978; Oski et al., 1971; Rice et al., 1975; Woodson et al., 1973; Zaroulis et al., 1979), whereas in hypoxia a lower P_{50} was beneficial (Eaton et al., 1974; Hall et al., 1936; Hebbel et al., 1978; Penney and Thomas, 1975). In general, lower P_{50} values are found in water breathers compared to air breathers, and this is supported by work on amphibians and air-breathing fishes. However, the fish data predominately come from studies of tropical species, such as from the Amazon, where aquatic hypoxia may confound the observed relationship (Johansen et al., 1978; Morris and Bridges, 1994; Powers et al., 1979; Wells et al.,

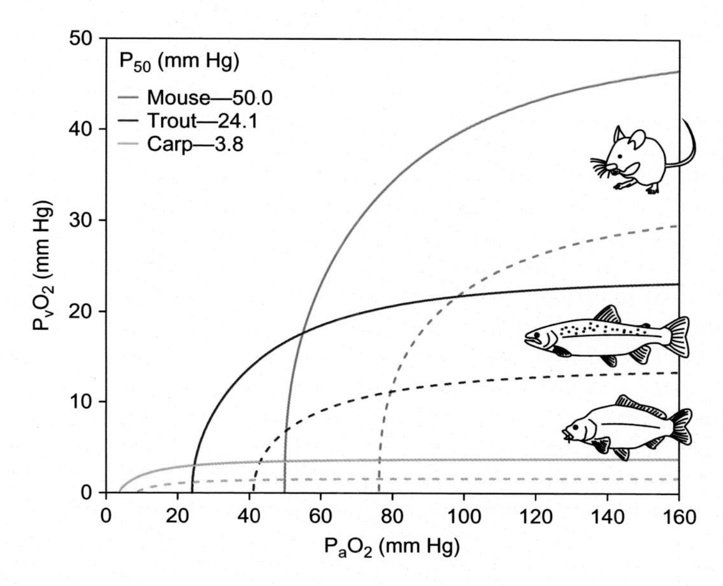

Fig. 3. Venous partial pressure of oxygen (P_vO_2), as a function of arterial PO_2 (P_aO_2), for the blood of mouse (*Mus musculus*, $P_{50}=50$ mm Hg, $n_H=2.6$; Newton and Peters, 1983), rainbow trout (*Oncorhynchus mykiss*, $P_{50}=24.1$ mm Hg, $n_H=2.05$; Tetens and Lykkeboe, 1981), and carp (*Cyprinus carpio*, $P_{50}=3.8$ mm Hg, $n_H=1.3$; Brauner et al., 2001), calculated using the equations described in Willford et al. (1982). P_{50} is the PO_2 at which hemoglobin (Hb) is 50% saturated, and n_H is the Hill coefficient. *Solid lines* are calculated at a constant Hb–O_2 unloading, expressed as the difference between arterial and venous Hb–O_2 saturations, $S_{a-v}O_2=50\%$, and *dashed lines* at $S_{a-v}O_2=75\%$.

2005; Yang et al., 1992). Graham (1997) provides an excellent synthesis on the subject, and his analysis supports the above conclusion that many water breathers have lower P_{50} values when compared to closely related air-breathing fishes. Nonetheless, this relationship is less consistent if applied more broadly, and with increasing phylogenetic distance. Based on limited evidence it has been suggested that species that are likely to encounter hypoxia have lower intrinsic Hb P_{50} values compared to species that inhabit O_2-rich environments, and that athletic species have higher intrinsic Hb P_{50} values as compared to more sluggish species (Jensen, 2004; Mairbäurl and Weber, 2012; Mislan et al., 2015). However, important exceptions prohibit us from making generalized predictions for a species (Herbert et al., 2006; Wells, 2009). Surprisingly, the relationship between *in vivo* blood P_{50} and hypoxia tolerance or maximal $\dot{M}O_2$ in fish has not been tested systematically, within or across fish species; but important mechanistic studies have improved our understanding of these relationships (Mandic et al., 2009; Regan et al., 2016; Speers-Roesch et al., 2012).

2.3. Erythrocyte Function

The intrinsic O_2 binding characteristics of Hb (including P_{50}) are determined by its amino acid sequence, and thus, are genetically coded. On top of this genotypic variation, Hb P_{50} is affected by temperature and the characteristics of the RBC intracellular environment. Some caution should be applied when inferring whole blood characteristics from results on isolated Hb alone, even if tested in buffer solutions that mimic *in vivo* conditions (see Berenbrink, 2006). Fish, like many other ectotherms, typically show Hb multiplicity, where functional heterogeneity between Hb isoforms may allow for adequate O_2 transport over a wide range of fluctuating conditions; and single isoforms may secure O_2 transport in some species or conditions (for reviews see Nikinmaa, 1990; Weber, 1990, 1996, 2000; Wells, 2009). RBCs in fish are nucleated and retain the ability to synthesize proteins including Hb until a late state of maturation (Speckner et al., 1989). The maturation of RBCs takes place in the circulation, and it is possible that *de novo* synthesis would favor Hb isoforms that suit the prevalent environmental conditions. This would allow the animal to change the intrinsic characteristics of circulating Hb (Marinsky et al., 1990; Nikinmaa, 2001), and represents an exciting area for further study.

2.3.1. Allosteric Modulators of Hemoglobin P_{50}

The RBC cytoplasm is isolated from the plasma by a lipid membrane, and the major factors that influence this microenvironment are the energy metabolism of the RBC and the transport of substrates and ions across the membrane (for review see Nikinmaa, 1990). Due to its intracellular location, Hb P_{50} can be effectively modulated by the concentration of a number of allosteric effectors within the RBC. The major allosteric effectors in fish are H^+ and organic phosphates (but also Cl^- and CO_2) that preferentially bind to Hb and stabilize the T-state. Thus, they delay the transition of Hb to the high affinity R-state and maintain high P_{50} values. Conversely, Hb oxygenation and the associated tertiary and quaternary conformational changes, place specific amino acids in different microenvironments, and thus, alter the affinity of binding sites for allosteric effectors (Benesch and Benesch, 1967; Jensen et al., 1998; Kilmartin and Rossi-Bernardi, 1973). This oxygenation-dependent binding and release of allosteric effectors, and the resulting modulation of Hb P_{50}, are a principal mechanism that fish use to fine-tune Hb–O_2 binding characteristics throughout the circulatory system and in response to environmental or metabolic challenges. The sensitivity of Hb–O_2 binding to allosteric effectors is sensitive to single amino acid substitutions at any of the involved binding sites, resulting in large interspecific differences among fishes. The inhibition of allosteric effector binding due to amino acid substitutions will typically decrease Hb P_{50} by shifting the equilibrium in favor of the higher affinity R-state (Storz and Moriyama, 2008).

Organic phosphates are common allosteric effectors in fish and their effect on Hb P_{50} has been reviewed extensively in the work of Val (2000) and others (Boutilier and Ferguson, 1989; Jensen, 2004; Jensen et al., 1993, 1998; Nikinmaa, 1990, 2001, 2003; Nikinmaa and Salama, 1998; Weber, 1982; Weber and Jensen, 1988; Wells, 2009). Briefly, the non-nucleated RBCs of mammals produce 2,3-DPG (2,3-diphosphoglycerate) by glycolysis, which is the primary organic phosphate involved in Hb P_{50} modulation. The nucleated RBCs of lower vertebrates, including fish, have mitochondria and also produce adenosine and guanosine triphosphates (ATP and GTP, collectively nucleotide triphosphates or NTP) by oxidative phosphorylation. Both NTPs bind to several sites on the Hb β-chain, whereas GTP forms an additional bond compared to ATP, which explains its greater effect on P_{50} (Lykkeboe et al., 1975; Weber et al., 1975). However, GTP is not an important modulator of P_{50} in all teleost species, due to its low concentration within RBCs (Gronenborn et al., 1984; Val et al., 1986; Weber, 1996; Weber et al., 1976b). Other organic phosphates, such as inositol pentaphosphate (IP_5), inositol diphosphate (IP_2), uridine nucleotides (UDP and UTP), and 2,3-DPG, have been reported in the RBCs of some fish species where they can exert allosteric effects on Hb. However, the importance of these phosphates as modulators of Hb P_{50} in fishes remains largely unstudied.

2.3.2. H^+ as Allosteric Effectors

Protons are important allosteric effectors of Hb, and the widespread and often significant pH sensitivity of fish Hbs warrants that this aspect be treated in more detail. As with other allosteric effectors discussed here, the binding of H^+ to Hb stabilizes the T-state and increases P_{50}. The resulting right shift of the OEC is termed the (alkaline) Bohr effect (Bohr et al., 1904) and describes H^+ binding to titratable groups on the Hb molecule above pH 6, whereas at lower pH values, many vertebrate Hbs show a reverse, or acid Bohr effect, which decreases P_{50} when additional H^+ are bound (for reviews see Berenbrink, 2006; Bonaventura et al., 2004; Giardina et al., 2004; Jensen, 2004; Jensen et al., 1998; Kilmartin and Rossi-Bernardi, 1973; Riggs, 1988). The physiological significance of the alkaline Bohr effect is apparent: CO_2 released into the blood by metabolically active tissues decreases blood pH at the capillaries, creating an arterial–venous pH difference (ΔpH_{a-v}). This decrease in pH increases Hb P_{50}, promoting the release of O_2 to the tissues. The situation is reversed at the gill, where low CO_2 tensions and high pH decrease P_{50} and promote Hb–O_2 binding. Mathematically, the Bohr effect is expressed as:

$$\Phi = -\frac{\Delta \log P_{50}}{\Delta pH}$$

Based on the principle of linked functions, the oxygenation of Hb decreases its affinity for the binding of allosteric effectors, including H^+, where the reciprocal effect is termed the Haldane effect (Christiansen et al., 1914). The molecular interactions between O_2 and H^+ binding were revealed by X-ray crystallography, largely on human Hb (Perutz, 1970, 1990; Perutz et al., 1960). Conformational changes during the T- to R-state transition, as Hb is oxygenated, alter the molecular environment surrounding titratable Bohr groups, decreasing their pK_a. Consequently, Hb–O_2 binding will lead to the release of H^+ from Hb, and reciprocally, the binding of H^+ to Bohr groups will stabilize the T-state, by introducing additional bonds, thereby delaying the transition to the high affinity R-state. Thermodynamic evidence that Bohr and Haldane effects are indeed mirror images of the same phenomenon was provided by Wyman (1964), who elegantly summarized the linkage between Bohr and Haldane coefficients for a symmetrical OEC and in the absence of additional allosteric effectors, using the equation:

$$-\frac{\Delta \log P_{50}}{\Delta pH} = \frac{1}{4}\Delta Z_H$$

In some teleost Hbs, the binding of H^+ results in negative cooperativity of Hb–O_2 binding (Yokoyama et al., 2004) and reduces Hb–O_2 carrying capacity, termed the Root effect (Root, 1931). This unique characteristic prevents Hb from becoming fully saturated even at 100 atm of pure O_2 (Scholander and Van Dam, 1954), and excellent reviews describe this phenomenon (Brittain, 2005, 2008; Jensen et al., 1998; Pelster and Decker, 2004; Pelster and Randall, 1998; Pelster and Weber, 1991; Root, 1931). In the past, the Root effect has been considered as simply an exaggerated Bohr effect. However, studies on Hb structure suggest different mechanistic and evolutionary origins of these two phenomena (Berenbrink et al., 2005; Brittain, 1987; Perutz and Brunori, 1982). Also, Root effect Hbs do not exhibit an acid Bohr effect, which contributes to the continued increase in P_{50} with decreasing pH.

Due to the high pH sensitivity of teleost Hb, RBC intracellular pH (pH_i) is a powerful modulator of blood P_{50}, and thus β_b, and has a large influence on cardiovascular O_2 transport and delivery in many fishes. RBC pH_i is determined by passive processes, such as diffusion and the Jacobs–Stewart cycle, and by active membrane transport. Importantly, any significant transfer of acid–base equivalents across the RBC membrane will be influenced by the activity of CA.

2.3.3. Physiology of Carbonic Anhydrase

Carbonic anhydrase, abundantly present in all vertebrate RBCs (Maren, 1967), is a zinc metalloenzyme that catalyzes the reversible CO_2 hydration/dehydration reaction (see below) that is central to CO_2 uptake from the tissues into the blood, transport through the circulatory system and excretion from the animal (for general reviews see Chegwidden and Carter, 2000; Chegwidden et al., 2000; Forster and Dodgson, 2000; Gilmour, 2012; Hewett-Emmett, 2000; Maren, 1967):

$$CO_2 + H_2O \overset{i}{\leftrightarrow} H_2CO_3 \overset{ii}{\leftrightarrow} HCO_3^- + H^+$$

The second part of this equation (ii), the ionization of H_2CO_3, is extremely fast (Eigen and DeMaeyer, 1963) and can safely be considered instantaneous under physiological conditions (Forster and Dodgson, 2000). But also term (i) will proceed at an appreciable rate even without a catalyst, albeit in a highly temperature-dependent manner. At 37°C the rate constants of the uncatalyzed hydration and dehydration reactions are 0.18 and 79 s^{-1}, respectively (Itada and Forster, 1977); see also the half-times measured in rainbow trout blood by Heming (1984) and values reported over a broader temperature range by Wang et al. (2009). In the presence of CA, as measured in human RBCs, both reactions are accelerated by up to 17,000-fold at 37°C (Forster and Dodgson, 2000).

The near ubiquity of CA in vertebrate tissues, including fishes, is rooted in its involvement in a wide range of physiological processes. Numerous excellent reviews summarize the importance of CA for CO_2 excretion in fishes (Esbaugh and Tufts, 2006; Evans et al., 2005; Henry and Heming, 1998; Henry and Swenson, 2000; Randall and Val, 1995; Tufts et al., 2003; Tufts and Perry, 1998), acid–base regulation (Gilmour, 2011; Gilmour and Perry, 2009; Haswell et al., 1980; Marshall and Grosell, 2006; Perry and Gilmour, 2006; Perry and Laurent, 1990) and ion-regulation (Evans et al., 2005; Gilmour and Perry, 2009; Haswell et al., 1980; Maetz, 1971; Maetz and Bornancin, 1975; Pelis and Renfro, 2004; Tresguerres et al., 2006). But more generally, the role of CA in physiological processes can be described as facilitated transport. Lipid membranes are relatively impermeable to charged ions, such as HCO_3^- and H^+, while CO_2, a small uncharged molecule, is highly lipid soluble and will rapidly diffuse across plasma membranes (Forster and Steen, 1969). In the presence of CA, $CO_2 - HCO_3^- - H^+$ species will rapidly equilibrate, both within a compartment and across lipid membranes, via the highly diffusible CO_2 (Forster and Dodgson, 2000). Therefore, the rate at which acid–base equivalents passively equilibrate across a membrane is a function of CA activity on either side of the membrane and the membrane permeability for the involved reactive species; in RBCs this process is termed the Jacobs–Stewart cycle (Jacobs and Stewart, 1942).

2.3.4. The Jacobs–Stewart Cycle

The phenomenon was first described in ammonium chloride-treated human RBCs, in which volume changes were faster upon addition of extracellular CA (Jacobs and Stewart, 1942). The individual steps of the cycle are depicted schematically in Fig. 4 (this sequence of events is reversed upon addition of an alkaline load): (1) When an acid load is added to a RBC suspension this creates a $CO_2 - HCO_3^- - H^+$ disequilibrium in the plasma and across the RBC membrane, as charged H^+ cannot easily move through lipid membranes (Forster and Dodgson, 2000). The Cl^-/HCO_3^- exchanger (or anion exchanger; AE) on the RBC membrane that quickly corrects the pH disequilibrium state across the membrane can only respond to changes in Cl^- or HCO_3^- gradients, but not to H^+, and the transporter will initially remain silent. (2) In the plasma, H^+ and HCO_3^- immediately reach equilibrium with H_2CO_3, which will dissociate to form CO_2 and water, notably at the slow, uncatalyzed rate. (3) This CO_2 will rapidly equilibrate across the cell membrane (half-time ~1 ms in human RBCs at 37°C; Forster, 1969; Wagner, 1977; Swenson and Maren, 1978), and within the RBC CO_2 will be hydrated to form HCO_3^- and H^+, a reaction that is catalyzed by the intracellular CA pool. The result is the transfer of H^+ into the RBC, that (4) will be buffered by intracellular buffers, chiefly Hb itself. (5) The formation of HCO_3^- within the RBC will activate the AE, exporting HCO_3^- and taking up Cl^-. The increase in intracellular [Cl^-] causes osmotic swelling, which allowed Jacobs and Stewart to make their initial observations. In the absence of extracellular CA, the transfer of H^+ into the RBC will proceed at the slow rate of uncatalyzed CO_2 formation in the plasma. Importantly, however, if extracellular CA is added the process will be accelerated and the AE will become rate limiting for acid–base equilibration (Jacobs and Stewart, 1942; Motais et al., 1989; Nikinmaa, 1990, 1992; Nikinmaa et al., 1990).

2.3.5. Steady-State Erythrocyte pH

Under steady-state conditions, when secondarily active transporters are silent, the RBC can be considered functionally impermeable to cations (for reviews see Nikinmaa, 1990, 1992). Exchangeable ions, such as HCO_3^- and Cl^-, are passively distributed across the RBC membrane and their concentration is determined by the overall net charge of the impermeable polyions within the cell, mainly Hb and organic phosphates, according to a Donnan-like equilibrium (Heinz, 1981; Hladky and Rink, 1977). Because

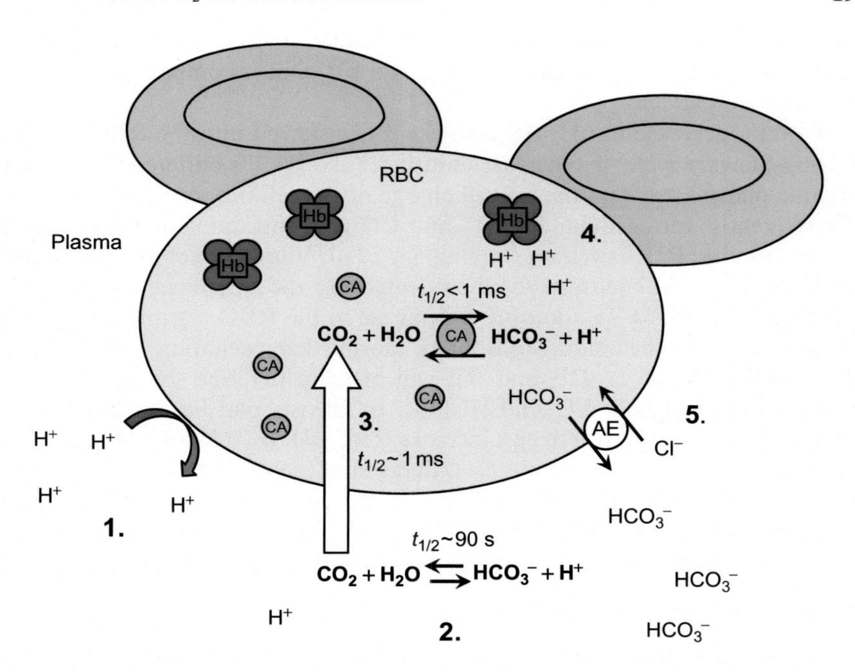

Fig. 4. The equilibration of an extracellular acid load across the red blood cell (RBC) membrane via the Jacobs–Stewart cycle (Jacobs and Stewart, 1942). (1) H^+ cannot easily move through lipid membranes and an extracellular acid load will create a $CO_2 - HCO_3^- - H^+$ disequilibrium in the plasma and across the RBC membrane. (2) H^+ and HCO_3^- react to form CO_2 and water at the slow, uncatalyzed rate ($t_{1/2} \sim 90$ s at pH 8 and 10°C; Heming, 1984). (3) CO_2 rapidly diffuses across the cell membrane ($t_{1/2} \sim 1$ ms in human RBCs at 37°C; Forster, 1969; Wagner, 1977; Swenson and Maren, 1978), and within the RBC carbonic anhydrase (CA) catalyzes its hydration into HCO_3^- and H^+ (in human RBCs CA accelerates CO_2–HCO_3^- reactions by 6500-fold, assuming a hydration rate constant, K_H, of 0.18 s^{-1} at 37°C; Forster and Crandall, 1975). (4) H^+ are buffered by intracellular buffers including hemoglobin (Hb), and (5) HCO_3^- is exported by the anion exchanger (AE) in exchange for Cl^-. In the absence of extracellular CA, the transfer of H^+ into the RBC will proceed at the slow rate of CO_2 formation in the plasma, however, if extracellular CA is added the process will accelerate and the AE will become rate limiting (Jacobs and Stewart, 1942; Motais et al., 1989; Nikinmaa, 1990, 1992; Nikinmaa et al., 1990).

the concentrations of HCO_3^- and H^+ are linked via the Jacobs–Stewart cycle, H^+ will also become passively distributed across the RBC membrane (Funder and Wieth, 1966; Heming et al., 1986; Hladky and Rink, 1977). The net charge of the intracellular polyions is negative under physiological conditions; thus pH_i is lower than extracellular pH ($pH_i < pH_e$) (Heming et al., 1986; Hladky and Rink, 1977; Jensen, 2004). The resulting transmembrane pH gradient is typically 0.2 in human RBCs (Hladky and Rink, 1977) and 0.7 pH units in the tench, a teleost (Jensen and Weber, 1982); this discrepancy is largely driven by a higher pH_e in tench at a pH_i that is comparable to other species.

Upon a decrease in pH_e the acidosis is transferred into the RBC via the Jacobs–Stewart cycle, the increase in intracellular [H^+] is buffered by Hb and organic phosphates, and the overall charge of the cell becomes less negative. Consequently, intracellular HCO_3^- and Cl^- concentrations increase, which causes osmotic RBC swelling (Nikinmaa, 1990; Van Slyke et al., 1923) and an increase in the anion ratio, thus decreasing the difference in pH_i–pH_e (Duhm, 1972, 1976). In addition, a decrease in the RBC organic phosphate concentration, or the binding of Bohr-H^+ during deoxygenation, will decrease the negative charge on Hb, and this will increase pH_i and the intracellular concentrations of both Cl^- and HCO_3^-. In teleosts that have a large Bohr effect, deoxygenation of Hb can increase RBC pH_i by 0.3–0.4 pH units, ten times more than in most mammals (Albers et al., 1983; Brauner et al., 1996; Jensen, 1986, 2004). The relationship between pH_i and pH_e is largely linear in oxygenated blood with a slope of 0.6–0.75, and interspecific differences in the position of this regression have been reviewed by Jensen (2004).

2.3.6. Active Regulation of Erythrocyte pH

The breadth of cellular transport processes in teleost RBCs has been extensively reviewed and will not be recapitulated here (Nikinmaa, 1990, 1992, 1997, 2003; Nikinmaa and Boutilier, 1995). For the present review, the most relevant transport processes are those that create an acid–base disequilibrium across the RBC membrane. Such may be the case for the K^+/Cl^- co-transporter (Jensen, 1990, 1995), a PO_2 sensitive transporter (Borgese et al., 1991; Gibson et al., 2000) involved in RBC volume recovery (Borgese et al., 1987; Cossins and Richardson, 1985; Lauf, 1982). However, whether the K^+/Cl^- cotransporter has a role in modulating Hb P_{50} via changes in pH_i or cell volume is still unclear (for review see Nikinmaa, 2003).

The vast majority of research on RBC pH_i regulation in fishes has focused on Na^+/H^+-exchangers (NHE). NHEs in teleost fish respond to osmotic shrinkage (Romero et al., 1996), intracellular acidification (Weaver et al., 1999) and catecholamines (Nikinmaa, 1982, 1986; Nikinmaa and Huestis, 1984). While NHEs in some species appear to respond to all stimuli, others respond to only one. For example, eel RBC NHEs are only activated by osmotic shrinkage and rainbow trout NHEs by catecholamines (Nikinmaa, 2003; Romero et al., 1996). Perhaps the best studied transporters in this group are β-adrenergically stimulated NHEs (β-NHE).

Upon a reduction in P_aO_2, catecholamines (adrenaline and noradrenaline) are released into the blood stream and bind to a β-adrenergic receptor on the RBC membrane (see Randall and Perry, 1992). An excitation cascade is activated that involves the intracellular accumulation of cyclic adenosine monophosphate (cAMP), which in turn activates the β-NHE (Mahe et al., 1985). The RBC adrenergic receptor of rainbow trout is more sensitive to noradrenaline than adrenaline (Tetens and Lykkeboe, 1988), while the transporter in carp may be responsive to noradrenaline only (Salama and Nikinmaa, 1990). The activity of β-NHE is strongly dependent on pH, with large interspecific variation that may be linked to the onset pH of the Root effect (Borgese et al., 1987; Salama and Nikinmaa, 1988, 1989), which is increased in the presence of NTPs (Jensen et al., 1998; Pelster and Weber, 1990). In rainbow trout, β-NHEs are quiescent at pH 8, even in the presence of catecholamines, but are activated as pH values decrease (see Nikinmaa, 1992). In addition, β-NHE activity in trout (Motais et al., 1987), flounder (Weaver et al., 1999) and carp (Salama and Nikinmaa, 1988) is inhibited at atmospheric O_2 tensions and it has been suggested that Hb may act as a signal transducer (Chetrite and Cassoly, 1985; Motais et al., 1987; Nikinmaa, 1992; Salama and Nikinmaa, 1988). Rainbow trout show a β-NHE response between 4 and 20°C, and activity increases about 10-fold with increasing temperature (Cossins and Kilbey, 1990); similar results were observed for Arctic char (Lecklin and Nikinmaa, 1999). The β-NHE response in rainbow trout and Arctic char appears to be highly seasonal with lower activity in the winter and highest activity in the spring (Cossins and Kilbey, 1989; Lecklin and Nikinmaa, 1999; Nikinmaa and Jensen, 1986).

Upon β-NHE activation, Na^+ moves down its concentration gradient (established by the Na^+/K^+-ATPase, NKA) and H^+ are displaced from equilibrium; the negative charge of intracellular polyions decreases, thus increasing pH_i. CA will rapidly correct this disequilibrium state within the RBC by hydrating CO_2 to HCO_3^- and H^+. As HCO_3^- accumulates within the RBC it is exported into the plasma in exchange for Cl^- and the intracellular accumulation of Na^+ and Cl^- causes osmotic swelling (Motais and Garcia-Romeu, 1988). The increase in RBC pH_i is related to the slower AE compared to NHE activity. This may seem surprising as the maximal rate of AE (5 mol kg cell dry weight^{-1} min^{-1}, in rainbow trout at 15°C; Romano and Passow, 1984; Nikinmaa and Boutilier, 1995) is ~200 × faster than the maximal initial activity of the β-NHE (30 mmol kg cell dry weight^{-1} min^{-1}, in rainbow trout at 20° C; Nikinmaa et al., 1990). Clearly, the rate of passive H^+ equilibration across the RBC membrane is not limited by the capacity of the AE, but by the uncatalyzed dehydration rate of HCO_3^- in the plasma (Jacobs and Stewart, 1942; Motais et al., 1989). This aspect is of pivotal importance for RBC pH_i control. It allows for a large H^+ extrusion, and thus large changes in pH_i, before HCO_3^- dehydration in the plasma produces a significant amount of CO_2 that will re-acidify the RBC via the Jacobs–Stewart cycle. As extracellular H^+ concentration increases, the uncatalyzed dehydration reaction accelerates, because the H_2CO_3 pool increases, while β-NHE activity slows as extracellular H^+ and intracellular Na^+ accumulate (Garcia-Romeu et al., 1988); the Na^+ gradient is also dependent on NKA activity, which increases during the β-adrenergic response (Bourne and Cossins, 1982; Ferguson and Boutilier, 1989; Palfrey and Greengard, 1981). At this point the apparent H^+ fluxes of the β-NHE and the Jacobs–Stewart cycle become equal, there is no net movement of H^+, and a new dynamic equilibrium is reached. The increase in pH_i will depend on the buffer capacity of Hb and organic phosphates within the RBC, where the low buffer capacity of teleost Hb drastically reduces the amount of H^+ that need to be extruded for a given change in pH_i (Nikinmaa, 1997). An elevated pH_i will decrease Hb P_{50} via the Bohr effect and safeguard O_2 carrying capacity of the blood during an extracellular acidosis (Cossins and Richardson, 1985; Nikinmaa, 1983), and in fact, O_2 transport in some fish species is compromised when the adrenergic response is inhibited during a blood acidosis (Nikinmaa et al., 1984; Perry and Kinkead, 1989; Vermette and Perry, 1988).

In addition, it appears that pH_i changes are influenced by RBC CA activity, which varies considerably between species. Carrie and Gilmour (2012) found that acetazolamide (AZ, a CA inhibitor) decreased the acidification of the extracellular space, presumably because the rate of β-NHE H^+ extrusion was limited by the intracellular production of H^+ from CO_2. Whether this affected RBC pH_i remains unclear. If the extracellular acidification was limited by the rate of intracellular H^+ production, this should have led to a depletion of H^+ within the RBC and an increased pH_i. The results of Nikinmaa et al. (1990) are not in line with these findings, but may have been confounded by the presence of extracellular CA activity from lysed RBCs, as the cells were resuspended in saline. The idea that interspecific differences in RBC CA activity may be related to the magnitude of the β-NHE response is intriguing (Carrie and Gilmour, 2012), and is consistent with previous findings (Dimberg, 1988, 1994; Forster and Steen, 1969; Maren and Swenson, 1980; Nikinmaa et al., 1990). These data indicate that Bohr shifts also depend on the high RBC CA activity to transfer an extracellular acidosis into the RBC.

3. REGULATION OF BLOOD O_2 CAPACITANCE IN VIVO

When tissues are confronted with a mismatch between O_2 demand and delivery, the maintenance of aerobic energy production requires corresponding adjustments to $\dot{M}O_2$ or the O_2 transport system according to the following equation.

$$\dot{M}O_2 = \dot{V}_b \times \beta_b(P_aO_2 - P_vO_2)$$

Increases in $\dot{V}_b$ and decreases in P_vO_2 are common responses to exercise or increases in temperature in fishes, but not during hypoxia where $\dot{Q}$ does not typically increase and P_aO_2 and P_vO_2 decrease proportionally (Fritsche and Nilsson, 1993; Randall and Daxboeck, 1984). As P_aO_2 is largely dictated by the environment, and changes in P_vO_2 are limited by the rate of diffusion into the tissues, increasing β_b is another effective mechanism to enhance O_2 delivery and maintain or even enhance $\dot{M}O_2$. The mechanisms that increase β_b can be classified as either: (i) capacity responses, in which [Hb] increases; or (ii) affinity responses due to changes in Hb P_{50}. However, since all of the variables in the above equation may change, experimental data can be difficult to interpret unless all the variables are measured simultaneously, and this has rarely been the case.

3.1. Hypoxia

During environmental hypoxia a mismatch between O_2 supply and demand potentially arises due to decreasing water PO_2, which decreases P_aO_2. The first response is often behavioral, and consists of either avoidance, aquatic surface respiration (ASR), or air-breathing in fishes that have such capacity. Obligatory water breathers will induce hypoxic-ventilatory, and cardiorespiratory reflexes, a suite of responses targeted at promoting diffusive and convective O_2 transport in the face of decreasing environmental PO_2 (see also Chapter 5, Volume 36B: Stecyk, 2017). If tissue hypoxemia cannot be avoided, the animal is faced with a metabolic challenge to energy (ATP) production, and may switch from oxidative phosphorylation at the mitochondria to substrate-level phosphorylation via glycolysis or phosphocreatine hydrolysis. Many fish species have a substantial capacity for anaerobic ATP production; however, these pathways are less efficient compared to oxidative phosphorylation, and are often limited by substrate availability and the accumulation of metabolic by-products. Therefore, sustained hypoxia may be endured by metabolic depression, a strategy that reduces $\dot{M}O_2$ but is only viable in habitats that allow the animal to disengage from interacting with its surroundings. The spectrum of hypoxic responses in fishes has been reviewed in a recent volume of the Fish Physiology series and other good reviews are available (Chapman and Mckenzie, 2009; Farrell and Richards, 2009; Gallaugher and Farrell, 1998; Nikinmaa, 1990; Perry et al., 2009; Richards, 2009; Val, 2000; Wells, 2009). The focus here is on the hematological responses that increase β_b, and that allow fishes to sustain aerobic energy production during hypoxia.

3.1.1. Capacity Response

Upon acute exposure to hypoxia, fish increase β_b by releasing additional RBC from the spleen (for review see Fänge and Nilsson, 1985; Gallaugher and Farrell, 1998; Nikinmaa, 1990; Nikinmaa and Salama, 1998; Nilsson, 1983; Perry and Wood, 1989; Randall and Perry, 1992). The spleen is one of several erythropoeitic sites in fish, and the one that has the largest store of RBCs sequestered from the circulation as well as immature RBCs. The hypoxia-induced release of catecholamines into the circulatory system of fish (see Randall and Perry, 1992) causes a muscular contraction of the spleen (Perry and Kinkead, 1989; Vermette and Perry, 1988; Yamamoto et al., 1985) that is mediated by α-adrenoreceptors (Fänge and Nilsson, 1985; Nilsson, 1984; Nilsson and Grove, 1974; Pearson et al., 1992). Splenic Hct is >90% in rainbow trout (Wells and Weber, 1990), or >20% of total RBCs (Pearson and Stevens, 1991), and the release of this RBC store can substantially increase β_b in the primary circulation by increasing Hct and [Hb] within 1 h of hypoxic exposure (Soivio et al., 1980; Tetens and Lykkeboe, 1985).

An increase in Hct during short-term hypoxia has been described in trout (Boutilier et al., 1988; Nikinmaa et al., 1980; Pearson and Stevens, 1991; Soivio et al., 1980; Swift and Lloyd, 1974; Tetens and Lykkeboe, 1981, 1985; Tun and Houston, 1986; Wells and Weber, 1990, 1991), but there is large variability in the magnitude of the Hct increase, and some studies (and species) show no response (Aota et al., 1990; Houston et al., 1996; Pan et al., 2017; Perry and Gilmour, 1996; Perry and Reid, 1994). Increased Hct during hypoxia was also observed in killifish (Greaney et al., 1980; Greaney and Powers, 1978), Atlantic cod (Petersen and Gamperl, 2011), sailfin molly (Timmerman and Chapman, 2004), sheephead minnow (Peterson, 1990), golden mullet (Soldatov and Parfenova, 2014), several Amazonian species (Affonso et al., 2002; Val et al., 1992), yellowtail (Yamamoto et al., 1985), and a notothenioid where Hct increased by a staggering 40% (Wells et al., 1989). However, hypoxia-tolerant tench (Jensen and Weber, 1982), plaice (Wood et al., 1975), carp (Lykkeboe and Weber, 1978), goldfish (Murad et al., 1990), and sturgeon (Baker et al., 2005) showed no increase in Hct during hypoxia. Hypoxia-tolerant species typically have low P_{50} (e.g., 4.8, 3.8, and 3.7 mm Hg in tench, Jensen and Weber, 1982; carp, Brauner et al., 2001; and goldfish, Regan et al., 2016, respectively), and their Hb will remain highly saturated unless PO_2 decreases to near anoxic levels (see Fig. 1). In all of the latter studies, water PO_2 during hypoxic exposure was likely well above the species' blood P_{50}, and may have been insufficient to induce hypoxemia and a compensatory hematological response.

Minor changes in Hct occur when plasma is skimmed from the primary circulation into the secondary circulation and interstitial space, at a rate defined by Starling's forces; mainly capillary blood pressure, colloid osmotic pressures, and vessel permeability (Aukland and Reed, 1993). In fishes, additional connections between arterial vessels of the primary circulation can selectively admit RBCs into the secondary circulation (Ishimatsu et al., 1988;

Milligan and Wood, 1986a; Sundin and Nilsson, 1992), which typically has a low Hct (<10%; for review see Gallaugher and Farrell, 1998; Olson, 1996; Perry and Wood, 1989; Steffensen and Lomholt, 1992). During hypoxia, Hct in the secondary circulation of the skin increases, which can be observed as a red blush, and may aid in cutaneous O_2 uptake during hypoxia; this response appears to be mediated by the release of nitric oxide (Jensen et al., 2009; Olson, 1984). Removing RBCs from the primary circulation may result in a decrease in β_b, but this effect is likely to be minor (Randall, 1985) and may be outweighed by the benefits of additional cutaneous O_2 uptake.

Other mechanisms of hemoconcentration include diuretic pathways (Swift and Lloyd, 1974; Vermette and Perry, 1987). Hypoxic rainbow trout show a doubling in urine flow rate that is mediated by an increase in plasma levels of cardiac peptides (Tervonen et al., 2006). While the exact stimulus is still unknown, it is likely that increased stretch of cardiomyocytes during hypoxia (where $\dot{Q}$ is maintained during bradycardia by an increase in stroke volume; Gamperl and Driedzic, 2009; Hughes, 1973) stimulates the release of cardiac natriuretic peptides (Arjamaa et al., 2014; Kokkonen et al., 2000; Tervonen et al., 1998, 2002; also see Chapter 5, Volume 36A: Imbrogno and Cerra, 2017). Cousins and Farrell (1996) and Cousins et al. (1997) have shown that the heart can sustain an increased release of cardiac natriuretic peptides for an appreciable period of time. As the levels of cardiac peptides were found to be different between the ventral and dorsal aorta, it appears that the gill scavenges the protein and plays a role in regulating systemic levels, an interesting area for future research (for a review see Arjamaa and Nikinmaa, 2009).

In rainbow trout exposed to chronic hypoxia, splenic contraction decreases after 24 h and the spleno-somatic index returns to pre-exposure values (Lai et al., 2006). However, erythropoietin levels in the kidney and the spleen were still elevated after 6 days of hypoxic exposure, which may indicate an upregulation of the erythropoietic pathway. In mammals, hypoxia-induced erythropoiesis is mediated by the transcription factor complex hypoxia-inducible factor-1 (HIF-1; Fandrey, 2004; Richards, 2009), which has been described in several hypoxia tolerant fish species (Heise et al., 2006; Law et al., 2006; Rissanen et al., 2006). These results are in line with other studies that report elevated Hct in chronically hypoxic rainbow trout (Tun and Houston, 1986), but exceptions exist (Bushnell et al., 1984; Wells and Weber, 1990). Hct has also been reported to be elevated during chronic hypoxia in Atlantic salmon (Härdig et al., 1978), goldfish (Murad et al., 1990), sailfin molly (Timmerman and Chapman, 2004), eel (Wood and Johansen, 1972), mudfish (Frey et al., 1998), Atlantic cod (Petersen and Gamperl, 2011), and a variety of Lake Nabugabo (Chapman et al., 2002) and Lake Victoria species (Rutjes et al., 2007). In contrast, chronic hypoxia had no effect on Hct in seabass, turbot (Pichavant et al., 2003), red drum (Pan et al., 2017), or steelhead trout (Motyka et al., 2017). Finally, in the notoriously hypoxia tolerant tambaqui, Hct was increased only transiently upon hypoxic exposure and levels normalized during chronic hypoxia (Affonso et al., 2002).

Fishes have a substantial scope to increase β_b by increasing [Hb], both acutely and chronically. The fact that many hypoxia-tolerant species do not increase Hct during hypoxia, or even decrease Hct after an initial peak, may indicate that the severity of the hypoxic challenge was not sufficient in these studies for these species. Alternatively, there is an elevated cost associated with circulating a more viscous blood, and cardiac work may be reduced by storing a substantial portion of the blood's O_2 carrying capacity in the spleen, which can be recruited when it is needed. This appears to be the strategy used by Antarctic fishes, which increase their Hct by 40%. Clearly, an increase in [Hb] is just one of many possible strategies to maintain $\dot{M}O_2$ in the face of hypoxia, and one that is typically paired with behavioral, ventilatory, cardiovascular and metabolic adjustments.

3.1.2. Affinity Response

In most vertebrates, including fish, the first response to unavoidable hypoxia is hyperventilation. Increased water flow over the gills decreases the difference in PO_2 between inspired and expired water, and therefore, increases the mean PO_2 gradient across the gill epithelium and P_aO_2. As a side effect, hyperventilation also lowers P_aCO_2 and causes a respiratory alkalosis, which decreases P_{50} via the Bohr effect and favors extraction of O_2 from the hypoxic water (for reviews see Malte and Weber, 1985; Perry and Gilmour, 2002; Perry and Wood, 1989; Perry et al., 2009; Piiper, 1998; Randall, 1990; Randall and Daxboeck, 1984).

In severe hypoxia, when P_aO_2 approximates P_{50}, many teleosts release catecholamines into the blood, which activates β-NHE on the RBC membrane (Randall and Perry, 1992; Thomas and Perry, 1992). The activation of β-NHE will decrease Hb P_{50}, and represents the most rapid form of active P_{50} modulation. The mode of action is two-fold: the largest effect is due to an elevation of RBC pH_i and a resulting reduction in P_{50} via the Bohr effect; in addition, cell swelling and the dilution of intracellular Hb and NTP will cause dissociation of the allosteric effectors from the Hb, and likewise decrease P_{50} (e.g., Nikinmaa, 1992). Activity of the β-NHE in all fishes is highly O_2-sensitive, and transporter activity increases at lower PO_2 (Bogdanova et al., 2009; Motais et al., 1987; Salama and Nikinmaa, 1988; Weaver et al., 1999). However, notable interspecific differences exist in the O_2 sensitivity of β-NHE. While a response is observed in normoxic rainbow trout, it is completely absent in normoxic carp, tench, and flounder (see Jensen, 2004;

Wells, 2009). In addition, hypoxia acclimation increases the number of adrenergic receptors on the RBC membrane in rainbow trout, potentially by externalizing additional receptors (Marttila and Nikinmaa, 1988; Reid et al., 1991; Reid and Perry, 1991).

During chronic hypoxia in fish, blood P_{50} is reduced by modulating the [NTP] within the RBC, a mechanism that has been reviewed in detail (Boutilier and Ferguson, 1989; Gallaugher and Farrell, 1998; Jensen, 2004; Jensen et al., 1993, 1998; Nikinmaa, 2001, 2002; Nikinmaa and Salama, 1998; Val, 2000; Weber, 1982; Weber and Jensen, 1988; Wells, 2009). In brief, when fish are exposed to hypoxia, [NTP] in the RBC and thus NTP binding to deoxygenated Hb decreases, which decreases P_{50} and enhances O_2 extraction from the hypoxic water (e.g., Frey et al., 1998; Gillen and Riggs, 1977; Greaney and Powers, 1977; Jensen and Weber, 1982; Johansen et al., 1976; Lykkeboe et al., 1975; Lykkeboe and Weber, 1978; Mandic et al., 2009; Peterson and Poluhowich, 1976; Soivio et al., 1980; Soldatov and Parfenova, 2014; Val, 2000; Weber et al., 1975, 1976a, 2000; Wells, 2009; Wells et al., 1989; Wood and Johansen, 1972, 1973); but notable exceptions exist (Pichavant et al., 2003). In addition, a decrease in total RBC [NTP] (but not hydrolysis of NTP) will increase RBC pH_i, due to a decrease in the charge of RBC polyions and a shift in the Donnan equilibrium (see Nikinmaa, 1990, 1992). When both ATP and GTP are present, the latter typically exerts the largest effect on P_{50}. The RBCs of rainbow trout contain almost exclusively ATP (Weber et al., 1976b), while RBCs in carp and eel contain about equal concentrations of ATP and GTP (Weber et al., 1976a; Weber and Lykkeboe, 1978). In these species, while total [NTP] decreases by approximately 50% in hypoxia, the higher sensitivity of Hb to GTP causes P_{50} to decrease more drastically in carp as compared to trout. Thus, hypoxia tolerant species may have an increased capacity to modulate blood P_{50} via [NTP] (Nikinmaa and Salama, 1998; Soivio et al., 1980).

Despite a substantial research effort, the specific mechanism by which [NTP] in fish RBCs is regulated is still unclear (Nikinmaa, 2001, 2002). Since ATP and GTP are produced by oxidative phosphorylation, it has been suggested that a decrease in PO_2 may affect mitochondrial NTP production, and thus, lower their concentration in the RBC (Ferguson and Boutilier, 1988; Ferguson et al., 1989; Greaney and Powers, 1978). However changes in [NTP] are observed at relatively high PO_2 values that are not expected to affect mitochondrial function directly (Soivio et al., 1980). A humoral regulation of RBC [NTP] is also plausible. Although catecholamines only change the ratio of ATP/ADP, instead of lowering total [NTP], cortisol is a potential candidate for modulating this response (Nikinmaa, 2002; Nikinmaa and Salama, 1998). Whether NTP are also transported across the RBC membrane is still unclear, and since [NTP] is often measured in whole blood, these techniques may be insensitive to this type of modulation; cAMP extrusion has been shown in avian RBCs (Heasley and Brunton, 1985; Heasley et al., 1985).

In addition to modulating the affinity of existing Hb isoforms, fish exposed to chronic hypoxia may differentially express isoforms that have a higher intrinsic affinity for O_2 or lower sensitivity to allosteric effectors. Changes in Hb isoform composition in response to hypoxia acclimation have been reported in rainbow trout (Marinsky et al., 1990), several Lake Victoria cichlids (Rutjes et al., 2007) and red drum (Pan et al., 2017), and are typically associated with a reduction in P_{50}. Pan et al. (2017) were able to link the differential expression of Hb isoforms in hypoxia acclimated red drum to a lower P_{50} of hemolysates, and to a higher *in vivo* hypoxia tolerance (by measuring critical O_2 tensions; P_{crit}); whether changes in RBC [NTP] also played a role in modulating Hb P_{50} was not assessed. The latter results clearly indicate plasticity in Hb isoform composition, and thus that Hb multiplicity may be an important element in the many strategies that fish use to respond to a variable environment.

Anemia can lead to a special case of "internal hypoxia" that imposes particular challenges on the animal. As opposed to environmental hypoxia, normoxic anemia does not decrease P_aO_2, but a decrease in β_b that may also lead to tissue hypoxemia. Conceptually, it is beneficial to increase P_{50} during anemia, which will increase P_vO_2 (Brauner and Wang, 1997). In fact, this response appears to be universal across mammals (Dhindsa et al., 1971; Eaton and Brewer, 1968; Woodson et al., 1978), amphibians (Brauner and Wang, 1997), and fish (Lane et al., 1981; Val et al., 1994), and is associated with increases in RBC [NTP]. However, reptilians may be an exception (Wang et al., 1999). Chronic anemia also induces erythropoiesis in order to correct the low β_b of the blood (Cameron and Wohlschlag, 1969; Houston and Murad, 1995; McLeod et al., 1978). But importantly, a large increase in $\dot{Q}$ is typically observed in anemic vertebrates, which is likely the largest contributor to sustaining O_2 transport and preventing tissue hypoxemia (Brauner and Wang, 1997; Wood et al., 1979a; Woodson et al., 1978).

3.2. Exercise

A mismatch between O_2 supply and demand that stems from an increase in $\dot{M}O_2$, as during exercise, is termed functional hypoxia (Farrell and Richards, 2009). The risk of functional hypoxia increases with the capacity to increase $\dot{M}O_2$ during exercise, which is reflected in a strong correlation between [Hb] and swimming performance (Fänge, 1992; Farrell, 1991, 1992; Wells, 2009; Wells and Baldwin, 1990). Also, arterial O_2 transport (the product of C_aO_2 and $\dot{Q}$) correlates well with maximum $\dot{M}O_2$ across species (Gallaugher et al., 2001; Thorarensen, 1994; Thorarensen et al., 1993). As during environmental hypoxia, exercise can induce the release of catecholamines, and an increase in β_b due to splenic RBC release. The conditions that induce catecholamine release in fish (Randall and Perry, 1992), and exercise induced changes in Hct, have been reviewed previously (Gallaugher and Farrell, 1998; Jones and Randall, 1978; Mairbäurl and Weber, 2012; Wells, 2009).

3.2.1. Capacity Response

In brief, sustained aerobic swimming in salmonids causes a graded increase in Hct that is dependent on swimming speed (Gallaugher, 1994; Gallaugher et al., 1992; Jensen, 1987; Nielsen and Lykkeboe, 1992; Thomas et al., 1987; Thorarensen et al., 1993, 1996; Yamamoto et al., 1980). Large changes in Hct have also been observed in aerobically swimming yellowtail (Yamamoto et al., 1980), Atlantic cod (Butler et al., 1989) and an Antarctic notothenioid (Franklin et al., 1993), and after burst exercise in rainbow trout (Milligan and Wood, 1986b, 1987; Pearson and Stevens, 1991; Primmett et al., 1986), starry flounder (Milligan and Wood, 1987), yellowtail (Yamamoto, 1991; Yamamoto et al., 1980), largemouth bass (Farlinger and Beamish, 1978), striped bass (Nikinmaa et al., 1984; Young and Cech, 1994b), several cyprinid species (Yamamoto, 1986), and three notothenioids (Egginton et al., 1991; Qvist et al., 1977).

In fish, the increase in Hct during exercise is typically a combined effect of the splenic release of RBCs (Gallaugher et al., 1992; Kita and Itazawa, 1989), RBC swelling, and a reduction in plasma volume (Yamamoto et al., 1980). However, increased gill ventilation during exercise can lead to osmotic disturbances (see Sardella and Bruner, 2007) such that swimming freshwater fish may gain water and plasma volume (Gallaugher et al., 1992; Wood and Randall, 1973) while seawater fish may become dehydrated (Gallaugher, 1994; Gallaugher et al., 2001). Besides the direct effects of hemo-dilution/concentration on Hct, changes in plasma osmolarity may result in changes in RBC volume (see Gallaugher and Farrell, 1998). Higher Hct due to cell swelling does not represent a significant increase in β_b since [Hb] stays constant, whereas those associated with plasma skimming or dehydration do increase β_b; whether these effects are large enough to affect O_2 transport and exercise performance is still unclear.

Exercise training may also increase β_b through erythropoiesis. In humans, erythropoiesis is stimulated by bouts of activity, and erythropoietin levels are typically elevated in the days following exercise (for reviews see Mairbäurl, 1994; Mairbäurl and Weber, 2012). In salmonids, there is some evidence of a small increase in β_b (via increased Hct) associated with exercise training (Dougan, 1993; Farlinger and Beamish, 1978; Gallaugher, 1994; Gallaugher et al., 2001; Hochachka, 1961; Thorarensen et al., 1993; Zbanyszek and

Smith, 1984). However, in other studies on rainbow trout and striped bass, this effect was absent (Davie et al., 1986; Gallaugher et al., 2001; Woodward and Smith, 1985; Young and Cech, 1993, 1994a). These inconsistencies may be related to different life stages, interspecific differences or methodology (particularly the intensity and duration of exercise training) between studies. Nonetheless, an increased swimming performance after training may not always be related to changes in β_b (Davie et al., 1986) as morphological and physiological adjustments to exercise training in fish are many fold (see Davison, 1997). Exercise-trained fish may have a greater capacity to increase Hct via splenic contraction, and therefore, may only elevate β_b transiently (Jørgensen and Jobling, 1993; Young and Cech, 1993). Perhaps an increased [Hb] is more important for CO_2 transport, pH homeostasis, or for safeguarding O_2 transport to tissues that receive restricted blood flow during exercise, such as the intestine (Thorarensen et al., 1993).

3.2.2. Affinity Response

Strenuous exercise in humans, as during maximal ergometer tests, decreases RBC 2,3-DPG levels, and thus, decreases blood P_{50}. This response appears to be independent of exercise duration, but is a direct consequence of the induced acidosis that inhibits the enzymes involved in glycolytic NTP production. The potential decrease in blood P_{50} is likely of minor physiological importance as it is negligible relative to the increase in P_{50} caused by a metabolic acidosis and the Bohr effect (see Mairbäurl, 1994; Mairbäurl et al., 1986).

The effect of exercise on blood P_{50} in fish has been covered in previous reviews (Nikinmaa, 1992; Nikinmaa and Boutilier, 1995; Weber, 2000; Weber and Jensen, 1988; Wood and Perry, 1985). Fish have nucleated RBCs that contain mitochondria, and NTP are largely produced by oxidative pathways (Ferguson and Boutilier, 1988). Changes in RBC [NTP], during exercise, will therefore depend on the balance between the production and consumption of NTP. Burst exercise in rainbow trout causes an increase in plasma catecholamine levels and activation of RBC β-NHE activity (Boutilier et al., 1986; Ferguson et al., 1989; Milligan and Wood, 1987; Nikinmaa, 1992; Primmett et al., 1986). Due to changes in the transmembrane Na^+ gradient, NKA activity increases and so do RBC ATP and O_2 consumption (Bourne and Cossins, 1982; Ferguson and Boutilier, 1989; Palfrey and Greengard, 1981). Under aerobic conditions *in vitro*, RBCs of Atlantic salmon are able to maintain constant [NTP]: [Hb] during adrenergic stimulation; however under anaerobic conditions [NTP]: [Hb] falls by ~20% (Ferguson et al., 1989; Ferguson and Boutilier, 1988). Other studies have observed a decrease in [NTP]: [Hb] after exhaustive exercise in rainbow trout (but not starry flounder; Milligan and

Wood, 1987), Atlantic salmon (Tufts et al., 1991), and several notothenioids (Qvist et al., 1977; Wells, 1978; Wells et al., 1984).

Many other studies on exercised teleosts show no absolute changes in RBC [NTP]: [Hb], but a decrease in [NTP] relative to RBC volume, as cells swell (Jensen et al., 1983; Lykkeboe and Weber, 1978; Nikinmaa, 1981; Nikinmaa et al., 1984; Soivio and Nikinmaa, 1981; Weber et al., 1976b; Wells et al., 2003). Surprisingly, a similar reduction in [NTP] due to cell swelling was observed in exhaustively exercised sandbar sharks that have no β-NHE response. Still, RBC pH_i and blood P_{50} were maintained during a severe acidosis. This indicates that at least some shark species may have mechanisms to defend RBC pH_i (Brill et al., 2008). A dilution of [NTP], due to RBC swelling, will cause the dissociation of NTP from Hb, and therefore, also decrease P_{50} (Nikinmaa, 1992). The observed changes in [NTP] are transient and energy status of the RBCs is typically restored within hours after exercise (Milligan and Wood, 1987). The correlational analysis of Jensen et al. (1990) indicates that [NTP] increases with decreasing blood pH in carp, and thus, that an acidosis may counteract some of the decrease in blood P_{50} observed during exercise or hypoxia, a possible negative feedback loop (Jensen et al., 1990). In addition, a negative correlation between the [NTP] and [Hb] of whole blood may indicate that RBCs released from the spleen have a higher [NTP]: [Hb] compared to circulating RBCs (Weber, 2000; Wells and Weber, 1990). These counteracting effects make it difficult to determine whether exercise leads to a net change in RBC [NTP] and blood P_{50}; overall the effects seem to be small, vary in direction, and are often completely absent.

Exhaustive exercise in salmonids typically leads to a marked acidosis due to CO_2 accumulation and glycolytic energy production in the muscle. In many teleosts (but not all; see Berenbrink et al., 2005), a release of catecholamines and the activation of β-NHE protect pH_i despite a reduction in plasma pH (Boutilier et al., 1986; Milligan and Wood, 1987; Primmett et al., 1986; Van den Thillart et al., 1983). However, there are many examples in which RBC pH_i protection during an acidosis is incomplete, even in species that have a β-NHE; e.g., during normoxia in tench (Jensen, 1987), striped bass (Nikinmaa et al., 1984), and starry flounder (Milligan and Wood, 1987), or when an adrenergic response is absent in salmonids (Ferguson and Boutilier, 1989; Nikinmaa and Jensen, 1986; Tufts et al., 1991). This may be due to seasonal variations in β-NHE activity or during specific life stages, such as spawning migrations. A reduction in RBC pH_i will lead to an increase in blood P_{50} via the Bohr effect (Jensen et al., 1983; Wood and Perry, 1985), which likely outweighs other allosteric effects on Hb (Duhm, 1972; Jensen, 2004; Jensen and Weber, 1982). Whether, the reduction in P_{50} via β-NHE activity, or the increase in P_{50} via the Bohr effect in the absence of pH_i

regulation, will increase β_b during exercise is difficult to assess. These changes would be either beneficial to O_2 loading at the gill or O_2 unloading at the tissue, depending on the conditions, and caution should be applied when assigning adaptive value to either observation (see Wells, 1990).

Exercise training in humans results in elevated levels of 2,3-DPG that essentially follow the induction of erythropoiesis. Mechanical destruction of senescent RBCs and their replacement with new RBC that contain higher 2,3-DPG levels can account for the observed effect (Mairbäurl, 1994). Few studies have focused on the effects of exercise training on blood P_{50} in fish (for reviews see Davison, 1997; Kieffer, 2000). There is presently no evidence to suggest that [NTP] in fish RBCs is altered by exercise training, but based on the observed response in humans this remains an intriguing area of study. Further, the response to adrenergic stimulation was enhanced in rainbow trout chased to exhaustion for seven consecutive days (apparently due to a higher sensitivity of the RBC adrenoreceptors; Perry et al., 1996a), and this may indirectly lead to a lower Hb P_{50}. Ultimately, the benefits of exercise training may be largely related to lessened metabolic and osmotic disturbances and a faster recovery from exercise, while swimming capacity or hematological variables may be unaffected (Gallaugher et al., 2001; Hernandez et al., 2002; Pearson et al., 1990; Young and Cech, 1993, 1994a). For example, a study by Davie et al. (1986) showed that exercise training in rainbow trout did not change β_b, but that muscle vascularization increased markedly.

3.3. Temperature

Temperature has profound effects on the relationship between O_2 supply and demand in fish. The routine metabolic rate of ectotherms increases with temperature due to thermodynamic effects. Therefore, a temperature-induced hypoxemia is best classified as a functional hypoxia, and may occur despite the maintenance of C_aO_2 (Steinhausen et al., 2008). Temperature-induced hypoxemia has received broad attention in recent years, mainly due to the imminent threats faced by many species as a result of climate change. Limitations to increasing tissue O_2 supply with rising temperatures appear to be set by the failure to induce a corresponding increase in $\dot{Q}$; the latter is related to a plateau and then collapse of cardiac function (Clark et al., 2008; Pörtner and Farrell, 2008; Pörtner and Lannig, 2009; see also Chapter 4, Volume 36B: Eliason and Anttila, 2017). However, the efficiency of oxidative phosphorylation in the mitochondria, which has been shown to decrease with temperature, thus limiting aerobic production of ATP, has also been proposed to play a role in determining limits in maximum metabolic rate and thermal tolerance (for reviews see; Guderley, 2004; Pörtner and Lannig, 2009; Pörtner et al., 2005).

3.3.1. Capacity Response

In general, small increases in Hct and [Hb] are observed with increasing acclimation temperature, but with substantial variability as summarized by Houston (1980, 1997) and commented on by Farrell (1997). Most studies have focused on rainbow trout and carp, and typically show some increase in Hct with increasing acclimation temperature (Albers et al., 1983; Anthony, 1961; Barron et al., 1987; Cameron, 1970; DeWilde and Houston, 1967; Houston and Cyr, 1974; Houston and DeWilde, 1968, 1969; Houston and Smeda, 1979; Lecklin et al., 1995; Martinez et al., 1994; Perry and Reid, 1994; Powers, 1974; Smeda and Houston, 1979; Weber et al., 1976b). However, an opposite response has also been reported (Nikinmaa et al., 1981; Taylor et al., 1993).

Indirect evidence indicates that erythropoiesis is induced by increasing temperature: in carp the concentration of young RBCs increased at higher temperatures, albeit without changes in Hct (Houston and Schrapp, 1994); increased temperature stimulated the recovery of Hct in anemic goldfish (Chudzik and Houston, 1983; Murad and Houston, 1992); and in tench, iron uptake into RBCs was increased at higher temperatures (Hevesy et al., 1964; Nikinmaa, 1990). *De novo* synthesis of Hb may also be associated with changes in isoform composition that safeguard O_2 transport at higher temperatures (i.e., by reducing the temperature sensitivity of Hb–O_2 binding). In fact, ample evidence indicates that isoform shifts are a common response to increased temperature in rainbow trout (Houston et al., 1996; Houston and Gingras-Bedard, 1994; Weber et al., 1976b), goldfish (Houston and Cyr, 1974; Houston and Gingras-Bedard, 1994; Houston and Rupert, 1976), and other teleost species (Houston et al., 1976). However, the functional significance of these changes remains untested.

Overall, it appears that only minor capacity-induced changes in β_b occur during exposure of fish to higher temperatures. The large variability in responses may be due to methodological inconsistencies in acclimation conditions, sampling, season, animal source, size, and general condition of the animals. In addition, changes in β_b may be of minor importance relative to ventilatory, cardiovascular or metabolic adjustments to increasing temperature. If O_2 transport at higher temperatures is largely limited by the ability to increase $\dot{Q}$ (e.g., Pörtner and Lannig, 2009), then higher Hct and the resulting higher viscosity of the blood may not always be advantageous. Thus, it appears that capacity responses to temperature-induced hypoxemia may be less important than modulating Hb–O_2 binding characteristics.

The Hct of temperate fish species is subjected to seasonal variations and tends to be lower in winter months (Powers, 1974), and in general, polar species have lower Hct as compared to temperate fishes (Scholander and

Van Dam, 1957; Wells, 1990). The Hct of Antarctic notothenioids species is, in fact, low (Everson and Ralph, 1968; Lecklin et al., 1995; Macdonald et al., 1987), but the reported range overlaps with that of temperate fish species. It has been suggested that a decrease in Hct may serve to lower the viscosity of the blood (Chien, 1987; Egginton, 1996), a parameter that increases with decreasing temperature, and may come at the expense of higher cardiac work. In addition, a reduction in [Hb] per unit of RBC volume increases RBC deformability by decreasing intracellular viscosity at low temperatures (for review see, Axelsson, 2005). Consistent with this idea, it appears that decreasing temperature has a larger effect on [Hb] compared to Hct, and results in a lower mean corpuscular [Hb] (Egginton, 1994, 1996; Hureau et al., 1977; Wells et al., 1980); however only a small effect was reported by others (Fletcher and Haedrich, 1987). Lecklin et al. (1995) showed that an acute decrease in temperature increased RBC membrane rigidity in rainbow trout, with a corresponding increase in the resistance of blood flow through narrow capillaries. Rainbow trout that had been acclimated to different temperatures showed the same temperature dependence of membrane rigidity, indicating a lack of plasticity in the rheological response of RBCs to temperature acclimation; but Hct and [Hb] decreased at lower temperature with no evident effects on mean corpuscular [Hb]. Further, the temperature dependence of membrane rigidity was similar between three species from vastly different thermal environments (rainbow trout, *T. mossambicus*, and *N. coriiceps*), perhaps indicating that temperature adaptation may not act on RBC rheology in these species (Lecklin et al., 1995).

At the lower extreme of an apparent continuum of decreasing [Hb] with reduced temperature are the icefishes of the family Channichthyidae (Notothenioidei), the only known vertebrates to lack Hb as adults (Ruud, 1954). These animals inhabit the southern ocean surrounding Antarctica, at temperatures of approximately −1.9°C and where the well-mixed water column ensures high O_2 tensions, creating the coldest and most thermally stable marine environment on earth (Sidell and O'Brien, 2006). As they lack Hb, O_2 in icefishes is transported entirely as physically dissolved O_2 in the plasma, and thus, the O_2 carrying capacity of their blood is <10% that of species with Hb (Holeton, 1970). An adequate O_2 supply to the tissues is facilitated by the higher solubility of O_2 in plasma at low temperatures and a low absolute metabolic rate (Hemmingsen, 1991). However, major cardiovascular adjustments are also necessary to compensate for the lack of Hb and the low O_2 carrying capacity of icefish blood. Compared to red-blooded notothenioids, icefish have a five-fold higher blood volume and correspondingly higher $\dot{Q}$ that is generated by enlarged hearts (see Chapter 4, Volume 36A: Farrell and Smith, 2017). Despite a decreased vascular resistance, due to larger capillaries, cardiac work in icefishes is higher than that in red-blooded species, and thus, it has been suggested that the loss of Hb was not adaptive, but was unimpeded by the cold-stable environment (Sidell and O'Brien, 2006; Wells, 1990). This refutes earlier ideas that the lower blood viscosity in the absence of RBCs may have represented an energetic advantage in terms of cardiac work (Cocca et al., 1997; di Prisco et al., 1991b; Zhao et al., 1998). In icefishes, the lack of Hb is the result of a single deletion event of the β-chain gene in the common ancestor of all extant species of the clade (Near et al., 2006). In contrast, the decrease in [Hb] in cold water red-blooded fishes is the result of changes in gene expression. Therefore, the complete lack of Hb in icefishes should be viewed as a separate case, and the selective pressures that brought about the temperature-dependent decrease in Hb expression in red-blooded species may not apply to icefishes (Sidell and O'Brien, 2006).

3.3.2. Affinity Response

Because Hb–O_2 binding at the gill is an exothermic reaction, higher water temperatures typically increase P_{50}, which could compromise O_2 uptake. However, the dissociation of allosteric effectors such as H^+, Cl^- or NTP upon Hb oxygenation are endothermic reactions, and this reduces the average heat of oxygenation, and therefore, the temperature sensitivity of Hb–O_2 binding (Jensen and Weber, 1987; Jensen et al., 1998). Accordingly, increased binding of allosteric effectors to Hb may reduce or even reverse the temperature effect of O_2 binding. Temperature insensitive Hbs are often found in species that experience large fluctuations in temperature, e.g., during foraging dives or vertical migrations through the water column. The lack of a temperature effect on Hb–O_2 binding may safeguard O_2 uptake at the gill over a wide range of temperatures (e.g., Brix et al., 1999; Fago et al., 1997; Hopkins and Cech, 1994; Lowe et al., 2000; Weber and Wells, 1989; Wells, 2009). A reversed temperature effect on Hb–O_2 binding is often associated with the presence of heat exchanging retes in regionally heterothermic fishes (Andersen et al., 1973; Carey and Gibson, 1983; Ikeda-Saito et al., 1983; Lilly et al., 2015; Skomal and Bernal, 2010), but exceptions exist (Clark et al., 2010). The functional significance of a reversed temperature effect on Hb–O_2 binding is still debated (for review see Morrison et al., 2015; Powers, 1980), but may involve a role in heat conservation at the gills that perhaps enabled the evolution of regional heterothermy in fishes (Weber and Fago, 2004; Weber et al., 2010).

As during environmental hypoxia, allosteric effectors can modulate blood P_{50} during temperature-induced functional hypoxia (for reviews see Jensen et al., 1993; Wells, 1999, 2009). During a normoxic increase in temperature, it can be beneficial to increase blood P_{50} by increasing RBC [NTP], and this has been observed in fish (Andersen et al., 1985; Dobson and Baldwin, 1982;

Frey et al., 1998; Laursen et al., 1985; Nikinmaa et al., 1980; Sollid et al., 2005; Tetens et al., 1984). However, when O_2 uptake becomes limiting at higher temperatures, and/or lower water PO_2, a decrease in P_{50} may be beneficial to safeguard O_2 uptake. This was observed in rainbow trout acclimated to temperatures of 18–24°C (Houston and Koss, 1984; Nikinmaa et al., 1980). However, other results indicate that in rainbow trout acclimated to 5, 10, or 15°C there was no effect on Hb P_{50} or on whole blood [ATP] (Weber et al., 1976b). This is in agreement with the data of Cameron (1971) in rainbow trout acclimated to 10, 15, or 20°C. A study by Laursen et al. (1985) reported that eel increased [NTP] at 17°C as compared to 2°C, but that a sharp decrease in [GTP] was observed when acclimating animals to 29°C. Therefore, blood P_{50} increased from 2 to 17°C ($Q_{10}=2.8$), but was largely unaffected by an increase in temperature from 17 to 29°C ($Q_{10}=1.2$). Similarly, in several catostomid species acclimated to temperatures ranging from 2 to 30°C, Hb P_{50} decreased due to a decrease in RBC [ATP] at the higher temperatures, with perhaps a confounding effect of decreased O_2 saturation in the water (Powers, 1974). Similar results were obtained in a subsequent study on killifish (Powers and Powers, 1975), where the decrease in Hb P_{50} was mediated by reduced RBC [ATP]. These effects persisted in a separate experiment, where the O_2 saturation of the water was controlled (Greaney and Powers, 1977). Bluefin tuna and yellowfin tuna acclimated to 17–24°C, likewise, showed a reduction in Hb P_{50} at higher temperatures. However, complex interactions of this effect with PCO_2 and assay temperature warrant a careful interpretation of these results (Lilly et al., 2015). Finally, acclimation of carp to 10 or 20°C resulted in a decrease in RBC [ATP] and P_{50} at higher temperatures (Albers et al., 1983); a finding similar to that observed in several notothenioids acclimated to −2 or >4°C (Qvist et al., 1977; Wells, 1978). Whether an increase or a decrease in blood P_{50} when fish are exposed to elevated temperatures enhances O_2 delivery to the tissues will be influenced by environmental O_2 availability in relation to Hb P_{50}. Thus, different responses in RBC [NTP] between studies may be due to interspecific differences and study conditions, especially acclimation conditions and the O_2 saturation of the water.

Less work has focused on the effects of cold acclimation on Hb P_{50}, but the data of Weber et al. (1976b) in rainbow trout acclimated to 5°C for over 4 months showed no effect on Hb P_{50} or RBC ATP concentration in whole blood. However, an acclimation effect on Hb–O_2 binding characteristics was detected in hemolysates, perhaps indicative of changes in the Hb isoform composition of the blood. Likewise, the data of Black et al. (1966) showed no differences in P_{50} between summer and winter brook trout (*Salvelinus fontinalis*). Whether salmonids are the best model to study cold acclimation in fish is debatable, and perhaps other, more eurythermal species, would show a stronger response of Hb P_{50} to cold acclimation. A recent study on killifish

(*Fundulus heteroclitus*) showed a significant effect of acclimation temperature on Hb P_{50} and a trend indicated a lower affinity at 5°C compared to higher temperatures. A significant interaction effect indicated that killifish from different subpopulations, stemming from different thermal habitats, responded differently to thermal acclimation (Chung et al., unpublished data). Whether changes in RBC [NTP] are part of the thermal response of Hb P_{50} in killifish remains to be tested. While Hb multiplicity is prevalent in temperate and tropical fish species, it appears to be less common in cold-adapted species. The well-studied notothenioid Hbs show low multiplicity, which has been linked to the stable Antarctic habitat (di Prisco, 1998; di Prisco et al., 1991a,b; Kunzmann, 1991; Wells et al., 1980). In species with a broader geographic distribution, habitat temperature correlates with polymorphism of some Hb isoforms, such as in turbot (Imsland et al., 2000) and Atlantic cod (Fyhn et al., 1994). However, large variability in the patterns of Hb multiplicity between species prohibits generalized conclusions (for review see Wells, 2005).

4. INTERACTION BETWEEN O_2 AND CO_2 TRANSPORT

While recruiting the venous reserve during either exercise (Brauner et al., 2000a; Kiceniuk and Jones, 1977; Stevens and Randall, 1967) or hypoxia (Holeton and Randall, 1967; Steffensen and Farrell, 1998) increases β_b, P_vO_2 must decrease if blood P_{50} is unchanged. The amount by which β_b can be increased without compromising O_2 diffusion to the tissues may be limited by the fish's critical minimum P_vO_2, which is greatly influenced by physical factors such as tissue capillary density and the location of the mitochondria within the cell relative to the capillary. However, this constraint can be released if the OEC is shifted to a higher P_{50} during capillary transit. Therefore, a Bohr effect that maintains a lower blood P_{50} at the gill and a higher P_{50} at the tissues will increase β_b and enhance Hb–O_2 unloading, and thus may support the same tissue $\dot{M}O_2$ at a lower [Hb] or $\dot{V}_b$, and importantly, without changing P_vO_2.

4.1. The Potential to Enhance Blood O_2 Capacitance via the Bohr-Effect

The magnitude of a Bohr shift is a function of the Bohr coefficient of Hb (Φ) and ΔpH_{a-v} during capillary transit. Based on these parameters the change in P_{50} that will result from a Bohr shift can be calculated as:

$$P_{v50} = P_{a50} \times \left(10^{(\Phi \times \Delta pH_{a-v})}\right)$$

where P_{v50} is the venous P_{50} after the Bohr shift and P_{a50} is the original arterial P_{50}. In rainbow trout with a $P_{a50} = 24.1$ mm Hg (Tetens and Lykkeboe, 1981), $\Phi = -0.91$ (Rummer and Brauner, 2015), and a $\Delta pH_{a-v} = -0.2$ (Kiceniuk and Jones, 1977), P_{v50} becomes 37 mm Hg and the OEC is right-shifted at the tissues compared to at the gill (Fig. 5).

When quantifying the effect of a Bohr shift at the tissues, one is faced with the uncertainty of the respective contributions of changes in PO_2 and SO_2. If the capillaries behave as a "closed" system, SO_2 is constant and PO_2 increases (Fig. 5, A–V_1), whereas in an "open" system PO_2 is constant and SO_2 decreases (Fig. 5, A–V_2). *In vivo* measurements indicate that neither is the case and that both PO_2 and SO_2 change during capillary transit (e.g., Brauner et al., 2000a; Kiceniuk and Jones, 1977), and thus, that a Bohr shift will result in a higher content of O_2 to be unloaded to the tissues at a higher PO_2. Nonetheless, calculating the contribution of either parameter exceeds the scope of the comprehensive models used here, as it requires insight into the diffusion kinetics of

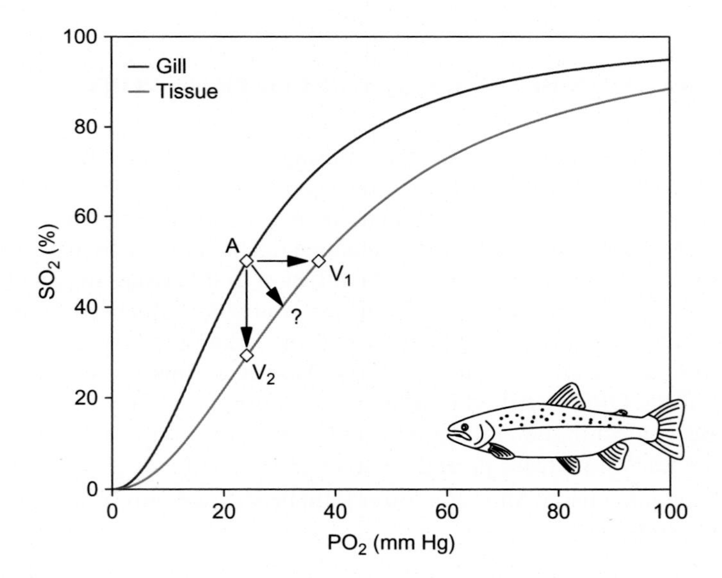

Fig. 5. Oxygen equilibrium curves (OEC) for rainbow trout blood (*Oncorhynchus mykiss*) calculated before (Gill) and after (Tissue) a Bohr shift during capillary transit. This analysis uses the Hill equation (Hill, 1910) and previously reported parameters for rainbow trout (Tetens and Lykkeboe, 1981), and assumes an arterial partial pressure of O_2 (P_aO_2) of 100 mm Hg, a Bohr coefficient (Φ) of -0.91 (Rummer and Brauner, 2015) and an arterial–venous pH difference (ΔpH_{a-v}) at the capillaries, of -0.2 (Kiceniuk and Jones, 1977). *Arrows* indicate the change in PO_2 at constant Hb–O_2 saturation (SO_2) in an "open system" (A–V_1); or in a "closed system" where SO_2 changes but PO_2 is constant (A–V_2). The *in vivo* situation at the capillaries (?) is likely intermediate.

O_2 in the microcirculation (see Egginton and Gaffney, 2010; Goldman, 2008; Secomb et al., 2004 for reviews, and Krogh, 1919 for one of the first attempts at modeling tissue O_2 supply). The small spatial and temporal scales at which these processes occur are still a major challenge for obtaining experimental data on microvascular O_2 transport; however, limited measurements are already possible (Ellis et al., 1990; Goldman, 2008; Golub et al., 1997, 2007; Japee et al., 2005).

Modeling of tissue O_2 supply is greatly simplified by assuming unloading at a constant PO_2 without limitations to O_2 diffusion (i.e., a fully "open" system, Fig. 5, A–V_2), and we will use this approach to illustrate a number of concepts. The change in $S_{a-v}O_2$ (%) can be calculated for both P_{50} values, i.e., before and after a Bohr shift, by rearranging the Hill equation:

$$S_{a-v}O_2(P_{v50}) = \left(\frac{P_aO_2^{n_H}}{(P_aO_2^{n_H} + P_{a50}^{n_H})}\right) - \left(\frac{P_vO_2^{n_H}}{(P_vO_2^{n_H} + P_{v50}^{n_H})}\right)$$

$$S_{a-v}O_2(P_{a50}) = \left(\frac{P_aO_2^{n_H}}{(P_aO_2^{n_H} + P_{a50}^{n_H})}\right) - \left(\frac{P_vO_2^{n_H}}{(P_vO_2^{n_H} + P_{a50}^{n_H})}\right)$$

and:

$$\Delta S_{a-v}O_2 = S_{a-v}O_2(P_{v50}) - S_{a-v}O_2(P_{a50})$$

The difference between $S_{a-v}O_2$ obtained at the two P_{50} values ($\Delta S_{a-v}O_2$) represents the potential for a Bohr shift to increase β_b and enhance Hb–O_2 unloading at the tissues (Fig. 6); assuming constant $\dot{V}_b$ and that $P_aO_2 = 100$ mm Hg, $P_vO_2 = 30$ mm Hg, $P_{a50} = 24.1$, and $P_{v50} = 37$ mm Hg. This model assumes that the change in P_{50}, and thus the Bohr shift, is an independent variable, which is not strictly accurate. The Bohr factor of Hb can be considered independent, while the ΔpH_{a-v} that determines the magnitude of the Bohr shift is clearly dependent on tissue $\dot{M}O_2$ and the respiratory quotient (RQ). Under resting conditions, $\dot{M}O_2$ in normoxic animals is independent of O_2 delivery and only at higher $\dot{M}O_2$ or when C_aO_2 decreases will $\dot{M}O_2$ become a function of O_2 delivery (see Wang and Malte, 2011). Clearly, a change in $\Delta S_{a-v}O_2$ that unloads more O_2 to the tissues must be preceded by a corresponding increase in $\dot{M}O_2$ at the tissues. The simplified model above ignores the physiological regulation of blood gases in response to $\dot{M}O_2$, but is a simplistic means of portraying the effects of a shift in P_{50} on Hb–O_2 unloading at the tissues. Importantly this will not lead to a higher tissue $\dot{M}O_2$, but will allow $\dot{M}O_2$ to be sustained at a lower $\dot{V}_b$ or C_aO_2.

As expected, higher Bohr coefficients and ΔpH_{a-v} values increase the potential for a Bohr shift to enhance O_2 unloading to the tissues, and in teleosts this effect is over 20% under the simulated conditions. Most mammals typically have lower Bohr coefficients compared to teleosts (Berenbrink et al., 2005),

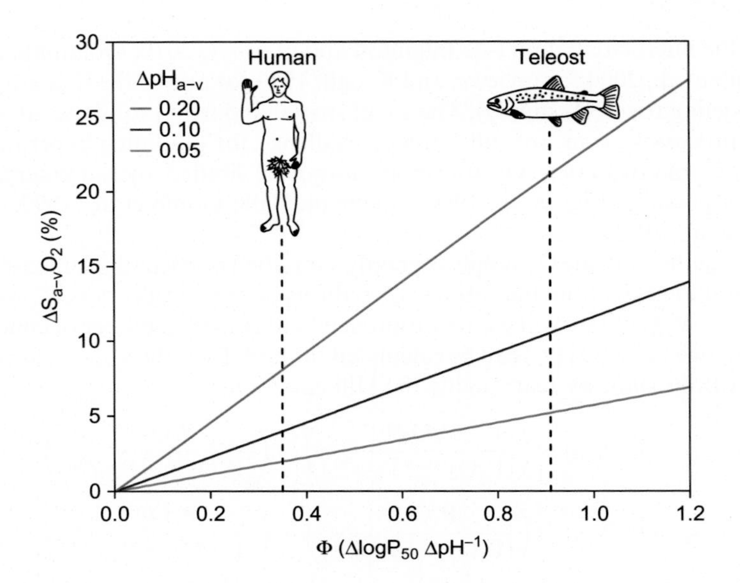

Fig. 6. The potential for a Bohr shift to enhance hemoglobin (Hb) oxygen (O_2) unloading at the tissues, expressed as the increase in arterial–venous Hb–O_2 saturation difference ($\Delta S_{a-v}O_2$) during capillary transit. $\Delta S_{a-v}O_2$ is plotted as a function of the Bohr coefficient (Φ) for three arterial–venous pH differences (ΔpH_{a-v}), and was calculated using the equations described in Willford et al. (1982). The Bohr coefficients for humans ($\Phi = -0.35$) and teleosts ($\Phi = -0.91$) are indicated by *dashed lines* (Rummer and Brauner, 2015) and are depicted as positive values. The maximal $\Delta S_{a-v}O_2$ at $\Delta pH_{a-v} = -0.2$, was 8% in humans and 21% in teleosts.

even though β_b is already higher due to a higher [Hb]. To allow for interspecific comparisons, Bartels (1972) introduced the term "effective Bohr effect," defined as the volume (mL) of O_2 unloaded from 100mL of blood at P_{50} and for a ΔpH_{a-v} of 0.2 units. This takes into account the higher C_aO_2 of human blood of 24mL O_2 100mL blood^{-1} (Dejours, 1981; Torrance et al., 1971) compared to 7.7mL O_2 100mL blood^{-1} (Perry and Gilmour, 1996) in rainbow trout (but values in some fish can exceed 15mL O_2 100mL blood^{-1} as in, e.g., bluefin tuna; Carey and Gibson, 1983). Therefore, effectively, a Bohr shift unloads a similar amount of O_2 in humans and rainbow trout (1.7 and 1.5mL O_2 100mL blood^{-1}, respectively) despite the higher Bohr coefficient of the teleost. However, in addition to a Bohr effect, many teleosts have a Root effect (Berenbrink et al., 2005; Brittain, 2005), which will further increase β_b and the amount of O_2 that is unloaded for a given ΔpH_{a-v}. Rummer and Brauner (2015) calculated that for a ΔpH_{a-v} of 0.2, such as used here, the Root effect Hb system in rainbow trout, may enhance β_b by over 70% relative to that without a Root shift.

The shape of the OEC and the position at which a given Bohr shift occurs will also influence its potential to increase β_b. To visualize this dependency, Fig. 7 plots $\Delta S_{a-v}O_2$ due to a Bohr shift (calculated using previous parameters) over a range of P_vO_2, and the OECs for rainbow trout, before and after the Bohr shift are shown. The maximal $\Delta S_{a-v}O_2$ is observed at the average P_{50} between both OECs, in this example 30 mm Hg. Based on the work of Kobayashi et al. (1994) on asymmetrical human OECs, this does not correspond with the point at which either OEC is steepest (typically below P_{50} and around $SO_2 = 0.38$) or with the point of maximal cooperativity (typically above P_{50}). Instead $\Delta S_{a-v}O_2$ is maximal at the point where both OECs have the same slope, half-way between the two P_{50} values. When moving away from this optimum, the relative change in β_b due to a Bohr shift will be smaller. Therefore, Bohr shifts at the gas exchange organ typically have little effect

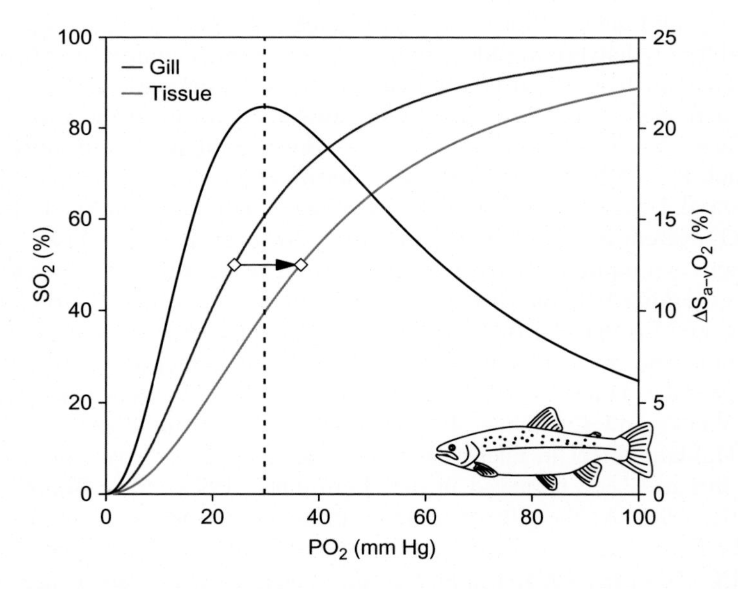

Fig. 7. The potential for a Bohr shift to enhance hemoglobin (Hb) oxygen (O_2) unloading at the tissues, expressed as the increase in the arterial–venous Hb–O_2 saturation difference ($\Delta S_{a-v}O_2$) during capillary transit. $\Delta S_{a-v}O_2$ (*black curve*) is plotted as a function of the partial pressure of O_2 (PO_2), and was calculated using the equations described in Willford et al. (1982). O_2 equilibrium curves (OEC) for rainbow trout (*Oncorhynchus mykiss*) were calculated for arterial (*red curve*) and venous (*blue curve*) blood using the Hill equation (Hill, 1910) and parameters for rainbow trout blood reported in Tetens and Lykkeboe (1981). P_{50} is the PO_2 at which Hb is 50% saturated; for this analysis arterial P_{a50} and venous P_{v50} were 24.1 and 37 mm Hg, respectively (connected by a *solid arrow*). $\Delta S_{a-v}O_2$ is maximal at ~30 mm Hg (*dashed line*), half-way between both P_{50} values.

on β_b, not because a Bohr effect is absent but because SO_2 is high and the slope of the OEC is small (Bartels, 1972). At the tissues, however, Bohr shifts occur closer to P_{50}, where the OEC is steep, thus maximizing the effect on β_b.

4.2. The Optimal Bohr-Coefficient

At the tissues, metabolically produced CO_2 acidifies the blood causing a ΔpH_{a-v}, and a large Bohr effect maximizes the change in P_{50}. On the other hand, a large Haldane effect buffers much of the acid load produced by the tissues as O_2 is unloaded, and thus, limits the ΔpH_{a-v}. The Bohr and Haldane effects are linked functions and numerically identical (Wyman, 1964), and thus favor either O_2 unloading or H^+ buffering, but not both. The analysis of Lapennas (1983) predicts that a Bohr–Haldane coefficient optimal for H^+ buffering and CO_2 transport would be one of similar magnitude to the RQ, as this will achieve perfect pH homeostasis and promote CO_2 hydration at the tissues and dehydration at the gill. Instead, a Bohr–Haldane coefficient optimal for O_2 delivery would be at half RQ, an optimal compromise between a Haldane coefficient that is small enough to allow for some ΔpH_{a-v} and a Bohr coefficient that is large enough to cause a significant Bohr shift.

Teleost fishes typically have large Bohr–Haldane coefficients that approach RQ (Albers et al., 1981; Berenbrink et al., 2005; Nikinmaa, 1983; Weber and Lykkeboe, 1978), and this suggests a function in H^+ buffering and CO_2 excretion rather than O_2 transport. However, Lapennas (1983) made two major assumptions based on the species that he investigated, which should not be generalized: (i) that there is no significant direct effect of CO_2 on Hb–O_2 affinity; and (ii) that the Bohr–Haldane coefficient is independent of Hb–O_2 saturation, and thus, constant over the OEC. The former assumption holds for most fishes (Farmer, 1979; Gillen and Riggs, 1973; Weber and Jensen, 1988; Weber and Lykkeboe, 1978). However, the latter does not, as the Bohr–Haldane effect in teleosts is non-linear over the *in vivo* range of the OEC; this has been reviewed in detail (Brauner, 1995, 1996; Brauner and Randall, 1996). At the gills of rainbow trout, tuna, and tench the majority of Bohr-H^+ are released between 50% and 100% SO_2, in the upper half of the OEC (Brauner, 1995; Brauner et al., 2000a; Jensen, 1986; Lowe et al., 1998). These are the same H^+ that are bound to Hb during deoxygenation at the tissues, and the same H^+ that elicit a Bohr shift (Wyman, 1964).

At rest or during moderate exercise in normoxia, in the upper half of the OEC, all Bohr-H^+ are available to drive gas exchange and the Bohr–Haldane coefficient is equal (or higher) than RQ. Consequently, pH homeostasis is complete and CO_2 excretion is largely supported by an oxygenation-dependent release of Bohr-H^+ at the gill; i.e., O_2 and CO_2 transport are coupled (Brauner, 1995). However, as the animal recruits the venous reserve during exercise, $S_{a-v}O_2$ increases without a corresponding increase in the availability of Bohr-H^+, and the Bohr–Haldane coefficient decreases to values that are close to half of RQ (Brauner and Randall, 1998); values optimal for O_2 transport, and that notably occur when $\dot{M}O_2$ is high (Brauner, 1995; Brauner et al., 2000b, 2001). During exercise, Hb still transports the maximal amount of Bohr-H^+; however, relatively more O_2 is consumed and CO_2 is produced at the tissues. O_2 and CO_2 transport become uncoupled as more acid enters the blood at the tissues. This amount of acid exceeds the buffering capacity of the Haldane effect, and enhances ΔpH_{a-v}, and therefore, the Bohr shift.

During hypoxia P_aO_2 and S_aO_2 decrease (Boutilier et al., 1988), and when P_aO_2 decreases to the P_{50} the uptake and release of Bohr-H^+ will be negligible, as only the lower part of the OEC is used for O_2 transport. An absence of Bohr-H^+ released at the gill causes an increase in RBC pH_i due to HCO_3^- dehydration, which may decrease P_{50} and benefit O_2 uptake during hypoxia (Brauner and Randall, 1996, 1998). At the tissues, however, using the lower part of the OEC may uncouple O_2 transport from H^+ buffering, and a Bohr shift may not be available to enhance O_2 unloading (Brauner and Randall, 1998). Nonetheless, as RBC [NTP] and P_{50} decrease during hypoxia (Jensen and Weber, 1982; Tetens and Lykkeboe, 1981; Weber and Lykkeboe, 1978; Wood and Johansen, 1972) the range of the OEC that is being used is shifted upward. This could make Bohr-H^+ binding sites accessible, although the Bohr coefficient itself in rainbow trout does not appear to be affected by changes in [NTP] (Bushnell et al., 1984). As conditions for aerobic energy production deteriorate, lactate and H^+ produced by substrate level phosphorylation enter the blood at the capillaries (Driedzic and Hochachka, 1978; Richards, 2009; Robergs et al., 2004). This metabolic acidosis overrides the optimal relationship between the Bohr–Haldane effect and RQ predicted by Lapennas (1983), and in teleosts large ΔpH_{a-v} are available to increase β_b, provided that some Bohr effect remains over the working range of the OEC. Whether an increase in RQ during hypoxia, due to changes in metabolic fuel use (switch from carbohydrate to lipids), can be of benefit to Hb–O_2 unloading via the Bohr effect in fish, remains to be studied.

5. A NOVEL MECHANISM OF ENHANCED Hb–O_2 UNLOADING IN TELEOSTS

Changes in blood P_{50} between the arterial and venous circulation can greatly enhance β_b via the Bohr effect, and this mechanism is likely important for enhancing O_2 delivery in most vertebrates. Teleosts, however, as opposed to many other fishes and mammals, have evolved Hb with higher pH sensitivity as reflected in a large Bohr effect and a Root effect, and specialized mechanisms that can generate a large ΔpH_{a-v}; in combination, these two factors induce large changes in blood P_{50} throughout the circulation, greatly increasing β_b, which may support a given $\dot{M}O_2$ at a lower [Hb] or $\dot{V}_b$.

In the teleost lineage, dedicated structures have evolved that create a large ΔpH_{a-v} locally, i.e., the *retia mirabilia* (singular *rete mirabile*). These are dense vascular counter-current exchangers found in the swim bladders and in the eyes of many teleosts species, and have been described in previous reviews (Pelster, 1997, 2004; Pelster and Randall, 1998; Scholander and Van Dam, 1954; Wittenberg and Haedrich, 1974; Wittenberg and Wittenberg, 1974). Briefly, the *rete* in the teleost swim bladder is closely associated with a gas gland consisting of specialized cells that produce large quantities of CO_2 from glucose via the pentose-phosphate shunt and of lactate and H^+ by glycolysis of glucose (see Pelster, 1995, 2004). The CO_2 that diffuses into the blood is rapidly converted into HCO_3^- and H^+ by RBC CA and a membrane-bound PACA isoform (Gervais and Tufts, 1998; Pelster, 1995; Würtz et al., 1999). The large ΔpH_{a-v} that results increases blood P_{50} via the Bohr effect and decreases Hb–O_2 carrying capacity via the Root effect, and this generates a high PO_2 within the *rete* (Kuhn et al., 1963). After this "single concentrating effect" O_2 and CO_2 diffuse from venous vessels into arterial vessels that run in parallel. This magnifies and "recycles" the acidosis, and multiplies PO_2 with every pass through the system. As a result, PO_2 of several hundred atmospheres can be generated, which is adequate to explain the presence of teleosts with air-filled swim bladders at depths of several thousand meters (Pelster, 1997). The *choroid rete* in the eye of most teleosts is, in many respects, analogous to the system in the swim bladder, although it is not nearly as well characterized (Wittenberg and Wittenberg, 1974). Importantly, the *choroid rete* is able to generate PO_2 values well in excess of atmospheric levels (Fairbanks et al., 1969; Wittenberg and Wittenberg, 1962) that drive O_2 across the large diffusion distances and meet the high O_2 demand of the largely avascular retina (Barnett, 1951). All fishes with well-developed *retes* have a Root effect, and this is a key element for generating PO_2 values well above atmospheric; without the Root effect, the increase in PO_2 generated by the single concentrating effect would simply be shunted into the venous system (Berenbrink et al., 2005).

The teleost *retes* are fascinating structures with tremendous capacity to generate ΔpH_{a-v}; however, the increase in β_b they achieve is strictly confined and available to just a few organs. Interestingly, accumulating evidence indicates that teleosts have evolved other mechanisms that can enhance ΔpH_{a-v} and increase β_b via a Bohr shift without the need for acidifying *retes*. A differential distribution of PACA in teleosts, an intrinsic property of the cardiovascular system, may facilitate and eliminate pH disequilibria throughout the circulation, enhance ΔpH_{a-v} locally and increase β_b for all tissues.

5.1. Short-Circuiting pH Disequilibria

5.1.1. Short-Circuiting pH Disequilibria Across the Erythrocyte Membrane

During an acidosis, such as may be experienced during hypoxia or exercise, many teleosts with pH-sensitive Hb protect O_2 transport by activating RBC β-NHE and uncoupling RBC pH_i from pH_e. The key element for displacing H^+ from the equilibrium across the RBC membrane is a Jacobs–Stewart cycle that is limited by the uncatalyzed production of CO_2 in the plasma, and thus, the absence of significant extracellular CA activity. There is accumulating evidence that teleosts, as opposed to all other vertebrates, in fact lack PACA at the respiratory surface (a more thorough discussion of PACA distribution in fishes will follow). An active β-NHE maintains RBC pH_i above equilibrium, and thus, creates a H^+ disequilibrium across the RBC membrane. However, if circulating RBCs encounter a site where PACA is anchored to the blood vessel wall, the H^+ disequilibrium across the RBC membrane will instantly be eliminated. Because the dehydration rate in the plasma is accelerated by PACA, H^+ and HCO_3^- that are extruded from the RBC will form CO_2, at a rate that exceeds H^+ efflux via β-NHE. Under these conditions, H^+ extrusion is futile and pH_i and pH_e once again become coupled via the Jacobs–Stewart cycle, effectively "short-circuiting" β-NHE activity. The result is a rapid transfer of H^+ from the plasma into the RBC, via diffusion of CO_2, which decreases pH_i and increases P_{50}. Short-circuiting RBC pH_i regulation will create a large ΔpH_{a-v} and increase β_b for every tissue that has PACA at its capillaries.

However, for this system to increase β_b, it is important that P_{50}, and thus, RBC pH_i are restored during venous transit and before the RBC reaches the gill; thus conceptually, this also requires the absence of functional PACA in the venous circulation of teleosts (see Fig. 8). Information on the transit times of a single RBC through the circulation in fish is surprisingly scarce. Best estimates are based on measurements of $\dot{Q}$ and the volume of the primary circulation, which yield an average transit time of blood through the entire cardiovascular system of ~2 min (Randall et al., 2014; Rummer and Brauner, 2011). In mammals the volume of the venous system has been estimated to be 70% of total blood volume (Pang, 2001; Rothe, 1993). Accurate data on fishes are still not available (Olson, 1992; Sandblom and Axelsson, 2007), but assuming a similar distribution of blood volume as in mammals, venous transit will be in excess of 1 min. Whether these venous transit times are sufficiently long for β-NHE activity to restore RBC pH_i after short-circuiting is yet to be tested.

5.1.2. Short-Circuiting pH Disequilibria in the Plasma

In the same way as abolishing H^+ disequilibria across the RBC membrane may increase β_b, abolishing pH disequilibria in the plasma may be of similar

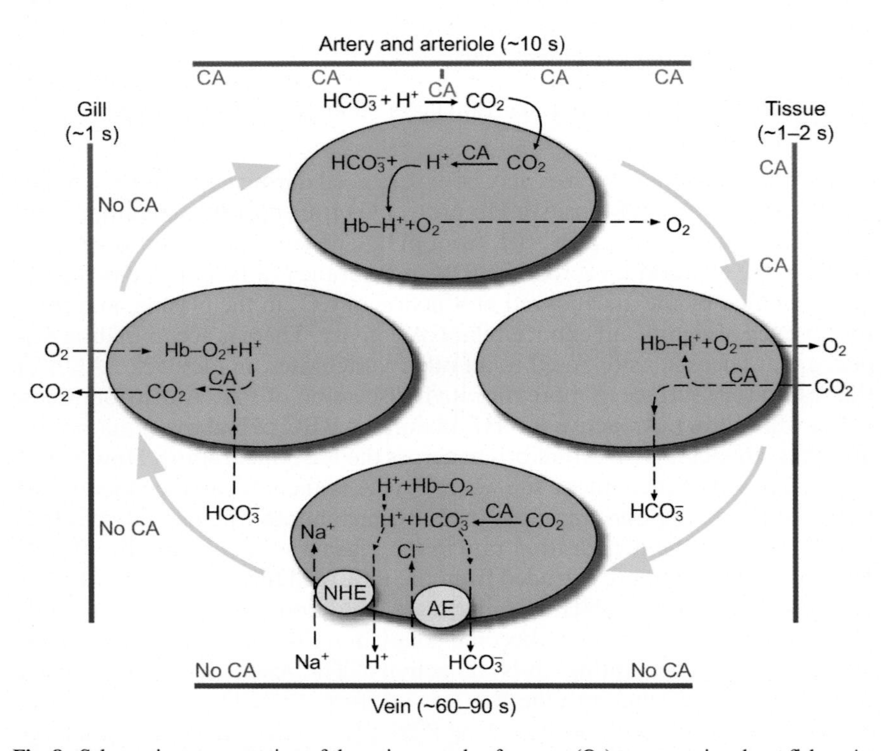

Fig. 8. Schematic representation of the unique mode of oxygen (O_2) transport in teleost fishes. At the gill, carbon dioxide (CO_2) excretion creates a $CO_2 - HCO_3^- - H^+$ disequilibrium in the absence of plasma-accessible carbonic anhydrase (PACA). When blood enters the arterioles and tissue capillaries, disequilibrium states are rapidly abolished in the presence of PACA. This transfers an acid load into the red blood cell (RBC), via the Jacobs–Steward cycle (Jacobs and Stewart, 1942) and enhances hemoglobin O_2 unloading to the tissues via a Bohr shift. Therefore, abolishing pH disequilibria across the RBC membrane or in the plasma may enhance the capacitance of the blood for O_2 (β_b). The absence of PACA in the venous system and the gills allows for the recovery of RBC intracellular pH via sodium–proton exchangers (NHE) on the RBC membrane that excrete H^+ into the plasma. Approximate transit times of blood through parts of the circulation are given in brackets. Reproduced from Randall, D., Rummer, J., Wilson, J., Wang, S. and Brauner, C., 2014. A unique mode of tissue oxygenation and the adaptive radiation of teleost fishes. J. Exp. Biol. 217, 1205–1214. With permission from the Journal of Experimental Biology.

benefit. There is ample evidence for such disequilibrium states at the gill of teleosts (Gilmour et al., 1994, 1997; Gilmour and Perry, 1994, 1996; Henry et al., 1988) and in the venous system (Perry et al., 1997). The various causes for pH disequilibria in blood have been reviewed (Gilmour, 1998a,b) and are always the result of a transfer of acid–base equivalents into or out of the blood, or between the RBC and the plasma (Bidani and Crandall, 1988).

In the absence of PACA, CO_2 excretion at the gill will create an excess of HCO_3^- and H^+ in the post-branchial blood and a pH that is lower than at equilibrium. If PACA is present in the post-branchial arterial system of teleosts, this pH disequilibrium will be quickly abolished, leading to a sudden increase in the partial pressure of CO_2 (PCO_2) and an acidification of the RBC via the Jacobs–Stewart cycle, increasing P_aO_2 (see Fig. 8). Forster and Steen (1969) provide strong evidence that the addition of CA ($0.02\,g\,L^{-1}$) to eel RBCs suspended in a solution that is in pH disequilibrium will quickly transfer the acidosis into the RBC and release O_2 from Hb (ΔPO_2 19 mm Hg, $t_{1/2}\sim 0.5\,s$ at 24°C; see original publication for exact conditions of the pH disequilibrium). This mechanism may be another way by which teleosts increase the ΔpH_{a-v} experienced by Hb, and since it does not require β-NHE activity, it may be operational even under routine conditions (Randall et al., 2014).

Creating and selectively abolishing pH disequilibria may be a fundamental aspect of the teleost O_2 transport system, and thus, relevant for nearly half of all vertebrates. This mechanism may not only increase β_b in fish that have Root effect Hbs and β-NHE activity, but may also function in animals with just a Bohr effect; but, in both cases, a heterogeneous distribution of PACA throughout the circulation is required. While Bohr effects, Root effects, and β-NHE activity have been extensively studied, the key aspect of this mechanism, the distribution of PACA, has not been thoroughly investigated in teleosts and will be discussed in detail in Section 5.2.

5.1.3. Evidence for RBC β-NHE Short-Circuiting

The first *in vitro* evidence for β-NHE short-circuiting by extracellular CA was provided by Motais et al. (1989). Their insightful study revealed that the regulation of RBC pH_i via β-NHE acidified the plasma and that the slow uncatalyzed rate of CO_2 production was the limiting step for re-equilibration via the Jacobs–Stewart cycle. When RBCs were incubated in the presence of extracellular CA, this limitation was removed and β-NHE stimulation did not decrease plasma pH significantly. However, surprisingly, RBC pH_i still increased in the presence of extracellular CA (Note: precise measurements of pH_i are challenging and analytical limitations may explain these results). In addition, blood was stimulated in a tonometer, where produced CO_2 is quickly washed out with the equilibration gas, instead of acidifying the RBCs. When Nikinmaa et al. (1990) stimulated RBCs that were incubated in a saline containing CA, the acidification of the plasma was small and no change in RBC pH_i was observed after addition of isoproterenol (ISO; a β-agonist), indicating that the β-NHE response was largely short-circuited. The difference in pH_i response between these two studies may be rooted in the six-fold higher concentration of CA used by Nikinmaa et al. (1990) and perhaps the more precise measurement of RBC pH_i from the distribution of radiolabeled DMO (5,5-dimethyl-2,4-oxazolidinedione).

The study of Rummer and Brauner (2011) elegantly avoided the challenges of measuring RBC pH_i, by loading the equilibrated blood into a closed vial and measuring PO_2 instead. In rainbow trout, that have pH-sensitive Hb, closed-system changes in PO_2 can be used as a proxy for changes in RBC pH_i and give direct insight into the effects of short-circuiting on Hb–O_2 binding. When extracellular CA was added to a closed system containing ISO stimulated RBCs, PO_2 increased by up to 25 mm Hg, reflecting a decrease in RBC pH_i. Because this was in a closed system, PCO_2 in the plasma was entirely available to be recycled through the RBC; this surely contributed to the pronounced effects on PO_2, but it is not representative of *in vivo* conditions. The effect of CA on PO_2 was completely abolished in the presence of the NHE inhibitor EIPA (ethyl isopropyl amiloride), thus substantiating the involvement of NHE activity in generating an H^+ disequilibrium. Interestingly, the addition of CA also caused an increase in PO_2 (up to 6 mm Hg) in the absence of β-NHE stimulation. It was concluded that another "house-keeping" NHE isoform may generate the H^+ disequilibrium that is short-circuited by CA. Evidence for an involvement of non-adrenergic NHE isoforms in hypoxic and hypercapnic RBC volume and pH_i regulation has been reported in sablefish (Rummer et al., 2010). However, the data of Tetens and Lykkeboe (1988), who measured a high sensitivity of rainbow trout RBC adrenoreceptors for noradrenaline, indicate that some stimulation of β-NHE is likely, even at the low catecholamine concentrations typical of resting fish; thus a background level of β-NHE activity should be considered in the results of Rummer and Brauner (2011). Nevertheless, the involvement of other NHE isoforms clearly warrants further studies, and their presence may indicate another potential way that β_b can be increased over a broader range of conditions.

Nikinmaa et al. (1984) measured arterial PO_2 in exercised striped bass that was higher than water and atmospheric PO_2. Given that hematological parameters clearly indicated β-NHE activity, this result may represent the first account of *in vivo* β-NHE short-circuiting by PACA in the post-branchial circulation. However, in the presence of propranolol (a β-adrenoreceptor antagonist) arterial PO_2 was still elevated above water PO_2, which points toward a cause other than β-NHE short-circuiting. If PACA was present in the dorsal aorta, a post-branchial disequilibrium in the plasma would be abolished, rapidly increasing PCO_2 and acidifying the RBC, causing an increase in P_aO_2 (Randall et al., 2014). Whether PACA activity is actually present in the dorsal aorta of teleosts is still unclear. However, the fact that numerous studies on teleosts were able to measure pH disequilibria in the dorsal aorta (Henry et al., 1988) or in the coeliac artery (which lays downstream of the dorsal aorta; Gilmour and Perry, 1994, 1996; Gilmour et al., 1994, 1997) may indicate that there is no, or insufficient, PACA activity in these vessels to fully correct a pH disequilibrium, and thus, to increase arterial PO_2. However, Randall et al. (2014) provide immunohistochemical support for the presence of PACA in the dorsal aortic endothelium of the glass catfish. Perhaps the large diameter of these vessels prohibits an effective exposure of RBC to PACA (i.e., significant effects may only be observed in smaller diameter arterioles and capillaries). Therefore, the situation in the post-branchial arterial system remains unresolved, but is clearly worthy of further investigation.

McKenzie et al. (2004) measured red muscle PO_2 values in rainbow trout that were consistently higher than those reported in mammalian studies, and unlike the mammalian measurements, were maintained (at >40 mm Hg) even during spontaneous struggling, exercise or hypoxia. This was despite large variations in P_aO_2, and is consistent with the idea that teleosts may enhance O_2 supply to the tissues via a Bohr shift. Whether this occurred specifically by the short-circuiting of β-NHE activity or pH disequilibria in the plasma was not tested. The finding that red muscle PO_2 is higher than previously reported values for mixed-venous PO_2 (Farrell and Clutterham, 2003; Stevens and Randall, 1967) is in line with a mechanism of enhanced Hb–O_2 unloading in rainbow trout. However, McKenzie et al. (2004) did not measure blood flow rates, and it is possible that elevated red muscle PO_2 resulted from an increased perfusion of the tissue relative to $\dot{M}O_2$. Mass-specific blood flow rates in rainbow trout may be double those in mammals (Egginton, 1987, 2002; Taylor et al., 1996), and diffusion distances in rainbow trout and striped bass are about 20% shorter as compared to those in rat and hamster *tibialis anterior* (Egginton, 2002), conditions that favor a higher muscle PO_2 in the first place. In rainbow trout, the P_{50} of myoglobin, an intracellular O_2 carrier molecule, is high compared to that of mammals, and this may have resulted in the higher red muscle PO_2 (Helbo and Fago, 2011). The role that myoglobin concentration, and its P_{50}, plays in supporting the higher tissue PO_2 observed in rainbow trout is still unclear (for reviews see Helbo et al., 2013; Wittenberg and Wittenberg, 2003).

The first study to specifically address PACA short-circuiting *in vivo* measured red muscle PO_2 in tubocurarine paralyzed rainbow trout that were force ventilated (Rummer et al., 2013). Exposure to hypercarbia (1.5% CO_2) caused a reduction in plasma pH and RBC pH_i and increased red muscle PO_2 by 30 mm Hg (65%), and this response was abolished by injection of the membrane impermeable CA inhibitor C18 (Scozzafava et al., 2000). The specificity of this CA inhibitor provides strong evidence for the involvement of PACA in enhancing red muscle PO_2 during periods of acidosis, but whether CA short-circuited a pH disequilibrium across the RBC membrane or in the plasma was not determined. The injection of C18 into normocarbic trout (in normoxia) had no effect on red muscle PO_2, indicating that this system of enhanced

Hb–O_2 unloading may not be operational under routine conditions, in the absence of an acidosis. Catecholamine levels were not elevated above resting levels throughout the entire experiment, but this may not preclude some low β-NHE activity (Tetens and Lykkeboe, 1988), especially in combination with a reduction in blood pH (Guizouarn et al., 1993; Nikinmaa, 1992; Weaver et al., 1999). Again, the involvement of other NHE isoforms is possible, but remains to be confirmed. Conceptually, during more stressful conditions, β-NHE activity may be sufficient to create the pH disequilibrium state that is required for short-circuiting, in which case an acidosis may not be required, a situation clearly worth addressing in future work. The conditions of an acidosis induced by hypercarbia, and the fact that the animals were paralyzed, and thus red muscle $\dot{M}O_2$ was likely low, are noteworthy and raise some concern about the relevance of these results for unrestrained fish. Nevertheless, Rummer et al. (2013) provide strong evidence for a mechanism that is able to enhance Hb–O_2 unloading at the red muscle during an acidosis, without changes in perfusion, and that relies on catalyzing the rate of $CO_2 - HCO_3^- - H^+$ reactions in the plasma.

Cooper et al. (2014) investigated the contribution of a Bohr shift to Hb–O_2 unloading in the intestinal tissue of European flounder. The osmoregulatory strategy of marine teleosts requires the secretion of large amounts of HCO_3^- into the intestinal lumen, a process that is matched by an extrusion of H^+ into the blood (Genz et al., 2008; Grosell and Genz, 2006; Grosell et al., 2005; Wilson, 2003; Wilson et al., 2002). This H^+ load creates a pH disequilibrium in the plasma that, in the presence of PACA, may be rapidly transferred into the RBC, creating a large ΔpH_{a-v} and increasing β_b at the intestine. Using combined data for seawater-acclimated rainbow trout and flounder, the benefit for Hb–O_2 unloading was estimated at over 42%. Intestinal HCO_3^- secretion is metabolically expensive (Taylor and Grosell, 2009) and a Bohr shift may not only be able to meet this high O_2 demand but also provide some autoregulation of β_b based on ΔpH_{a-v}, and thus, HCO_3^- secretion; a fascinating idea that calls for measurements of tissue PO_2 and the localization of PACA in the intestinal capillaries of marine teleosts.

Hypoxia-acclimated Atlantic cod showed a decreased $\dot{Q}$ during rest or exercise, compared to normoxia acclimated fish, however reached similar $\dot{M}O_{2max}$ and U_{crit} values (Petersen and Gamperl, 2010). Given that blood O_2 carrying capacity was only slightly elevated in hypoxia-acclimated fish, and that P_{50} and n_H were unchanged, this indicates that the greater $\dot{M}O_2/\dot{Q}$ in hypoxia-acclimated fish was due to an increase in β_b (Petersen and Gamperl, 2011). Similar results were observed in hypoxia-acclimated steelhead trout that were unable to increase $\dot{Q}$ during a temperature challenge to the same degree as normoxia-acclimated fish, but were able to elevate their

$\dot{M}O_2$ to the same extent; i.e., their aerobic scope did not differ (Motyka et al., 2017). This increase in $\dot{M}O_2/\dot{Q}$ during different aerobic challenges (temperature and exercise) is in line with a system in which RBC β-NHE short-circuiting in the presence of PACA increases Hb–O_2 unloading at the tissues. The result is a larger $S_{a-v}O_2$ at the tissues, and thus, an increase in β_b. Because hypoxia-acclimated fish appear unable to increase $\dot{Q}$ to the same degree as normoxia-acclimated conspecifics, they may rely heavily on increased β_b in order to sustain similar $\dot{M}O_2$. Thus, hypoxia-acclimated fish may be a powerful model in which to study the effects of β-NHE short-circuiting by PACA on O_2 transport in teleosts.

Clearly additional studies are required to increase our, still rudimentary, understanding of the complex kinetics of blood gases and acid–base balance that occur during β-NHE short-circuiting. If β-NHE activity protects Hb P_{50} during an acidosis (Nikinmaa et al., 1984; Perry and Kinkead, 1989; Vermette and Perry, 1988), conceptually, short-circuiting this mechanism by the injection of soluble CA *in vivo* should decrease C_aO_2. However, two studies that injected CA into rainbow trout, exposed to hypoxia (Lessard et al., 1995) or swimming at U_{crit} (Wood and Munger, 1994), found no effect on C_aO_2. Whether CA concentration was not increased sufficiently to short-circuit β-NHE activity or whether the low buffer capacity of teleost plasma limited extracellular CO_2 production that is required for short-circuiting, is still unclear (Desforges et al., 2001; Gilmour and Desforges, 2000; Gilmour et al., 2004). In addition, the time course of pH_e recovery after β-NHE stimulation in tonometers (Carrie and Gilmour, 2012; Motais et al., 1989) has remarkable resemblance with the time course measured *in vivo* (Carrie and Gilmour, 2012; Claireaux et al., 1988; Fievet et al., 1988; Motais et al., 1989; Thomas et al., 1988); pH_e typically recovers within 20–60 min of stimulation, which is slower than expected based on the uncatalyzed reaction rates in the plasma. Tonometers are open systems, and thus, acid–base equivalents are titrated out over this time course and the same may occur at the fish gill *in vivo*. The recovery of pH_e may greatly reduce any pH disequilibrium across the RBC membrane and the benefits of β-NHE short-circuiting may be limited to the initial phase of a β-NHE response in the presence of a plasma acidosis. This idea remains to be thoroughly tested, and whether a complete state of equilibrium is ever achieved *in vivo* is questionable (Gilmour, 1998a).

5.2. The Distribution of Plasma-Accessible Carbonic Anhydrase in the Circulation of Fishes

The model of enhanced teleost Hb–O_2 unloading proposed here requires that PACA is functionally absent at the gills and in the venous system, but present at the tissue capillaries where O_2 unloading occurs and potentially in arterioles where an increase in PO_2 may "prime" the system for subsequent diffusion of O_2 to the mitochondria. Thus, teleosts are predicted to have a heterogeneous distribution of PACA in the circulatory system, whereas other fishes without pH-sensitive Hb or β-NHE activity may possess PACA that is homogenously distributed throughout the circulation; this may be the case in Chondrichthyes, hagfishes, and perhaps basal actinopterygians (see below). Despite the potential implications that the distribution of PACA may have for our understanding of O_2 transport in teleosts, this aspect has not been studied systematically.

Like most other vertebrates, fishes transport the vast majority of CO_2 produced by the tissues as HCO_3^- in the plasma. This mode of CO_2 transport inevitably requires CA activity to catalyze the, otherwise slow, spontaneous formation of CO_2 at the gas exchange organ. The study of CO_2 transport and excretion in vertebrates, is therefore, inevitably intertwined with the localization and characterization of CA pools in the blood and at the gas exchange surface, and numerous reviews have been written on the subject (Esbaugh and Tufts, 2006; Evans et al., 2005; Geers and Gros, 2000; Gilmour, 1998a,b, 2011, 2012; Gilmour and Perry, 2009, 2010; Henry and Heming, 1998; Henry and Swenson, 2000; Maren, 1967; Morrison et al., 2015; Perry, 1986; Randall and Val, 1995; Swenson, 2000; Tufts et al., 2003; Tufts and Perry, 1998).

Today, 16 α-CAs (the vertebrate gene family of CA isozymes) have been identified (Chegwidden and Carter, 2000; Hewett-Emmett, 2000), of which only 13 isoforms possess catalytic activity. Here the focus will be on isoforms that are putatively accessible to the blood plasma, therefore mainly membrane-bound isoforms and to a lesser extent soluble CA isoforms. There are five membrane-bound CA isoforms in mammals (Hilvo et al., 2005; Sly, 2000), three with transmembrane domains (IX, XII, and XIV), and two (IV and XV), which are anchored to the membrane by the common linkage molecule glycophosphatidylinositol (GPI; Cross, 1987; Low et al., 1988). It appears that all membrane-bound CAs have a high catalytic activity, perhaps with the exception of CAXV (see Esbaugh and Tufts, 2006). In the research on PACA distribution in fishes, CAIV has received the most attention, and very little information is available for other membrane-bound isoforms.

The catalytic activities of CA are among the highest enzymatic rates known today (human CAII $k_{cat}=10^6 s^{-1}$), which puts exceptional requirements on methods for the measurement of enzyme kinetics and characteristics (Khalifah and Silverman, 1991). Consequently, a broad spectrum of techniques are used to assess CA distribution in tissues, and these are discussed in the following section with attention to methodological constraints. Taken together, the evidence from different methods paints a picture of PACA distribution in the vasculature of fishes that is somewhat conclusive, at least for a few model species.

5.2.1. Plasma-Accessible Carbonic Anhydrase at the Gill

5.2.1.1. Biochemical Studies. A first piece of evidence for the existence of PACAs in teleosts was provided by studies that measured the CA activity in cellular fractions of homogenized gill tissue, mostly by using the electrometric ΔpH assay (Henry, 1991; Henry and Kormanik, 1985; Maren and Couto, 1979). After a final step of ultracentrifugation (100,000 g), a microsomal pellet is obtained that contains the intra- and extracellular membranes, and thus, putative PACA isoforms. Cytosolic CA isoforms are recovered from the supernatant and typically represent the largest CA pool, as high CA activities are found in the cytosol of the major gill cell types; i.e., pavement-, chloride-, mucous-, and pillar cells (Conley and Mallatt, 1988; Dimberg et al., 1981; Flügel et al., 1991; Lacy, 1983; Rahim et al., 1988; Sender et al., 1999; Wilson et al., 2000). In comparison, microsomal CA activity is typically low, and large intra and interspecific variability is observed across studies. This variability is largely due to the difficulty of clearing the gills of RBCs that contain large amounts of CA, and can contaminate all fractions. Therefore, absolute values of microsomal CA activity should be considered with caution and no clear interspecific trends can be identified based on this metric alone. However, a number of treatments can be applied to the isolated cellular fractions to help identify their CA isoform distribution. (i) Washing the microsomal pellet will transfer entrained cytosolic CA into the supernatant, leaving mostly membrane-associated CA activity in the pellet. (ii) Microsomal pellets can be treated with phosphatidylinositol specific phospholipase C (PI-PLC), an enzyme that cleaves GPI anchors and releases membrane-bound CAIV or CAXV into solution (Bottcher et al., 1994; Zhu and Sly, 1990). (iii) In mammals, membrane bound CAIV is less susceptible to denaturation by sodium dodecyl sulfate (SDS) due to the presence of two stabilizing disulfide bonds (Waheed et al., 1992a, 1996; Whitney and Briggle, 1982). However, some uncertainty exists whether SDS is an equally reliable identifier of CAIV in lower vertebrates (Bottcher et al., 1994; Gervais and Tufts, 1998; Gilmour et al., 2002; Maffia et al., 1996). (iv) The inhibition constant (k_i) for commonly used sulfonamide CA inhibitors, such as AZ, can differentiate between CA isoforms, and the inhibition kinetics are well studied in mammalian CA isoforms (Baird et al., 1997; Maren, 1967; Maren et al., 1993).

The findings of biochemical studies that assessed CA isoform distribution in the gills of 19 fish species are summarized in Table 2. Overall, these results indicate that teleosts may have no, or at least low amounts of, membrane-bound CA activity at the gills, the exception among teleosts being the icefishes. Scarce biochemical data from single studies likewise indicate an absence of PACA at the gills of lamprey and *Amia* (or bowfin, *Amia calva)*. In contrast,

Table 2

Summary of studies that used differential centrifugation and biochemical measurements of carbonic anhydrase (CA) activity in gill homogenates of 19 fish species

Species	Microsomal CA activity (% of total)[a]	N	Sensitive to washing	Sensitive to PI-PLC[b]	Sensitive to SDS[c]	Difference in AZ k_i[d] between microsomal and cytosolic fraction (nM)	Reference
Teleosts							
Channel catfish (*Ictalurus punctatus*)	3.5[e]	4	n				Henry et al. (1988)
Brown bullhead (*Ameiurus nebulosus*)	1.4	4	y	n	n		Gilmour et al. (2002)
Rainbow trout (*Oncorhynchus mykiss*)	3.8	4					Henry et al. (1993)
	0.7	4					Henry et al. (1997a)
		4	y	n			Gilmour et al. (2001)
					y		Stabenau and Heming (2003)
Steelhead trout (*Oncorhynchus mykiss*)					y		Stabenau and Heming (2003)
Brown trout (*Salmo trutta*)					y		Stabenau and Heming (2003)
Smallmouth bass (*Micropterus dolomieu*)					y		Stabenau and Heming (2003)
European flounder (*Platichthys flesus*)	0.19	20					Sender et al. (1999)
European eel (*Anguilla anguilla*)[f]	~10						Maffia et al. (2001)
Bowfin (*Amia calva*)	3	6	y		n		Gervais and Tufts (1998)
Red-blooded notothenioids							
N. coriiceps	1.3	6	y	y	y	y	Tufts et al. (2002)
N. rossii		6	y	y	y		(Harter et al., unpublished data)
T. bernacchii[f]	~10						Maffia et al. (2001)

White-blooded notothenioids							
C. aceratus	2.3	6	y	y	n	n	Tufts et al. (2002)
C. gunnari		6	y	y	n	y	Harter et al. (unpublished data)
C. hamatus[f]	~10						Maffia et al. (2001)
Agnathans							
Sea lamprey (*Petromyzon marinus*)	2.2	4					Henry et al. (1993)
Pacific hagfish (*Eptatretus stouti*)	5	10	n	y	n		Esbaugh et al. (2009)
Chondrichthyes							
Spiny dogfish (*Squalus acanthias*)	2.7	4					Henry et al. (1997a)
		4	n	y		y	Gilmour et al. (2001)
Longnose skate (*Raja rhina*)	5.1	4	n	y	n		Gilmour et al. (2002)
Spotted ratfish (*Hydrolagus colliei*)	3.3	4	y	y	n		Gilmour et al. (2002)

[a]Membrane-bound CA isoforms are typically found in the microsomal fraction (containing the cell membranes), whereas soluble CA isoforms are found in the cytosolic fraction (typically the largest CA pool in fish gills).

[b]PI-PLC, phosphatidylinositol specific phospholipase C, an enzyme that cleaves glycophosphatidylinositol anchors and releases membrane-bound CAIV or CAXV into solution.

[c]SDS, sodium dodecyl sulfate, an anionic surfactant.

[d]AZ k_i, inhibition constant for acetazolamide, a CA inhibitor.

[e]Henry et al. (1988) report that microsomal CA activity is 9% of total gill CA activity when standardized by weight, and 11% when standardized to protein content. However, when recalculated as % of total CA activity (expressed as μmol CO_2 mL^{-1} min^{-1} × fraction volume in mL), we obtain a value of 3.5%; this nomenclature was adopted by the same authors in subsequent publications (Henry et al., 1993) and is comparable to other studies.

[f]The study of Maffia et al. (2001) does not provide the actual data on microsomal CA activity, but mentions that it was ~10% of total gill CA activity in the three species studied.

hagfish and Chondrichthyes consistently show biochemical markers for the presence of a membrane-bound CAIV isoform at the gill.

5.2.1.2. Inhibitor Studies. Perhaps the most conclusive studies in determining the availability of PACA in the vasculature are those that assessed CA activity *in vivo* by measuring pH disequilibrium states or CO_2 excretion rates, and by using specific CA inhibitors. Perry et al. (1982) measured CO_2 excretion in a spontaneously ventilating, perfused, rainbow trout head preparation and in perfused gills of trout and coho salmon. CO_2 excretion was dependent on Hct and in no preparation was there significant CO_2 excretion when perfusing with plasma or saline alone; thus HCO_3^- dehydration was almost entirely due to RBC CA activity. A number of studies on rainbow trout have described an increase in CO_2 excretion after the injection of soluble CA to the circulation (Currie et al., 1995; Desforges et al., 2001, 2002; Julio et al., 2000; Wood and Munger, 1994), and *in vitro* work has identified the AE as the rate limiting step in CO_2 excretion (for review see Perry and Gilmour, 2002). In addition, CO_2 excretion was unaffected by inhibition of the extracellular CA pool (with benzolamide, Bz; for inhibitor limitations see Supuran and Scozzafava, 2004), but decreased when RBC CA was inhibited (with AZ) in channel catfish (Henry et al., 1988) and rainbow trout (Gilmour et al., 1997). Therefore, if HCO_3^- access to RBC CA is the rate-limiting step in branchial CO_2 excretion, then the dehydration of HCO_3^- in the plasma must occur largely at the uncatalyzed rate.

The presence of a post-branchial pH disequilibrium in rainbow trout indicates the absence of sufficient CA activity to ensure equilibrium conditions (Gilmour and Perry, 1994; Gilmour et al., 1994, 1997; Perry et al., 1997). The disequilibrium pH technique warrants some careful considerations when applied to whole blood, as transfer of acid–base equivalents between the two blood compartments may confound results (for review see Gilmour, 1998b). However, the magnitude and direction of the post-branchial disequilibrium observed in rainbow trout, and the response to selective inhibitors, is indicative of negligible extracellular CA activity, even when considering the caveats of the disequilibrium pH technique.

A study on African lungfish (*Protopterus dolloi*) found no effect of BZ on CO_2 excretion rates or acid–base status, even though the gills were identified as the major site of gas exchange for both CO_2 and O_2 (Perry et al., 2005). This result does not exclude the presence of PACA in lungfish gills, but like in other air breathers, PACA does not appear to be a major contributor to CO_2 excretion perhaps due to limited H^+ availability in the poorly buffered plasma (Bidani, 1991; Bidani and Heming, 1991; Cardenas et al., 1998; Henry and Swenson, 2000). More puzzling, however, is the finding that AZ did not significantly reduced CO_2 excretion, which would indicate that CO_2 excretion at the gill or the skin in *P. dolloi* may not depend on the catalyzed dehydration of HCO_3^-. As suggested for *A. gigas*, this situation may be facilitated by the low metabolic rates and high internal PCO_2 in this air-breathing species (Brauner and Val, 1996).

In contrast, a number of studies that selectively inhibited CA pools in the dogfish gill found evidence for the presence of a membrane-bound PACA that contributes to the dehydration of HCO_3^-, and therefore, plays a role in CO_2 excretion (Gilmour and Perry, 2004; Gilmour et al., 1997, 2001; Patel et al., 1997; Swenson et al., 1995b, 1996; Wilson et al., 2000). A possible confounding factor in the latter studies is the presence of soluble CA activity in the plasma of elasmobranchs (Gilmour et al., 1997, 2002; Henry et al., 1997a; Patel et al., 1997; Wood et al., 1994). However, perfusing dogfish gills with saline and excluding the confounding CA activity in real plasma, confirmed the presence of a membrane-bound PACA pool (Wilson et al., 2000).

A negative post-branchial pH disequilibrium state was detected in the gills of dogfish, which was smaller in magnitude and in the opposite direction to the one observed in rainbow trout (Gilmour et al., 1997; Wilson, 1995). Consistent with the previously mentioned constraints of the disequilibrium pH technique, this observation may have been caused by continued HCO_3^- uptake into RBCs after the blood had left the gill, a situation that has been described in mammals (Crandall and Bidani, 1981), or by the presence of anatomical or functional shunts (Gilmour et al., 1997). Wilson et al. (2000) reported a similar disequilibrium in dogfish gills perfused with saline, i.e., without RBCs and transmembrane AE, which may point toward shunting as a cause. Regardless, this result does not exclude the presence of PACA at the gill of dogfish, but indicates that circulating CA is insufficient to establish equilibrium conditions, consistent with the findings of Henry et al. (1997a).

Overall, the data on dogfish provide strong evidence for a membrane bound PACA isoform at the gill and circulating CA activity in the plasma. In contrast, functional studies on rainbow trout, coho and the channel catfish found undetectable levels of PACA activity at the gills. But H^+ availability in the poorly buffered plasma of teleosts needs to be considered as a potential factor limiting extracellular CA activity (Desforges et al., 2001; Gilmour et al., 2004). Therefore, a complete absence of PACA activity in teleosts cannot be concluded from functional studies alone.

5.2.1.3. Imaging and Molecular Studies. Interesting insights are also gained from studies that localized CA distribution in fish gills by means of histochemical-, immunohistochemical staining or *in-situ* hybridization. Rahim et al. (1988) used a combination of immunofluorescence with light microscopy, and immunogold labelling with electron microscopy, on the gills of rainbow trout and carp. In trout, antibodies were raised against a CA isoform purified from gill tissue homogenates and collected from the supernatant after ultracentrifugation. No staining was detected in gill endothelial cells facing the blood plasma. However, the antibody used may have been specific to a cytosolic isoform of CA (as it was recovered from the supernatant and not the microsomal fraction), and unreactive with membrane-bound isoforms. The same applies to the study of Sender et al. (1999) that investigated the cellular CA distribution in flounder gills, by raising an antibody against CA isolated from the supernatant after ultracentrifugation. However, Georgalis et al. (2006) cloned a CAIV isoform in rainbow trout, and detected no CAIV mRNA expression or protein at the gill; further evidence for an absence of PACA at the gill of rainbow trout. A membrane-bound CAXV isoform is expressed in the skin and gills of developing zebrafish (Lin et al., 2008), that may play a role in Na^+ uptake via NHE (Ito et al., 2013; Yan et al., 2007). However, it has only been localized on the apical membrane of ionocytes, facing the water; whether CAXV in fish also has a plasma-accessible orientation remains to be tested. Recent work from our lab on Antarctic notothenioids provides immunohistochemical evidence for the presence of PACA at the gills of icefish. Reactivity to an anti-trout CAIV antibody was absent in the gills of the red-blooded *N. rossii*, but clearly stained the luminal vascular endothelium in the Hb-less *C. gunnari* (Harter et al., unpublished data).

Dogfish showed staining at the gills for a mammalian CAII, but not for a mammalian CAIV antibody (Wilson et al., 2000). However, in western blots of gill membrane preparations, the same CAIV antibody reacted with a band at 48 kDa, indicating that the results in fixed gill tissue may have been confounded by epitope masking (Werner et al., 1996). More conclusive results were obtained by Gilmour et al. (2007) who detected a CAIV isoform associated with the plasma membrane of gill pillar cells, using an antibody raised against trout CAIV. These results are strengthened by homology cloning of the dogfish CAIV isozyme paired with *in situ* hybridization and mRNA expression profiles of gill tissue. Therefore, Gilmour et al. (2007) convincingly demonstrate the presence of a CAIV isoform in the dogfish gill, in a location that is consistent with its plasma-accessible orientation.

In hagfish two high-activity membrane-bound CA isoforms were detected (Esbaugh et al., 2009). Using homology cloning and phylogenetic analysis, these were identified as a CAIV and CAXV isoform. CAIV is expressed primarily at the gill, and is likely the major isoform available to plasma reactions driving branchial CO_2 excretion, although some CAXV expression was detected as well. The cellular orientation of both isoforms was not determined and whether they are plasma-accessible remains to be confirmed.

In line with biochemical and functional data, molecular evidence points toward the presence of a membrane bound CAIV isoform at the gills of dogfish and possibly elasmobranchs in general; the same appears to be the case in hagfish. On the other hand, molecular techniques also indicate that PACA is likely absent at the gills of teleosts, although icefish seem to be an exception. While this combined evidence allows for some general patterns to be recognized, clearly more work on other species is required before any overarching statements can be made.

5.2.2. Plasma-Accessible Carbonic Anhydrase at the Tissue

In mammals, a number of studies provide strong evidence for the presence of PACA activity in the capillaries of skeletal muscle (Bruns et al., 1986; Dermietzel et al., 1985; Effros and Weissman, 1979; Geers and Gros, 1984; Geers et al., 1985; Gros and Dodgson, 1988; Hemptinne et al., 1987; Lönnerholm, 1980; Ridderstrale, 1979; Riley et al., 1982; Waheed et al., 1992b; Wetzel and Gros, 1990, 1998; Wetzel et al., 1990; Zborowska-Sluis et al., 1974), the brain (Ghandour et al., 1992), and the heart (Decker et al., 1996; Knüppel-Ruppert et al., 2000; Sender et al., 1994, 1998; Vandenberg et al., 1996), which is located on the luminal surface of the capillary endothelium (Decker et al., 1996; Sender et al., 1994).

While much less work has been done on fish tissues, it appears that white muscle in rainbow trout has a CA distribution similar to that in mammals, including a membrane-bound PACAIV isoform (Henry and Heming, 1998; Henry et al., 1997b; Sanyal et al., 1982, 1984; Siffert and Gros, 1982; Wang et al., 1998). The physiological functions of membrane-bound extracellular CA at the tissues are diverse, and a role has been shown in cellular CO_2 (Geers and Gros, 1988), ammonia (Henry et al., 1997b; Wang et al., 1998) and lactate excretion (Wetzel et al., 2001), and in facilitating rapid Ca^{2+} movements across the sarcoplasmic reticulum (Wetzel et al., 1990) during muscle contraction (for review see Geers and Gros, 2000; Sly, 2000).

A pH disequilibrium has been described in the venous system of cannulated rainbow trout (Perry et al., 1997), results that are in line with mammalian studies that used a similar stop-flow approach (O'Brasky and Crandall, 1980; O'Brasky et al., 1979). Based on previous considerations regarding pH disequilibria (Gilmour, 1998b), the latter findings do not exclude the presence of PACA activity in the capillary beds of fish skeletal muscle, but may indicate that PACA activity is absent in the venous system of teleosts, and almost certainly as a soluble form in the plasma. Whether a venous pH disequilibrium can function to enhance Hb–O_2 unloading to the teleost heart in the presence of PACA is still unclear. However, Perry et al. (1997) have observed similar pH disequilibria reaching as far as the bulbus arteriosus, perhaps indicating a promising avenue for future studies. These results are in line with the immunohistochemical localization of PACA in the dorsal aorta of glass catfish and its absence in the ventral aorta (Randall et al., 2014). However, the fact that the antibody used was raised against a human RBC CAII, and not a membrane-bound CA isoform, is of concern. In addition, siluriforms have secondarily lost a Root effect and β-NHE activity (Berenbrink et al., 2005), and the proposed mode of enhanced Hb–O_2 unloading may not apply to these species.

Georgalis et al. (2006) investigated the CA isoforms involved in branchial and renal acid–base regulation in rainbow trout and cloned two cytosolic isoforms that had been identified previously (Esbaugh et al., 2004, 2005) and one additional CAIV isoform. Interestingly CAIV was detected in the brains and the heart, which is in line with the results of mammalian studies where PACA was reported in the capillaries of these tissues. In addition, evidence for the expression of a CAIV isoform was found in the contractile elements of the coho salmon heart (Alderman et al., 2016). Non-specific histochemical staining localized CA activity in the atrial lumen, on cells and trabeculae of the spongy myocardium. The cellular orientation of this CA isoform was assessed by a novel *in situ* electrometric ΔpH assay, providing strong support for a plasma-accessible orientation. The hydration rate of CO_2 within the atrial lumen was reduced significantly upon either the addition of C18 or pretreatment of the atria with PI-PLC, indicating that the observed catalytic activity was due to a GPI membrane-bound CA isoform, and is consistent with the observed expression patterns of CAIV. Due to a surface area to volume ratio that is less favorable than in capillaries, RBCs circulating through the heart may only experience the effects of PACA in the proximity of trabeculae or the chamber's walls. However, these are the microenvironments where O_2 release is most relevant. In light of the great increase in β_b that is dependent on PACA in the red muscle of rainbow trout (Rummer et al., 2013), it is also possible that the hearts of salmonids may benefit from this mode of enhanced Hb–O_2 unloading (Alderman et al., 2016); particularly in situations of low P_vO_2, which generally coincides with the release of catecholamines and the activation of RBC β-NHE. However, a potential role of PACA in functions related to cellular waste excretion or muscle contraction must be considered as well.

A similar observation was made in the hearts of lamprey, where 26% of homogenate CA activity was found in the microsomal fraction, indicating a large pool of membrane-associated CA activity (Esbaugh and Tufts, 2004). In general, biochemical markers indicate a membrane-bound isoform that is distinct from cytosolic CA, but that is not susceptible to PI-PLC, thus ruling out the GPI-linked CAIV or CAXV isoforms. In hagfish, CAXV is expressed widely throughout the tissues and high mRNA levels for this CA were also detected in RBCs (Esbaugh et al., 2009). The cellular orientation of the membrane-bound CA isoforms in agnathans has not been studied, and their accessibility to the plasma is undetermined.

Like in mammals, it appears that membrane-bound CA isoforms in fish are present in many tissues and are likely to be involved in a variety of processes; i.e., facilitating membrane transport of ions and metabolites. Previous studies indicate that PACA is present in the vasculature of mammals, including the capillary beds of several organs. However, in fishes, the cellular orientation of putative PACA isoforms is largely unstudied outside of the gill. Thus far, data indicate that: (i) glass catfish may have PACA in the dorsal aorta, but not in the ventral aorta; (ii) CA activity appears to be absent in the venous system of rainbow trout; and (iii) a PACA isoform was detected in the heart lumen of coho salmon. Thus, it appears that PACA has a heterogeneous distribution in the vasculature of teleosts, but data on other fishes are unavailable. If the presence of PACA in the gills of hagfish and dogfish is an indicator, these species may show a homogenous presence of PACA throughout the vasculature, which is the situation in invertebrates and may represent the ancestral state (Henry, 1984; Stabenau and Heming, 2003). Clearly, this is an idea that needs to be substantiated by future work.

5.2.3. Plasma Carbonic Anhydrase Inhibitors

The presence of CA in most tissues and gas exchange surfaces is evidence that there is a physiological need to accelerate the slow uncatalyzed reaction rates and provide rapid equilibrium between $CO_2 - HCO_3^- - H^+$ species. In stark contrast to this, stands the absence of CA activity in the plasma of most vertebrates. Henry and Swenson (2000) rightfully argue that plasma is an unfavorable environment to support significant CA activity (Bidani et al., 1983; Desforges et al., 2001; Gilmour et al., 2004), which may explain the lack of CA activity in plasma. However, it does raise questions about the presence of plasma CA inhibitors (PICA).

These PICA have been reported widely across vertebrates, including mammals (Booth, 1938a,b; Rispens et al., 1985; Hill, 1986; Roush and Fierke, 1992) and teleosts (Dimberg, 1994; Gervais and Tufts, 1998; Haswell et al., 1983; Heming, 1984; Heming and Watson, 1986; Henry et al., 1997a; Maetz, 1956; Peters et al., 2000). However, notable exceptions are found in most groups (for a review in fishes see Henry and Heming, 1998). Among fishes, the Chondrichthyes and agnathans seem to lack PICA, and notably it is these groups where gill PACA and circulating CA activity are found (Esbaugh et al., 2009; Esbaugh and Tufts, 2004; Gilmour et al., 1997, 2001, 2002; Henry et al., 1997a; Wood et al., 1994). In dogfish, circulating CA activity correlates with the concentration of iron and Hb in the plasma, suggesting that it is mostly derived from RBC lysis (Henry et al., 1997a). Heming et al. (1986) reported the absence of PICA in *Amia*, but its presence was later confirmed at low concentrations (Gervais and Tufts, 1999). Also, the channel catfish has low PICA levels (Heming, 1984; Henry et al., 1997a) and PICA may be completely absent in the brown bullhead (Gilmour et al., 2002).

Interestingly, PICA are highly specific, and their inhibitory capacity decreases markedly when applied to the RBC CA of other species. It appears that salmonids and eel possess the most potent PICA, as it will significantly inhibit CA activity of other species as well; but their cross-reactivity decreases with phylogenetic distance (Henry and Heming, 1998; Henry et al., 1997a; Maetz, 1956) and no fish PICA studied to date has been shown to inhibit bovine RBC CA (Dimberg, 1994; Henry et al., 1997a). The high specificity of PICA is reflected in their inhibition kinetics for different CA isoforms, where CAIV is much less sensitive to PICA as compared to cytosolic isoforms (Gervais and Tufts, 1998; Heming et al., 1993; Roush and Fierke, 1992). The physiological role of PICA is still debated, not only in fishes. The fact that their inhibition is strongest for cytosolic CA isoforms may suggest that they have some function in inactivating or scavenging CA released from RBC or tissue lysis (Booth, 1938a; Henry and Heming, 1998; Henry et al., 1997a).

Due to its small molecular size, circulating CA may be filtered from the plasma at the kidney. Superoxide dismutase, a molecule of similar size (32 vs 29 kDa for CA), was cleared from rat plasma with a halftime of minutes and localized in the urine (Bayati et al., 1988; Odlind et al., 1988). The binding of a larger molecule to CA from cell lysis may slow or prevent glomerular filtration by the kidney and conserve the enzyme for uptake into other tissues for the recycling of Zn (Appelgren et al., 1989; Henry and Heming, 1998; Henry and Swenson, 2000); the structural similarity of porcine PICA with transferrin supports this role (Henry and Heming, 1998).

In teleosts with pH-sensitive Hb, it has been proposed that PICA may enable RBC pH_i protection via β-NHE, keeping extracellular CA activity from lysis low (Nikinmaa, 2003; Nikinmaa et al., 1984; Primmett et al., 1986). Based on a few species there appears to be a correlation between the presence of PICA and the pH sensitivity of Hb–O_2 binding. Salmonids have the most potent PICA found in fishes, and also show a pronounced Root effect and β-NHE activity, whereas elasmobranchs, catfishes, and hagfish that lack PICA have Hbs with low pH sensitivity and no β-NHE (Berenbrink et al., 2005). However, the susceptibility of β-NHE activity to low levels of extracellular CA activity is questionable (Lessard et al., 1995; Wood and Munger, 1994). Surprisingly, the Hb-less icefish also have a PICA (Harter et al., unpublished data; Maffia et al., 2001; Tufts et al., 2002) that inhibits the RBC CA of closely related notothenioids. While PICA in these species may be an evolutionary relic from a red-blooded ancestry, it may also point toward a function of PICA other than the protection of pH sensitive O_2 transport (Lessard et al., 1995; Nikinmaa et al., 1990; Tufts et al., 2002). Eels have a potent PICA, but show no adrenergic β-NHE response (Hyde and Perry, 1990; Perry and Reid, 1992a,b), and this may also be the situation in starry flounder (Milligan and Wood, 1987; Wood and Milligan, 1987). Nonetheless, these species likely have RBC NHE that respond to other stimuli. Lamprey have pH-sensitive Hb and actively maintain elevated RBC pH_i via NHE (Tufts and Perry, 1998), but lack PICA, and in fact, show some circulating CA activity (Esbaugh and Tufts, 2004).

Because PICA are ineffective in inhibiting membrane-bound CAIV, their presence or absence is not a predictor of the presence of PACA in the vasculature. On the other hand, the absence of a PICA is permissive for, but not always indicative of, the presence of circulating CA activity (Gilmour et al., 2002). Differences in the expression of PICA may be rooted in the susceptibility of RBCs to lysis, and the selective pressures for the evolution of PICA may include RBC fragility (Henry et al., 1997a). Clearly, and despite substantial research effort, the physiological significance of PICA has not been fully resolved and remains a worthwhile area for future studies.

5.2.4. The Evolution of Differential Plasma-Accessible Carbonic Anhydrase Distribution in Fishes

Despite the information above, it is still difficult to draw a complete picture of PACA distribution in the vasculature across fish species. The use of a variety of methods provides pieces of evidence that are only conclusive if taken together. Stabenau and Heming (2003) provide, perhaps, the broadest survey for the presence of PACA across vertebrate gas exchange organs; importantly using comparable methodology (some of the original data was collected by Heming et al., 1994; Stabenau and Heming, 1999; Stabenau and Vietti, 2002; Stabenau et al., 1996). It appears that among vertebrates, teleost fishes stand out in their absence of PACA at the gas exchange surface, a trait that is highly conserved across vertebrate phyla. However, data are only available for a few fish species, and this makes it challenging to place the distribution of PACA in a phylogenetic context.

In an attempt to resolve this uncertainty, Table 3 summarizes the findings on the distribution of PACA in fishes in a conceptual context, and in relation to other known blood characteristics (see previous sections for details and references). In every fish species studied, the presence of RBC NHE activity occurs in the absence of PACA at the gill (this includes both teleost and lamprey), which is in line with a functional constraint of PACA activity on RBC pH_i regulation. In addition, PACA is often absent at the gill of species that have lowered their plasma buffer capacity, which typically coincides with a low Hb buffer capacity and the presence of a significant Bohr–Haldane effect, teleost traits that first evolved in the basal actinopterygians.

Table 3
Conceptual summary of blood characteristics in select fish lineages, relative to the distribution of carbonic anhydrase (CA) activity at the gas exchange surface

Species	Circul. CA[a]	PICA[b]	PACA[c]	β plasma[d]	β Hb[e]	Bohr-Hb[f]	NHE[g]
Chondrichthyes	+	–	+	+	+	–	–
Hagfish	+	–	+	+	+	–	–
Lamprey	+	–	–	–	–	+	+
Amia	–	+	–	–	–	+	–
Salmonids	–	+	–	–	–	+	+
Flounder	–	+	–	–	–	+	+
Siluriforms	–	+	–	+	+	–	–
Red-blooded notothenioids	–	+	–	–	–	+	+
Icefish	–	+	+	+			

[a]Circulating CA activity in the plasma; soluble isoforms (+ indicates presence, – absence).
[b]Plasma inhibitor of CA (+ indicates presence, – absence).
[c]Plasma-accessible CA at the gill; membrane-bound isoforms (+ indicates presence, – absence).
[d]Non-bicarbonate buffer capacity of plasma (+ indicates relatively high, – relatively low).
[e]Hemoglobin buffer capacity (+ indicates relatively high, – relatively low).
[f]Bohr–Haldane effect of hemoglobin (+ indicates relatively high, – relatively low).
[g]Sodium proton exchanger activity on the red blood cell membrane; adrenergic β-NHE in teleosts and non-adrenergic NHE in lamprey (+ indicates presence, – absence).

If one accepts that a homogenous distribution of PACA in the circulatory system represents the ancestral state, the question arises as to why some fish lineages (possibly independently) have lost PACA at the gills. The selective advantage of a β-NHE that allowed for the Root effect to be expressed in the physiological pH range may in itself have driven the loss of PACA at the gill of basal actinopterygians (Randall et al., 2014). If this was the case, one may predict that the loss of PACA at the gill coincided with a shift of the Root effect into the physiological pH range and with the appearance of a β-NHE. However, based on the results of a single study (Gervais and Tufts, 1998), and without support from molecular or functional data, it appears that *Amia* do not possess PACA at the gills, despite lacking a Root effect in the physiological pH range (Regan and Brauner, 2010b) or a β-NHE (Berenbrink et al., 2005). This may indicate that the loss of PACA at the gill was not associated with the need to regulate RBC pH_i, but may still have released the functional constraint on the evolution of β-NHE activity. This hypothesis needs to be thoroughly substantiated with additional data on *Amia*, on other basal actinopterygians, and on teleosts. It is also possible that PACA activity at the gill decreased much earlier with the evolution of a significant Bohr effect. The lack of CA activity in the plasma uncouples RBC

pH_i from plasma pH_e, and may protect blood P_{50} during an acidosis (Randall et al., 2014); an idea that is supported by the data of Rummer et al. (2013). The latter study reported no effect of hypercarbia on RBC pH_i in the absence of plasma CA activity. If the initial function of the Bohr–Haldane effect was to maintain pH homeostasis, it seems possible that PACA at the gill would be lost to shift HCO_3^- dehydration entirely into the RBC, where large amounts of H^+ need to be buffered upon Hb oxygenation. Clearly, there are a number of testable hypotheses as to why teleosts may have decreased (or lost) PACA activity at the gill, and these are exciting avenues for future studies. The following sections are likewise tentative, and remain to be substantiated experimentally; however, they are based on important findings from previous studies.

It is possible that levels of branchial PACA gradually decreased in concert with plasma non-bicarbonate buffer capacity in the basal actinopterygians. In mammals, it has been suggested that the low plasma buffer capacity limits H^+ availability to PACA at the lung endothelium, and thus, that the contribution of this CA pool to CO_2 excretion is less than 10% (Bidani and Heming, 1991; Heming and Bidani, 1992; Henry and Swenson, 2000; Swenson et al., 1993a,b). Experimentally changing plasma buffer capacity in rainbow trout (three-fold with HEPES) had no effect on arterial PCO_2 of CA vs saline-injected fish, indicating that H^+ availability in the plasma is not limiting to HCO_3^- dehydration (Desforges et al., 2001). However, in the presence of RBC AE, plasma HCO_3^- has functional access to the large buffer capacity of the Bohr–Haldane effect (Brauner, 1996; Perry and Gilmour, 1993; Perry et al., 1996b). Therefore, increasing plasma buffer capacity three-fold may be negligible considering H^+ availability within the RBC, and that CO_2 excretion will still largely occur via the RBC (Desforges et al., 2001; Gilmour and Desforges, 2000; Gilmour et al., 2004). When AE was inhibited with DIDS (4,4′-diisothiocyano-2,2′-stilbenedisulfonic acid) *in vitro* (Gilmour and Desforges, 2000) or *in vivo* (Gilmour et al., 2004), CO_2 excretion decreased markedly and became a function of plasma buffer capacity. This indicates that, like in mammals, the low buffer capacity of the plasma in rainbow trout is unable to support significant HCO_3^- dehydration rates, even with extracellular CA activity. In the presence of a fast AE (Nikinmaa and Boutilier, 1995), a large Haldane effect (Jensen, 1989) and abundant RBC CA activity (Maren, 1967), PACA at the gill may be largely inconsequential. Therefore, the reduction of PACA at the gill may have been a by-product of the reduction in plasma buffer capacity and may in itself not have been selected for. The reason why mammals retained PACA at the lung may be related to the control of respiration in air breathers, which hinges on the acid–base status of the blood leaving the lung. Having PACA at the lung endothelium will abolish disequilibrium states in the plasma that may confound the sensing of respiratory set-points (Swenson, 2000; Swenson et al., 1995a, 1998), a functional constraint that does not apply to fishes.

The non-bicarbonate buffer capacity of fish plasma decreases from about 50%–70% of whole blood buffer capacity in elasmobranchs to 20%–40% in teleosts (Gilmour et al., 2002; Szebedinszky and Gilmour, 2002; Tufts and Perry, 1998). Like the reduction in Hb buffer capacity, it is likely that a reduced plasma buffer capacity evolved in concert with the increase in the Bohr effect in basal actinopterygians, perhaps to maximize ΔpH_{a-v} (Berenbrink et al., 2005; Regan and Brauner, 2010a). However, no systematic data on plasma buffer capacity are available for this group. Gonzalez et al. (2001) measured the non-bicarbonate buffer capacity of separated plasma in *Amia* and reported values that are several-fold lower compared to dogfish (Lenfant and Johansen, 1966), and less than half those in rainbow trout (Wood et al., 1982). These results are consistent with PACA activity at the gill decreasing in concert with plasma buffer capacity due to a limitation in H^+ availability. Whether these changes occurred gradually throughout the evolution of basal actinopterygians is unknown, but Hb-buffer values, the Bohr effect, and the onset pH of the Root effect changed substantially in the common ancestor of teleosts and *Amia*, compared to the rather constant levels in more ancestral species (Regan and Brauner, 2010a,b); a similar pattern may be observed with plasma buffer capacity and the distribution of PACA. However, at present, all available data are based on two studies (Gervais and Tufts, 1998; Gonzalez et al., 2001) and on a single species, which in addition is a facultative air-breather a circumstance that may well influence the observed relationships (Brauner and Berenbrink, 2007; Graham, 1997). More studies on gill PACA activity and plasma buffer capacities are clearly required, and these should incorporate a broader range of species including other basal actinopterygians and non-air-breathing representatives of this group.

Siluriforms have secondarily lost both the *retia* at the eye and the swim bladder, a trait that correlates with the loss of a significant Root effect and β-NHE activity (Berenbrink et al., 2005). In addition, siluriforms have a high plasma buffer capacity (Cameron and Kormanik, 1982; Gilmour et al., 2002), which stems from a high concentration of histidine-rich albumins in the plasma (Szebedinszky and Gilmour, 2002). Despite the absence of a Root effect and β-NHE activity, siluriforms do not have PACA at the gill (Gilmour et al., 2002; Henry et al., 1988). Whether this absence is a vestigial trait, dictated by the teleost ancestry of siluriforms rather than their physiology, is still unclear. However, based on the work of Desforges et al. (2001) it appears that regaining branchial PACA may be of little benefit to CO_2 excretion in the presence of a rapid AE and sufficient H^+ availability within the RBC (Berenbrink et al., 2005). A high plasma buffer capacity has also been reported to correlate with hypoxia and hypercarbia tolerance in three cyprinids and rainbow trout (Cech et al., 1994), whether the same relationship holds for a wider range of fish species including the siluriforms remains to be tested.

Two species of Antarctic icefish (Channichthyidae), *C. gunnari* and *C. aceratus*, are the only known teleosts to possess PACA at the gills (Harter et al., unpublished data; Tufts et al., 2002). This clade derives from a perciform ancestor (Eastman, 2005; Near et al., 2003, 2004) that likely had the full extent of teleost blood characteristics. These include a Root effect, a *choroid rete*, β-NHE activity, and low Hb and plasma buffer capacities, and many of these traits are still found in the closely related red-blooded notothenioids (Davison et al., 1988; Eastman and Lannoo, 2003, 2004; Near et al., 2004; Wells, 2005; Wells et al., 1984, 1989; Wujcik et al., 2007). It is worthwhile noting that many red-blooded notothenioids reportedly have a *choroid rete*, whereas this structure is lost in the icefishes due to the absence of a Root effect, and O_2 is supplied to the eyes by a dense network of arterial vessels (Eastman and Lannoo, 2003, 2004; Sidell and O'Brien, 2006; Wells, 2005; Wujcik et al., 2007). If the common ancestor of notothenioids lacked PACA at the gill, then it must have been secondarily re-acquired in the icefish. This suggests that teleosts have the capacity to express this trait when physiological requirements exist, unlike the loss of Hb that is rooted in a deletion of the β-globin gene (Near et al., 2006). Morphological and physiological adaptations enabled icefishes to thrive without Hb and RBCs (see review by Kock, 2005). However, because CO_2 is mainly transported as HCO_3^- in the plasma (Hemmingsen and Douglas, 1972), CA activity is required to catalyze the dehydration reaction at the gill, especially at low temperatures. A membrane-bound PACAIV isoform in the icefish gill likely catalyzes plasma reactions in the absence of RBC CA (Harter et al., unpublished data; Tufts et al., 2002). Previous studies indicate, in agreement with the latter hypothesis, that icefish have exceptionally high plasma buffer capacities (Acierno et al., 1997; Feller and Gerday, 1997; Feller et al., 1994; Harter et al. unpublished data).

Hemoglobin and plasma non-bicarbonate buffer capacity are also high in hagfish, and in this species CO_2 excretion takes place largely in the plasma, catalyzed by a PACAIV isoform at the gill and circulating CA activity (Esbaugh et al., 2009; Jensen, 1999; Tufts and Perry, 1998). However, the species of lamprey that have been studied, much like teleosts, have a low plasma buffer capacity, and thus, CO_2 formation in the plasma may be negligible (Bidani and Heming, 1991; Desforges et al., 2001; Gilmour et al., 2004). Lampreys also have a pronounced Bohr–Haldane effect (Ferguson et al., 1992; Jensen, 1999; Nikinmaa et al., 1995; Nikinmaa and Mattsoff, 1992) and regulate RBC pH_i above equilibrium by active H^+ extrusion via NHE (Nikinmaa, 1986; Nikinmaa et al., 1986; Tufts et al., 1992). These characteristics allow lamprey to transport CO_2 almost entirely as HCO_3^- within the RBCs (Tufts and Boutilier, 1989); a CO_2 transport strategy that differs from that of teleosts and elasmobranchs and is related to the lack of a functional RBC AE (Cameron et al., 1996; Cameron and Tufts, 1994; Nikinmaa and Railo, 1987; Ohnishi and Asai, 1985; Tufts and Boutilier, 1989). Lamprey RBC NHE respond to intracellular acidification (Nikinmaa et al., 1993; Virkki and Nikinmaa, 1994), hypercarbia (Tufts et al., 1992), and hypoxia (Ferguson et al., 1992), but not to adrenergic stimulation (Tufts, 1991; Virkki and Nikinmaa, 1994). Therefore, lamprey NHEs may function continuously to maintain RBC pH_i within tolerable limits, and in fact, transport of CO_2 in the RBCs is compromised when RBC NHEs are disabled (Cameron and Tufts, 1994; Tufts, 1992). Much like the β-NHE in teleosts, lamprey NHE function is likely dependent upon the uncatalyzed rate of HCO_3^- dehydration in the plasma to establish an H^+ disequilibrium across the RBC membrane.

It appears that PACA is absent at the gill of lamprey. However, this is based on scarce evidence from a single study (Henry et al., 1993). In addition, Esbaugh et al. (2004) reported low CA activity in lamprey plasma with no evidence for a PICA. These findings stand in contrast to the need for RBC pH_i regulation. However, RBC lysis during sampling may cause elevated plasma CA activity, and *in vivo* levels may be low enough for active RBC pH_i regulation. Lamprey and teleosts appear to have converged on a number of blood characteristics (low Hb and plasma buffer capacity, pronounced Bohr–Haldane effects, and RBC pH_i regulation) by distinctly different mechanisms (monomeric vs tetrameric Hb and non-adrenergic vs β-NHE; Nikinmaa et al., 1995; Tufts and Perry, 1998) that may have required the loss of PACA at the gill. Whether this convergence on many fundamental blood characteristics correlates with similarities in habitat and life history between lamprey and teleosts is still unclear, and the early divergence of these groups (Arnason et al., 2004; Janvier, 2007) may represent a unique opportunity to study the evolution of these traits.

Another pattern of PACA distribution that stands out is that all fish species studied to date that possess PACA at the gill are osmoconforming marine species (icefishes being the notable exception), whereas all fishes that have lost PACA at the gill are osmoregulators. However, the current models of ionoregulation in fish do not reveal any evident mechanistic constraints for having PACA at the gill (see Evans et al., 2005). Representatives of both freshwater and marine teleosts (flounder and seawater-acclimated rainbow trout; Gilmour et al., 1997; Sender et al., 1999) appear to lack PACA at the gills, and this may indicate that this trait is compatible with either ionoregulatory strategy. As all extant marine actinopterygians may have a freshwater ancestry, similar patterns in PACA distribution across teleosts may be evolutionarily constrained (Vega and Wiens, 2012). Thus, it will be revealing to examine PACA distribution in freshwater rays (e.g., Potamotrygon) which have independently transitioned from an osmoconforming marine to an osmoregulating freshwater strategy; studies that are currently being performed in our lab. If the presence of PACA at the gill conflicts with the transport mechanisms required to uptake ions from freshwater, then PACA may have been lost during freshwater invasion in lamprey and actinopterygians (Janvier, 2007); and Potamotrygon rays may display a similar pattern despite being elasmobranchs.

The previous discussion summarizes the extent of our knowledge on branchial PACA in fishes, but due to the evident lack of data it is clearly speculative. Nonetheless, if nothing else, it will hopefully provide a different view on O_2 transport in fishes, a research field that overall has received considerable attention. The distribution of CA, an enzyme usually implicated in acid–base regulation, may have shaped the evolution of the unique mode of O_2 transport in teleosts and played an important part in the extensive adaptive radiation of this most speciose group of vertebrates (Helfman et al., 2009; Randall et al., 2014). The following section outlines the putative series of events that may have led to the differential distribution of PACA in the circulatory system of teleosts. It is a tentative extension of the elegant analysis of Berenbrink et al. (2005), and considers the recent perspectives provided by Randall et al. (2014):

(1) In the ancestral state, plasma buffer capacities were high, Hb had a low Bohr effect, no Root effect, and a high buffer capacity, RBCs had no β-NHE, and PACA was present throughout the circulatory system (Henry, 1984; Stabenau and Heming, 2003); thus HCO_3^- dehydration took place in the plasma and in the RBC. This ancestral condition is conserved in hagfish and Chondrichthyes (Esbaugh et al., 2009; Gilmour and Perry, 2010).

(2) The magnitude of the Bohr effect increased in basal actinopterygians and it has been suggested that the Root effect may initially have been a molecular by-product of this process; thus, the evolution of the Root effect may have been driven by the selective advantages of the Bohr effect (Berenbrink et al., 2005). At a time where acidifying *retes* had not yet evolved, the selective advantage of the Bohr effect was likely an increase in β_b, perhaps not only enhancing Hb–O_2 unloading in all tissues but also enhancing the coupling between O_2 and CO_2 transport that promotes pH homeostasis via the Haldane effect.

(3) In the basal actinopterygians, Hb buffer capacity decreased in concert with an increase in the Bohr effect, and this reduced the amount of acid required to elicit a given ΔpH_{a-v} and maximized the resulting Bohr shift.

(4) The onset pH of the Root effect in basal actinopterygians was likely outside the range of physiological pH values, even in extreme scenarios of exercise and hypoxia (Regan and Brauner, 2010b). In contrast, more derived teleosts have a Root effect onset pH that is well within the pH range that may be experienced *in vivo* (Berenbrink, 2007; Herbert et al., 2006; Pelster and Weber, 1990; Waser and Heisler, 2005).

(5) The evolution of functional *retes* was only possible in the presence of a Root effect, as otherwise O_2 would have been shunted away from the tissues in the counter-current exchangers. A marked increase in the onset pH of the Root effect occurred in the common ancestor of *Amia* and teleosts, which coincides with the first occurrence of a *choroid rete* (Berenbrink et al., 2005). While the Root effect in *Amia* is still unlikely to become activated in the general circulation (Regan and Brauner, 2010b), in the *choroid rete* blood is acidified sufficiently to reach the threshold pH. Therefore, with the evolution of a *choroid rete*, the Root effect increased β_b at the eyes and came under positive selection.

(6) To maximize β_b via the Root effect within the *choroid rete*, plasma and Hb buffer capacities continued to decrease in basal teleosts.

(7) The reduction of plasma buffer capacity limited the availability of H^+ in the plasma, which is required to sustain high CA catalytic activities (Gilmour et al., 2004). Furthermore, the presence of rapid AE and the release of oxygenation-dependent H^+ due to the Bohr–Haldane effect may have obviated the need for PACA at the gill, and thus activity decreased before the divergence of *Amia* and teleosts (but alternative hypotheses need to be considered and are discussed in Section 5.2.4). The absence of any significant HCO_3^- dehydration in the plasma drives CO_2 excretion entirely through the RBC, and thus strengthens the coupling between O_2 and CO_2 transport in teleosts.

(8) The ΔpH_{a-v} required to elicit the Root effect was reduced by shifting the onset pH further into the physiological pH range. However, a Root effect that is activated in the physiological pH range poses a risk to O_2 uptake at the gills, for example, in situations where blood pH decreases systemically (Nikinmaa, 1992). Therefore, a higher onset pH of the Root effect likely required the active regulation of RBC pH_i during an acidosis to protect general O_2 transport. A functional β-NHE is first observed in the Elopomorpha (Berenbrink, 2007; Berenbrink et al., 2005), but whether this correlates with a shift of the Root effect into the physiological pH range has not been tested.

(9) The activity of a β-NHE requires a slow uncatalyzed rate of HCO_3^- dehydration in the plasma, and thus, the absence of significant branchial PACA activity (Motais et al., 1989; Nikinmaa et al., 1990). The reduction of PACA activity in basal actinopterygians released this constraint and allowed for the evolution of a β-NHE that subsequently allowed for the Root effect to be expressed in the physiological pH range, maximizing its potential for increasing β_b. Due to the smaller ΔpH_{a-v} required to trigger the Root effect, its benefits were then available to enhance β_b for all tissues, without the need for acidifying *retes*, and with ΔpHa–v confined to sites with PACA; this may be the case in rainbow trout red muscle today (Rummer et al., 2013).

Clearly the distribution of PACA in the vasculature of fishes is a vastly understudied area. Since the work of Berenbrink et al. (2005), it is evident that the distribution of PACA is a missing component in the sequence of events that led to the unique mode of O_2 and CO_2 transport in modern teleosts. The lack of PACA at the gills of teleosts and lampreys may have released a functional constraint on RBC pH_i regulation, and may have been driven by a reduction in plasma buffer capacity that rendered branchial PACA activity obsolete. The absence of PACA at the gills of osmoregulating fish species vs osmoconformers leaves ample room for alternative hypotheses. There is a clear need for reliable data on PACA activity in more fish species, including other basal actinopterygians and teleosts, but also fishes occupying different habitats (e.g., freshwater rays and notothenioids) and with different life histories and ecologies (e.g., air-breathing fishes). The heterogeneous distribution of PACA in the cardiovascular system of teleosts is pivotal to a mode of O_2 transport that may greatly enhance β_b and support higher $\dot{M}O_2$ without increasing [Hb] or $\dot{V}_b$. Clearly, this is an area ripe for further investigation, and that may have consequences for our understanding of O_2 transport in nearly half of all extant vertebrates.

REFERENCES

Acierno, R., Maffia, M., Rollo, M., Storelli, C., 1997. Buffer capacity in the blood of the hemoglobinless Antarctic fish *Chionodraco hamatus*. Comp. Biochem. Physiol. A Physiol. 118, 989–992.

Affonso, E., Polez, V., Correa, C., Mazon, A., Araujo, M., Moraes, G., Rantin, F., 2002. Blood parameters and metabolites in the teleost fish *Colossoma macropomum* exposed to sulfide or hypoxia. Comp. Biochem. Physiol. C 133, 375–382.

Albers, C., Götz, K., Welbers, P., 1981. Oxygen transport and acid-base balance in the blood of the sheatfish, *Silurus glanis*. Respir. Physiol. 46, 223–236.

Albers, C., Goetz, K.-H., Hughes, G.M., 1983. Effect of acclimation temperature on intraerythrocytic acid-base balance and nucleoside triphosphates in the carp, *Cyprinus carpio*. Respir. Physiol. 34, 145–159.

Alderman, S.L., Harter, T.S., Wilson, J.M., Supuran, C.T., Farrell, A.P., Brauner, C.J., 2016. Evidence for a plasma-accessible carbonic anhydrase in the lumen of salmon heart that may enhance oxygen delivery to the myocardium. J. Exp. Biol. 219, 719–724.

Andersen, M.E., Olson, J.S., Gibson, Q.H., Carey, F.G., 1973. Studies on ligand binding to hemoglobins from teleosts and elasmobranchs. J. Biol. Chem. 248, 331–341.

Andersen, N., Laursen, J., Lykkeboe, G., 1985. Seasonal variations in hematocrit, red cell hemoglobin and nucleoside triphosphate concentrations, in the European eel *Anguilla anguilla*. Comp. Biochem. Physiol. A Comp. Physiol. 81, 87–92.

Anthony, E., 1961. The oxygen capacity of goldfish (*Carassius auratus* L.) blood in relation to thermal environment. J. Exp. Biol. 38, 93–107.

Aota, S., Holmgren, K.D., Gallaugher, P., Randall, D.J., 1990. A possible role for catecholamines in the ventilatory responses associated with internal acidosis or external hypoxia in rainbow trout *Oncorhynchus mykiss*. J. Exp. Biol. 151, 57–70.

Appelgren, L.E., Odlind, B., Wistr, P., 1989. Tissue distribution of ^{125}I-labelled carbonic anhydrase isozymes I, II and III in the rat. Acta Physiol. Scand. 137, 449–456.

Arjamaa, O., Nikinmaa, M., 2009. Natriuretic peptides in hormonal regulation of hypoxia responses. Am. J. Phys. Regul. Integr. Comp. Phys. 296, R257–R264.

Arjamaa, O., Vuolteenaho, O., Kivi, E., Nikinmaa, M., 2014. Hypoxia increases the release of salmon cardiac peptide (sCP) from the heart of rainbow trout (*Oncorhynchus mykiss*) under constant mechanical load *in vitro*. Fish Physiol. Biochem. 40, 67–73.

Arnason, U., Gullberg, A., Janke, A., Joss, J., Elmerot, C., 2004. Mitogenomic analyses of deep gnathostome divergences: a fish is a fish. Gene 333, 61–70.

Aukland, K., Reed, R., 1993. Interstitial-lymphatic mechanisms in the control of extracellular fluid volume. Physiol. Rev. 73, 1–78.

Axelsson, M., 2005. The circulatory system and its control. In: Steffensen, J.F., Farrel, A.P. (Eds.), The Physiology of Polar Fishes. In: Fish Physiology, vol. 22. Academic Press, New York, pp. 239–280.

Bailey, J., Sephton, D.H., Driedzic, W.R., 1990. Oxygen uptake by isolated perfused fish hearts with differing myoglobin concentrations under hypoxic conditions. J. Mol. Cell. Cardiol. 22, 1125–1134.

Baird, T.T., Waheed, A., Okuyama, T., Sly, W.S., Fierke, C.A., 1997. Catalysis and inhibition of human carbonic anhydrase IV. Biochemist 36, 2669–2678.

Baker, D.W., Wood, A.M., Kieffer, J.D., 2005. Juvenile Atlantic and shortnose sturgeons (Family: Acipenseridae) have different hematological responses to acute environmental hypoxia. Physiol. Biochem. Zool. 78, 916–925.

Bakker, J., Gortmaker, G., Vrolijk, A., Offerijns, F., 1976. The influence of the position of the oxygen dissociation curve on oxygen-dependent functions of the isolated perfused rat liver. Pflugers Arch. 362, 21–31.

Barnett, C., 1951. The structure and function of the choroidal gland of teleostean fish. J. Anat. 85, 113.

Barron, M.G., Tarr, B.D., Hayton, W.L., 1987. Temperature-dependence of cardiac output and regional blood flow in rainbow trout, *Salmo gairdneri* Richardson. J. Fish Biol. 31, 735–744.

Bartels, H., 1972. In: Rorth, M., Astrup, P. (Eds.), The biological significance of the Bohr effect. Proceedings of the Alfred Benzon Symposium IV: Oxygen Affinity of Hemoglobin and Red Cell Acid Base Status. Munksgaard, Copenhagen, pp. 717–735.

Bayati, A., Källskog, Ö., Odlind, B., Wolgast, M., 1988. Plasma elimination kinetics and renal handling of copper/zinc superoxide dismutase in the rat. Acta Physiol. Scand. 134, 65–74.

Benesch, R., Benesch, R.E., 1967. The effect of organic phosphates from the human erythrocyte on the allosteric properties of hemoglobin. Biochem. Biophys. Res. Commun. 26, 162–167.

Berenbrink, M., 2006. Evolution of vertebrate haemoglobins: histidine side chains, specific buffer value and Bohr effect. Respir. Physiol. Neurobiol. 154, 165–184.

Berenbrink, M., 2007. Historical reconstructions of evolving physiological complexity: O_2 secretion in the eye and swimbladder of fishes. J. Exp. Biol. 210, 1641–1652.

Berenbrink, M., Koldkjaer, P., Kepp, O., Cossins, A.R., 2005. Evolution of oxygen secretion in fishes and the emergence of a complex physiological system. Science 307, 1752–1757.

Bidani, A., 1991. Analysis of abnormalities of capillary CO_2 exchange *in vivo*. J. Appl. Physiol. 70, 1686–1699.

Bidani, A., Crandall, E.D., 1988. Velocity of CO_2 exchange in the lungs. Annu. Rev. Physiol. 50, 639–652.

Bidani, A., Heming, T.A., 1991. Effects of perfusate buffer capacity on capillary CO_2-HCO_3^--H^+ reactions: theory. J. Appl. Physiol. 71, 1460–1468.

Bidani, A., Mathew, S., Crandall, E., 1983. Pulmonary vascular carbonic anhydrase activity. J. Appl. Physiol. 55, 75–83.

Black, E.C., Kirkpatrick, D., Tucker, H.H., 1966. Oxygen dissociation curves of the blood of brook trout (*Salvelinus fontinalis*) acclimated to summer and winter temperatures. Can. J. Fish. Aquat. Sci. 23, 1–13.

Bogdanova, A., Berenbrink, M., Nikinmaa, M., 2009. Oxygen-dependent ion transport in erythrocytes. Acta Physiol. 195, 305–319.

Bohr, C., Hasselbalch, K., Krogh, A., 1904. About a new biological relation of high importance that the blood carbonic acid tension exercises on its oxygen binding. Skand. Arch. Physiol. 16, 402–412.

Bonaventura, C., Crumbliss, A.L., Weber, R.E., 2004. New insights into the proton-dependent oxygen affinity of Root effect haemoglobins. Acta Physiol. Scand. 182, 245–258.

Booth, V., 1938a. The carbonic anhydrase inhibitor in serum. J. Physiol. 91, 474–489.

Booth, V., 1938b. Carbonic anhydrase activity inside corpuscles. Enzyme-substrate accessibility factors. J. Physiol. 93, 117–128.

Booth, J.H., 1978. The distribution of blood flow in the gills of fish: application of a new technique to rainbow trout (*Salmo gairdneri*). J. Exp. Biol. 73, 119–129.

Borgese, F., Garcia-Romeu, F., Motais, R., 1987. Ion movements and volume changes induced by catecholamines in erythrocytes of rainbow trout: effect of pH. J. Physiol. 382, 145–157.

Borgese, F., Motais, R., Garcia-Romeu, F., 1991. Regulation of Cl^--dependent K^+ transport by oxy-deoxyhemoglobin transitions in trout red cells. Biochim. Biophys. Acta Biomembr. 1066, 252–256.

Bottcher, K., Waheed, A., Sly, W.S., 1994. Membrane-associated carbonic anhydrase from the crab gill: purification, characterization, and comparison with mammalian CAs. Arch. Biochem. Biophys. 312, 429–435.

Bourne, P.K., Cossins, A.R., 1982. On the instability of K^+ influx in erythrocytes of the rainbow trout, *Salmo gairdneri*, and the role of catecholamine hormones in maintaining *in vivo* influx activity. J. Exp. Biol. 101, 93–104.

Boutilier, R.G., Ferguson, R.A., 1989. Nucleated red cell function: metabolism and pH regulation. Can. J. Zool. 67, 2986–2993.

Boutilier, R.G., Heming, T.A., Iwama, G.K., 1984. Physicochemical parameters for use in fish respiratory physiology. In: Hoar, W.S., Randall, D.J. (Eds.), In: Fish Physiology, vol. 10A. Academic Press, New York, pp. 403–426.

Boutilier, R., Iwama, G., Randall, D., 1986. The promotion of catecholamine release in rainbow trout, *Salmo gairdneri*, by acute acidosis: interactions between red cell pH and haemoglobin oxygen-carrying capacity. J. Exp. Biol. 123, 145–157.

Boutilier, R., Dobson, G., Hoeger, U., Randall, D., 1988. Acute exposure to graded levels of hypoxia in rainbow trout (*Salmo gairdneri*): metabolic and respiratory adaptations. Respir. Physiol. 71, 69–82.

Brauner, C.J., 1995. The interaction between O_2 and CO_2 movements during aerobic exercise in fish. Braz. J. Med. Biol. Res. 28, 1185–1189.

Brauner, C.J., 1996. An analysis of the transport of oxygen and carbon dioxide in fish. In: Zoology. UBC, Vancouver (Ph.D.).

Brauner, C., Berenbrink, M., 2007. Gas transport and exchange. In: Farrell, A.P., Brauner, C.J. (Eds.), In: Fish Physiology, vol. 26. Academic Press, New York, pp. 213–282.

Brauner, C.J., Randall, D.J., 1996. The interaction between oxygen and carbon dioxide movements in fishes. Comp. Biochem. Physiol. 113A, 83–90.

Brauner, C.J., Randall, D., 1998. The linkage between oxygen and carbon dioxide transport. In: Perry II, S.F., Tufts, B.L. (Eds.), Fish Respiration. In: Fish Physiology, vol. 17. Academic press, New York, pp. 283–320.

Brauner, C.J., Val, A.L., 1996. The interaction between O_2 and CO_2 exchange in the obligate air breather, *Arapaima gigas*, and the facultative air breather, *Lipossarcus pardalis*. In: Val, A.L., Almeida-Val, V.M.F., Randall, D.J. (Eds.), Physiology and Biochemistry of the Fishes of the Amazon. INPA, Manaus, Brazil, pp. 101–110.

Brauner, C.J., Val, A.L., 2005. Oxygen transfer. In: Farrell, A.P., Brauner, C.J. (Eds.), The Physiology of Tropical Fishes. In: Fish Physiology, vol. 21. Academic Press, New York, pp. 277–306.

Brauner, C.J., Wang, T., 1997. The optimal oxygen equilibrium curve: a comparison between environmental hypoxia and anemia. Am. Zool. 37, 101–108.

Brauner, C.J., Gilmour, K.M., Perry, S.F., 1996. Effect of haemoglobin oxygenation on Bohr proton release and CO2 excretion in the rainbow trout. Respir. Physiol. 106, 65–70.

Brauner, C., Thorarensen, H., Gallaugher, P., Farrell, A., Randall, D., 2000a. The interaction between O_2 and CO_2 exchange in rainbow trout during graded sustained exercise. Respir. Physiol. 119, 83–96.

Brauner, C., Thorarensen, H., Gallaugher, P., Farrell, A., Randall, D., 2000b. CO_2 transport and excretion in rainbow trout (*Oncorhynchus mykiss*) during graded sustained exercise. Respir. Physiol. 119, 69–82.

Brauner, C., Wang, T., Val, A., Jensen, F.B., 2001. Non-linear release of Bohr protons with haemoglobin-oxygenation in the blood of two teleost fishes; carp (*Cyprinus carpio*) and tambaqui (*Colossoma macropomum*). Fish Physiol. Biochem. 24, 97–104.

Brill, R., Bushnell, P., Schroff, S., Seifert, R., Galvin, M., 2008. Effects of anaerobic exercise accompanying catch-and-release fishing on blood-oxygen affinity of the sandbar shark (*Carcharhinus plumbeus*, Nardo). J. Exp. Mar. Biol. Ecol. 354, 132–143.

Brittain, T., 1987. The root effect. Comp. Biochem. Physiol. 86B, 473–481.

Brittain, T., 2005. Root effect hemoglobins. J. Inorg. Biochem. 99, 120–129.

Brittain, T., 2008. Extreme pH sensitivity in the binding of oxygen to some fish hemoglobins: the root effect. In: Ghosh, A. (Ed.), The Smallest Biomolecules: Diatomics and Their Interactions with Heme Protein. Elsevier, Amsterdam, p. 219.

Brix, O., Clements, K.D., Wells, R.M.G., 1999. Haemoglobin components and oxygen transport in relation to habitat distribution in triplefin fishes (Tripterygiidae). J. Comp. Physiol. B. 169, 329–334.

Bruns, W., Dermietzel, R., Gros, G., 1986. Carbonic anhydrase in the sarcoplasmic reticulum of rabbit skeletal muscle. J. Physiol. 371, 351–364.

Bushnell, P.G., Steffensen, J.F., Johansen, K., 1984. Oxygen consumption and swimming performance in hypoxia acclimated rainbow trout *Salmo gairdneri*. J. Exp. Biol. 113, 225–235.

Butler, P.J., Metcalfe, J.D., 1989. Cardiovascular and respiratory systems. In: Shuttleworth, T.J. (Ed.), Physiology of Elasmobranch Fishes. Springer-Verlag, Berlin, pp. 1–47.

Butler, P.J., Axelsson, M., Ehrenstrom, F., Metcalfe, J.D., Nilsson, S., 1989. Circulating catecholamines and swimming performance in the atlantic cod, *Gadus morhua*. J. Exp. Biol. 141, 377–387.

Cameron, J.N., 1970. The influence of environmental variables on the hematology of pinfish (*Lagodon rhomboides*) and striped mullet (*Mugil cephalus*). Comp. Biochem. Physiol. 32, 175–192.
Cameron, J.N., 1971. Oxygen dissociation characteristics of the blood of the rainbow trout, *Salmo gairdneri*. Comp. Biochem. Physiol. 38, 699–704.
Cameron, J.N., Kormanik, G.A., 1982. Intracellular and extracellular acid-base status as a function of temperature in the fresh-water channel catfish, *Ictalurus punctatus*. J. Exp. Biol. 99, 127–142.
Cameron, J.N., Polhemus, J.A., 1974. Theory of CO_2 exchange in trout gills. J. Exp. Biol. 60, 183–194.
Cameron, B., Tufts, B., 1994. In vitro investigation of the factors contributing to the unique CO2 transport properties of blood in the sea lamprey (Petromyzon marinus). J. Exp. Biol. 197, 337–348.
Cameron, J.N., Wohlschlag, D.E., 1969. Respiratory response to experimentally induced anaemia in the pinfish (*Lagodon rhomboides*). J. Exp. Biol. 50, 307–317.
Cameron, B.A., Perry II, S.F., Wu, C., Ko, K., Tufts, B.L., 1996. Bicarbonate permeability and immunological evidence for an anion exchanger-like protein in the red blood cells of the sea lamprey, *Petromyzon marinus*. J. Comp. Physiol. B. 166, 197–204.
Cardenas, V., Heming, T.A., Bidani, A., 1998. Kinetics of CO_2 excretion and intravascular pH disequilibria during carbonic anhydrase inhibition. J. Appl. Physiol. 84, 683–694.
Carey, F.G., Gibson, Q.H., 1983. Heat and oxygen exchange in the *rete mirabile* of the bluefin tuna, *Thunnus thynnus*. Comp. Biochem. Physiol. A Physiol. 74, 333–342.
Carrie, D.W., Gilmour, K.M., 2012. Intracellular carbonic anhydrase contributes to the red blood cell adrenergic response in rainbow trout *Oncorhynchus mykiss*. Respir. Physiol. Neurobiol. 184, 60–64.
Cech, J.J., Castleberry, D.T., Hopkins, T.E., 1994. Temperature and CO_2 effects on blood O_2 equilibria in northern squawfish, *Ptychocheilus oregonensis*. Can. J. Fish. Aquat. Sci. 51, 13–19.
Chapman, L.J., Mckenzie, D.J., 2009. Behavioral responses and ecological consequences. In: Farrel, A.P., Brauer, C.J. (Eds.), Hypoxia. In: Fish Physiology, vol. 27. Academic Press, New York, pp. 25–77.
Chapman, L.J., Chapman, C.A., Nordlie, F.G., Rosenberger, A.E., 2002. Physiological refugia: swamps, hypoxia tolerance and maintenance of fish diversity in the Lake Victoria region. Comp. Biochem. Physiol. A Mol. Integr. Physiol. 133, 421–437.
Chegwidden, W.R., Carter, N.D., 2000. Introduction to the carbonic anhydrases. In: Chegwidden, W.R., Carter, N.D., Edwards, Y.H. (Eds.), The Carbonic Anhydrases: New Horizons. Springer, Berlin, pp. 13–28.
Chegwidden, W.R., Dodgson, S.J., Spencer, I.M., 2000. The roles of carbonic anhydrase in metabolism, cell growth and cancer in animals. In: Chegwidden, W.R., Carter, N.D., Edwards, Y.H. (Eds.), The Carbonic Anhydrases: New Horizons. Springer, Berlin, pp. 343–363.
Chetrite, G., Cassoly, R., 1985. Affinity of hemoglobin for the cytoplasmic fragment of human erythrocyte membrane band 3. J. Mol. Biol. 185, 639–644.
Chien, S., 1987. Physiological and pathophysiological significance of hemorheology. In: Chien, S., Dormandy, J., Ernst, E., Matrai, A. (Eds.), Clinical Hemorheology: Applications in Cardiovascular and Hematological Disease, Diabetes, Surgery and Gynecology. Springer, Dordrecht, pp. 125–164.
Christiansen, J., Douglas, C., Haldane, J., 1914. The absorption and dissociation of carbon dioxide by human blood. J. Physiol. 48, 244–271.
Christoforides, C., Hedley-Whyte, J., 1969. Effect of temperature and hemoglobin concentration on solubility of O_2 in blood. J. Appl. Physiol. 27, 592–596.

Chudzik, J., Houston, A., 1983. Temperature and erythropoiesis in goldfish. Can. J. Zool. 61, 1322–1325.

Clack, J.A., 2007. Devonian climate change, breathing, and the origin of the tetrapod stem group. Integr. Comp. Biol. 47, 510–523.

Claireaux, G., Thomas, S., Fievet, B., Motais, R., 1988. Adaptive respiratory responses of trout to acute hypoxia. II. Blood oxygen carrying properties during hypoxia. Respir. Physiol. 74, 91–98.

Clark, T.D., Sandblom, E., Cox, G.K., Hinch, S.G., Farrell, A.P., 2008. Circulatory limits to oxygen supply during an acute temperature increase in the Chinook salmon (*Oncorhynchus tshawytscha*). Am. J. Phys. Regul. Integr. Comp. Phys. 295, R1631–R1639.

Clark, T.D., Rummer, J., Sepulveda, C., Farrell, A., Brauner, C., 2010. Reduced and reversed temperature dependence of blood oxygenation in an ectothermic scombrid fish: implications for the evolution of regional heterothermy? J. Comp. Physiol. B. 180, 73–82.

Cocca, E., Ratnayake-Lecamwasam, M., Parker, S.K., Camardella, L., Ciaramella, M., di Prisco, G., Detrich, H.W., 1997. Do the hemoglobinless icefishes have globin genes? Comp. Biochem. Physiol. 118, 1027–1030.

Conley, D.M., Mallatt, J., 1988. Histochemical localization of Na^{+}-K^{+} ATPase and carbonic anhydrase activity in gills of 17 fish species. Can. J. Zool. 66, 2398–2405.

Cooper, C.A., Regan, M.D., Brauner, C.J., De Bastos, E.S.R., Wilson, R.W., 2014. Osmoregulatory bicarbonate secretion exploits H^{+}-sensitive haemoglobins to autoregulate intestinal O_2 delivery in euryhaline teleosts. J. Comp. Physiol. B. 184, 865–876.

Cossins, A.R., Kilbey, R.V., 1989. The seasonal modulation of Na^{+}/H^{+} exchanger activity in trout erythrocytes. J. Exp. Biol. 144, 463–478.

Cossins, A.R., Kilbey, R.V., 1990. The temperature dependence of the adrenergic Na^{+}/H^{+} exchanger of trout erythrocytes. J. Exp. Biol. 148, 303–312.

Cossins, A.R., Richardson, P.A., 1985. Adrenaline-induced Na^{+}/H^{+} exchange in trout erythrocytes and its effects upon oxygen carrying capacity. J. Exp. Biol. 118, 229–246.

Cousins, K., Farrell, A., 1996. Stretch-induced release of atrial natriuretic factor from the heart of rainbow trout (*Oncorhynchus mykiss*). Can. J. Zool. 74, 380–387.

Cousins, K., Farrell, A., Sweeting, R., Vesely, D., Keen, J., 1997. Release of atrial natriuretic factor prohormone peptides 1-30, 31-67 and 99-126 from freshwater-and seawater-acclimated perfused trout (*Oncorhynchus mykiss*) hearts. J. Exp. Biol. 200, 1351–1362.

Crandall, E.D., Bidani, A., 1981. Effects of red blood cell HCO_3^{-}/Cl^{-} exchange kinetics on lung CO_2 transfer: theory. J. Appl. Physiol. 50, 265–271.

Cross, G.A., 1987. Eukaryotic protein modification and membrane attachment via phosphatidylinositol. Cell 48, 179–181.

Currie, S., Kieffer, J.D., Tufts, B.L., 1995. The effects of blood CO_2 reaction rates on CO_2 removal from muscle in exercised trout. Respir. Physiol. 100, 261–269.

Davie, P.S., Farrell, A.P., 1991. Cardiac performance of an isolated heart preparation from the dogfish (*Squalus acanthias*): the effects of hypoxia and coronary artery perfusion. Can. J. Zool. 69, 1822–1828.

Davie, P.S., Wells, R.M.G., Tetens, V., 1986. Effects of sustained swimming on rainbow trout muscle structure, blood oxygen transport, and lactate dehydrogenase isozymes: evidence for increased aerobic capacity of white muscle. J. Exp. Zool. 237, 159–171.

Davison, W., 1997. The effects of exercise training on teleost fish, a review of recent literature. Comp. Biochem. Physiol. A Physiol. 117, 67–75.

Davison, W., Forster, M.E., Franklin, C.E., Taylor, H.H., 1988. Recovery from exhausting exercise in an Antarctic fish, *Pagothenia borchgrevinki*. Polar Biol. 8, 167–171.

Decker, B., Sender, S., Gros, G., 1996. Membrane-associated carbonic anhydrase IV in skeletal muscle: subcellular localization. Histochem. Cell Biol. 106, 405–411.

Dejours, P., 1981. Principles of Comparative Respiratory Physiology. Elsevier/North-Holland Biomedical Press, Amsterdam.

Dermietzel, R., Leibstein, A., Siffert, W., Zamboglou, N., Gros, G., 1985. A fast screening method for histochemical localization of carbonic anhydrase. Application to kidney, skeletal muscle, and thrombocytes. J. Histochem. Cytochem. 33, 93–98.

Desforges, P.R., Gilmour, K., Perry II, S.F., 2001. The effects of exogenous extracellular carbonic anhydrase on CO_2 excretion in rainbow trout (*Oncorhynchus mykiss*): role of plasma buffering capacity. J. Comp. Physiol. B. 171, 465–473.

Desforges, P.R., Harman, S.S., Gilmour, K.M., Perry, S.F., 2002. Sensitivity of CO_2 excretion to blood flow changes in trout is determined by carbonic anhydrase availability. Am. J. Phys. Regul. Integr. Comp. Phys. 282, R501–R508.

DeWilde, M.A., Houston, A., 1967. Hematological aspects of the thermoacclimatory process in the rainbow trout, *Salmo gairdneri*. Can. J. Fish. Aquat. Sci. 24, 2267–2281.

Dhindsa, D.S., Hoversland, A.S., Neill, W.A., Metcalfe, J., 1971. Changes in blood oxygen affinity and hemodynamics in anemic dogs. Respir. Physiol. 11, 346–353.

di Prisco, G., 1998. Molecular adaptations in Antarctic fish hemoglobins. In: di Prisco, G., Pisano, E., Clark, M. (Eds.), Fishes of Antarctica. A Biological Overview. Springer, Milan, pp. 339–353.

di Prisco, G., Condo, S.G., Tamburrini, M., Giardina, B., 1991a. Oxygen transport in extreme environments. Trends Biochem. Sci. 16, 471–474.

di Prisco, G., D'Avino, R., Caruso, C., Tamburini, M., Camardella, L., Rutigliano, B., Carratore, V., Romano, M., 1991b. The biochemistry of oxygen transport in the red-blooded Antarctic fish. In: di Prisco, G., Maresca, B., Tota, B. (Eds.), Biology of Antarctic Fish. Springer, Berlin, pp. 263–281.

Dimberg, K., 1988. Inhibition of carbonic anhydrase *in vivo* in the freshwater-adapted rainbow trout: differential effects of acetazolamide and metazolamide on blood CO_2 levels. Comp. Biochem. Physiol. A Physiol. 91, 253–258.

Dimberg, K., 1994. The carbonic anhydrase inhibitor in trout plasma: purification and its effect on carbonic anhydrase activity and the Root effect. Fish Physiol. Biochem. 12, 381–386.

Dimberg, K., Höglund, L., Knutsson, P., Ridderstråle, Y., 1981. Histochemical localization of carbonic anhydrase in gill lamellae from young salmon (*Salmo salar* L) adapted to fresh and salt water. Acta Physiol. Scand. 112, 218–220.

Dobson, G., Baldwin, J., 1982. Regulation of blood oxygen affinity in the Australian blackfish *Gadopsis Marmoratus*: II. Thermal acclimation. J. Exp. Biol. 99, 245–254.

Dougan, M.C, 1993. Growth and development of chinook salmon, Oncorhynchus tshawytscha: effects of exercise training and seawater transfer. In: Zoology, vol. PhD. University of Canterbury, Christchurch, NZ.

Driedzic, W.R., Hochachka, P.W., 1978. Metabolism in fish during exercise. In: Hoar, W.S., Randall, D. (Eds.), Locomotion. In: Fish Physiology, vol. 7. Academic press, New York, pp. 503–544.

Dudley, R., Gans, C., 1991. A critique of symmorphosis and optimality models in physiology. Physiol. Zool. 64, 627–637.

Duhm, J., 1972. The effect of 2, 3-DPG and other organic phosphates on the Donnan equilibrium and the oxygen affinity of human blood. In: Oxygen Affinity of Hemoglobin and Red Cell Acid Base Status. Academic Press, New York, pp. 583–594.

Duhm, J., 1976. Influence of 2, 3-diphosphoglycerate on the buffering properties of human blood. Pflugers Arch. 363, 61–67.

Eastman, J.T., 2005. The nature of the diversity of Antarctic fishes. Polar Biol. 28, 93–107.

Eastman, J.T., Lannoo, M.J., 2003. Diversification of brain and sense organ morphology in Antarctic dragonfishes (Perciformes: Notothenioidei: Bathydraconidae). J. Morphol. 258, 130–150.

Eastman, J.T., Lannoo, M.J., 2004. Brain and sense organ anatomy and histology in hemoglobinless Antarctic icefishes (Perciformes: Notothenioidei: Channichthyidae). J. Morphol. 260, 117–140.

Eaton, J.W., Brewer, G.J., 1968. The relationship between red cell 2,3-Diphosphoglycerate and levels of hemoglobin in the human. PNAS 61, 756–760.

Eaton, J.W., Skelton, T., Berger, E., 1974. Survival at extreme altitude: protective effect of increased hemoglobin-oxygen affinity. Science 183, 743–744.

Effros, R.M., Weissman, M.L., 1979. Carbonic anhydrase activity of the cat hind leg. J. Appl. Physiol. 47, 1090–1098.

Egginton, S., 1987. Effects of an anabolic hormone on aerobic capacity of rat striated muscle. Pflugers Arch. 410, 356–361.

Egginton, S., 1994. Stress response in two Antarctic teleosts (*Notothenia coriiceps* Richardson and *Chaenocephalus aceratus* Lönnberg) following capture and surgery. J. Comp. Physiol. B. 164, 482–491.

Egginton, S., 1996. Blood rheology of Antarctic fishes: viscosity adaptations at very low temperatures. J. Fish Biol. 48, 513–521.

Egginton, S., 2002. Temperature and angiogenesis: the possible role of mechanical factors in capillary growth. Comp. Biochem. Physiol. A Mol. Integr. Physiol. 132, 773–787.

Egginton, S., Gaffney, E., 2010. Tissue capillary supply–it's quality not quantity that counts! Exp. Physiol. 95, 971–979.

Egginton, S., Taylor, E.W., Wilson, R.W., Johnston, I.A., Moon, T.W., 1991. Stress response in the Antarctic teleosts (*Notothenia neglecta* and *N. rossi* Richardson). J. Fish Biol. 38, 225–235.

Eigen, M., DeMaeyer, L., 1963. Relaxation methods. In: Friess, S.L., Lewis, E.S., Weissberger, A. (Eds.), Investigation of Rates and Mechanisms of Reactions. In: Techniques of Organic Chemistry, vol. 8. Interscience, New York, pp. 895–1054.

Eliason, E.J., Anttila, K., 2017. Temperature and the cardiovascular system. In: Gamperl, A.K., Gillis, T.E., Farrell, A.P., Brauner, C.J. (Eds.), The Cardiovascular System: Development, Plasticity and Physiological Responses, Fish Physiology. vol. 36B. Academic Press, San Diego. pp. 235–297. Chapter 4.

Ellis, C.G., Ellsworth, M.L., Pittman, R., 1990. Determination of red blood cell oxygenation *in vivo* by dual video densitometric image analysis. Am. J. Phys. 258, H1216–H1223.

Esbaugh, A., Tufts, B., 2004. Evidence for a membrane-bound carbonic anhydrase in the heart of an ancient vertebrate, the sea lamprey (*Petromyzon marinus*). J. Comp. Physiol. B. 174, 399–406.

Esbaugh, A.J., Tufts, B.L., 2006. The structure and function of carbonic anhydrase isozymes in the respiratory system of vertebrates. Respir. Physiol. Neurobiol. 154, 185–198.

Esbaugh, A., Lund, S., Tufts, B.L., 2004. Comparative physiology and molecular analysis of carbonic anhydrase from the red blood cells of teleost fish. J. Comp. Physiol. B. 174, 429–438.

Esbaugh, A.J., Perry, S.F., Bayaa, M., Georgalis, T., Nickerson, J., Tufts, B.L., Gilmour, K.M., 2005. Cytoplasmic carbonic anhydrase isozymes in rainbow trout *Oncorhynchus mykiss*: comparative physiology and molecular evolution. J. Exp. Biol. 208, 1951–1961.

Esbaugh, A.J., Gilmour, K., Perry, S., 2009. Membrane-associated carbonic anhydrase in the respiratory system of the Pacific hagfish (*Eptatretus stouti*). Respir. Physiol. Neurobiol. 166, 107–116.

Evans, D.H., Piermarini, P.M., Choe, K.P., 2005. The multifunctional fish gill: dominant site of gas exchange, osmoregulation, acid–base regulation, and excretion of nitrogenous waste. Physiol. Rev. 85, 97–177.

Everson, I., Ralph, R., 1968. Blood analyses of some Antarctic fish. Br. Antarctic Surv. Bull. 15, 59–62.

Fago, A., Wells, R.M.G., Weber, R.E., 1997. Temperature-dependent enthalpy of oxygenation in Antarctic fish hemoglobins. Comp. Biochem. Physiol. 118, 319–326.
Fairbanks, M.B., Hoffert, J.R., Fromm, P.O., 1969. The dependence of the oxygen-concentrating mechanism of the teleost eye (*Salmo gairdneri*) on the enzyme carbonic anhydrase. J. Gen. Physiol. 54, 203–211.
Fandrey, J., 2004. Oxygen-dependent and tissue-specific regulation of erythropoietin gene expression. Am. J. Phys. Regul. Integr. Comp. Phys. 286, R977–R988.
Fänge, R., 1992. Fish blood cells. In: Hoar, W.S., Randall, D., Farrell, A. (Eds.), The Cardiovascular System. Fish Physiology, vol. 7B. Academic press, New York.
Fänge, R., Nilsson, S., 1985. The fish spleen: structure and function. Experientia 41, 152–158.
Farlinger, S., Beamish, F., 1978. Changes in blood chemistry and critical swimming speed of largemouth bass, *Micropterus salmoides*, with physical conditioning. Trans. Am. Fish. Soc. 107, 523–527.
Farmer, M., 1979. The transition from water to air breathing: effects of CO_2 on hemoglobin function. Comp. Biochem. Physiol. 62A, 109–114.
Farrell, A.P., 1991. From hagfish to tuna: a perspective on cardiac function in fish. Physiol. Zool. 64, 1137–1164.
Farrell, A., 1992. Cardiac output in fish: regulation and limitations. In: Eduardo, J., Bicudo, P.W. (Eds.), The Vertebrate Gas Transport Cascade: Adaptions to Environmet and Mode of Life. CRC Press, Boca Raton, pp. 208–214.
Farrell, A., 1997. Effects of temperature on cardiovascular performance. In: Wood, C.M., McDonald, D.G. (Eds.), Global Warming: Implications for Freshwater and Marine Fish. In: Society for Experimental Biology Seminar Series, vol. 61. Cambridge University Press, Cambridge, pp. 135–158.
Farrell, A., 2007. Cardiorespiratory performance during prolonged swimming tests with salmonids: a perspective on temperature effects and potential analytical pitfalls. Philos. Trans. R. Soc. B 362, 2017–2030.
Farrell, A., Clutterham, S., 2003. On-line venous oxygen tensions in rainbow trout during graded exercise at two acclimation temperatures. J. Exp. Biol. 206, 487–496.
Farrell, A., Jones, D.R., 1992. The heart. In: Hoar, W.S., Randall, D., Farrell, A. (Eds.), The Cardiovascular System. In: Fish Physiology, vol. 7A. Academic press, New York, pp. 1–88.
Farrell, A.P., Richards, J.G., 2009. Defining hypoxia: an integrative synthesis of the responses of fish to hypoxia. In: Farrell, A.P., Brauner, C.J. (Eds.), Hypoxia. In: Fish Physiology, vol. 27. Academic Press, New York, pp. 487–503.
Farrell, A.P., Smith, F., 2017. Cardiac form, function and physiology. In: Gamperl, A.K., Gillis, T.E., Farrell, A.P., Brauner, C.J. (Eds.), The Cardiovascular System: Morphology, Control and Function. Fish Physiology, vol. 36A. Academic Press, San Diego, pp. 155–264.
Farrell, A.P., Eliason, E.J., Clark, T.D., Steinhausen, M.F., 2014. Oxygen removal from water versus arterial oxygen delivery: calibrating the Fick equation in Pacific salmon. J. Comp. Physiol. B. 184, 855–864.
Feller, G., Gerday, C., 1997. Adaptations of the hemoglobinless Antarctic icefish (Channichthyidae) to hypoxia tolerance. Comp. Biochem. Physiol. A Physiol. 118, 981–987.
Feller, G., Poncin, A., Aittaleb, M., Schyns, R., Gerday, C., 1994. The blood proteins of the Antarctic icefish *Channichthys rhinoceratus*: biological significance and purification of the two main components. Comp. Biochem. Physiol. B 109, 89–97.
Ferguson, R., Boutilier, R., 1988. Metabolic energy production during adrenergic pH regulation in red cells of the Atlantic salmon, *Salmo salar*. Respir. Physiol. 74, 65–75.
Ferguson, R.A., Boutilier, R.G., 1989. Metabolic-membrane coupling in red blood cells of trout: effects of anoxia and adrenergic stimulation. J. Exp. Biol. 143, 149–164.

Ferguson, R., Tufts, B., Boutilier, R., 1989. Energy metabolism in trout red cells: consequences of adrenergic stimulation *in vivo* and *in vitro*. J. Exp. Biol. 143, 133–147.

Ferguson, R.A., Sehdev, N., Bagatto, B., Tufts, B.L., 1992. *In vitro* interactions between oxygen and carbon dioxide transport in the blood of the sea lamprey (*Petromyzon marinus*). J. Exp. Biol. 173, 25–41.

Fick, A., 1870. Ueber die Messung des Blutquantums in den Herzventrikeln. Sitz. Physik. Med. Ges 2, 16.

Fievet, B., Claireaux, G., Thomas, S., Motais, R., 1988. Adaptive respiratory responses of trout to acute hypoxia. III. Ion movements and pH changes in the red blood cell. Respir. Physiol. 74, 99–113.

Fletcher, G.L., Haedrich, R.T., 1987. Rheological properties of rainbow trout blood. Can. J. Zool. 65, 879–883.

Flügel, C., Lütjen-Drecoll, E., Zadunaisky, J.A., 1991. Histochemical demonstration of carbonic anhydrase in gills and opercular epithelium of seawater-and freshwater-adapted killyfish (*Fundulus heteroclitus*). Acta Histochem. 91, 67–75.

Forster, R.E., 1969. The rate of CO_2 equilibration between red cells and plasma. In: Forster, R.E., Edsall, J.T., Otis, A.B., Roughton, F.J.W. (Eds.), In: CO2, Chemical, Biological and Physiological Aspects, vol. 188. NASA Special Publication, Washington, pp. 275–284.

Forster, R.E., Crandall, E.D., 1975. Time course of exchanges between red cells and extracellular fluid during CO_2 uptake. J. Appl. Physiol. 38, 710–718.

Forster, R.E., Dodgson, S.J., 2000. Membrane transport and provision of substrates for carbonic anhydrase: in vertebrates. In: Dodgeson, S.J., Tashian, R.E., Gros, G., Carter, N.D. (Eds.), The Carbonic Anhydrases. Springer, New York, pp. 263–280.

Forster, R.E., Steen, J.B., 1969. The rate of the 'Root shift' in eel red cells and eel haemoglobin solutions. J. Physiol. 204, 259–282.

Franklin, C.E., Davison, W., McKenzie, J., 1993. The role of the spleen during exercise in the Antarctic teleost, *Pagothenia borchgrevinki*. J. Exp. Biol. 174, 381–386.

Frey, B.J., Weber, R.E., Van, A.W.J., Fago, A., 1998. The haemoglobin system of the mudfish, *Labeo capensis*: adaptations to temperature and hypoxia. Comp. Biochem. Physiol. B 120, 735–742.

Fritsche, R., Nilsson, S., 1993. Cardiovascular and ventilatory control during hypoxia. In: Rankin, J.C., Jensen, F.B. (Eds.), Fish Ecophysiology. Chapman & Hall, London, pp. 180–206.

Funder, J., Wieth, J.O., 1966. Chloride and hydrogen ion distribution between human red cells and plasma. Acta Physiol. Scand. 68, 234–245.

Fyhn, U., Brix, O., Naevdal, G., Johansen, T., 1994. In: New variants of the haemoglobins of Atlantic cod: a tool for discriminating between coastal and Arctic cod populations. ICES Marine Science Symposia. vol. 198. International Council for the Exploration of the Sea, Copenhagen, Denmark, pp. 666–670.

Gallaugher, P.E., 1994. The Role of Haematocrit in Oxygen Transport and Swimming in Salmonid Fishes (Ph.D.). Simon Fraser University, Vancouver.

Gallaugher, P., Farrell, A.P., 1998. Hematocrit and blood oxygen-carrying capacity. In: Perry II, S.F., Tufts, B.L. (Eds.), Fish Respiration. In: Fish Physiology, vol. 17. Academic press, New York, pp. 185–227.

Gallaugher, P., Axelsson, M., Farrell, A.P., 1992. Swimming performance and haematological variables in splenectomized rainbow trout, *Oncorhynchus mykiss*. J. Exp. Biol. 171, 301–314.

Gallaugher, P., Thorarensen, H., Farrell, A.P., 1995. Hematocrit in oxygen transport and swimming in rainbow trout (*Oncorhynchus mykiss*). Respir. Physiol. 102, 279–292.

Gallaugher, P.E., Thorarensen, H., Kiessling, A., Farrell, A.P., 2001. Effects of high intensity exercise training on cardiovascular function, oxygen uptake, internal oxygen transport and

osmotic balance in chinook salmon (*Oncorhynchus tshawytscha*) during critical speed swimming. J. Exp. Biol. 204, 2861–2872.
Gamperl, A.K., Driedzic, W., 2009. Cardiovascular function and cardiac metabolism. Fish Physiol. 27, 301–360.
Garcia-Romeu, F., Motais, R., Borgese, F., 1988. Desensitization by external Na^+ of the cyclic AMP-dependent Na^+/H^+ antiporter in trout red blood cells. J. Gen. Physiol. 91, 529–548.
Geers, C., Gros, G., 1984. Inhibition properties and inhibition kinetics of an extracellular carbonic anhydrase in perfused skeletal muscle. Respir. Physiol. 56, 269–287.
Geers, C., Gros, G., 1988. Carbonic anhydrase inhibition affects contraction of directly stimulated rat soleus. Life Sci. 42, 37–45.
Geers, C., Gros, G., 2000. Carbon dioxide transport and carbonic anhydrase in blood and muscle. Physiol. Rev. 80, 681–715.
Geers, C., Gros, G., Gartner, A., 1985. Extracellular carbonic anhydrase of skeletal muscle associated with the sarcolemma. J. Appl. Physiol. 59, 548–558.
Genz, J., Taylor, J.R., Grosell, M., 2008. Effects of salinity on intestinal bicarbonate secretion and compensatory regulation of acid-base balance in *Opsanus beta*. J. Exp. Biol. 211, 2327–2335.
Georgalis, T., Gilmour, K., Yorston, J., Perry, S.F., 2006. Roles of cytosolic and membrane-bound carbonic anhydrase in renal control of acid-base balance in rainbow trout, *Oncorhynchus mykiss*. Am. J. Phys. 291, F407–F421.
Gervais, M.R., Tufts, B.L., 1998. Evidence for membrane-bound carbonic anhydrase in the air bladder of bowfin (*Amia calva*), a primitive air-breathing fish. J. Exp. Biol. 201, 2205–2212.
Gervais, M.R., Tufts, B.L., 1999. Characterization of carbonic anhydrase and anion exchange in the erythrocytes of bowfin (*Amia calva*), a primitive air-breathing fish. Comp. Biochem. Physiol. A Mol. Integr. Physiol. 123, 343–350.
Ghandour, M., Langley, O., Zhu, X., Waheed, A., Sly, W., 1992. Carbonic anhydrase IV on brain capillary endothelial cells: a marker associated with the blood-brain barrier. PNAS 89, 6823–6827.
Giardina, B., Mosca, D., De Rosa, M.C., 2004. The Bohr effect of haemoglobin in vertebrates: an example of molecular adaptation to different physiological requirements. Acta Physiol. Scand. 182, 229–244.
Gibson, J., Cossins, A., Ellory, J., 2000. Oxygen-sensitive membrane transporters in vertebrate red cells. J. Exp. Biol. 203, 1395–1407.
Gillen, R.G., Riggs, A., 1973. Structure and function of the isolated hemoglobins of the American eel, (*Anguilla rostrata*). J. Biol. Chem. 248, 1961–1969.
Gillen, R.G., Riggs, A., 1977. Enhancement of alkaline bohr effect of some fish hemoglobins with adenosine-triphosphate. Arch. Biochem. Biophys. 183, 678–685.
Gilmour, K., 1998a. Causes and consequences of acid-base disequilibria. In: Perry II, S.F., Tufts, B.L. (Eds.), Fish Respiration. In: Fish Physiology, vol. 17. Academic Press, New York, pp. 321–348.
Gilmour, K.M., 1998b. The disequilibrium pH: a tool for the localization of carbonic anhydrase. Comp. Biochem. Physiol. A Mol. Integr. Physiol. 119, 243–254.
Gilmour, K.M., 2011. Perspectives on carbonic anhydrase. Comp. Biochem. Physiol. A Mol. Integr. Physiol. 157, 193–197.
Gilmour, K.M., 2012. New insights into the many functions of carbonic anhydrase in fish gills. Respir. Physiol. Neurobiol. 184, 223–230.
Gilmour, K., Desforges, P., 2000. HCO_3^- dehydration in the plasma of rainbow trout: the role of buffering capacity. In: Ion Transfer Across Fish Gills. Proceedings of International Fish Physiology Symposium, pp. 23–27.

Gilmour, K.M., Perry, S.F., 1994. The effects of hypoxia, hypercapnia on the acid-base disequilibrium in the arterial blood of rainbow trout. J. Exp. Biol. 192, 269–284.
Gilmour, K.M., Perry, S.F., 1996. The effects of experimental anaemia on CO_2 excretion *in vitro* in rainbow trout, *Oncorhynchus mykiss*. Fish Physiol. Biochem. 15, 83–94.
Gilmour, K.M., Perry, S.F., 2004. Branchial membrane-associated carbonic anhydrase activity maintains CO_2 excretion in severely anemic dogfish. Am. J. Phys. Regul. Integr. Comp. Phys. 286, R1138–R1148.
Gilmour, K.M., Perry, S.F., 2009. Carbonic anhydrase and acid-base regulation in fish. J. Exp. Biol. 212, 1647–1661.
Gilmour, K.M., Perry, S.F., 2010. Gas transfer in dogfish: a unique model of CO_2 excretion. Comp. Biochem. Physiol. A Mol. Integr. Physiol. 155, 476–485.
Gilmour, K.M., Randall, D., Perry, S.F., 1994. Acid-base disequilibrium in the arterial blood of rainbow trout. Respir. Physiol. 96, 259–272.
Gilmour, K.M., Henry, R.P., Wood, C.M., Perry, S.F., 1997. Extracellular carbonic anhydrase and an acid-base disequilibrium in the blood of the dogfish (*Squalus acanthias*). J. Exp. Biol. 200, 173–183.
Gilmour, K.M., Perry, S.F., Bernier, N.J., Henry, R.P., Wood, C.M., 2001. Extracellular carbonic anhydrase in the dogfish, *Squalus acanthias*: a role in CO_2 excretion. Physiol. Biochem. Zool. 74, 477–492.
Gilmour, K.M., Shah, B., Szebedinszky, C., 2002. An investigation of carbonic anhydrase activity in the gills and blood plasma of brown bullhead (*Ameiurus nebulosus*), longnose skate (*Raja rhina*), and spotted ratfish (*Hydrolagus colliei*). J. Comp. Physiol. B. 172, 77–86.
Gilmour, K.M., Desforges, P.R., Perry, S.F., 2004. Buffering limits plasma HCO_3^- dehydration when red blood cell anion exchange is inhibited. Respir. Physiol. Neurobiol. 140, 173–187.
Gilmour, K., Bayaa, M., Kenney, L., McNeill, B., Perry, S., 2007. Type IV carbonic anhydrase is present in the gills of spiny dogfish (*Squalus acanthias*). Am. J. Phys. Regul. Integr. Comp. Phys. 292, R556–R567.
Goldman, D., 2008. Theoretical models of microvascular oxygen transport to tissue. Microcirculation 15, 795–811.
Golub, A.S., Popel, A.S., Zheng, L., Pittman, R.N., 1997. Analysis of phosphorescence in heterogeneous systems using distributions of quencher concentration. Biophys. J. 73, 452.
Golub, A.S., Barker, M.C., Pittman, R.N., 2007. PO_2 profiles near arterioles and tissue oxygen consumption in rat mesentery. Am. J. Phys. 293, H1097–H1106.
Gonzalez, R.J., Milligan, L., Pagnotta, A., McDonald, D., 2001. Effect of air breathing on acid-base and ion regulation after exhaustive exercise and during low pH exposure in the bowfin, *Amia calva*. Physiol. Biochem. Zool. 74, 502–509.
Graham, J.B., 1997. Air-Breathing Fishes: Evolution, Diversity, and Adaptation. Academic Press, New York.
Graham, M.S., Fletcher, G.L., 1983. Blood and plasma viscosity of winter flounder: influence of temperature, red cell concentration, and shear rate. Can. J. Zool. 61, 2344–2350.
Greaney, G.S., Powers, D.A., 1977. Cellular regulation of an allosteric modifier of fish haemoglobin. Nature 270, 73–74.
Greaney, G.S., Powers, D.A., 1978. Allosteric modifiers of fish hemoglobins: *in vitro* and *in vivo* studies of the effect of ambient oxygen and pH on erythrocyte ATP concentrations. J. Exp. Zool. 203, 339–349.
Greaney, G.S., Place, A.R., Cashon, R.E., Smith, G., Powers, D.A., 1980. Time course of changes in enzyme activities and blood respiratory properties of killifish during long-term acclimation to hypoxia. Physiol. Zool. 53, 136–144.

Gronenborn, A.M., Clore, G.M., Brunori, M., Giardina, B., Falcioni, G., Perutz, M.F., 1984. Stereochemistry of ATP and GTP bound to fish haemoglobins: a transferred nuclear overhauser enhancement, ^{31}P-nuclear magnetic resonance, oxygen equilibrium and molecular modelling study. J. Mol. Biol. 178, 731–742.

Gros, G., Dodgson, S., 1988. Velocity of CO_2 exchange in muscle and liver. Annu. Rev. Physiol. 50, 669–694.

Grosell, M., Genz, J., 2006. Ouabain-sensitive bicarbonate secretion and acid absorption by the marine teleost fish intestine play a role in osmoregulation. Am. J. Phys. 291, R1145–R1156.

Grosell, M., Wood, C.M., Wilson, R.W., Bury, N.R., Hogstrand, C., Rankin, J.C., Jensen, F.B., 2005. Bicarbonate secretion plays a role in chloride and water absorption of the European flounder intestine. Am. J. Phys. 288, R936–R946.

Guderley, H., 2004. Metabolic responses to low temperature in fish muscle. Biol. Rev. 79, 409–427.

Guizouarn, H., Borgese, F., Pellissier, B., Garcia-Romeu, F., Motais, R., 1993. Role of protein phosphorylation and dephosphorylation in activation and desensitization of the cAMP-dependent Na^+/H^+ antiport. J. Biol. Chem. 268, 8632–8639.

Hall, F., Dill, D., Barron, E., 1936. Comparative physiology in high altitudes. J. Cell. Comp. Physiol. 8, 301–313.

Härdig, J., Olsson, L., Höglund, L.B., 1978. Autoradiography on erythrokinesis and multihemoglobins in juvenile *Salmo salar* L. at various respiratory gas regimes. Acta Physiol. Scand. 103, 240–251.

Harken, A.H., Woods, M., 1976. The influence of oxyhemoglobin affinity on tissue oxygen consumption. Ann. Surg. 183, 130.

Haswell, M.S., Randall, D.J., Perry, S.F., 1980. Fish gill carbonic anhydrase: acid-base regulation or salt transport? Am. J. Phys. Regul. Integr. Comp. Phys. 238, R240–R245.

Haswell, M.S., Raffin, J.-P., Leray, C., 1983. An investigation of the carbonic anhydrase inhibitor in eel plasma. Comp. Biochem. Physiol. A Comp. Physiol. 74, 175–177.

Heasley, L.E., Brunton, L.L., 1985. Prostaglandin A1 metabolism and inhibition of cyclic AMP extrusion by avian erythrocytes. J. Biol. Chem. 260, 11514–11519.

Heasley, L., Watson, M., Brunton, L., 1985. Putative inhibitor of cyclic AMP efflux: chromatography, amino acid composition, and identification as a prostaglandin A1-glutathione adduct. J. Biol. Chem. 260, 11520–11523.

Hebbel, R.P., Eaton, J.W., Kronenberg, R.S., Zanjani, E.D., Moore, L.G., Berger, E.M., 1978. Human llamas. Adaptation to altitude in subjects with high hemoglobin oxygen affinity. J. Clin. Invest., 593–600.

Heinz, E., 1981. Electrical Potentials in Biological Membrane Transport. Springer, Berlin.

Heise, K., Puntarulo, S., Nikinmaa, M., Lucassen, M., Pörtner, H.-O., Abele, D., 2006. Oxidative stress and HIF-1 DNA binding during stressful cold exposure and recovery in the North Sea eelpout (*Zoarces viviparus*). Comp. Biochem. Physiol. A Mol. Integr. Physiol. 143, 494–503.

Helbo, S., Fago, A., 2011. Allosteric modulation by S-nitrosation in the low-O_2 affinity myoglobin from rainbow trout. Am. J. Phys. Regul. Integr. Comp. Phys. 300, R101–R108.

Helbo, S., Weber, R.E., Fago, A., 2013. Expression patterns and adaptive functional diversity of vertebrate myoglobins. Biochim. Biophys. Acta, Proteins Proteomics 1834, 1832–1839.

Helfman, G., Collette, B.B., Facey, D.E., Bowen, B.W., 2009. The Diversity of Fishes: Biology, Evolution, and Ecology. Wiley-Blackwell, Oxford.

Heming, T.A., 1984. The role of fish erythrocytes in transport and excretion of carbon dioxide. In: Zoology. UBC, Vancouver (Ph.D.).

Heming, T.A., Bidani, A., 1992. Influence of proton availability on intracapillary CO_2-HCO_3^--H^+ reactions in isolated rat lungs. J. Appl. Physiol. 72, 2140–2148.
Heming, T., Watson, T., 1986. Activity and inhibition of carbonic anhydrase in *Amia calva*, a bimodal-breathing holostean fish. J. Fish Biol. 28, 385–392.
Heming, T.A., Randall, D.J., Boutilier, R.G., Iwama, G.K., Primmett, D., 1986. Ionic equilibria in red blood cells of rainbow trout (*Salmo gairdneri*): Cl^-, HCO_3^- and H^+. Respir. Physiol. 65, 223–234.
Heming, T.A., Vanoye, C.G., Stabenau, E.K., Roush, E.D., Fierke, C.A., Bidani, A., 1993. Inhibitor sensitivity of pulmonary vascular carbonic-anhydrase. J. Appl. Physiol. 75, 1642–1649.
Heming, T.A., Stabenau, E.K., Vanoye, C.G., Moghadasi, H., Bidani, A., 1994. Roles of intra- and extracellular carbonic anhydrase in alveolar-capillary CO_2 equilibration. J. Appl. Physiol. 77, 697–705.
Hemmingsen, E.A., 1991. Respiratory and cardiovascular adaptations in hemoglobin-free fish: resolved and unresolved problems. In: DiPrisco, G., Maresca, B., Tota, B. (Eds.), Biology of Antarctic fish, Vol. 2. Springer, New York, pp. 191–203.
Hemmingsen, E., Douglas, E., 1972. Respiratory and circulatory responses in a hemoglobin-free fish, *Chaenocepahlus aceratus*, to changes in temperature and oxygen tension. Comp. Biochem. Physiol. A Comp. Physiol. 43, 1031–1043.
Hemptinne, A.d., Marrannes, R., Vanheel, B., 1987. Surface pH and the control of intracellular pH in cardiac and skeletal muscle. Can. J. Physiol. Pharmacol. 65, 970–977.
Henry, R.P., 1984. The role of carbonic anhydrase in blood ion and acid-base regulation. Am. Zool. 24, 241–251.
Henry, R.P., 1991. Techniques for measuring carbonic anhydrase activity *in vitro*. In: Dogdson, S., Tashian, R.E., Gros, G., Carter, N.D. (Eds.), The Carbonic Anhydrases. Springer, New York, pp. 119–125.
Henry, R.P., Heming, T., 1998. Carbon anhydrase and respiratory gas exchange. In: Perry II, S.F., Tuffs, B.L. (Eds.), Fish Respiration. In: Fish Physiology, vol. 17. Academic Press, New York, pp. 75–112.
Henry, R.P., Kormanik, G.A., 1985. Carbonic anhydrase activity and calcium deposition during the molt cycle of the blue crab *Callinectes sapidus*. J. Crustac. Biol. 5, 234–241.
Henry, R.P., Swenson, E.R., 2000. The distribution and physiological significance of carbonic anhydrase in vertebrate gas exchange organs. Respir. Physiol. 121, 1–12.
Henry, R.P., Smatresk, N.J., Cameron, J.N., 1988. The distribution of branchial carbonic anhydrase and the effects of gill and erythrocyte carbonic anhydrase inhibition in the channel catfish, *Ictalurus punctatus*. J. Exp. Biol. 134, 201–218.
Henry, R.P., Tufts, B.L., Boutilier, R.G., 1993. The distribution of carbonic anhydrase type I and II isozymes in lamprey and trout: possible co-evolution with erythrocyte chloride/bicarbonate exchange. J. Comp. Physiol. B. 163, 380–388.
Henry, R.P., Gilmour, K.M., Wood, C.M., Perry, S.F., 1997a. Extracellular carbonic anhydrase activity and carbonic anhydrase inhibitors in the circulatory system of fish. Physiol. Zool. 70, 650–659.
Henry, R.P., Wang, Y., Wood, C.M., 1997b. Carbonic anhydrase facilitates CO_2 and NH_3 transport across the sarcolemma of trout white muscle. Am. J. Phys. 272, 1754–1761.
Herbert, N., Skov, P., Wells, R., Steffensen, J., 2006. Whole blood-oxygen binding properties of four cold-temperate marine fishes: blood affinity is independent of pH-dependent binding, routine swimming performance, and environmental hypoxia. Physiol. Biochem. Zool. 79, 909–918.
Hernandez, M., Mendiola, P., de Costa, J., Zamora, S., 2002. Effects of intense exercise training on rainbow trout growth, body composition and metabolic responses. J. Physiol. Biochem. 58, 1–7.

Hevesy, G., Lockner, D., Sletten, K., 1964. Iron metabolism and erythrocyte formation in fish. Acta Physiol. Scand. 60, 256–266.
Hewett-Emmett, D., 2000. Evolution and distribution of the carbonic anhydrase gene families. In: Chegwidden, W.R., Carter, N.D., Edwards, Y.H. (Eds.), The Carbonic Anhydrases: New Horizons. Springer, Berlin, pp. 29–76.
Hill, A.V., 1910. The possible effects of the aggregation of the molecules of haemoglobin on its dissociation curves. J. Physiol. 40, iv–vii.
Hill, E.P., 1986. Inhibition of carbonic anhydrase by plasma of dogs and rabbits. J. Appl. Physiol. 60, 191–197.
Hilvo, M., Tolvanen, M., Clark, A., Shen, B., Shah, G., Waheed, A., Halmi, P., Hanninen, M., Hamalainen, J., Vihinen, M., 2005. Characterization of CA XV, a new GPI-anchored form of carbonic anhydrase. Biochem. J. 392, 83–92.
Hladky, S., Rink, T., 1977. pH equilibrium across the red cell membrane. In: Membrane Transport in Red Cells. Academic Press, London, pp. 115–135.
Hochachka, P., 1961. The effect of physical training on oxygen debt and glycogen reserves in trout. Can. J. Zool. 39, 767–776.
Holeton, G.F., 1970. Oxygen uptake and circulation by a hemoglobinless Antarctic fish (*Chaenocephalus aceratus* Lonnberg) compared with three red-blooded Antarctic fish. Comp. Biochem. Physiol. 34, 457–471.
Holeton, G.F., Randall, D.J., 1967. The effect of hypoxia upon the partial pressure of gases in the blood and water afferent and efferent to the gills of rainbow trout. J. Exp. Biol. 46, 317–327.
Hopkins, T.E., Cech, J.J., 1994. Temperature effects on blood-oxygen equilibria in relation to movements of the bat ray, *Myliobatis Californica* in tomales bay, California. Mar. Behav. Physiol. 24, 227–235.
Houston, A.H., 1980. Components of the hematological response of fishes to environmental temperature change: a review. In: Ali, M.A. (Ed.), Environmental Physiology of Fishes. Springer, New York, pp. 241–298.
Houston, A.H., 1997. Review: are the classical hematological variables acceptable indicators of fish health? Trans. Am. Fish. Soc. 126, 879–894.
Houston, A., Cyr, D., 1974. Thermoacclimatory variation in the haemoglobin systems of goldfish (*Carassius auratus*) and rainbow trout (*Salmo gairdneri*). J. Exp. Biol. 61, 455–461.
Houston, A., DeWilde, M.A., 1968. Thermoacclimatory variations in the haematology of the common carp, *Cyprinus carpio*. J. Exp. Biol. 49, 71–81.
Houston, A., DeWilde, M.A., 1969. Environmental temperature and the body fluid system of the fresh-water teleost—III. Hematology and blood volume of thermally acclimated brook trout, *Salvelinus fontinalis*. Comp. Biochem. Physiol. 28, 877–885.
Houston, A., Gingras-Bedard, J.H., 1994. Variable versus constant temperature acclimation regimes: effects on hemoglobin isomorph profile in goldfish, *Carassius auratus*. Fish Physiol. Biochem. 13, 445–450.
Houston, A., Koss, T., 1984. Erythrocytic haemoglobin, magnesium and nucleoside triphosphate levels in rainbow trout exposed to progressive heat stress. J. Therm. Biol. 9, 159–164.
Houston, A.H., Murad, A., 1995. Erythrodynamics in fish: recovery of the goldfish *Carassius auratus* from acute anemia. Can. J. Zool. 73, 411–418.
Houston, A., Rupert, R., 1976. Immediate response of the hemoglobin system of the goldfish, *Carassius auratus*, to temperature change. Can. J. Zool. 54, 1737–1741.
Houston, A.H., Schrapp, M.P., 1994. Thermoacclimatory hematological response: have we been using appropriate conditions and assessment methods? Can. J. Zool. 72, 1238–1242.
Houston, A.H., Smeda, J.S., 1979. Thermoacclimatory changes in the ionic microenvironment of haemoglobin in the stenothermal rainbow trout (*Salmo gairdneri*) and eurythermal carp (*Cyprinus carpio*). J. Exp. Biol. 80, 317–340.

Houston, A., Mearow, K., Smeda, J., 1976. Further observations upon the hemoglobin systems of thermally-acclimated freshwater teleosts: pumpkinseed (*Lepomis gibbosus*), white sucker (*Catostomus commersoni*), carp (*Cyprinus carpio*), goldfish (*Carassius auratus*) and carp-goldfish hybrids. Comp. Biochem. Physiol. A Comp. Physiol. 54, 267–273.

Houston, A., Dobric, N., Kahurananga, R., 1996. The nature of hematological response in fish. Fish Physiol. Biochem. 15, 339–347.

Hughes, G.M., 1973. Respiratory responses to hypoxia in fish. Am. Zool. 13, 475–489.

Hughes, G., Horimoto, M., Kikuchi, Y., Kakiuchi, Y., Koyama, T., 1981. Blood flow velocity in microvessels of the gill filaments of the goldfish (*Carassius auratus* L.). J. Exp. Biol. 90, 327–331.

Hureau, J.C., Petit, D., Fine, J.M., Marneux, M., 1977. New cytological, biochemical, and physiological data on the colorless blood of the channichthyidae (pisces, teleosteans, perciformes). In: Llano, G.A. (Ed.), Adaptations Within Antarctic Ecosystems. Smithsonian Institution, Washington, D. C, pp. 459–477.

Hyde, D.A., Perry, S.F., 1990. Absence of adrenergic red cell pH and oxygen content regulation in American eel (*Anguilla rostrata*) during hypercapnic acidosis *in vivo* and *in vitro*. J. Comp. Physiol. 159, 687–693.

Ikeda-Saito, M., Yonetani, T., Gibson, Q.H., 1983. Oxygen equilibrium studies on hemoglobin from the bluefin tuna (*Thunnus thynnus*). J. Mol. Biol. 168, 673–686.

Imbrogno, S., Cerra, M.C., 2017. Hormonal and autacoid control of cardiac function. In: Gamperl, A.K., Gillis, T.E., Farrell, A.P., Brauner, C.J. (Eds.), The Cardiovascular System: Morphology, Control and Function. Fish Physiology. vol. 36A. Academic Press, San Diego, pp. 265-316.

Imsland, A., Foss, A., Stefansson, S., Nævdal, G., 2000. Hemoglobin genotypes of turbot (*Scophthalmus maximus*): consequences for growth and variations in optimal temperature for growth. Fish Physiol. Biochem. 23, 75–81.

Ishimatsu, A., Iwama, G.K., Heisler, N., 1988. *In vivo* analysis of partitioning of cardiac output between systemic and central venous sinus circuits in rainbow trout: a new approach using chronic cannulation of the branchial vein. J. Exp. Biol. 137, 75–88.

Itada, N., Forster, R.E., 1977. Carbonic anhydrase activity in intact red blood cells measured with ^{18}O exchange. J. Biol. Chem. 252, 3881–3890.

Ito, Y., Kobayashi, S., Nakamura, N., Miyagi, H., Esaki, M., Hoshijima, K., Hirose, S., 2013. Close association of carbonic anhydrase (CA2a and CA15a), Na^+/H^+ exchanger (Nhe3b), and ammonia transporter Rhcg1 in zebrafish ionocytes responsible for Na^+ uptake. Front. Physiol. 4, 1–17.

Jacobs, M., Stewart, D.R., 1942. The role of carbonic anhydrase in certain ionic exchanges involving the erythrocyte. J. Gen. Physiol. 25, 539–552.

Janvier, P., 2007. Living primitive fishes and fishes from deep time. In: McKenzie, D.J., Farrell, A.P., Brauner, C.J. (Eds.), Primitive Fishes. In: Fish Physiology, vol. 26. Academic Press, New York, pp. 1–51.

Japee, S.A., Pittman, R.N., Ellis, C.G., 2005. A new video image analysis system to study red blood cell dynamics and oxygenation in capillary networks. Microcirculation 12, 489–506.

Jensen, F.B., 1986. Pronounced influence of Hb-O_2 saturation on red cell pH in tench blood *in vivo* and *in vitro*. J. Exp. Zool. 238, 119–124.

Jensen, F.B., 1987. Influences of exercise-stress and adrenaline upon intra-and extracellular acid-base status, electrolyte composition and respiratory properties of blood in tench (*Tinca tinca*) at different seasons. J. Comp. Physiol. B. 157, 51–60.

Jensen, F.B., 1989. Hydrogen ion equilibria in fish haemoglobins. J. Exp. Biol. 143, 225–234.

Jensen, F.B., 1990. Nitrite and red cell function in carp: control factors for nitrite entry, membrane potassium ion permeation, oxygen affinity and methaemoglobin formation. J. Exp. Biol. 152, 149–166.

Jensen, F.B., 1991. Multiple strategies in oxygen and carbon dioxide transport by haemoglobin. In: Woakes, A.J., Greishaber, M.K., Bridges, C.R. (Eds.), Physiological Strategies for Gas Exchange and Metabolism. Cambridge University Press, Cambridge, pp. 55–78.

Jensen, F.B., 1995. Regulatory volume decrease in carp red blood cells: mechanisms and oxygenation-dependency of volume-activated potassium and amino acid transport. J. Exp. Biol. 198, 155–165.

Jensen, F.B., 1999. Haemoglobin H^+ equilibria in lamprey (*Lampetra fluviatilis*) and hagfish (*Myxine glutinosa*). J. Exp. Biol. 202, 1963–1968.

Jensen, F.B., 2004. Red blood cell pH, the Bohr effect, and other oxygenation-linked phenomena in blood O_2 and CO_2 transport. Acta Physiol. Scand. 182, 215–227.

Jensen, F.B., Weber, R.E., 1982. Respiratory properties of tench blood and hemoglobin. Adaptation to hypoxic-hypercapnic water. Mol. Phys. 2, 235–250.

Jensen, F.B., Weber, R.E., 1987. Thermodynamic analysis of precisely measured oxygen equilibria of tench (*Tinca tinca*) hemoglobin and their dependence on ATP and protons. J. Comp. Physiol. 157, 137–143.

Jensen, B.J., Nikinmaa, M., Weber, R.E., 1983. Effects of exercise stress on acid-base balance and respiratory function in blood of the teleost *Tinca tinca*. Respir. Physiol. 51, 291–301.

Jensen, F.B., Andersen, N.A., Heisler, N., 1990. Interrelationships between red cell nucleoside triphosphate content, and blood pH, O_2-tension and haemoglobin concentration in the carp, *Cyprinus carpio*. Fish Physiol. Biochem. 8, 459–464.

Jensen, F.B., Nikinmaa, M., Weber, R.E., 1993. Environmental perturbations of oxygen transport in teleost fishes: causes, consequences and compensations. In: Rankin, J.C., Jensen, F.B. (Eds.), Fish Ecophysiology. Chapman & Hall, London, pp. 161–179.

Jensen, F.B., Fago, A., Weber, R.E., 1998. Hemoglobin structure and function. In: Perry II, S.F., Tufts, B.L. (Eds.), Fish Respiration. In: Fish Physiology, vol. 17. Academic Press, New York, pp. 1–40.

Jensen, L.D.E., Cao, R., Hedlund, E.-M., Söll, I., Lundberg, J.O., Hauptmann, G., Steffensen, J.F., Cao, Y., 2009. Nitric oxide permits hypoxia-induced lymphatic perfusion by controlling arterial-lymphatic conduits in zebrafish and glass catfish. PNAS 106, 18408–18413.

Johansen, K., Lykkeboe, G., Weber, R.E., Maloiy, G.M.O., 1976. Respiratory properties of blood in awake and estivating lungfish, *Protopterus amphibius*. Respir. Physiol. 27, 335–345.

Johansen, K., Mangum, C.P., Lykkeboe, G., 1978. Respiratory properties of the blood of amazon fishes. Can. J. Zool. 56, 891–897.

Jones, D.R., 1971. The effect of hypoxia and anaemia on the swimming performance of rainbow trout (*Salmo gairdneri*). J. Exp. Biol. 55, 541–551.

Jones, D.R., Randall, D.J., 1978. The respiratory and circulatory systems during exercise. In: Hoar, W.S., Randall, D.J. (Eds.), Locomotion. In: Fish Physiology, vol. 7. Academic Press, New York, pp. 425–501.

Jørgensen, E.H., Jobling, M., 1993. The effects of exercise on growth, food utilisation and osmoregulatory capacity of juvenile Atlantic salmon, *Salmo salar*. Aquaculture 116, 233–246.

Julio, A.E., Desforges, P.R., Perry, S.F., 2000. Apparent diffusion limitations for CO_2 excretion in rainbow trout are relieved by injections of carbonic anhydrase. Respir. Physiol. 121, 53–64.

Khalifah, R.G., Silverman, D.N., 1991. Carbonic anhydrase kinetics and molecular function. In: Dodgeson, S.J., Tashian, R.E., Gros, G., Carter, N.D. (Eds.), Carbonic Anhydrases: Cellular Physiology and Molecular Genetics. Springer Science and Business, New York, pp. 49–70.

Kiceniuk, J.W., Jones, D.R., 1977. The oxygen transport system in trout (*Salmo gairdneri*) during sustained exercise. J. Exp. Biol. 69, 247–260.

Kieffer, J.D., 2000. Limits to exhaustive exercise in fish. Comp. Biochem. Physiol. 136, 161–179.

Kilmartin, J., Rossi-Bernardi, L., 1973. Interaction of hemoglobin with hydrogen ions, carbon dioxide, and organic phosphates. Physiol. Rev. 53, 836–890.

Kita, J., Itazawa, Y., 1989. Release of erythrocytes from the spleen during exercise and splenic constriction by adrenaline infusion in the rainbow trout. Jpn. J. Ichthyol. 36, 48–52.

Knüppel-Ruppert, A.S., Gros, G., Harringer, W., Kubis, H.-P., 2000. Immunochemical evidence for a unique GPI-anchored carbonic anhydrase isozyme in human cardiomyocytes. Am. J. Phys. 278, H1335–H1344.

Kobayashi, M., Ishigaki, K., Kobyashi, M., Imai, K., 1994. Shape of the haemoglobin-oxygen equilibrium curve and oxygen transport efficiency. Respir. Physiol. 95, 321–328.

Kock, K.-H., 2005. Antarctic icefishes (*Channichthyidae*): a unique family of fishes. A review, Part I. Polar Biol. 28, 862–895.

Kokkonen, K., Vierimaa, H., Bergström, S., Tervonen, V., Arjamaa, O., Ruskoaho, H., Järvilehto, M., Vuolteenaho, O., 2000. Novel salmon cardiac peptide hormone is released from the ventricle by regulated secretory pathway. Am. J. Phys. 278, E285–E292.

Krogh, A., 1919. The number and distribution of capillaries in muscles with calculations of the oxygen pressure head necessary for supplying the tissue. J. Physiol. 52, 409–415.

Kuhn, W., Ramel, A., Kuhn, H.J., Marti, E., 1963. The filling mechanism of the swim bladder. Generation of high gas pressures through hairpin countercurrent multiplication. Experientia 19, 497–511.

Kunzmann, A., 1991. Blood physiology and ecological consequences in Weddell Sea fishes (Antarctica). In: Mathematisch-Naturwissenschaftliche Fakultaet. Christian-Albrechts-Universitaet zu Kiel, Kiel (Ph.D.).

Lacy, E.R., 1983. Histochemical and biochemical studies of carbonic anhydrase activity in the opercular epithelium of the euryhaline teleost, *Fundulus heteroclitus*. Am. J. Anat. 166, 19–39.

Lai, J.C.C., Kakuta, I., Mok, H.O.L., Rummer, J.L., Randall, D., 2006. Effects of moderate and substantial hypoxia on erythropoietin levels in rainbow trout kidney and spleen. J. Exp. Biol. 209, 2734–2738.

Lane, H., Rolfe, A., Nelson, J., 1981. Changes in the nucleotide triphosphate/haemoglobin and nucleotide triphosphate/red cell ratios of rainbow trout, *Salmo gairdneri* Richardson, subjected to prolonged starvation and bleeding. J. Fish Biol. 18, 661–668.

Lapennas, G.N., 1983. The magnitude of the Bohr coefficient: optimal for oxygen delivery. Respir. Physiol. 54, 161–172.

Lauf, P.K., 1982. Evidence for chloride dependent potassium and water transport induced by hyposmotic stress in erythrocytes of the marine teleost, *Opsanus tau*. J. Comp. Physiol. 146, 9–16.

Laursen, J.S., Andersen, N.A., Lykkeboe, G., 1985. Temperature acclimation and oxygen binding properties of blood of the European eel (*Anguilla anguilla*). Comp. Biochem. Physiol. 81A, 79–86.

Law, S.H., Wu, R.S., Ng, P.K., Richard, M., Kong, R.Y., 2006. Cloning and expression analysis of two distinct HIF-alpha isoforms–gcHIF-1alpha and gcHIF-4alpha–from the hypoxia-tolerant grass carp, *Ctenopharyngodon idellus*. BMC Mol. Biol. 7, 1.

Lecklin, T., Nikinmaa, M., 1999. Seasonal and temperature effects on the adrenergic responses of Arctic char (*Salvelinus alpinus*) erythrocytes. J. Exp. Biol. 202, 2233–2238.

Lecklin, T., Nash, G., Egginton, S., 1995. Do fish acclimated to low temperature improve microcirculatory perfusion by adapting red cell rheology? J. Exp. Biol. 198, 1801–1808.

Lenfant, C., Johansen, K., 1966. Respiratory function in the elasmobranch *Squalus suckleyi* G. Respir. Physiol. 1, 13–29.

Lessard, J., Val, A., Aota, S., Randall, D., 1995. Why is there no carbonic anhydrase activity available to fish plasma? J. Exp. Biol. 198, 31–38.

Lilly, L.E., Bonaventura, J., Lipnick, M.S., Block, B.A., 2015. Effect of temperature acclimation on red blood cell oxygen affinity in Pacific bluefin tuna (*Thunnus orientalis*) and yellowfin tuna (*Thunnus albacares*). Comp. Biochem. Physiol. A Mol. Integr. Physiol. 181, 36–44.

Lin, T.-Y., Liao, B.-K., Horng, J.-L., Yan, J.-J., Hsiao, C.-D., Hwang, P.-P., 2008. Carbonic anhydrase 2-like a and 15a are involved in acid-base regulation and Na^+ uptake in zebrafish H^+-ATPase-rich cells. Am. J. Phys. Cell Physiol. 294, C1250–C1260.

Lönnerholm, G., 1980. Carbonic anhydrase in rat liver and rabbit skeletal muscle: further evidence for the specificity of the histochemical cobalt-phosphate method of Hansson. J. Histochem. Cytochem. 28, 427–433.

Low, M.G., Stiernberg, J., Waneck, G.L., Flavell, R.A., Kincade, P.W., 1988. Cell-specific heterogeneity in sensitivity of phosphatidylinositol-anchored membrane antigens to release by phospholipase C. J. Immunol. Methods 113, 101–111.

Lowe, T.E., Brill, R.W., Cousins, K.L., 1998. Responses of the red blood cells from two high-energy-demand teleosts, yellowfin tuna(*Thunnus albacares*) and skipjack tuna (*Katsuwonus pelamis*), to catecholamines. J. Comp. Physiol. B. 168, 405–418.

Lowe, T., Brill, R., Cousins, K., 2000. Blood oxygen-binding characteristics of bigeye tuna (*Thunnus obesus*), a high-energy-demand teleost that is tolerant of low ambient oxygen. Mar. Biol. 136, 1087–1098.

Lykkeboe, G., Weber, R.E., 1978. Changes in the respiratory properties of the blood in the carp, *Cyprinus carpio*, induced by diurnal variation in ambient oxygen tension. J. Comp. Physiol. 128, 117–125.

Lykkeboe, G., Johansen, K., Maloiy, G.M.O., 1975. Functional properties of hemoglobins in the teleost *Tilapia grahami*. J. Comp. Physiol. 104, 1–11.

Macdonald, J.A., Montgomery, J.C., Wells, R.M.G., 1987. Comparative physiology of Antarctic fishes. Adv. Mar. Biol. 24, 321–388.

Maetz, M., 1956. Le dosage de l'anhydrase carbonique etude de quelques substances inhibitrices et activatrices. Bull. Soc. Chim. Biol. 38, 288–289. Paris: Elsevier.

Maetz, J., 1971. Fish gills: mechanisms of salt transfer in fresh water and sea water. Philos. Trans. R. Soc. B 262, 209–249.

Maetz, J., Bornancin, M., 1975. Biochemical and biophysical aspects of salt excretion by chloride cells in teleosts. Fortschr. Zool. 23, 322.

Maffia, M., Trischitta, F., Lionetto, M., Storelli, C., Schettino, T., 1996. Bicarbonate absorption in eel intestine: evidence for the presence of membrane-bound carbonic anhydrase on the brush border membranes of the enterocyte. J. Exp. Zool. 275, 365–373.

Maffia, M., Rizzello, A., Acierno, R., Rollo, M., Chiloiro, R., Storelli, C., 2001. Carbonic anhydrase activity in tissues of the icefish *Chionodraco hamatus* and of the red-blooded teleosts *Trematomus bernacchii* and *Anguilla anguilla*. J. Exp. Biol. 204, 3983–3992.

Mahe, Y., Garciaromeu, F., Motais, R., 1985. Inhibition by amiloride of both adenylate-cyclase activity and the Na^+/H^+ antiporter in fish erythrocytes. Eur. J. Pharmacol. 116, 199–206.

Mairbäurl, H., 1994. Red blood cell function in hypoxia at altitude and exercise. Int. J. Sports Med. 15, 51–63.

Mairbäurl, H., Weber, R.E., 2012. Oxygen transport by hemoglobin. Compr. Physiol. 2, 1463–1489.

Mairbäurl, H., Schobersberger, W., Hasibeder, W., Schwaberger, G., Gaesser, G., Tanaka, K., 1986. Regulation of red cell 2, 3-DPG and Hb-O_2-affinity during acute exercise. Eur. J. Appl. Physiol. 55, 174–180.

Malte, H., Weber, R.E., 1985. A mathematical model for gas exchange in the fish gill based on non-linear blood gas equilibrium curves. Respir. Physiol. 62, 359–374.

Malte, H., Weber, R.E., 1987. The effect of shape and position of the oxygen equilibrium curve on extraction and ventilation requirement in fishes. Respir. Physiol. 70, 221–228.
Malte, H., Weber, R.E., 1989. Gas exchange in fish gills with parallel inhomogeneities. Respir. Physiol. 76, 129–137.
Mandic, M., Todgham, A.E., Richards, J.G., 2009. Mechanisms and evolution of hypoxia tolerance in fish. Proc. R. Soc. B 276, 735–744.
Maren, T.H., 1967. Carbonic anhydrase: chemistry, physiology, and inhibition. Physiol. Rev. 47, 595–781.
Maren, T.H., Couto, E.O., 1979. The nature of anion inhibition of human red cell carbonic anhydrases. Arch. Biochem. Biophys. 196, 501–510.
Maren, T.H., Swenson, E.R., 1980. A comparative study of the kinetics of the Bohr effect in vertebrates. J. Physiol. 303, 535.
Maren, T., Wynns, G., Wistrand, P., 1993. Chemical properties of carbonic anhydrase IV, the membrane-bound enzyme. Mol. Pharmacol. 44, 901–905.
Marinsky, C., Houston, A., Murad, A., 1990. Effect of hypoxia on hemoglobin isomorph abundances in rainbow trout, *Salmo gairdneri*. Can. J. Zool. 68, 884–888.
Marshall, W.S., Grosell, M., 2006. Ion transport, osmoregulation and acid–base balance. In: Evans, D., Claiborne, J.B. (Eds.), The Physiology of Fishes. CRC Press, Boca Raton, pp. 177–230.
Martinez, F., Garcia-Riera, M., Ganteras, M., De Costa, J., Zamora, S., 1994. Blood parameters in rainbow trout (*Oncorhynchus mykiss*): simultaneous influence of various factors. Comp. Biochem. Physiol. A Physiol. 107, 95–100.
Marttila, O.N.T., Nikinmaa, M., 1988. Binding of β-adrenergic antagonists ^{3}H-DHA and ^{3}H-CGP 12177 to intact rainbow trout (*Salmo gairdneri*) and carp (*Cyprinus carpio*) red blood cells. Gen. Comp. Endocrinol. 70, 429–435.
McKenzie, D.J., Wong, S., Randall, D.J., Egginton, S., Taylor, E.W., Farrell, A.P., 2004. The effects of sustained exercise and hypoxia upon oxygen tensions in the red muscle of rainbow trout. J. Exp. Biol. 207, 3629–3637.
McLeod, T.F., Sigel, M.M., Yunis, A.A., 1978. Regulation of erythropoiesis in the Florida gar, *Lepisosteus platyrhincus*. Comp. Biochem. Physiol. A Physiol. 60, 145–150.
Milligan, C.L., Wood, C.M., 1986a. Tissue intracellular acid-base status and the fate of lactate after exhaustive exercise in the rainbow trout. J. Exp. Biol. 123, 123–144.
Milligan, C.L., Wood, C.M., 1986b. Intracellular and extracellular acid-base status and H^+ exchange with the environment after exhaustive exercise in the rainbow trout. J. Exp. Biol. 123, 93–121.
Milligan, C.L., Wood, C.M., 1987. Regulation of blood-oxygen transport and red-cell pH_i after exhaustive activity in rainbow-trout (*Salmo gairdneri*) and starry flounder (*Platichthys stellatus*). J. Exp. Biol. 133, 263–282.
Mislan, K., Dunne, J.P., Sarmiento, J.L., 2015. The fundamental niche of blood oxygen binding in the pelagic ocean. Oikos 125, 938–949.
Monod, J., Wyman, J., Changeux, J.-P., 1965. On the nature of allosteric transitions: a plausible model. J. Mol. Biol. 12, 88–118.
Moores, W., Willford, D., Crum, J., Neville, J., Weiskopf, R., Dembitsky, W., 1978. Alteration of myocardial-function resulting from changes in hemoglobin oxygen-affinity. In: Circulation, vol. 58. Ames Heart Assoc, Dallas, Tx, p. 225.
Morris, S., Bridges, C.R., 1994. Properties of respiratory pigments in bimodal breathing animals: air and water breathing by fish and crustaceans. Am. Zool. 34, 216–228.
Morrison, P.R., Gilmour, K.M., Brauner, C.J., 2015. Oxygen and carbon dioxide transport in elasmobranchs. In: Shadwick, R.E., Farrell, A.P., Brauner, C.J. (Eds.), In: Fish Physiology, vol. 34B. Academic Press, London, pp. 127–219.

Motais, R., Garcia-Romeu, F., 1988. Effects of catecholamines and cyclic nucleotides on Na^+/H^+ exchange. In: Grinstein, S. (Ed.), Na^+/H^+ Exchange. CRC Press, Boca Raton, FL, pp. 255–270.
Motais, R., Garcia-Romeu, F., Borgese, F., 1987. The control of Na^+/H^+ exchange by molecular oxygen in trout erythrocytes. A possible role of hemoglobin as a transducer. J. Gen. Physiol. 90, 197–207.
Motais, R., Fievet, B., Garcia-Romeu, F., Thomas, S., 1989. Na^+-H^+ exchange and pH regulation in red blood cells: role of uncatalyzed $H_2CO_3^-$ dehydration. Am. J. Phys. 256, C728–C735.
Motyka, R., Norin, T., Petersen, L.H., Huggett, D.B., Gamperl, A.K., 2017. Long-term hypoxia exposure alters the cardiorespiratory physiology of steelhead trout (*Oncorhynchus mykiss*), but does not affect their upper thermal tolerance. J. Therm. Biol. 68, 149–161.
Murad, A., Houston, A.H., 1992. Maturation of the goldfish (*Carassius auratus*) erythrocyte. Comp. Biochem. Physiol. 102A, 107–110.
Murad, A., Houston, A., Samson, L., 1990. Haematological response to reduced oxygen-carrying capacity, increased temperature and hypoxia in goldfish, *Carassius auratus* L. J. Fish Biol. 36, 289–305.
Nagai, K., Perutz, M.F., Poyart, C., 1985. Oxygen binding properties of human mutant hemoglobins synthesized in *Escherichia coli.* PNAS 82, 7252–7255.
Near, T.J., Pesavento, J.J., Cheng, C.-H.C., 2003. Mitochondrial DNA, morphology, and the phylogenetic relationships of Antarctic icefishes (Notothenioidei: Channichthyidae). Mol. Phylogenet. Evol. 28, 87–98.
Near, T.J., Pesavento, J.J., Cheng, C.-H.C., 2004. Phylogenetic investigations of Antarctic notothenioid fishes (Perciformes: Notothenioidei) using complete gene sequences of the mitochondrial encoded 16S rRNA. Mol. Phylogenet. Evol. 32, 881–891.
Near, T.J., Parker, S.K., Detrich, H.W., 2006. A genomic fossil reveals key steps in hemoglobin loss by the Antarctic icefishes. Mol. Biol. Evol. 23, 2008–2016.
Nelson, J.S., Grande, T.C., Wilson, M.V., 2006. Fishes of the World. John Wiley & Sons, Hoboken, NJ.
Newton, M., Peters, J., 1983. Physiological variation of mouse haemoglobins. Proc. R. Soc. Lond. B 218, 443–453.
Nielsen, O.B., Lykkeboe, G., 1992. Changes in plasma and erythrocyte K^+ during hypercapnia and different grades of exercise in trout. J. Appl. Physiol. 72, 1285–1290.
Nikinmaa, M., 1981. Respiratory adjustments of rainbow trout (*Salmo gairdneri* Richardson) to changes in environmental temperature and oxygen availability. In: Zoology. Helsinki University, Helsinki (Ph.D.).
Nikinmaa, M., 1982. Effects of adrenaline on red-cell volume and concentration gradient of protons across the red-cell membrane in the rainbow-trout, *Salmo-gairdneri.* Mol. Phys. 2, 287–297.
Nikinmaa, M., 1983. Adrenergic regulation of hemoglobin oxygen-affinity in rainbow-trout red-cells. J. Comp. Physiol. 152, 67–72.
Nikinmaa, M., 1986. Red cell pH of lamprey (*Lampetra fluviatilis*) is actively regulated. J. Comp. Physiol. B. 156, 747–750.
Nikinmaa, M., 1990. Vertebrate Red Blood Cells. Adaptations of Function to Respiratory Requirements. Springer, Berlin.
Nikinmaa, M., 1992. Membrane transport and control of hemoglobin-oxygen affinity in nucleated erythrocytes. Physiol. Rev. 72, 301–321.
Nikinmaa, M., 1997. Oxygen and carbon dioxide transport in vertebrate erythrocytes: an evolutionary change in the role of membrane transport. J. Exp. Biol. 200, 369–380.
Nikinmaa, M., 2001. Haemoglobin function in vertebrates: evolutionary changes in cellular regulation in hypoxia. Respir. Physiol. 128, 317–329.

Nikinmaa, M., 2002. Oxygen-dependent cellular functions—why fishes and their aquatic environment are a prime choice to study. Comp. Biochem. Physiol. 133, 1–16.
Nikinmaa, M., 2003. Gas transport. In: Bernhardt, I., Ellory, J.C. (Eds.), Red Cell Membrane Transport in Health and Disease. Springer, Berlin, pp. 489–509.
Nikinmaa, M., Boutilier, R., 1995. Adrenergic control of red cell pH, organic phosphate concentrations and haemoglobin function in teleost fish. In: Heisler, N. (Ed.), In: Mechanisms of Systemic Regulation: Respiration and Circulation, vol. 21. Springer, Berlin, pp. 107–133.
Nikinmaa, M., Huestis, W.H., 1984. Adrenergic swelling of nucleated erythrocytes - cellular mechanisms in a bird, domestic goose, and 2 teleosts, striped bass and rainbow-trout. J. Exp. Biol. 113, 215–224.
Nikinmaa, M., Jensen, F.B., 1986. Blood oxygen transport and acid-base status of stressed trout (*Salmo gairdneri*): pre-and postbranchial values in winter fish. Comp. Biochem. Physiol. A Physiol. 84, 391–396.
Nikinmaa, M., Mattsoff, L., 1992. Effects of oxygen saturation on the CO_2 transport properties of *Lampetra* red cells. Respir. Physiol. 87, 219–230.
Nikinmaa, M., Railo, E., 1987. Anion movements across lamprey (*Lampetra fluviatilis*) red cell membrane. Biochim. Biophys. Acta Biomembr. 899, 134–136.
Nikinmaa, M., Salama, A., 1998. Oxygen transport in fish. In: Perry II, S.F., Tuffs, B.L. (Eds.), Fish Respiration. In: Fish Physiology, vol. 17. Academic press, New York, pp. 141–184.
Nikinmaa, M., Tuurala, H., Soivio, A., 1980. Thermoacclimatory changes in blood oxygen binding properties and gill secondary lamellar structure of *Salmo gairdneri*. J. Comp. Physiol. 140, 255–260.
Nikinmaa, M., Soivio, A., Railo, E., 1981. Blood volume of *Salmo gairdneri*: influence of ambient temperature. Comp. Biochem. Physiol. A Physiol. 69, 767–769.
Nikinmaa, M., Cech, J.J.J., McEnroe, M., 1984. Blood oxygen transport in stressed striped bass (*Morone saxatilis*): role of β-adrenergic responses. J. Comp. Physiol. 154, 365–369.
Nikinmaa, M., Kunnamo-Ojala, T., Railo, E., 1986. Mechanisms of pH regulation in lamprey (*Lampetra fluviatilis*) red blood cells. J. Exp. Biol. 122, 355–367.
Nikinmaa, M., Tiihonen, K., Paajaste, M., 1990. Adrenergic control of red-cell pH in Salmonid fish—roles of the sodium proton-exchange, jacobs-stewart cycle and membrane-potential. J. Exp. Biol. 154, 257–271.
Nikinmaa, M., Tufts, B., Boutilier, R., 1993. Volume and pH regulation in agnathan erythrocytes: comparisons between the hagfish, *Myxine glutinosa*, and the lampreys, *Petromyzon marinus* and *Lampetra fluviatilis*. J. Comp. Physiol. B. 163, 608–613.
Nikinmaa, M., Airaksinen, S., Virkki, L.V., 1995. Haemoglobin function in intact lamprey erythrocytes: interactions with membrane function in the regulation of gas transport and acid-base balance. J. Exp. Biol. 198, 2423–2430.
Nilsson, S., 1983. Autonomic Nerve Function in the Vertebrates. Springer Science & Business Media, Berlin.
Nilsson, S., 1984. Adrenergic control systems in fish. Mar. Biol. Lett. 5, 127–146.
Nilsson, S., Grove, D.J., 1974. Adrenergic and cholinergic innervation of the spleen of the Cod: *Gadus morhua*. Eur. J. Pharmacol. 28, 135–143.
Nilsson, G.E., Östlund-Nilsson, S., 2005. Hypoxia tolerance in coral reef fishes. In: Farrell, A.P., Brauner, C.J. (Eds.), The Physiology of Tropical Fishes. In: Fish Physiology, vol. 21. Academic Press, New York, pp. 583–596.
O'Brasky, J.E., Crandall, E.D., 1980. Organ and species differences in tissue vascular carbonic anhydrase activity. J. Appl. Physiol. 49, 211–217.
O'Brasky, J., Mauro, T., Crandall, E., 1979. Postcapillary pH disequilibrium after gas exchange in isolated perfused liver. J. Appl. Physiol. 47, 1079–1083.

Odlind, B., Appelgren, L.E., Bayati, A., Wolgast, M., 1988. Tissue distribution of ^{125}I-labelled bovine superoxide dismutase (SOD) in the rat. Pharmacol. Toxicol. 62, 95–100.
Ohnishi, S.T., Asai, H., 1985. Lamprey erythrocytes lack glycoproteins and anion transport. Comp. Biochem. Physiol. B 81, 405–407.
Olson, K., 1984. Distribution of flow and plasma skimming in isolated perfused gills of three teleosts. J. Exp. Biol. 109, 97–108.
Olson, K.R., 1992. Blood and extracellular fluid volume regulation: role of the renin-angiotensin, kallikrein-kinin systems and atrial natriuretic peptides. In: Hoar, W.S., Randall, D.J., Farrell, A.P. (Eds.), The Cardiovascular System. In: Fish Physiology, vol. 12B. Academic Press, New York, pp. 135–254.
Olson, K.R., 1996. Secondary circulation in fish: anatomical organization and physiological significance. J. Exp. Zool. A Ecol. Integr. Physiol. 275, 172–185.
Oski, F.A., Marshall, B.E., Cohen, P.J., Sugerman, H.J., Miller, L.D., 1971. Exercise with anemia: the role of the left-shifted or right-shifted oxygen-hemoglobin equilibrium curve. Ann. Intern. Med. 74, 44–46.
Palfrey, H.C., Greengard, P., 1981. Hormone-sensitive ion transport systems in erythrocytes as models for epithelial ion pathways. Ann. N. Y. Acad. Sci. 372, 291–308.
Pan, Y.K., Ern, R., Morrison, P.R., Brauner, C.J., Esbaugh, A.J., 2017. Acclimation to prolonged hypoxia alters hemoglobin isoform expression and increases hemoglobin oxygen affinity and aerobic performance in a marine fish. Sci Rep 7, 7834.
Pang, C.C., 2001. Autonomic control of the venous system in health and disease: effects of drugs. Pharmacol. Ther. 90, 179–230.
Patel, C.B., Maren, T.H., Mills, J., Swenson, E.R., 1997. Effects of a high molecular weight carbonic anhydrase (CA) inhibitor, F3500, on respiratory acidosis in the shark, *Squalus acanthias*. Bull. Mt. Desert Isl. Biol. Lab. 36, 65–68.
Pearson, M.P., Stevens, E.D., 1991. Size and hematological impact of the splenic erythrocyte reservoir in rainbow trout, *Oncorhynchus mykiss*. Fish Physiol. Biochem. 9, 39–50.
Pearson, M., Spriet, L., Stevens, E., 1990. Effect of sprint training on swim performance and white muscle metabolism during exercise and recovery in rainbow trout (*Salmo gairdneri*). J. Exp. Biol. 149, 45–60.
Pearson, M., Van Der Kraak, G., Stevens, E.D., 1992. *In vivo* pharmacology of spleen contraction in rainbow trout. Can. J. Zool. 70, 625–627.
Pelis, R.M., Renfro, J.L., 2004. Role of tubular secretion and carbonic anhydrase in vertebrate renal sulfate excretion. Am. J. Phys. Regul. Integr. Comp. Phys. 287, R491–R501.
Pelster, B., 1995. Mechanisms of acid release in isolated gas gland cells of the European eel *Anguilla anguilla*. Am. J. Phys. Regul. Integr. Comp. Phys. 269, R793–799.
Pelster, B., 1997. Buoyancy at depth. In: Randall, D., Farrell, A. (Eds.), Deep-Sea Fishes. In: Fish Physiology, vol. 16. Academic Press, New York, pp. 195–238.
Pelster, B., 2004. pH regulation and swimbladder function in fish. Respir. Physiol. Neurobiol. 144, 179–190.
Pelster, B., Bagatto, B., 2010. Respiration. In: Perry, S.F., Ekker, M.E., Farrell, A.P., Brauner, C.J. (Eds.), Zebrafish. In: Fish Physiology, vol. 29. Academic Press, New York, pp. 289–309.
Pelster, B., Decker, H., 2004. The root effect—a physiological perspective. Micron 35, 73–74.
Pelster, B., Randall, D.J., 1998. Physiology of the root effect. In: Perry, S.F., Tufts, B.L. (Eds.), Fish Respiration. In: Fish physiology, vol. 17. Academic Press, New York, pp. 113–140.
Pelster, B., Weber, R.E., 1990. Influence of organic phosphates on the root effect of multiple fish haemoglobins. J. Exp. Biol. 149, 425–437.
Pelster, B., Weber, R.E., 1991. The physiology of the root effect. In: Gilles, R., Butler, P.J., Greger, R., Mangum, C.P., Somero, G.N., Takahashi, K., Weber, R.E. (Eds.), Advances in Comparative and Environmental Physiology. Springer, Heidelberg/New York, pp. 51–77.

Penney, D., Thomas, M., 1975. Hematological alterations and response to acute hypobaric stress. J. Appl. Physiol. 39, 1034–1037.

Perry, S.F., 1986. Carbon dioxide excretion in fishes. Can. J. Zool. 64, 565–572.

Perry, S.F., Gilmour, K., 1993. An evaluation of factors limiting carbon dioxide excretion by trout red blood cells *in vitro*. J. Exp. Biol. 180, 39–54.

Perry, S.F., Gilmour, K.M., 1996. Consequences of catecholamine release on ventilation and blood oxygen transport during hypoxia and hypercapnia in an elasmobranch (*Squalus acanthias*) and a teleost (*Oncorhynchus mykiss*). J. Exp. Biol. 199, 2105–2118.

Perry, S.F., Gilmour, K.M., 2002. Sensing and transfer of respiratory gases at the fish gill. J. Exp. Zool. 293, 249–263.

Perry, S.F., Gilmour, K.M., 2006. Acid-base balance and CO_2 excretion in fish: unanswered questions and emerging models. Respir. Physiol. Neurobiol. 154, 199–215.

Perry, S.F., Kinkead, R., 1989. The role of catecholamines in regulating arterial oxygen content during acute hypercapnic acidosis in rainbow trout (*Salmo gairdneri*). Respir. Physiol. 77, 365–377.

Perry, S.F., Laurent, P., 1990. The role of carbonic anhydrase in carbon dioxide excretion, acid-base balance and ionic regulation in aquatic gill breathers. In: Truchot, J.P., Lahlou, B. (Eds.), In: Animal Nutrition and Transport Processes 2. Transport, Respiration and Excretion: Comparative and Environmental Aspects, vol. 6. Karger, Basel, pp. 39–57.

Perry, S.F., Reid, S.D., 1992a. The relationship between *beta*-adrenoceptors and adrenergic responsiveness in trout (*Oncorhynchus mykiss*) and eel (*Anguilla rostrata*) erythrocytes. J. Exp. Biol. 167, 235–250.

Perry, S.F., Reid, S.D., 1992b. Relationship between blood O_2-content and catecholamine levels during hypoxia in rainbow trout and American eel. Am. J. Phys. 263, R240–R249.

Perry, S., Reid, S., 1994. The effects of acclimation temperature on the dynamics of catecholamine release during acute hypoxia in the rainbow trout *Oncorhynchus mykiss*. J. Exp. Biol. 186, 289–307.

Perry, S.F., Wood, C.M., 1989. Control and coordination of gas transfer in fishes. Can. J. Zool. 67, 2961–2970.

Perry, S.F., Davie, P.S., Daxboeck, C., Randall, D.J., 1982. A comparison of CO_2 excretion in a spontaneously ventilating blood-perfused trout preparation and saline-perfused gill preparations: contribution of the branchial epithelium and red blood cell. J. Exp. Biol. 101, 47–60.

Perry, S., Reid, S., Salama, A., 1996a. The effects of repeated physical stress on the b-adrenergic response of the rainbow trout red blood cell. J. Exp. Biol. 199, 549–2497.

Perry, S.F., Wood, C.M., Walsh, P.J., Thomas, S., 1996b. Fish red blood cell carbon dioxide excretion *in vitro*. A comparative study. Comp. Biochem. Physiol. 113A, 121–130.

Perry, S., Brauner, C., Tufts, B., Gilmour, K., 1997. Acid-base disequilibrium in the venous blood of rainbow trout (*Oncorhynchus mykiss*). Exp. Biol. Online 2, 1–10.

Perry, S.F., Gilmour, K.M., Swenson, E.R., Vulesevic, B., Chew, S.F., Ip, Y.K., 2005. An investigation of the role of carbonic anhydrase in aquatic and aerial gas transfer in the African lungfish *Protopterus dolloi*. J. Exp. Biol. 208, 3805–3815.

Perry, S., Jonz, M., Gilmour, K., 2009. Oxygen sensing and the hypoxic ventilatory response. In: Farrel, A.P., Brauer, C.J. (Eds.), Hypoxia. In: Fish Physiology, vol. 27. Academic Press, New York, pp. 193–253.

Perutz, M.F., 1970. Stereochemistry of cooperative effects in haemoglobin. Nature 228, 726–734.

Perutz, M., 1990. Mechanisms of Cooperativity and Allosteric Regulation in Proteins. Cambridge University Press, Cambridge.

Perutz, M.F., Brunori, M., 1982. Stereochemistry of cooperative effects in fish and amphibian haemoglobins. Nature 299, 421–426.

Perutz, M.F., Rossmann, M.G., Cullis, A.F., Muirhead, H., Will, G., North, A.C.T., 1960. Structure of haemoglobin - three-dimensional fourier synthesis at 5.5-Å resolution, obtained by X-ray analysis. Nature 185, 416–422.
Perutz, M., Fermi, G., Luisi, B., Shaanan, B., Liddington, R., 1987. Stereochemistry of cooperative mechanisms in hemoglobin. Acc. Chem. Res. 20, 309–321.
Peters, T., Papadopoulos, F., Kubis, H.-P., Gros, G., 2000. Properties of a carbonic anhydrase inhibitor protein in flounder serum. J. Exp. Biol. 203, 3003–3009.
Petersen, L., Gamperl, A., 2010. Effect of acute and chronic hypoxia on the swimming performance, metabolic capacity and cardiac function of Atlantic cod (*Gadus morhua*). J. Exp. Biol. 213, 808–819.
Petersen, L., Gamperl, A.K., 2011. Cod (*Gadus morhua*) cardiorespiratory physiology and hypoxia tolerance following acclimation to low-oxygen conditions. Physiol. Biochem. Zool. 84, 18–31.
Peterson, M.S., 1990. Hypoxia-induced physiological changes in two mangrove swamp fishes: sheepshead minnow, *Cyprinodon variegatus* Lacepede and sailfin molly, *Poecilia latipinna* Lesueur. Comp. Biochem. Physiol. A Physiol. 97, 17–21.
Peterson, A.J., Poluhowich, J.J., 1976. The effects of organic phosphates on the oxygenation behavior of eel multiple hemoglobins. Comp. Biochem. Physiol. A Physiol. 55, 351–354.
Pichavant, K., Maxime, V., Soulier, P., Boeuf, G., Nonnotte, G., 2003. A comparative study of blood oxygen transport in turbot and sea bass: effect of chronic hypoxia. J. Fish Biol. 62, 928–937.
Piiper, J., 1998. Branchial gas transfer models. Comp. Biochem. Physiol. 119, 125–130.
Pörtner, H.O., Farrell, A.P., 2008. Ecology: physiology and climate change. Science 322, 690–692.
Pörtner, H.O., Lannig, G., 2009. Oxygen and capacity limited thermal tolerance. In: Richards, J.G., Farrell, A.P., Brauner, C.J. (Eds.), Hypoxia. In: Fish Physiology, vol. 27. Academic Press, New York, pp. 143–191.
Pörtner, H.-O., Lucassen, M., Storch, D., 2005. Metabolic biochemistry: its role in thermal tolerance and in the capacities of physiological and ecological function. In: Farrell, A.P., Steffensen, J.F. (Eds.), The Physiology of Polar Fishes. In: Fish Physiology, vol. 22. Academic Press, New York, pp. 79–154.
Powers, D.A., 1974. Structure, function, and molecular ecology of fish hemoglobins. Ann. N. Y. Acad. Sci. 241, 472–490.
Powers, D.A., 1980. Molecular ecology of teleost fish hemoglobins: strategies for adapting to changing environments. Am. Zool. 162, 139–162.
Powers, D.A., Powers, D., 1975. Predicting gene frequencies in natural populations: a testable hypothesis. In: Markert, C.L. (Ed.), In: Isozymes. Genetics and evolution, vol. IV. Academic Press, New York, pp. 63–84.
Powers, D.A., Fyhn, H.J., Fyhn, U.E.H., Martin, J.P., Garlick, R.L., Wood, S.C., 1979. A comparative study of the oxygen equilibria of blood from 40 genera of Amazonian fishes. Comp. Biochem. Physiol. 62, 67–85.
Primmett, D.R.N., Randall, D.J., Mazeaud, M., Boutilier, R.G., 1986. The role of catecholamines in erythrocyte pH regulation and oxygen-transport in rainbow-trout (*Salmo-gairdneri*) during exercise. J. Exp. Biol. 122, 139–148.
Qvist, J., Weber, R.E., DeVries, A., Zapol, W., 1977. pH and haemoglobin oxygen affinity in blood from the Antarctic cod *Dissostichus mawsoni*. J. Exp. Biol. 67, 77–88.
Rahim, S., Delaunoy, J.-P., Laurent, P., 1988. Identification and immunocytochemical localization of two different carbonic anhydrase isoenzymes in teleostean fish erythrocytes and gill epithelia. Histochemistry 89, 451–459.
Randall, D., 1982a. The control of respiration and circulation in fish during exercise and hypoxia. J. Exp. Biol. 100, 275–288.

Randall, D.J., 1982b. Blood flow through gills. In: Houlihan, D.F., Rankin, J.C., Shuttleworth, T.J. (Eds.), In: Gills: Society for Experimental Biology Seminar Series, vol. 16. Cambridge University Press, Cambridge, pp. 173–191.

Randall, D., 1985. In: Johansen, K., Burggren, W. (Eds.), Shunts in fish gills. In Cardiovascular Shunts, Alfred Benzon Symposium. In: vol. 21. Munksgaard, Copenhagen, pp. 71–87.

Randall, D., 1990. Control and co-ordination of gas exchange in water breathers. In: Boutilier, R.G. (Ed.), In: Advances in Comparative and Environmental Physiology, vol. 6. Springer, Berlin, pp. 253–278.

Randall, D., Daxboeck, C., 1984. Oxygen and carbon dioxide transfer across fish gills. In: Hoar, W.S., Randall, D.J. (Eds.), Gills. In: Fish Physiology, vol. 10A. Academic Press, New York, pp. 263–314.

Randall, D.J., Perry, S.F., 1992. Catecholamines. In: Hoar, W.S., Randall, D.J., Farrell, A.P. (Eds.), The Cardiovascular System. In: Fish Physiology, vol. 12B. Academic Press, New York, pp. 255–300.

Randall, D., Val, A., 1995. The role of carbonic anhydrase in aquatic gas exchange. In: Heisler, N., Boutilier, R.G. (Eds.), Mechanisms of Systemic Regulation. Springer, Berlin, pp. 25–39.

Randall, D., Rummer, J., Wilson, J., Wang, S., Brauner, C., 2014. A unique mode of tissue oxygenation and the adaptive radiation of teleost fishes. J. Exp. Biol. 217, 1205–1214.

Regan, M., Brauner, C., 2010a. The transition in hemoglobin proton-binding characteristics within the basal actinopterygian fishes. J. Comp. Physiol. B. 180, 521–530.

Regan, M.D., Brauner, C.J., 2010b. The evolution of Root effect hemoglobins in the absence of intracellular pH protection of the red blood cell: insights from primitive fishes. J. Comp. Physiol. B. 180, 695–706.

Regan, M.D., Gill, I., Richards, J.G., 2016. Calorespirometry reveals that goldfish prioritize aerobic metabolism over metabolic rate depression in all but near-anoxic environments. J. Exp. Biol. 220, 564–572.

Reid, S., Perry, S., 1991. The effects and physiological consequences of raised levels of cortisol on rainbow trout (*Oncorhynchus mykiss*) erythrocyte beta-adrenoreceptors. J. Exp. Biol. 158, 217–240.

Reid, S., Moon, T., Perry, S., 1991. Characterization of beta-adrenoreceptors of rainbow trout (*Oncorhynchus mykiss*) erythrocytes. J. Exp. Biol. 158, 199–216.

Rice, C.L., Herman, C.M., Kiesow, L.A., Homer, L.D., John, D.A., Valeri, R., 1975. Benefits from improved oxygen delivery of blood in shock therapy. J. Surg. Res. 19, 193–198.

Richards, J.G., 2009. Metabolic and molecular responses of fish to hypoxia. In: Farrell, A.P., Brauner, C.J. (Eds.), Hypoxia. In: Fish Physiology, vol. 27. Academic Press, New York, pp. 443–485.

Richardson, T.Q., Guyton, A.C., 1959. Effects of polycythemia and anemia on cardiac output and other circulatory factors. Am. J. Phys. 197, 1167–1170.

Richardson, S.L., Swietach, P., 2016. Red blood cell thickness is evolutionarily constrained by slow, hemoglobin-restricted diffusion in cytoplasm. Sci Rep 6, 36018.

Ridderstrale, Y., 1979. Observations on the localization of carbonic anhydrase in muscle. Acta Physiol. Scand. 106, 239–240.

Riggs, A., 1970. Properties of fish hemoglobins. In: Hoar, W.S., Randall, D. (Eds.), The Nervous System, Circulation, and Respiration. In: Fish Physiology, vol. 4. Academic Press, New York, pp. 209–252.

Riggs, A., 1976. Factors in the evolution of hemoglobin function. Fed. Proc. 35, 2115–2118.

Riggs, A.F., 1988. The Bohr effect. Annu. Rev. Physiol. 50, 181–204.

Riley, D., Ellis, S., Bain, J., 1982. Carbonic anhydrase activity in skeletal muscle fiber types, axons, spindles, and capillaries of rat soleus and extensor digitorum longus muscles. J. Histochem. Cytochem. 30, 1275–1288.

Rispens, P., Hessels, J., Zwart, A., Zijlstra, W., 1985. Inhibition of carbonic anhydrase in dog plasma. Pflugers Arch. 403, 344–347.

Rissanen, E., Tranberg, H.K., Sollid, J., Nilsson, G.E., Nikinmaa, M., 2006. Temperature regulates hypoxia-inducible factor-1 (HIF-1) in a poikilothermic vertebrate, crucian carp (*Carassius carassius*). J. Exp. Biol. 209, 994–1003.

Robergs, R.A., Ghiasvand, F., Parker, D., 2004. Biochemistry of exercise-induced metabolic acidosis. Am. J. Phys. Regul. Integr. Comp. Phys. 287, R502–R516.

Romano, L., Passow, H., 1984. Characterization of anion transport system in trout red blood cell. Am. J. Phys. 246, C330–C338.

Romero, M.G., Guizouarn, H., Pellissier, B., GarciaRomeu, F., Motais, R., 1996. The erythrocyte Na^+/H^+ exchangers of eel (*Anguilla anguilla*) and rainbow trout (*Oncorhynchus mykiss*): a comparative study. J. Exp. Biol. 199, 415–426.

Root, R.W., 1931. The respiratory function of the blood of marine fishes. Biol. Bull. 61, 427–456.

Rothe, C.F., 1993. Mean circulatory filling pressure: its meaning and measurement. J. Appl. Physiol. 74, 499–509.

Roush, E.D., Fierke, C.A., 1992. Purification and characterization of a carbonic anhydrase II inhibitor from porcine plasma. Biochemist 31, 12536–12542.

Rummer, J.L., Brauner, C.J., 2011. Plasma-accessible carbonic anhydrase at the tissue of a teleost fish may greatly enhance oxygen delivery: *in vitro* evidence in rainbow trout, *Oncorhynchus mykiss*. J. Exp. Biol. 214, 2319–2328.

Rummer, J.L., Brauner, C.J., 2015. Root effect haemoglobins in fish may greatly enhance general oxygen delivery relative to other vertebrates. PLoS One 10. e0139477.

Rummer, J.L., Roshan-Moniri, M., Balfry, S.K., Brauner, C.J., 2010. Use it or lose it? Sablefish, *Anoplopoma fimbria*, a species representing a fifth teleostean group where the beta NHE associated with the red blood cell adrenergic stress response has been secondarily lost. J. Exp. Biol. 213, 1503–1512.

Rummer, J.L., McKenzie, D.J., Innocenti, A., Supuran, C.T., Brauner, C.J., 2013. Root effect hemoglobin may have evolved to enhance general tissue oxygen delivery. Science 340, 1327–1329.

Rutjes, H.A., Nieveen, M.C., Weber, R.E., Witte, F., Van den Thillart, G.E.E.J.M., 2007. Multiple strategies of Lake Victoria cichlids to cope with lifelong hypoxia include hemoglobin switching. Am. J. Phys. Regul. Integr. Comp. Phys. 293, R1376–1383.

Ruud, J.T., 1954. Vertebrates without erythrocytes and blood pigment. Nature 173, 848–850.

Salama, A., Nikinmaa, M., 1988. The adrenergic responses of carp (*Cyprinus carpio*) red cells: effects of PO_2 and pH. J. Exp. Biol. 136, 405–416.

Salama, A., Nikinmaa, M., 1989. Species differences in the adrenergic responses of fish red cells: studies on whitefish, pikeperch, trout and carp. Fish Physiol. Biochem. 6, 167–173.

Salama, A., Nikinmaa, M., 1990. Effect of oxygen tension on catecholamine-induced formation of cAMP and on swelling of carp red blood cells. Am. J. Phys. 259, C723–C726.

Sandblom, E., Axelsson, M., 2007. The venous circulation: a piscine perspective. Comp. Biochem. Physiol. A Mol. Integr. Physiol. 148, 785–801.

Sanyal, G., Swenson, E., Pessah, N., Maren, T., 1982. The carbon dioxide hydration activity of skeletal muscle carbonic anhydrase. Inhibition by sulfonamides and anions. Mol. Pharmacol. 22, 211–220.

Sanyal, G., Swenson, E.R., Maren, T.H., 1984. The isolation of carbonic anhydrase from the muscle of *Squalus acanthias* and *Scomber scombrus*: inhibition studies. Bull. Mt. Desert Isl. Biol. Lab. 24, 66–68.

Sardella, B.A., Bruner, C.J., 2007. The osmo-respiratory compromise in fish. In: Fernandes, M.N., Rantin, F.T., Glass, M.L., Kapoor, B.G. (Eds.), Fish Respiration and Environment. Science Publishers, Enfield, pp. 147–165.

Scholander, P.F., Van Dam, L., 1954. Secretion of gases against high pressures in the swimbladder of deep sea fishes. I. Oxygen dissociation in blood. Biol. Bull. 107, 247–259.

Scholander, P.F., Van Dam, L., 1957. The concentration of hemoglobin in some cold water Arctic fishes. J. Cell. Physiol. 49, 1–4.

Scozzafava, A., Briganti, F., Ilies, M.A., Supuran, C.T., 2000. Carbonic anhydrase inhibitors: synthesis of membrane-impermeant low molecular weight sulfonamides possessing *in vivo* selectivity for the membrane-bound versus cytosolic isozymes. J. Med. Chem. 43, 292–300.

Secomb, T.W., Hsu, R., Park, E.Y., Dewhirst, M.W., 2004. Green's function methods for analysis of oxygen delivery to tissue by microvascular networks. Ann. Biomed. Eng. 32, 1519–1529.

Sender, S., Gros, G., Waheed, A., Hageman, G.S., Sly, W.S., 1994. Immunohistochemical localization of carbonic anhydrase IV in capillaries of rat and human skeletal muscle. J. Histochem. Cytochem. 42, 1229–1236.

Sender, S., Decker, B., Fenske, C.D., Sly, W.S., Carter, N.D., Gros, G., 1998. Localization of carbonic anhydrase IV in rat and human heart muscle. J. Histochem. Cytochem. 46, 855–861.

Sender, S., Böttcher, K., Cetin, Y., Gros, G., 1999. Carbonic anhydrase in the gills of seawater- and freshwater-acclimated flounders *Platichthys flesus*: purification, characterization, and immunohistochemical localization. J. Histochem. Cytochem. 47, 43–50.

Sidell, B.D., O'Brien, K.M., 2006. When bad things happen to good fish: the loss of hemoglobin and myoglobin expression in Antarctic icefishes. J. Exp. Biol. 209, 1791–1802.

Siffert, W., Gros, G., 1982. Carbonic anhydrase C in white-skeletal-muscle tissue. Biochem. J. 205, 559–566.

Skomal, G., Bernal, D., 2010. Physiological responses to stress in sharks. In: Carrier, J.C., Musick, J.A., Heithaus, M.R. (Eds.), Sharks and Their Relatives II: Biodiversity, Adaptive Physiology, and Conservation. CRC Press, Boca Raton, pp. 459–490.

Sly, W.S., 2000. The membrane carbonic anhydrases: from CO_2 transport to tumor markers. In: -Chegwidden, W.R., Carter, N.D., Edwards, Y.H. (Eds.), The Carbonic Anhydrases: New Horizons. Springer, Berlin, pp. 95–104.

Smeda, J.S., Houston, A., 1979. Evidence of weight dependent differential hematological response to increased environmental temperature by carp, *Cyprinus carpio*. Environ. Biol. Fish 4, 89–92.

Smith, F.M., Jones, D.R., 1982. The effect of changes in blood oxygen-carrying capacity on ventilation volume in the rainbow trout (*Salmo gairdneri*). J. Exp. Biol. 97, 325–334.

Soivio, A., Nikinmaa, M., 1981. The swelling of erythrocytes in relation to the oxygen affinity of the blood of the rainbow trout, *Salmo gairdneri* Richardson. In: Pickering, A.D. (Ed.), Stress and Fish. Academic Press, New York, pp. 103–119.

Soivio, A., Nikinmaa, M., Westman, K., 1980. The blood oxygen binding properties of hypoxic *Salmo gairdneri*. J. Comp. Physiol. 136, 83–87.

Soldatov, A., Parfenova, I., 2014. Hemoglobin system of golden mullet (*Liza aurata*, Risso) at adaptation to conditions of outer hypoxia. J. Evol. Biochem. Physiol. 50, 81–87.

Sollid, J., Weber, R.E., Nilsson, G.E., 2005. Temperature alters the respiratory surface area of crucian carp *Carassius carassius* and goldfish *Carassius auratus*. J. Exp. Biol. 208, 1109–1116.

Speckner, W., Schindler, J.F., Albers, C., 1989. Age-dependent changes in volume and haemoglobin content of erythrocytes in the carp (*Cyprinus carpio* L.). J. Exp. Biol. 141, 133–149.

Speers-Roesch, B., Richards, J.G., Brauner, C.J., Farrell, A.P., Hickey, A.J., Wang, Y.S., Renshaw, G.M., 2012. Hypoxia tolerance in elasmobranchs. I. Critical oxygen tension as a measure of blood oxygen transport during hypoxia exposure. J. Exp. Biol. 215, 93–102.

Stabenau, E.K., Heming, T.A., 1999. CO_2 excretion and postcapillary pH equilibration in blood-perfused turtle lungs. J. Exp. Biol. 202, 965–975.
Stabenau, E.K., Heming, T., 2003. Pulmonary carbonic anhydrase in vertebrate gas exchange organs. Comp. Biochem. Physiol. A Mol. Integr. Physiol. 136, 271–279.
Stabenau, E.K., Vietti, K.R., 2002. Pulmonary carbonic anhydrase in the garter snake, *Thamnophis sirtalis*. Physiol. Biochem. Zool. 75, 83–89.
Stabenau, E.K., Bidani, A., Heming, T.A., 1996. Physiological characterization of pulmonary carbonic anhydrase in the turtle. Respir. Physiol. 104, 187–196.
Stecyk, J.A.W., 2017. Cardiovascular responses to limiting oxygen levels. In: Gamperl, A.K., Gillis, T.E., Farrell, A.P., Brauner, C.J. (Eds.), The Cardiovascular System: Development, Plasticity and Physiological Responses. Fish Physiology, vol. 36B. Academic Press, San Diego. pp. 299–371. Chapter 5.
Steffensen, J.F., 2005. Respiratory systems and metabolic rates. Fish Physiol. 22, 203.
Steffensen, J.F., Farrell, A., 1998. Swimming performance, venous oxygen tension and cardiac performance of coronary-ligated rainbow trout, *Oncorhynchus mykiss*, exposed to progressive hypoxia. Comp. Biochem. Physiol. A Mol. Integr. Physiol. 119, 585–592.
Steffensen, J., Lomholt, J.P., 1992. The secondary vascular system. In: Hoar, W.S., Randall, D.J., Farrell, A.P. (Eds.), The Cardiovascular System. In: Fish Physiology, vol. 12. Academic Press, New York, pp. 185–213.
Steinhausen, M.F., Sandblom, E., Eliason, E.J., Verhille, C., Farrell, A.P., 2008. The effect of acute temperature increases on the cardiorespiratory performance of resting and swimming sockeye salmon (*Oncorhynchus nerka*). J. Exp. Biol. 211, 3915–3926.
Stevens, E.D., Randall, D.J., 1967. Changes of gas concentrations in blood and water during moderate swimming activity in rainbow trout. J. Exp. Biol. 46, 329–337.
Storz, J.F., Moriyama, H., 2008. Mechanisms of hemoglobin adaptation to high altitude hypoxia. High Alt. Med. Biol. 9, 148–157.
Sundin, L., Nilsson, S., 1992. Arterio-venous branchial blood flow in the Atlantic cod *Gadus morhua*. J. Exp. Biol. 165, 73–84.
Supuran, C.T., Scozzafava, A., 2004. Benzolamide is not a membrane-impermeant carbonic anhydrase inhibitor. J. Enzyme Inhib. Med. Chem. 19, 269–273.
Swenson, E.R., 2000. Respiratory and renal roles of carbonic anhydrase in gas exchange and acid-base regulation. In: Chegwidden, W.R., Carter, N.D., Edwards, Y.H. (Eds.), The Carbonic Anhydrases: New Horizons. Springer, Berlin, pp. 281–341.
Swenson, E.R., Maren, T.H., 1978. A quantitative analysis of CO_2 transport at rest and during maximal exercise. Respir. Physiol. 35, 129–159.
Swenson, E.R., Gronlund, J., Ohlsson, J., Hlastala, M.P., 1993a. *In vivo* quantitation of carbonic anhydrase and band 3 protein contributions to pulmonary gas exchange. J. Appl. Physiol. 74, 838–848.
Swenson, E.R., Robertson, H.T., Hlastala, M., 1993b. Effects of carbonic anhydrase inhibition on ventilation-perfusion matching in the dog lung. J. Clin. Invest. 92, 702.
Swenson, E.R., Graham, M.M., Hlastala, M.P., 1995a. Acetazolamide slows VA/Q matching after changes in regional blood flow. J. Appl. Physiol. 78, 1312–1318.
Swenson, E., Lippincott, L., Maren, T., 1995b. Effect of gill membrane-bound carbonic anhydrase inhibition on branchial bicarbonate excretion in the dogfish shark, Squalus acanthias. Bull. Mt. Desert Isl. Biol. Lab. 34, 94–95.
Swenson, E., Taschner, B., Maren, T., 1996. Effect of membrane-bound carbonic anhydrase (CA) inhibition on bicarbonate excretion in the shark, *Squalus acanthias*. Bull. Mt. Desert Isl. Biol. Lab. 35, 35.
Swenson, E., Brogan, T., Hedges, R., Deem, S., 1998. Acetazolamide slows hypoxic pulmonary vasoconstriction (HPV). Am. J. Respir. Crit. Care Med. 157, A380.

Swift, D., Lloyd, R., 1974. Changes in urine flow rate and haematocrit value of rainbow trout *Salmo gairdneri* (Richardson) exposed to hypoxia. J. Fish Biol. 6, 379–387.

Szebedinszky, C., Gilmour, K., 2002. The buffering power of plasma in brown bullhead (*Ameiurus nebulosus*). Comp. Biochem. Physiol. 131, 171–183.

Taylor, J.R., Grosell, M., 2009. The intestinal response to feeding in seawater gulf toadfish, *Opsanus beta*, includes elevated base secretion and increased epithelial oxygen consumption. J. Exp. Biol. 212, 3873–3881.

Taylor, S., Egginton, S., Taylor, E., 1993. Respiratory and cardiovascular-responses in rainbow-trout (*Oncorhynchus mykiss*) to aerobic exercise over a range of acclimation temperatures. J. Physiol. 459, P19.

Taylor, S., Egginton, S., Taylor, E., 1996. Seasonal temperature acclimatisation of rainbow trout: cardiovascular and morphometric influences on maximal sustainable exercise level. J. Exp. Biol. 199, 835–845.

Tenney, S.M., 1995. In: Sutton, J.R., Houston, C.S., Coates, G. (Eds.), Hypoxia and the brain: functional significance of differences in mammalian hemoglobin affinity for oxygen. Proceeding of the 9th International Hypoxia Symposium. Queen city printers, Burlington, Vt., Lake Louise, Canada, pp. 57–68.

Tervonen, V., Arjamaa, O., Kokkonen, K., Ruskoaho, H., Vuolteenaho, O., 1998. A novel cardiac hormone related to A-, B-, and C-type natriuretic peptides. Endocrinology 139, 4021–4025.

Tervonen, V., Ruskoaho, H., Lecklin, T., Ilves, M., Vuolteenaho, O., 2002. Salmon cardiac natriuretic peptide is a volume-regulating hormone. Am. J. Phys. 283, E353–E361.

Tervonen, V., Vuolteenaho, O., Nikinmaa, M., 2006. Haemoconcentration via diuresis in short-term hypoxia: a possible role for cardiac natriuretic peptide in rainbow trout. Comp. Biochem. Physiol. A Mol. Integr. Physiol. 144, 86–92.

Tetens, V., Lykkeboe, G., 1981. Blood respiratory properties of rainbow trout, *Salmo gairdneri*: responses to hypoxia acclimation and anoxic incubation of blood *in vitro*. J. Comp. Physiol. 145, 117–125.

Tetens, V., Lykkeboe, G., 1985. Acute exposure of rainbow trout to mild and deep hypoxia: O_2 affinity and O_2 capacitance of arterial blood. Respir. Physiol. 61, 221–235.

Tetens, V., Lykkeboe, G., 1988. Potency of adrenaline and noradrenaline for b-adrenergic proton extrusion from red cells of rainbow trout, *Salmo gairdneri*. J. Exp. Biol. 134, 267–280.

Tetens, V., Wells, R.M., Devries, A.L., 1984. Antarctic fish blood: respiratory properties and the effects of thermal acclimation. J. Exp. Biol. 109, 265–279.

Thomas, S., Perry, S.F., 1992. Control and consequences of adrenergic activation of red blood cell Na^+/H^+ exchange on blood oxygen and carbon dioxide transport in fish. J. Exp. Zool. 263, 160–175.

Thomas, S., Poupin, J., Lykkeboe, G., Johansen, K., 1987. Effects of graded exercise on blood gas tensions and acid-base characteristics of rainbow trout. Respir. Physiol. 68, 85–97.

Thomas, S., Fievet, B., Claireaux, G., Motais, R., 1988. Adaptive respiratory responses of trout to acute hypoxia. I. Effects of water ionic composition on blood acid-base status response and gill morphology. Respir. Physiol. 74, 77–89.

Thorarensen, H., 1994. Gastrointestinal blood flow in chinook salmon (Oncorhynchus tshawytscha). In: Biological Sciences. Simon Fraser University, Burnaby, BC, Canada (Ph.D.).

Thorarensen, H., Gallaugher, P.E., Kiessling, A.K., Farrell, A.P., 1993. Intestinal blood-flow in swimming chinook salmon *Oncorhynchus tshawytscha* and the effects of hematocrit on blood flow distribution. J. Exp. Biol. 179, 115–129.

Thorarensen, H., Gallaugher, P., Farrell, A.P., 1996. Cardiac output in swimming rainbow trout, *Oncorhynchus mykiss*, acclimated to seawater. Physiol. Zool. 69, 139–153.

Timmerman, C.M., Chapman, L.J., 2004. Behavioral and physiological compensation for chronic hypoxia in the sailfin molly (*Poecilia latipinna*). Physiol. Biochem. Zool. 77, 601–610.
Torrance, J., Lenfant, C., Cruz, J., Marticorena, E., 1971. Oxygen transport mechanisms in residents at high altitude. Respir. Physiol. 11, 1–15.
Tresguerres, M., Katoh, F., Orr, E., Parks, S.K., Goss, G.G., 2006. Chloride uptake and base secretion in freshwater fish: a transepithelial ion-transport metabolon? Physiol. Biochem. Zool. 79, 981–996.
Tufts, B.L., 1991. Acid-base regulation and blood gas transport following exhaustive exercise in an Agnathan the sea lamprey *Petromyzon marinus*. J. Exp. Biol. 159, 371–386.
Tufts, B.L., 1992. *In vitro* evidence for sodium-dependent pH regulation in sea lamprey (*Petromyzon marinus*) red blood cells. Can. J. Zool. 70, 411–416.
Tufts, B.L., Boutilier, R.G., 1989. The absence of rapid chloride/bicarbonate exchange in lamprey erythrocytes: implications for CO_2 transport and ion distributions between plasma and erythrocytes in the blood of *Petromyzon Marinus*. J. Exp. Biol. 144, 565–576.
Tufts, B.L., Perry II, S.F., 1998. Carbon dioxide transport and excretion. In: Perry II, S.F., Tufts, B.L. (Eds.), Fish Respiration. In: Fish Physiology, vol. 17. Academic Press, New York, pp. 229–282.
Tufts, B.L., Tang, Y., Tufts, K., Boutilier, R.G., 1991. Exhaustive exercise in "wild" Atlantic Salmon (*Salmo salar*): acid-base regulation and blood gas transport. Can. J. Fish. Aquat. Sci. 48, 868–874.
Tufts, B.L., Bagatto, B., Cameron, B., 1992. *In vivo* analysis of gas transport in arterial and venous blood of the sea lamprey *Petromyzon marinus*. J. Exp. Biol. 169, 105–119.
Tufts, B., Gervais, M., Staebler, M., Weaver, J., 2002. Subcellular distribution and characterization of gill carbonic anhydrase and evidence for a plasma carbonic anhydrase inhibitor in Antarctic fish. J. Comp. Physiol. B. 172, 287–295.
Tufts, B.L., Esbaugh, A., Lund, S.G., 2003. Comparative physiology and molecular evolution of carbonic anhydrase in the erythrocytes of early vertebrates. Comp. Biochem. Physiol. A Mol. Integr. Physiol. 136, 259–269.
Tun, N., Houston, A., 1986. Temperature, oxygen, photoperiod, and the hemoglobin system of the rainbow trout, *Salmo gairdneri*. Can. J. Zool. 64, 1883–1888.
Turek, Z., Kreuzer, F., Hoofd, L.J.C., 1973. Advantage or disadvantage of a decrease of blood oxygen affinity for tissue oxygen supply at hypoxia. Pflugers Arch. 342, 185–197.
Val, A.L., 2000. Organic phosphates in the red blood cells of fish. Comp. Biochem. Physiol. A Mol. Integr. Physiol. 125A, 417–435.
Val, A.L., Schwantes, A.R., Almeida-Val, V.M.F., 1986. Biological aspects of Amazonian fishes-VI. Hemoglobins and whole blood properties of *Semaprochilodus* species (Prochilodontidae) at two phases of migration. Comp. Biochem. Physiol. 83B, 659–667.
Val, A.L., Affonso, E.G., Almeida-Val, V.M.F., 1992. Adaptive features of Amazon fishes: blood characteristics of Curimata (*Prochilodus* cf. *Inigricans*, Osteichthyes). Physiol. Zool. 65, 832–843.
Val, A.L., Mazur, C.F., De Salvo-Souza, R.H., Iwama, G.K., 1994. Effects of experimental anaemia on intra-erythrocytic phosphate levels in rainbow trout, *Oncorhynchus mykiss*. J. Fish Biol. 45, 269–277.
Van den Thillart, G., Randall, D., Hoaren, L., 1983. CO_2 and H^+ excretion by swimming coho salmon, *Oncorhynchus kisutch*. J. Exp. Biol. 107, 169–180.
Van Slyke, D.D., Wu, H., McLean, F.C., 1923. Studies of gas and electrolyte equilibria in the blood V. Factors controlling the electrolyte and water distribution in the blood. J. Biol. Chem. 56, 765–849.

Vandenberg, J.I., Carter, N.D., Bethell, H., Nogradi, A., Ridderstrale, Y., Metcalfe, J.C., Grace, A.A., 1996. Carbonic anhydrase and cardiac pH regulation. Am. J. Phys. 271, C1838–C1846.

Vega, G.C., Wiens, J.J., 2012. Why are there so few fish in the sea? Proc. R. Soc. B Biol. Sci. 279, 2323–2329.

Vermette, M., Perry, S., 1987. The effects of prolonged epinephrine infusion on the physiology of the rainbow trout, *Salmo gairdneri*. III. Renal ionic fluxes. J. Exp. Biol. 128, 269–285.

Vermette, M., Perry, S., 1988. Effects of prolonged epinephrine infusion on blood respiratory and acid-base states in the rainbow trout: alpha and beta effects. Fish Physiol. Biochem. 4, 189–202.

Virkki, L., Nikinmaa, M., 1994. Activation and physiological role of Na^+/H^+ exchange in lamprey (*Lampetra fluviatilis*) erythrocytes. J. Exp. Biol. 191, 89–105.

Vornanen, M., Stecyk, J.A.W., Nilsson, G.E., 2009. The anoxia-tolerant crucian carp (*Carassius Carassius* L.). In: Farrell, A.P., Brauner, C.J. (Eds.), Hypoxia. In: Fish Physiology, vol. 27. Academic Press, New York, pp. 397–441.

Wagner, P.D., 1977. Diffusion and chemical reaction in pulmonary gas exchange. Physiol. Rev. 57, 257–312.

Waheed, A., Zhu, X., Sly, W., 1992a. Membrane-associated carbonic anhydrase from rat lung. Purification, characterization, tissue distribution, and comparison with carbonic anhydrase IVs of other mammals. J. Biol. Chem. 267, 3308–3311.

Waheed, A., Zhu, X.L., Sly, W.S., Wetzel, P., Gros, G., 1992b. Rat skeletal muscle membrane associated carbonic anhydrase is 39-kDa, glycosylated, GPI-anchored CA IV. Arch. Biochem. Biophys. 294, 550–556.

Waheed, A., Okuyama, T., Heyduk, T., Sly, W.S., 1996. Carbonic anhydrase IV: purification of a secretory form of the recombinant human enzyme and identification of the positions and importance of its disulfide bonds. Arch. Biochem. Biophys. 333, 432–438.

Wang, T., Malte, H., 2011. O2 uptake and transport: the optimal P50. In: Farrell, A. (Ed.), In: Encycopedia of Fish Physiology, vol. 2. Academic Press, New York, pp. 893–898.

Wang, Y., Henry, R.P., Wright, P.M., Heigenhauser, J.F., Wood, C.M., 1998. Respiratory and metabolic functions of carbonic anhydrase in exercised white muscle of trout. Am. J. Phys. 275, 1766–1779.

Wang, T., Brauner, C.J., Milsom, W.K., 1999. The effect of isovolemic anaemia on blood O_2 affinity and red cell triphosphate concentrations in the painted turtle (*Chrysemys picta*). Comp. Biochem. Physiol. A Mol. Integr. Physiol. 122, 341–346.

Wang, X., Conway, W., Burns, R., McCann, N., Maeder, M., 2009. Comprehensive study of the hydration and dehydration reactions of carbon dioxide in aqueous solution. J. Phys. Chem. A 114, 1734–1740.

Waser, W., Heisler, N., 2005. Oxygen delivery to the fish eye: root effect as crucial factor for elevated retinal PO_2. J. Exp. Biol. 208, 4035–4047.

Weaver, Y.R., Kiessling, K., Cossins, A.R., 1999. Responses of the Na^+/H^+ exchanger of European flounder red blood cells to hypertonic, beta-adrenergic and acidotic stimuli. J. Exp. Biol. 202, 21–32.

Weber, R.E., 1982. Intraspecific adaptation of hemoglobin function in fish to oxygen availability. In: Addink, A.D.F., Spronk, N. (Eds.), Exogenous and Endogenous Influences on Metabolic and Neural Control. Pergamon Press, Oxford, pp. 87–102.

Weber, R.E., 1990. Functional significance and structural basis of multiple hemoglobins with special reference to ectothermic vertebrates. In: Truchot, J.P., Lahlou, B. (Eds.), In: Comparative Physiology, vol. 6. Karger, Basel, Switzerland, pp. 58–75.

Weber, R.E., 1996. Hemoglobin adaptations in Amazonian and temperate fish with special reference to hypoxia, allosteric effectors and functional heterogeneity. In: Val, A.L.,

Almeida-Val, V.M.F., Randall, D.J. (Eds.), Physiology and Biochemistry of the Fishes of the Amazon. INPA, Manaus, Amazonas, Brazil, pp. 75–90.

Weber, R.E., 2000. Adaptations for oxygen transport: lessons from fish hemoglobins. In: di Prisco, G., Giardina, B., Weber, R.E. (Eds.), Hemoglobin Function in Vertebrates: Molecular Adaptation in Extreme and Temperate Environments. Springer, Berlin, pp. 23–37.

Weber, R.E., Fago, A., 2004. Functional adaptation and its molecular basis in vertebrate hemoglobins, neuroglobins and cytoglobins. Respir. Physiol. Neurobiol. 144, 141–159.

Weber, R.E., Jensen, F.B., 1988. Functional adaptations in hemoglobins from ectothermic vertebrates. Annu. Rev. Physiol. 50, 161–179.

Weber, R.E., Lykkeboe, G., 1978. Respiratory adaptations in carp blood influences of hypoxia, red cell organic phosphates, divalent cations and CO_2 on hemoglobin-oxygen affinity. J. Comp. Physiol. 128, 127–137.

Weber, R.E., Wells, R.M., 1989. Hemoglobin structure and function. In: Wood, S.C. (Ed.), Lung Biology in Health and Disease. Comparative Pulmonary Physiology, Current Concepts. Marcel Dekker, Inc., New York, pp. 279–310.

Weber, R.E., Lykkeboe, G., Johansen, K., 1975. Biochemical aspects of the adaptation of hemoglobin-oxygen affinity of eels to hypoxia. Life Sci. 17, 1345–1349.

Weber, R.E., Lykkeboe, G., Johansen, K., 1976a. Physiological properties of eel haemoglobin: hypoxic acclimation, phosphate effects and multiplicity. J. Exp. Biol. 64, 75–88.

Weber, R.E., Wood, S.C., Lomholt, J.P., 1976b. Temperature acclimation and oxygen-binding properties of blood and multiple haemoglobins of rainbow trout. J. Exp. Biol. 65, 333–345.

Weber, R., Fago, A., Val, A., Bang, A., Van, H.M., Dewilde, S., Zal, F., Moens, L., 2000. Isohemoglobin differentiation in the bimodal-breathing amazon catfish *Hoplosternum littorale*. J. Biol. Chem. 275, 297–305.

Weber, R.E., Campbell, K.L., Fago, A., Malte, H., Jensen, F.B., 2010. ATP-induced temperature independence of hemoglobin-O_2 affinity in heterothermic billfish. J. Exp. Biol. 213, 1579–1585.

Weinberg, S.R., LoBue, J., Siegel, C.D., Gordon, A.S., 1976. Hematopoiesis of the kissing gourami (*Helostoma temmincki*). Effects of starvation, bleeding, and plasma-stimulating factors on its erythropoiesis. Can. J. Zool. 54, 1115–1127.

Wells, R., 1978. Respiratory adaptation and energy metabolism in Antarctic nototheniid fishes. N. Z. J. Zool. 5, 813–815.

Wells, R.M.G., 1990. Hemoglobin physiology in vertebrate animals: a cautionary approach to adaptationist thinking. In: Boutilier, R.G. (Ed.), In: Advances in Comparative and Environmental Physiology: Vertebrate Gas Exchange From Environment to Cell, vol. 6. Springer, Berlin, pp. 143–161.

Wells, R.M.G., 1999. Haemoglobin function in aquatic animals: molecular adaptations to environmental challenge. Mar. Freshw. Res. 50, 933–939.

Wells, R.M.G., 2005. Blood-gas transport and hemoglobin function in polar fishes: does low temperature explain physiological characters? In: Steffensen, J.F., Farrel, A.P. (Eds.), The Physiology of Polar Fishes. In: Fish Physiology, vol. 22. Academic Press, New York, pp. 281–316.

Wells, R.M., 2009. Blood-gas transport and hemoglobin function: adaptations for functional and environmental hypoxia. In: Farrell, A.P., Brauner, C.J. (Eds.), Hypoxia. In: Fish Physiology, vol. 27. Academic Press, New York, pp. 255–299.

Wells, R., Baldwin, J., 1990. Oxygen transport potential in tropical reef fish with special reference to blood viscosity and haematocrit. J. Exp. Mar. Biol. Ecol. 141, 131–143.

Wells, R., Weber, R., 1990. The spleen in hypoxic and exercised rainbow trout. J. Exp. Biol. 150, 461–466.

Wells, R., Weber, R.E., 1991. Is there an optimal haematocrit for rainbow trout, *Oncorhynchm mykiss* (Walbaum)? An interpretation of recent data based on blood viscosity measurements. J. Fish Biol. 38, 53–65.

Wells, R., Ashby, M., Duncan, S., Macdonald, J., 1980. Comparative study of the erythrocytes and haemoglobins in nototheniid fishes from Antarctica. J. Fish Biol. 17, 517–527.

Wells, R., Tetens, V., Devries, A., 1984. Recovery from stress following capture and anaesthesia of Antarctic fish: haematology and blood chemistry. J. Fish Biol. 25, 567–576.

Wells, R.M.G., Grigg, G., Beard, L., Summers, G., 1989. Hypoxic responses in a fish from a stable environment: blood oxygen transport in the antartic fish *Pagothenia Borchgrevinki*. J. Exp. Biol. 141, 97–111.

Wells, R., Baldwin, J., Seymour, R., Baudinette, R., Christian, K., Bennett, M., 2003. Oxygen transport capacity in the air-breathing fish, *Megalops cyprinoides*: compensations for strenuous exercise. Comp. Biochem. Physiol. A Mol. Integr. Physiol. 134, 45–53.

Wells, R.M.G., Baldwin, J., Seymour, R.S., Christian, K., Brittain, T., 2005. Red blood cell function and haematology in two tropical freshwater fishes from Australia. Comp. Biochem. Physiol. A Mol. Integr. Physiol. 141, 87–93.

Werner, M., von Wasielewski, R., Komminoth, P., 1996. Antigen retrieval, signal amplification and intensification in immunohistochemistry. Histochem. Cell Biol. 105, 253–260.

West, J.B., Wagner, P.D., 1980. Predicted gas exchange on the summit of Mt. Everest. Respir. Physiol. 42, 1–16.

Wetzel, P., Gros, G., 1990. Sarcolemmal carbonic anhydrase in red and white rabbit skeletal muscle. Arch. Biochem. Biophys. 279, 345–354.

Wetzel, P., Gros, G., 1998. Inhibition and kinetic properties of membrane-bound carbonic anhydrases in rabbit skeletal muscles. Arch. Biochem. Biophys. 356, 151–158.

Wetzel, P., Liebner, T., Gros, G., 1990. Carbonic anhydrase inhibition and calcium transients in soleus fibers. FEBS Lett. 267, 66–70.

Wetzel, P., Hasse, A., Papadopoulos, S., Voipio, J., Kaila, K., Gros, G., 2001. Extracellular carbonic anhydrase activity facilitates lactic acid transport in rat skeletal muscle fibres. J. Physiol. 531, 743–756.

Whitney, P., Briggle, T.V., 1982. Membrane-associated carbonic anhydrase purified from bovine lung. J. Biol. Chem. 257, 12056–12059.

Willford, D.C., Hill, E.P., Moores, W.Y., 1982. Theoretical analysis of optimal P_{50}. J. Appl. Physiol. 52, 1043–1048.

Wilson, J.M., 1995. The localization of branchial carbonic anhydrase. In: Zoology. The University of British Columbia, Vancouver. vol. MSc.

Wilson, R., 2003. Intestinal bicarbonate secretion in marine teleost fish—source of bicarbonate, pH sensitivity, and consequences for whole animal acid–base and calcium homeostasis. Biochim. Biophys. Acta Biomembr. 1618, 163–174.

Wilson, J., Randall, D., Vogl, A., Harris, J., Sly, W., Iwama, G., 2000. Branchial carbonic anhydrase is present in the dogfish, *Squalus acanthias*. Fish Physiol. Biochem. 22, 329–336.

Wilson, R.W., Wilson, J.M., Grosell, M., 2002. Intestinal bicarbonate secretion by marine teleost fish—why and how? Biochim. Biophys. Acta 1566, 182–193.

Wittenberg, J.B., Haedrich, R.L., 1974. Choroid *rete mirabile* of fish eye. 2. Distribution and relation to pseudobranch and to swimbladder *rete mirabile*. Biol. Bull. 146, 137–156.

Wittenberg, J.B., Wittenberg, B.A., 1962. Active secretion of oxygen into the eye of fish. Nature 194, 106–107.

Wittenberg, J.B., Wittenberg, B.A., 1974. Choroid *rete mirabile* of fish eye. 1. Oxygen secretion and structure-comparison with swimbladder *rete mirabile*. Biol. Bull. 146, 116–136.

Wittenberg, J.B., Wittenberg, B.A., 2003. Myoglobin function reassessed. J. Exp. Biol. 206, 2011–2020.

Wood, S.C., Johansen, K., 1972. Adaptation to hypoxia by increased Hb-O_2 affinity and decreased red cell ATP concentration. Nature 237, 278–279.
Wood, S., Johansen, K., 1973. Organic phosphate metabolism in nucleated red cells: influence of hypoxia on eel Hb-O_2 affinity. Neth. J. Sea Res. 7, 328–338.
Wood, C.M., Milligan, C.L., 1987. Adrenergic Analysis of Extracellular and Intracellular Lactate and H^+ Dynamics after Strenuous Exercise in the Starry Flounder *Platichthys stellatus*. Physiol. Zool, 69–81.
Wood, C.M., Munger, R.S., 1994. Carbonic anhydrase injection provides evidence for the role of blood acid-base status in stimulating ventilation after exhaustive exercise in rainbow trout. J. Exp. Biol. 194, 225–253.
Wood, C.M., Perry, S.F., 1985. Respiratory, circulatory, and metabolic adjustments to exercise in fish. In: Gilles, R. (Ed.), Circulation, Respiration and Metabolism. Current Comparative Approaches. Springer, Berlin, pp. 1–22.
Wood, C.M., Randall, D., 1973. The influence of swimming activity on water balance in the rainbow trout (*Salmo gairdneri*). J. Comp. Physiol. A. 82, 257–276.
Wood, S.C., Kjell, J., Weber, R.E., 1975. Effects of ambient PO_2 on hemoglobin-oxygen affinity and red cell ATP concentrations in a benthic fish, *Pleuronectes platessa*. Respir. Physiol. 25, 259–267.
Wood, C.M., McMahon, B.R., McDonald, D.G., 1979a. Respiratory, ventilatory, and cardiovascular responses to experimental anaemia in the starry flounder, *Platichthys stellatus*. J. Exp. Biol. 82, 139–162.
Wood, C.M., McMahon, B.R., McDonald, D.G., 1979b. Respiratory gas exchange in the resting starry flounder, *Platichthys stellatus*: a comparison with other teleosts. J. Exp. Biol. 78, 167–179.
Wood, C.M., McDonald, D.G., McMahon, B.R., 1982. The influence of experimental anemia on blood acid-base regulation *in vivo* and *in vitro* in the starry flounder (*Platichthys stellatus*) and the rainbow trout (*Salmo gairdneri*). J. Exp. Biol. 96, 221–237.
Wood, C.M., Perry, S.F., Walsh, P.J., Thomas, S., 1994. HCO_3^-- dehydration by the blood of an elasmobranch in the absence of a Haldane effect. Respir. Physiol. 98, 319–337.
Woodson, R.D., Wranne, B., Detter, J.C., 1973. Effect of increased blood oxygen affinity on work performance of rats. J. Clin. Invest. 52, 2717.
Woodson, R.D., Wills, R.E., Lenfant, C., 1978. Effect of acute and established anemia on O_2 transport at rest, submaximal and maximal work. J. Appl. Physiol. 44, 36–43.
Woodward, J., Smith, L., 1985. Exercise training and the stress response in rainbow trout, *Salmo gairdneri* Richardson. J. Fish Biol. 26, 435–447.
Wujcik, J.M., Wang, G., Eastman, J.T., Sidell, B.D., 2007. Morphometry of retinal vasculature in Antarctic fishes is dependent upon the level of hemoglobin in circulation. J. Exp. Biol. 210, 815–824.
Würtz, J., Salvenmoser, W., Pelster, B., 1999. Localization of carbonic anhydrase in swimbladder of European eel (*Anguilla anguilla*) and perch (*Perca fluviatilis*). Acta Physiol. Scand. 165, 219–224.
Wyman, J., 1964. Linked functions and reciprocal effects in haemoglobin: a second look. Adv. Protein Chem. 19, 223–286.
Yamamoto, K., 1986. Contraction of spleen in exercised cyprinid. Comp. Biochem. Physiol. A Comp. Physiol. 87, 1083–1087.
Yamamoto, K., 1991. Increase of arterial O_2 content in exercised yellowtail (*Seriola quinqueradiata*). Comp. Biochem. Physiol. A Physiol. 98, 43–46.
Yamamoto, K.I., Itazawa, Y., Kobayashi, H., 1980. Supply of erythrocytes into the circulating blood from the spleen of exercised fish. Comp. Biochem. Physiol. 65, 5–11.
Yamamoto, K., Itazawa, Y., Kobayashi, H., 1985. Direct observation of fish spleen by an abdominal window method and its application to exercised and hypoxic yellowtail (*Seriola quinqueradiata*). Gyoruigaku Zasshi 31, 427–433.

Yan, J.-J., Chou, M.-Y., Kaneko, T., Hwang, P.-P., 2007. Gene expression of Na^+/H^+ exchanger in zebrafish H^+-ATPase-rich cells during acclimation to low-Na^+ and acidic environments. Am. J. Phys. 293, C1814–C1823.

Yang, T., Lai, N., Graham, J., Somero, G., 1992. Respiratory, blood, and heart enzymatic adaptations of *Sebastolobus alascanus* (Scorpaenidae; Teleostei) to the oxygen minimum zone: a comparative study. Biol. Bull. 183, 490–499.

Yokoyama, T., Chong, K.T., Miyazaki, G., Morimoto, H., Shih, D.T.-B., Unzai, S., Tame, J.R., Park, S.-Y., 2004. Novel mechanisms of pH sensitivity in tuna hemoglobin: a structural explanation of the Root effect. J. Biol. Chem. 279, 28632–28640.

Young, P.S., Cech, J.J., 1993. Effects of exercise conditioning on stress responses and recovery in cultured and wild young-of-the-year striped bass, *Morone saxatilis*. Can. J. Fish. Aquat. Sci. 50, 2094–2099.

Young, P.S., Cech, J.J., 1994a. Effects of different exercise conditioning velocities on the energy reserves and swimming stress responses in young-of-the-year striped bass (*Morone saxatilis*). Can. J. Fish. Aquat. Sci. 51, 1528–1534.

Young, P.S., Cech, J.J., 1994b. Optimum exercise conditioning velocity for growth, muscular development, and swimming performance in young-of-the-year striped bass (*Morone saxatilis*). Can. J. Fish. Aquat. Sci. 51, 1519–1527.

Zaroulis, C., Pivacek, L., Lowrie, G., Valeri, C., 1979. Lactic acidemia in baboons after transfusion of red blood cells with improved oxygen transport function and exposure to severe arterial hypoxemia. Transfusion 19, 420–425.

Zbanyszek, R., Smith, L.S., 1984. Changes in carbonic anhydrase activity in coho salmon smolts resulting from physical training and transfer into seawater. Comp. Biochem. Physiol. A Comp. Physiol. 79, 229–233.

Zborowska-Sluis, D.T., L'Abbate, A., Klassen, G.A., 1974. Evidence of carbonic anhydrase activity in skeletal muscle: a role for facilitative carbon dioxide transport. Respir. Physiol. 21, 341–350.

Zhao, Y., Ratnayake-Lecamwasam, M., Parker, S.K., Cocca, E., Camardella, L., di Prisco, G., Detrich, H.W., 1998. The major adult α-globin gene of Antarctic teleosts and its remnants in the hemoglobinless icefishes calibration of the mutational clock for nuclear genes. J. Biol. Chem. 273, 14745–14752.

Zhu, X., Sly, W., 1990. Carbonic anhydrase IV from human lung. Purification, characterization, and comparison with membrane carbonic anhydrase from human kidney. J. Biol. Chem. 265, 8795–8801.

2

CARDIOVASCULAR DEVELOPMENT IN EMBRYONIC AND LARVAL FISHES

WARREN W. BURGGREN[1]
BENJAMIN DUBANSKY
NAIM M. BAUTISTA

University of North Texas, Denton, TX, United States
[1]Corresponding author: burggren@unt.edu

The Cardiovascular System: Development, Plasticity and Physiological Responses, Volume 36B
FISH PHYSIOLOGY

DOI: https://doi.org/10.1016/bs.fp.2017.09.002

Embryonic, larval, and juvenile fish develop in environments that frequently present severe challenges, not just to maintain homeostasis, but to survive. Key to survival is a functional cardiovascular system, which transports respiratory gases, nutrients, and wastes in response to varying tissue needs. This review begins with consideration of how the heart and circulation progressively develop to replace respiration by simple diffusion across the surface area of young fishes, and how the circulation may function initially to aid angiogenesis rather than transport. The morphology and regulation of heart formation, including the heart tube, cardiac chambers and valves, and the cardiac conduction systems is discussed, as is myocardial differentiation. Similarly, the process of angiogenesis and the formation of vascular beds are outlined, with brief mention of the secondary circulation and the lymphatic system in developing fishes. A focus of the chapter is the ontogeny of cardiovascular regulation, including regulation of heart rate, blood pressure, stroke volume, cardiac output, and the peripheral vasculature. The cardiovascular system must respond to environmental variation, so the effects of temperature, oxygenation, and toxicants on the functioning of the cardiovascular system are explored. The chapter concludes with a discussion of ongoing and needed technological advances, and our emerging understanding of potential epigenetic influences on developing fishes.

1. INTRODUCTION: DEVELOPMENT IN A CHALLENGING ENVIRONMENT

The early development of the cardiovascular system of fishes follows a basic pattern not unlike that of all vertebrates. During the earliest phases of embryonic development, cells—including those ultimately forming the cardiovascular system—divide, specify, move, interact, and reach a highly differentiated state. Cardiac organogenesis then proceeds with the coordinated migration of cells from the different germ layers, cell assembly into various structures/locations, and increased cell interaction via cell signaling. Collectively, these processes lead to appropriately formed and differentially functioning cardiovascular tissues. Finally, to ensure appropriate function and the continued survival of the organism as body size and complexity increases, there must be an integration and interaction of individual cardiovascular tissues and organs with the other systems within the body (Burggren et al., 2014; Stainier et al., 1993; Yelon, 2001). All of these developmental processes play out similarly for other organ systems, but in vertebrates occur first in the cardiovascular system (Burggren and Bagatto, 2008).

Despite the overall commonality of cardiovascular development in fishes, there are significant developmental variations (especially in cardiac function) that derive from inherent taxonomic differences; although few systematic studies of interspecific variation have been carried out. Complicating any assessment of the patterns of overall cardiac development is the mode of reproduction, which, in turn, can shape overall development as well as that of the cardiovascular system. Consequently, to help understand apparent variation in the cardiovascular development between taxa, we first briefly review reproductive modes in fishes, and the possible implications that they may have for physiological systems that regulate various processes in early life-stage fishes.

1.1. Modes of Reproduction and Development in Fishes

A wide variety of reproductive modes have evolved in fishes (Avise et al., 2002). *Oviparity*, where eggs are externally deposited and fertilized, is the most common form of reproduction in fishes and is found in ~90% of teleost fishes and nearly half of cartilaginous fishes (Reading and Sullivan, 2011). Nutrition for the eggs in oviparous species is strictly provided by the initial maternal provisioning of substrates within the oocyte, which can vary greatly with genetically fixed traits such as egg size or semelparity/iteroparity (Crespi and Teo, 2002; Einum and Fleming, 2007), or even with parental experience (Kindsvater et al., 2016). Once the eggs are laid, the degree of parental guarding ranges from nothing—i.e., simple dispersal of eggs on a substrate or in the water column—to subsequent intensive guarding by the male and/or female (more common in freshwater than marine species) (Mank et al., 2005; Reynolds et al., 2002).

A much smaller proportion of fishes exhibit *viviparity* or live birth. This reproductive mode has evolved in more than 50 families, including 40 families of chondrichthyans. Viviparity has long been a source of fascination to fish biologists—and aspects of this mode of reproduction have been reviewed by several authors (see Hoar and Randall, 1988; Schultz, 1961; Uribe and Grier, 2010; Wourms, 1981). The simplicity of the term "viviparity" belies several quite distinct modes of reproduction. In truly viviparous fishes, the eggs are fertilized internally, and a direct connection to the mother's gonoducts is established shortly thereafter[1]—such fishes are categorized as *matrotrophic*. As that term suggests, the developing embryos receive nutrition from the

[1]We use here the general term "gonoduct" rather than "oviduct," because female teleost fishes lack mullerian ducts (Campuzano-Caballero and Uribe, 2014).

mother, often through a placental analogue (Kwan et al., 2015; Panhuis et al., 2011; Schindler, 2015). Such fishes are typically born as relatively well developed, free swimming, individuals and their cardiovascular systems are similarly well developed and actively perfusing systemic tissues. In a subcategory of viviparity termed *ovoviviparity,* eggs are also internally fertilized and there is live birth as in true viviparity, but during their development the eggs have no direct nutrient-providing placental connection to the mother's gonoducts. Instead, in ovoviviparous fish, the developing embryos receive all nutrition from the yolk laid down at the time of oocyte creation (i.e., are *lecithotrophic*). The eggs hatch within the oviduct prior to birth and, as in other viviparous fishes, they are born as well developed, free-swimming, larvae.[2]

Given the enormous diversity within fishes, it is of little surprise that there is, in fact, a gradient between matrotrophy and lecithotrophy (Trexler and DeAngelis, 2003; Wourms, 1981), particularly in the family Poeciliidae (e.g., guppies, mollies, platies, and swordtails) (Kwan et al., 2015) and in the Selachii (sharks and rays) (Frazer et al., 2012). Discussing variations on the theme of viviparity in fishes is beyond the scope of this review. However, what is relevant is that to a lesser or greater extent viviparity binds the physiology of the mother (i.e., osmoregulation, respiration, nutrition and immunology, and circulation) to that of her internally developing offspring (Wourms, 1981). This relationship, in addition, affords opportunities beyond egg provisioning and extends to epigenetic influences on the cardiovascular and other systems of the offspring (Kindsvater et al., 2016).

1.2. Physiological Consequences of Reproduction and Development Mode

Advantages exist for each form of reproduction in fishes. Oviparity, as the most common form of reproduction in fishes, typically takes the form of a mass reproduction event, with tens of thousands or even millions of eggs being released in some species—the ultimate example of r-selected reproduction. Survival rates to adulthood are near zero (i.e., commonly <1%) in such species, with much of that mortality occurring in the first days following hatching (Browman, 1989; Burggren and Bagatto, 2008; McCasker et al., 2014). With such high mortality levels, cardiovascular function in the early life-history

[2]A long-standing debate exists over the terminology of fish life stages. There is a wide spectrum of ontogenic plans in fish, reflected in several classification scenarios that propose to subdivide embryonic, larval, and juvenile life phases, but it is beyond the scope of this chapter to review this literature. In this chapter, we use the terms "larva" and "larval" broadly to describe fish that have hatched and are free-existing in the environment, but we also strive to give ages or additional developmental details where applicable. See Section 1.3 and Balon (1999) or Fuiman (2002) for an introduction on the varied ontogenic plans in fish.

stages may contribute little to survival compared to stochastic aspects of where the larvae are distributed posthatch.

Viviparity, on the other hand, allows for larger and more developmentally advanced fish at birth/hatch. This larger size places added transport demands on the developing cardiovascular system of these young fishes. However, it also has a number of advantages. The fish's larger size at birth/hatch: (1) presumably reduces predation by other fishes and, in freshwater habitats, by amphibians and predatory aquatic insects; (2) it helps young fishes to overcome the physical limitations to locomotion and hydraulic feeding due to the high kinematic viscosity of water (China and Holzman, 2014; Danos, 2012; Fuiman and Batty, 1997); and (3) the developing individuals of both viviparous and ovoviviparous fishes receive some degree of benefit from the presumably relatively stable environment of the oviduct. Even in the case of ovoviviparous, lecithotropic larvae, the gonoduct will contribute to larval homeostasis through physicochemical constancy with respect to the exchange of respiratory gases and waste product elimination.[3] Little is known, however, about the contribution of the maternal circulation to the developing embryo in such species.

Viviparity also has some potential disadvantages for the developing fish. For example, the very act of labor may present a short, but highly vulnerable, period of predation risk for the mother, and so too for the offspring remaining within her. Moreover, fewer young can be accommodated in the gonoducts of viviparous fishes, and all offspring are indirectly at risk from any stressors that affect the mother at any time during development. An additional potential disadvantage is that, while the larger body size of viviparous animals may help avoid predation, it also results in a lower surface-to-volume ratio, which potentially places limits on the diffusional exchange of nutrients, ions, and waste products across a nonspecialized body surface (see later). Thus, large viviparous young are not only more mature at birth, but they may need a larger branchial surface area to ensure adequate ion and gas exchange for their larger body (discussed below in Section 5).

Irrespective of mode of reproduction—oviparous, viviparous, or ovoviviparous—young fishes upon hatching or birth are thrust into an environment that is replete with potential stressors. Whether in a marine or freshwater setting, the ambient environment is a source of major osmotic stress; e.g., either through the challenges related to excessive osmotic water loss in seawater or osmotic water gain from fresh water. Additionally, microenvironments may be low in oxygen, especially in fresh water bodies. These and other stressors (e.g., acidification, pollution, nonoptimal temperatures) impact early

[3]To our knowledge, chemical characterization (e.g., PO_2, pH, osmolality) of the internal environment where the developing larvae reside has not yet been measured.

life-stage fishes that may have varying degrees of cardiovascular development at the time of hatch/birth, depending on reproductive mode.

Compounding the impact of environmental stressors on many early life-stage fishes is a very high surface-to-volume ratio due to their small size (Burggren and Bagatto, 2008) and/or presence of a large yolk sac (whose outer surface is typically highly vascularized). Relatively high surface areas aid in direct respiratory gas exchange and waste elimination, which is primarily cutaneous in early development (see Section 2). However, it is this large surface area across which poorly controlled fluxes of water, ions, and even nutrients can present challenges to the regulation of internal homeostasis. This is, in part, because in many developing fishes physiological regulatory mechanisms for the cardiovascular and other systems are lacking or are relatively immature. This is particularly true in the case for oviparous species, which in many instances hatch from the egg only hours after the initiation of the heart beat, and before cardiac or ventilatory reflexes occur (see Section 5.2). The limited ability to regulate their own internal milieu in the face of environmental stressors, which may vary sharply in space and time, may contribute to the extraordinarily high early mortality in most species of fishes.

1.3. Developmental Rate and Timing

In many fish species, especially those in which parents guard or otherwise interact with their offspring, larval development may occur at a relatively slow pace while the yolk sac is absorbed. In other species, development is quite rapid, beginning with cardiovascular function. These contrasts in developmental timing and rate parallel the classification of newly hatched birds as either precocial (immediately able to thermoregulate, forage, and generally fend for themselves soon after hatching) or altricial (eyes remaining closed, very dependent upon material care for weeks after hatching). While this formal classification has not been extended to fishes and their offspring, the young of various species of fish indeed fall upon a similar gradient between precocial and altricial (Balon, 1999). A more precocial life history has been associated with greater swimming activity and more efficient dispersal (Kopf et al., 2014), and this presumably increases early demands on the cardiovascular system. Thus, in about 25% or more of all fish species, the young benefit from some form of parental care and protection—e.g., in the stickleback the male guards the eggs for about a week while they hatch, then ensures that the embryos/larvae do not stray far from the nest for another week. In many other species, however, the larvae immediately disperse from the site of hatching and more quickly acquire juvenile characteristics. Of course, young fishes that are released as a result of live birth are often ready to fend for themselves. Such precocial lifestyle strategies in either viviparous or ovoviviparous fishes may be accelerated in species that inhabit very warm tropical environments. In the air breathing tropical gar, *Atractosteus tropicus*, for example, the yolk sac is absorbed with two days of hatching, and feeding, air-breathing and the regulation of gill ventilation and heart rate (f_H) start as early as 2.5 days (Burggren et al., 2016).

The concept of precocial and altricial larval forms will prove useful for fish. Yet, such generalizations made about the developmental program of fish, or even lineages of fishes (including those in this chapter), should retain an unspoken caveat—that the timing of the appearance of individual traits between species is, in fact, not always in accordance with the demands of the environment. Rather, after hatch/birth, fish may be poorly equipped to deal with niche-specific stressors. For example, although a precocial fish may transition to exogenous feeding more quickly and be better suited to compete for resources, altricial forms are more likely to have higher fecundity with increased dispersal (Balon, 1999; Belanger et al., 2010; Fuiman, 2002). This means that an altricial species will likely remain less active for a longer period [and have delayed maturation of the cardiovascular (and other) systems], because survival relies more on reproductive strategy (i.e., high clutch size, parental care, a supportive microenvironment, camouflage, buoyancy, etc.,) rather than the timing of physiological traits. This dichotomy between altricial and precocial early life-stage fishes, and the gradient of their readiness to deal with extremes, belies generalizations about the development of physiological phenotypes in fish. In doing so, it necessitates comparative studies that unite similarities in physiological mechanisms within the context of a species' developmental time.

Against this backdrop of varied challenges to internal homeostasis in early life-stage fishes, that may vary with reproductive mode, we now turn to the role of the cardiovascular system in the earliest developmental stages.

2. A FRAMEWORK FOR UNDERSTANDING CARDIOVASCULAR DEVELOPMENT

2.1. The Changing Roles of the Embryonic and Larval Heart and Circulation

The cardiovascular system has traditionally been seen as the servant of the body, pumping blood to the whole organism and ensuring the delivery of nutrients, regulators (hormones and molecular chaperons), respiratory gases (O_2 and CO_2) and pressure, and in endothermic vertebrates, heat (Burggren, 2013; Gilbert, 2014). This perspective is interpolated back to the circulation of the embryo and larva. Yet, in many instances, there is little to no evidence to support this assumption, and in some case evidence exists to the contrary (Burggren, 2004, 2013). To understand the changing role of the circulation during development, we must first understand the characteristics of the general body integument, and how cutaneous exchange may contribute to early physiological processes in developing fish.

2.2. Cutaneous Materials Exchange

The skin of larval fishes immediately after hatching is very thin, less than 10 μm in many species, and even as thin as 2 μm (Le Guellec et al., 2004; Schreiber, 2001). This potentially makes the cutaneous surface of the body (excluding external or internal gills) a highly efficient route for the diffusive exchange of materials between the ambient environment and the larva's internal tissues. Indeed, using average values of metabolic rate and gradients for oxygen diffusion, a hypothetical animal that is spherical can attain a diameter of $\sim \leq 1$ mm without needing a circulation for internal convection (Burggren, 2013; Crossley et al., 2016). In fact, many otherwise complex metazoans that lack an internal circulatory system for bulk transport reach or even exceed this size. Most early stage fishes are well under this diameter at hatch. Moreover, upon hatching almost all of their functional surface area is in the skin rather than associated with their rudimentary gills. For example, an estimated 95% of the total surface area of newly hatched Atlantic salmon, *Salmo salar*, is localized in the skin (including the yolk sac covering), with the poorly developed gills providing the remaining 5% (Wells and Pinder, 1996a). Even in the anabantid air breathing fish *Trichogaster trichopterus*, the skin is ~80% of potential respiratory surface area as late as 35 days posthatch, even with ongoing development of both gills and the labyrinth organ (Blank and Burggren, 2014). Thus, while diffusion itself is a relatively slow and inefficient process for moving materials when distances are large, the typically very small size (and thus internal diffusion distances) of most larval fishes upon hatching, coupled with their very thin skin, enables them to efficiently capture oxygen and eliminate wastes across the general body surface. Interestingly, at least in some species, cutaneous exchange appears to be enhanced by pectoral fin movements that create active water currents over the skin and disrupt the boundary layer next to the skin (Hale, 2014; Liem, 1981). In summary, numerous authors have concluded that in the first hours to days following hatching, larval fishes (both marine and freshwater) do not require convective blood flow for exchange of materials between the environment and the internal tissues (Hale, 2014; Jacob et al., 2002; Jonz and Nurse, 2005; Kopp et al., 2005; Mirkovic and Rombough, 1998; Pelster and Burggren, 1996; Rombough, 1998; Rombough and Moroz, 1997; Wells and Pinder, 1996b).

Finally, in a discussion of cutaneous respiration, it is important to note that the skin of many fishes continues to play a significant role in respiratory gas exchange into adulthood (e.g., see Chapter 4, Volume 36A: Farrell and Smith, 2017), especially in those fishes with a well-vascularized cuticle overlying their scales, or with highly vascularized fin surfaces. Cutaneous gas exchange may be particularly important in air breathing fish during emersion, when some degree of gill collapse in air reduces branchial surface area. In such fishes, cutaneous O_2 uptake (often supplemented by O_2 uptake associated with an air breathing organ of various forms) can help avoid hypoxia in the short term, with CO_2 excretion generally remaining largely restricted to aquatic branchial routes (Cooper et al., 2012; Feder and Burggren, 1985; Graham, 1997; Jonz et al., 2016; Martin, 2014; Randall et al., 1981; Sacca and Burggren, 1982; Urbina et al., 2014).

2.3. The Cutaneous-Branchial Transition and Role of Circulatory Bulk Flow

Most larval fishes grow rapidly through a combination of hyperplasia and hypertrophy, and this includes the developing circulation (see Section 3.1). As the skin grows thicker with development, the internal gills proliferate and begin to be efficiently ventilated. Consequently, the exchange of both respiratory gases and ions switches from predominantly cutaneous to predominantly branchial (Brauner and Rombough, 2012; Hale, 2014; Rombough, 1998; Rombough and Moroz, 1997; Varsamos et al., 2005; Wells and Pinder, 1996b). Eventually, however, the circulatory system becomes well developed, and convective transport by the blood of respiratory gases, nutrients, and wastes to and from the gills largely replaces simple diffusive exchange across the skin. Importantly, in many larval fishes, the yolk sac (the exterior surface of which typically is well vascularized) persists for several days, and doubtlessly acts as an additional respiratory surface.

Fig. 1 shows the transition from tissue gas exchange by direct diffusion, in embryos and early larval stages, to tissue gas exchange by internal convection.

2.4. Early Functions of the Beating Heart

The heart of most fishes begins to beat before hatching/birth, producing convective flow of blood, as evident from the easily observed motion of red blood cells through the transparent body wall (see Section 5.2). Yet, cutaneous diffusion, not convection, appears to be the primary mode of exchange of respiratory gases, ions and waste products for days after the heart starts beating. This holds true not only for fishes but for amphibian embryos/larvae and bird embryos as well (Burggren, 2013; Burggren et al., 2000, 2004; Crossley et al., 2016; Fransen and Lemanski, 1989; Hale, 2014). The "early" beating of the heart has been termed "prosynchronotropy," and

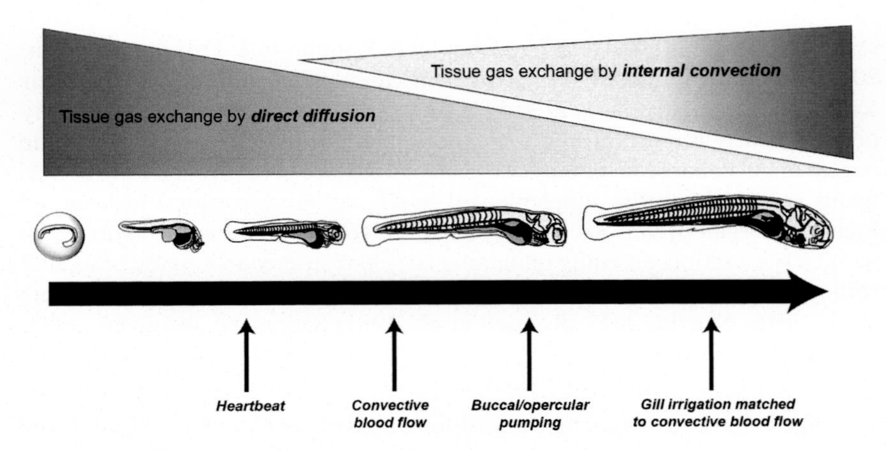

Fig. 1. Schematic diagram of the changing modes of tissue gas exchange during early embryonic and larval development in a teleost fish.

appears characteristic of every vertebrate embryo examined in this context (Burggren, 2004; Burggren and Territo, 1995; Crossley et al., 2016; Territo and Burggren, 1998).

Why, then, does the heart form, begin to beat, and develop pressure and flow in advance of its function as an organ of bulk transport? Various hypotheses have been suggested, but most revolve around the continuing development of the cardiovascular system itself, rather than its eventual role in the transport of various substances and gases. Details of angiogenesis, and the putative role for the heart in helping to expand the growing vasculature through the generation of pulsatile pressure, are discussed in Section 4.

While the earliest role(s) of the circulation remain unclear, the ubiquitous statement that "the heart is the first organ to function in the embryo" is true, though the *purpose* of this early heart beat remains uncertain. We now turn to the actual development of the heart of fishes.

3. CARDIAC DEVELOPMENT

Studies of heart development in fish have considered numerous species, but most recently have focused primarily on model species such as the zebrafish (*Danio rerio*). The same reasons that make the zebrafish a model for so many different types of developmental studies apply to investigations focused on the cardiovascular system. Essentially, zebrafish are easily reared and bred, and produce large numbers of transparent eggs. Although their eggs are smaller than those of many fishes, their transparency enables prehatch development to be visualized at the single-cell level with high resolution (Glickman and Yelon, 2002; Huo et al., 2015; Langenbacher et al., 2011; Poon and Brand, 2013; Sharmili and Ahila, 2015; Stainier and Fishman, 1994; Yelon, 2001). Consequently, most of what is described below is derived from zebrafish studies, though other species are incorporated as appropriate.[4]

3.1. Formation and Development of the Heart

3.1.1. Heart Tube Formation

Heart formation in fishes follows the same general pattern as in all vertebrates up to the point of cardiac chamber formation, after which different structures associated with different vertebrate classes emerge (Jensen et al., 2013). In brief, after fertilization, the fish egg remains at the one-cell stage (i.e., as a zygote) until the first cleavage occurs. Although the timing of cleavage and its synchronization is variable among (and within) species, and affected by temperature (Desnitskiy, 2015; Francisco Simão et al., 2007; Kimmel et al., 1995) in the zebrafish, the first cleavage occurs around 40 min postfertilization at 28 ± 0.5°C (Kimmel et al., 1995). The zygote period is followed by the cleavage period, which lasts, approximately until the 64-cell stage, which typically takes ~2 h at optimal temperature in the zebrafish. In turn, cleavage is followed by the blastula period. Yolk syncytial layer formation and the beginning of epiboly occur during the mid-blastula transition (Gilbert, 2014). During this phase, the first cardiac progenitor cells (CPCs) have been tracked (Bakkers, 2011). Specifically, in the early blastula (~512 cells), the myocardial progenitors are located in the marginal zone (Bakkers, 2011; Lee et al., 1994; Stainier et al., 1993; Yelon, 2001). However, it is not until the late blastula stage (5 h postfertilization—40% epiboly in the zebrafish), when the myocardial progenitors become located bilaterally, that three layers of cells are formed close to the embryonic margin of the marginal zone (Yelon, 2001). At this time, myocardial ventricular progenitors are located dorsal to the atrial progenitors (a position that is retained through gastrulation). In comparison to the myocardial cells, atrial and ventricular endocardial progenitor cells appear to be distributed randomly (Keegan

[4]The zebrafish as a model organism is reaching near-venerable status. Yet, animals become "model" organisms in part because they are very well understood, not necessarily because they are truly representative (Burggren, 2000). Certainly, any traits regarding length of development, physiological rates, etc., will vary in fish embryos that develop at temperatures other than 25–28°C, the typical range for zebrafish development. Though a heretical thought to many, we urge caution in wholesale extrapolation of findings specifically in zebrafish to the general piscine condition, until independent verification is made.

et al., 2004; Staudt and Stainier, 2012). Interestingly, injection of a trace dye into single blastomeres at different times during the blastula period reveals that, in contrast with early blastula stage where cells contribute to both major cardiac chambers (atrium and ventricle), during mid-blastula these cells only contribute to the ventricle (Stainier et al., 1993).

During the late blastula (40% epiboly) and early gastrula (50% epiboly to bud) periods (Kimmel et al., 1995; Yelon, 2001), the cardiac progenitor cells appear at either side of the central axis between 60 and 120 degrees and between 240 and 300 degrees (considering the embryonic margin as 0 degrees) of the body plan, and are among the first to involute. As the CPCs move inward, they turn and change direction of movement, reaching an area between 30 and 60 degrees posterior to the animal pole (Stainier et al., 1993; Stainier and Fishman, 1992; Yelon, 2001).

At the 8-somite stage (the segmentation period) the first CPCs start arriving at each side of the embryonic axis (Fig. 2A). By the 15-somite stage (Fig. 2B), progenitor cells arising from the primordial splanchnic mesoderm (in the lateral plate) have arranged to form the two tubular primordia (Burggren and Bagatto, 2008; Gilbert, 2014; Rombough, 1997; Stainier et al., 1993). By the 21-somite stage (Fig. 2C), both cardiac tubes are formed and begin to migrate to the center of the body plan (Stainier et al., 1993). The two cardiac tubes make contact at the embryonic midline by 18 h postfertilization (hpf) in the zebrafish (Fig. 2D). The contact between tubes develops from posterior to anterior (Staudt and Stainier, 2012), forming the heart cone which will subsequently become the primary heart tube (Rohr et al., 2008).

Irregular contraction of the first myocardial cells occurs at the 22-somite stage in zebrafish (about 19–20 hpf), and at approximately the 26-somite stage (~22 hpf) the definitive heart tube forms (Fig. 2D; Burggren and Bagatto, 2008; Rohr et al., 2008). The zebrafish heart starts contracting in a coordinated manner by the 26-somite stage with a frequency of 25 bpm at 28°C (Schwerte, 2009). By 24 hpf f_H has almost quadrupled (i.e., to 90 bpm), and limited blood circulation begins (Pelster and Burggren, 1996; Rohr et al., 2008; Schwerte, 2009; Stainier et al., 1993). During this stage the myocardial tissue involutes and rotates to the left. This asymmetric leftward rotation positions the endocardial tissue within the tube. The myocardium then surrounds the endocardial tissue (Bakkers, 2011). At 30 hpf (the pharyngula period in the zebrafish), the f_H increases to 140 bpm (Schwerte, 2009).

3.1.2. Cardiac Chamber Formation

Elasmobranchs and teleosts have four specialized chambers arranged in series: *sinus venosus*, atrium, ventricle, and most distal to the heart the outflow

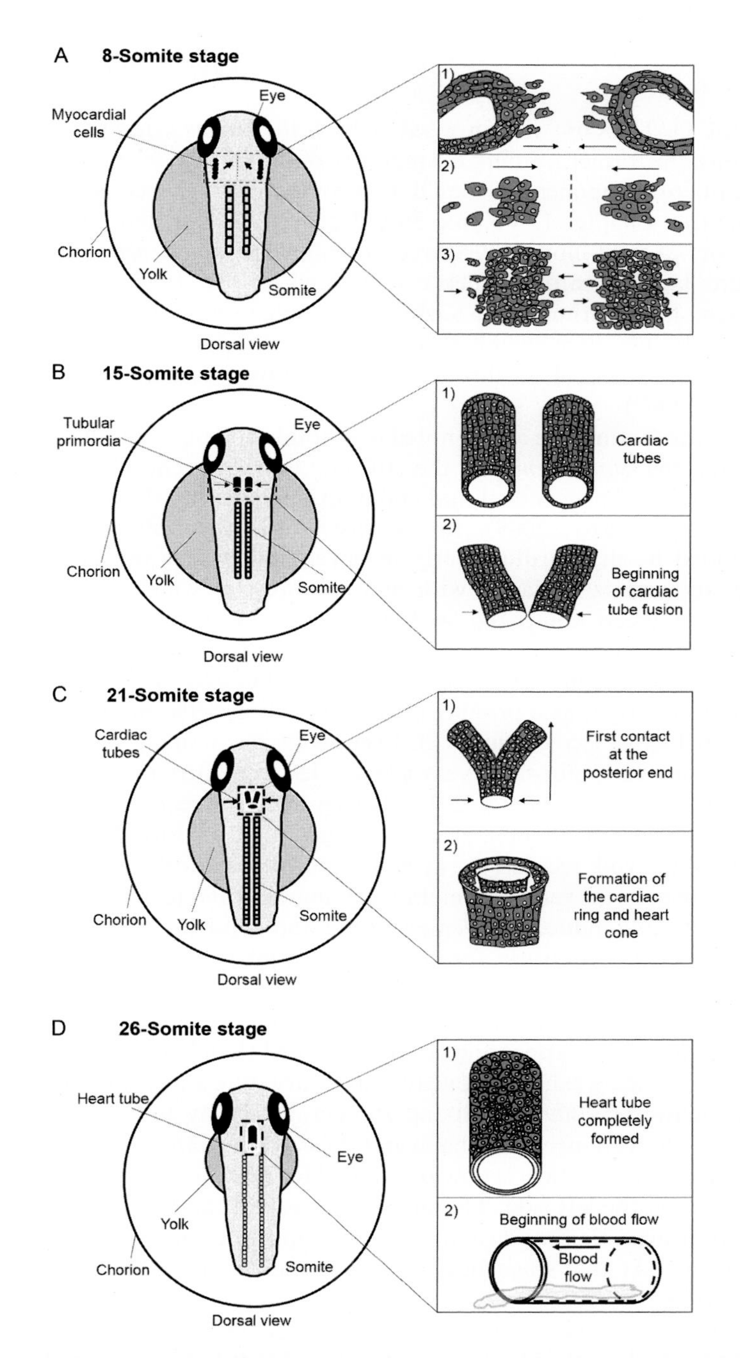

Fig. 2. Heart tube formation in the zebrafish. (A) At the 8-somite stage, myocardial progenitors from the splanchnic lateral plate mesoderm begin their migration (1). At this stage they start to aggregate at either side of the embryonic axis (2) and then, get arranged to form the cardiac tubes (3). (B) By the 15-somite stage the two primordial cardiac tubes have formed (1), and migration to the embryonic axis starts (2). (C) During the 21-somite stage, contact between the primordial cardiac tubes takes place at the posterior end (1), subsequently forming the cardiac ring and the heart cone (2). (D) By the 26-somite stage, the cardiac tube has been completely formed (1) and a weak blood flow has begun (2).

tract (OFT). The OFT is comprised of both the *bulbus arteriosus* and *conus arteriosus*; the former forming the majority of the OFT in teleosts and the latter the major component of the OFT in some primitive species and elasmobranchs (see Chapter 1, Volume 36A: Icardo, 2017). In the lungfishes, the *conus arteriosus* is followed by a structure called the *truncus arteriosus*, a vessel structure lacking cardiac muscle (Burggren et al., 1997; Burggren and Johansen, 1986; Grimes and Kirby, 2009; Johansen and Burggren, 1980). In all fishes, it is these chambers in which the blood is received, pressurized, and ejected to circulate throughout the entire vasculature (Fisher and Burggren, 2007).

Notwithstanding the above noted taxonomic and semantic differences, in all fishes the *sinus venosus*, the atrium and the ventricle are derived from myocardial progenitors and maintain their identity through adulthood (Grimes and Kirby, 2009). In contrast, while the bulbus arteriosus is surrounded by myocardium early in development, this myocardial tissue disappears and is replaced with smooth muscle, which provides more compliance when compared with the conus arteriosus (Moorman and Christoffels, 2003).

At 30 hpf (the pharyngula period in the zebrafish) the heart tube starts looping to the right as a product of the interaction between mechanical and molecular factors (see Section 3.2). Eventually, this looping process leads to the asymmetrical (right-sided ventricle and left-sided atrium) arrangement of the heart (Staudt and Stainier, 2012). However, before reaching this asymmetrical positioning, the looping process changes heart polarity (Fig. 3A and B), wherein as the yolk recedes, the atrium is pulled posteriorly, and cephalic differentiation pulls the ventricle anteriorly (Langeland and Kimmel, 1997). These morphologically induced movements twist and constrict the region between the ventricle and the atrium, forming the atrioventricular canal. A ballooning (Fig. 3C) of the chambers follows as cardiac looping continues (Staudt and Stainier, 2012). At this point in development the atrium and the ventricle can be readily identified, appearing as "bean-shaped chambers" (Bakkers, 2011), and both chambers exhibit a concave inner curvature and a convex outer curvature (Schwerte, 2009; Staudt and Stainier, 2012). By 36 hpf, the heart has completely looped into an S-shape and the cardiac chambers are clearly differentiated. The ventricle is now positioned to the right side of the body with the atrium on the left side. The sinus venosus and the bulbus arteriosus, or conus arteriosus, thus reside at the posterior and anterior regions of the heart, respectively. At this stage heart rate has increased to 180 bpm (Stainier et al., 1993).

By the fifth day (120 hpf), the heart of the zebrafish has attained its adult arrangement where the atrium is located dorsal to the ventricle, and heart rate is at its peak of 230–250 bpm (Schwerte, 2009; Stainier et al., 1993).

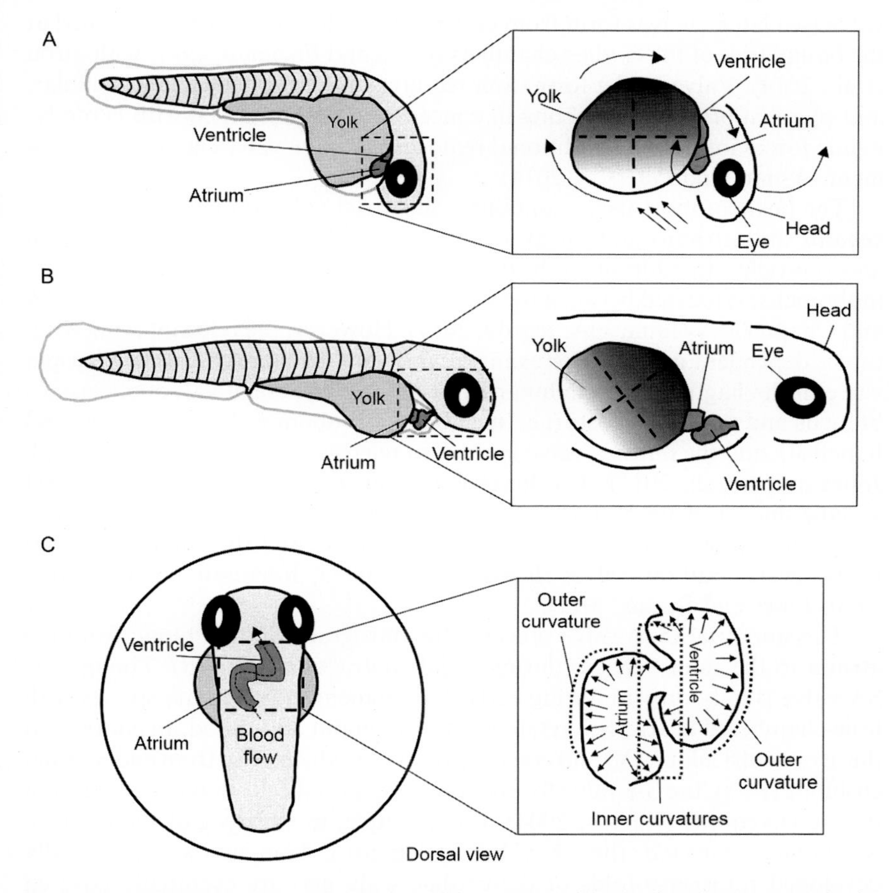

Fig. 3. Change in heart polarity, and the ballooning process of the cardiac chambers. (A) As the fish embryo develops, the yolk sac recedes back and up, pulling the atrium to a position above and slightly posterior to the ventricle. (B) The ventricle, in turn, is pulled anteriorly as the head attains a straight position. (C) The ballooning process of the chambers takes place after the looping process and the change in polarity, leading to the formation of an inner curvature and outer curvature.

3.1.3. Cardiac Valve Formation

Cardiac valves allow for unidirectional blood flow through the heart, and their improper function leads to valvular pathologies or eventually death (Brown et al., 2016; Martin and Bartman, 2009; Peal et al., 2011; Staudt and Stainier, 2012). Indeed, the correct formation of cardiac valves is a critical period for the developing vertebrate. The development of the cardiac valves occurs immediately after cardiac looping. With the exception of the sinoatrial valve (see later), valves form from endocardial cells specifically positioned at the boundaries of the cardiac chambers (Chen and Fishman, 1997; Kalogirou et al., 2014). Valve leaflet formation requires molecular, epigenetic, cellular, and physical processes all acting in concert (see Section 3.1.3) with hemodynamic forces influencing additional remodeling of the valves later in development (Asnani and Peterson, 2014).

The heart of most fishes contains a sinoatrial valve (in between the *sinus venosus* and atrium) and an atrioventricular valve (in between the atrium and ventricle). In addition, a third valve (or several additional valves in elasmobranchs) is located between the ventricle and the ventral aorta (see Fig. 4A and Chapter 1, Volume 36A: Icardo, 2017). However, the relevant nomenclature is dependent on taxa. For example, this valve(s) is/are called the bicuspid valve in the hagfish, the semilunar valve in lamprey, conal valves in elasmobranchs and ancient teleosts (i.e., members of Elopomorpha such as eels and bonefish), and the outflow valve in modern teleosts (Grimes and Kirby, 2009; Jones and Braun, 2011). The lungfish appear as exceptions to this general scheme since, instead of having an atrioventricular valve, they have an atrioventricular plug between the atrium and ventricle, and the conus arteriosus bears several outflow valves (Icardo et al., 2005a; Johansen and Burggren, 1980; Jones and Braun, 2011).

The sinoatrial (SA) valve prevents the retrograde flow of blood from the atrium to the sinus venosus during atrial contraction (Fig. 4B). Though the SA valve is only present during early development in vertebrate species with four-chambered hearts, fish retain a SA throughout adulthood. In contrast to the atrioventricular (AV) and conal valves (CV), which arise from endocardial cushions (ECs), the SA valve forms from myocardial cells that originate from the sinus venosus (Ramos, 2004). For example, in embryos of the dogfish, *Scyliorhinus canicula*, the SA valve leaflets arise from two asymmetrically developed transverse folds of the cardiac wall, and are eventually covered by epicardial tissue (Gallego et al., 1997). Notably, while the morphogenetic processes of valve formation are generally regarded as conserved across vertebrates, in dogfish, only the right sinoatrial fold that forms the right SA valve leaflet (but not the left) has been described as homologous to that of higher vertebrates (Gallego et al., 1997).

The heart is comprised of a thicker myocardial cell layer that surrounds a thinner endothelial layer (which faces the tubular cavity). The two major layers are separated by the extracellular matrix (ECM, also called *cardiac jelly*), which among other compounds contains proteoglycans, hyaluronic acid and collagen (Armstrong and Bischoff, 2004; Peal et al., 2011). The atrioventricular valves (Fig. 4C), which are located in the atrioventricular canal, begin to arise from endocardial cushions that develop in the wall of the heart at this

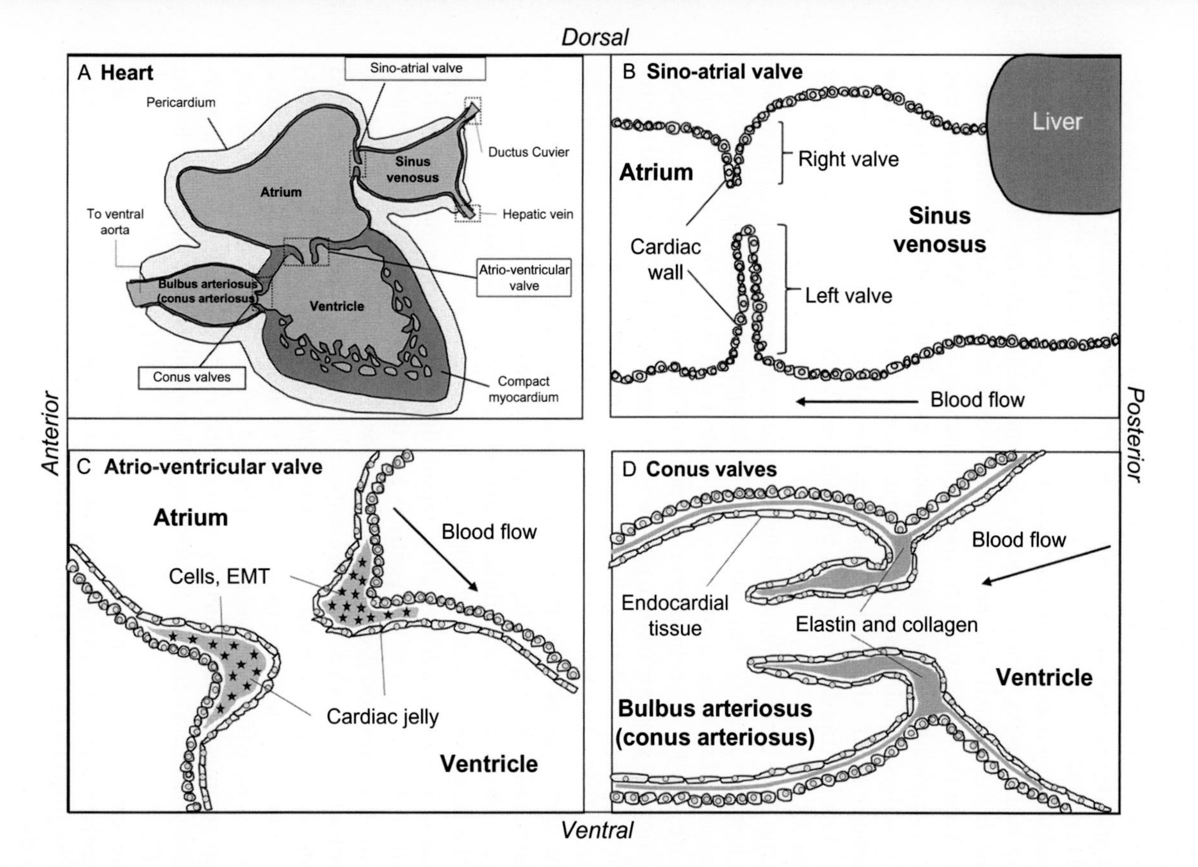

Fig. 4. Fish cardiac valves (nomenclature is taxa-specific). (A) General scheme of a fish heart and cardiac valve location. (B) Sino-atrial valve of a 17 mm dogfish (*Scyliorhinus canicula*), the left valve is larger than the right valve. (C) Atrio-ventricular valve in zebrafish (*Danio rerio*) at 48–96 hpf, cells in endothelial-to-mesenchymal transition (EMT) are shown. (D) Conus valve of an adult gilthead seabream (*Sparus auratus*), this valve is attached to the cardiac wall by a core of collagen and elastin.

junction (Beis et al., 2005). The formation of the endocardial cushions presents as a thickening at the AV canal, and first requires the migration of endothelial cells that delaminate from the endothelial wall and enter the extracellular matrix (Peal et al., 2011). This movement expands the space between the endothelial and myocardial layers, causing swelling that pushes the thinner endothelial layer into the lumen within the AV ring. Concurrently, endothelial cells remodel as they undergo an endothelial-to-mesenchymal transition or EMT (Scherz et al., 2008), a process where the endothelial cells de-differentiate to become mesenchymal and become thicker and more rigid (Butcher et al., 2007). After the EMT and subsequent thickening of the AV canal, primitive leaflets are visible. These leaflets subsequently develop into true atrioventricular valves by elongation and maturation, during which a change in appearance occurs that is dependent on animal size (Martin and Bartman, 2009). Leaflet transformation is also shaped by hemodynamic forces (see Section 3.1.3).

In the three extant genera of lungfishes, an atrioventricular plug is present in the heart, in lieu of a flap-like valve. Developmentally, the AV plug arises from myocardial progenitors from the ventral and atrial chambers (Grimes and Kirby, 2009). Two functional purposes of the plug have been proposed: to functionally separate oxygenated and deoxygenated blood flow; and to prevent retrograde blood flow during ventricular contraction (Burggren and Johansen, 1986). In adult *Protopterus dolloi*, the AV plug is formed by a dense hyaline cartilaginous core, derived from neural crest-derived chondrocytes, and surrounded by a thick layer of connective tissue composed primarily of collagen, elastin and extracellular matrix (Icardo et al., 2005a).

Little is known about development of the heart and cardiac valves in hagfishes. Studies in adult hagfish report the presence of a bicuspid valve, located at the junction of the ventricle and the ventral aorta (Wright, 1984). This valve is composed of two leaflets, positioned to the left and right of the ventricle cranial edge, that are composed of a collagenous core surrounded by endocardial tissue (Grimes and Kirby, 2009; Icardo et al., 2016). Interestingly, although connective tissue serves as an insertion area for the leaflets, differences were found between the right and left side junctions. For example, the valvar sinus wall is formed by myocardial tissue on the left side, whereas it is composed primarily of connective tissue on the right side—see Icardo et al. (2016).

The semilunar valves in lampreys are supported by myocardial tissue of the ventricular walls and also consist of two leaflets, similar to the semilunar valves of hagfish (Grimes and Kirby, 2009). These leaflets lie in the arterial pole, where the myocardial tissue connects to the smooth muscle of the arterial system, and they appear to be transitionary as the leaflets are composed of both myocardium and smooth muscle (Grimes and Kirby, 2009; Richardson et al.,

2010). Little is known of valve development in the embryos and larvae of either hagfish or lamprey.

In teleosts the leaflets of the outflow valves, also called *conus valves* (Fig. 4D) or (erroneously) bulboventricular valves, arise from the walls of the conus arteriosus (Icardo et al., 2003). However, their location varies between species depending of the morphology of the conus and its transitional pattern into the bulbus (Icardo, 2006). The leaflets' flap-like thinner distal end is composed of luminal fibrosa (collagen and elastin) covered by endocardial tissue. In contrast, the proximal thicker side of the leaflets has a dense cellular core with a collagen-rich region and elongated fibroblasts; this likely provides the valves with mechanical stiffening during ventricular diastole (Icardo, 2012; Icardo et al., 2003; Lorenzale et al., 2017).

Finally, the proximal portion of the outflow tract in the lungfishes (also called *conus arteriosus* due to its well-vascularized compact myocardium) bears several rows of valves situated on both the ventral and dorsal walls of the tract. However, even in adult specimens of *Protopterus dolloi*, these valves seem to be rudimentary and poorly developed (Icardo et al., 2005b).

3.1.4. Myocardial Differentiation

Myocardial differentiation has been studied for decades, in a variety of animal models, due to its importance for the correct form and function of the heart and the biomedical significance of this process in congenital heart defects in humans (Meilhac et al., 2014). Myocardial differentiation appears highly conserved across vertebrate groups (Evans et al., 2010). In zebrafish, the first myocardial progenitor cells can be mapped in the early blastula stage. They are located in the marginal zone on either side of the embryo, and lateral progenitor cells are more likely to be involved in heart formation than medial cells (Bakkers, 2011; Lee et al., 1994; Stainier et al., 1993; Staudt and Stainier, 2012). Blastomeres are located at 90 degrees longitude in early blastula stages (from a lateral perspective, they are located in the midline between the future ventral and dorsal sides of the embryo), and contribute solely to myocardial cells. In contrast, ventral blastomeres also give rise to endothelial, endocardial and blood cells later in development (Stainier et al., 1993).

Myocardial formation begins during the mid-blastula stage, when deep layer cell linages first become motile. At the onset of the gastrula period, the hypoblast (mesendoderm) is formed by involution of cells from the blastoderm. In zebrafish at ~27°C, 7–8 h later (i.e., during the early somite stages of the segmentation period) myocardial progenitors form the lateral plate mesoderm. These cells have undergone a few divisions, and continue to migrate to the ventral zone of the tubular primordia as a part of the anterior lateral plate mesoderm (Bakkers, 2011; De Pater et al., 2012; Stainier et al., 1993; Yelon, 2001).

Although cells can be mapped in early stages of development, they are not committed to becoming cardiac cells (Meilhac et al., 2014). In fact, cell fate is not fixed until they reach the anterior lateral plate mesoderm (between the 12- and 15-somite stages) during the mid-blastula stage. This is when deep layer cell lineages first become motile, and when several signaling pathways become active and myocardial differentiation begins (Bakkers, 2011; De Pater et al., 2012; Evans et al., 2010). Specifically, the initiation of differentiation is marked by the expression of sarcomere proteins (myosin light chain polypeptide 7—cmlc1, cmlc2) and, notably, by the regionalization of a population of myocardial precursors in the myocardial field in a medial to lateral direction; the medial cells express myosin heavy chain isoforms that indicate that these cells are ventricular and that the lateral cells are atrial (Bakkers, 2011; Yelon, 2001).

Finally, myocardial differentiation continues during the pharyngula period (de Pater et al., 2009). At this point, a second lineage of cardiac progenitor cells begins to contribute to the construction of the heart. This second lineage of CPCs is called the second heart field and has been, until this stage, less active at differentiation and segregated from the other CPCs (i.e., first heart field) (Van Vliet et al., 2012). Now, new cardiomyocytes arise from the second heart field's CPCs, and from neural crest cells from pharyngeal arches one and two. These latter cells add cells to the arterial and venous poles in this second wave of cell differentiation, which coincides with looping (Bakkers, 2011; Cavanaugh et al., 2015; Kelly, 2012; Paige et al., 2015; Van Vliet et al., 2012).

3.1.5. Development of the Cardiac Conduction System

The cardiac conduction system of fishes is a specialized network of cardiomyocytes in charge of generation and propagation of the electrical impulses required to trigger cardiac contractions (Staudt and Stainier, 2012; van Weerd and Christoffels, 2016). While nervous control of the heart is functionally similar in fish and other vertebrates, distinct anatomical differences exist (Burggren and Bagatto, 2008; Zaccone et al., 2009). Specific His bundles and Purkinje cells have not been identified in zebrafish, but a trabecular band provides the functional continuity required for electrical transmission (see Section 5; Gamperl and Shiels, 2014; Sedmera et al., 2011).

The cardiac conduction system is derived from cardiogenic progenitors (Chi et al., 2008). Specifically, they arise from posterior CPCs from the second heart field, which also give rise to the venous pole (inflow tract) of the heart (Paige et al., 2015; Van Vliet et al., 2012). In developing zebrafish the first contracting cardiac cells have been identified during the 22-somite stage at ~19 hpf (see Sections 3.1.1 and 3.1.4). These weak contractions are myogenic and follow an irregular pattern, likely because of immature development of the sarcomeres and sarcoplasmic reticulum (van Weerd and Christoffels, 2016). During the subsequent hours, as the heart tube continues to elongate, the cardiac cells continue to mature until the 26-somite stage (~22–24 hpf). This is when coordinated and regulated beats begin (Stainier et al., 1993), which can be attributed to the emergence of pacemaker cells (Staudt and Stainier, 2012).

The majority of the Osteichthyes have pacemaker cells in the sino-atrial node (Bakkers, 2011; Gamperl and Shiels, 2014). Consequently, contractions are initiated at the inner curvature of the venous pole of the heart tube and follow a unidirectional path along the heart tube in a slow, peristaltic fashion (Arrenberg et al., 2010; Bakkers, 2011). Coincident with heart chamber formation around 36 hpf in the zebrafish, electrical transmission begins to exhibit an atrio-ventricular delay (Staudt and Stainier, 2012). This delay is likely caused by the ring form of the atrio-ventricular constriction (Moorman et al., 1998), and is characterized by the slowing of calcium activation at the atrio-ventricular constriction, followed by an increase in conduction rate in the lateral areas of the ventricle (Chi et al., 2008). Functionally, this delay may help to prevent the retrograde flow of blood from ventricle to atrium and from outflow tract to ventricle, prior to maturation of functional valves (Chi et al., 2008). Eventually, it is this conduction delay, combined with concurrent valve formation, that results in coordinated contraction of the atrium then ventricle (Bakkers, 2011).

Cardiac impulse conduction continues to be optimized as development continues. For example, myocardial trabecular bands of the ventricle are visible at 100 hpf in the zebrafish (Peshkovsky et al., 2011). The presence of the conductive trabecular bands suggests an improvement in signal transmission (Gamperl and Shiels, 2014; Sedmera et al., 2003), and this is confirmed by the presence of a distinct calcium wave that is detectable in the atrioventricular canal (Chi et al., 2008). By the time the heart has attained its adult configuration (5 dpf in the zebrafish), the trabecular network of the conduction system is present within the ventricle (Peshkovsky et al., 2011; Stainier et al., 1993).

Finally, in both embryonic and adult zebrafish, the trabecula increase conduction velocity in the heart's conduction system (Staudt and Stainier, 2012). Due to the configuration of the heart chambers in adult fishes, electrical activation is conducted unidirectionally from the apex of the ventricle to its base (also see Section 5).

3.2. Regulation of Heart Formation

As evident from the earlier discussion, the formation and development of the fish heart is a complex process, with each step directed and coordinated to ensure a mature, functional organ. Remarkably, this process must occur even as the heart is beginning to function, a process similar to climbing on a floating log with a saw and nails in hand, and making a wooden boat from the log beneath your feet while floating down the river. Each step in heart formation is orchestrated and dictated by the interaction of mechanical (i.e., hemodynamic), molecular (i.e., signaling pathways) and genetic factors.

3.2.1. Mechanical Factors

Early studies of vertebrate heart development assumed that blood flow and the mechanical stresses/strains it produced on the growing heart had only a subsidiary role in actual heart formation (Manasek, 1981; Manasek and Monroe, 1972). These conclusions were based on gross anatomical observations in amphibians showing that the general makeup of the heart (e.g., cardiac looping) appeared to proceed normally even in the absence of cardiac contractions—e.g., in cardiac mutants without a heart beat (Fransen and Lemanski, 1989; Lemanski et al., 1995). However, when linear heart tubes of the zebrafish were cultured ex situ, they displayed dextral looping (Noel et al., 2013). These data are consistent with the assumption that, whereas the dextral looping (C-looping phase) of the heart is orchestrated by intrinsic (i.e., endogenous cytoskeletal strain) factors, the S-looping phase is dictated by both intrinsic and extrinsic factors (Lindsey et al., 2014). However, more recent, finer scale cellular and molecular observations, in association with biomechanical engineering models, have indicated that heart tissue is, indeed, subtly molded and influenced by internal hemodynamic forces in the embryos of fishes and birds (Kowalski et al., 2013, 2014; Wang et al., 2009; Yang et al., 2014).

The mechanical factors involved in cardiac formation include intrinsic factors such as cytoskeletal tension (isometric tension, cell stiffness), and extrinsic factors such as blood flow through the developing heart [fluid pressure and wall shear stress (Fig. 5A; Gjorevski and Nelson, 2010)]. Adhesion forces between cells and the extracellular matrix, among other processes, influence cell

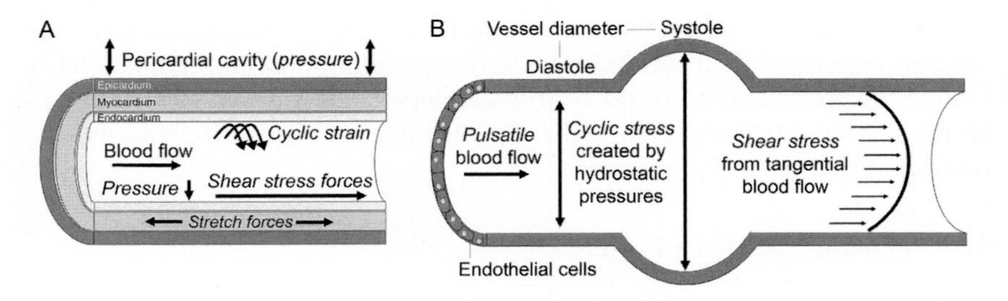

Fig. 5. Mechanical factors influencing formation and development of: (A) the cardiac tube, and (B) the vasculature.

proliferation, migration, differentiation, growth and apoptosis (see Haase et al., 2014; Mammoto et al., 2013).

During heart tube formation, myocardial cells migrate and eventually fuse to form a hollow tube. Cell migration takes place within the extracellular matrix fluid, which affects cell autonomous motility, and also applies elastic resistance forces on these cells (Buxboim et al., 2010; Zamir et al., 2006). Cell adhesion and connective tissue motions within the extracellular matrix act as mechanical stimuli for movement (Lindsey et al., 2014). An increase in stiffness of the substrate, for example, increases the number and size of focal adhesions between cells and substrate—see Haase et al. (2014). Cells also exert forces upon their surrounding matrix and neighboring cells (Buxboim et al., 2010). Based on computational modeling experiments employing avian embryos, active contraction of the endoderm apparently pulls the bilateral cardiac tubes toward the embryo midline, enabling their fusion and the subsequent formation of the hollow cardiac tube (Taber, 1998; Varner and Taber, 2012).

Once the heart tube has been formed, the cardiomyocytes start to contract and rudimentary blood flow begins (see Section 3.1.3). The onset of pulsatile pressures with the lumen of the cardiac tube leads to wall shear stress, cyclic strain, transmural pressure oscillations across the cardiac wall, and circumferential stress when the walls stretch (Andrés-Delgado and Mercader, 2016). These mechanical forces directly influence heart development. Using zebrafish embryos during the linear heart tube stage, blood flow was blocked at the sino-atrial boundary or at the ventricle–bulbus arteriosus boundary, and cardiac development was assessed 20 h later. The bulbus arteriosus failed to develop, heart looping was inhibited, and the walls of the inflow and outflow tracts collapsed and fused; these observations all suggest that shear forces are important in remodeling cardiac tissues in this fish (Hove et al., 2003).

After it forms, the heart tube loops and the cardiac chambers form. The ballooning process of the bean-shaped cardiac chambers creates a concave inner and a convex outer curvature (see Section 3.1.2). The inner concave curvature is composed of cells that maintain their cuboidal shape, while the outer curvature is composed of cells which elongate and flatten (Lindsey et al., 2014). Both enlargement and elongation are promoted by the shear forces against the walls exerted by blood flow, which themselves are tempered by the intrinsic tissue tension (Auman et al., 2007).

Valve formation is also affected by mechanical forces. For example, blood flow exerts differential wall sheer and strain forces on each curvature of the cardiac chambers. Research with the avian model has shown that these forces are smaller in the outer curvature than in the inner curvature (Yalcin et al., 2011). Larger shear forces may promote both the expansion of the extracellular matrix and the endothelial-to-mesenchymal transition, whereas lower forces may assist in the formation of trabeculations in the avian model (Yalcin et al., 2011). Further, it is important to note that valvulogenesis occurs as the endocardial cushions begin to arise perpendicularly to the direction of blood flow once the inflow reaches its peak velocity (Lindsey et al., 2014). This is probably due to the induction of morphogenetic expression related to mechanical forces. Valvulogenesis fails after blood flow blockage in the zebrafish, suggesting that a reduction of the shear forces by $\sim$10-fold has a negative effect on heart development (Hove et al., 2003).

Collectively, these data suggest that the integration of intrinsic and extrinsic mechanical factors is critical for heart development in fishes. All of these stimuli are felt by mechanosensors (cilia, integrin signaling components, the cell membrane, among others) and are mechano-transduced (see Haase et al., 2014; Stoppel et al., 2016). This latter process involves the activation of molecular cues acting together with biophysical cues during heart development.

3.2.2. Molecular and Genetic Factors

Our understanding of cardiac development has benefited greatly from the rapidly expanding approaches and tools available for molecular investigations. The molecular factors involved in cardiac morphogenesis are nearly innumerable, with new genes and their expression products discovered daily. Many of these genes and gene products are highly conserved, and apply to development of the heart of fishes as well as mammals and other vertebrates. Reviewing these advances is beyond the scope of this chapter, and the reader is invited to refer to the literature cited below for an introduction into this voluminous literature. Finally, the reader is again reminded that the wide exploitation of the zebrafish in particular, especially for studying molecular aspects of development, leaves the following discussion highly skewed toward this species.

Early cardiac progenitors are located near the marginal zone before involution, and migrate to the midline during gastrulation of the embryo to form the cardiac tube (see Sections 3.1.1 and 3.1.4). This process is directed by the expression of the gene *miles apart* and lysosphingolipid siphogosin-1-phosphate (S1P) via its cognate G protein-coupled receptor S1pr2 and its downstream effector Gα13 (Burggren and Bagatto, 2008; Ye et al., 2015). The expression of the spinster homologue *toh* (within the yolk syncytial layer) regulates the transport of S1P (from the yolk to the embryonic tissue) and the expression of *fibronectin* within the cellular matrix and *sdc2* within the extra-embryonic yolk, and so *toh* is essential for cell migration (Bakkers, 2011). In addition, heart tube formation is inhibited by the loss of cell apicobasal polarity within the myocardial progenitors. This inhibition is associated with the expression of the protein mpp5 encoded by protein kinase C and *nok* as a result of a mutation in the gene *has* (Rohr et al., 2008).

During early gastrulation (between the 1- and 3-somite stage in zebrafish), the transcription factor *Nkx2.5* is required for cardiogenic differentiation. *Nkx2.5* expression is indirectly regulated by bone morphogenetic proteins (BMPs) and Nodal signaling pathways (members of the Tgf-B superfamily), which in turn induce *gata5* expression (Bakkers, 2011; Yelon, 2001). Between the 6th and 9th somite stage, the myocardial progenitors are migrating to the posterior part of the anterior lateral plate mesoderm (see Section 3.1.4), where *hand2* (which is also required for cardiogenic differentiation) and *gata4* are expressed (De Pater et al., 2012; Staudt and Stainier, 2012). Within the anterior lateral plate mesoderm, growth factors, Nodal signaling pathway members, and BMP are asymmetrically expressed (BMP expressed more on the left side), and this results in the asymmetrical morphogenesis of the heart (Bakkers, 2011; Evans et al., 2010). In addition, the expression of *spaw* (Nodal-related gene) and its effects on the expression of *pitx2, lefty1,* and *lefty2* are critical for cardiac left–right asymmetry (Rohr et al., 2008).

By the 12th to 15th somite stage of zebrafish, the population of myocardial precursors is regionalized in a medial to lateral direction, as shown by the expression of *vmhc* in the medial section (i.e., the ventricular section which is also regulated by *gata5, fgf8,* and *Zoep*), and *myh6lamhc* in the lateral section (atrium). The proliferation of myocardial precursor cells is regulated by several signaling pathways including retinoic acid (RA), Wnt (a portmanteau of the historical gene names wingless and int-1), Hedgehog (HH), fibroblast growth factor (FGF), BMP, and Nodal pathways (Bakkers, 2011; Li et al., 2016; Staudt and Stainier, 2012; Yelon, 2001).

As mentioned earlier, cardiac chamber formation is highly dependent on both intrinsic and extrinsic mechanical factors. However, a number of genes have also been linked with chamber morphogenesis that function independently of mechanical factors. An example is the lack of sarcomere formation in *half-hearted* mutants, which leads to the formation of enlarged ventricles in zebrafish (Staudt and Stainier, 2012). Additionally, in *weak atrium* mutants, the lack of atrium contractility leads to a failure of ventricular cells to increase their surface area and elongate (Staudt and Stainier, 2012).

Several signaling pathways have been identified in relation to valvulogenesis, including Notch, BMP, TGF, Wnt (which also regulates the regenerative process in many systems), NFATc, Erb2/4, and *has2*, as well as microRNAs (Bakkers, 2011; Kalogirou et al., 2014). Specifically, for the atrioventricular region, in mammals Wnt/B-catenin signaling is required for both endocardial proliferation and endothelial-to-mesenchymal transition in endocardial cells, and induces the expression of *has2* during the formation of the atrioventricular valve (Bakkers, 2011). In the zebrafish, the expression of *bmp4* and *versican* are restricted to the atrioventricular canal when constriction takes place (Staudt and Stainier, 2012). Similar to chamber development, valvulogenesis is closely correlated with mechanical forces. Of particular interest for atrioventricular canal formation is the gene *klf2*, since it is a shear-sensitive gene and also acts during the invagination process in the endocardial cushions (Bakkers, 2011; Staudt and Stainier, 2012). In addition, SMAD proteins (a portmanteau of homologs of *Drosophila* and *Caenorhabditis elegans*) and *notch1b,* which are involved in signaling between cellular layers and with adjacent cells, are important for cushion and valve development, and for proper myocardial differentiation in the atrioventricular canal (De Pater et al., 2012; Peal et al., 2011).

The development of the cardiac conduction system is also influenced by molecular factors including Nkx2.5 and Tbx5. The absence of either leads to a failure in maturation and maintenance of atrioventricular conduction (Chi et al., 2008). Expression of *bmp4* occurs in the atrioventricular canal, where it, in turn, regulates the expression of *tbx2b*. This factor performs two distinct functions. On one hand, *tbx2b* is required for the conduction delay (see Section 3.1.5) observed during cardiac myocardial activation. However, it also inhibits the expression of *nppa,* a marker for the cells that form the outer curvature of the chambers (Bakkers, 2011). Similarly, inhibition of *neuregulin* or *notch1b* results in AV conduction delay (Staudt and Stainier, 2012).

In addition to genetic factors, development and maturation of the cardiac system in fish (as in other vertebrates including humans) are modulated by epigenetic processes such as histone modification, DNA methylation and ATP-dependent chromatin remodeling (Han et al., 2011; Lu et al., 2016; Van Weerd et al., 2011). For example, combined blockage of SmyD1a and SmyD1b (two histone methyltransferases) in the zebrafish results in inhibition of muscle contraction and myofibril alignment (Tan et al., 2006). Similarly, while separate morpholino injections of *Smyd3* or *Setd7* lead to cardiac edema, knock-down of both of these factors (Smyd3 and Setd7) synergistically affects looping of the heart tube in zebrafish (Kim et al., 2015).

Chromatin remodeling complexes play nodal roles during cardiac development in vertebrates—reviewed in Han et al. (2011). In the zebrafish, for example, based on their contribution to the cardiac region in the anterior lateral plate mesoderm, Smarcd3 (transcriptional factor of the BAF complex) and Gata5 are suggested to induce cardiac fate in cells and differentiation in cardiac lineages (Lu et al., 2016). In addition, the Polymerase Associated Factor 1 (PAF1) complex consists of five components (Leo1, Ctr9, Cdc, Paf1, and Rtf1) and is involved in chamber morphogenesis, cardiac differentiation, and elongation of the heart tube in the piscine model (Langenbacher et al., 2011; Lu et al., 2016).

Finally, it is clear that the formation of a pumping organ is not enough to ensure the delivery of nutrients and oxygen throughout the body, and the removal of waste products, and for this reason the creation of a vascular network is needed to reach every region of the body.

4. VASCULAR DEVELOPMENT

4.1. Vascular Structure and Pattern

4.1.1. The Embryo

Circulation of blood begins around 22–24 hpf in the zebrafish. At this point, blood is being pumped by the heart into the ventral aorta, which perfuses the right and left first aortic branches, which in turn merge to form the dorsal artery that runs posteriorly within the organism and becomes the caudal artery (Rombough, 1997; Fig. 6A). From the caudal artery, blood passes to the caudal vein, which runs cranially becoming the posterior cardinal vein. This vessel then splits into the *ducti of Cuvier* before, finally, connecting with the inflow tract of the heart (Fig. 6B). During this early period of development of the circulation, the pattern and growth of cranial vessels take place, but no vitelline arterioles are observed (Isogai et al., 2001). By two days postfertilization, near the onset of hatching (Fig. 6C), the intersegmental blood vessels have distinct lumens and blood movement within them is evident. Also appearing at this stage are the supraintestinal artery, which eventually will supply blood to the intestine, and the subintestinal vein which empties into the posterior cardinal vein. Finally, by 48 hpf, the third and fourth aortic arches appear, the opercular artery replaces the hyoid aortic arch, the lateral head vein begins to drain the head region, and the first endothelial cells of the eye are situated close to the lens (Isogai et al., 2001; Mably and Childs, 2010).

4.1.2. The Larva

By the third day postfertilization (Fig. 6D), the axial and intersegmental vessels elongate as the girth and height of the fish increase. Dorsal branches have their arterial root at the dorsal tip of intersegmental arteries, whereas the ventral intercostal vessels are connected with the dorsal aorta (Isogai et al., 2001). The right posterior cardinal vein carries most of the venous return from the trunk at this point, and four mesenteric arteries branch from the

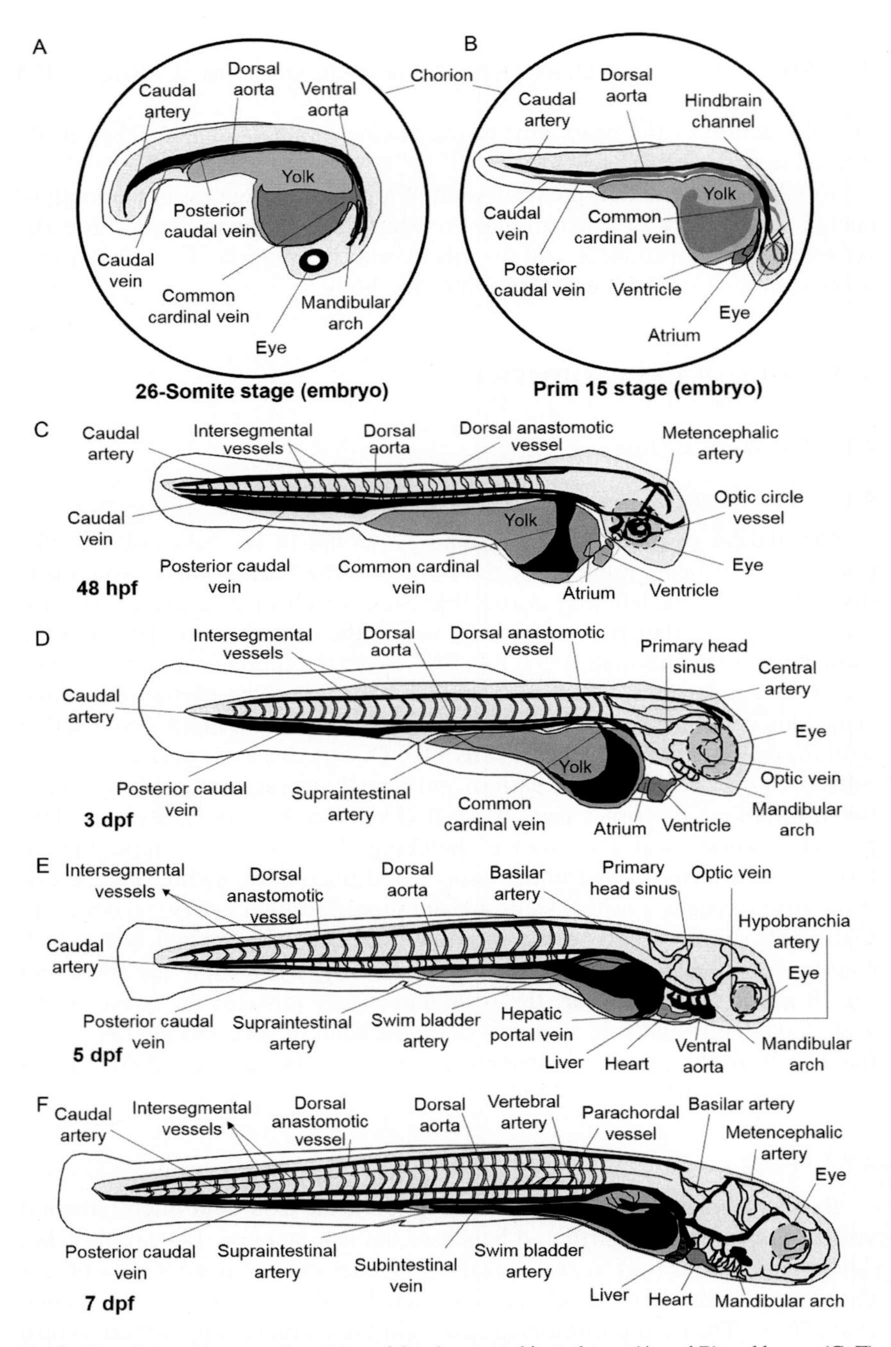

Fig. 6. Vascular structures and patterns of developmental in embryos (A and B) and larvae (C–F) of the zebrafish. Note that the appearance of intersegmental vessels occurs within 49 hpf. Early in ontogeny, vascular development is conserved evolutionarily, but as development proceeds there may be divergence from the zebrafish pattern. See text for additional details.

dorsal aorta to form the caudal part of the supraintestinal artery. Interestingly, both the right and left subintestinal veins now form a continuous vascular network across most of the dorsal and lateral sections of the yolk. The anterior part of this network, in particular, will become the hepatic portal vein (Isogai et al., 2001). By 4–5 dpf (Fig. 6E), as the yolk is progressively absorbed, the left subintestinal vein starts to degenerate. By the fifth day postfertilization, the afferent and efferent branchial arteries have completely differentiated, and by 7 dpf (Fig. 6F) they begin to perfuse the gills in synch with ventilation.

4.2. Vasculogenesis/Angiogenesis

4.2.1. Formation of the Circulatory Network

The formation of new blood vessels in fish embryos and larvae involves *vasculogenesis* and/or *angiogenesis* (Ellertsdóttir et al., 2010). Vasculogenesis involves de novo formation of blood vessels by the coalescence of hemangioblasts (differentiated in situ). When new vessels arise (i.e., sprout) from preexisting vessels by endothelial cell division this is termed angiogenesis (Burggren, 2013; Ellertsdóttir et al., 2010).

Vasculogenesis (Fig. 7) takes place early in embryonic and larval development in fishes. Around the 12-somite stage in zebrafish, the first migration of angioblasts is required for the formation of primary axial vessels of the trunk (Fig. 7A and B); this occurs before the initiation of convective blood flow (Burggren and Bagatto, 2008; Stratman et al., 2015). Briefly, once aggregates of angioblasts have reached the embryonic midline (beneath the hypochord and the mesoderm) around the 14-somite stage in the zebrafish, segregation and ventral sprouting take place (Fig. 7C and D; Herbert et al., 2009). This process is partially dictated by expression at around 17 hpf in the zebrafish of arteriovenous markers such as *ephrin-b2a* and *ephb4a*, within future arterial and venous cells, respectively (Ellertsdóttir et al., 2010). From 20 to 30 hpf, venous cells migrate ventrally within the zebrafish embryo. Just as the arterial cells form a dorsal aorta, these venous cells acquire a tubular arrangement around emerging blood cells to become the postcardinal vein (Fig. 7E) (Ellertsdóttir et al., 2010).

In contrast to venous tissue differentiation, future arterial angioblasts are found near tissues expressing *sonic hedgehog* (*shh*) and are exposed to vascular endothelial growth factor (VEGF); VEGF induces subsequent differentiation into arterial cells (Simons and Eichmann, 2015). Later, angiopoietins and their tyrosine kinase receptors are important for stabilizing the vessel walls in zebrafish by means of interactions between endothelial cells and the smooth muscle layers (Stratman et al., 2015; Weinstein and Lawson, 2002). For a more detailed review of the molecular control of arteriogenesis see Simons and

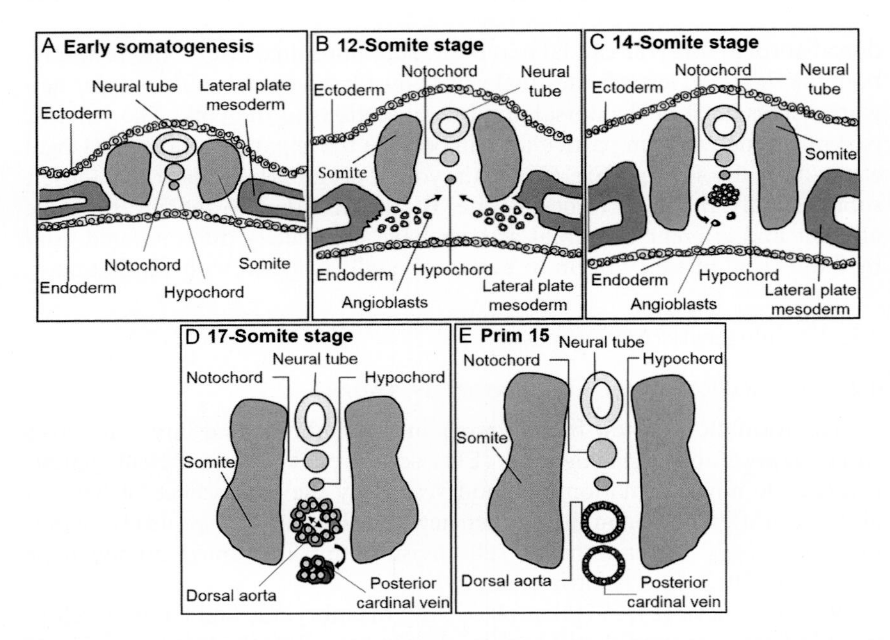

Fig. 7. Vasculogenesis in fishes. (A) During early somatogenesis, the lateral plate mesoderm moves to reach both sides of the embryonic axis. (B) By the 12-somite stage, angioblasts from the lateral plate mesoderm begin migration to the embryonic axis where the future dorsal aorta will be formed. (C) At the 14-somite stage, angioblasts aggregate beneath the hypochord and some of them start migrating ventrally where the posterior cardinal vein will be located. (D) By the 17-somite stage the angioblasts aggregate and start their arrangement. (E) By the Prim 15 stage the dorsal aorta and the posterior cardinal vein have been formed.

Eichmann (2015). As for the formation of the heart (see Section 3.2.1), arterial–venous specification is dependent on blood flow during angiogenesis, at least in developing birds (le Noble et al., 2004).

Angiogenesis occurs in one of two ways. Sprouting refers to the branching of new vessels from preexisting ones, and includes migration of endothelial cells toward angiogenic stimuli, alignment into a bipolar arrangement, the formation of a lumen, division of cells distant to the "bud," and the connection between different sprouts to initiate circulation (Ellertsdóttir et al., 2010). Secondly, although not described in the piscine model, intussusception refers to the process by which intervascular tissue structures, or transluminal tissue pillars, develop into secondary vessels, and subsequently fuse to form complete vascular entities (Burggren, 2013; Makanya et al., 2009).

The formation of secondary vessels is influenced by mechanical and molecular factors, just as in the formation of the heart tube and in vasculogenesis.

Whereas cyclic stress and shear forces are important mechanical factors (see Section 3.2.1) influencing angiogenesis, molecular factors such as VEGF, BMP, Semaphorin–plexin, Chemokines (Cxcr4), and such genes as *c1qrl/cd93* and *c1qrl/clec14a* appear to be the master molecular guides for vascular patterning (Schuermann et al., 2014; Stratman et al., 2015). Taken together, these factors support the *synangiotropy* hypothesis, where the initiation of the heartbeat and its pulsatile blood flow early in development occurs synchronously with the need for angiogenesis in the peripheral circulation.

Unfortunately, few experiments have investigated hemodynamic forces within the cardiovascular system during the early development of fish. However, several hypotheses, and their testing involving other embryonic vertebrates, provide some insight. One hypothesis is that the heart begins to beat as early as it does—prior to the absolute need for internal convective blood flow—to assist in forming the shape and even the tissue composition and distribution within the heart itself. In this scenario, both centrifugal forces generated by blood flow through the heart, as well as the size changes and myocardial stretching associated with systole and diastole, help create the intricate form of the normally developing heart—for an introduction to this literature see Burggren (2013), Collins and Stainier (2016), Kowalski et al. (2013) and Peralta et al. (2013).

Another suggested hypothesis for the early beating of the heart—compatible with the hydraulic shaping hypothesis outlined earlier—is that the early generation of pressure and pulsatile flow actually participates in angiogenesis in the central and peripheral circulation. In this hypothesis, the pressurized central circulation helps stimulate expansion of the peripheral vessels, just like a long, narrow balloon is blown up by the air pressure within the balloon. Further, it is proposed that the sheer/strain forces on the vessel walls resulting from the pulsatile blood flow stimulate the release of VEGF and numerous other compounds (Fig. 8). Then, through a paracrine action, these compounds stimulate vascular endothelial cell growth and division (Burggren, 2013; Chappell and Bautch, 2010; Chung and Ferrara, 2011; Craig and Sumanas, 2016; Ferrara, 2001). Evidence for the angiogenesis hypothesis is, as yet, circumstantial. In one study on early chicken embryos, f_H and cardiac output ($\dot{Q}$) were pharmacologically decreased, which increased arterial pulse pressure in the vessels of the chorioallantoic membrane (CAM), and thus, presumably increased the sheer–strain in its developing vessels. This bradycardia caused complex changes in CAM vascular development (decreased vessel density, but increased vessel length) (Branum et al., 2013). How changes in arterial pulsatility might affect developing fish embryos and larvae has not been determined, although mutations leading to slow and/or irregular heart rate are available for investigation (Kopp et al., 2005, 2014; Warren et al., 2001).

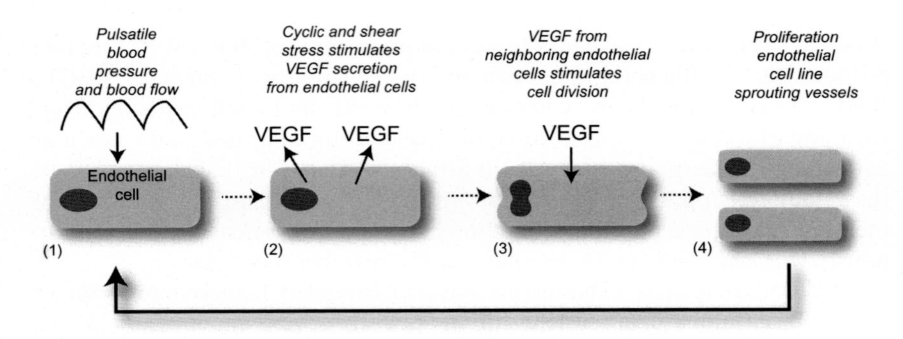

Fig. 8. A schematic diagram showing how pulsatile blood pressure and flow may lead to endothelial cell division and blood vessel sprouting in early vertebrate embryos. After Burggren, W.W., 2013. Cardiovascular development and angiogenesis in the early vertebrate embryo. Cardiovasc. Eng. Technol. 4, 234–245.

Numerous studies have probed the molecular, genetic and physiological basis of pulsatile blood flow in the process of both heart formation and angiogenesis. However, very few have considered the relative timing of angiogenesis, heart development, and the onset of the need for internal convective blood flow for bulk transport. Collectively, the data suggest that the heart transitions from a focus on building itself and the vascular system to becoming an organ for creating bulk materials transport. This transition occurs "just in time," as the rapid growth of vertebrate embryos and larvae increasingly make the simple diffusive delivery of materials across the skin to the underlying tissues inadequate.

The formation of the vascular network requires morphogenetic and mechanical mechanisms, which induce vessel pruning, tissue perfusion, the addition of segments, anastomosis, and remodeling (Herwig et al., 2011; Zakrzewicz et al., 2002). In the zebrafish, the development of the intersegmental blood vessels is the first evidence of vascular bed formation (Ellertsdóttir et al., 2010; Stratman et al., 2015). Briefly, around 22 hpf in zebrafish, the first sprouts formed by endothelial cells emerge bilaterally from the dorsal aorta. These sprouts grow dorsally, aligned with the somite boundaries, and reach the dorsal–lateral roof of the neural tube. By 28–30 hpf, the dorsally located cells branch, sending extensions toward the anterior (cranial) and posterior (caudal). Subsequent interconnections (anastomoses) take place around 34 hpf. Following the formation of the first interconnections, a secondary wave of angiogenic sprouts arises from the posterior cardinal vein or from the dorsal aorta (around 40 hpf), and connect with the basal portion of the primary segments. These connections become intersegmental veins or intersegmental arteries or form parachordal lymphangioblasts (see later). During this process, *Netrin, Slits, Semaphorin, Ephrin,* and VEGF signaling play an important role as attractants or repellents for both endothelial and neural cells (Ellertsdóttir et al., 2010; Isogai et al., 2001; Stratman et al., 2015). Subsequent formation of anastomoses is thought to take place in a similar manner. In addition tubulogenesis of angiogenic sprouts may occur via either budding, cord hollowing, or cell hollowing (Ellertsdóttir et al., 2010).

4.2.2. Vascular Remodeling During Development

Vascular remodeling, the development of structural changes in response to mechanical stresses to meet functional demands, is a normal process in a developing animal. For example, a change in the mean cross-sectional area of the major arteries and veins is evident during early development of the zebrafish (Bagatto and Burggren, 2006). In addition, mechanical and molecular factors act together during vascular remodeling. For example, mechanical factors elicited by blood flow and pressure (shear force and cyclic stress, respectively) act on the capillary walls of mice, causing the release of nitric oxide (NO, a regulator of vascular diameter). Nitric oxide then acts as a downstream mediator of VEGF, FGF, and ANG-1 factors. In contrast, the absence of NO causes an increase in vessel wall thickness (by doubling the proliferation of smooth muscle) and impairs luminal remodeling of the vessels (Rudic et al., 1998; Stewart and Langille, 2004).

Vascular remodeling is important in attenuating potential stressors that occur during development (Pries et al., 2005). For example, the so-called angioadaptation in response to a stressor could be evident as an increase in the number of sprouting vessels and intussusceptions along a vessel. Alternatively, angioadaptation could be evident as a decrease or increase in pruning, or an increase or decrease in the diameter and wall mass of arteries and veins (see Zakrzewicz et al., 2002).

4.3. The Secondary Circulation

A distinct secondary circulation exists in teleost fish. Part of this circulation is associated with the systemic circulation, arising from the dorsal aorta and segmental arteries, and perfuses external tissues such as the skin, scales, fins, peritoneum, and the buccal cavity (Gamperl and Shiels, 2014; Olson, 1996; Olson and Farrell, 2011). The second region of the secondary circulation is associated with the gills and is composed of nutrient arteries and the intralamellar vasculature. Interestingly, blood can follow two pathways; it can perfuse the intralamellar vasculature, subsequently entering the branchial veins and returning to the venous side, or blood can enter the nutrient arteries, feeding the supporting tissue and the adductor and abductor muscles of the gill arch (Gamperl and Shiels, 2014; Chapter 7, Volume 36A: Sandblom and Gräns, 2017).

The secondary circulation forms an independent parallel circulatory system, with a volume ranging between 10% and 50% of the volume of the primary circulation. The adult secondary circulation has quite low pressures (~0.4 kPa) and apparently has little or no role in gas exchange (Dewar et al., 1994; Olson, 1996; Rummer et al., 2014).

4.4. Lymphatic System

The lymphatic system's main functions involve the maintenance of fluid balance, fat absorption, and the drainage of fluids and macromolecules from interstitial spaces within the organism. The lymphatic system is also an essential component of immune responses (Karpanen and Mäkinen, 2006; Yaniv et al., 2006).

Unlike the secondary vasculature, which is connected with the venous system, the lymphatic system is composed of blind-ended capillaries that lack connections with the blood vascular system. In fishes, the lymphatic system arises early in development from a subset of venous endothelial cells that sprout from the major veins (Yaniv et al., 2006). This lymphatic endothelial differentiation (lymphanogenesis) is orchestrated by the expression of the transcription factor *Prox-1* (Karpanen and Mäkinen, 2006). The expression of *VEGFc* is also essential for lymphatic formation in teleosts (Yaniv et al., 2006), as in mammals (Tammela et al., 2005). In addition, equivalence in function (i.e., the collection and draining of fluids and macromolecules from interstitial spaces) and anatomy (i.e., the exhibition of facial, pectoral, and lateral lymphatic vessels that merge before draining into the cardinal vein) of the lymphatic system between developing zebrafish and other vertebrates has been demonstrated employing angiography and lymphangiography techniques (Yaniv et al., 2006).

5. ONTOGENY OF CARDIOVASCULAR REGULATION

5.1. Introduction

With the great diversity in fish species, we find among them appropriate models to study heart development. Most frequently studied in the context of cardiac regulation have been zebrafish, medaka, salmonids (such as rainbow trout) and, to a lesser extent, killifish. However, these popular models fall far short of representing the diversity of habitats and respective phenotypes within fishes. The genetic determination of phenotype produces a standard body plan, but the environment clearly affects the phenotype. As such, epigenetic factors can intercede the genotype beyond pleiotropy, where gene regulation can remaster the phenotype. This adds complexity to the variability that is part of the evolutionary process. Indeed, we find that this is well represented during the ontogeny of the fish cardiovascular system, where we see extensive modifications in response to physiochemical effectors (Incardona et al., 2015; Incardona and Scholz, 2016; Johnson et al., 2015; Sorhus et al., 2016; Xu et al., 2016). This can result in an alteration of the developmental trajectory of various cardiovascular phenotypes associated with environmental pressures. Such environmental pressures within an animal's niche, be they transient or sustained, will dictate the effectiveness of reproductive strategies and are reflected in the development of cardiovascular regulation and its plasticity. In this context, the ability to regulate components of cardiovascular function emerges at different points in development in different species, with synchronicity to match the needs of growth and development.

In addition to the combined difficulties and opportunities presented by species diversity, there are also practical matters associated with studying larval fish. Recording physiologic data from early life-stage fish is difficult and time-consuming. Doing so generally requires either anesthesia or creative methods for restraining the larvae, and invariably requires the use of magnification (Burggren and Blank, 2009; Burggren and Fritsche, 1995; Schwerte and Fritsche, 2003). So, in contrast to the wide variety of straightforward methods for detecting differences in molecular phenotype, there are fewer options for evaluating the physiological phenotype of the cardiovascular system in early life-stage fish. High throughput systems for evaluating embryonic and larval fish health are available to study metabolic and cardiac function during development (Akagi et al., 2013; Brown et al., 2016; Letamendia et al., 2012; Lisa et al., 2014; Oziolor et al., 2016; Spomer et al., 2012). Through such techniques, it is possible to measure physiological, behavioral, developmental and morphological endpoints of large numbers of animals nearly simultaneously. Still, few advances have been made in the past decade that enable in-depth investigations of the functional properties of the developing fish heart, although technologies exist that will enable further understanding of the ontogeny of cardiovascular regulation.

5.2. Ontogeny of Heart Function

5.2.1. Conduction System of the Heart

Teleost fish and older evolutionary classes such as cyclostomes and Chondrichthyes (and other ectothermal vertebrates) retain a sinus venosus (see Section 3.1.2). The sinus venosus houses the dominant pacemaker of the heart, and contracts ahead of the atrium in some species—contributing to atrial filling (Jensen et al., 2014). In hagfish and lungfish, the sinus venosus is a distinct chamber (a sinus) proximal to the atrium. A distinct pause in conduction from the sinus to the atrium is often evident in electrocardiograms (Jensen et al., 2014). This sinoatrial signature event has not been well characterized in more derived teleost fish, which possesses a more reduced sinus venosus, as described below. In teleost fish, the sinus venosus is a separate chamber during early development, and contraction is pronounced. However, during continuing embryonic development, the sinus venosus of many teleost fish becomes greatly reduced, and is all but lost in notable species such as zebrafish, rainbow trout, and killifish, in contrast with the meaning of the term "sinus."

In zebrafish and trout, the dominant pacemaker of the heart is a ring of spindle-shaped cells that encircle the junction of the *sinus venosus* and the atrium; analogous to the sinoatrial node in mammals (Haverinen and Vornanen, 2007; Jensen et al., 2014; Tessadori et al., 2012). Though there is little myocardium in the sinus venosus of most fish, there is some myocardial tissue between the sinus venosus and atrium, where the sinus venosus pacemaker is located and likely contracts to some degree (Haverinen and Vornanen, 2007). Still, despite the differences in morphology and apparent function during development and across evolutionary time, there is a consensus that the origin of the conduction signal begins at the sinus venosus, near the junction with the atrium in fish (Burggren and Bagatto, 2008; Chapter 4, Volume 36A: Farrell and Smith, 2017).

During early development of the zebrafish, conduction of electrical activation is propagated along the heart tube, but as the atrium and ventricle become distinct, the wavelike conduction signal desynchronizes and becomes bimodal (Boselli et al., 2015). Conduction tissue at least somewhat analogous to the AV node in mammals and birds is found at the atrioventricular junction of fishes (Šolc, 2007). Like the mammalian AV node, the AV node in fish creates a delay in the signal coming from the atrium, prior to propagation of the signal through the ventricle from apex to base (Šolc, 2007). This prevents tachyarrhythmia in the atrium from reaching the ventricle (Burggren and Bagatto, 2008; Irisawa, 1978). The development of the conduction delay is concomitant with the formation of the atrioventricular valve, and likely represents a coordination of the electrophysiological and morphological components necessary for proper blood flow (Boselli et al., 2015). Like in the mammalian heart, AV node cells exhibit stretch-induced tachycardia (Burggren and Bagatto, 2008), and are thought to function similarly in all vertebrates. No histologically defined conducting tissue has been described in the atria of developing fish, but cells that have a phenotype of conducting cells are present in the adult atrium and sinus venosus (Šolc, 2007).

In adult ectothermic vertebrates, ventricular contractions occur from apex to base, and are conducted through ventricular trabeculae, *in lieu* of the

His-Purkinje system found in mammals (Jensen et al., 2012; Sedmera et al., 2003). Many teleost species do not have completely trabeculated ventricles (Icardo, 2006; Chapters 1 and 4, Volume 36A: Icardo, 2017 and Farrell and Smith, 2017, respectively). In these species, the ventricle is composed of an external compacta layer and an internal spongiosa layer (hereafter referred to as a "compact ventricle"). Signals are conducted through the spongy myocardium of fish with compact ventricles; this tissue thus serves both contractile and conductive functions in these fish, and it is likely that the activation signal is conducted across the atrium in the same way (Jensen et al., 2012; Šolc, 2007). There is no consensus on the functional or evolutionary significance of the interspecific variation in ventricular morphology among fishes, though more derived species have a more compact ventricle and the morphology of the outflow tract appears to be linked to the morphology of the ventricle (see later) (Icardo, 2006, 2012).

Another conduction node may exist between the ventricle and OFT of adult fishes (Burggren and Bagatto, 2008), though no investigations have identified candidate pacemaker cells in this region using histological surveys or molecular markers in fish. However, a nerve plexus that encircles the conus valves has origins in both the bulbus arteriosus and the ventral aorta in adult goldfish (*Carassius auratus*) (Newton et al., 2014). The adult zebrafish also has a well innervated cardiac outflow tract, with axons that originate from the atrioventricular plexus and spread along the wall of the bulbus arteriosus. Additionally, there is a branchiocardiac nerve trunk that appears to originate from the fourth branchial arch (Stoyek et al., 2015). This branchiocardiac nerve is likely a well-timed feature, arising during larval development in time to coincide with the requirement for convective blood flow and gill respiration that occurs after hatch in larval fishes.

There is little information on conduction in, or activation of, the conus arteriosus of adult fish, let alone larvae, though fish represent a tractable model to study the development of the vertebrate outflow tract (Grimes and Kirby, 2009; Hutson and Kirby, 2007). In more ancient fish species, there is a muscular conus arteriosus *and* an elastic bulbus arteriosus (see Section 3.1.2). This is also the case in less derived teleost species, where the presence of both a conus and bulbus arteriosus coincides with a fully trabeculated ventricle (i.e., no compactum layer); whereas more derived teleost species lack a substantial conus arteriosus, but do have a compact ventricle (Icardo, 2006; Chapter 1, Volume 36A: Icardo, 2017). The conus arteriosus of these derived teleost species is generally constructed of vascularized compact myocardium similar to that of a compact ventricle (Icardo, 2006). However, it is conceivable that the compact myocardium of the conus arteriosus conducts the signal from the ventricle to the bulbus arteriosus in a similar way that the compact myocardium conducts signals in more derived fish species. If this is the case, it is likely that a distal conduction node is located at the junction with the bulbus arteriosus, where a conductive compact myocardium ends and the elastic tissue of the bulbus begins.

5.2.2. Heart Rate

5.2.2.1. Control of Heart Rate. The heart starts beating early in development and may be irregular, prior to settling into a rhythm as development of autonomic control mechanisms advances (Burggren and Bagatto, 2008). Each heart beat is initiated by the pacemaker in the myocardium at the sinoatrial junction (see Section 3.1.5), but the overall heart rate is controlled by excitatory and inhibitory signals from the autonomic nervous system, as discussed below (Taylor et al., 2014: Chapter 4, Volume 36A: Farrell and Smith, 2017).

Heart rate generally increases during the early embryonic development of vertebrates, then slows down prior to or just after hatching (Burggren and Warburton, 1994), and this trend holds for developing fishes (Fig. 9). The progressive increase in f_H appears to be created, at least in part, by increasing concentrations of circulating catecholamines that establish beta adrenergic

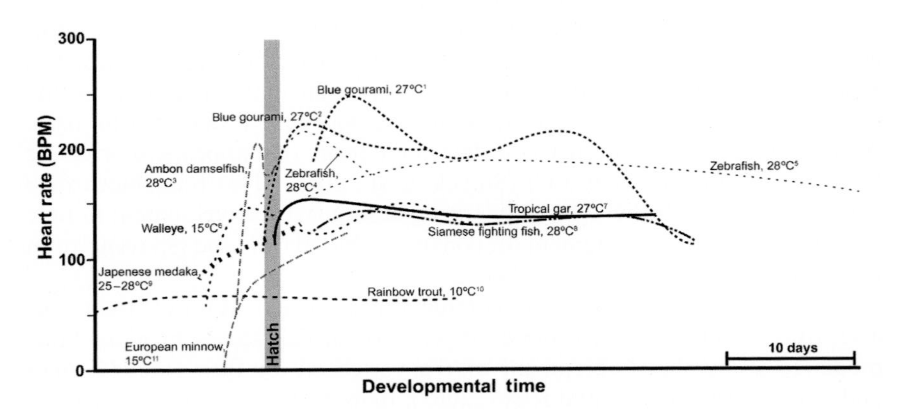

Fig. 9. Developmental changes in heart rate in several teleost fish. Heart rate data were digitized and replotted as polynomial trend lines. Trend lines were adjusted along the x-axis according to original plot coordinates to coincide with the presumed hatching time indicated in the source literature. Data were collected at the given temperature for each species. 1. Blue gourami, *Trichopterus trichopodus* (Mendez-Sanchez, 2015); 2. Blue gourami, *Trichopterus trichopodus* (Blank, 2009); 3. Ambon damselfish, *Pomacentrus amboinensis* (McCormick and Nechaev, 2002); 4. Zebrafish, *Danio rerio* (Jacob et al., 2002); 5. Zebrafish, *Danio rerio* (Barrionuevo and Burggren, 1999); 6. Walleye, *Sander vitreus* (McElman and Balon, 1979); 7. Tropical gar, *Atractosteus tropicus* (Burggren et al., 2016); 8. Siamese fighting fish, *Betta splendens* (Mendez-Sanchez, 2015); 9. Japanese medaka, *Oryzias latipes* (Colman et al., 2005); 10. Rainbow trout, *Oncorhynchus mykiss* (Miller et al., 2011); 11. European minnow, *Pomacentrus phoxinus* (Schonweger et al., 2000).

tone—the stimulatory input to the circulatory system (Miller et al., 2011; Taylor et al., 2014). With this single stimulatory input, it is not surprising that stressors may have little to no effect on f_H in some species of embryonic fish, since the inhibitory control mechanisms appear later in development, depending on the developmental plan of the species (Miller et al., 2011; Schwerte, 2009). Prior to the development of inhibitory control of f_H, the cause of bradycardia in embryos is unclear. In zebrafish, the first rise in f_H coincides with an increase in oxygen consumption (Schwerte, 2009), which is commonly used as a proxy for measuring metabolic rate in fish. However, despite the close link between metabolic rate and cardiac function in adult fish, there is some uncoupling between heart rate and metabolic rate in embryonic fish (discussed below in Section 5.3).

Cardiac innervation and its role in the control of heart function are well documented in adult fish (e.g., see Chapter 4, Volume 36A: Farrell and Smith, 2017), but we know little about the ontogeny of innervation of the fish heart and the extent of central control. Most information is inferred from the first appearance of physiological responses, like changes in gill ventilation and f_H in response to hypoxia or activity (Barrionuevo and Burggren, 1999; Burggren et al., 2016; Holeton, 1971; Jacob et al., 2002; Liem, 1981; McDonald and McMahon, 1977; Pelster et al., 2003; Vulesevic et al., 2006). Like other vertebrates, the fish heart receives innervation from spinal (sympathetic) and cranial (parasympathetic) limbs of the autonomic nervous system, which converge in a network surrounding the sino-atrial region of the heart (Stoyek et al., 2015). Developing rainbow trout follow a common vertebrate pattern, where the early onset of stimulatory adrenergic tone eventually becomes tempered by an inhibitory cholinergic tone that appears just after hatching (Miller et al., 2011). This causes a slowing of the heart rate of these larvae that is likely due to a newly acquired innervation by the vagus (X) nerve; functional receptors for both cholinergic and adrenergic agonists are likely present before innervation is in place (Taylor et al., 2014). Not surprisingly, the appearance of cholinergic tone at hatch in rainbow trout is well timed for the onset of gill ventilation and swimming activity, and the challenges associated with direct exposure to the environment. Interestingly, there appear to be considerable differences in the rate of development of the various regulatory components controlling the circulation that likely reflects the altricial or precocial nature of the posthatch larvae. In the zebrafish, the balance exerted by vagal tone is not set until well over a week posthatch ($\sim$ 12 dpf), around the time that the gills start to serve a dominant role in respiration and NMDA receptors begin to play a role in regulating gill ventilation (Schwerte et al., 2006). NMDA receptors and their involvement in ventilatory responses have been described, but it is unclear if they play a role in cardiovascular regulation in developing fishes (Burggren and Bagatto, 2008). In the very rapidly developing, and relative to zebrafish, quite precocial tropical gar (*Atractosteus tropicus*), both cardiac and ventilatory responses become apparent as early as 2.5 days after hatching (Burggren et al., 2016). Also linked to the timing and onset of the ventilatory response is functional innervation of neuroepithelial cells (NECs), an important part of the oxygen-sensing system (Jonz et al., 2015).

5.2.2.2. Emerging Techniques for Electrocardiology in Fish Larvae. Future research on the ontogeny of cardiovascular regulation in larval fishes is well poised to advance rapidly. Pharmacological studies using adrenergic and cholinergic agonists and antagonists, and receptor blockade combined with fine measurements of cardiovascular function, can illustrate the ontogeny of cardiac regulation in fish as it has in mammals (Jonz et al., 2015; Taylor et al., 2014). A number of functional assays employ image analysis of the circulation (Bagatto and Burggren, 2006; Bark et al., 2017; Bromnimann et al., 2016; Hove, 2006; Johnson et al., 2013; Schwerte and Pelster, 2000; Zeng et al., 2014). However, there is also much promise in electrophysiological techniques for deriving data recordings of heart function that are translatable to the developing fish condition. Indeed, electrocardiogram (ECG) surface recordings are now being performed on fish employing both single and dual electrode techniques (Dhillon et al., 2013; Liu et al., 2016; Fig. 10). Such recordings can be performed noninvasively to produce ECG traces that show fine details of the conduction system in embryonic and larval fish (Vornanen and Hassinen, 2016). Though needle electrodes or electrode arrays can be used for adult fish (Kermorgant et al., 2015; Yu et al., 2012), the most translatable data for embryonic or larval fish have been produced using an electrolyte-filled glass micropipette as an electrode, connected to a silver chloride-coated wire lead (Dhillon et al., 2013). Using these or similar methodologies, noninvasive ECG recordings can reveal a distinguishable P wave that represents atrial depolarization, a QRS complex that is indicative of ventricular depolarization and a T wave corresponding to ventricular repolarization in embryonic and larval fish (Fig. 10B; Dhillon et al., 2013). From these recordings, we can study variables such as the QT interval, and changes in P–R, R–T and T–P; the latter is the main component that influences changes in f_H and enables fine scale evaluation of arrhythmias in embryonic and larval fish (Arnaout et al., 2007; Burggren and Bagatto, 2008; Dhillon et al., 2013; Leong et al., 2010; Verkerk and Remme, 2012; Vornanen and Hassinen, 2016).

Few studies have utilized ECG data in early life-stage fish, despite the widespread use of heart function as a “master” physiological variable. Visually derived data on heart function are valuable and image recording analyses that use pixel data to derive f_H offer insights into variations in f_H and are

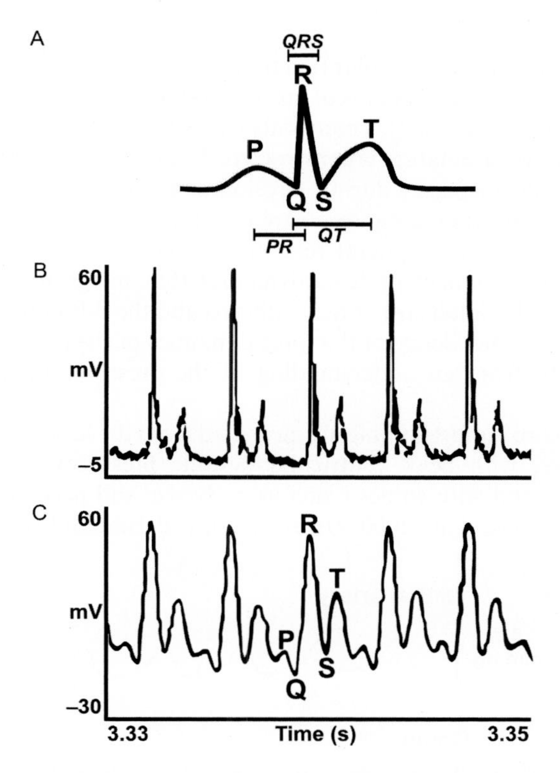

Fig. 10. Electrocardiogram of larval zebrafish. (A) Diagram of an electrocardiogram illustrating the P wave (atrial polarization), the QRS complex (ventricular polarization), and the T wave (ventricular repolarization). Time intervals of Q–T, P–R, and R–T are commonly used as indicators of heart dysfunction. T–P is often used to establish changes in heart rate and arrhythmias. (B) ECG recording from a larval zebrafish. (C) ECG recording after implementing a low pass filter to reduce background noise to establish a signal for analysis. Figure modified from Dhillon, S.S., Dóró, E., Magyary, I., Egginton, S., Sík, A., Müller, F., 2013. Optimisation of embryonic and larval ECG measurement in zebrafish for quantifying the effect of QT prolonging drugs. PLoS One 8, e60552.

discussed below. However, these recordings have limitations since blood flow and changes in morphology are used as a proxy for signal generation that could otherwise be recorded as an electrical impulse in an ECG. An understanding of the development of cardiac control in fishes, and the environmental stressors that affect it, will be useful in a variety of contexts, including aquaculture management. Further, the popularity and utility of using developing fish as a model for the vertebrate condition will likely drive the production of commercial systems specifically designed for high-throughput ECG measurements of embryonic and larval fish.

5.2.3. Blood Pressure

Development of cardiovascular function generally proceeds as cardiac output increases to meet the demands of growth and differentiation in vertebrates (Schwerte and Fritsche, 2003). There is also a decrease in peripheral resistance that occurs as new vasculature is constructed. Thus, even though systolic pressure progressively increases during development to meet growth demands (Hu et al., 2000), the increase in pressure is not as great as might be expected due to the decrease in peripheral resistance. There is little information for fishes on the development of hemodynamics that integrates structure and function, due to the small size of fish embryos and the difficulties in data collection. However, knowledge of the force dynamics of the cardiovascular system will benefit from an understanding of the pressures in the heart and vasculature.

Blood pressure is most commonly measured from the heart or outflow tract of larval fishes, with peak ventricular systolic pressures in the range of 0.5–2.0 mmHg, and with values rising as body size and peripheral resistance increase (Fig. 11; Hu et al., 2000; Hu et al., 2001; Pelster and Burggren, 1996).

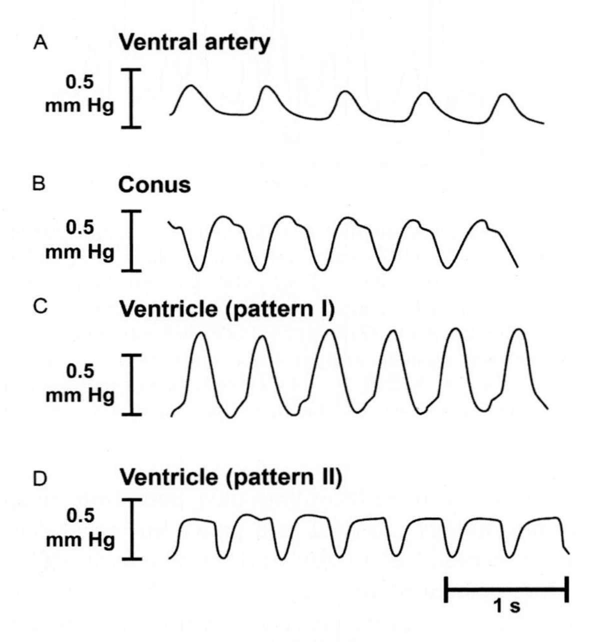

Fig. 11. Blood pressure in 84–96 hpf larval zebrafish. Representative blood pressure traces recorded from (A) ventral artery, (B) conus arteriosus, and (C) ventricle. Modified from Schwerte, T., Uberbacher, D., Pelster, B., 2003. Non-invasive imaging of blood cell concentration and blood distribution in zebrafish Danio rerio incubated in hypoic conditions in vivo. J. Exp. Biol. 206, 1299–1307.

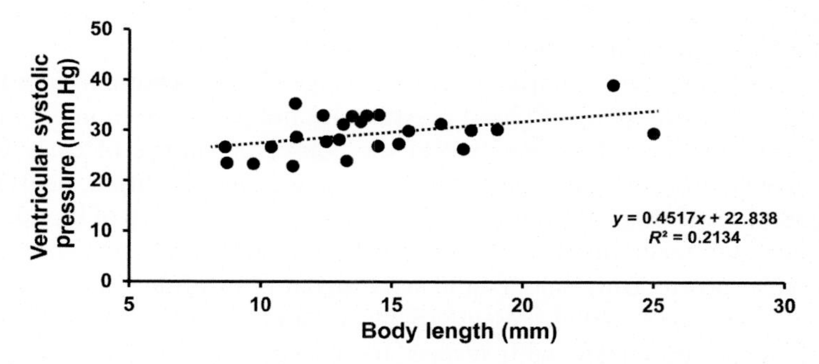

Fig. 12. Systolic ventricular pressure in larval Gulf killifish (*Fundulus grandis*) (two to six weeks postfertilization) as a function of growth and differentiation (B. Dubansky, unpublished).

These values, which are comparable to those measured in larval amphibians (Hou and Burggren, 1995), are much lower than those of adult fishes which could be higher than 25 mmHg (and more typically 40 mmHg), even in small fish like the stickleback (Jones et al., 2005; McDonald et al., 2010; Sandblom et al., 2010; Chapter 4, Volume 36A: Farrell and Smith, 2017). Outflow pressure in larval fishes increases linearly with size as the time required for systolic contraction shortens (Fig. 12), despite the decrease in heart rate that occurs as vagal tone develops (Burggren and Bagatto, 2008; Hu et al., 2001). Still, pressures generated by the heart of embryos and larvae are very low, not surprising given that the myocardium is only a few cell layers thick. Despite the increase in ventricular pressure described during ontogeny, it is unclear if this continues to increase linearly with body size in adults and between species (Burggren and Bagatto, 2008).

To integrate cardiac output into a finer-scale metric of cardiac function, blood pressure must be considered. Hydrostatic pressure in very small larval fishes can be measured using a servo-null micropipette system (Hu et al., 2001; Pelster and Bemis, 1992; Pelster and Burggren, 1996). These systems can be inaccurate at very low pressures, but low pressure transducers are being integrated into existing systems so that more precise measurements can be at the pressures seen in fish larvae. Using these servo-null micropipette systems, it is conceivable that pressures could be measured in vessels with diameters as small as 7 μm, and thus, used to measure vascular resistance, among other things (Pelster and Burggren, 1996; Schwerte and Fritsche, 2003). Though this technique is time-consuming and requires more operator training than other higher-throughput methods for measuring cardiac function in small animals, it offers significant advantages. In addition to obtaining physiological values for blood pressure, one can simultaneously collect data on f_H. Further, if coupled with image analysis of stroke volume (V_S), one can also calculate peripheral resistance and obtain pressure–volume traces to evaluate contractility, and thus, the efficiency of cardiac function.

There are only a few reports of early life-stage blood pressure in zebrafish, and data on the ontogeny of blood pressure are not available to make multispecies comparisons. However, due to the conserved nature of early heart function in vertebrates, it can be assumed that since cardiac output, heart rate, and general physiology are similar in fish species, the ontogeny of blood pressure progresses concomitantly, though this might vary with precocial vs altricial species. So, to truly understand hemodynamics during ontogeny, it is necessary to measure blood pressure (Schwerte and Fritsche, 2003). This will be particularly important with respect to the variability in environmental adaptations that occur during development, and comparative studies of blood pressure ontogeny in multiple species (and populations) will contribute to our understanding of hemodynamics and cardiovascular development.

5.2.4. Stroke Volume and Cardiac Output

After hatch, when vagal tone develops and f_H stabilizes, heart chamber size continues to increase such that cardiac output continues to increase to support growth and activity after hatch. As development and integration of the autonomic nervous system progresses, the role of the latter in controlling f_H and stroke volume (V_S) will determine the ability of the fish to rapidly respond with alterations in cardiac output.

Data derived from video capture of the heart, through the typically transparent body wall of larval fishes, are generally used to calculate ventricular stroke volume and cardiac output (Fig. 13). Such data are most commonly derived by measuring changes in pixelation and/or chamber volume at ventricular diastole and systole (Bagatto and Burggren, 2006; Burggren and Blank, 2009; Kalogirou et al., 2014; Perrichon and Burggren, 2017; Shin et al., 2010). This becomes more difficult as pigmentation develops, obscuring the chamber margins and other internal structures. Zebrafish mutants with defects in pigmentation such as *crystal, casper, golden, albino, rose,* and *panther* result in fish with different degrees of enhanced and maintained transparency (Antinucci and Hindges, 2016; Bark et al., 2017; White et al., 2008). Some of these strains are particularly useful (i.e., *crystal*) for characterizing developmental changes in V_S and cardiac output without interference from developing pigmentation (Antinucci and Hindges, 2016). Crossing the *casper* zebrafish line with the *Tg* (*cm1c2:nuDsRed*) line, which has a red fluorescent heart (Hoage et al., 2012), may hold promise for new and more accurate measurements of changes in pixel values based on fluorescent energy. However, pixel values can also be used to calculate blood velocity, kinematics of the chamber walls, flow rates, and pumping efficiency (Denvir et al., 2008; Johnson et al., 2013; Kopp et al., 2005; Schwerte and Pelster, 2000), which may add context

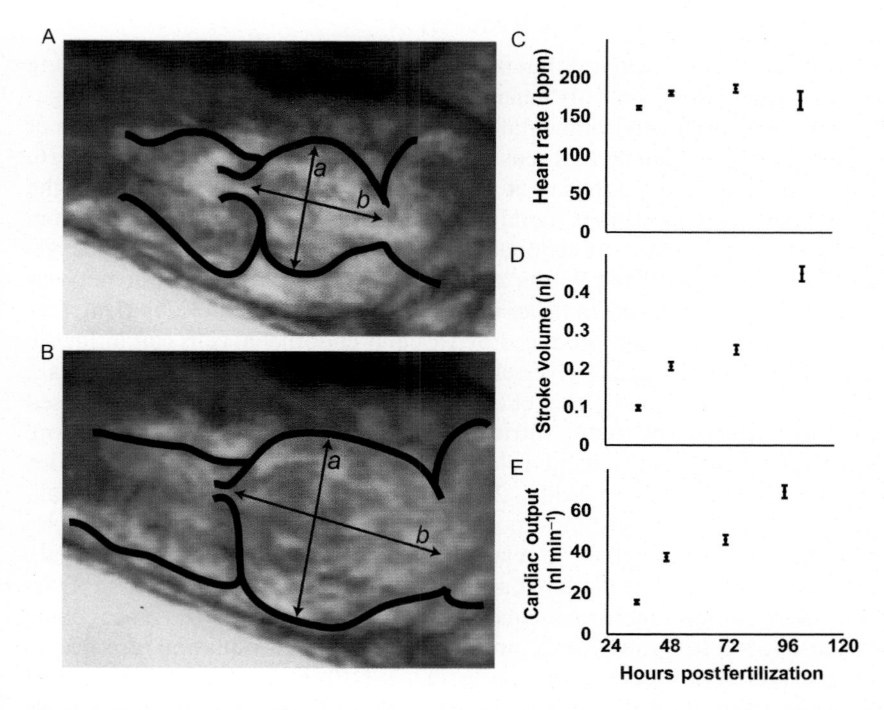

Fig. 13. Visually derived cardiac output in Mahi–Mahi (*Coryphaena hippurus*) larvae at 96 hpf. Ventricular volume was calculated assuming that the chamber was a prolate spheroid, and V_S was determined by subtracting the ventricular volume at end systole (A) from the volume at end diastole (B). Heart rate (C) can then be multiplied by the calculated stroke volume (D) to determine cardiac output (E) (i.e., heart rate × stroke volume).

to image-driven analysis of cardiac output. Still, data from video capture present some difficulties in obtaining accurate calculations of cardiac output, in part because the shape of the heart (and thus the model used to calculate volume) changes during the early development of fish larvae (Perrichon and Burggren, 2017).

Taking advantage of more transparent zebrafish mutants will be a boon to studying the general cardiovascular physiology of larval fishes. However, a few other fishes are either naturally transparent or have transparent strains. For example, the glass catfish, *Kryptopterus vitreolus* (previously *Kryptopterus bicirrhis* or *K. minor*), has been exploited for examining patterns of blood and lymph flow (Dahl Ejby Jensen et al., 2009; Rummer et al., 2014; Tripathi et al., 2003). The crucian carp (*Carassius auratus*) also has a strain that remains transparent for several weeks after hatching, and has been studied in the context of the genetics and development of body pigmentation (Xu et al., 2015).

5.2.5. Ontogeny of Vascular Regulation

The pattern of vascular differentiation and growth through the developing tissues, presumably, is mostly determined by genetics, but blood flow is regulated in response to environmental cues and influences virtually all aspects of vasculature development, as discussed in Section 4 (Broemnimann et al., 2016; Goenezen et al., 2012; Johnson et al., 2013). As in the heart, receptors in the vasculature and endogenous signaling molecules become effective modulators of vascular tone, prior to functional innervation of the peripheral nervous system (Fritsche et al., 2000; Pelster et al., 2005). Although autonomic nervous control of the vasculature in fishes is not active until later in development, several chemical cell signals produced by vascular endothelial cells can influence vasculature tone by causing vessel dilation and contraction (Fritsche et al., 2000; Pelster et al., 2005; Sykes et al., 2016). Most prominently (at least based upon its frequency of study), nitric oxide (NO) is present as vessels develop, and likely plays a prominent role in the control of vascular tone in fish embryos (Bagatto, 2005; Eddy, 2005; Fritsche et al., 2000; Pelster et al., 2005). Nitric oxide contributes to vascular tone by signaling smooth muscles in the vascular endothelium to relax, causing vasodilation (Gutterman et al., 2016; Heiss et al., 2015; Pelster et al., 2005). In early stage zebrafish (5 dpf larvae) there is immunohistochemical evidence of endothelial nitric oxide synthase (eNOS), a nitric oxide producing enzyme. Vasodilation also occurs when a NO donor (sodium nitroprusside—SNP) is injected into the dorsal artery and vein, and this affects the vasculature branching off these vessels and increases blood flow (Fritsche et al., 2000; Pelster et al., 2005). The increase in blood flow, and the associated sheer stress on the vasculature, contributes to vasculogenesis and angiogenesis (Hove, 2006; Hove et al., 2003; Pelster et al., 2005). NO also modulates VEGF-induced angiogenesis and vascular permeability, adding to its role in development of the vasculature (Pelster et al., 2005; Schwerte et al., 2003; Schwerte and Fritsche, 2003; Schwerte and Pelster, 2000; Sykes et al., 2016). Although NO does not change the appearance and structure of the vascular bed, it can influence the timing of development of the trunk vasculature in zebrafish, and this suggests that there is potential for plasticity in vascular bed development (Isogai et al., 2001, 2003; Pelster et al., 2005). NO is also thought to be responsible for the vasodilatory tone that develops soon after hatching (Clough, 2015). In contrast, vasoconstriction is potentially linked to the arachidonic acid metabolite, thromboxane A_2, generated by cyclooxygenase (COX) and thromboxane synthase (Watkins et al., 2012). This may indicate that inflammatory pathways can influence vascular tone early in development since the COX-thromboxane pathway is involved in circulatory failure of early stage zebrafish exposed to toxicants (Teraoka et al., 2009, 2014). Toxicants can alter vascular tone by the disruption of calcium handling, by decreasing NO concentrations, and by increasing oxidative stress (Cypher et al., 2015; Gao and Wang, 2014 in Cypher et al., 2015). Such autoregulation probably develops early in ontogeny in most fish species, since localized responses of the microcirculation do not require the development of innervation or the secretory capacity of the endocrine glands, but are critical for development of the circulatory system and its response to stressors. In some fish species, full autoregulation may be complete just after formation of the heart chambers is achieved (Bagatto, 2005; Balashov et al., 1991; Balashov and Soltitskij, 1991; Burggren and Bagatto, 2008), although this is poorly understood. Studies of the peripheral vascular system's role throughout cardiovascular development in fishes will be valuable for understanding the mechanisms responsible for regulating vascular tone in response to stressors, and the development of the ventilatory control between the gill vasculature and heart. In adult fishes, the muscle displays considerable plasticity, and this also extends to larvae and juveniles. For example, exercise (swim training) increases the capillarization of the trunk musculature of zebrafish at 21–32 dpf, but not in larvae only trained from 9 to 15 dpf (Pelster et al., 2003). These data suggest that the plasticity of the vascular system is reduced in earlier stage fish, an idea that is supported by data which shows that the timing (i.e., dpf) of the constriction and dilation associated with the injection of phenylephrine (PE) and SNP, respectively, were not affected by swim training (Bagatto, 2005). However, the lack of an effect of training on muscle capillarization from 9 to 15 dpf is not mirrored in other effects of swim training [e.g., decreased oxygen consumption during exertion, increased survival in hypoxia (Bagatto et al., 2001; Gore and Burggren, 2006)], and in zebrafish incubated at lower temperatures (20°C vs 25°C) the characteristic dilation in response to SNP was delayed by two days and the effects of PE delayed by three days. These latter data suggest that plasticity in the developmental program for vasculature tone does exist. Interestingly, in fish exposed to hypoxia, although the response to SNP was similarly delayed (not appearing until 10 dpf), there was no effect of hypoxia on vasoconstriction by PE (Bagatto, 2005).

Importantly, since a response to these (and many other) stressors causes an overall delay in development, developmental stage should be taken into account. As such, if developmental stage is considered *in lieu* of developmental time (i.e., dpf), the onset of PE-mediated vasoconstriction was actually accelerated in response to hypoxia. Additionally, there was no difference in the onset of SNP-mediated vasodilation or either response when early stage zebrafish larvae were exposed to lower temperature or swim training (this is also true for the adrenergic response) (Bagatto, 2005). Consequently, it appears that differences in the timing of emergence of vasodilation and vasoconstriction were likely caused by delays in development, as a secondary effect of the stressor. Thus, although environmental pressures can slow overall development, they can also accelerate aspects of development, reflecting developmental heterokairy (Andrewartha et al., 2011; Mueller et al., 2015a; Spicer et al., 2011).

There is little information on the ontogeny of peripheral vascular regulation in fish. This is surprising because endothelial dysfunction is clearly a contributing factor to the observed effects of environmental stressors on heart function in vertebrates (Watkins et al., 2012). Increased peripheral resistance causes a substantial increase in workload on the heart, and hemodynamic and mechanical forces can severely affect the function of the cardiovascular system (Pelster et al., 2010). Since proper development of the microvasculature at the tissue level (e.g., for the delivery of nutrients, gases and metabolites) is a crucial component of the capacity for organisms to respond to physiological stressors later in life, understanding the vascular network and its regulation and control is also critical for understanding the pathology of metabolic disease (Clough, 2015). Furthermore, knowledge of the patterning of the vasculature and its remodeling during growth in vertebrates has largely been gained by studying fish models, in which vascular development is sensitive to dysglycemia, dyslipidemia and redox state, in addition to the stressors mentioned earlier (Clough, 2015). Since environmental stressors during development can cause such diverse and complex cardiovascular phenotypes that organisms are significantly affected later in life (Patti, 2013), fish models will continue to advance our understanding of developmental plasticity and metabolic diseases (Brown et al., 2016; Dahme et al., 2009).

5.3. Cardiac Regulation of Whole Body Metabolism

The clear link between cardiovascular and metabolic activity is illustrated by the increasing use of developing fish as a biomedical model for metabolic disease, where insights from whole-animal physiological measurements of metabolic activity hold great promise (Schlegel and Gut, 2015). Whole body metabolic rate increases during development, at a rate dependent on requisite energy demands, and concurrent with the development of the functional capacity of the heart. However, cardiac output and the convective transport of oxygen are not crucial for the metabolic demands of early stage fish (see Section 2); though embryos and larvae still respond to low oxygen by increasing cardiac output and redirecting blood flow (Pelster et al., 2010). A significant physiological link exists between cardiac activity and blood supply to the tissues in response to changes in oxygen demand, but the relationship between cardiac activity and whole-body metabolic regulation during early development in larval fishes remains unclear. For example, although increases in temperature are linked to higher values of cardiac output and metabolic rate, the temperature-induced metabolic rate far exceeds the increase observed in cardiac output. This indicates that early heart development is not as well linked to metabolism as it is in adults (Mirkovic and Rombough, 1998; Pelster, 2002). This may be due to the underdeveloped vasculature of larval fish, and the fact that autonomic control of f_H and cardiac output is tightly linked to the high efficiency of gas exchange at the gills of adult fish (Taylor et al., 2014). Balancing of autonomic cholinergic and adrenergic tone appears to begin after hatching when the gills begin to become functional (as described earlier). Thus, the onset of cardiac metabolic regulation is likely also coupled to hatching/birth, and the subsequent onset of gill ventilation. It thus follows that the various components of the cardiovascular system, when established, respond to metabolic requirements on an "as needed" basis.

In zebrafish, metabolic rate continues to increase until after hatch, when it begins to settle to a value that becomes consistent with body size (Fig. 14; Barrionuevo and Burggren, 1999). This pattern and timing is similar to the change in heart rate during development. Although it is generally true that metabolic rates follow a well-defined relationship with mass in adult fish, we see contrary data on the mass-specific metabolic rate of fish during early development—for example Barrionuevo and Burggren (1999) and Mendez-Sanchez and Burggren (2017). This is not surprising given that differences in energetic demand between fish in early development might differ from that

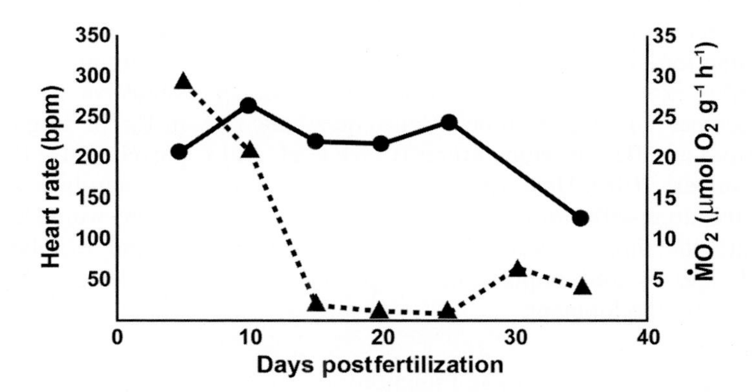

Fig. 14. Heart rate (*solid line*) and metabolic rate ($\dot{M}O_2$, *dashed line*) during ontogeny in larval blue gourami (*Trichopterus trichopodus*). Note that $\dot{M}O_2$ and heart rate are not positively coupled during early life stages, as seen in adult fish. Data from Blank, T.M., 2009. Cardio-respiratory ontogeny and the transition to bimodal respiration in an air-breathing fish, the blue gourami (Trichogaster trichopterus): morphological and physiological development in normoxia and hypoxia. In: Biological Sciences. University of North Texas, Denton, Texas, USA, Ph.D. and Mendez-Sanchez, J.F., 2015. Environmental Modulation of the Onset of Air-Breathing of the Siamese Fighting Fish and the Blue Gourami. University of North Texas, Denton, Texas, USA, Ph.D.

of adults (Rombough, 2011; Rombough and Moroz, 1997). The increase in energetic demand during development is also present in species that are more precocious at hatch. These species have a higher capacity to increase oxygen consumption to meet the presumed increase in physiological demand, as compared to altricial species that are less developed and have less active lifestyles at hatch (Killen et al., 2007).

Heart rate can be used for estimating metabolic rate in adult fish in situ, and also for fish species that are not good sustained swimmers (Norin and Clark, 2016). However, using f_H as a proxy for metabolic rate is unlikely to be extendable to fish embryos and larvae, especially given their early dependence upon cutaneous respiration (see Section 2.2). Moreover, the developmental trajectory of aerobic scope in fish with different lifestyles is indeed different (Killen et al., 2007). Since environmental pressures are often specific to a taxonomic group, a stressor could necessitate earlier regulatory capacity prior to hatch, depending on the fish's niche. For example, a highly hypoxia-tolerant species might develop a stronger link between metabolic rate and heart functions earlier than a less-tolerant species, based on the demands of the environment. Such questions would be very worthy of further investigation.

Despite a loose correlation between whole body metabolic rate and heart rate in developing fish, measurements of metabolic rate are well suited as a proxy for physical fitness. Knowing maximum metabolic rate ($\dot{M}O_{2max}$) provides insights into the aerobic capacity of a fish, and relates to the efficiency of the cardiovascular system. Adult fishes are widely used for such determinations, i.e., where aerobic scope is determined as maximum metabolic (i.e., after exhaustive exercise in a swim flume or exhaustive chasing—discussed below) minus routine or basal metabolic rate, or by charting metabolic rates at different temperatures (Clark et al., 2013; Farrell, 2013; Powell and Gamperl, 2016). However, a developmental series where aerobic scope is estimated along with cardiac development throughout ontogeny would help to delineate the link between cardiovascular development and metabolism. Indeed, species and population comparisons will likely depict adaptive responses in development.

5.4. Environment and Cardiac Function

Although the generation of physiological data on cardiovascular ontogeny in fish lags behind that of molecular data, there exists a firm physiological foundation which can be used in future studies to improve our understanding of environmental effects on cardiovascular development and performance. Development is altered by both biotic and abiotic environmental factors, and stressors can be dynamic, or fluctuate stochastically in type, number, intensity and duration. So too can stressors differentially affect different life stages. As such, the developing cardiovascular system becomes tuned to regulate critical functions in response to expected stressors, and as the need arises to meet physiological challenges. Beyond this, we see that novel stressors (at least in evolutionary terms) such as anthropogenic contaminants affect the development of the cardiovascular system, though mechanisms are in place to deal with some of these stressors as well.

Against this backdrop, we now consider the effects of temperature and oxygen availability, and water quality, on the developing heart.

5.4.1. Temperature

Temperature affects the timing of development and developmental success, not just in fishes but in a wide variety of organisms. In terms of cardiac function and its underlying metabolism, temperature is positively correlated with f_H and a variety of other cardiac parameters (e.g., contractility, morphology, etc.,) within the thermal limits of a species during development (Barrionuevo and Burggren, 1999; Denvir et al., 2008). Although increased temperature can decrease hatching time, interestingly, this shortened incubation period can be associated with a smaller adult (Lopez-Olmeda and Sanchez-Vazquez, 2011), and the response to temperature likely affects a variety of developing systems (Barrionuevo and Burggren, 1999). The temperature regime to which a fish is exposed is likely a critical variable to consider with regard to development, and strongly influences fish physiology in addition to behavior and species' distribution (Brown and Green, 2014; Lopez-Olmeda and Sanchez-Vazquez, 2011; Mueller et al., 2015b; Schnurr et al., 2014). Indeed, there is a Q_{10}[5] for virtually everything from body mass to energy assimilation in developing fish, and many aspects of cardiac function (Barrionuevo and Burggren, 1999; Lee et al., 2016; Mueller et al., 2015b; Schnurr et al., 2014; Somero, 2010). Compounding the situation, aspects of the fish's phenotype are differentially affected by temperature, and this results in different Q_{10} values for different traits. As an example, in larval and juvenile fishes the Q_{10} for somatogenesis is typically ~2.8, but ~2 for heart development (Crossley and Burggren, 2009). Further, the Q_{10} for stroke volume in zebrafish larvae prior to swim bladder inflation may be near 1–1.2, though heart rate and cardiac output may be nearer to 1.8–2.2 (Jacob et al., 2002). In both zebrafish and rainbow trout larvae, stroke volume is nearly temperature-independent, with a Q_{10} near 1 (Jacob et al., 2002; Mirkovic and Rombough, 1998; Pelster, 2002). As such, although heart rate is temperature-dependent, the resultant change in cardiac output may not be significantly affected due to a relatively stable stroke volume during early development

[5]Q_{10} is a measure of temperature sensitivity as defined by the change of a physiological variable over a 10°C change in temperature.

(Jacob et al., 2002). Beyond such effects of temperature on developmental time and physiological functions, exceeding upper thermal tolerance limits can cause morphological cardiovascular defects such as edema, hematoma, and delayed vessel growth, and result in mortality (Burggren and Bagatto, 2008; Rodrigues-Galdino et al., 2010).

Temperature appears to have an effect on vasculogenesis that goes beyond Q_{10}. At low temperatures slow arrhythmic heart rates can occur during early developmental stages that are critical for establishing normal blood circulation (Watanabe-Asaka et al., 2014). In southern-adapted species of Japanese medaka, colder incubation temperatures cause a bradycardia and an arrhythmic heart rate that affects normal cardiovascular development (Watanabe-Asaka et al., 2014). Compared to northern-adapted species, cold exposed southern medaka embryos also show severe retrograde blood flow, a reduction in hatching success, and severe edema and hematomas (Watanabe-Asaka et al., 2014). Finally, the cold-exposed southern medaka that hatched showed moderate retrograde blood flow, the degeneration of pectoral and tail fins presumably due to poor blood flow, and were unable to swim and died soon after hatching. Whether vascular development in other larval fishes is similarly affected by low (or high) temperatures remains to be determined.

Acclimation temperature is highly influential in setting the upper and lower temperature tolerances of larval and juvenile fishes, and in addition to genetic components, it can be definitive of the boundaries of eurythermal and stenothermal species traits (Lopez-Olmeda and Sanchez-Vazquez, 2011; McBryan et al., 2013). Not surprisingly, metabolic scope at temperature extremes is used as a proxy for understanding aerobic capacity. This link between metabolic performance and temperature tolerance is strongly correlated with cardiac performance (as described earlier), though little work has been done on developing fish in this respect. Complicating the collection of data on the effects of temperature on developing fish is that developmental plasticity itself can be temperature-dependent (Burggren and Blank, 2009; Burggren and Reyna, 2011). Further, stenothermal fish have been shown to have a increased sensitivity to temperature early in development, but become more stenothermal as they mature, though f_H sensitivity to temperature does not always parallel the connection between metabolic rate and temperature (Barrionuevo and Burggren, 1999). Additionally, since temperature is effective at altering developmental timing of so many aspects of physiology, comparisons between studies and even between treatment groups in the same study suffer from difficulties imposed by a dissociation of chronological and developmental age (Barrionuevo and Burggren, 1999; Mueller et al., 2015a). Still, the effects of temperature on development are profound, and plasticity evoked by temperature stress can result in varied outcomes depending on the timing and duration of the exposure. This results in differential effects on the development of cardiac and metabolic performance (Eme et al., 2015; Mueller et al., 2015b).

5.4.2. Water Chemistry

The water chemistry of aquatic environments is highly variable and differs greatly between bodies of water, and such properties are definitive of an ecological niche. Thus, physiochemical properties in aquatic environments are relative to a population. However, anthropogenic changes to the aquatic ecosystem are artificial selective pressures and unequivocally degrade water quality. Runoff from industrial and agricultural fields and river pollution cause hypoxic conditions by contributing to algal blooms, and climate change causes increases in temperature and acidity. Further, and even more prevalent, may be the more acute effects of pollution on water chemistry (Celander, 2011; Dubansky et al., 2013). We now consider specific types of water quality degradation, and their cardiovascular effects.

5.4.2.1. Hypoxia. Hypoxic exposure during development affects numerous aspects of physiology, leading to alterations in developmental trajectory and associated compensatory responses. While expression of some genes related to cardiac development becomes altered in response to hypoxic incubation (see Section 3.2.2), embryonic exposure to hypoxia in fishes has long-term negative effects on physiologic function that impact future swim performance and growth (Johnston et al., 2013; Miller et al., 2008, 2011). Chronic hypoxia also lowers f_H throughout development in multiple teleost species, aside from a general developmental delay and reduction in metabolic rate. Interestingly, rainbow trout embryos can show a slight tachycardia in response to acute hypoxia, and when exposed to chronic hypoxia they show an increase in the chronotropic response to beta-adrenergic stimulation early in development and a delay in the development of cholinergic control of the heart (Miller et al., 2011). In this case, it could be reasoned that tachycardia augments diffusive oxygen delivery by increasing convective oxygen transport capacity (Miller et al., 2011).

Hypoxia also has a significant effect on the onset of cardiovascular regulation in fishes by accelerating the onset of adrenergic and vasoconstrictive responses relative to the developmental program. For example, in zebrafish, the first vessel responses to phenylephrine occur earlier in larvae exposed to hypoxia during development compared to normoxic controls (Bagatto, 2005; Burggren and Bagatto, 2008). This effect of hypoxia-induced plasticity during growth is not surprising, since we see complex interactions that influence physiological function, along with the slow growth and depressed metabolism generally described as the effect of hypoxia on development.

The response to hypoxia at the cellular level is mediated by hypoxia-inducible factor-1 (HIF-1). The HIF-1 pathway is active as early as 24 h postfertilization in zebrafish, and early upregulation in response to hypoxia exposure has effects on hypoxia tolerance in larvae (Lu et al., 2014). Early parental hypoxia can actually increase hypoxia tolerance in the F1 generation of zebrafish (Strick et al., 1991). Thus, in a species that is only moderately tolerant to low oxygen, early hypoxic exposure may equip fish with an increased ability to handle hypoxia later in life. These findings align with the HIF-1 pathway's connection with the cardiovascular system, where increased HIF-1 pathway activity leads to enhanced cardiac activity, vascularization, erythrocyte number, and aerobic capacity at the tissue level (Dzijan-Horn et al., 2014).

In early life stages, it is thought that the metabolic demands of larval fishes are not tightly coupled with cardiac activity, and little change is seen in cardiac activity in response to hypoxia (as mentioned earlier) (Jacob et al., 2002). Indeed, we see a steadily decreasing oxygen carrying capacity at 3, 7, and 15 days postfertilization (dpf) in normoxic zebrafish (Dzijan-Horn et al., 2014). In contrast, 7 and 15 dpf larvae exposed to hypoxia showed significant enhancement of oxygen carrying capacity due to a combination of increased blood cell concentration and cardiac output (largely due to an increase in stroke volume) (Dzijan-Horn et al., 2014). Such increases in convective oxygen carrying capacity likely occur at earlier stages as well. Although early stage larvae may not rely heavily on convective oxygen transport, an increase in convective oxygen carrying capacity could represent a compensatory response to low PO_2 when faced with a smaller gradient for inward oxygen diffusion; as mentioned earlier in reference to rainbow trout showing a tachycardic response to chronic hypoxia (Miller et al., 2011). In zebrafish larvae, cardiac output is increased under hypoxic conditions, indicating an increased requirement/ or ability for convective delivery of oxygen at or near hatching (Jacob et al., 2002). Efficient convective transport of oxygen could enhance internal transport oxygen by diffusion (Cypher et al., 2015; Rombough, 2002, 2007). In *Fundulus heteroclitus*, a relatively hypoxia-tolerant fish, though early growth is affected only by severe hypoxia (<3 kPa), growth rate recovers in later posthatch stages indicating that some unknown compensatory mechanism(s) exist in these fish (Rees et al., 2012). Future studies should examine the ontogeny of multiple aspects of cardiovascular development in such species that are highly tolerant of hypoxia, and precocious at hatch, to investigate the timing of cardiovascular regulation as an adaptive response to low oxygen environments.

5.4.2.2. Toxicants. We know little about how anthropogenic environmental perturbations alter the relative timing of the development of cardiovascular regulation of fish (e.g., heterokairy), despite the known cardiotoxicity of fish to numerous chemicals including metals, endocrine disruptors, pesticides, and organics (Di Giulio and Clark, 2015; Jezierska et al., 2009; King-Heiden et al., 2012; Kopf and Walker, 2009; Santos et al., 2014; Weis and Weis, 1997: Chapter 6, Volume 36B: Incardona and Scholz, 2017). Cardiovascular responses to toxicant exposure include altered f_H, arrhythmia, loss of contractility, ischemia, hematoma and ascites, in addition to gross defects in morphology (Carney et al., 2006; Weis and Weis, 1974; Weis and Weis, 1977; Yang et al., 2009: also see Chapter 6, Volume 36B: Incardona and Scholz, 2017). Interestingly, similar cardiovascular perturbations appear after exposure to a wide range of toxicants, suggesting basic patterns of disruption of cardiovascular development. Indeed, zebrafish with mutations in the von Hippel–Lindau tumor suppressor can exhibit cardiac dilation, cardiomegaly, and stretched cardiomyocytes along with severe edema (Pelster et al., 2010), and fish incubated at high temperatures and hypoxia also display these cardiac defects (as described earlier in Section 5.4.2.1). Further, cardiotoxicity may appear concurrently with skeletal defects, osmoregulatory dysfunction, neural defects, and decreased growth and survival (Carney et al., 2006). This broad suite of responses to multiple toxicants potentially complicates the specific cause–effect relationship, requiring future detailed studies following the path from gene to organismal phenotype.

Bioassays that evaluate embryotoxicity in fish have become popular for evaluating the toxicity of drugs and pollutants, a process that started in the 1970s (Wells et al., 1997). Indeed, the scientific literature on this subject is voluminous in both the type and nature of toxicant studied and the experimental methods used. From an environmental health perspective, we pay particular attention to the effects of ubiquitous dioxin-like compounds (DLCs) such as polycyclic aromatic hydrocarbons (PAHs), polychlorinated biphenols (PCBs), and polychlorinated dibenzodioxins and furans (PCDD/Fs) on the development of the piscine cardiovascular system. We focus on these compounds, as this fairly diverse group of chemicals is especially cardiotoxic and well characterized with respect to the ontogeny of the cardiovascular system of fishes, relative to other groups of toxicants (metals, pesticides, etc.). As such, the timeline of cardiotoxicity from DLCs has been extensively reviewed (Carney et al., 2006; Kopf and Walker, 2009).

The effects of DLC toxicity on the cardiovascular system become obvious at the onset of organogenesis. Numerous morphological defects occur in the developing cardiovascular system of fish exposed to DLCs, including improper looping of the heart, reduced ventricle size, thinning of chamber walls (Carney et al., 2004), altered chamber shape and orientation (Incardona et al., 2015), arrhythmias, and a reduction of contractility (Brette et al., 2014). The most obvious first symptoms of DLC toxicity in fish development are bradycardia and reduced blood flow (King-Heiden et al., 2012). Vascular leakage from the medial yolk vein and a decrease in vascular growth are among the earliest obvious effects to appear, following DNA degradation and apoptosis in the embryonic vasculature (Kopf and Walker, 2009). Blood is commonly seen outside of the vasculature and often described as hemorrhages (Carney et al., 2006; Jung et al., 2013; Raimondo et al., 2014). Cell–cell junctions play a crucial role in regulating vascular activity, contributing greatly to both structural integrity and vascular homeostasis, as defects in the junction proteins are thought to lead to hemorrhaging (Gore et al., 2012).

Dioxin like compounds also lead to defects in cardiac physiology early in development. A decrease in blood flow precedes the characteristic yolk sac and pericardial edema seen in virtually all vertebrates exposed to DLCs (Kopf and Walker, 2009). Decreased blood perfusion obviously plays a critical role in the developmental toxicity of DLCs (King-Heiden et al., 2012). However, reduced cardiac output occurs before a reduction in blood flow (Carney et al., 2006; Incardona et al., 2011), and defects in cardiac morphology generally precede edema (Antkiewicz et al., 2005; Incardona et al., 2004; Incardona and Scholz, 2016). Knockdown of cardiac troponin T by morpholino injection during embryogenesis eliminates the heartbeat, which leads to the same suite of defects associated with cardiotoxicity of DLCs and other toxicants, indicating that disruption of cardiac function is enough to cause DLC-induced cardiotoxicity (Incardona et al., 2004). Collectively, these data suggest that altered heart performance is chiefly responsible for the gross defects in overall cardiovascular function, though the molecular foundations of these effects are multifaceted and as yet to be determined.

Dioxin like compounds are ligands for intracellular aryl hydrocarbon receptors (AHR), and upon binding result in a battery of genes being transcribed that generate proteins involved in the detoxification of DLCs. This pathway is thought to be a key mediator of cardiovascular effects of DLCs. Accordingly, evidence of AHR pathway induction is seen at the onset of organogenesis and circulation in both zebrafish and killifish (Andreasen et al., 2002; Powell et al., 2000), coinciding with the onset of DLC-induced cardiovascular defects. Activation of the AHR pathway in vertebrates exposed to DLCs is universal, and results in the production of a battery of proteins involved in DLC metabolism and excretion. As such, increased induction of the AHR and its downstream products in the vascular endothelium are hallmarks of exposure to DLCs (Dubansky et al., 2013; Sarasquete and Segner, 2000; Woodin et al., 1997). Due to the conserved nature of this pathway and the link between AHR in cardiotoxicity and exposure to DLCs, this pathway is the primary molecular target with regards to the study of the toxic effects of DLCs in fishes. However, the role of the AHR is not clear since some DLCs exert toxicity through AHR-independent mechanisms (Incardona et al., 2005; Incardona and Scholz, 2016). For example, killifish populations on the Atlantic and Gulf of Mexico coasts have acquired resistance to DLC-induced cardiac defects that coincide with a decrease in AHR activity, and some of these fish are found to be cross-resistant to compounds that are not metabolized by the AHR pathway, indicating that other mechanisms are also strongly influential in cardiotoxicity (Di Giulio and Clark, 2015; Jia and Burggren, 1997; Oziolor et al., 2014, 2016). Still, it is clear that exposure to toxicants causes cardiac dysfunction, and it is not surprising that even low exposure concentrations of DLCs during development can exert intragenerational epigenetic effects on cardiovascular morphology and performance later in life (Corrales et al., 2014; Incardona et al., 2015). This points to lasting effects caused by disruption in the developmental success of the cardiovascular system.

6. CONCLUSION AND SUGGESTED DIRECTIONS FOR FUTURE RESEARCH

Development of cardiovascular form and function in larval fishes is a highly dynamic process. Some aspects of development of the general circulation are highly conserved among vertebrates. Yet, other aspects (e.g., the rapidity of development, physiological rates) are highly variable even between closely related species due to both taxonomic differences and environmental factors (especially temperature). Despite these differences, some general principles, opportunities, and challenges emerge.

6.1. Key Importance of the Cardiovascular System

Successful formation and onset of circulatory function is absolutely fundamental to subsequent larval development and growth, whether its role is in shaping the circulation or in actually providing convective transport of respiratory gases, nutrients, wastes, and chemical signals. Alteration of the form and function of the cardiovascular system during early development can potentially translate into long-term impacts to physical fitness, and thus, survivorship (see later). Unfortunately, much of information in this arena is correlational rather than causative. A prime example is toxicological studies. Pericardial edema is a frequent symptom of exposure to toxicants, and is also a harbinger of ensuing death. Yet, it is unclear whether pericardial edema causes death, or is merely symptomatic of developing failure of osmoregulation and circulation. Future experiments should begin to explore causation even as correlations continue to be explored.

6.2. Assessment of Fitness

Ultimately, changes in fitness best indicate the impact of adjustments to the cardiovascular system of fishes. While ecologists would use successful reproduction as a measure of fitness, if a larval fish does not survive it obviously cannot subsequently reproduce. Thus, survival and factors related to survival are good measures of ultimate fitness in larval fishes. In this regard, aerobic performance and the behavior of early stage fish have often been examined in the context of survival and fitness. Critical swimming speed assays are tractable and can be used to establish such a link between the cardiovascular system and aerobic performance. Long established for adult fish, such experimental trials are difficult to complete for small larval and juvenile fish, which may not be as prone to sustained swimming in experimental conditions. Several techniques have been used to gauge maximal cardiac function, and include inducing cardiac work through pharmaceutical manipulation to achieve a maximal f_H, forced swimming, and temperature extremes. As the heart reaches maximal working capacity, f_H, cardiac output, metabolic rate, and other variables are measured to generate a working relationship between physical activity and potential aerobic capacity, and thus ultimately a relationship with fitness. Future studies that focus on the performance of larval fishes in controlled active states, will thus, help establish the role and importance of the cardiovascular system in larval fishes.

6.3. Vulnerability to Environment

Early life experiences affect morphological and physiological development of fish later in life. As an example, although larval swim training had no significant effects on cardiac output, it did enhance the mitochondrial density and capillarizationof red and intermediate muscle fibers in zebrafish (Pelster et al., 2003). Though no change was found in heart activity in that study, the results indicate an increase in the ability for aerobic capacity. Indeed, early swim training is also correlated with enhanced survival in hypoxia in zebrafish larvae, and a decrease in mass-specific oxygen consumption in swim trained free-swimming larvae (Bagatto et al., 2001). Interestingly, swimming stamina in zebrafish is heritable, with larvae bred from fish that have higher swimming stamina having a lower resting heart rate compared to larvae bred from lower stamina fish (Gore and Burggren, 2012). Conversely, exposure to even sublethal concentrations of toxicants can result in a decrease in cardiac performance and dysmorphogenesis of the heart later in life (Hicken et al., 2011; Mager et al., 2014). As such, swimming ability is an effective indicator of a fish's ability to perform physically, and will be a useful parameter to focus on in future studies; though the reader is cautioned when comparing between species and life stages that may be reflective of specific abilities and developmental effects.

6.4. Technological Advances

Specialized instrumentation is often required to measure cardiac activity in developing fish, especially to track variables that would otherwise be routine in larger animals. The growing popularity of using early life-stage fish in biomedical research has driven the establishment of methodologies to measure a variety of cardiac endpoints (as mentioned earlier in Section 5.2), where microtechniques have enabled measurements of pressure and electrical signals in the heart of embryonic and larval fish. Such data not only help to measure effects in experimental treatments, but also continue to establish links between what can be learned in fish and the application of that knowledge to other vertebrate groups. Further, image analysis is time-consuming when comparing large numbers of early stage fish. Automation of imaging enables large numbers of individual fish to be processed both robotically for sorting and orientation (Pardo-Martin et al., 2010), the analysis of vasculature function, f_H, morphology, and cardiac output when combined with software-driven techniques (Bagatto and Burggren, 2006; Burggren and Fritsche, 1995; Denvir et al., 2008; Schwerte and Pelster, 2000; Watkins et al., 2012), and microsurgery and genetic or pharmaceutical screening (Pardo-Martin et al., 2010; Spomer et al., 2012). Future studies that more actively incorporate microtechniques, automation, and software-driven techniques into assessments of phenotypic variations in embryonic and larval fish will be well rewarded.

6.5. Need for "Vertical Integration" From Molecules to Physiology, Behavior, and Ecology

The field of larval fish physiology suffers from—or, perhaps more appropriately is not benefitting from—an all-too-common stratification of disciplines. A frequent pattern is that one lab studies genes, another studies morphology, and yet another physiology. Future studies by collaborating laboratories will be well advised to follow specific phenomena from the gene to whole organism. The need for such collaboration is underscored by the fact that substantial changes in molecular phenotype (say, as a result of an environmental stressor) may or may not translate into adjustments in cellular phenotype. And cellular phenotypic changes may not translate into tissue phenotypes, and so on all the way up to organ system modifications. Even if there are morphological phenotypic modifications, easily deployed changes in physiological performance or behavior may partially or fully mitigate phenotypic changes at the most basic molecular level. Such a suggestion is not novel—e.g., Noble (2015)—but demands repeating as we consider future experimentation on larval fishes.

6.6. Epigenetic Influences

Though beyond the scope of this review, epigenetic influences both within and across generations are increasingly being viewed as a source of variation especially in physiological studies (Burggren, 2014). Larval fishes not only show considerable phenotypic plasticity because of intragenerational changes in gene expression, but also show transgenerational epigenetic inheritance of physiological characteristics and toxicant susceptibility/resistance (Corrales et al., 2014; Di Giulio and Clark, 2015; Ho and Burggren, 2012). Such outcomes, underpinned by epigenetic mechanisms of modified gene expression and nongenetic inheritance, may have an important role in survival, speciation, and evolution by helping organisms temporarily "bridge" unfavorable conditions that may last for a few generations (Burggren, 2016). At a minimum, future studies would be well advised to understand, whenever possible, the previous history (going back multiple generations) of the animals they experiment upon.

A final reflection is that our observations and understanding of the physiology of larval fishes are confined to an extraordinarily small fraction—probably <0.005%—of all known fish species. Given the wonderful yet confounding diversity of fishes, we have many fascinating aspects of larval fish physiology yet to be discovered.

REFERENCES

Akagi, J., Khoshmanesh, K., Hall, C.J., Cooper, J.M., Crosier, K.E., Crosier, P.S., Wlodkowic, D., 2013. Fish on chips: microfluidic living embryo array for accelerated in vivo angiogenesis assays. Sensors Actuators B Chem. 189, 11–20.

Andreasen, E.A., Spitsbergen, J.M., Tanguay, R.L., Stegeman, J.J., Heideman, W., Peterson, R.E., 2002. Tissue-specific expression of AHR2, ARNT2, and CYP1A in zebrafish embryos and larvae: effects of developmental stage and 2,3,7,8-tetrachlorodibenzo-p-dioxin exposure. Toxicol. Sci. 68, 403–419.

Andrés-Delgado, L., Mercader, N., 2016. Interplay between cardiac function and heart development. Biochim. Biophys. Acta (BBA) Mol. Cell Res. 1863, 1707–1716.

Andrewartha, S.J., Tazawa, H., Burggren, W.W., 2011. Hematocrit and blood osmolality in developing chicken embryos (Gallus gallus): in vivo and in vitro regulation. Respir. Physiol. Neurobiol. 179, 142–150.

Antinucci, P., Hindges, R., 2016. A crystal-clear zebrafish for in vivo imaging. Sci Rep 6. 29490
Antkiewicz, D.S., Burns, C.G., Carney, S.A., Peterson, R.E., Heideman, W., 2005. Heart malformation is an early response to TCDD in embryonic zebrafish. Toxicol. Sci. 84, 368–377.
Armstrong, E.J., Bischoff, J., 2004. Heart valve development: endothelial cell signaling and differentiation. Circ. Res. 95, 459–470.
Arnaout, R., Ferrer, T., Huisken, J., Spitzer, K., Stainier, D.Y.R., Tristani-Firouzi, M., Chi, N.C., 2007. Zebrafish model for human long QT syndrome. Proc. Natl. Acad. Sci. U. S. A. 104, 11316–11321.
Arrenberg, A.B., Stainier, D.Y., Baier, H., Huisken, J., 2010. Optogenetic control of cardiac function. Science 330, 971–974.
Asnani, A., Peterson, R.T., 2014. The zebrafish as a tool to identify novel therapies for human cardiovascular disease. Dis. Model. Mech. 7, 763–767.
Auman, H.J., Coleman, H., Riley, H.E., Olale, F., Tsai, H.J., Yelon, D., 2007. Functional modulation of cardiac form through regionally confined cell shape changes. PLoS Biol. 5, 0604–0615.
Avise, J.C., Jones, A.G., Walker, D., DeWoody, J.A., 2002. Genetic mating systems and reproductive natural histories of fishes: lessons for ecology and evolution. Annu. Rev. Genet. 36, 19–45.
Bagatto, B., 2005. Ontogeny of cardiovascular control in zebrafish (Danio rerio): effects of developmental environment. Comp. Biochem. Physiol. A Mol. Integr. Physiol. 141, 391–400.
Bagatto, B., Burggren, W., 2006. A three-dimensional functional assessment of heart and vessel development in the larva of the zebrafish (Danio rerio). Physiol. Biochem. Zool. 79, 194–201.
Bagatto, B., Pelster, B., Burggren, W.W., 2001. Growth and metabolism of larval zebrafish: effects of swim training. J. Exp. Biol. 204, 4335–4343.
Bakkers, J., 2011. Zebrafish as a model to study cardiac development and human cardiac disease. Cardiovasc. Res. 91, 279–288.
Balashov, N.V., Soltitskij, V.V., 1991. On the development of cardiac muscle reactivity in Black Sea mullets at early stages of ontogenesis. In: Kulikova, N.G. (Ed.), Mullet Culture in the Azov and Black Sea Basins. VNIRO, Moscow, pp. 51–59.
Balashov, N.V., Soltitskii, V.V., Makukhina, L.I., 1991. The effect of noradrenaline and acetylcholine on cardiac muscle of the golden mullett Liza auratus at early stages of ontogenesis. Zh. Evol. Biokhim. Fiziol. 27, 255–257.
Balon, E.K., 1999. Alternative ways to become a juvenile or a definitive phenotype (and on some persisting linguistic offenses). Environ. Biol. Fish 56, 17–38.
Bark Jr., D.L., Johnson, B., Garrity, D., Dasi, L.P., 2017. Valveless pumping mechanics of the embryonic heart during cardiac looping: pressure and flow through micro-PIV. J. Biomech. 50, 50–55.
Barrionuevo, W.R., Burggren, W.W., 1999. O_2 consumption and heart rate in developing zebrafish (Danio Rerio): influence of temperature and ambient O2. Am. J. Phys. 276, R505–13.
Beis, D., Bartman, T., Jin, S.-W., Scott, I.C., D'Amico, L.A., Ober, E.A., Verkade, H., Frantsve, J., Field, H.a., Wehman, A., et al., 2005. Genetic and cellular analyses of zebrafish atrioventricular cushion and valve development. Development 132, 4193–4204.
Belanger, S.E., Balon, E.K., Rawlings, J.M., 2010. Saltatory ontogeny of fishes and sensitive early life stages for ecotoxicology tests. Aquat. Toxicol. 97, 88–95.
Blank, T.M., 2009. Cardio-respiratory ontogeny and the transition to bimodal respiration in an air-breathing fish, the blue gourami (Trichogaster trichopterus): morphological and physiological development in normoxia and hypoxia. In: Biological Sciences. University of North Texas, Denton, Texas, USA (Ph.D.).

Blank, T., Burggren, W.W., 2014. Hypoxia-induced developmental plasticity of the gills and air-breathing organ of the air-breathing fish blue gourami (*Trichopodus trichopterus*). J. Fish Biol. 84, 808–826.

Boselli, F., Freund, J.B., Vermot, J., 2015. Blood flow mechanics in cardiovascular development. Cell. Mol. Life Sci. 72, 2545–2559.

Branum, S.R., Yamada-Fisher, M., Burggren, W., 2013. Reduced heart rate and cardiac output differentially affect angiogenesis, growth, and development in early chicken embryos (Gallus domesticus). Physiol. Biochem. Zool. 86, 370–382.

Brauner, C.J., Rombough, P.J., 2012. Ontogeny and paleophysiology of the gill: new insights from larval and air-breathing fish. Respir. Physiol. Neurobiol. 184, 293–300.

Brette, F., Machado, B., Cros, C., Incardona, J.P., Scholz, N.L., Block, B.A., 2014. Crude oil impairs cardiac excitation-contraction coupling in fish. Science 343, 772–776.

Broemnimann, D., Djukic, T., Triet, R., Dellenbach, C., Saveljic, I., Rieger, M., Rohr, S., Filipovic, N., Djonov, V., 2016. Pharmacological modulation of hemodynamics in adult zebrafish in vivo. PLoS One 11, e0150948.

Browman, H.I., 1989. Embryology, ethology and ecology of ontogenetic critical periods in fish. Brain Behav. Evol. 34, 5–12.

Brown, C.A., Green, C.C., 2014. Metabolic and embryonic responses to terrestrial incubation of Fundulus grandis embryos across a temperature gradient. J. Fish Biol. 84, 732–747.

Brown, D.R., Samsa, L.A., Qian, L., Liu, J., 2016. Advances in the study of heart development and disease using zebrafish. J. Cardiovasc. Dev. Dis. 3, 13.

Burggren, W.W., 2000. Developmental physiology, animal models, and the August Krogh principle. Zool. Anal. Complex Syst. 102, 148–156.

Burggren, W.W., 2004. What is the purpose of the embryonic heart beat? Or how facts can ultimately prevail over physiological dogma. Physiol. Biochem. Zool. 77, 333–345.

Burggren, W.W., 2013. Cardiovascular development and angiogenesis in the early vertebrate embryo. Cardiovasc. Eng. Technol. 4, 234–245.

Burggren, W.W., 2014. Epigenetics as a source of variation in comparative animal physiology—or—Lamarck is lookin' pretty good these days. J. Exp. Biol. 217, 682–689.

Burggren, W.W., 2016. Epigenetic inheritance and its role in evolutionary biology: re-evaluation and new perspectives. Biology 5, 24.

Burggren, W., Bagatto, B., 2008. Cardiovascular anatomy and physiology of Larval fishes. In: Finn, N., Kapoor, B.G. (Eds.), Fish Larval Physiology. New Delhi, Oxford & IBH Publishing Co. Pvt. Ltd.

Burggren, W., Blank, T., 2009. Physiological study of larval fishes: challenges and opportunities. Sci. Mar. 2009, 99–110.

Burggren, W., Fritsche, R., 1995. Cardiovascular measurements in animals in the milligram range. Braz. J. Med. Biol. Res. 28, 1291–1305.

Burggren, W.W., Johansen, K., 1986. Circulation and respiration in lungfishes (Dipnoi). J. Morphol. 190 (Suppl. S1), 217–236.

Burggren, W.W., Reyna, K.S., 2011. Developmental trajectories, critical windows and phenotypic alteration during cardio-respiratory development. Respir. Physiol. Neurobiol. 178, 13–21.

Burggren, W.W., Territo, P.R., 1995. Early development of blood oxygen transport. In: Houston, J., Coates, J. (Eds.), Hypoxia and Brain. Queen City Printer, Burlington, VT, pp. 45–56.

Burggren, W.W., Warburton, S.J., 1994. Patterns of form and function in developing hearts: contributions from non-mammalian vertebrates. Cardioscience 5, 183–191.

Burggren, W., Farrell, A.P., Lillywhite, H., 1997. Vertebrate cardiovascular systems. In: Dantzler, W. (Ed.), Handbook of Comparative Physiology. Oxford University Press, Oxford, pp. 215–308.

Burggren, W.W., Warburton, S.J., Slivkoff, M.D., 2000. Interruption of cardiac output does not affect short-term growth and metabolic rate in day 3 and 4 chick embryos. J. Exp. Biol. 203, 3831–3838.

Burggren, W.W., Khorrami, S., Pinder, A., Sun, T., 2004. Body, eye, and chorioallantoic vessel growth are not dependent on cardiac output level in day 3-4 chicken embryos. Am. J. Phys. Regul. Integr. Comp. Phys. 287, R1399–406.

Burggren, W.W., Christoffels, V.M., Crossley 2nd, D.A., Enok, S., Farrell, A.P., Hedrick, M.S., Hicks, J.W., Jensen, B., Moorman, A.F., Mueller, C.A., et al., 2014. Comparative cardiovascular physiology: future trends, opportunities and challenges. Acta Physiol (Oxford) 210, 257–276.

Burggren, W.W., Martinez Bautista, G., Camarillo Coop, S., Marquez Couturier, G., Paramo Delgadillo, S., Alvarez Gonzalez, C.A., 2016. Developmental cardiorespiroatory physiology of the air breathing tropical gar, Atractosteus tropicus, (in press). Am. J. Phys. Regul. Integr. Comp. Phys. 311 (4), R689–R701.

Butcher, J.T., McQuinn, T.C., Sedmera, D., Turner, D., Markwald, R.R., 2007. Transitions in early embryonic atrioventricular valvular function correspond with changes in cushion biomechanics that are predictable by tissue composition. Circ. Res. 100, 1503–1511.

Buxboim, A., Ivanovska, I.L., Discher, D.E., 2010. Matrix elasticity, cytoskeletal forces and physics of the nucleus: how deeply do cells 'feel' outside and in? J. Cell Sci. 123, 297–308.

Campuzano-Caballero, J.C., Uribe, M.C., 2014. Structure of the female gonoduct of the viviparous teleost Poecilia reticulata (Poeciliidae) during nongestation and gestation stages. J. Morphol. 275, 247–257.

Carney, S.A., Peterson, R.E., Heideman, W., 2004. 2,3,7,8-tetrachlorodibenzo-p-dioxin activation of the aryl hydrocarbon receptor/aryl hydrocarbon receptor nuclear translocator pathway causes developmental toxicity through a CYP1A-independent mechanism in zebrafish. Mol. Pharmacol. 66, 512–521.

Carney, S.A., Prasch, A.L., Heideman, W., Peterson, R.E., 2006. Understanding dioxin developmental toxicity using the zebrafish model. Birth Defects Res. Part A Clin. Mol. Teratol. 76, 7–18.

Cavanaugh, A.M., Huang, J., Chen, J.N., 2015. Two developmentally distinct populations of neural crest cells contribute to the zebrafish heart. Dev. Biol. 404, 103–112.

Celander, M.C., 2011. Cocktail effects on biomarker responses in fish. Aquat. Toxicol. 105, 72–77.

Chappell, J.C., Bautch, V.L., 2010. Vascular development: genetic mechanisms and links to vascular disease. Curr. Top. Dev. Biol. 90, 43–72.

Chen, J.-N., Fishman, C.M., 1997. Genetic dissection of heart development. In: Burggren, W.W., Keller, B.B. (Eds.), Development of Cardiovascular Systems: Molecules to Organisms. Cambridge University Press, Cambridge.

Chi, N.C., Shaw, R.M., Jungblut, B., Huisken, J., Ferrer, T., Arnaout, R., Scott, I., Beis, D., Xiao, T., Baier, H., et al., 2008. Genetic and physiologic dissection of the vertebrate cardiac conduction system. PLoS Biol. 6, 1006–1019.

China, V., Holzman, R., 2014. Hydrodynamic starvation in first-feeding larval fishes. Proc. Natl. Acad. Sci. U. S. A. 111, 8083–8088.

Chung, A.S., Ferrara, N., 2011. Developmental and pathological angiogenesis. Annu. Rev. Cell Dev. Biol. 27, 563–584.

Clark, T.D., Sandblom, E., Jutfelt, F., 2013. Aerobic scope measurements of fishes in an era of climate change: respirometry, relevance and recommendations. J. Exp. Biol. 216, 2771–2782.

Clough, G.F., 2015. Developmental conditioning of the vasculature. Compr. Physiol. 5, 397–438.

Collins, M.M., Stainier, D.Y., 2016. Organ function as a modulator of organ formation: lessons from zebrafish. Curr. Top. Dev. Biol. 117, 417–433.

Colman, J.R., Twiner, M.J., Hess, P., McMahon, T., Satake, M., Yasumoto, T., Doucette, G.J., Ramsdell, J.S., 2005. Teratogenic effects of azaspiracid-1 identified by microinjection of Japanese medaka (Oryzias latipes) embryos. Toxicon 45, 881–890.

Cooper, C.A., Litwiller, S.L., Murrant, C.L., Wright, P.A., 2012. Cutaneous vasoregulation during short- and long-term aerial acclimation in the amphibious mangrove rivulus, Kryptolebias marmoratus. Comp. Biochem. Physiol. B Biochem. Mol. Biol. 161, 268–274.

Corrales, J., Thornton, C., White, M., Willett, K.L., 2014. Multigenerational effects of benzo[a] pyrene exposure on survival and developmental deformities in zebrafish larvae. Aquat. Toxicol. 148, 16–26.

Craig, M.P., Sumanas, S., 2016. ETS transcription factors in embryonic vascular development. Angiogenesis 19, 275–285.

Crespi, B.J., Teo, R., 2002. Comparative phylogenetic analysis of the evolution of semelparity and life history in salmonid fishes. Evolution 56, 1008–1020.

Crossley 2nd, D.A., Burggren, W.W., 2009. Development of cardiac form and function in ectothermic sauropsids. J. Morphol. 270, 1400–1412.

Crossley, D.A.I., Burggren, W.W., Reiber, C., Altimiras, J., Rodnick, K.J., 2016. Mass transport: circulatory system with emphasis on non endothermic species. Compr. Physiol. https://doi.org/10.1002/cphy.c150010 (in press).

Cypher, A.D., Ickes, J.R., Bagatto, B., 2015. Bisphenol A alters the cardiovascular response to hypoxia in Danio rerio embryos. Comp. Biochem. Physiol. C Toxicol. Pharmacol. 174, 39–45.

Dahl Ejby Jensen, L., Cao, R., Hedlund, E.M., Soll, I., Lundberg, J.O., Hauptmann, G., Steffensen, J.F., Cao, Y., 2009. Nitric oxide permits hypoxia-induced lymphatic perfusion by controlling arterial-lymphatic conduits in zebrafish and glass catfish. Proc. Natl. Acad. Sci. U. S. A. 106, 18408–18413.

Dahme, T., Katus, H.A., Rottbauer, W., 2009. Fishing for the genetic basis of cardiovascular disease. Dis. Model. Mech. 2, 18–22.

Danos, N., 2012. Locomotor development of zebrafish (Danio rerio) under novel hydrodynamic conditions. J. Exp. Zool. A Ecol. Genet. Physiol. 317, 117–126.

de Pater, E., Clijsters, L., Marques, S.R., Lin, Y.-F., Garavito-Aguilar, Z.V., Yelon, D., Bakkers, J., 2009. Distinct phases of cardiomyocyte differentiation regulate growth of the zebrafish heart. Development 136, 1633–1641.

De Pater, E., Ciampricotti, M., Priller, F., Veerkamp, J., Strate, I., Smith, K., Lagendijk, A.K., Schilling, T.F., Herzog, W., Abdelilah-Seyfried, S., et al., 2012. Bmp signaling exerts opposite effects on cardiac differentiation. Circ. Res. 110, 578–587.

Denvir, M.A., Tucker, C.S., Mullins, J.J., 2008. Systolic and diastolic ventricular function in zebrafish embryos: influence of norepenephrine, MS-222 and temperature. BMC Biotechnol. 8, 1–8.

Desnitskiy, A.G., 2015. On the features of embryonic cleavage in diverse fish species. Russ. J. Dev. Biol. 46, 326–332.

Dewar, H., Brill, R.W., Olson, K.R., 1994. Secondary circulation of the vascular heat exchangers in skipjack tuna, Katsuwonus pelamis. J. Exp. Zool. 269, 566–570.

Dhillon, S.S., Dóró, É., Magyary, I., Egginton, S., Sík, A., Müller, F., 2013. Optimisation of embryonic and larval ECG measurement in zebrafish for quantifying the effect of QT prolonging drugs. PLoS One 8, e60552.

Di Giulio, R.T., Clark, B.W., 2015. The Elizabeth River story: a case study in evolutionary toxicology. J. Toxicol. Environ. Health B Crit. Rev. 18, 259–298.

Dubansky, B., Whitehead, A., Miller, J., Rice, C.D., Galvez, F., 2013. Multi-tissue molecular, genomic, and developmental effects of the Deepwater horizon oil spill on resident Gulf killifish (*Fundulus grandis*). Environ. Sci. Technol. 47, 5074–5082.

Dzijan-Horn, M., Langwieser, N., Groha, P., Bradaric, C., Linhardt, M., Bottiger, C., Byrne, R.A., Steppich, B., Koppara, T., Godel, J., et al., 2014. Safety and efficacy of a potential treatment algorithm by using manual compression repair and ultrasound-guided thrombin injection for the management of iatrogenic femoral artery pseudoaneurysm in a large patient cohort. Circ. Cardiovasc. Interv. 7, 207–215.

Eddy, F.B., 2005. Role of nitric oxide in larval and juvenile fish. Comp. Biochem. Physiol. A Mol. Integr.Physiol. 142, 221–230.

Einum, S., Fleming, I.A., 2007. Of chickens and eggs: diverging propagule size of iteroparous and semelparous organisms. Evolution 61, 232–238.

Ellertsdóttir, E., Lenard, A., Blum, Y., Krudewig, A., Herwig, L., Affolter, M., Belting, H.G., 2010. Vascular morphogenesis in the zebrafish embryo. Dev. Biol. 341, 56–65.

Eme, J., Mueller, C.A., Manzon, R.G., Somers, C.M., Boreham, D.R., Wilson, J.Y., 2015. Critical windows in embryonic development: shifting incubation temperatures alter heart rate and oxygen consumption of Lake Whitefish (Coregonus clupeaformis) embryos and hatchlings. Comp. Biochem. Physiol. A Mol. Integr. Physiol. 179, 71–80.

Evans, S.M., Yelon, D., Conlon, F.L., Kirby, M.L., 2010. Myocardial lineage development. Circ. Res. 107, 1428–1444.

Farrell, A.P., 2013. Aerobic scope and its optimum temperature: clarifying their usefulness and limitations—correspondence on J. Exp. Biol. 216, 2771-2782. J. Exp. Biol. 216, 4493–4494.

Farrell, A.P., Smith, F., 2017. Cardiac form, function and physiology. In: Gamperl, A.K., Gillis, T.E., Farrell, A.P., Brauner, C.J. (Eds.), The Cardiovascular System: Morphology. In: Control and Function. Fish Physiology, vol. 36A. Academic Press, San Diego, pp. 155–264.

Feder, M.E., Burggren, W.W., 1985. Cutaneous gas exchange in vertebrates: design, patterns, control and implications. Biol. Rev. Camb. Philos. Soc. 60, 1–45.

Ferrara, N., 2001. Role of vascular endothelial growth factor in regulation of physiological angiogenesis. Am. J. Phys. Cell Physiol. 280, C1358–66.

Fisher, S.A., Burggren, W.W., 2007. Role of hypoxia in the evolution and development of the cardiovascular system. Antioxid. Redox Signal. 9, 1339–1352.

Francisco Simão, M., Pérez Camps, M., García-Ximénez, F., 2007. Short communication. Zebrafish embryo development can be reversibly arrested at the MBT stage by exposure to a water temperature of 16°C. Span. J. Agric. Res. 5, 181–185.

Fransen, M.E., Lemanski, L.F., 1989. Studies of heart development in normal and cardiac lethal mutant axolotls: a review. Scanning Microsc. 3, 1101–1115.

Frazer, H.A., Ellis, M., Huveneers, C., 2012. Can a threshold value be used to classify chondrichthyan reproductive modes: systematic review and validation using an oviparous species. PLoS One 7, e50196.

Fritsche, R., Schwerte, T., Pelster, B., 2000. Nitric oxide and vascular reactivity in developing zebrafish, Danio rerio. Am. J. Phys. Regul. Integr. Comp. Phys. 279, R2200–R2207.

Fuiman, L.A., 2002. Special considerations of fish eggs and larvae. In: Fuiman, L.A., Werner, R.G. (Eds.), Fishery Science: The Unique Contributions of Early Life Stages. Blackwell Science, Oxford, pp. 1–32.

Fuiman, L., Batty, R., 1997. What a drag it is getting cold: partitioning the physical and physiological effects of temperature on fish swimming. J. Exp. Biol. 200, 1745–1755.

Gallego, A., Durán, A.C., De Andrés, A.V., Navarro, P., Muñoz-Chápuli, R., 1997. Anatomy and development of the sinoatrial valves in the dogfish (Scyliorhinus canicula). Anat. Rec. 248, 224–232.

Gamperl, K.A., Shiels, A.H., 2014. Cardiovascular system. In: Evans, D.H., Claiborne, B.J., Currie, S. (Eds.), The Physiology of Fishes. vol. IV. CRC Press, Boca Raton.

Gilbert, S.F., 2014. Developmental Biology. Sinauer Associates Inc., Sunderland, MA.

Gjorevski, N., Nelson, C.M., 2010. The mechanics of development: models and methods for tissue morphogenesis. Birth Defects Res. C Embryo Today 90, 193–202.

Glickman, N.S., Yelon, D., 2002. Cardiac development in zebrafish: coordination of form and function. Semin. Cell Dev. Biol. 13, 507–513.

Goenezen, S., Rennie, M.Y., Rugonyi, S., 2012. Biomechanics of early cardiac development. Biomech. Model. Mechanobiol. 11, 1187–1204.

Gore, M.R., Burggren, W.W., 2006. Parental swimming performance influences the ontogeny of cardiac and metabolic performance in larval zebrafish (Danio rerio). Integr. Comp. Biol. 46, E199.

Gore, M., Burggren, W.W., 2012. Cardiac and metabolic physiology of early larval zebrafish (Danio rerio) reflects parental swimming stamina. Front. Physiol. 3, 35.

Gore, A.V., Monzo, K., Cha, Y.R., Pan, W.J., Weinstein, B.M., 2012. Vascular development in the zebrafish. Cold Spring Harb. Perspect. Med. 2 (5), a006684. https://doi.org/10.1101/cshperspect.a006684.

Graham, J.B., 1997. Air-Breathing Fishes: Evolution, Diversity and Adaptation. Academic Press, San Diego, CA.

Grimes, A.C., Kirby, M.L., 2009. The outflow tract of the heart in fishes: anatomy, genes and evolution. J. Fish Biol. 74, 983–1036.

Gutterman, D.D., Chabowski, D.S., Kadlec, A.O., Durand, M.J., Freed, J.K., Ait-Aissa, K., Beyer, A.M., 2016. The human microcirculation regulation of flow and beyond. Circ. Res. 118, 157–172.

Haase, K., Al-Rekabi, Z., Pelling, A.E., 2014. Mechanical cues direct focal adhesion dynamics. Prog. Mol. Biol. Transl. Sci. 126, 103–134.

Hale, M.E., 2014. Developmental change in the function of movement systems: transition of the pectoral fins between respiratory and locomotor roles in zebrafish. Integr. Comp. Biol. 54, 238–249.

Han, P., Hang, C.T., Yang, J., Chang, C.P., 2011. Chromatin remodeling in cardiovascular development and physiology. Circ. Res. 108, 378–396.

Haverinen, J., Vornanen, M., 2007. Temperature acclimation modifies sinoatrial pacemaker mechanism of the rainbow trout heart. Am. J. Phys. Regul. Integr. Comp. Phys. 292, R1023–R1032.

Heiss, C., Rodriguez-Mateos, A., Kelm, M., 2015. Central role of eNOS in the maintenance of endothelial homeostasis. Antioxid. Redox Signal. 22, 1230–1242.

Herbert, S.P., Huisken, J., Kim, T.N., Feldman, M.E., Houseman, B.T., Wang, R.A., Shokat, K.M., Stainier, D.Y., 2009. Arterial-venous segregation by selective cell sprouting: an alternative mode of blood vessel formation. Science 326, 294–298.

Herwig, L., Blum, Y., Krudewig, A., Ellertsdottir, E., Lenard, A., Belting, H.G., Affolter, M., 2011. Distinct cellular mechanisms of blood vessel fusion in the zebrafish embryo. Curr. Biol. 21, 1942–1948.

Hicken, C.E., Linbo, T.L., Baldwin, D.H., Willis, M.L., Myers, M.S., Holland, L., Larsen, M., Stekoll, M.S., Rice, S.D., Collier, T.K., et al., 2011. Sublethal exposure to crude oil during embryonic development alters cardiac morphology and reduces aerobic capacity in adult fish. Proc. Natl. Acad. Sci. 108, 7086–7090.

Ho, D.H., Burggren, W.W., 2012. Parental hypoxic exposure confers offspring hypoxia resistance in zebrafish (Danio rerio). J. Exp. Biol. 215, 4208–4216.

Hoage, T., Ding, Y., Xu, X., 2012. Quantifying cardiac functions in embryonic and adult zebrafish. Methods Mol. Biol. (Clifton, N.J.) 843, 11–20.

Hoar, W.S., Randall, D.J., 1988. The Physiology of Developing Fish. Part B. Viviparity and Post-hatching Juveniles. Academic Press, New York, NY.

Holeton, G.F., 1971. Respiratory and circulatory responses of rainbow trout larvae to carbon monoxide and to hypoxia. J. Exp. Biol. 55, 683–694.

Hou, P.C., Burggren, W.W., 1995. Blood pressures and heart rate during larval development in the anuran amphibian Xenopus laevis. Am. J. Phys. 269, R1120–5.

Hove, J.R., 2006. Quantifying cardiovascular flow dynamics during early development. Pediatr. Res. 60, 6–13.

Hove, J.R., Köster, R.W., Forouhar, A.S., Acevedo-Bolton, G., Fraser, S.E., Gharib, M., 2003. Intracardiac fluid forces are an essential epigenetic factor for embryonic cardiogenesis. Nature 421, 172–177.

Hu, N., Sedmera, D., Yost, H.J., Clark, E.B., 2000. Structure and function of the developing zebrafish heart. Anat. Rec. 260, 148–157.

Hu, N., Yost, H.J., Clark, E.B., 2001. Cardiac morphology and blood pressure in the adult zebrafish. Anat. Rec. 264, 1–12.

Huo, Z.R., Marshall, L., Zhou, W., Ruan, Z.S., Xu, B., He, B., Xu, X.I., 2015. Zebrafish models of heart development and cardiovascular diseases. 2, 1.

Hutson, M.R., Kirby, M.L., 2007. Model systems for the study of heart development and disease—cardiac neural crest and conotruncal malformations. Semin. Cell Dev. Biol. 18, 101–110.

Icardo, J.M., 2006. Conus arteriosus of the teleost heart: dismissed, but not missed. Anat. Rec. A Discov. Mol. Cell. Evol. Biol. 288, 900–908.

Icardo, J.M., 2012. The teleost heart: a morphological approach. In: Sedmera, D., Wang, T. (Eds.), Ontogeny and Phylogeny of the Vertebrate Heart. Springer New York, New York, NY, pp. 35–53.

Icardo, J.M., 2017. Heart morphology and anatomy. In: Gamperl, A.K., Gillis, T.E., Farrell, A.P., Brauner, C.J. (Eds.), The Cardiovascular System: Morphology, Control and Function, Fish Physiology. In: vol. 36A. Academic Press, San Diego, pp. 1–54.

Icardo, J.M., Brunelli, E., Perrotta, I., Colvee, E., Wong, W.P., Ip, Y.K., 2005a. Ventricle and outflow tract of the African lungfish Protopterus dolloi. J. Morphol. 265, 43–51.

Icardo, J.M., Ojeda, J.L., Colvee, E., Tota, B., Wong, W.P., Ip, Y.K., 2005b. Heart inflow tract of the African lungfish *Protopterus dolloi*. J. Morphol. 263, 30–38.

Icardo, J.M., Schib, J.L., Ojeda, J.L., Durán, A.C., Guerrero, A., Colvee, E., Amelio, D., Sans-Coma, V., 2003. The conus valves of the adult gilthead seabream (Sparus auratus). J. Anat. 202, 537–550.

Icardo, J.M., Colvee, E., Schorno, S., Lauriano, E.R., Fudge, D.S., Glover, C.N., Zaccone, G., 2016. Morphological analysis of the hagfish heart. I. The ventricle, the arterial connection and the ventral aorta. J. Morphol. 277, 326–340.

Incardona, J.P., Scholz, N.L., 2016. The influence of heart developmental anatomy on cardiotoxicity-based adverse outcome pathways in fish. Aquat. Toxicol. 177, 515–525.

Incardona, J.P., Scholz, N.L., 2017. Environmental Pollution and the Fish Heart. In: Gamperl, A.K., Gillis, T.E., Farrell, A.P., Brauner, C.J. (Eds.), In: The Cardiovascular System: Development, Plasticity and Physiological Responses, Fish Physiology, vol. 36B. Academic Press, San Diego. pp. 373–433. Chapter 6.

Incardona, J.P., Collier, T.K., Scholz, N.L., 2004. Defects in cardiac function precede morphological abnormalities in fish embryos exposed to polycyclic aromatic hydrocarbons. Toxicol. Appl. Pharmacol. 196, 191–205.

Incardona, J.P., Carls, M.G., Teraoka, H., Sloan, C.A., Collier, T.K., Scholz, N.L., 2005. Aryl hydrocarbon receptor-independent toxicity of weathered crude oil during fish development. Environ. Health Perspect. 113, 1755–1762.

Incardona, J.P., Linbo, T.L., Scholz, N.L., 2011. Cardiac toxicity of 5-ring polycyclic aromatic hydrocarbons is differentially dependent on the aryl hydrocarbon receptor 2 isoform during zebrafish development. Toxicol. Appl. Pharmacol. 257, 242–249.

Incardona, J.P., Carls, M.G., Holland, L., Linbo, T.L., Baldwin, D.H., Myers, M.S., Peck, K.A., Tagal, M., Rice, S.D., Scholz, N.L., 2015. Very low embryonic crude oil exposures cause lasting cardiac defects in salmon and herring. Sci Rep 5, 13499.

Irisawa, H., 1978. Comparative physiology of the cardiac pacemaker mechanism. Physiol. Rev. 58, 461–498.

Isogai, S., Horiguchi, M., Weinstein, B.M., 2001. The vascular anatomy of the developing zebrafish: an atlas of embryonic and early larval development. Dev. Biol. 230, 278–301.

Isogai, S., Lawson, N.D., Torrealday, S., Horiguchi, M., Weinstein, B.M., 2003. Angiogenic network formation in the developing vertebrate trunk. Development 130, 5281–5290.

Jacob, E., Drexel, M., Schwerte, T., Pelster, B., 2002. Influence of hypoxia and of hypoxemia on the development of cardiac activity in zebrafish larvae. Am. J. Phys. Regul. Integr. Comp. Phys. 283, R911–R917.

Jensen, B., Boukens, B.J.D., Postma, A.V., Gunst, Q.D., van den Hoff, M.J.B., Moorman, A.F.M., Wang, T., Christoffels, V.M., 2012. Identifying the evolutionary building blocks of the cardiac conduction system. PLoS One 7, e44231.

Jensen, B., Wang, T., Christoffels, V.M., Moorman, A.F., 2013. Evolution and development of the building plan of the vertebrate heart. Biochim. Biophys. Acta 1833, 783–794.

Jensen, B., Boukens, B., Wang, T., Moorman, A., Christoffels, V., 2014. Evolution of the sinus Venosus from fish to human. J. Cardiovasc. Dev. Dis. 1, 14.

Jezierska, B., Lugowska, K., Witeska, M., 2009. The effects of heavy metals on embryonic development of fish (a review). Fish Physiol. Biochem. 35, 625–640.

Jia, X., Burggren, W., 1997. Developmental changes in chemoreceptive control of gill ventilation in larval bullfrogs (Rana catesbeiana). II. Sites of O2-sensitive chemoreceptors. J. Exp. Biol. 200, 2237–2248.

Johansen, K., Burggren, W.W., 1980. Cardiovascular function in lower vertebrates. In: Bourne, G. (Ed.), Hearts and Heart-Like Organs. Academic Press, New York, NY.

Johnson, B.M., Garrity, D.M., Dasi, L.P., 2013. Quantifying function in the early embryonic heart. J. Biomech. Eng. 135, 041006.

Johnson, B., Bark, D., Van Herck, I., Garrity, D., Dasi, L.P., 2015. Altered mechanical state in the embryonic heart results in time-dependent decreases in cardiac function. Biomech. Model. Mechanobiol. 14, 1379–1389.

Johnston, E.F., Alderman, S.L., Gillis, T.E., 2013. Chronic hypoxia exposure of trout embryos alters swimming performance and cardiac gene expression in larvae. Physiol. Biochem. Zool. 86, 567–575.

Jones, D.R., Braun, M.H., 2011. The outflow tract from the heart. In: Farrell, A.P. (Ed.), Encyclopedia of Fish Physiology, from Genome to Environment. Academic Press, London, UK, pp. 1015–1029.

Jones, D.R., Perbhoo, K., Braun, M.H., 2005. Necrophysiological determination of blood pressure in fishes. Naturwissenschaften 92, 582–585.

Jonz, M.G., Nurse, C.A., 2005. Development of oxygen sensing in the gills of zebrafish. J. Exp. Biol. 208, 1537–1549.

Jonz, M.G., Zachar, P.C., Da Fonte, D.F., Mierzwa, A.S., 2015. Peripheral chemoreceptors in fish: a brief history and a look ahead. Comp. Biochem. Physiol. A Mol. Integr. Physiol. 186, 27–38.

Jonz, M.G., Buck, L.T., Perry, S.F., Schwerte, T., Zaccone, G., 2016. Sensing and surviving hypoxia in vertebrates. Ann. N. Y. Acad. Sci. 1365, 43–58.

Jung, J.-H., Hicken, C.E., Boyd, D., Anulacion, B.F., Carls, M.G., Shim, W.J., Incardona, J.P., 2013. Geologically distinct crude oils cause a common cardiotoxicity syndrome in developing zebrafish. Chemosphere 91, 1146–1155.

Kalogirou, S., Malissovas, N., Moro, E., Argenton, F., Stainier, D.Y.R., Beis, D., 2014. Intracardiac flow dynamics regulate atrioventricular valve morphogenesis. Cardiovasc. Res. 104, 49–60.

Karpanen, T., Mäkinen, T., 2006. Regulation of lymphangiogenesis—from cell fate determination to vessel remodeling. Exp. Cell Res. 312, 575–583.

Keegan, B.R., Meyer, D., Yelon, D., 2004. Organization of cardiac chamber progenitors in the zebrafish blastula. Development 131, 3081–3091.

Kelly, R.G., 2012. The second heart field. Curr. Top. Dev. Biol. 100, 33–65.

Kermorgant, M., Lancien, F., Mimassi, N., Le Mevel, J.C., 2015. Central actions of serotonin and fluoxetine on the QT interval of the electrocardiogram in trout. Comp. Biochem. Physiol. C Toxicol. Pharmacol. 167, 190–199.

Killen, S.S., Costa, I., Brown, J.A., Gamperl, A.K., 2007. Little left in the tank: metabolic scaling in marine teleosts and its implications for aerobic scope. Proc. R. Soc. B Biol. Sci. 274, 431–438.

Kim, S., Sundaramoorthi, H., Jagadeeswaran, P., 2015. Dioxin-induced thrombocyte aggregation in zebrafish. Blood Cells Mol. Dis. 54, 116–122.

Kimmel, C.B., Ballard, W.W., Kimmel, S.R., Ullmann, B., Schilling, T.F., 1995. Stages of embryonic development of the zebrafish. Dev. Dyn. 203, 253–310.

Kindsvater, H.K., Braun, D.C., Otto, S.P., Reynolds, J.D., 2016. Costs of reproduction can explain the correlated evolution of semelparity and egg size: theory and a test with salmon. Ecol. Lett. 19, 687–696.

King-Heiden, T.C., Mehta, V., Xiong, K.M., Lanham, K.A., Antkiewicz, D.S., Ganser, A., Heideman, W., Peterson, R.E., 2012. Reproductive and developmental toxicity of dioxin in fish. Mol. Cell. Endocrinol. 354, 121–138.

Kopf, P.G., Walker, M.K., 2009. Overview of developmental heart defects by dioxins, PCBs, and pesticides. J. Environ. Sci. Health C Environ. Carcinog. Ecotoxicol. Rev. 27, 276–285.

Kopf, S.M., Humphries, P., Watts, R.J., 2014. Ontogeny of critical and prolonged swimming performance for the larvae of six Australian freshwater fish species. J. Fish Biol. 84, 1820–1841.

Kopp, R., Schwerte, T., Pelster, B., 2005. Cardiac performance in the zebrafish breakdance mutant. J. Exp. Biol. 208, 2123–2134.

Kopp, R., Bauer, I., Ramalingam, A., Egg, M., Schwerte, T., 2014. Prolonged hypoxia increases survival even in zebrafish (*Danio rerio*) showing cardiac arrhythmia. PLoS One 9, e89099.

Kowalski, W.J., Dur, O., Wang, Y., Patrick, M.J., Tinney, J.P., Keller, B.B., Pekkan, K., 2013. Critical transitions in early embryonic aortic arch patterning and hemodynamics. PLoS One 8, e60271.

Kowalski, W.J., Teslovich, N.C., Menon, P.G., Tinney, J.P., Keller, B.B., Pekkan, K., 2014. Left atrial ligation alters intracardiac flow patterns and the biomechanical landscape in the chick embryo. Dev. Dyn. 243, 652–662.

Kwan, L., Fris, M., Rodd, F.H., Rowe, L., Tuhela, L., Panhuis, T.M., 2015. An examination of the variation in maternal placentae across the genus Poeciliopsis (Poeciliidae). J. Morphol. 276, 707–720.

Langeland, A.J., Kimmel, B.C., 1997. Fishes. In: Gilbert, F.S., Raunio, M.A. (Eds.), Embryology. Contructing the Organism. Sinauer Associates, Sunderland, MA.

Langenbacher, A.D., Nguyen, C.T., Cavanaugh, A.M., Huang, J., Lu, F., Chen, J.N., 2011. The PAF1 complex differentially regulates cardiomyocyte specification. Dev. Biol. 353, 19–28.

Le Guellec, D., Morvan-Dubois, G., Sire, J.Y., 2004. Skin development in bony fish with particular emphasis on collagen deposition in the dermis of the zebrafish (Danio rerio). Int. J. Dev. Biol. 48, 217–231.

le Noble, F., Moyon, D., Pardanaud, L., Yuan, L., Djonov, V., Matthijsen, R., Bréant, C., Fleury, V., Eichmann, A., le, N.F., et al., 2004. Flow regulates arterial-venous differentiation in the chick embryo yolk sac. Development 131, 361–375.

Lee, R.K., Stainier, D.Y., Weinstein, B.M., Fishman, M.C., 1994. Cardiovascular development in the zebrafish. II. Endocardial progenitors are sequestered within the heart field. Development 120, 3361–3366.

Lee, L., Genge, C.E., Cua, M., Sheng, X.Y., Rayani, K., Beg, M.F., Sarunic, M.V., Tibbits, G.F., 2016. Functional assessment of cardiac responses of adult zebrafish (Danio rerio) to acute and chronic temperature change using high-resolution echocardiography. PLoS One 11, e0145163.

Lemanski, L.F., La France, S.M., Erginel-Unaltuna, N., Luque, E.A., Ward, S.M., Fransen, M.E., Mangiacapra, F.J., Nakatsugawa, M., Lemanski, S.L., Capone, R.B., et al., 1995. The cardiac mutant gene c in axolotls: cellular, developmental, and molecular studies. Cell. Mol. Biol. Res. 41, 293–305.

Leong, I.U.S., Skinner, J.R., Shelling, A.N., Love, D.R., 2010. Zebrafish as a model for long QT syndrome: the evidence and the means of manipulating zebrafish gene expression. Acta Physiol. 199, 257–276.

Letamendia, A., Quevedo, C., Ibarbia, I., Virto, J.M., Holgado, O., Diez, M., Izpisua Belmonte, J.C., Callol-Massot, C., 2012. Development and validation of an automated high-throughput system for zebrafish in vivo screenings. PLoS One 7, e36690.

Li, J., Yue, Y., Zhao, Q., 2016. Retinoic acid signaling is essential for Valvulogenesis by affecting endocardial cushions formation in zebrafish embryos. Zebrafish 13, 9–18.

Liem, K.F., 1981. Larvae of air-breathing fishes as countercurrent-flow devices in hypoxic environments. Science 211, 1177–1179.

Lindsey, S.E., Butcher, J.T., Yalcin, H.C., 2014. Mechanical regulation of cardiac development. Front. Physiol. 5, 318.

Lisa, T., Reif, D.M., St Mary, L., Geier, M.C., Truong, H.D., Tanguay, R.L., 2014. Multidimensional in vivo hazard assessment using zebrafish. Toxicol. Sci. 137, 212–233.

Liu, C.C., Li, L., Lam, Y.W., Siu, C.W., Cheng, S.H., 2016. Improvement of surface ECG recording in adult zebrafish reveals that the value of this model exceeds our expectation. Sci Rep 6, 25073.

Lopez-Olmeda, J.F., Sanchez-Vazquez, F.J., 2011. Thermal biology of zebrafish (Danio rerio). J. Therm. Biol. 36, 91–104.

Lorenzale, M., Lopez-Unzu, M.A., Fernandez, M.C., Duran, A.C., Fernandez, B., Soto-Navarrete, M.T., Sans-Coma, V., 2017. Anatomical, histochemical and immunohistochemicalcharacterisation of the cardiac outflow tract of the silver arowana, *Osteoglossum bicirrhosum* (Teleostei: Osteoglossiformes). Zoology 120, 15–23.

Lu, W., Zhang, Y., McDonald, D.O., Jing, H., Carroll, B., Robertson, N., Zhang, Q., Griffin, H., Sanderson, S., Lakey, J.H., et al., 2014. Dual proteolytic pathways govern glycolysis and immune competence. Cell 159, 1578–1590.

Lu, F., Langenbacher, A., Chen, J.-N., 2016. Transcriptional regulation of heart development in zebrafish. J. Cardiovasc. Dev. Dis. 3, 14.

Mably, D.J., Childs, J.S., 2010. Developmental physiology of the zebrafish cardiovascular system. In: Perry, S., Ekker, M., Farrell, A., Braune, C. (Eds.), Zebrafish, Fish Physiology, vol. 29. Academic Press, New York, pp. 249–286.

Mager, E.M., Esbaugh, A.J., Stieglitz, J.D., Hoenig, R., Bodinier, C., Incardona, J.P., Scholz, N.L., Benetti, D.D., Grosell, M., 2014. Acute embryonic or juvenile exposure to Deepwater horizon crude oil impairs the swimming performance of Mahi-Mahi (Coryphaena hippurus). Environ. Sci. Technol. 48, 7053–7061.

Makanya, A.N., Hlushchuk, R., Djonov, V.G., 2009. Intussusceptive angiogenesis and its role in vascular morphogenesis, patterning, and remodeling. Angiogenesis 12, 113–123.

Mammoto, T., Mammoto, A., Ingber, D.E., 2013. Mechanobiology and developmental control. Annu. Rev. Cell Dev. Biol. 29, 27–61.

Manasek, F.J., 1981. Determinants of heart shape in early embryos. Fed. Proc. 40, 2011–2016.

Manasek, F.J., Monroe, R.G., 1972. Early cardiac morphogenesis is independent of function. Dev. Biol. 27, 584–588.

Mank, J.E., Promislow, D.E., Avise, J.C., 2005. Phylogenetic perspectives in the evolution of parental care in ray-finned fishes. Evolution 59, 1570–1578.

Martin, K.L., 2014. Theme and variations: amphibious air-breathing intertidal fishes. J. Fish Biol. 84, 577–602.

Martin, R.T., Bartman, T., 2009. Analysis of heart valve development in larval zebrafish. Dev. Dyn. 238, 1796–1802.

McBryan, T.L., Anttila, K., Healy, T.M., Schulte, P.M., 2013. Responses to temperature and hypoxia as interacting stressors in fish: implications for adaptation to environmental change. Integr. Comp. Biol. 53, 648–659.

McCasker, N., Humphries, P., Meredith, S., Klomp, N., 2014. Contrasting patterns of larval mortality in two sympatric riverine fish species: a test of the critical period hypothesis. PLoS One 9, e109317.

McCormick, M.I., Nechaev, I.V., 2002. Influence of cortisol on developmental rhythms during embryogenesis in a tropical damselfish. J. Exp. Zool. 293, 456–466.

McDonald, D.G., McMahon, B.R., 1977. Respiratory development in Arctic char Salvelinus Alpinus under conditions of normoxia and chronic hypoxia. Can. J. Zool. 55, 1461–1467.

McDonald, M.D., Gilmour, K.M., Walsh, P.J., Perry, S.F., 2010. Cardiovascular and respiratory reflexes of the gulf toadfish (Opsanus beta) during acute hypoxia. Respir. Physiol. Neurobiol. 170, 59–66.

McElman, J.F., Balon, E.K., 1979. Early ontogeny of walleye, Stizostedion vitreum, with steps of saltatory development. Environ. Biol. Fish 4, 309–348.

Meilhac, S.M., Lescroart, F., Blanpain, C., Buckingham, M.E., 2014. Cardiac cell lineages that form the heart. Cold Spring Harb. Perspect. Med. 4, 1–14.

Mendez-Sanchez, J.F., 2015. Environmental Modulation of the Onset of Air-Breathing of the Siamese Fighting Fish and the Blue Gourami (Ph.D.). University of North Texas, Denton, Texas, USA.

Mendez-Sanchez, J.F., Burggren, W.W., 2017. Cardio-respiratory physiological phenotypic plasticity in developing air breathing anabantid fishes (*Betta splendens* and *Trichopodus trichopterus*). Phys. Rep. e13359 (in press).

Miller, S.C., Reeb, S.E., Wright, P.A., Gillis, T.E., 2008. Oxygen concentration in the water boundary layer next to rainbow trout (Oncorhynchus mykiss) embryos is influenced by hypoxia exposure time, metabolic rate, and water flow. Can. J. Fish. Aquat. Sci. 65, 2170–2177.

Miller, S.C., Gillis, T.E., Wright, P.A., 2011. The ontogeny of regulatory control of the rainbow trout (*Oncorhynchus mykiss*) heart and how this is influenced by chronic hypoxia exposure. J. Exp. Biol. 214, 2065–2072.

Mirkovic, T., Rombough, P., 1998. The effect of body mass and temperature on the heart rate, stroke volume, and cardiac output of larvae of the rainbow trout, Oncorhynchus mykiss. Physiol. Zool. 71, 191–197.

Moorman, A.F.M., Christoffels, V.M., 2003. Cardiac chamber formation: development, genes, and evolution. Physiol. Rev. 83, 1223–1267.

Moorman, A.F., de Jong, F., Denyn, M.M., Lamers, W.H., 1998. Development of the cardiac conduction system. Circ. Res. 82, 629–644.

Mueller, C.A., Burggren, W.W., Tazawa, H., 2015a. The physiology of the avian embryo. In: Whittow, G.C. (Ed.), Sturkie's Avian Physiology. Elsevier, New York, NY, pp. 739–766.
Mueller, C.A., Eme, J., Manzon, R.G., Somers, C.M., Boreham, D.R., Wilson, J.Y., 2015b. Embryonic critical windows: changes in incubation temperature alter survival, hatchling phenotype, and cost of development in lake whitefish (Coregonus clupeaformis). J. Comp. Physiol. B, Biochem. Syst. Environ. Physiol. 185, 315–331.
Newton, C.M., Stoyek, M.R., Croll, R.P., Smith, F.M., 2014. Regional innervation of the heart in the goldfish, Carassius auratus: a confocal microscopy study. J. Comp. Neurol. 522, 456–478.
Noble, D., 2015. Evolution beyond neo-Darwinism. J. Exp. Biol. 218, 7–13.
Noel, E.S., Verhoeven, M., Lagendijk, A.K., Tessadori, F., Smith, K., Choorapoikayil, S., den Hertog, J., Bakkers, J., 2013. A nodal-independent and tissue-intrinsic mechanism controls heart-looping chirality. Nat. Commun. 4, 1–9.
Norin, T., Clark, T.D., 2016. Measurement and relevance of maximum metabolic rate in fishes. J. Fish Biol. 88, 122–151.
Olson, K.R., 1996. Secondary circulation in fish: anatomical organization and physiological significance. J. Exp. Zool. 275, 172–185.
Olson, K.R., Farrell, A.P., 2011. Design and physiology of capillaries and secondary circulation: secondary circulation and lymphatic anatomy. In: Farrell, A.P. (Ed.), Encyclopedia of Fish Physiology. In: vol. 2. Elsevier, The Netherlands, pp. 1161–1168.
Oziolor, E.M., Bigorgne, E., Aguilar, L., Usenko, S., Matson, C.W., 2014. Evolved resistance to PCB- and PAH-induced cardiac teratogenesis, and reduced CYP1A activity in gulf killifish (*Fundulus grandis*) populations from the Houston Ship Channel, Texas. Aquat. Toxicol. 150, 210–219.
Oziolor, E.M., Dubansky, B., Burggren, W.W., Matson, C.W., 2016. Cross-resistance in gulf killifish (*Fundulus grandis*) populations resistant to dioxin-like compounds. Aquat. Toxicol. 175, 222–231.
Paige, S.L., Plonowska, K., Xu, A., Wu, S.M., 2015. Molecular regulation of cardiomyocyte differentiation. Circ. Res. 116, 341–353.
Panhuis, T.M., Broitman-Maduro, G., Uhrig, J., Maduro, M., Reznick, D.N., 2011. Analysis of expressed sequence tags from the placenta of the live-bearing fish Poeciliopsis (Poeciliidae). J. Hered. 102, 352–361.
Pardo-Martin, C., Chang, T.-Y., Koo, B.K., Gilleland, C.L., Wasserman, S.C., Yanik, M.F., 2010. High-throughput in vivo vertebrate screening. Nat. Methods 7, 634–636.
Patti, M.E., 2013. Intergenerational programming of metabolic disease: evidence from human populations and experimental animal models. Cell. Mol. Life Sci. 70, 1597–1608.
Peal, D.S., Lynch, S.N., Milan, D.J., 2011. Patterning and development of the atrioventricular canal in zebrafish. J. Cardiovasc. Transl. Res. 4, 720–726.
Pelster, B., 2002. Developmental plasticity in the cardiovascular system of fish, with special reference to the zebrafish. Comp Biochem. Physiol. A Mol. Integr. Physiol. 133, 547–553.
Pelster, B., Bemis, W.E., 1992. Structure and function of the external gill filaments of embryonic skates (Raja-Erinacea). Respir. Physiol. 89, 1–13.
Pelster, B., Burggren, W.W., 1996. Disruption of hemoglobin oxygen transport does not impact oxygen-dependent physiological processes in developing embryos of zebra fish (Danio rerio). Circ. Res. 79, 358–362.
Pelster, B., Sanger, A.M., Siegele, M., Schwerte, T., 2003. Influence of swim training on cardiac activity, tissue capillarization, and mitochondrial density in muscle tissue of zebrafish larvae. Am. J. Phys. Regul. Integr. Comp. Phys. 285, R339–R347.
Pelster, B., Grillitsch, S., Schwerte, T., 2005. NO as a mediator during the early development of the cardiovascular system in the zebrafish. Comp. Biochem. Physiol. A Mol. Integr. Physiol. 142, 215–220.

Pelster, B., Gittenberger-de Groot, A.C., Poelmann, R.E., Rombough, P., Schwerte, T., Thompson, M.B., 2010. Functional plasticity of the developing cardiovascular system: examples from different vertebrates. Physiol. Biochem. Zool. 83, 775–791.

Peralta, M., Steed, E., Harlepp, S., Gonzalez-Rosa, J.M., Monduc, F., Ariza-Cosano, A., Cortes, A., Rayon, T., Gomez-Skarmeta, J.L., Zapata, A., et al., 2013. Heartbeat-driven pericardiac fluid forces contribute to epicardium morphogenesis. Curr. Biol. 23, 1726–1735.

Perrichon, P., Burggren, W.W., 2017. Heart shape and volume modeling in larval fish for estimation of cardiac output. Front. Physiol. 8, 464.

Peshkovsky, C., Totong, R., Yelon, D., 2011. Dependence of cardiac trabeculation on neuregulin signaling and blood flow in zebrafish. Dev. Dyn. 240, 446–456.

Poon, K.L., Brand, T., 2013. The zebrafish model system in cardiovascular research: a tiny fish with mighty prospects. Glob. Cardiol. Sci. Pract. 2013, 9–28.

Powell, M.D., Gamperl, A.K., 2016. Effects of *Loma morhua* (Microsporidia) infection on the cardiorespiratory performance of Atlantic cod *Gadus morhua* (L). J. Fish Dis. 39 (2), 189–204.

Powell, W.H., Bright, R., Bello, S.M., Hahn, M.E., 2000. Developmental and tissue-specific expression of AHR1, AHR2, and ARNT2 in dioxin-sensitive and -resistant populations of the marine fish Fundulus heteroclitus. Toxicol. Sci. 57, 229–239.

Pries, A.R., Reglin, B., Secomb, T.W., 2005. Remodeling of blood vessels: responses of diameter and wall thickness to hemodynamic and metabolic stimuli. Hypertension 46, 725–731.

Raimondo, S., Jackson, C.R., Krzykwa, J., Hemmer, B.L., Awkerman, J.A., Barron, M.G., 2014. Developmental toxicity of Louisiana crude oil-spiked sediment to zebrafish. Ecotoxicol. Environ. Saf. 108, 265–272.

Ramos, C., 2004. The structure and ultrastructure of the sinus venosus in the mature dogfish (Scyliorhinus canicula): the endocardium, the epicardium and the subepicardial space. Tissue Cell 36, 399–407.

Randall, D.J., Burggren, W.W., Farrell, A.P., Haswell, M.S., 1981. The Evolution of Air Breathing in Vertebrates. Cambridge University Press, Cambridge, UK.

Reading, B.J., Sullivan, C.V., 2011. The Reproductive Organs and Processes: Vitellogenesis in Fishes. Esevier Inc., Amsterdam, The Netherlands.

Rees, B.B., Targett, T.E., Ciotti, B.J., Tolman, C.A., Akkina, S.S., Gallaty, A.M., 2012. Temporal dynamics in growth and white skeletal muscle composition of the mummichog Fundulus heteroclitus during chronic hypoxia and hyperoxia. J. Fish Biol. 81, 148–164.

Reynolds, J.D., Goodwin, N.B., Freckleton, R.P., 2002. Evolutionary transitions in parental care and live bearing in vertebrates. Philos. Trans. R. Soc. Lond. Ser. B Biol. Sci. 357, 269–281.

Richardson, M.K., Admiraal, J., Wright, G.M., 2010. Developmental anatomy of lampreys. Biol. Rev. 85, 1–33.

Rodrigues-Galdino, A.M., Maiolino, C.V., Forgati, M., Donatti, L., Mikos, J.D., Carneiro, P.C.F., Rios, F.S., 2010. Development of the neotropical catfish Rhamdia quelen (Siluriformes, Heptapteridae) incubated in different temperature regimes. Zygote 18, 131–144.

Rohr, S., Otten, C., Abdelilah-Seyfried, S., 2008. Asymmetric involution of the myocardial field drives heart tube formation in zebrafish. Circ. Res. 102 (2), e12–e9.

Rombough, P., 1997. Piscine cardiovascular development. In: Burggren, W.W., Keller, B. (Eds.), Development of Cardiovascular Systems Molecules to Organisms, Cambridge University Press, Cambridge

Rombough, P.J., 1998. Partitioning of oxygen uptake between the gills and skin in fish larvae: a novel method for estimating cutaneous oxygen uptake. J. Exp. Biol. 201, 1763–1769.

Rombough, P., 2002. Gills are needed for ionoregulation before they are needed for O-2 uptake in developing zebrafish, Danio rerio. J. Exp. Biol. 205, 1787–1794.

Rombough, P., 2007. The functional ontogeny of the teleost gill: which comes first, gas or ion exchange? Comp. Biochem. Physiol. A Mol. Integr. Physiol. 148, 732–742.

Rombough, P., 2011. The energetics of embryonic growth. Respir. Physiol. Neurobiol. 178, 22–29.

Rombough, P.J., Moroz, B.M., 1997. The scaling and potential importance of cutaneous and branchial surfaces in respiratory gas exchange in larval and juvenile walleye Stizostedion vitreum. J. Exp. Biol. 200, 2459–2468.

Rudic, R.D., Shesely, E.G., Maeda, N., Smithies, O., Segal, S.S., Sessa, W.C., 1998. Direct evidence for the importance of endothelium-derived nitric oxide in vascular remodeling. J. Clin. Investig. 101, 731–736.

Rummer, J.L., Wang, S., Steffensen, J.F., Randall, D.J., 2014. Function and control of the fish secondary vascular system, a contrast to mammalian lymphatic systems. J. Exp. Biol. 217, 751–757.

Sacca, R., Burggren, W., 1982. Oxygen uptake in air and water in the air-breathing reedfish Calamoichthys calabaricus: role of skin, gills and lungs. J. Exp. Biol. 97, 179–186.

Sandblom, E., Gräns, A., 2017. Form, function, and control of the vasculature. In: Gamperl, A.K., Gillis, T.E., Farrell, A.P., Brauner, C.J. (Eds.), The Cardiovascular System: Morphology, Control and Function, Fish Physiology. In: vol. 36A. Academic Press, San Diego, pp. 369–434.

Sandblom, E., Olsson, C., Davison, W., Axelsson, M., 2010. Nervous and humoral catecholaminergic control of blood pressure and cardiac performance in the Antarctic fish Pagothenia borchgrevinki. Comp. Biochem. Physiol. A Mol. Integr. Physiol. 156, 232–236.

Santos, D., Matos, M., Coimbra, A.M., 2014. Developmental toxicity of endocrine disruptors in early life stages of zebrafish, a genetic and embryogenesis study. Neurotoxicol. Teratol. 46, 18–25.

Sarasquete, C., Segner, H., 2000. Cytochrome P4501A (CYP1A) in teleostean fishes. A review of immunohistochemical studies. Sci. Total Environ. 247, 313–332.

Scherz, P.J., Huisken, J., Sahai-Hernandez, P., Stainier, D.Y.R., 2008. High-speed imaging of developing heart valves reveals interplay of morphogenesis and function. Development 135, 1179–1187.

Schindler, J.F., 2015. Structure and function of placental exchange surfaces in goodeid fishes (Teleostei: Atheriniformes). J. Morphol. 276, 991–1003.

Schlegel, A., Gut, P., 2015. Metabolic insights from zebrafish genetics, physiology, and chemical biology. Cell. Mol. Life Sci. 72, 2249–2260.

Schnurr, M.E., Yin, Y., Scott, G.R., 2014. Temperature during embryonic development has persistent effects on metabolic enzymes in the muscle of zebrafish. J. Exp. Biol. 217, 1370–1380.

Schonweger, G., Schwerte, T., Pelster, B., 2000. Temperature-dependent development of cardiac activity in unrestrained larvae of the minnow Phoxinus phoxinus. Am. J. Phys. Regul. Integr. Comp. Phys. 279, R1634–40.

Schreiber, A.M., 2001. Metamorphosis and early larval development of the flatfishes (Pleuronectiformes): an osmoregulatory perspective. Comp. Biochem. Physiol. B Biochem. Mol. Biol. 129, 587–595.

Schuermann, A., Helker, C.S.M., Herzog, W., 2014. Angiogenesis in zebrafish. Semin. Cell Dev. Biol. 31, 106–114.

Schultz, R.J., 1961. Reproductive mechanism of unisexual and bisexual strains of the viviparous fish *Poeciiopsis*. Evolution 15, 302–325.

Schwerte, T., 2009. Cardio-respiratory control during early development in the model animal zebrafish. Acta Histochem. 111, 230–243.

Schwerte, T., Fritsche, R., 2003. Understanding cardiovascular physiology in zebrafish and Xenopus larvae: the use of microtechniques. Comp. Biochem. Physiol. A Mol. Integr. Physiol. 135, 131–145.

Schwerte, T., Pelster, B., 2000. Digital motion analysis as a tool for analysing the shape and performance of the circulatory system in transparent animals. J. Exp. Biol. 203, 1659–1669.

Schwerte, T., Uberbacher, D., Pelster, B., 2003. Non-invasive imaging of blood cell concentration and blood distribution in zebrafish *Danio rerio* incubated in hypoic conditions *in vivo*. J. Exp. Biol. 206, 1299–1307.

Schwerte, T., Prem, C., Mairosl, A., Pelster, B., 2006. Development of the sympatho-vagal balance in the cardiovascular system in zebrafish (Danio rerio) characterized by power spectrum and classical signal analysis. J. Exp. Biol. 209, 1093–1100.

Sedmera, D., Reckova, M., deAlmeida, A., Sedmerova, M., Biermann, M., Volejnik, J., Sarre, A., Raddatz, E., McCarthy, R.A., Gourdie, R.G., et al., 2003. Functional and morphological evidence for a ventricular conduction system in zebrafish and Xenopus hearts. Am. J. Physiol. Heart Circ. Physiol. 284, H1152–H1160.

Sedmera, D., Reckova, M., Sedmerova, M., Volejnik, J., Sarre, A., Raddatz, E., Mccarthy, R.A., Gourdie, R.G., Thompson, R.P., Liu, J., et al., 2011. Functional and morphological evidence for a ventricular conduction system in zebrafish and Xenopus hearts functional and morphological evidence for a ventricular conduction system in zebrafish and Xenopus hearts. Society 29425, 1152–1160.

Sharmili, S., Ahila, A.J., 2015. Stages of embryonic development of the zebrafish. Eur. J. Biotechnol. Sci. 3, 6–11.

Shin, J.T., Pomerantsev, E.V., Mably, J.D., MacRae, C.A., 2010. High-resolution cardiovascular function confirms functional orthology of myocardial contractility pathways in zebrafish. Physiol. Genomics 42, 300–309.

Simons, M., Eichmann, A., 2015. Molecular controls of arterial morphogenesis. Circ. Res. 116, 1712–1724.

Šolc, D., 2007. The heart and heart conducting system in the kingdom of animals: a comparative approach to its evolution. Exp. Clin. Cardiol. 12, 113–118.

Somero, G.N., 2010. The physiology of climate change: how potentials for acclimatization and genetic adaptation will determine 'winners' and 'losers'. J. Exp. Biol. 213, 912–920.

Sorhus, E., Incardona, J.P., Furmanek, T., Jentoft, S., Meier, S., Edvardsen, R.B., 2016. Developmental transcriptomics in Atlantic haddock: illuminating pattern formation and organogenesis in non-model vertebrates. Dev. Biol. 411, 301–313.

Spicer, J.I., Rundle, S.D., Tills, O., 2011. Studying the altered timing of physiological events during development: it's about time...or is it? Respir. Physiol. Neurobiol. 178, 3–12.

Spomer, W., Pfriem, A., Alshut, R., Just, S., Pylatiuk, C., 2012. High-throughput screening of zebrafish embryos using automated heart detection and imaging. J. Lab. Autom. 17, 435–442.

Stainier, D.Y.R., Fishman, M.C., 1992. Patterning the zebrafish heart tube: acquisition of anteroposterior polarity. Dev. Biol. 153, 91–101.

Stainier, D.Y., Fishman, M.C., 1994. The zebrafish as a model system to study cardiovascular development. Trends Cardiovasc. Med. 4, 207–212.

Stainier, D.Y., Lee, R.K., Fishman, M.C., 1993. Cardiovascular development in the zebrafish. I. Myocardial fate map and heart tube formation. Development 119, 31–40.

Staudt, D., Stainier, D., 2012. Uncovering the molecular and cellular mechanisms of heart development using the zebrafish. Annu. Rev. Genet. 46, 397–418.

Stewart, D.J., Langille, B.L., 2004. Tied down by shear force: role for Tie1 in postnatal vascular remodeling? Circ. Res. 94, 271–272.

Stoppel, W.L., Kaplan, D.L., Black, L.D., 2016. Electrical and mechanical stimulation of cardiac cells and tissue constructs. Adv. Drug Deliv. Rev. 96, 135–155.

Stoyek, M.R., Croll, R.P., Smith, F.M., 2015. Intrinsic and extrinsic innervation of the heart in zebrafish (Danio rerio). J. Comp. Neurol. 523, 1683–1700.

Stratman, A.N., Yuja, J.A., Mulligan, T.S., Butler, M.G., Cause, E.T., Weinstein, B.M., 2015. Blood vessel formation. In: Moody, S. (Ed.), Principles of Developmental Genetics. Academic Press, New York, NY, pp. 421–449.

Strick, D.M., Waycaster, R.L., Montani, J.P., Gay, W.J., Adair, T.H., 1991. Morphometric measurements of chorioallantoic membrane vascularity: effects of hypoxia and hyperoxia. Am. J. Phys. 260, H1385–9.

Sykes, B.G., Van Steyn, P.M., Vignali, J.D., Winalski, J., Lozier, J., Bell, W.E., Turner, J.E., 2016. The relationship between estrogen and nitric oxide in the prevention of cardiac and vascular anomalies in the developing zebrafish (Danio Rerio). Brain Sci. 6, E51.

Taber, L.A., 1998. Mechanical aspects of cardiac development. Prog. Biophys. Mol. Biol. 69, 237–255.

Tammela, T., Petrova, T.V., Alitalo, K., 2005. Molecular lymphangiogenesis: new players. Trends Cell Biol. 15, 434–441.

Tan, X., Rotllant, J., Li, H., De Deyne, P., DeDeyne, P., Du, S.J., 2006. SmyD1, a histone methyltransferase, is required for myofibril organization and muscle contraction in zebrafish embryos. Proc. Natl. Acad. Sci. U. S. A. 103, 2713–2718.

Taylor, E.W., Leite, C.A.C., Sartori, M.R., Wang, T., Abe, A.S., Crossley, D.A., 2014. The phylogeny and ontogeny of autonomic control of the heart and cardiorespiratory interactions in vertebrates. J. Exp. Biol. 217, 690–703.

Teraoka, H., Kubota, A., Dong, W., Kawai, Y., Yamazaki, K., Mori, C., Harada, Y., Peterson, R.E., Hiraga, T., 2009. Role of the cyclooxygenase 2–thromboxane pathway in 2,3,7,8-tetrachlorodibenzo-p-dioxin-induced decrease in mesencephalic vein blood flow in the zebrafish embryo. Toxicol. Appl. Pharmacol. 234, 33–40.

Teraoka, H., Okuno, Y., Nijoukubo, D., Yamakoshi, A., Peterson, R.E., Stegeman, J.J., Kitazawa, T., Hiraga, T., Kubota, A., 2014. Involvement of COX2-thromboxane pathway in TCDD-induced precardiac edema in developing zebrafish. Aquat. Toxicol. 154, 19–26.

Territo, P.R., Burggren, W.W., 1998. Cardio-respiratory ontogeny during chronic carbon monoxide exposure in the clawed frog Xenopus laevis. J. Exp. Biol. 201, 1461–1472.

Tessadori, F., van Weerd, J.H., Burkhard, S.B., Verkerk, A.O., de Pater, E., Boukens, B.J., Vink, A., Christoffels, V.M., Bakkers, J., 2012. Identification and functional characterization of cardiac pacemaker cells in zebrafish. PLoS One 7, e47644.

Trexler, J.C., DeAngelis, D.L., 2003. Resource allocation in offspring provisioning: an evaluation of the conditions favoring the evolution of matrotrophy. Am. Nat. 162, 574–585.

Tripathi, S., Sahu, D.B., Kumar, R., Kumar, A., 2003. Effect of acute exposure of sodium arsenite (Na3 Aso3) on some haematological parameters of Clarias Batrachus (common Indian cat fish) in vivo. Indian J. Environ. Health 45, 183–188.

Urbina, M.A., Meredith, A.S., Glover, C.N., Forster, M.E., 2014. The importance of cutaneous gas exchange during aerial and aquatic respiration in galaxiids. J. Fish Biol. 84, 759–773.

Uribe, M.C., Grier, H.J., 2010. Vivaparous Fishes II. New Life Publications, Homestead, FL.

Van Vliet, P., Wu, S.M., Zaffran, S., Puceat, M., 2012. Early cardiac development: a view from stem cells to embryos. Cardiovasc. Res. 96, 352–362.

van Weerd, J.H., Christoffels, V.M., 2016. The formation and function of the cardiac conduction system. Development 143, 197–210.

Van Weerd, J.H., Koshiba-Takeuchi, K., Kwon, C., Takeuchi, J.K., 2011. Epigenetic factors and cardiac development. Cardiovasc. Res. 91, 203–211.

Varner, V.D., Taber, L.A., 2012. Not just inductive: a crucial mechanical role for the endoderm during heart tube assembly. Development 139, 1680–1690.

Varsamos, S., Nebel, C., Charmantier, G., 2005. Ontogeny of osmoregulation in postembryonic fish: a review. Comp. Biochem. Physiol. A Mol. Integr. Physiol. 141, 401–429.

Verkerk, A.O., Remme, C.A., 2012. Zebrafish: a novel research tool for cardiac (patho)electrophysiology and ion channel disorders. Front. Physiol. 3, 255.

Vornanen, M., Hassinen, M., 2016. Zebrafish heart as a model for human cardiac electrophysiology. Channels 10, 101–110.

Vulesevic, B., McNeill, B., Perry, S.F., 2006. Chemoreceptor plasticity and respiratory acclimation in the zebrafish Danio rerio. J. Exp. Biol. 209, 1261–1273.

Wang, Y., Dur, O., Patrick, M.J., Tinney, J.P., Tobita, K., Keller, B.B., Pekkan, K., 2009. Aortic arch morphogenesis and flow modeling in the chick embryo. Ann. Biomed. Eng. 37, 1069–1081.

Warren, K.S., Baker, K., Fishman, M.C., 2001. The slow mo mutation reduces pacemaker current and heart rate in adult zebrafish. Am. J. Physiol. Heart Circ. Physiol. 281, H1711–9.

Watanabe-Asaka, T., Sekiya, Y., Wada, H., Yasuda, T., Okubo, I., Oda, S., Mitani, H., 2014. Regular heartbeat rhythm at the heartbeat initiation stage is essential for normal cardiogenesis at low temperature. BMC Dev. Biol. 14, 12.

Watkins, S.C., Maniar, S., Mosher, M., Roman, B.L., Tsang, M., St Croix, C.M., 2012. High resolution imaging of vascular function in zebrafish. PLoS One 7, e44018.

Weinstein, B.M., Lawson, N.D., 2002. Arteries, veins, Notch, and VEGF. Cold Spring Harb. Symp. Quant. Biol. 67, 155–162.

Weis, P., Weis, J.S., 1974. Cardiac-malformation and other effects due to insecticides in embryos of killifish, *Fundulus heteroclitus*. Teratology 10, 263–267.

Weis, P., Weis, J.S., 1977. Methylmercury teratogenesis in killifish, *Fundulus heteroclitus*. Teratology 16, 317–325.

Weis, J.S., Weis, P., 1997. Aquatic testing with early life stages of killifish. In: Wells, P.G., Lee, K., Blaise, C. (Eds.), Microscale Testing in Aquatic Toxicology. CRC Press, Boca Roton, FL, pp. 479–490.

Wells, P., Pinder, A., 1996a. The respiratory development of Atlantic salmon. I. Morphometry of gills, yolk sac and body surface. J. Exp. Biol. 199, 2725–2736.

Wells, P., Pinder, A., 1996b. The respiratory development of Atlantic salmon. II. Partitioning of oxygen uptake among gills, yolk sac and body surfaces. J. Exp. Biol. 199, 2737–2744.

Wells, P.G., Lee, K., Blaise, C., 1997. Microscale Testing in Aquatic Toxicology: Advances, Technques, and Practice. CRC Press, Boca Raton, FL.

White, R.M., Sessa, A., Burke, C., Bowman, T., LeBlanc, J., Ceol, C., Bourque, C., Dovey, M., Goessling, W., Burns, C.E., et al., 2008. Transparent adult zebrafish as a tool for in vivo transplantation analysis. Cell Stem Cell 2, 183–189.

Woodin, B.R., Smolowitz, R.M., Stegeman, J.J., 1997. Induction of cytochrome P4501A in the intertidal fish *Anoplarchus purpurescens* by Prudhoe Bay crude oil and environmental induction in fish from Prince William Sound. Environ. Sci. Technol. 31, 1198–1205.

Wourms, J., 1981. Viviparity: the materenal-fetal relationship in fishes. Am. Zool. 21, 473–515.

Wright, G.M., 1984. Structure of the conus arteriosus and ventral aorta in the sea lamprey, Petromyzon marinus, and the Atlantic hagfish, Myxine glutinosa: microfibrils, a major component. Can. J. Zool. 62, 2445–2456.

Xu, W., Tong, G.X., Geng, L.W., Jiang, H.F., 2015. Body color development and genetic analysis of hybrid transparent crucian carp (Carassius auratus). Genet. Mol. Res. 14, 4399–4407.

Xu, E.G., Mager, E.M., Grosell, M., Pasparakis, C., Schlenker, L.S., Stieglitz, J.D., Benetti, D., Hazard, E.S., Courtney, S.M., Diamante, G., et al., 2016. Time- and oil-dependent transcriptomic and physiological responses to Deepwater horizon oil in Mahi-Mahi (Coryphaena hippurus) embryos and larvae. Environ. Sci. Technol. 50, 7842–7851.

Yalcin, H.C., Shekhar, A., McQuinn, T.C., Butcher, J.T., 2011. Hemodynamic patterning of the avian atrioventricular valve. Dev. Dyn. 240, 23–35.

Yang, L., Ho, N.Y., Alshut, R., Legradi, J., Weiss, C., Reischl, M., Mikut, R., Liebel, U., Mueller, F., Straehle, U., 2009. Zebrafish embryos as models for embryotoxic and teratological effects of chemicals. Reprod. Toxicol. 28, 245–253.

Yang, J., Hartjes, K.A., Nelson, T.J., Xu, X., 2014. Cessation of contraction induces cardiomyocyte remodeling during zebrafish cardiogenesis. Am. J. Physiol. Heart Circ. Physiol. 306, H382–95.

Yaniv, K., Isogai, S., Castranova, D., Dye, L., Hitomi, J., Weinstein, B.M., 2006. Live imaging of lymphatic development in the zebrafish. Nat. Med. 12, 711–716.
Ye, D., Xie, H., Hu, B., Lin, F., 2015. Endoderm convergence controls subduction of the myocardial precursors during heart-tube formation. Development 142, 2928–2940.
Yelon, D., 2001. Cardiac patterning and morphogenesis in zebrafish. Dev. Dyn. 222, 552–563.
Yu, F., Zhao, Y., Gu, J., Quigley, K.L., Chi, N.C., Tai, Y.-C., Hsiai, T.K., 2012. Flexible microelectrode arrays to interface epicardial electrical signals with intracardial calcium transients in zebrafish hearts. Biomed. Microdevices 14, 357–366.
Zaccone, G., Mauceri, A., Maisano, M., Giannetto, A., Parrino, V., Fasulo, S., 2009. Distribution and neurotransmitter localization in the heart of the ray-finned fish, bichir (Polypterus Bichir Bichir Geoffroy St. Hilaire, 1802). Acta Histochem. 111, 93–103.
Zakrzewicz, A., Secomb, T.W., Pries, A.R., 2002. Angioadaptation: keeping the vascular system in shape. News Physiol. Sci. 17, 197–201.
Zamir, E.A., Czirók, A., Cui, C., Little, C.D., Rongish, B.J., 2006. Mesodermal cell displacements during avian gastrulation are due to both individual cell-autonomous and convective tissue movements. Proc. Natl. Acad. Sci. U. S. A. 103, 19806–19811.
Zeng, Y., Yan, B., Sun, Q., He, S., Jiang, J., Wen, Z., Qu, J.Y., 2014. In vivo micro-vascular imaging and flow cytometry in zebrafish using two-photon excited endogenous fluorescence. Biomed. Opt. Express 5, 653–663.

3

CARDIAC PRECONDITIONING, REMODELING AND REGENERATION

TODD E. GILLIS[1]
ELIZABETH F. JOHNSTON

University of Guelph, Guelph, ON, Canada
[1]Corresponding author: tgillis@uoguelph.ca

The Cardiovascular System: Development, Plasticity and Physiological Responses, Volume 36B
FISH PHYSIOLOGY

DOI: https://doi.org/10.1016/bs.fp.2017.09.004

This chapter describes the molecular and cellular mechanisms that enable the heart of some fish species to maintain function during acute changes in physiological and environmental conditions, as well as to remodel in response to chronic stressors. Specifically, we cover hypoxic preconditioning in rainbow trout, as well as the remodeling response of the heart from a number of fish species to hypoxia, anoxia, and thermal acclimation. This remodeling response represents a highly integrated and regulated process across multiple levels of biological organization. In addition, we examine the capacity of the zebrafish heart to regenerate following cardiac injury, the mechanisms involved in this response, and how this ability differs from that seen in mammalian species.

1. INTRODUCTION

The function of the vertebrate heart is extremely sensitive to changes in physiological conditions and workload. For fish living in waters where environmental conditions can vary both spatially and temporally, the adjustment of cardiac output ($\dot{Q}$) to support physiological requirements can be a potential challenge. For example, temperate freshwater fishes live in waters where temperature changes with depth as well as season, while tide pool sculpins and small-bodied fishes, such as zebrafish, can experience significant daily changes in oxygen availability and temperature (Gillis and Tibbits, 2002; Mandic et al., 2009; Todgham et al., 2005). In addition, there can be significant variation in the demands placed on the heart over the lifetime of a fish, such as in Pacific salmon species, due to arduous terminal spawning migrations and reproductive requirements (Eliason et al., 2011, 2013). Not surprisingly, numerous studies over the last 30 years have demonstrated that the function and morphology of many fish hearts are highly plastic, and that these changes result from responses that can occur across multiple levels of biological organization and timescales (Farrell et al., 1988, 1991; Graham and Farrell, 1989; Johnson et al., 2014; Keen et al., 1993, 2016a; Klaiman et al., 2011, 2014; Rodnick and Sidell, 1997). The benefit of such responses is the maintenance of blood delivery to the tissues, and as a result, the continued survival and function of the animal.

A change in physiological conditions puts a stress on the heart as a result of alterations in the biochemical and biophysical conditions under which it is functioning. The result can be a change in contractile function and/or an altered ability to meet the oxygen requirements of the animal. For example, a decrease in temperature affects the function of the Ca^{2+} handling proteins, the metabolic capacity of the heart, and decreases the Ca^{2+} sensitivity of the contractile elements (see Chapter 4, Volume 36B: Eliason and Anttila, 2017).

For the heart to continue to function effectively under these conditions, short-term and long-term solutions may be required, which represent a highly integrated and regulated response (Keen et al., 2016b).

Cardiac remodeling can be described as changes to the size, shape, and/or function of the heart that result from modifications in the molecular, cellular, and/or connective tissue composition of the tissue caused by changes in genome expression (Cohn et al., 2000). The ability of the heart from a number of fish species to remodel in response to a physiological stressor makes them useful models to study the phenotypic plasticity of the vertebrate heart. The following chapter examines the mechanisms that underlie the remodeling response of the hearts from select fish to physiological stressors such as low oxygen or changes in temperature. In addition, we will look at immediate or short-term responses that preclude a transcriptional response and cannot, therefore, be classified as cardiac remodeling. Such responses include hypoxic preconditioning, a phenomenon that protects the heart from a subsequent prolonged period of hypoxia exposure. In fish, the preconditioning response helps to ensure survival in environments where oxygen levels can be variable. We then describe the development and consequences of more chronic stressors, such as prolonged limitations in oxygen and seasonal changes in temperature that lead to significant remodeling of the structure and function of the heart, and the role played by key hormones (e.g., 11-ketotestosterone and angiotensin II). Throughout this discussion, we draw parallels between what is known of the remodeling response of the fish heart as compared to what occurs in the mammalian heart, highlighting conservation of the underlying molecular and cellular pathways that regulate cardiac function. Finally, we discuss the regenerative capacity of the zebrafish (*Danio rerio*) heart following cardiac injury, and how this response differs from that in mammals.

2. CARDIAC RESPONSE TO HYPOXIA, ANOXIA AND ANEMIA

2.1. Hypoxia Preconditioning

Exposure of the vertebrate heart to a period of extreme hypoxia/anoxia followed by reoxygenation causes necrosis and apoptosis of the myocardium and leads to permanent cardiac damage (Cohen et al., 2000). The primary cause of this injury is mitochondrial dysfunction that leads to ion imbalances, the formation of reactive oxygen species (ROS), and the activation of proapoptotic proteins (Cohen et al., 2000). The effect of oxygen deprivation/reoxygenation on the heart is relevant to fish as in their natural environment they can be exposed to varying concentrations of dissolved oxygen, including extreme hypoxia or anoxia. Such conditions can be caused by eutrophication, an increase in water temperature, ice cover in winter, or diurnal conditions within tidal pools (Lefevre et al., 2014; Mandic et al., 2009; Stecyk et al., 2011; Todgham et al., 2005). Such conditions can have significant consequences on both the immediate functional capacity of the fish, as well as their long-term survival. However, Rees et al. (2001) demonstrated that exposing zebrafish to 48 h of non-lethal hypoxia (10% O_2) increased the survival rate upon exposure of the fish to a normally lethal level of hypoxia (5% O_2). This work suggested that the initial period of hypoxia had a protective effect against successive, more severe, periods of hypoxic insult (Rees et al., 2001). Subsequent studies have illustrated that an initial period of hypoxia exposure helps to protect the heart of rainbow trout (*Oncorhynchus mykiss*; Gamperl et al., 2001), and the brain of epaulet sharks, (*Hemiscyllium ocellatum*; Renshaw et al., 2002), from a second hypoxic exposure. Gamperl et al. (2001) first demonstrated cardiac preconditioning in fish in an experiment with a strain of hypoxia sensitive rainbow trout. These authors, using an *in situ* preparation, found that 5 min of severe hypoxia (perfusate $PO_2 < 5$ mm Hg) exposure protected cardiac function during a subsequent 15 min exposure. Without such preconditioning, 15 min of severe hypoxia caused maximum cardiac output ($\dot{Q}_{max}$) to decrease by approximately 25% (Gamperl et al., 2001) (Fig. 1). Interestingly, however, work with a strain of hypoxia insensitive trout demonstrated that cardiac function was not as affected by hypoxia exposure (Faust et al., 2004), and that even a comparatively intense preconditioning stimuli (20 min of severe hypoxia) did not induce a cardioprotective/preconditioning response (Overgaard et al., 2004). The work of Overgaard et al. (2004) also suggested that the ability of the heart from hypoxia insensitive trout to recover function following severe hypoxic exposure may be related to high levels of lactate efflux during oxygen limitation. Therefore, the noted differences in hypoxia sensitivity between species, and between strains of rainbow trout, may be related to differences in the ability to transport lactate out of the myocytes. Further work is required to test this hypothesis. Hypoxia preconditioning has now been demonstrated for the hearts of goldfish, (*Carassius auratus*, Chen et al., 2005), yellow flounder (*Limanda ferruginea* and epaulet sharks Rytkonen et al., 2010), but not Atlantic cod (*Gadus morhua*: Gamperl and Farrell, 2004; MacCormack and Driedzic, 2002). This clearly suggests that there are differences in the ability of cardiac myocytes from different species, and strains of rainbow trout, to respond to severe hypoxia. Such differences may reflect the consequences of different environmental pressures on these fishes during evolutionary history.

In mammalian hearts, ATP-sensitive K^+ (K_{ATP}) channels in the mitochondrial membrane are critical to the hypoxia/ischemic preconditioning response (Gross, 2000). For an extensive review of this subject, please see Foster and Coetzee (2016). Briefly, K_{ATP} channels are present in the

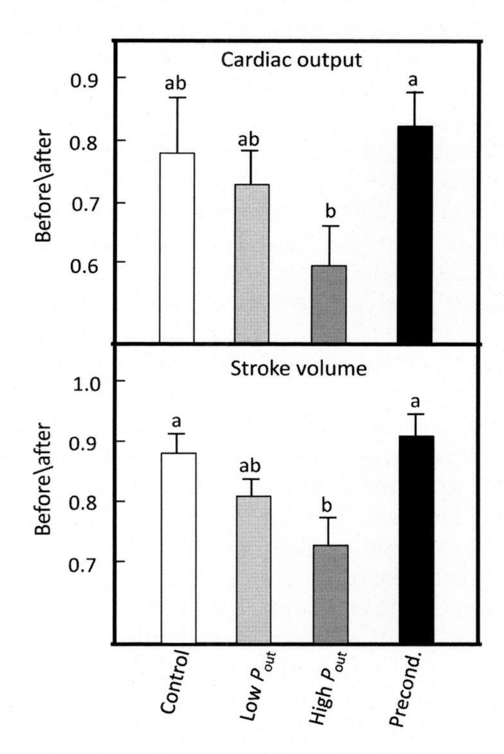

Fig. 1. Influence of hypoxic preconditioning on maximum performance of *in situ* working rainbow trout hearts measured before and after exposure to severe hypoxia (perfusate $PO_2 < 5$ mmHg). Control = oxygenated perfusion; low output pressure (P_{out}; 10 cm H_2O), 15 min severe hypoxic exposure at low P_{out}; high P_{out}, 15 min of severe hypoxic exposure at high P_{out} (50 cm H_2O), Precond., 5 min of severe hypoxia at a P_{out} of 10 cm H_2O, followed after 20 min of oxygenated perfusion, before the protocol described for the high P_{out} group. *Dissimilar letters* indicate before/after values that were significantly different from each other. Cardiac output was measured as: $mL\,min^{-1}\,kg^{-1}$ and stroke volume as: $mL\,kg^{-1}$. *Figure modified from Gamperl, A.K., Todgham, A.E., Parkhouse, W.S., Dill, R., Farrell, A.P., 2001. Recovery of trout myocardial function following anoxia: preconditioning in a non-mammalian model. Am. J. Physiol. Regul. Integr. Comp. Physiol. 281, R1755–1763.*

sarcolemmal (SL) membrane and the inner mitochondrial membrane, and open when the ATP/ADP ratio decreases. In the mammalian heart, ischemic preconditioning is initiated by the activation of multiple isoforms of protein kinase C (PKC) including cPKCα (Bouwman et al., 2007; Hassouna et al., 2004) and nPKCε (Farah and Sossin, 2012), and these target the K_{ATP} channels in the SL and inner mitochondrial membrane. The opening of K_{ATP} channels in the SL membrane decreases the metabolic requirements of the cell by inhibiting contractile function (Ala-Rami et al., 2003; Das et al., 2002; Debska et al., 2002). This effect is due to a decrease in the ion gradients required to move Ca^{2+} through the membrane (Ala-Rami et al., 2003; Das et al., 2002; Debska et al., 2002). The phosphorylation of K_{ATP} channels in the inner mitochondrial membrane results in the loss of the K^+ gradient across the membrane, and a decrease in mitochondrial activity (Gross and Fryer, 1999; Murry et al., 1990). As a result, the cardiac myocytes are less susceptible to damage during reoxygenation because the damaging processes described above, including the release of ROS and the activation of apoptotic proteins, are avoided (Foster and Coetzee, 2016; Gross, 2000; Gross and Fryer, 1999).

Experiments by MacCormack and Driedzic (2002) have established the presence of mitochondrial K_{ATP} (mK_{ATP}) channels in the hearts of yellow flounder *L. ferruginea.* These authors demonstrated that different K_{ATP} channel antagonists [diazoxide, glibenclamide, and sodium 5-hydroxydecanoic acid (5-HD)], alter isometric force production by ventricular preparations (MacCormack and Driedzic, 2002). More specifically, 5-HD, a mitochondrial K_{ATP} (mK_{ATP})-specific blocker, improved force production while the specific mK_{ATP} agonist diazoxide preserved resting tension and eliminated anoxic force potentiation (MacCormack and Driedzic, 2002). Interestingly, similar ventricular preparations from the Atlantic cod were not affected by the K_{ATP} channel antagonists indicating that there are differences in the K_{ATP} channels expressed in different fish species. Work by Cameron et al. (2013) has also demonstrated that the open probability of sarcolemmal K_{ATP} channels is increased when goldfish, *C. auratus*, are exposed to moderate hypoxia, and thus, that there is active modification of membrane ion gradients in this hypoxia tolerant species. Additional research is clearly required to establish the role of K_{ATP} channels in the preconditioning response in fish hearts, and if differences in channel function relate to differences in hypoxia tolerance between species.

2.2. Cardiac Remodeling in Response to Hypoxia

The following section describes the response of the fish heart to hypoxic exposure. It is, however, important to note that in the natural environment, changes in dissolved oxygen can occur concurrently with variations in other environmental conditions (e.g., temperature or pH). Owing to the complexity of designing experiments that simultaneously manipulate several environmental variables, there have only been a few studies where the consequences of multiple aquatic stressors on the function of the cardiovascular system of fishes have been examined (Anttila et al., 2013, 2015; McBryan et al., 2013). Nonetheless, the results of these studies clearly demonstrate that such work is critical to fully understanding the response of animals to changes in environmental conditions, and thus, the impacts of global climate change.

With regards to the effects of hypoxia alone, recent work by Parente et al. (2013) showed that exposure of adult zebrafish to severe hypoxia (5% O_2) for 15 min, followed by recovery in normoxic water, resulted in significant changes in the molecular, biochemical, and physiological function of the heart. These changes included evidence of enhanced oxidative stress, an inflammatory response, activation of hypoxia inducible factor 1-α (HIF 1-α)-dependent genes, cellular apoptosis, and necrosis, and depressed ventricular function, and were primarily due to a marked, and rapid, increase in reactive ROS production caused by reoxygenation (Parente et al., 2013). These effects have the potential to result in long-term, potentially permanent, cardiac damage, and thus, these authors also tracked the progression of cellular apoptosis and proliferation in the heart following this treatment (Parente et al., 2013). At 18 h (h) after the severe hypoxic exposure, approximately 10% of myocyte nuclei were positive for TUNEL (terminal deoxynucleotidyl transferase dUTP nick-end labeling) indicating apoptosis (Fig. 2) (Parente et al., 2013). However, after 30 days (d) of recovery in normoxia, the hearts had regained a complete compliment of functional cardiomyocytes (Parente et al., 2013). This is an interesting result, as it demonstrates that the zebrafish heart does not necessarily tolerate oxygen deprivation,

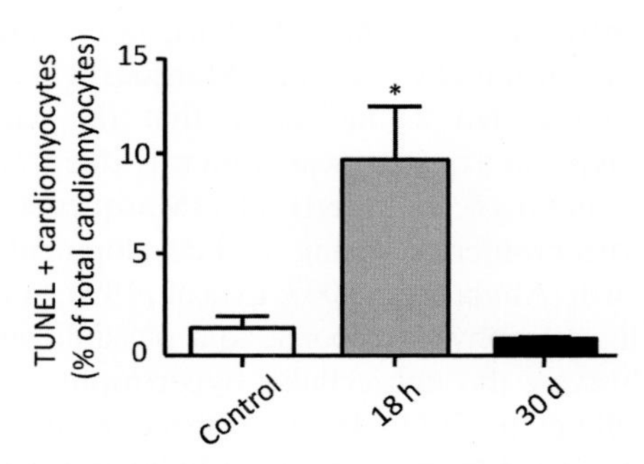

Fig. 2. Apoptosis and regeneration of cardiomyocytes in zebrafish hearts in response to acute severe hypoxia. Zebrafish were exposed to 5% dissolved oxygen for 15 min. This figure represents the number of cells that were positive for TUNEL (marker of cellular apoptosis) in the control fish, and hypoxia-exposed fish after after 18 hours (h) and 30 days (d) of recovery. After 30 d, the number of TUNEL+ cells decreased to a point where it was not significant from the control group. This indicates that the hearts had removed the apoptotic cells and replaced them with new cardiomyocytes. Figure modified from Parente, V., Balasso, S., Pompilio, G., Verduci, L., Colombo, G.I., Milano, G., Guerrini, U., Squadroni, L., Cotelli, F., Pozzoli, O., Capogrossi, M.C., 2013. Hypoxia/reoxygenation cardiac injury and regeneration in zebrafish adult heart. PLoS One 8, e53748.

instead, cells die and are replaced at a later time to ensure the long-term maintenance of heart function.

HIF 1-α is one of the primary mediators of the cellular response to hypoxic exposure in the vertebrate heart (Nikinmaa and Rees, 2005; Semenza, 2002; Zhang et al., 2012). This transcriptional factor regulates the expression of gene transcripts that encode for proteins involved in angiogenesis, erythropoiesis, glucose transport, glycolysis, cell proliferation, and cell survival (Nikinmaa and Rees, 2005; Semenza, 2002; Zhang et al., 2012). The expression of these proteins, and the resultant changes in cellular and physiological function, increase oxygen delivery, and carbohydrate metabolism in the heart (Nikinmaa and Rees, 2005). Rissanen et al. (2006) demonstrated that the exposure of crucian carp, *Carrassius carrassius*, to hypoxia for 45 min at 18°C caused an increase in HIF 1-α protein in the heart. One consequence of increased HIF 1-α production is an increase in the expression of CD73 an enzyme that converts adenosine monophosphate to adenosine (Semenza, 2011). Such a response may be relevant for preserving cardiac integrity, as adenosine is a key regulator of ischemic preconditioning (Semenza, 2011).

Hypoxia exposure of zebrafish and the cichlid, *Haplochromis piceatus*, also causes morphological changes to the heart (Marques et al., 2008). In the study by Marques et al. (2008), fish were exposed to increasing levels of hypoxia over 4 d and then kept at 10% O_2 for 21 d. Histological analyses showed that chronic hypoxia exposure resulted in a decrease in the size of the ventricular outflow tract, a decrease in the size of the lacunae in the central cavity of the ventricle (i.e., the ventricular cavity had filled in), and an increase in the number of nuclei in the myocardium by over 50% (Marques et al., 2008) (Fig. 3). The latter finding was interpreted as indicating that the cardiac hypertrophy (growth) caused by hypoxia exposure was primarily due to myocyte proliferation (hyperplasia) and not myocyte hypertrophy (Marques et al., 2008). In mammals, cardiac myocytes proliferate during fetal development, and then exit the cell cycle following birth (Ahuja et al., 2007; Li et al., 1996; Soonpaa et al., 1996). Therefore, cardiac hypertrophy in response to a physiological or pathological stressor occurs exclusively through cellular hypertrophy (Ahuja et al., 2007; Li et al., 1996; Porrello et al., 2011). In the Marques et al. (2008) study, transcriptomic analysis was used to examine the cellular response of the zebrafish heart to hypoxia exposure and the resultant remodeling (Marques et al., 2008). These authors reported that there were changes in the expression levels of 2.5% of the 15,000 transcripts (375 transcripts) analyzed in the heart, with 69% of these increasing in abundance (Marques et al., 2008). When organized by cellular function, the 260 transcripts that increased were associated with cardiac hypertrophy, protection from ROS, angiogenesis, metabolism, transcription, and tissue growth and development (Marques et al., 2008). Together, these changes would enable the growth, and maintenance, of new myocardium.

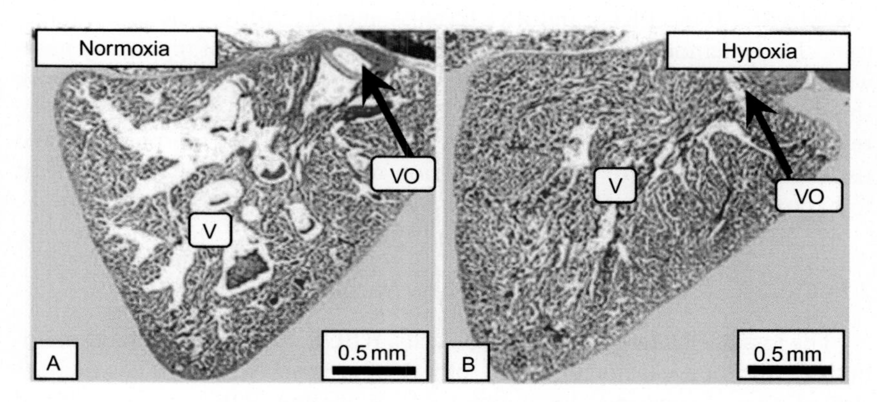

Fig. 3. Chronic hypoxia exposure induces cardiac hypertrophy in the cichlid, *Haplochromis piceatus*. (A) A micrograph of a heart from a fish maintained in normoxia. (B) A micrograph of the heart from a fish maintained in hypoxia (10% air saturation) for 21 d. Hearts were fixed and then stained with *Azan blue*. V, ventricle, VO, ventricular outflow tract. Figure modified from Marques, I.J., Leito, J.T., Spaink, H.P., Testerink, J., Jaspers, R.T., Witte, F., van den Berg, S., Bagowski, C.P., 2008. Transcriptome analysis of the response to chronic constant hypoxia in zebrafish hearts. J. Comp. Physiol. B 178, 77–92.

In addition, the increased expression of six transcripts associated with protection from ROS indicates that these hearts were well prepared for reoxygenation. For example, previous work with rats suggests that the product of one of the transcripts, ceruloplasmin, plays a role in cardiac protection due to its antioxidant function (Kennedy et al., 2014).

One common response of the vertebrate heart to an increase in workload is the activation of cell-signaling pathways that regulate gene transcription pathways associated with cardiac hypertrophy or changes in the collagen composition of the heart (Katsumi et al., 2004). These pathways involve the activation of a number of protein kinases, through protein phosphorylation, including mitogen-activated protein kinases (MAPK), extracellular signal-regulated protein kinase (ERK), c-JUN NH_2 terminal kinase (JNK), p38-MAPK, PKCε, and PKCδ (Churchill and Mochly-Rosen, 2007; Katsumi et al., 2004; Kolosova et al., 2011; Komuro and Yazaki, 1993). The activation of some of these kinases (PKCε, PKCδ, MAPK, and p38-MAPK) is also associated with the preconditioning response of the heart to hypoxia (Bouwman et al., 2007; Churchill and Mochly-Rosen, 2007; Farah and Sossin, 2012; Gaitanaki et al., 2007, 2008; Hassouna et al., 2004). Interestingly, recent work by Nilsson et al. (2015) demonstrated that exposure of crucian carp, *Carassius carassius*, to anoxia does not affect the levels of these kinases in the heart, but it does result in a sustained increase in their phosphorylation. The proteins that were found to be more phosphorylated with anoxia exposure were ERK 1/2, p38-MAPK, JNK, PKCε and PKCδ (Nilsson et al., 2015). This study, therefore, clearly demonstrated that there is active management of regulatory pathways in the heart of the crucian carp during anoxia exposure. Future work should aim to identify the tissue level consequences of the activation of these regulatory pathways, including changes to heart function, morphology, and tissue composition. For further discussion of the response of the fish heart to hypoxia/anoxia, see Chapter 5, Volume 36B (Stecyk, 2017).

2.3. Cardiac Remodeling in Response to Anemia

The loss of erythrocytes (red blood cells, RBCs) due to anemia has a number of effects on cardiorespiratory physiology. Work with rats has demonstrated that decreasing the number of circulating RBCs reduces blood viscosity, and as a result, vascular resistance (Olivetti et al., 1992). This leads to vasodilation of muscular blood vessels and greater systemic blood volume (Olivetti et al., 1992). In addition, anemia (lower than normal RBCs) reduces the oxygen carrying capacity of the blood, and in concert with the above factors, results in an increase in $\dot{Q}$. This increase in $\dot{Q}$ is not accomplished with a higher f_H, instead, the elevated $\dot{Q}$ is achieved through an increase in ventricular end-diastolic volume and stroke volume (V_S), and this leads to pressure-induced biventricular hypertrophy in mammals (Olivetti et al., 1992). In fish, there are a number of hematologic disorders that can result in anemia, including toxicant exposure and renal or splenic diseases (Clauss et al., 2008). As in mammals, chronic anemia has been shown to induce morphological and functional changes in the fish heart. For example, Simonot and Farrell (2007) repeatedly injected warm- and cold-acclimated (~17.6 and 6.4°C, respectively) rainbow trout with phenylhydrazine for 8 weeks to reduce hematocrit levels. This treatment causes erythrocytes to lyse, and thus, results in hemolytic anemia. When hematocrit levels fell below 10%, $\dot{Q}$ increased significantly (Simonot and Farrell, 2007). In the anemic, warm-acclimated, fish, relative ventricular mass (RVM) started to increase at 2 weeks post-injection and was 41% greater than control fish by 8 weeks of treatment (Simonot and Farrell, 2007). In the anemic, cold-acclimated rainbow trout, RVM was 18.6% greater than controls by 8 weeks, and this change was due primarily to an approximately 28% increase in the dry mass of the compact myocardium (Simonot and Farrell, 2007). Such a change in cardiac morphology would enable greater force to be generated by ventricular contraction. Work by McClelland et al. (2005) suggests that cardiac hypertrophy in rainbow trout caused by anemia is due to cellular hyperplasia. These authors demonstrated that anemia caused by phenylhydrazine triggered a 35% increase in the DNA content of the ventricle (McClelland et al., 2005). This result was interpreted as an increase in cell nuclei, and therefore, an increase in the number of cardiomyocytes (McClelland et al., 2005). As in rainbow trout, anemia has also been shown to cause cardiac hypertrophy in Atlantic halibut, *Hippoglossus hippoglossus* (Powell et al., 2012) as well as zebrafish (Sun et al., 2009). This suggests that cardiac hypertrophy is a common response of fishes to anemia.

3. RESPONSE OF THE HEART TO ACUTE AND CHRONIC CHANGES IN TEMPERATURE

3.1. The Initial Response to a Temperature Challenge

Temperate fish species live in environments where temperatures can change by more than 20°C during the year (Gillis and Tibbits, 2002), while many fish, such as tide pool species, can be exposed to rapid daily changes in temperature (Todgham et al., 2006). Each of these scenarios requires an active response so that the structural and functional characteristics of the heart are preserved. The initial cellular response to an acute change in temperature protects biological components such as proteins and cellular membranes from damage (Basu et al., 2002). Subsequent to this response, the tissue may begin to remodel in order to maintain function. As has been discussed in previous reviews (Basu et al., 2002; Kiang and Tsokos, 1998), heat shock proteins (HSPs) are essential in helping to protect biological tissues from damage caused by an acute physiological stressor, including temperature change. These proteins function as molecular chaperones, and protect newly expressed proteins as well as serve as templates for the refolding of damaged proteins (Basu et al., 2002; Kiang and Tsokos, 1998). Multiple studies with different fish species have demonstrated that the transcripts for HSPs, specifically HSP30, HSP70 and HSP90, increase in fish tissues in response to acute changes in temperature, as well as hypoxia and anoxia (Basu et al., 2002; Currie et al., 2000; Todgham et al., 2005, 2006). For example, Currie et al. (2000) demonstrated that a 15°C heat shock caused transcripts for HSP70 and HSP30 to be expressed in the heart of 10°C acclimated rainbow trout, when they were not detected prior to the heat shock. More recent work by Stenslokken et al., (2010) has demonstrated that anoxia exposure of the crucian carp, *C. carrassius*, caused HSP70 to increase in the hearts of fish acclimated to 8 and 13°C, but that the increase at 13°C was much larger. This suggests that acclimation temperature influences the response of the animal to anoxia (Stenslokken et al., 2010). This is of obvious significance when considering the consequence of hypoxia on animals in their natural environment, where they are exposed to other multiple stressors simultaneously.

3.2. Response of the Heart to a Chronic Temperature Stressor

As reviewed in (Chapter 4, Volume 36B: Eliason and Anttila, 2017), changes in temperature have a significant impact on cardiac function. There are, however, several temperate species that are able to maintain function throughout the year (Keen et al., 2016b). One reason for this is that the heart of some fish species can undergo significant remodeling in order to alter its physiological capabilities (Keen et al., 2016b). Laboratory studies have demonstrated that this remodeling response includes changes in the expression of gene transcripts for troponin I (Alderman et al., 2012) and troponin C (Genge et al., 2013), the phosphorylation state of the contractile proteins (Klaiman et al., 2011, 2014), the expression level of sarco(endo)plasmic reticulum Ca^{2+}-ATPase (SERCA) (Korajoki and Vornanen, 2013), cardiac hypertrophy (Graham and Farrell, 1989; Klaiman et al., 2011), changes in the thickness of the compact myocardium (Johnson et al., 2014; Keen et al., 2016a; Klaiman et al., 2011, 2014), as well as significant modifications in the amount and composition of cardiac connective tissue, specifically collagen (Johnson et al., 2014; Keen et al., 2016a; Klaiman et al., 2011). Characterization of the functional properties of the rainbow trout heart also reveals that cold acclimation increases the passive stiffness of the whole ventricle (Keen et al., 2016a), the Ca^{2+} sensitivity of the myofilaments (Fig. 4), as well as the rate and magnitude of force generation by the intact ventricle (Fig. 4) (Klaiman et al., 2014). The increase in passive stiffness is thought to be due to the changes in collagen composition (Keen et al., 2016a), while the increased contractile function may be due, at least in part, to a change in the phosphorylation state of the contractile proteins (Klaiman et al., 2014).

The experiments that characterized the above changes used acclimation periods of between 4 and 8 weeks. This indicates that the molecular pathways that underpin these responses are activated relatively quickly and translate into changes in the structure and function of the heart. It is not surprisingly, therefore, that Vornanen et al. (2005) found that 4 weeks of cold (4°C) acclimation in rainbow trout had a significant effect on the cardiac transcriptome, and that the expression of 9% of the 1380 genes monitored was affected; the majority (82%) of these responses an increase in transcript abundance (Vornanen et al., 2005) (Fig. 5). These changes in the transcriptome were accompanied by cardiac hypertrophy. This suggests that the transcriptomic changes were actively being translated into the proteins that were responsible for cellular growth and tissue remodeling. The fact that hypoxia exposure and cold acclimation both cause cardiac hypertrophy, but the mechanism responsible for the tissue growth is not the same (cellular hyperplasia *vs* cellular hypertrophy), is quite interesting. Future experiments should identify how

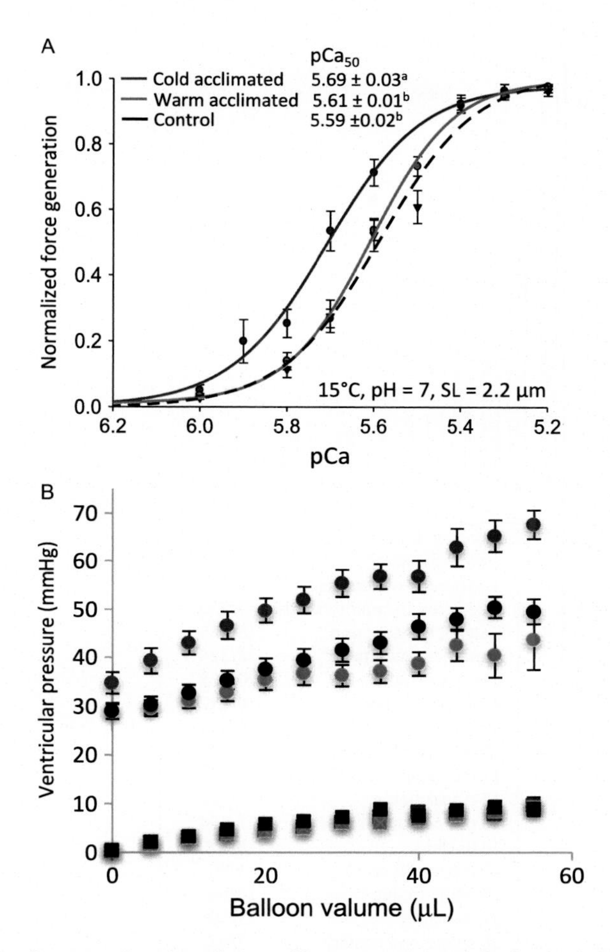

Fig. 4. Contractile properties of cardiac myofilaments and ventricles from rainbow trout acclimated to 4, 11 (control) and 17°C for 8 weeks. (A) Relative Ca^{2+}-activated force generated by cardiac myofilaments measured at 15°C. pCa_{50} is the pCa at half-maximum force. *Different superscript letters* denote a significant difference between values ($P < 0.05$). (B) Pressure development by ventricles measured at 15°C using a Langendorff preparation. *Circles* indicate ventricular developed pressures while *squares* indicate diastolic pressures. The balloon volume of "0" is the volume at baseline conditions. Developed pressures at balloon volumes greater than baseline were higher for the 4°C acclimated (*blue symbols*) fish than those for the 11°C (*black symbols*) and 17°C (*red symbols*) acclimated fish. pCa is the negative log of the Ca^{2+} concentration. SL = sarcomere length. Figures modified from Klaiman, J.M., Pyle, W.G., Gillis, T.E., 2014. Cold acclimation increases cardiac myofilament function and ventricular pressure generation in trout. J. Exp. Biol. 217, 4132–4140.

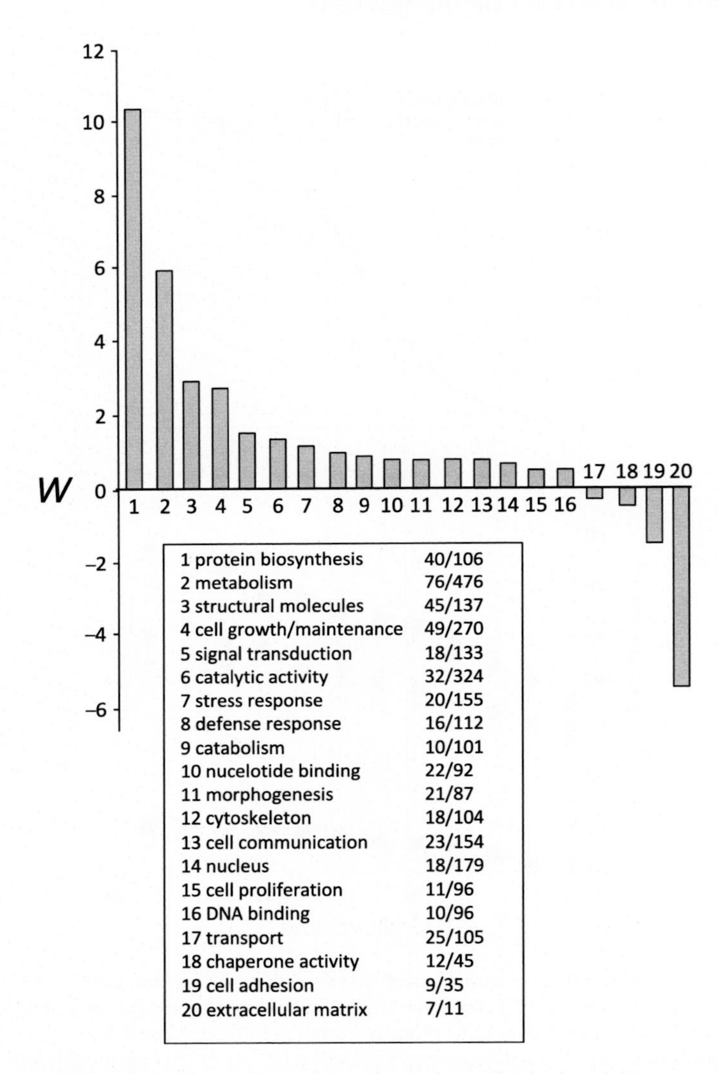

Fig. 5. Effect of temperature acclimation on transcript expression in the rainbow trout ventricle. Ventricular transcriptomes of cold-acclimated (4°C) and warm-acclimated (18°C) rainbow trout heart are compared. The number on the *X*-axis corresponds to 1 of the 20 functional categories [gene ontology (GO) terms] described. *W* (weight) is the relative contribution of the specific functional category to transcriptomic differences between hearts from warm- and cold-acclimated fish. The ratio beside each category indicates the number of genes in that category that changed with cold acclimation. Figure modified from Vornanen, M., Hassinen, M., Koskinen, H., Krasnov, A., 2005. Steady-state effects of temperature acclimation on the transcriptome of the rainbow trout heart. Am. J. Physiol. Regul. Integr. Comp. Physiol. 289, R1177–R1184.

the signaling mechanisms responsible for these responses differ between the two conditions.

The 2005 study by Vornanen and colleagues demonstrated that the transcripts most affected by cold acclimation in the rainbow trout heart are involved in protein synthesis, cellular integrity, cell growth, and the maintenance of cellular metabolism (Fig. 5). More specifically, there was a significant increase in the gene transcripts that encode for ribosomal proteins, and transcriptional factors for the initiation and maintenance of protein synthesis (Vornanen et al., 2005). Such changes would support cellular growth. There was also a significant decrease in the abundance of transcripts associated with the collagen matrix (Vornanen et al., 2005). While Vornanen et al. (2005) did not measure changes in collagen content in the heart with cold acclimation, this result does suggest that there is active regulation of collagen composition during thermal acclimation. Transcriptional changes with cold acclimation also indicated an increased reliance on glycolytic substrates for ATP production (i.e., an increase in transcripts for glycogen phosphorylase and malate dehydrogenase) and a decrease in mitochondrial content (i.e., a decrease in transcripts for ATP/ADP translocases and cytochrome *c*) (Vornanen et al., 2005). However, this response of the rainbow trout heart is quite different from what has been reported for other species. For example, measurement of the activities of a number of enzymes in the hearts of cold-acclimated chain pickerel (*Esox niger*: Kleckner and Sidell, 1985), white perch (*Morone americana*; Sephton and Driedzic, 1991) and carp (*Cyprinus carpio*; Cai and Adelman, 1990) suggests an increased reliance on fatty acids as substrates for aerobic metabolism (Driedzic et al., 1996). In addition, cold acclimation of Pacific bluefin tuna, *Thunnus orientalis*, has been shown to result in an increase in mitochondrial density within the cardiac myocytes (Shiels et al., 2011), as well as a transcriptional response that would support an increase in lipid metabolism within the heart (Jayasundara et al., 2013). It is important to remember, that while Pacific bluefin tuna do display regional endothermy, the temperature of the heart is the same as the environment. The variation in metabolic changes in response to cold acclimation between fish species, reiterates the fact that there is considerable variation in the response of the fish hearts from different species to an environmental stressor. For more information on the effects of cold (both acute and chronic) on cardiac metabolism, see Chapter 6, Volume 36A (Rodnick and Gesser, 2017).

The trout heart, like that of other active fish species, contains a spongy myocardium and a compact myocardium (Klaiman et al., 2011). However, the hypertrophy of the rainbow trout heart with cold acclimation is due to the growth of the spongy myocardium, and not the compact myocardium (Klaiman et al., 2011). In fact, experiments with rainbow trout (Keen et al., 2016a; Klaiman et al., 2011, 2014) and zebrafish (Johnson et al., 2014) indicate that cold acclimation causes a decrease in compact myocardium. As the spongy layer makes up a much greater proportion of the ventricle, this explains the increase in overall heart size with cold acclimation in some species (Klaiman et al., 2011). Hypertrophy of the trout myocardium with cold acclimation would help compensate for the loss of force generating capacity of the muscle caused by the decrease in temperature, while a change in compact myocardium and collagen composition would modify the passive properties of the heart (Keen et al., 2016b). While the rainbow trout heart can grow via cellular hyperplasia (McClelland et al., 2005), the tissue hypertrophy caused by cold acclimation is thought to be primarily due to cellular hypertrophy (Keen et al., 2016a; Klaiman et al., 2011). Opposite to the effects of acclimation to low temperatures, acclimation of male rainbow trout to warmer temperatures results in an increase in the thickness of the compact myocardium, a decrease in the amount of spongy myocardium, and a decrease in connective tissue content of both myocardial layers (Klaiman et al., 2011). Together, the opposing effects of warm and cold acclimation on the rainbow trout heart, suggest that the hearts of these temperate fishes reversibly remodel as environmental temperatures cycle between winter and summer (Klaiman et al., 2011).

While cold acclimation causes cardiac hypertrophy in some fishes (green sunfish, *Lepomis cyanellus*: Kent et al., 1988; channel catfish, *Ictalurus punctatus*: Kent et al., 1988; smallmouth bass, *Micropterus dolomieui*: Sephton and Driedzic, 1991; striped bass, *Morone saxatilis*: Rodnick and Sidell, 1997; and rainbow trout), this response is not seen in white perch (*M. americana*) or yellow perch (*Perca flavescens*) (Sephton and Driedzic, 1990). In addition, as mentioned above, cold acclimation of zebrafish was found to only cause a decrease in the thickness of the compact myocardium without hypertrophy of the spongy myocardium. There are also differences between fish species in the effect of cold acclimation on the metabolic capacity of the heart. For example, Tschantz et al. (2002) have demonstrated that there is variation in the effect of cold acclimation on the activities of hexokinase, lactate dehydrogenase, and cytochrome oxidase in the hearts of largemouth bass, *Micropterus salmoides*; green sunfish, *L. cyanellus*; bluegill, *Lepomis macrochirus*; white crappie, *Pomonix annularis*; and black crappie, *Pomoxis nigromaculatus* (Tschantz et al., 2002). For example, the activity of cytochrome oxidase was increased in the hearts of largemouth bass and white crappie with cold acclimation, but the activity of this enzyme was not affected in the hearts of the other three species (Tschantz et al., 2002). There are multiple factors that can potentially explain these interspecific differences in the response of the fish heart to thermal acclimation. These include differences in overwintering strategies, native environmental conditions, and behavioral responses to seasonal changes in environmental temperature. Clearly, however, this variability represents a fertile ground to begin testing hypotheses regarding the molecular basis of thermal plasticity of the fish heart.

3.3. Remodeling of Connective Tissue With Thermal Acclimation

While cold acclimation can cause cardiac hypertrophy in fish, it can also induce changes in the connective tissue content of the heart (Keen et al., 2016a, b; Klaiman et al., 2011). For example, Klaiman et al. (2011) showed that cold acclimation increased connective tissue content in the rainbow trout heart (Fig. 6), while Keen et al. (2016a) demonstrated that cold acclimation increases the collagen content of this species' heart (Fig. 6). Collagen is the major component of connective tissue. The relative amount and type of connective tissue present in the extracellular matrix (ECM) of biological tissues is partially regulated by matrix metalloproteinases (MMPs) (Visse and Nagase, 2003). MMPs are zinc-dependent endopeptidases that catabolize ECM proteins. In fish, MMP13 degrades collagen into gelatin by catalyzing the hydrolysis process (Hillegass et al., 2007). MMP2 and MMP9 digest this hydrolyzed collagen (gelatin) into waste products for removal (Kubota et al., 2003; Li et al., 2002). The activity of all MMPs is regulated, in part, by tissue inhibitors of metalloproteinase (TIMP), which bind to the MMP proforms and can ultimately reduce the rate of connective tissue degradation (Willenbrock et al., 1993). Using quantitative PCR, Keen et al. (2016a) found that cold acclimation of rainbow trout decreased the expression of MMP2 and MMP13 mRNA, and increased transcript levels of TIMP2 in the heart. As mentioned above, these authors also demonstrated that cold acclimation increases the connective tissue content of the rainbow trout heart, suggesting that the changes in transcript abundance for MMPs and TIMP translate into an increase in collagen content.

Cold acclimation of zebrafish appears to have the opposite effect on the collagen content of the heart. Johnson et al. (2014) reported that the hearts of cold-acclimated zebrafish had less collagen, significantly thinner collagen fibers, and a decreased ratio of Type I:Type III collagen (Fig. 6). Type I collagen is thicker and stiffer than Type III (Jalil et al., 1988, 1989; Nelson et al., 2008; Pauschinger et al., 1999). In human hearts, increased collagen fiber thickness and increased Type I:Type III collagen correlate with stiffening of the myocardium and changes in the biomechanical properties of the muscle (Jalil et al., 1988, 1989; Nelson et al., 2008; Pauschinger et al., 1999). Studies by Keen et al. (2016a) with rainbow trout showed that cold-induced fibrosis and hypertrophy are associated with increased passive stiffness of the whole ventricle and with increased micromechanical stiffness of tissue sections. Changes in collagen composition of a tissue are due to changes in the production as well as the turnover of collagen isoforms

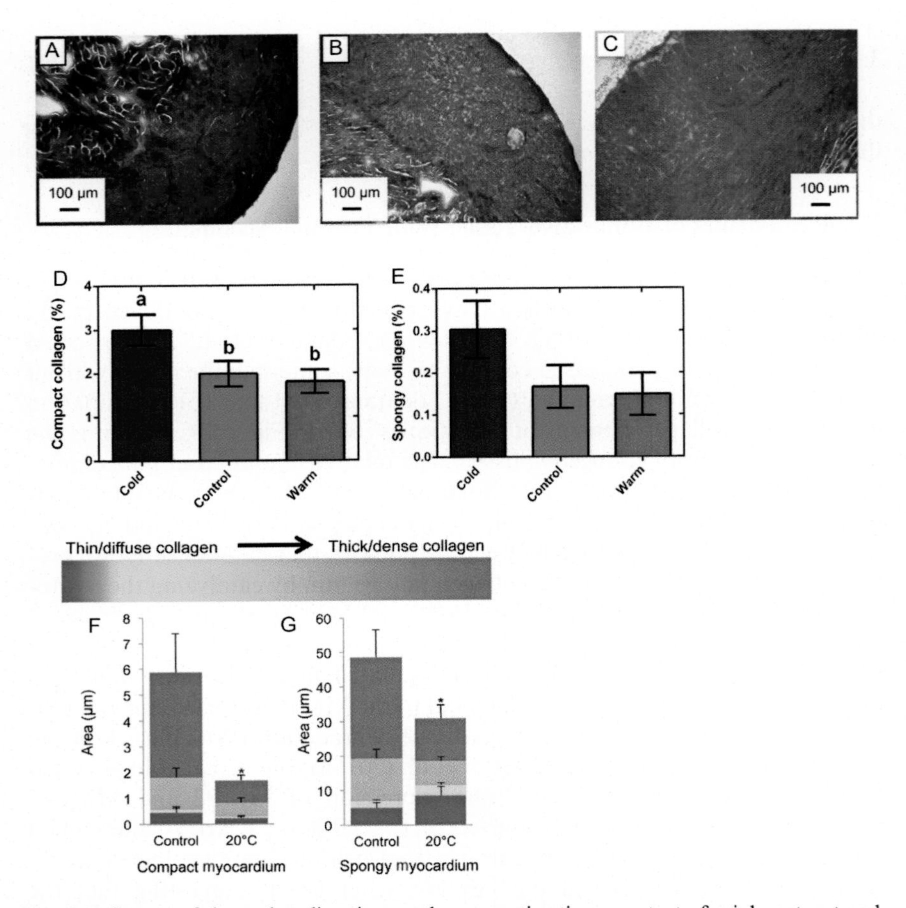

Fig. 6. Influence of thermal acclimation on the connective tissue content of rainbow trout and zebrafish hearts. (A)–(C) Masson's trichrome stained sections of ventricular compact layer from thermally acclimated rainbow trout. (A) Cold-acclimated, 4°C, (B) control, 12°C, (C) warm-acclimated, 17°C; where *pink/purple* is muscle, *blue* is connective tissue, and *white* or *very pale pink* is "extrabundular" space. (D) The hearts of cold-acclimated male rainbow trout had significantly more connective tissue in the spongy layer than that of either control or warm-acclimated male fish. (E) Cold acclimation of male rainbow trout caused an increase in connective tissue content in the compact layer compared to controls, while warm acclimation of male rainbow trout caused a decrease in connective tissue content compared to controls. (D) and (E) The amount of connective tissue present in the spongy layer and compact layer is presented as arbitrary units (AU), and represents the ratio of connective tissue present in the compartment in relation to muscle. Values are mean ± SEM. (F) and (G) Cold acclimation decreases collagen content in the spongy (F) and compact myocardium (G) of zebrafish hearts. Collagen content of ventricular tissue sections from zebrafish hearts was quantified using picro-sirus red staining as in Johnson et al. (2014). (F) and (G) The area occupied by each of the four collagen fiber types in the compact myocardium was calculated within the middle cross-section of hearts from control and cold-acclimated zebrafish. Panels (C)–(E): Modified from Klaiman, J.M., Fenna, A.J., Shiels, H.A., Macri, J., Gillis, T.E., 2011. Cardiac remodeling in fish: strategies to maintain heart function during temperature change. PLoS One 6, e24464; Panels (F) and (G): modified from Johnson, A.C., Turko, A.J., Klaiman, J.M., Johnston, E.F., Gillis, T.E., 2014. Cold acclimation alters the connective tissue content of the zebrafish (Danio rerio) heart. J. Exp. Biol. 217, 1868–1875.

(Visse and Nagase, 2003). Indeed, Johnson et al. (2014) found that changes in the collagen composition of the zebrafish heart with cold acclimation occurred in conjunction with an increase in the expression of gene transcripts for MMP2, MMP9, COL1A1, and TIMP2. The increased expression of the transcripts for MMP2 and MMP9 suggests that the heart is increasing its capacity to remove collagen from the ECM (Johnson et al., 2014). The observed decrease in cardiac collagen content with cold acclimation supports this idea. However, the increase in the expression of gene transcripts for TIMP2 suggests that MMP activity may also be regulated (Johnson et al., 2014). This idea is supported by the work of Li et al. (2002) which suggests that an increase in the production of MMPs, without a counterbalancing increase in TIMPs, leads to functional defects of the myocardial collagen network, and as a result, myocarditis. While the changes in collagen content in response to cold acclimation are opposite in rainbow trout and zebrafish, the results suggest that similar strategies are being used to regulate collagen content. That is, manipulation of the expression of MMPs and TIMPs.

It is not clear why the hearts of rainbow trout and zebrafish differ in their response to cold acclimation. One possible explanation is related to a difference in blood (ventricular) pressures between the two species (Keen et al., 2016b). While there are no reliable measurements for adult zebrafish (see Chapter 4, Volume 36A: Farrell and Smith, 2017), the ventricular systolic pressure of 2.5 cm Gulf killifish (*Fundulus grandis*), a fish similar in size to zebrafish, is approximately 30 mm Hg (Chapter 2, Volume 36B: Burggren et al., 2017). This is lower than the ventral aortic systolic pressure in trout weighing ~750 g (35–60 mm Hg, depending on RVM; Clark and Rodnick, 1999). If this difference in arterial blood pressure (blood pressure regulation) also results in a decrease in central venous blood pressure (i.e., cardiac filling pressure), an increase in the stiffness of the zebrafish myocardium with cold acclimation could make it particularly difficult for the ventricular lacunae of this species to fill with blood during diastole (Keen et al., 2016b). Further work is obviously required to determine why cold acclimation has opposite effects on the collagen content of zebrafish and trout hearts, and to examine how cold acclimation influences the passive stiffness of trout and zebrafish hearts.

3.4. Initiation of Cardiac Remodeling With Thermal Acclimation

One critical question regarding the remodeling response in the fish heart with thermal acclimation is, what is the signal that initiates it? Graham and Farrell (1989) have suggested that cold-induced cardiac hypertrophy occurs in response to the increase in blood viscosity caused by a decrease in environmental temperature. An increase in blood viscosity, due to a decrease in the fluidity of the erythrocyte membranes as well as an increase in plasma viscosity, would increase vascular resistance, and therefore, increase the amount of work performed by the heart (Farrell, 1984). The resultant hypertrophy, would thus, be an adaptive response to increased workload (Keen et al., 2016b). However, an increase in blood viscosity can also induce hypertrophy and fibrosis by increasing shear stress within the heart (Devereux et al., 1984). Human patients exhibiting pathological left-ventricular hypertrophy associated with hypertension (high blood pressure) have increased blood viscosity (Devereux et al., 1984). In contrast, Devereux et al. (1984) demonstrated that patients with hypertension, but normal blood viscosity, have a normal left ventricular mass suggesting that mechanical stimulation of cardiac growth by rheology is more dependent upon blood viscosity than blood pressure. Both increased blood pressure and viscosity lead to mechanical overload as a result of stretching of the cardiomyocytes. Studies by Komuro and Yazaki (1993) have demonstrated that mechanical overload can induce the expression of gene transcripts involved in the immediate–early response associated with the activation of mechanosensitive receptors, such as *c-fos*, *c-jun*, *c-myc*, and *Erg-1* (Komuro and Yazaki, 1993).[a] The activity of MAPKs that initiate protein synthesis pathways is also increased in response to stretch and shows similarities in signaling patterns as when activated by growth factors or hormones (Komuro and Yazaki, 1993). Therefore, the increase in blood viscosity caused by a reduction in temperature could directly activate hypertrophic pathways in the myocardium by causing an increase in the mechanical forces acting on the myocytes (Keen et al., 2016b). In addition, one of the growth factors released within cardiac myocytes in response to stretch is the cytokine transforming growth factor-beta 1 (TGF-β1) (Dobaczewski et al., 2011). Not only does this cytokine trigger muscle hypertrophy (Dobaczewski et al., 2011), but it also plays a role in cardiac fibrosis as will be discussed in Section 5.5. Recently, Johnston and Gillis (2017) demonstrated that treatment of trout cardiac fibroblasts with physiologically relevant concentrations of TGF-β1 increased total collagen levels at 48 and 72 h post-treatment, and an increase in collagen type I protein after 7 d (Fig. 7). Cells treated with TGF-β1 also had lower levels of the gene transcript for *mmp-2* after 48 h, and higher levels of the gene transcript for collagen type Iα1 (*col1a1*) after 72 h (Johnston and Gillis, 2017). These changes in gene expression, similar to that seen in the hearts of cold-acclimated trout (Keen et al., 2016a), suggest that the measured increases in collagen deposition were due to a decrease in the activity of MMPs and an increase in collagen synthesis (Johnston and Gillis, 2017). Together, these results indicate that TGF-β1 is a regulator of ECM composition in cultured trout cardiac fibroblasts and suggest that this cytokine may play a role in

[a]Going forward gene transcripts are indicated using italicized lowercase letters while proteins are indicated in non-italicized uppercase letters.

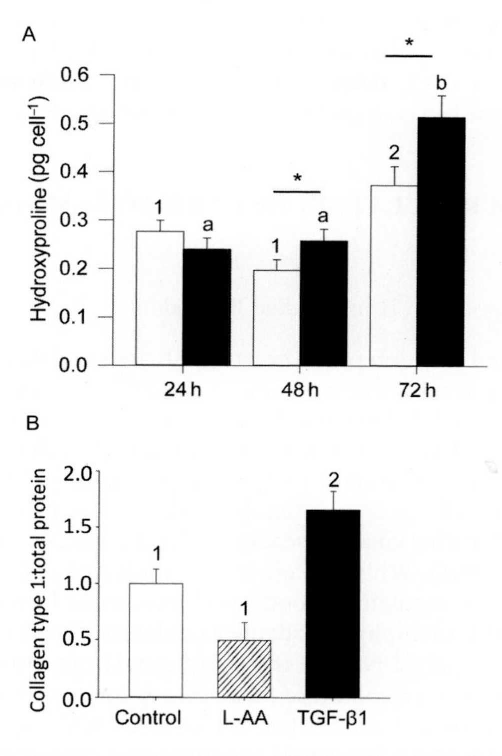

Fig. 7. The effect of TGF-β1 on collagen production by cultured trout cardiac fibroblasts. (A) The average amount of hydroxyproline produced per cell (pg cell^{-1}) after 24, 48, and 72h of TGF-β1 treatment. Hydroxyproline is an amino acid that occurs regularly (typically every third residue) in the helices of the collagen chains, and can thus, be used as an indirect measure of total collagen (Kafienah and Sims, 2004). *Numbers* indicate a significant effect of time on the amount of hydroxyproline produced within control cells ($P<0.05$). *Letters* indicate a significant effect of time on the amount of hydroxyproline produced per cell in the TGF-β1-treated group ($P<0.05$). *, significant effect of TGF-β1 on hydroxyproline concentration between control and TGF-β1-treated cells ($P<0.05$). $n=5$, where each n represents a separate cell line established from a single heart, and each n contains 8–15 technical replicates. (B) Mean collagen type 1 levels measured using Western blotting were quantified by densitometry and then standardized to total protein in the lane. The standardized collagen measurement for the control group was set to 1 and values presented for the treatment groups are relative to this value. Cardiac fibroblast cultures were treated with medium alone (control), or medium containing either L-ascorbic acid or TGF-β1 for 7d. L-ascorbic acid (L-AA) induces mammalian cultured fibroblasts to synthesize and secrete collagen type I into the ECM, and aids in the formation of structured triple helix filaments (Murad et al., 1981). Differences in collagen abundance between treatments were determined with a one-way ANOVA and *post hoc* Bonferroni tests ($n=3$, $P<0.06$). *Bars* indicated by different numbers are different from each other. Figure modified from Johnston, E.F., Gillis, T.E., 2017. Transforming growth factor beta-1 (TGF-beta1) stimulates collagen synthesis in cultured rainbow trout cardiac fibroblasts. J. Exp. Biol. 220, 2645–2653.

regulating collagen content in the trout heart during thermal acclimation. Future work should: (1) characterize changes in TGF-β1 production during cold acclimation and (2) determine how the production of this cytokine changes at different levels of preload (volume) and afterload (pressure).

4. ENDOCRINE REGULATION OF CARDIAC REMODELING IN FISH

4.1. Role of Angiotensin II in Cardiac Remodeling

The growth and function of the heart in fish are regulated, in part, by circulating levels of protein hormones (Guo et al., 2001). For example, angiotensin II (AngII) is a peptide hormone that regulates blood pressure by exerting vasoconstrictive effects on vascular smooth muscle cells via the plasma membrane-bound angiotensin II receptor, type 1 (AT_1) (Guo et al., 2001). Activation of the AT_1 receptor stimulates G-coupled proteins resulting in the activation of tyrosine kinase cascades, and as a result, vasculature contraction (Guo et al., 2001). While the effects of AngII and AT_1 stimulation are most well known for regulating blood vessel tone, these factors can also play a role in myocyte hypertrophy in both the vasculature (Griffin et al., 1991) and heart (Sadoshima et al., 1993). In the heart, AngII can either be the initial signal mediating mechanical stress (discussed below), or can bind directly to AT_1 receptors present on cardiac myocytes (Griffin et al., 1991). For example, activation of AT_1 by AngII in cultured rat cardiac myocytes causes a rapid increase in gene transcripts that promote cardiac growth (*c-fos*, *c-jun*, *jun B*, *Egr-1*, and *c-myc*), and in the expression of protein markers of cardiac hypertrophy including atrial natriuretic factor (ANP) (Sadoshima and Izumo, 1993). In addition, vasoconstriction stimulated by chronic levels of AngII causes cardiac hypertrophy as it leads to elevated blood pressure, increased mechanical stress on the heart, and the release of cytokines (Baker et al., 1990). These cytokines then activate intercellular receptors and initiate the hypertrophic response and fibrotic signaling cascades (Ocaranza and Jalil, 2010).

AngII influences cardiac function in fish in a similar fashion as it does in mammals (Butler and Oudit, 1995; Le Mevel et al., 2008). For a review, see Le Mevel et al. (2008). Experiments in fish also demonstrate that chronic elevated levels of AngII cause cardiac hypertrophy (Imbrogno et al., 2013). For example, chronic elevation of AngII via peritoneal injection results in hypertrophy of the European eel (*Anguilla anguilla*) ventricle and an increase in V_S, as measured using a Langendorff preparation (Imbrogno et al., 2013). In the hearts of eels treated with AngII, there was an increase in proteins that regulate cell growth including C-KIT and ARC (Imbrogno et al., 2013). C-KIT is a tyrosine kinase receptor that, when activated, initiates signaling pathways that are involved in growth and proliferation. ARC acts as an apoptosis repressor in cardiac muscle (Koseki et al., 1998). Together, elevated concentrations of C-KIT and ARC contribute to cardiac hypertrophy by protecting cardiomyocytes and supporting cellular growth. The indirect effects of AngII on the heart are caused by vasoconstriction leading to increased blood pressure, increased mechanical stress, and the release of cytokines (Baker et al., 1990). These cytokines then activate intercellular receptors and initiate the hypertrophic response and fibrotic signaling cascades (Ocaranza and Jalil, 2010). For more information on how AngII affects the fish cardiovascular system, see Chapter 5, Volume 36A (Imbrogno and Cerra, 2017).

4.2. Role of 11-Ketotestosterone in Cardiac Remodeling

Elevated levels of circulating androgens can cause cardiac hypertrophy in male rainbow trout during the spawning season (Thorarensen et al., 1996). Sexually mature male rainbow trout have increased RVM compared to females and sexually immature males (Davie and Thorarensen, 1997; Franklin and Davie, 1992; Graham and Farrell, 1992; West and Driedzic, 1999), and this enhanced cardiac mass is associated with increases in maximum V_S and cardiac power output (CPO) (Franklin and Davie, 1992). These changes are hypothesized to support the increased functional demands placed upon the hearts of male fish during spawning, when they are competing with other males for access to females (Davie and Thorarensen, 1997; Franklin and Davie, 1992; Graham and Farrell, 1992; West and Driedzic, 1999). This response is a likely consequence of increased 11-ketotestosterone production as Davie and Thorarensen (1997) demonstrated that the injection of male rainbow trout with testosterone caused a 1.7-fold increase in ventricle mass. As suggested by Clark and Rodnick (1999), the increased levels of this steroid in sexually mature trout may be caused by an increase in hemodynamic load due to an increase in arterial blood pressure (afterload) and volume (preload). The increased activity levels associated with the spawning season could induce these changes in cardiovascular characteristics. In mammals, testosterone initiates the signaling cascade that leads to cardiac hypertrophy by stimulating cellular growth (Altamirano et al., 2009). It specifically binds to a receptor called extracellular signal-regulated protein kinases 1 and 2 (ERK1/2) (Altamirano et al., 2009). ERK1/2 is an extracellular signal-regulated kinase that targets a protein complex called mechanistic target of rapamycin complex 1 (mTORC1) (Altamirano et al., 2009). This protein complex is involved in cellular survival, proliferation, and growth (Richardson et al., 2004), and integral to cardiac hypertrophy *in vitro* (Boluyt et al., 1997; Kenessey and Ojamaa,

2006; Takano et al., 1996) and *in vivo* (Kemi et al., 2008; McMullen et al., 2004; Shioi et al., 2003). The end result is an increase in protein production and ribosome biosynthesis that lead to cellular hypertrophy (Altamirano et al., 2009). Further work is required to determine if this pathway, which has been characterized in mammalian models, is also present in all fishes.

5. THE ZEBRAFISH HEART AS A MODEL FOR CARDIAC REGENERATION

5.1. Summary of the Regenerative Process

The zebrafish heart can regenerate following injury (Jopling et al., 2010; Poss et al., 2002a, b; Raya et al., 2003; Sun and Weber, 2000). This response involves the initial replacement of the injured tissue with a fibrin clot, which is then replaced with a temporary collagen-based scar, and then by normal myocardial tissue (Fig. 8) (Jopling et al., 2010). The regeneration/repair of the damaged tissue is associated with inflammation, wound healing, angiogenesis, dedifferentiation of terminally differentiated cells, cellular proliferation, cellular differentiation, and neovascularization (Jopling et al., 2010; Kikuchi et al., 2011b; Talman and Ruskoaho, 2016). This differs from the response of the mammalian heart to injury, where the remaining cardiac myocytes hypertrophy in an effort to compensate for the lost tissue and the damaged tissue is replaced with a permanent fibrotic scar generated by fibroblasts and myofibroblasts (Eghbali et al., 1991; Li et al., 1996; Soonpaa et al., 1996; van den Borne et al., 2010; Whittaker et al., 1989). While this initial response is needed to prevent further damage to the myocardium, there is also a reactive fibrotic response in the tissues surrounding the injured area that increases the impact of the injury on the heart, and can lead to further dysfunction and eventual heart failure (Weber et al., 2013).

The ability of zebrafish to regenerate their heart at all life stages, makes these animals a powerful model for characterizing the cellular mechanisms involved in repopulating the heart with functional myocytes following injury (Kikuchi, 2014; Yin, 2013). This process is dependent upon a multitude of coordinated signaling events, with gene targets that direct the development of newly produced cells to form myocytes, the vasculature, fibroblasts, etc. For detailed reviews, see Major and Poss (2007), Yin (2013), and Kikuchi (2014). Here, we will focus on more recent work describing the cellular signaling and transcriptional regulation of the regeneration response in the hearts of adult zebrafish. In these experiments, a portion of the ventricle is amputated or damaged using a cryoprobe, and this leads to 20%–30% of the myocardium, near the apex, being removed (Poss et al., 2002a, b; Raya et al., 2003; Sun and Weber, 2000). Experiments utilizing genetically modified zebrafish are also

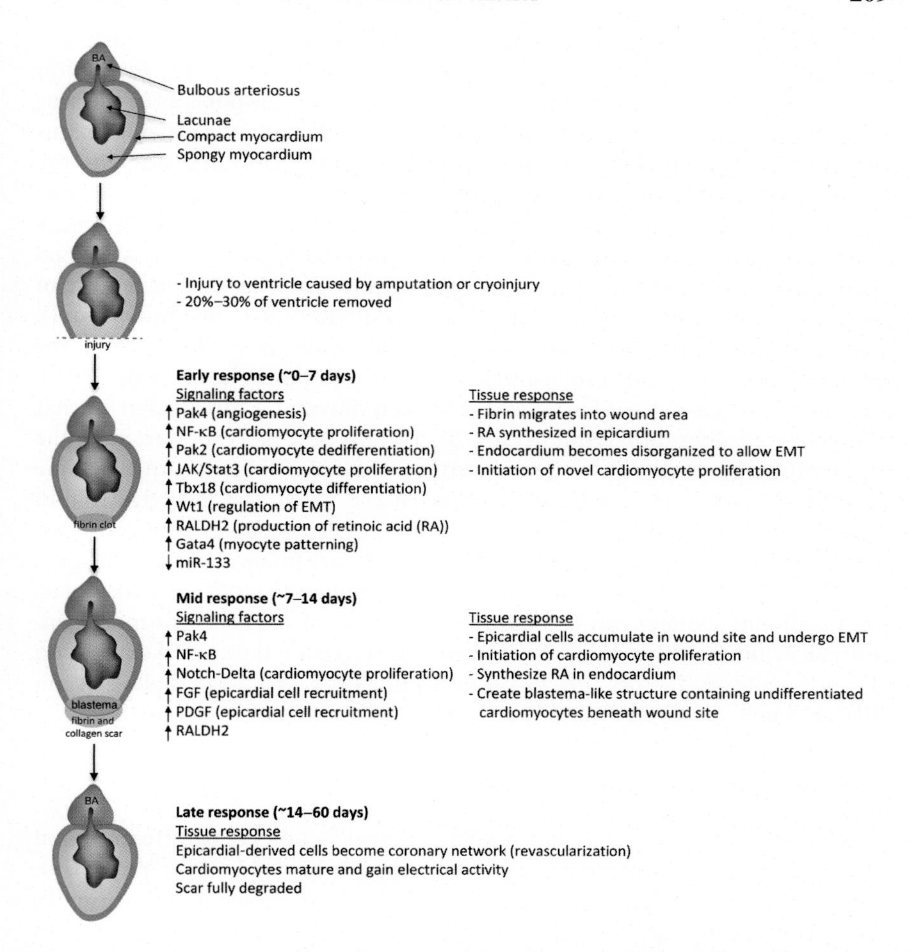

Fig. 8. Summary of cardiac regeneration in zebrafish. Regrowth of the heart has been broken down into three critical time periods that represent common sampling times in various literature sources. The signaling factors that have been shown to appear during these time points, and their relative change compared to control hearts, have been highlighted. The response of the tissue associated with these signaling factors has been included and represents the gross observed changes during these critical times. EMT, endothelial-to-mesenchyme transition; RA, retinoic acid; miR-133, microRNA-133; FGF, fibroblast growth factor; PDGF, platelet-derived growth factor.

used extensively to investigate the molecular regulation of this response. To help facilitate the discussion of cardiac regeneration in the zebrafish heart, the information provided is organized into a number of subsections that focus on particular cardiac tissues and processes. However, we will begin with an overview of the process of regeneration.

Much like the wound healing process in the mammalian heart, the zebrafish heart responds to injury immediately with fibrin clot formation followed soon after by a scar (Yin, 2013) (Fig. 8). The components of the scar are one factor that sets the zebrafish and mammalian hearts apart. The mammalian scar is composed mainly of collagen types I and III, whereas the zebrafish scar is rich in fibrin (Poss et al., 2002b; Sun and Weber, 2000). During the regenerative process, this fibrin scar will be broken down and disappear (Poss et al., 2002b; Sun and Weber, 2000). This is the opposite to what occurs in mammals, where the collagen scar is permanent. Another unique feature of the injured zebrafish heart is that at approximately 7 d post-injury (dpi), cardiomyocytes begin to proliferate, and will continue to do so until the regenerated heart is indistinguishable from a non-injured heart (Yin, 2013). During this process, the fibrotic clot is broken down at a rate similar to that of myocyte proliferation (Sallin et al., 2015). This process is supported by the regenerating epicardium that grows around the degrading clot, in the same direction as myocyte proliferation—toward the apical wound edge (Sallin et al., 2015). The gradual deposition of structural extracellular connective tissue, as well as the formation of new blood vessels, are integral parts of the process (Yin, 2013). The signaling factors that regulate these large-scale tissue compositional changes can be unique to one layer of the heart, or more generally distributed as suggested by studies characterizing the spatial expression of key regulatory factors (discussed in Sections 5.2 and 5.3) (Jopling et al., 2010; Kikuchi et al., 2010, 2011b).

5.2. Role of the Myocardium in Regeneration

Myocardial regeneration in zebrafish is dependent on the dedifferentiation and proliferation of existing myocytes (reviewed in Kikuchi, 2014; Yin, 2013). Using genetic fate mapping, Kikuchi et al. (2010) have shown that the primary source of these myocytes in the adult zebrafish is the compact myocardium. In these studies, the expression of *gata4*, a transcription factor gene in the developing embryonic heart and essential for normal cardiac patterning, was localized to the compact layer at 3 dpi. It is these cells, that express *gata4*, that will proliferate and fill the wound (Kikuchi et al., 2010). Once generated, the new myocytes migrate to the wound (Jopling et al., 2010; Kikuchi et al., 2010, 2011b) (Fig. 8). Another potential source of cardiac myocytes following injury to the zebrafish ventricle is the atrium (Kikuchi et al., 2011b). Recent studies have demonstrated that, following injury to the ventricle, atrial cardiomyocytes will dedifferentiate into cardiac progenitor cells and then migrate to the site of injury (Kikuchi et al., 2011b; Zhang et al., 2013). These cells then play a significant role in repairing the injury (Kikuchi et al., 2011b; Zhang et al., 2013).

A regulator of myocyte proliferation is the nuclear factor kappa light-chain enhancer of activated B cells (NF-κB) (Baldwin, 1996). This nuclear transcription factor binds to a response element that is located within the promoter region of multiple genes and is induced by a broad range of stressors including mediators of immune function, UV irradiation, growth factors, and viral infections (Baldwin, 1996). Following cardiac injury NF-κB rapidly gains transcriptional activity, which is essential for its role as a regulator of the early response to cellular stress (Fig. 9) (Karra et al., 2015). In the regenerating zebrafish heart, NF-κB is localized to proliferating cardiomyocytes in the compact myocardium by 1 dpi and will be expressed until at least 14 dpi (Karra et al., 2015) (Fig. 9). The importance of NF-κB is evidenced by the downstream activation of a wide variety of target genes (reviewed in Dabek et al., 2010). In the zebrafish heart, NF-κB inhibition impairs regeneration after 30 d by decreasing cardiomyocyte proliferation and epicardial responses, a result that confirms its importance as one of the critical long-term regulators of cardiac regeneration (Karra et al., 2015) (Fig. 9). Regeneration of the zebrafish myocardium is regulated in part by the expression of p-21-activated kinase 2 (PAK2). p-21-Activated kinases (PAKs) are a family of serine/threonine protein kinases involved in the proliferation and dedifferentiation of cardiomyocytes, as well as angiogenesis. Peng et al. (2016) demonstrated that PAK2 was concentrated in the cardiomyocytes at the apical edge of the wound at 3 and 7 dpi (Fig. 10). The signaling of PAK2 was also found to be dependent upon the activity of its upstream effectors, cell division control protein 42 (CDC42), and Ras-related C3 botulinum toxin substrate 1 (RAC1). While the inhibition of PAK4 leads to muted angiogenesis in the myocardium, blocking the activity of PAK2 causes cardiomyocyte proliferation to cease. The authors also determined that PAK2 plays a role in directing the dedifferentiation of preexisting cardiomyocytes, as suggested by the inhibition of the early cardiomyocyte markers *gata4* and N2.261 (Peng et al., 2016). These data support the importance of PAK signaling in the regenerating heart and also suggest that activation of the G protein-coupled receptors associated with CDC42 and RAC1 are critical to this process.

The repopulation of the injured myocardium with new cardiac myocytes in the zebrafish heart is regulated by neuregulin 1 (NRG1) (Gemberling et al., 2015). NRG1 is a paracrine signaling factor that binds to tyrosine kinase receptors and regulates cardiac cell processes throughout the life of the animal; including the maintenance of cellular integrity (Odiete et al., 2012). In an *in situ* experiment by Gemberling et al. (2015), the highest level of *nrg1* expression in regenerating zebrafish hearts was measured in the compact myocardium after 7 dpt (days post-treatment) (Fig. 11A). This time point in cardiac regeneration is important as it correlates with the time of peak cardiomyocyte proliferation. Inhibition of *nrg1* in zebrafish following

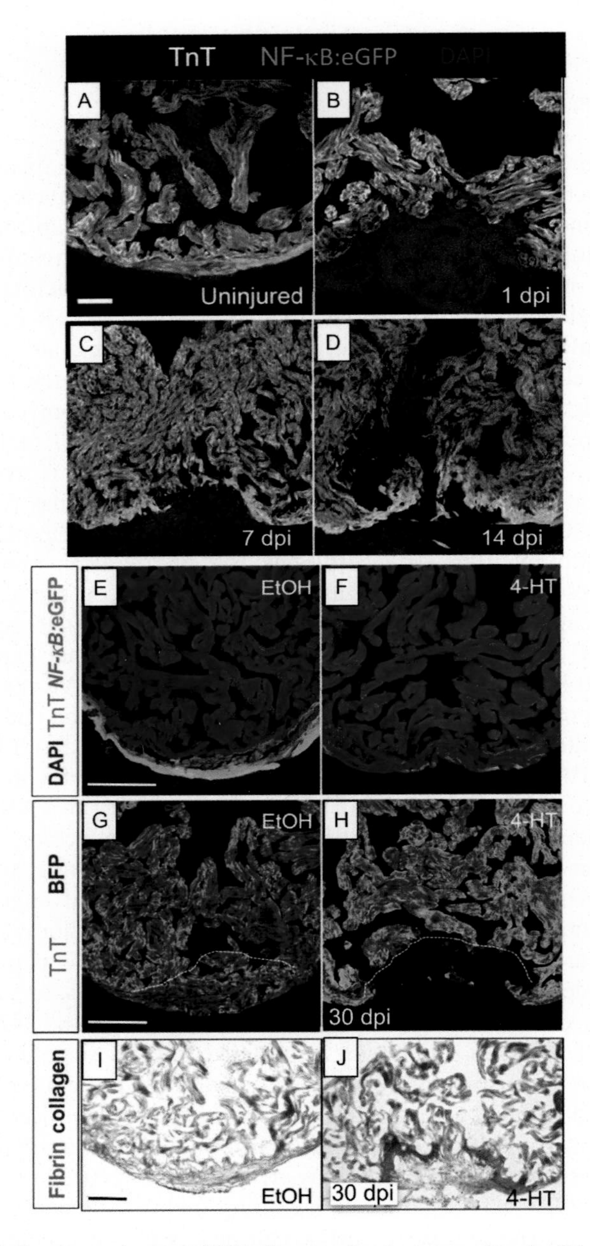

Fig. 9. Immunofluorescent images of NF-κB expression in transgenic zebrafish hearts. (A) An uninjured heart with regular NF-κB expression. Seven to 14dpi (B), NF-κB is upregulated and localized to the wound site (C) and (D). Cardiac troponin T (TnT) staining denotes cardiomyocytes, DAPI stains cell nuclei. To demonstrate that NF-κB plays a critical role in cardiac regeneration, NF-κB was inhibited in an injured heart. Green fluorescence shows that NF-κB is clearly present in normal hearts treated with an ethanol carrier (E), and lacking in fish treated with 4-hydroxytamoxifen (4-HT) (F). Regenerated untreated hearts have fully restored cardiomyocytes (G), whereas hearts where NF-κB was inhibited do not regenerate cardiomyocytes (H). Untreated hearts also do not demonstrate significant scarring (I), but hearts with NF-κB inhibition show a scar containing fibrin and collagen after 30dpi (J). Modified from Karra, R., Knecht, A.K., Kikuchi, K., Poss, K.D., 2015. Myocardial NF-kappaB activation is essential for zebrafish heart regeneration. Proc. Natl. Acad. Sci. U.S.A. 112, 13255–13260.

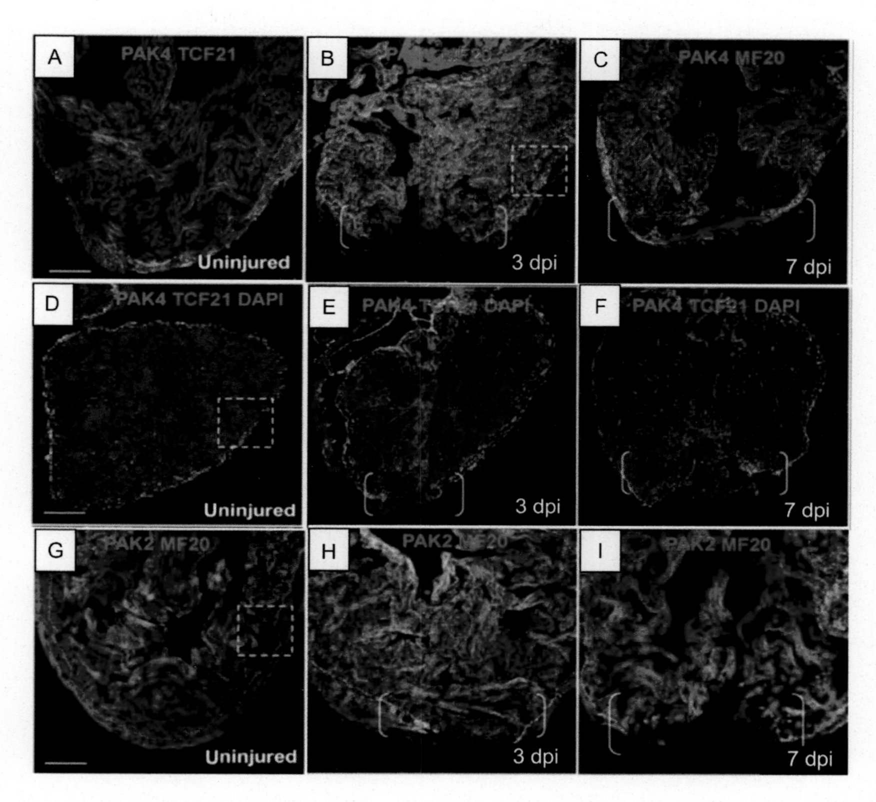

Fig. 10. Differential localized expressions of PAK2 and PAK4 in the uninjured and injured zebrafish heart after 3 and 7 dpi. In the uninjured heart, PAK4 is expressed in the epicardium as well as in the vasculature of the compact myocardium (A) and (D). PAK2 is also expressed in the uninjured heart (G), but is restricted to the cardiomyocytes within the spongy myocardium. In the injured heart, PAK4 appears in the rest of the heart between 3 and 7 dpi (B), (C), (E), and (F). PAK2 is expressed in the rest of the heart after 3 dpi and is localized to the wound site after 7 dpi (H) and (I). Modified from Peng, X., He, Q., Li, G., Ma, J., Zhong, T.P., 2016. RAC1–PAK2 pathway is essential for zebrafish heart regeneration. Biochem. Biophys. Res. Commun. 472, 637–642.

cardiac injury decreased cardiomyocyte proliferation, while reactivation of *nrg1* in healthy non-injured zebrafish caused the heart to hypertrophy (Gemberling et al., 2015). This data support a key role for *nrg1* in cardiac regeneration.

A second signaling pathway that is crucial to cardiomyocyte repopulation is the Janus kinase 1-STAT3 pathway (Fang et al., 2013). Janus kinase 1 (JAK1) is a tyrosine kinase receptor associated with a transmembrane

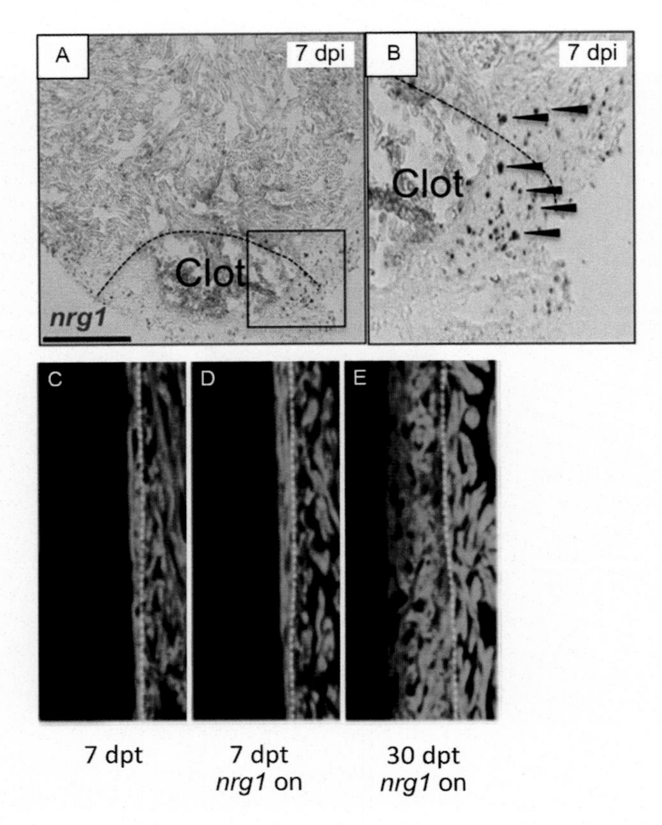

Fig. 11. Expression of *nrg1* (NRGI transcript) during cardiac regeneration in zebrafish. (A) The increased expression and localization of *nrg1* at 7 dpi, after initial clot formation. (B) A close-up of panel (A). *Arrowheads* indicate areas of RNAscope signals detecting *nrg1* expression via *in situ* hybridization. To determine if *nrg1* plays a role in cardiomyocyte growth, overexpression experiments were employed in transgenic fish. (C) A region of compact and spongy myocardium (compact myocardium to the left of the *dotted line*) in fish that did not have *nrg1* overexpression whereas panel (D) is myocardium where there was *nrg1* overexpression (Nrg1 on). *nrg1* overexpression was activated by bathing the transgenic fish in 5 μM tamoxifen. Controls were wild-type fish exposed to tamoxifen (7 dpt = 7 days post-treatment). (E) Hyperplastic myocardium growth in zebrafish after 30 d of *nrg1* overexpression stimulated by tamoxifen treatment. Control fish (no tamoxifen treatment) did not experience significant myocardium growth (not shown). Modified from Gemberling, M., Karra, R., Dickson, A.L., Poss, K.D., 2015. Nrg1 is an injury-induced cardiomyocyte mitogen for the endogenous heart regeneration program in zebrafish. eLife 4.

domain that phosphorylates different protein isoforms of signal transducers and activators of transcription (STAT) in response to the binding of a ligand (Schindler and Darnell, 1995). STAT phosphorylation results in its dimerization and movement into the nucleus where it functions as a transcription factor (Murray, 2007). STAT3 activation is known to play a role in cytoprotection, hypertrophy, angiogenesis, and ECM regulation within the mammalian heart (Hilfiker-Kleiner et al., 2005). In the regenerating zebrafish heart, inhibition of STAT3 reduced cardiomyocyte proliferation (Fang et al., 2013). Interestingly, inhibition of STAT3 during zebrafish development does not restrict cardiac growth (Fang et al., 2013), highlighting that some processes during cardiac regeneration are specific to adult gene regulation pathways.

5.3. Role of the Epicardium in Regeneration

The epicardium is a serous, nonmuscular, membrane that surrounds the heart and is located adjacent to the compact myocardium in the zebrafish heart. Following cardiac injury, the embryonic developmental program is reactivated in the epicardium, and this leads to the production of transcriptional factors and retinoic acid (RA). These molecules are critical to driving, and directing, the regenerative process (Yin, 2013) (Fig. 8). The developmental program activated in the epicardium includes the induction of T-box transcription factor-18 (TBX18) (Lepilina et al., 2006), Wilms' tumor protein (WT1) (Kikuchi et al., 2011a), and the production of retinaldehyde dehydrogenase 2 (RALDH2) (Kikuchi et al., 2011a; Lepilina et al., 2006), a RA synthesizing enzyme. TBX18 transduction plays a role in directing the differentiation of cardiac progenitor cells into cardiomyocytes by enabling a process known as epithelial-to-mesenchymal transition (EMT), where cells in an epithelial layer detach and migrate and/or differentiate (Cai et al., 2008). This can occur multiple times within tissues during development, enabling growth of the tissue (Cai et al., 2008). WT1 is also involved in regulating EMTs (Martinez-Estrada et al., 2010). RA signaling is essential for normal cardiac development during embryogenesis, where it directs patterning of the heart and has a mitogenic function (Kastner et al., 1994; Ryckebusch et al., 2008; Sucov et al., 1994; Waxman et al., 2008). Cells from the epicardium also accumulate in the wound area 7–14 dpi, where their recruitment is dependent on fibroblast growth factor (FGF) and platelet-derived growth factor (PDGF) (Kikuchi et al., 2011b). During this process, a subset of the epicardial cells undergoes an EMT to generate the vasculature for the new cardiomyocytes (Lepilina et al., 2006). It is not known if the epicardial region of the heart is directly responsible for producing new myocytes. Studies in mouse and chick embryos maintain that epicardial-derived cells (EPDCs) give rise to coronary vascular smooth muscle cells, but there is some evidence that EPDCs can differentiate into myocytes under specific conditions (Masters and Riley, 2014). While future research should aim to address the direct contribution of the epicardium in myocyte renewal, there is still strong evidence that the epicardium participates in important signaling events that are required for *de novo* myocyte synthesis in the regenerating zebrafish heart.

Recent work has demonstrated that the epicardium is also the location where specific PAKs are first expressed in the heart following cardiac injury (Peng et al., 2016). In mammals, PAK4 is highly expressed in early stages of development and tapers off with age (Liu et al., 2008), and this early expression is critical to heart formation as *pak4* knockout causes fatal cardiac abnormalities in mice (Nekrasova and Minden, 2012). In the cardiac regeneration experiments, a portion of the ventricle of adult zebrafish was surgically amputated then the phosphorylation activity of PAKs in the epicardium and myocardium were characterized at 3 and 7 dpi. Using immunofluorescence, the authors found that PAK4 expression is localized to the epicardium in the uninjured heart (Peng et al., 2016). After injury, this expression begins to expand to the rest of the heart, and occurs throughout the myocardium between 3 and 7 dpi (Peng et al., 2016). Functional inhibition of PAK4 in this injured heart model did not affect overall cardiac regeneration, but it did prevent angiogenesis (Peng et al., 2016). Together, the results from the Peng et al. (2016) study suggest that there is a "reservoir" of PAK4 in the epicardium of zebrafish, and that this is a requirement for heart regeneration due to the role of PAK4 in regulating angiogenesis as new cardiomyocytes emerge.

The epicardium is also a source of fibronectin, a key component of the ECM, during the initial phases of wound healing and regeneration. Studies by Wang et al. (2013) demonstrate that the entire epicardium expresses fibronectin by 1 dpi, and that after 7 dpi the fibronectin is primarily localized to the wound site. It is likely that this mechanism contributes to the initial clot which is composed mainly of mature fibrin; which disappears several days into regeneration (Poss et al., 2002a, b). At this stage of regeneration, the scar area in an injured heart fills in with a collagen matrix due to TGF-β1 signaling (Chablais and Jazwinska, 2012) as described in Section 5.5.

5.4. Role of the Endocardium in Regeneration

Our current understanding of the role of the endocardium during regeneration is limited compared to that of the myocardium and epicardium. Recent work has demonstrated that damage to the zebrafish ventricle caused by amputation induces the endocardium to undergo significant morphological changes, as well as become a site for RA production (Kikuchi et al., 2011b). Following injury, cells in the endocardium near the site of damage become rounded in appearance and detach from the underlying myofibers, resulting in a "disorganized" appearance (Kikuchi et al., 2011b). There is also an increase in the expression of RALDH2 indicative of RA production

(Kikuchi et al., 2011b). Interestingly, damage to the heart of *Polypterus senegalus*, an ancient fish, also causes proliferation of cardiac myocytes and the expression of RALDH2 in the endocardium and epicardium. Similar injury to the mouse heart does not cause proliferation of cardiac myocytes or stimulate the expression of RALDH2 in the heart (Kikuchi et al., 2011b). This result suggests a direct linkage between RALDH2 expression and the ability to regenerate cardiac tissue following an injury.

Two other signaling molecules that play a role in regulating the response of the endocardium to cardiac injury are the notch membrane receptors and the delta transmembrane ligand (Niessen and Karsan, 2008; Zhou and Liu, 2014). Notch receptors interact with the delta ligand on an adjacent cell (Niessen and Karsan, 2008; Zhou and Liu, 2014). When activated, the delta ligand initiates a signaling cascade that results in the production of gene transcripts that are involved in regulating cardiac development, myocyte proliferation, differentiation, and cell fate (Niessen and Karsan, 2008; Zhou and Liu, 2014). Cardiac injury in zebrafish results in an increase in the expression of *notch1b* and *deltaC* throughout the endocardium (Raya et al., 2003). A more recent study found that cardiac injury increases the expression of *notch1a*, *notch1b* and *notch2* in the area of the endocardium adjacent to the wound site (Zhao et al., 2014). Interestingly, it was also found that blocking or overactivating notch receptor expression leads to the dysregulation of cardiomyocyte proliferation (Zhao et al., 2014). Together, these results suggest that cell–cell interactions are important during cardiac regeneration, and that future work should identify the downstream targets of notch signaling during cardiac regeneration.

5.5. Cardiac Fibrosis Pathways

When the mammalian heart is injured, the wound is infiltrated with collagen types I and III, and a scar is formed. This scar is permanent and can lead to significant contractile dysfunction and eventual cardiac failure (reviewed in Sadoshima and Izumo, 1997). The process of scar formation is regulated, in part, by stretching of the myocardium, the latter caused by an increase in blood pressure that results from increased vascular resistance (Carver et al., 1991; Sadoshima et al., 1992; Yamazaki et al., 1995). The stretching of cardiac myocytes causes the release of paracrine and autocrine factors that initiate signaling cascades that result in the upregulation of genes controlling connective tissue deposition (Carver et al., 1991; Sadoshima et al., 1992; Yamazaki et al., 1995). These connective tissue proteins are deposited into the ECM by fibroblasts, and can be degraded by MMPs (discussed in Section 3.3) as part of normal cellular maintenance. However, when deposition occurs at a faster rate than degradation, this can lead to a pathological fibrotic state. This response is also exacerbated by pathological factors such as high blood flow or high blood pressure. The stretching of cardiac fibroblasts influences multiple cellular processes that are activated through various pathways including: the activation of stretch-sensitive ion channels (Li et al., 2009), the disruption of the glycocalyx (Gupta and Grande-Allen, 2006), the stimulation of integrins (MacKenna et al., 1998), and the activation of G protein-coupled receptors (Gudi et al., 1998). In addition, the stretching of fibroblasts beyond their typical physiological range causes the release of TGF-β1 (Ruwhof et al., 2000). This cytokine initiates a signaling cascade that results in the deposition of collagen in the myocardium and can lead to fibrosis (Eghbali et al., 1991; Wright et al., 1991). Collagen deposition occurs because TGF-β1 release leads to an increase in the expression of TIMP and an increase in the expression of collagen genes (Eghbali et al., 1991; Wright et al., 1991). TGF-β1 also plays a role in the formation of the permanent scar that results from cardiac injury or myocardial infarction in mammalian hearts (Bujak and Frangogiannis, 2007). Comparatively, while TGF-β1 plays a role in the initial scar formation following cardiac injury in the zebrafish heart, it is also involved in cardiomyocyte regeneration (Chablais and Jazwinska, 2012). More specifically, Chablais and Jazwinska (2012) found that three types of TGF-β are expressed in the zebrafish heart within the first week post-injury, including TGF-β1 at 7 dpi when cardiomyocyte proliferation is initiated, and that these are not present in the uninjured heart. Thus, it appears as though the zebrafish heart is able to regulate the concentration of TGF-β subtypes depending on the type of connective tissue deposition required. It is not known if TGF-β1 plays a role in the connective tissue remodeling of other fish species. An important next step would, therefore, be to determine if TGF-β1 induces cardiac fibrosis in other species of fish or if it is implicated in the remodeling response to the aforementioned stressors (Table 1). Knowing the role of this cytokine in the regulation of collagen deposition in zebrafish would be of significant interest since TGF-β 1 is known to be a stimulator of muscle hypertrophy, as discussed in Section 3.4 and reviewed in Dobaczewski et al. (2011).

5.6. Role of microRNAs in Cardiac Regeneration in Zebrafish

MicroRNAs (miRNAs) are small non-coding strands of RNA that play a role in regulating gene expression by binding to the 3′-untranslated region (UTR) of their target RNAs, thereby inhibiting translation. Recent work has suggested that miRNAs regulate the expression of at least 60% of human proteins (Yin, 2013). In addition, one miRNA can inhibit the translation of more than one protein, indicating that the regulatory control exerted by microRNAs is far more complex than previously thought (Yin et al., 2012). Studies to understand the role of miRNAs in regulating cardiac regeneration are just beginning. However, recent work on the role of miRNA-133 in cardiac

Table 1
Summary of remodeling response of the hearts from several fish species to different chronic stressors and stimuli

Stressor/stimuli	Chronic response	Species
Cold acclimation	Spongy myocardium hypertrophy[a,d] Spongy myocardium fibrosis[a] Decrease in compact myocardium thickness[a,c] Increase in ventricular pressure generation[b] Increased ratio of thin: thick collagen[c] Increased TIMP expression[e] Decreased MMP2 and MMP13 expression[e]	Rainbow trout[a,b,d,e] Zebrafish[c]
Warm acclimation	Spongy myocardium atrophy[a] Increase in compact layer thickness[a,e] Decreased collagen in compact myocardium[a,e]	Rainbow trout[a,e]
Hypoxia	Cardiomyocyte apoptosis[f] Cardiomyocyte regeneration[f] Cardiomyocyte growth and proliferation[g]	Zebrafish[f,g]
Endocrine		Rainbow trout[h,i,j,k] European eel[l]
Testosterone	Ventricular hypertrophy[h,i,j,k,l] Increased stroke volume[h]	
Angiotensin II	Ventricular hypertrophy[l] Increase in C-KIT, ARC, HSP-90, "eNOS-like" enzyme protein expression[l] Increased stroke volume[l]	
Anemia	Ventricular hypertrophy[k,l,m,n,o] Myocyte hyperplasia[n,p]	Rainbow trout[k,l,m,n] Zebrafish[p] Atlantic halibut[o]

[a]Klaiman et al. (2011); [b]Klaiman et al. (2014); [c]Johnson et al. (2014); [d]Graham and Farrell (1989); [e]Keen et al. (2016a, b); [f]Parente et al. (2013); [g]Marques et al. (2008); [h]Franklin and Davie (1992); [i]Davie and Thorarensen (1997); [j]Graham and Farrell (1992); [k]West and Driedzic (1999); [l]Imbrogno et al. (2013); [m]Simonot and Farrell (2007); [n]McClelland et al. (2005); [o]Powell et al. (2012); [p]Sun et al. (2009).

regeneration in zebrafish (Yin et al., 2012) demonstrated that the expression of this miRNA was suppressed following cardiac injury. This is relevant as miRNA-133 plays a role in cardiac development (Yin et al., 2012). Subsequent experiments found that an increase in miRNA-133 expression following injury inhibited myocardial regeneration, and that the inhibition of miRNA-133 enhanced cardiomyocyte proliferation (Yin, 2013). These results, therefore, suggest that miRNA-133 plays an important role in regulating the regeneration of the myocardium following injury. Recent work by Beauchemin et al. (2015) has demonstrated that miRNA-101a also plays a role in the regulation of cardiomyocyte proliferation in the zebrafish heart. This effect was demonstrated when the suppression of this miRNA, following injury to the ventricle by amputation, caused an increase in cardiomyocyte proliferation but reduced subsequent removal of the scar tissue (Beauchemin et al., 2015). This study also showed that miRNA-101a was downregulated in the early stages of regeneration, but upregulated between 7 and 14dpi. This indicates that a balance between miRNAs and their target genes is essential for the multiple cellular processes involved in cardiac regeneration to be effectively regulated (Beauchemin et al., 2015).

5.7. Heart Regeneration Across Species

Most of the research on the cellular mechanisms responsible for cardiac regeneration in fish has used zebrafish as a model. Experiments with the giant danio (*Danio aequipinnatus*) have shown that the heart of this animal regenerates in a manner similar to that of the zebrafish, including novel myocyte synthesis and epicardial cell proliferation (Lafontant et al., 2012). However, the regenerative processes described in zebrafish and giant danio are not conserved across all fish species. For instance, work by Grivas et al. (2014) found that removal of a similar proportion of the goldfish ventricle, as was done for zebrafish and giant danio (i.e., 20%–25%), led to very low survival rates of experimental fish. However, in the same study, removing only one-ninth of the goldfish heart allowed successful regeneration (Grivas et al., 2014). This suggests that there are differences between species in how much of the heart can be regenerated. In addition, Ito et al. (2014) demonstrated that the injured medaka heart retains a fibrotic scar and does not incorporate new myocytes after ventricular resection. Grivas et al. (2014) suggest that the variation in regenerative potential across fish species demonstrates an ecological, and therefore evolutionary, consequence. Interestingly, cardiac regeneration following injury in an adult vertebrate was first described in urodele amphibians (Becker et al., 1974; McDonnell and Oberpriller, 1983; Oberpriller et al., 1995; Oberpriller and Oberpriller, 1974), and the general strategy employed to regrow the heart in these animals is similar to that seen in the zebrafish (Kikuchi et al., 2011b; Witman et al., 2011; Zhang et al., 2013). This indicates that the capacity to regenerate the heart in adult vertebrates is an ancestral trait that was lost in an ancestor of the mammals after the divergence of the amphibians from the vertebrate lineage. For a review of tissue repair in fish and amphibians, see Jewhurst and McLaughlin (2016).

5.8. Heart *vs* Limb Regeneration

Interestingly, there are similarities between limb regeneration in salamanders and heart regeneration in zebrafish that may provide some insight into the fundamental requirements for tissue regeneration that are absent in mammals.

Limb regeneration in salamanders involves a highly regulated regrowth of bone, cartilage, nerves, muscle, and connective tissue to replace the lost limb, or a portion of the limb (Jazwinska and Sallin, 2016). This process involves the dedifferentiation of terminally differentiated cells, cellular proliferation, and then cellular differentiation (Jazwinska and Sallin, 2016). A key component of this regenerative process is the formation of a blastema between the original uninjured tissue and the apical ectodermal cap that forms over the wound site (Jazwinska and Sallin, 2016). A blastema is a mass of dedifferentiated cells that gives rise to the specific cell types that comprise the various tissues within the new limb. In the salamander limb, the origin of the blastema is primarily dedifferentiated cells from tissues adjacent to the wound (Jazwinska and Sallin, 2016). The formation of the blastema during limb regeneration has similarities to the regeneration of the zebrafish heart. Studies by Sallin et al. (2015) showed that the regenerative response in the zebrafish heart involves two key processes that allow for cardiomyocyte repopulation. The first is the ability of mature cardiomyocytes to undergo hyperplasia, and the second is for a transient population of undifferentiated cells to form a blastema (Fig. 8). Together, it is thought that these two processes allow for new cardiomyocytes to fill in where a portion of the ventricle was amputated or injured via cryoinjury (Sallin et al., 2015). In the salamander limb, FGF secreted by the apical ectodermal cap, is a key signaling factor during regeneration. Similarly, in the zebrafish heart, FGF is critical as it directs the migration of epicardial cells into the wound site to support neovascularization (Lepilina et al., 2006). The blood vessels created in the wound area support the continuity of progenitor cell division (Lepilina et al., 2006). A second signaling factor common to cardiac regeneration and limb regeneration is RA (Cunningham and Duester, 2015). As discussed above, RA is required for heart regeneration, while in limb regeneration RA is involved in directing the development of the proximal—distal patterning of the limb. It is thought that increasing concentrations of RA at the site of injury along the proximal–distal axis of the limb regulate HOX gene expression for growth and development of new tissue (Cunningham and Duester, 2015). Future work should further investigate the role of RA in directing patterning of the regenerating heart.

One interesting question to consider regarding tissue regeneration is, why the zebrafish heart and salamander limb are able to completely regenerate following an injury while the mammalian heart cannot? One potential explanation is related to the thick fibrous scar that forms following cardiac injury in the mammalian heart. This scar protects the wound as well as helps to maintain the structural integrity of the myocardium during contraction and pressure generation. A similar scar is formed following amputation of a mammalian limb, but not a salamander limb (Jazwinska and Sallin, 2016). One possible reason why a tough scar is not formed in the fish heart is that it is not needed. In mice, systolic blood pressure reaches approximately 120 mmHg (Henry et al., 1967), while a salamanders systolic blood pressure is approximately 22 mmHg (Shelton and Jones, 1968) and in small fish, such as 2.5 cm larval Gulf killifish, peak ventricular pressure is approximately 30 mmHg (Chapter 2, Volume 36B: Burggren et al., 2017). Injury to a mouse heart, therefore, has a greater potential to cause a rupture to the chamber wall during contraction and pressure generation leading to massive blood loss and death. The relatively low pressure of the salamander and fish cardiovascular system reduces the potential for such a problem, suggesting that the need for a tough scar over a site of injury is not needed. The absence of true scar formation with salamander limb regeneration is thought to be one reason why limb regeneration is possible (Jazwinska and Sallin, 2016). The same may, therefore, be true for the fish heart following an injury.

6. PERSPECTIVES

The ability of some fish species to remodel their hearts in response to a change in physiological conditions, or to protect their hearts from damage caused by an acute stressor, provides benefit to animals living in conditions that can be highly variable. This chapter has discussed how the cellular and molecular mechanisms underlying these responses are highly integrated, function across levels of biological organization and are conserved across species. However, it is also clear that significant work is required to fully understand the molecular mechanisms that underpin these responses. This includes full characterization of the mechanisms that drive the dedifferentiation and proliferation of cardiomyocytes, as well as the mechanisms that control the fibrotic response, immediately following injury. The application of what is learned by such studies may contribute to the development of strategies to prevent cardiac damage caused by ischemia reperfusion injury, the treatment of cardiac pathologies, as well as the repair of damaged cardiac muscle. For example, there are increasing efforts to control the fibrotic scar formation in the mammalian heart following an injury, so as to increase the potential effectiveness of cardiac regeneration strategies (Talman and Ruskoaho, 2016). Recent work has demonstrated that the application of NRG1 (discussed in Section 5.3) following cardiac injury in swine acts to suppress fibrosis post-infarction and improve cardiac function (Galindo et al., 2014). However, for regeneration to occur, the heart must replace the damaged myocardium with functioning myocytes. By utilizing knowledge gained in studies of how cardiomyocytes dedifferentiate and proliferate in the zebrafish heart following injury it may become possible to reprogram mammalian cardiomyocytes to behave in a similar manner following injury.

ACKNOWLEDGMENTS

E.F.J. is supported by an Ontario Graduate Scholarship, and research in the Gillis Lab is supported by operating grants from Natural Sciences and Engineering Research Council of Canada (NSERC), the Department of Fisheries and Oceans, Canada, and the Canadian Foundation for Innovation. The authors thank Dr. S.L. Alderman, as well as two anonymous reviewers, for comments on earlier drafts of this chapter.

REFERENCES

Ahuja, P., Sdek, P., MacLellan, W.R., 2007. Cardiac myocyte cell cycle control in development, disease, and regeneration. Physiol. Rev. 87, 521–544.

Ala-Rami, A., Ylitalo, K.V., Hassinen, I.E., 2003. Ischaemic preconditioning and a mitochondrial KATP channel opener both produce cardioprotection accompanied by F1F0-ATPase inhibition in early ischaemia. Basic Res. Cardiol. 98, 250–258.

Alderman, S.L., Klaiman, J.M., Deck, C.A., Gillis, T.E., 2012. Effect of cold acclimation on troponin I isoform expression in striated muscle of rainbow trout. Am. J. Physiol. Regul. Integr. Comp. Physiol. 303, R168–176.

Altamirano, F., Oyarce, C., Silva, P., Toyos, M., Wilson, C., Lavandero, S., Uhlen, P., Estrada, M., 2009. Testosterone induces cardiomyocyte hypertrophy through mammalian target of rapamycin complex 1 pathway. J. Endocrinol. 202, 299–307.

Anttila, K., Dhillon, R.S., Boulding, E.G., Farrell, A.P., Glebe, B.D., Elliott, J.A., Wolters, W.R., Schulte, P.M., 2013. Variation in temperature tolerance among families of Atlantic salmon (Salmo salar) is associated with hypoxia tolerance, ventricle size and myoglobin level. J. Exp. Biol. 216, 1183–1190.

Anttila, K., Lewis, M., Prokkola, J.M., Kanerva, M., Seppanen, E., Kolari, I., Nikinmaa, M., 2015. Warm acclimation and oxygen depletion induce species-specific responses in salmonids. J. Exp. Biol. 218, 1471–1477.

Baker, K.M., Chernin, M.I., Wixson, S.K., Aceto, J.F., 1990. Renin–angiotensin system involvement in pressure-overload cardiac hypertrophy in rats. Am. J. Physiol. 259, H324–332.

Baldwin Jr., A.S., 1996. The NF-kappa B and I kappa B proteins: new discoveries and insights. Annu. Rev. Immunol. 14, 649–683.

Basu, N., Todgham, A.E., Ackerman, P.A., Bibeau, M.R., Nakano, K., Schulte, P.M., Iwama, G.K., 2002. Heat shock protein genes and their functional significance in fish. Gene 295, 173–183.

Beauchemin, M., Smith, A., Yin, V.P., 2015. Dynamic microRNA-101a and Fosab expression controls zebrafish heart regeneration. Development 142, 4026–4037.

Becker, R.O., Chapin, S., Sherry, R., 1974. Regeneration of the ventricular myocardium in amphibians. Nature 248, 145–147.

Boluyt, M.O., Zheng, J.S., Younes, A., Long, X., O'Neill, L., Silverman, H., Lakatta, E.G., Crow, M.T., 1997. Rapamycin inhibits alpha 1-adrenergic receptor-stimulated cardiac myocyte hypertrophy but not activation of hypertrophy-associated genes. Evidence for involvement of p70 S6 kinase. Circ. Res. 81, 176–186.

Bouwman, R.A., Musters, R.J., van Beek-Harmsen, B.J., de Lange, J.J., Lamberts, R.R., Loer, S.A., Boer, C., 2007. Sevoflurane-induced cardioprotection depends on PKC-alpha activation via production of reactive oxygen species. Br. J. Anaesth. 99, 639–645.

Bujak, M., Frangogiannis, N.G., 2007. The role of TGF-beta signaling in myocardial infarction and cardiac remodeling. Cardiovasc. Res. 74, 184–195.

Burggren, W., Dubansky, B., Martinez Bautista, N., 2017. Cardiovascular development in embryonic and larval fishes. In: Gamperl, A.K., Gillis, T.E., Farrell, A.P., Brauner, C.J. (Eds.), The Cardiovascular System: Development, Plasticity and Physiological Responses. In: Fish Physiology, vol. 36B. Elsevier, Amsterdam. pp. 107–184. Chapter 2.

Butler, D.G., Oudit, G.Y., 1995. Angiotensin-I- and -III-mediated cardiovascular responses in the freshwater North American eel, Anguilla rostrata: effect of Phe8 deletion. Gen. Comp. Endocrinol. 97, 259–269.

Cai, Y.J., Adelman, I.R., 1990. Temperature-acclimation in respiratory and cytochrome-C oxidase activity in common carp (Cyprinus carpio). Comp. Biochem. Physiol. A Physiol. 95, 139–144.

Cai, C.L., Martin, J.C., Sun, Y., Cui, L., Wang, L., Ouyang, K., Yang, L., Bu, L., Liang, X., Zhang, X., Stallcup, W.B., Denton, C.P., McCulloch, A., Chen, J., Evans, S.M., 2008. A myocardial lineage derives from Tbx18 epicardial cells. Nature 454, 104–108.

Cameron, J.S., DeWitt, J.P., Ngo, T.T., Yajnik, T., Chan, S., Chung, E., Kang, E., 2013. Cardiac K(ATP) channel alterations associated with acclimation to hypoxia in goldfish (Carassius auratus L.). Comp. Biochem. Physiol. A Mol. Integr. Physiol. 164, 554–564.

Carver, W., Nagpal, M.L., Nachtigal, M., Borg, T.K., Terracio, L., 1991. Collagen expression in mechanically stimulated cardiac fibroblasts. Circ. Res. 69, 116–122.

Chablais, F., Jazwinska, A., 2012. The regenerative capacity of the zebrafish heart is dependent on TGFbeta signaling. Development 139, 1921–1930.

Chen, J., Zhu, J.X., Wilson, I., Cameron, J.S., 2005. Cardioprotective effects of K ATP channel activation during hypoxia in goldfish Carassius auratus. J. Exp. Biol. 208, 2765–2772.

Churchill, E.N., Mochly-Rosen, D., 2007. The roles of PKCdelta and epsilon isoenzymes in the regulation of myocardial ischaemia/reperfusion injury. Biochem. Soc. Trans. 35, 1040–1042.

Clark, R.J., Rodnick, K.J., 1999. Pressure and volume overloads are associated with ventricular hypertrophy in male rainbow trout. Am. J. Physiol. 277, R938–946.

Clauss, T.M., Dove, A.D., Arnold, J.E., 2008. Hematologic disorders of fish. Vet. Clin. North Am. Exot. Anim. Pract. 11, 445–462.

Cohen, M.V., Baines, C.P., Downey, J.M., 2000. Ischemic preconditioning: from adenosine receptor to KATP channel. Annu. Rev. Physiol. 62, 79–109.

Cohn, J.N., Ferrari, R., Sharpe, N., 2000. Cardiac remodeling—concepts and clinical implications: a consensus paper from an international forum on cardiac remodeling. Behalf of an International forum on cardiac remodeling. J. Am. Coll. Cardiol. 35, 569–582.

Cunningham, T.J., Duester, G., 2015. Mechanisms of retinoic acid signalling and its roles in organ and limb development. Nat. Rev. Mol. Cell Biol. 16, 110–123.

Currie, S., Moyes, C.D., Tufts, B.L., 2000. The effects of heat shock and acclimation temperature on hsp70 and hsp30 mRNA expression in rainbow trout: in vivo and in vitro comparisons. J. Fish Biol. 56, 398–408.

Dabek, J., Kulach, A., Gasior, Z., 2010. Nuclear factor kappa-light-chain-enhancer of activated B cells (NF-kappaB): a new potential therapeutic target in atherosclerosis? Pharmacol. Rep. 62, 778–783.

Das, B., Sarkar, C., Karanth, K.S., 2002. Selective mitochondrial K(ATP) channel activation results in antiarrhythmic effect during experimental myocardial ischemia/reperfusion in anesthetized rabbits. Eur. J. Pharmacol. 437, 165–171.

Davie, P.S., Thorarensen, H., 1997. Heart growth in rainbow trout in response to exogenous testosterone and 17-alpha methyltestosterone. Comp. Biochem. Phys. A 117, 227–230.

Debska, G., Kicinska, A., Skalska, J., Szewczyk, A., May, R., Elger, C.E., Kunz, W.S., 2002. Opening of potassium channels modulates mitochondrial function in rat skeletal muscle. Biochim. Biophys. Acta 1556, 97–105.

Devereux, R.B., Drayer, J.I., Chien, S., Pickering, T.G., Letcher, R.L., DeYoung, J.L., Sealey, J.E., Laragh, J.H., 1984. Whole blood viscosity as a determinant of cardiac hypertrophy in systemic hypertension. Am. J. Cardiol. 54, 592–595.

Dobaczewski, M., Chen, W., Frangogiannis, N.G., 2011. Transforming growth factor (TGF)-beta signaling in cardiac remodeling. J. Mol. Cell. Cardiol. 51, 600–606.

Driedzic, W.R., Bailey, J.R., Sephton, D.H., 1996. Cardiac adaptations to low temperature non-polar teleost fish. J. Exp. Zool. 275, 186–195.

Eghbali, M., Tomek, R., Sukhatme, V.P., Woods, C., Bhambi, B., 1991. Differential effects of transforming growth factor-beta 1 and phorbol myristate acetate on cardiac fibroblasts. Regulation of fibrillar collagen mRNAs and expression of early transcription factors. Circ. Res. 69, 483–490.

Eliason, E.J., Anttila, K., 2017. Temperature and the cardiovascular system. In: Gamperl, A.K., Gillis, T.E., Farrell, A.P., Brauner, C.J. (Eds.), The Cardiovascular System: Development, Plasticity and Physiological Responses. In: Fish Physiology, vol. 36B. Elsevier, Amsterdam. pp. 235–297. Chapter 4.

Eliason, E.J., Clark, T.D., Hague, M.J., Hanson, L.M., Gallagher, Z.S., Jeffries, K.M., Gale, M.K., Patterson, D.A., Hinch, S.G., Farrell, A.P., 2011. Differences in thermal tolerance among sockeye salmon populations. Science 332, 109–112.

Eliason, E.J., Wilson, S.M., Farrell, A.P., Cooke, S.J., Hinch, S.G., 2013. Low cardiac and aerobic scope in a coastal population of sockeye salmon Oncorhynchus nerka with a short upriver migration. J. Fish Biol. 82, 2104–2112.

Fang, Y., Gupta, V., Karra, R., Holdway, J.E., Kikuchi, K., Poss, K.D., 2013. Translational profiling of cardiomyocytes identifies an early Jak1/STAT3 injury response required for zebrafish heart regeneration. Proc. Natl. Acad. Sci. U.S.A. 110, 13416–13421.

Farah, C.A., Sossin, W.S., 2012. The role of C2 domains in PKC signaling. Adv. Exp. Med. Biol. 740, 663–683.

Farrell, A.P., 1984. A review of cardiac-performance in the teleost heart—intrinsic and humoral regulation. Can. J. Zool. 62, 523–536.

Farrell, A.P., Smith, F., 2017. Cardiac form, function and physiology. In: Gamperl, A.K., Gillis, T.E., Farrell, A.P., Brauner, C.J. (Eds.), The Cardiovascular System: Morphology, Control and Function. Fish Physiology, vol. 36A. Elsevier, Amsterdam. pp. 155–264. Chapter 4.

Farrell, A.P., Hammons, A.M., Graham, M.S., Tibbits, G.F., 1988. Cardiac growth in rainbow-trout, Salmo gairdneri. Can. J. Zool. 66, 2368–2373.

Farrell, A.P., Johansen, J.A., Suarez, R.K., 1991. Effects of exercise-training on cardiac performance and muscle enzymes in rainbow trout, Oncorhynchus mykiss. Fish Physiol. Biochem. 9, 303–312.

Faust, H.A., Gamperl, A.K., Rodnick, K.J., 2004. All rainbow trout (Oncorhynchus mykiss) are not created equal: intra-specific variation in cardiac hypoxia tolerance. J. Exp. Biol. 207, 1005–1015.

Foster, M.N., Coetzee, W.A., 2016. KATP channels in the cardiovascular system. Physiol. Rev. 96, 177–252.

Franklin, C.E., Davie, P.S., 1992. Sexual maturity can double heart mass and cardiac power output in male rainbow trout. J. Exp. Biol. 171, 139–148.

Gaitanaki, C., Kalpachidou, T., Aggeli, I.K., Papazafiri, P., Beis, I., 2007. CoCl2 induces protective events via the p38-MAPK signalling pathway and ANP in the perfused amphibian heart. J. Exp. Biol. 210, 2267–2277.

Gaitanaki, C., Mastri, M., Aggeli, I.K., Beis, I., 2008. Differential roles of p38-MAPK and JNKs in mediating early protection or apoptosis in the hyperthermic perfused amphibian heart. J. Exp. Biol. 211, 2524–2532.

Galindo, C.L., Kasasbeh, E., Murphy, A., Ryzhov, S., Lenihan, S., Ahmad, F.A., Williams, P., Nunnally, A., Adcock, J., Song, Y., Harrell, F.E., Tran, T.L.,

Parry, T.J., Iaci, J., Ganguly, A., Feoktistov, I., Stephenson, M.K., Caggiano, A.O., Sawyer, D.B., Cleator, J.H., 2014. Anti-remodeling and anti-fibrotic effects of the neuregulin-1beta glial growth factor 2 in a large animal model of heart failure. J. Am. Heart Assoc. 3. e000773.

Gamperl, A.K., Farrell, A.P., 2004. Cardiac plasticity in fishes: environmental influences and intraspecific differences. J. Exp. Biol. 207, 2539–2550.

Gamperl, A.K., Todgham, A.E., Parkhouse, W.S., Dill, R., Farrell, A.P., 2001. Recovery of trout myocardial function following anoxia: preconditioning in a non-mammalian model. Am. J. Physiol. Regul. Integr. Comp. Physiol. 281, R1755–1763.

Gemberling, M., Karra, R., Dickson, A.L., Poss, K.D., 2015. Nrgl is an injury-induced cardiomyocyte mitogen for the endogenous heart regeneration program in zebrafish. eLife 4., 1–17, https://doi.org/10.7554/eLife.05871.

Genge, C.E., Davidson, W.S., Tibbits, G.F., 2013. Adult teleost heart expresses two distinct troponin C paralogs: cardiac TnC and a novel and teleost-specific ssTnC in a chamber- and temperature-dependent manner. Physiol. Genomics 45, 866–875.

Gillis, T.E., Tibbits, G.F., 2002. Beating the cold: the functional evolution of troponin C in teleost fish. Comp. Biochem. Physiol. A Mol. Integr. Physiol. 132, 763–772.

Graham, M.S., Farrell, A.P., 1989. The effect of temperature acclimation and adrenaline on the performance of a perfused trout heart. Physiol. Zool. 62, 38–61.

Graham, M.S., Farrell, A.P., 1992. Environmental-influences on cardiovascular variables in rainbow-trout, Oncorhynchus mykiss (Walbaum). J. Fish Biol. 41, 851–858.

Griffin, S.A., Brown, W.C., MacPherson, F., McGrath, J.C., Wilson, V.G., Korsgaard, N., Mulvany, M.J., Lever, A.F., 1991. Angiotensin II causes vascular hypertrophy in part by a non-pressor mechanism. Hypertension 17, 626–635.

Grivas, J., Haag, M., Johnson, A., Manalo, T., Roell, J., Das, T.L., Brown, E., Burns, A.R., Lafontant, P.J., 2014. Cardiac repair and regenerative potential in the goldfish (Carassius auratus) heart. Comp. Biochem. Physiol. C Toxicol. Pharmacol. 163, 14–23.

Gross, G.J., 2000. The role of mitochondrial KATP channels in cardioprotection. Basic Res. Cardiol. 95, 280–284.

Gross, G.J., Fryer, R.M., 1999. Sarcolemmal versus mitochondrial ATP-sensitive K+ channels and myocardial preconditioning. Circ. Res. 84, 973–979.

Gudi, S.R., Lee, A.A., Clark, C.B., Frangos, J.A., 1998. Equibiaxial strain and strain rate stimulate early activation of G proteins in cardiac fibroblasts. Am. J. Physiol. 274, C1424–1428.

Guo, D.F., Sun, Y.L., Hamet, P., Inagami, T., 2001. The angiotensin II type 1 receptor and receptor-associated proteins. Cell Res. 11, 165–180.

Gupta, V., Grande-Allen, K.J., 2006. Effects of static and cyclic loading in regulating extracellular matrix synthesis by cardiovascular cells. Cardiovasc. Res. 72, 375–383.

Hassouna, A., Matata, B.M., Galinanes, M., 2004. PKC-epsilon is upstream and PKC-alpha is downstream of mitoKATP channels in the signal transduction pathway of ischemic preconditioning of human myocardium. Am. J. Physiol. Cell Physiol. 287, C1418–1425.

Henry, J.P., Meehan, J.P., Stephens, P.M., 1967. The use of psychosocial stimuli to induce prolonged systolic hypertension in mice. Psychosom. Med. 29, 408–432.

Hilfiker-Kleiner, D., Hilfiker, A., Drexler, H., 2005. Many good reasons to have STAT3 in the heart. Pharmacol. Ther. 107, 131–137.

Hillegass, J.M., Villano, C.M., Cooper, K.R., White, L.A., 2007. Matrix metalloproteinase-13 is required for zebra fish (Danio rerio) development and is a target for glucocorticoids. Toxicol. Sci. 100, 168–179.

Imbrogno, S., Cerra, M.C., 2017. Hormonal and autacoid control of cardiac function. In: Gamperl, A.K., Gillis, T.E., Farrell, A.P., Brauner, C.J. (Eds.), The Cardiovascular System: Morphology, Control and Function. In: Fish Physiology, vol. 36A. Elsevier, Amsterdam. pp. 265–315. Chapter 5.

Imbrogno, S., Garofalo, F., Amelio, D., Capria, C., Cerra, M.C., 2013. Humoral control of cardiac remodeling in fish: role of Angiotensin II. Gen. Comp. Endocrinol. 194, 189–197.

Ito, K., Morioka, M., Kimura, S., Tasaki, M., Inohaya, K., Kudo, A., 2014. Differential reparative phenotypes between zebrafish and medaka after cardiac injury. Dev. Dyn. 243, 1106–1115.

Jalil, J.E., Doering, C.W., Janicki, J.S., Pick, R., Clark, W.A., Abrahams, C., Weber, K.T., 1988. Structural vs. contractile protein remodeling and myocardial stiffness in hypertrophied rat left ventricle. J. Mol. Cell. Cardiol. 20, 1179–1187.

Jalil, J.E., Doering, C.W., Janicki, J.S., Pick, R., Shroff, S.G., Weber, K.T., 1989. Fibrillar collagen and myocardial stiffness in the intact hypertrophied rat left ventricle. Circ. Res. 64, 1041–1050.

Jayasundara, N., Gardner, L.D., Block, B.A., 2013. Effects of temperature acclimation on Pacific bluefin tuna (Thunnus orientalis) cardiac transcriptome. Am. J. Physiol. Regul. Integr. Comp. Physiol. 305, R1010–1020.

Jazwinska, A., Sallin, P., 2016. Regeneration versus scarring in vertebrate appendages and heart. J. Pathol. 238, 233–246.

Jewhurst, K., McLaughlin, K.A., 2016. Beyond the mammalian heart: fish and amphibians for cardiac repair and regeneration. J. Dev. Biol. 4, 1.

Johnson, A.C., Turko, A.J., Klaiman, J.M., Johnston, E.F., Gillis, T.E., 2014. Cold acclimation alters the connective tissue content of the zebrafish (Danio rerio) heart. J. Exp. Biol. 217, 1868–1875.

Johnston, E.F., Gillis, T.E., 2017. Transforming growth factor beta-1 (TGF-beta1) stimulates collagen synthesis in cultured rainbow trout cardiac fibroblasts. J. Exp. Biol. 220, 2645–2653.

Jopling, C., Sleep, E., Raya, M., Marti, M., Raya, A., Izpisua Belmonte, J.C., 2010. Zebrafish heart regeneration occurs by cardiomyocyte dedifferentiation and proliferation. Nature 464, 606–609.

Kafienah, W., Sims, T.J., 2004. Biochemical methods for the analysis of tissue-engineered cartilage. Methods Mol. Biol. 238, 217–230.

Karra, R., Knecht, A.K., Kikuchi, K., Poss, K.D., 2015. Myocardial NF-kappaB activation is essential for zebrafish heart regeneration. Proc. Natl. Acad. Sci. U.S.A. 112, 13255–13260.

Kastner, P., Grondona, J.M., Mark, M., Gansmuller, A., LeMeur, M., Decimo, D., Vonesch, J.L., Dolle, P., Chambon, P., 1994. Genetic analysis of RXR alpha developmental function: convergence of RXR and RAR signaling pathways in heart and eye morphogenesis. Cell 78, 987–1003.

Katsumi, A., Orr, A.W., Tzima, E., Schwartz, M.A., 2004. Integrins in mechanotransduction. J. Biol. Chem. 279, 12001–12004.

Keen, J.E., Vianzon, D.M., Farrell, A.P., Tibbits, G.F., 1993. Thermal-acclimation alters both adrenergic sensitivity and adrenoceptor density in cardiac tissue of rainbow-trout. J. Exp. Biol. 181, 27–47.

Keen, A.N., Fenna, A.J., McConnell, J.C., Sherratt, M.J., Gardner, P., Shiels, H.A., 2016a. The dynamic nature of hypertrophic and fibrotic remodeling of the fish ventricle. Front. Physiol. 6, 427.

Keen, A.N., Klaiman, J.M., Shiels, H.A., Gillis, T.E., 2016b. Temperature-induced cardiac remodeling in fish. J. Exp. Biol. 220, 147–160.

Kemi, O.J., Ceci, M., Wisloff, U., Grimaldi, S., Gallo, P., Smith, G.L., Condorelli, G., Ellingsen, O., 2008. Activation or inactivation of cardiac Akt/mTOR signaling diverges physiological from pathological hypertrophy. J. Cell. Physiol. 214, 316–321.

Kenessey, A., Ojamaa, K., 2006. Thyroid hormone stimulates protein synthesis in the cardiomyocyte by activating the Akt-mTOR and p70S6K pathways. J. Biol. Chem. 281, 20666–20672.

Kennedy, D.J., Fan, Y., Wu, Y., Pepoy, M., Hazen, S.L., Tang, W.H., 2014. Plasma ceruloplasmin, a regulator of nitric oxide activity, and incident cardiovascular risk in patients with CKD. Clin. J. Am. Soc. Nephrol. 9, 462–467.

Kent, J., Koban, M., Prosser, C.L., 1988. Cold-acclimation-induced protein hypertrophy in channel catfish and green sunfish. J. Comp. Physiol. B 158, 185–198.

Kiang, J.G., Tsokos, G.C., 1998. Heat shock protein 70 kDa: molecular biology, biochemistry, and physiology. Pharmacol. Ther. 80, 183–201.

Kikuchi, K., 2014. Advances in understanding the mechanism of zebrafish heart regeneration. Stem Cell Res. 13, 542–555.

Kikuchi, K., Holdway, J.E., Werdich, A.A., Anderson, R.M., Fang, Y., Egnaczyk, G.F., Evans, T., Macrae, C.A., Stainier, D.Y., Poss, K.D., 2010. Primary contribution to zebrafish heart regeneration by gata4(+) cardiomyocytes. Nature 464, 601–605.

Kikuchi, K., Gupta, V., Wang, J., Holdway, J.E., Wills, A.A., Fang, Y., Poss, K.D., 2011a. tcf21 + epicardial cells adopt non-myocardial fates during zebrafish heart development and regeneration. Development 138, 2895–2902.

Kikuchi, K., Holdway, J.E., Major, R.J., Blum, N., Dahn, R.D., Begemann, G., Poss, K.D., 2011b. Retinoic acid production by endocardium and epicardium is an injury response essential for zebrafish heart regeneration. Dev. Cell 20, 397–404.

Klaiman, J.M., Fenna, A.J., Shiels, H.A., Macri, J., Gillis, T.E., 2011. Cardiac remodeling in fish: strategies to maintain heart function during temperature change. PLoS One 6 . e24464.

Klaiman, J.M., Pyle, W.G., Gillis, T.E., 2014. Cold acclimation increases cardiac myofilament function and ventricular pressure generation in trout. J. Exp. Biol. 217, 4132–4140.

Kleckner, N.W., Sidell, B.D., 1985. Comparison of maximal activities of enzymes from tissues of thermally acclimated and naturally acclimatized chain pickerel (Esox niger). Physiol. Zool. 58, 18–28.

Kolosova, I., Nethery, D., Kern, J.A., 2011. Role of Smad2/3 and p38 MAP kinase in TGF-beta1-induced epithelial–mesenchymal transition of pulmonary epithelial cells. J. Cell. Physiol. 226, 1248–1254.

Komuro, I., Yazaki, Y., 1993. Control of cardiac gene expression by mechanical stress. Annu. Rev. Physiol. 55, 55–75.

Korajoki, H., Vornanen, M., 2013. Temperature dependence of sarco(endo)plasmic reticulum Ca2+ ATPase expression in fish hearts. J. Comp. Physiol. B 183, 467–476.

Koseki, T., Inohara, N., Chen, S., Nunez, G., 1998. ARC, an inhibitor of apoptosis expressed in skeletal muscle and heart that interacts selectively with caspases. Proc. Natl. Acad. Sci. U.S.A. 95, 5156–5160.

Kubota, M., Kinoshita, M., Takeuchi, K., Kubota, S., Toyohara, H., Sakaguchi, M., 2003. Solubilization of type I collagen from fish muscle connective tissue by matrix metalloproteinase-9 at chilled temperature. Fish. Sci. 69, 1053–1059.

Lafontant, P.J., Burns, A.R., Grivas, J.A., Lesch, M.A., Lala, T.D., Reuter, S.P., Field, L.J., Frounfelter, T.D., 2012. The giant danio (D. aequipinnatus) as a model of cardiac remodeling and regeneration. Anat. Rec. 295, 234–248.

Le Mevel, J.C., Lancien, F., Mimassi, N., 2008. Central cardiovascular actions of angiotensin II in trout. Gen. Comp. Endocrinol. 157, 27–34.

Lefevre, S., Damsgaard, C., Pascale, D.R., Nilsson, G.E., Stecyk, J.A., 2014. Air breathing in the Arctic: influence of temperature, hypoxia, activity and restricted air access on respiratory physiology of the Alaska blackfish Dallia pectoralis. J. Exp. Biol. 217, 4387–4398.

Lepilina, A., Coon, A.N., Kikuchi, K., Holdway, J.E., Roberts, R.W., Burns, C.G., Poss, K.D., 2006. A dynamic epicardial injury response supports progenitor cell activity during zebrafish heart regeneration. Cell 127, 607–619.

Li, F., Wang, X., Capasso, J.M., Gerdes, A.M., 1996. Rapid transition of cardiac myocytes from hyperplasia to hypertrophy during postnatal development. J. Mol. Cell. Cardiol. 28, 1737–1746.

Li, J., Schwimmbeck, P.L., Tschope, C., Leschka, S., Husmann, L., Rutschow, S., Reichenbach, F., Noutsias, M., Kobalz, U., Poller, W., Spillmann, F., Zeichhardt, H., Schultheiss, H.P., Pauschinger, M., 2002. Collagen degradation in a murine myocarditis model: relevance of matrix metalloproteinase in association with inflammatory induction. Cardiovasc. Res. 56, 235–247.

Li, G.R., Sun, H.Y., Chen, J.B., Zhou, Y., Tse, H.F., Lau, C.P., 2009. Characterization of multiple ion channels in cultured human cardiac fibroblasts. PLoS One 4. e7307.

Liu, Y., Xiao, H., Tian, Y., Nekrasova, T., Hao, X., Lee, H.J., Suh, N., Yang, C.S., Minden, A., 2008. The pak4 protein kinase plays a key role in cell survival and tumorigenesis in athymic mice. Mol. Cancer Res. 6, 1215–1224.

MacCormack, T.J., Driedzic, W.R., 2002. Mitochondrial ATP-sensitive K+ channels influence force development and anoxic contractility in a flatfish, yellowtail flounder Limanda ferruginea, but not Atlantic cod Gadus morhua heart. J. Exp. Biol. 205, 1411–1418.

MacKenna, D.A., Dolfi, F., Vuori, K., Ruoslahti, E., 1998. Extracellular signal-regulated kinase and c-Jun NH2-terminal kinase activation by mechanical stretch is integrin-dependent and matrix-specific in rat cardiac fibroblasts. J. Clin. Invest. 101, 301–310.

Major, R.J., Poss, K.D., 2007. Zebrafish heart regeneration as a model for cardiac tissue repair. Drug Discov. Today Dis. Models 4, 219–225.

Mandic, M., Sloman, K.A., Richards, J.G., 2009. Escaping to the surface: a phylogenetically independent analysis of hypoxia-induced respiratory behaviors in sculpins. Physiol. Biochem. Zool. 82, 730–738.

Marques, I.J., Leito, J.T., Spaink, H.P., Testerink, J., Jaspers, R.T., Witte, F., van den Berg, S., Bagowski, C.P., 2008. Transcriptome analysis of the response to chronic constant hypoxia in zebrafish hearts. J. Comp. Physiol. B 178, 77–92.

Martinez-Estrada, O.M., Lettice, L.A., Essafi, A., Guadix, J.A., Slight, J., Velecela, V., Hall, E., Reichmann, J., Devenney, P.S., Hohenstein, P., Hosen, N., Hill, R.E., Munoz-Chapuli, R., Hastie, N.D., 2010. Wt1 is required for cardiovascular progenitor cell formation through transcriptional control of Snail and E-cadherin. Nat. Genet. 42, 89–93.

Masters, M., Riley, P., 2014. The epicardium signals the way towards heart regeneration. Stem Cell Res. 13, 683–692.

McBryan, T.L., Anttila, K., Healy, T.M., Schulte, P.M., 2013. Responses to temperature and hypoxia as interacting stressors in fish: implications for adaptation to environmental change. Integr. Comp. Biol. 53, 648–659.

McClelland, G.B., Dalziel, A.C., Fragoso, N.M., Moyes, C.D., 2005. Muscle remodeling in relation to blood supply: implications for seasonal changes in mitochondrial enzymes. J. Exp. Biol. 208, 515–522.

McDonnell, T.J., Oberpriller, J.O., 1983. The atrial proliferative response following partial ventricular amputation in the heart of the adult newt. A light and electron microscopic autoradiographic study. Tissue Cell 15, 351–363.

McMullen, J.R., Sherwood, M.C., Tarnavski, O., Zhang, L., Dorfman, A.L., Shioi, T., Izumo, S., 2004. Inhibition of mTOR signaling with rapamycin regresses established cardiac hypertrophy induced by pressure overload. Circulation 109, 3050–3055.

Murad, S., Grove, D., Lindberg, K.A., Reynolds, G., Sivarajah, A., Pinnell, S.R., 1981. Regulation of collagen synthesis by ascorbic acid. Proc. Natl. Acad. Sci. U.S.A. 78, 2879–2882.

Murray, P.J., 2007. The JAK-STAT signaling pathway: input and output integration. J. Immunol. 178, 2623–2629.

Murry, C.E., Richard, V.J., Reimer, K.A., Jennings, R.B., 1990. Ischemic preconditioning slows energy metabolism and delays ultrastructural damage during a sustained ischemic episode. Circ. Res. 66, 913–931.

Nekrasova, T., Minden, A., 2012. Role for p21-activated kinase PAK4 in development of the mammalian heart. Transgenic Res. 21, 797–811.

Nelson, O.L., Robbins, C.T., Wu, Y., Granzier, H., 2008. Titin isoform switching is a major cardiac adaptive response in hibernating grizzly bears. Am. J. Physiol. Heart Circ. Physiol. 295, H366–371.

Niessen, K., Karsan, A., 2008. Notch signaling in cardiac development. Circ. Res. 102, 1169–1181.

Nikinmaa, M., Rees, B.B., 2005. Oxygen-dependent gene expression in fishes. Am. J. Physiol. Regul. Integr. Comp. Physiol. 288, R1079–R1090.

Nilsson, G.E., Vaage, J., Stenslokken, K.O., 2015. Oxygen- and temperature-dependent expression of survival protein kinases in crucian carp (Carassius carassius) heart and brain. Am. J. Physiol. Regul. Integr. Comp. Physiol. 308, R50–61.

Oberpriller, J.O., Oberpriller, J.C., 1974. Response of the adult newt ventricle to injury. J. Exp. Zool. 187, 249–253.

Oberpriller, J.O., Oberpriller, J.C., Matz, D.G., Soonpaa, M.H., 1995. Stimulation of proliferative events in the adult amphibian cardiac myocyte. Ann. N.Y. Acad. Sci. 752, 30–46.

Ocaranza, M.P., Jalil, J.E., 2010. Mitogen-activated protein kinases as biomarkers of hypertension or cardiac pressure overload. Hypertension 55, 23–25.

Odiete, O., Hill, M.F., Sawyer, D.B., 2012. Neuregulin in cardiovascular development and disease. Circ. Res. 111, 1376–1385.

Olivetti, G., Quaini, F., Lagrasta, C., Ricci, R., Tiberti, G., Capasso, J.M., Anversa, P., 1992. Myocyte cellular hypertrophy and hyperplasia contribute to ventricular wall remodeling in anemia-induced cardiac hypertrophy in rats. Am. J. Pathol. 141, 227–239.

Overgaard, J., Stecyk, J.A., Gesser, H., Wang, T., Gamperl, A.K., Farrell, A.P., 2004. Preconditioning stimuli do not benefit the myocardium of hypoxia-tolerant rainbow trout (Oncorhynchus mykiss). J. Comp. Physiol. B 174, 329–340.

Parente, V., Balasso, S., Pompilio, G., Verduci, L., Colombo, G.I., Milano, G., Guerrini, U., Squadroni, L., Cotelli, F., Pozzoli, O., Capogrossi, M.C., 2013. Hypoxia/reoxygenation cardiac injury and regeneration in zebrafish adult heart. PLoS One 8. e53748.

Pauschinger, M., Knopf, D., Petschauer, S., Doerner, A., Poller, W., Schwimmbeck, P.L., Kuhl, U., Schultheiss, H.P., 1999. Dilated cardiomyopathy is associated with significant changes in collagen type I/III ratio. Circulation 99, 2750–2756.

Peng, X., He, Q., Li, G., Ma, J., Zhong, T.P., 2016. RAC1-PAK2 pathway is essential for zebrafish heart regeneration. Biochem. Biophys. Res. Commun. 472, 637–642.

Porrello, E.R., Mahmoud, A.I., Simpson, E., Hill, J.A., Richardson, J.A., Olson, E.N., Sadek, H.A., 2011. Transient regenerative potential of the neonatal mouse heart. Science 331, 1078–1080.

Poss, K.D., Nechiporuk, A., Hillam, A.M., Johnson, S.L., Keating, M.T., 2002a. Mps1 defines a proximal blastemal proliferative compartment essential for zebrafish fin regeneration. Development 129, 5141–5149.

Poss, K.D., Wilson, L.G., Keating, M.T., 2002b. Heart regeneration in zebrafish. Science 298, 2188–2190.

Powell, M.D., Burke, M.S., Dahle, D., 2012. Cardiac remodelling of Atlantic halibut Hippoglossus induced by experimental anaemia with phenylhydrazine. J. Fish Biol. 81, 335–344.

Raya, A., Koth, C.M., Buscher, D., Kawakami, Y., Itoh, T., Raya, R.M., Sternik, G., Tsai, H.J., Rodriguez-Esteban, C., Izpisua-Belmonte, J.C., 2003. Activation of Notch signaling pathway precedes heart regeneration in zebrafish. Proc. Natl. Acad. Sci. U.S.A. 100 (Suppl. 1), 11889–11895.

Rees, B.B., Sudradjat, F.A., Love, J.W., 2001. Acclimation to hypoxia increases survival time of zebrafish, Danio rerio, during lethal hypoxia. J. Exp. Zool. 289, 266–272.

Renshaw, G.M., Kerrisk, C.B., Nilsson, G.E., 2002. The role of adenosine in the anoxic survival of the epaulette shark, Hemiscyllium ocellatum. Comp. Biochem. Physiol. B Biochem. Mol. Biol. 131, 133–141.

Richardson, C.J., Schalm, S.S., Blenis, J., 2004. PI3-kinase and TOR: PIKTORing cell growth. Semin. Cell Dev. Biol. 15, 147–159.

Rissanen, E., Tranberg, H.K., Sollid, J., Nilsson, G.E., Nikinmaa, M., 2006. Temperature regulates hypoxia-inducible factor-1 (HIF-1) in a poikilothermic vertebrate, crucian carp (Carassius carassius). J. Exp. Biol. 209, 994–1003.

Rodnick, K.J., Gesser, H., 2017. Cardiac energy metabolism. In: Gamperl, A.K., Gillis, T.E., Farrell, A.P., Brauner, C.J. (Eds.), The Cardiovascular System: Morphology, Control and Function. Fish Physiology, vol. 36A. Elsevier, Amsterdam. pp. 317–367. Chapter 6.

Rodnick, K.J., Sidell, B.D., 1997. Structural and biochemical analyses of cardiac ventricular enlargement in cold-acclimated striped bass. Am. J. Physiol. 273, R252–258.

Ruwhof, C., van Wamel, A.E., Egas, J.M., van der Laarse, A., 2000. Cyclic stretch induces the release of growth promoting factors from cultured neonatal cardiomyocytes and cardiac fibroblasts. Mol. Cell Biochem. 208, 89–98.

Ryckebusch, L., Wang, Z., Bertrand, N., Lin, S.C., Chi, X., Schwartz, R., Zaffran, S., Niederreither, K., 2008. Retinoic acid deficiency alters second heart field formation. Proc. Natl. Acad. Sci. U.S.A. 105, 2913–2918.

Rytkonen, K.T., Renshaw, G.M., Ashton, K.J., Williams-Pritchard, G., Leder, E.H., Nikinmaa, M., 2010. Elasmobranch qPCR reference genes: a case study of hypoxia preconditioned epaulet sharks. BMC Mol. Biol. 11, 27.

Sadoshima, J., Izumo, S., 1993. Molecular characterization of angiotensin II-induced hypertrophy of cardiac myocytes and hyperplasia of cardiac fibroblasts. Critical role of the AT1 receptor subtype. Circ. Res. 73, 413–423.

Sadoshima, J., Izumo, S., 1997. The cellular and molecular response of cardiac myocytes to mechanical stress. Annu. Rev. Physiol. 59, 551–571.

Sadoshima, J., Jahn, L., Takahashi, T., Kulik, T.J., Izumo, S., 1992. Molecular characterization of the stretch-induced adaptation of cultured cardiac cells. An in vitro model of load-induced cardiac hypertrophy. J. Biol. Chem. 267, 10551–10560.

Sadoshima, J., Xu, Y., Slayter, H.S., Izumo, S., 1993. Autocrine release of angiotensin II mediates stretch-induced hypertrophy of cardiac myocytes in vitro. Cell 75, 977–984.

Sallin, P., de Preux Charles, A.S., Duruz, V., Pfefferli, C., Jazwinska, A., 2015. A dual epimorphic and compensatory mode of heart regeneration in zebrafish. Dev. Biol. 399, 27–40.

Schindler, C., Darnell Jr., J.E., 1995. Transcriptional responses to polypeptide ligands: the JAK–STAT pathway. Annu. Rev. Biochem. 64, 621–651.

Semenza, G., 2002. Signal transduction to hypoxia-inducible factor 1. Biochem. Pharmacol. 64, 993–998.

Semenza, G.L., 2011. Hypoxia-inducible factor 1: regulator of mitochondrial metabolism and mediator of ischemic preconditioning. Biochim. Biophys. Acta 1813, 1263–1268.

Sephton, D.H., Driedzic, W.R., 1990. Effects of acute and chronic temperature transition on enzymes of cardaic metabolism in white perch (*Morone americana*), yellow perch (*Perca flavescens*), and smallmouth bass (Micropterus dolomieui). Can. J. Zool. 69, 258–262.

Sephton, D.H., Driedzic, W.R., 1991. Effect of acute and chronic temperature transition on enzymes of cardiac metabolism in white perch (Morone americana), yellow perch

(Perca flavescens), and smallmouth bass (Micropterus dolomieui). Can. J. Zool. 69, 258–262.

Shelton, G., Jones, D.R., 1968. A comparative study of central blood pressures in five amphibians. J. Exp. Biol. 49, 631–643.

Shiels, H.A., Di Maio, A., Thompson, S., Block, B.A., 2011. Warm fish with cold hearts: thermal plasticity of excitation–contraction coupling in bluefin tuna. Proc. Biol. Sci. 278, 18–27.

Shioi, T., McMullen, J.R., Tarnavski, O., Converso, K., Sherwood, M.C., Manning, W.J., Izumo, S., 2003. Rapamycin attenuates load-induced cardiac hypertrophy in mice. Circulation 107, 1664–1670.

Simonot, D.L., Farrell, A.P., 2007. Cardiac remodeling in rainbow trout *Oncorhynchus mykiss* Walbaum in response to phenylhydrazine-induced anemia. J. Exp. Biol. 210, 2574–2584.

Soonpaa, M.H., Kim, K.K., Pajak, L., Franklin, M., Field, L.J., 1996. Cardiomyocyte DNA synthesis and binucleation during murine development. Am. J. Physiol. 271, H2183–2189.

Stecyk, J.A.W., 2017. Cardiovascular responses to limiting oxygen levels. In: Gamperl, A.K., Gillis, T.E., Farrell, A.P., Brauner, C.J. (Eds.), The Cardiovascular System: Development, Plasticity and Physiological Responses. Fish Physiology, vol. 36B. Elsevier, Amsterdam. pp. 299–371. Chapter 5.

Stecyk, J.A., Larsen, B.C., Nilsson, G.E., 2011. Intrinsic contractile properties of the crucian carp (*Carassius carassius*) heart during anoxic and acidotic stress. Am. J. Physiol. Regul. Integr. Comp. Physiol. 301, R1132–1142.

Stenslokken, K.O., Ellefsen, S., Larsen, H.K., Vaage, J., Nilsson, G.E., 2010. Expression of heat shock proteins in anoxic crucian carp (*Carassius carassius*): support for cold as a preparatory cue for anoxia. Am. J. Physiol. Regul. Integr. Comp. Physiol. 298, R1499–1508.

Sucov, H.M., Dyson, E., Gumeringer, C.L., Price, J., Chien, K.R., Evans, R.M., 1994. RXR alpha mutant mice establish a genetic basis for vitamin A signaling in heart morphogenesis. Genes Dev. 8, 1007–1018.

Sun, Y., Weber, K.T., 2000. Infarct scar: a dynamic tissue. Cardiovasc. Res. 46, 250–256.

Sun, X., Hoage, T., Bai, P., Ding, Y., Chen, Z., Zhang, R., Huang, W., Jahangir, A., Paw, B., Li, Y.G., Xu, X., 2009. Cardiac hypertrophy involves both myocyte hypertrophy and hyperplasia in anemic zebrafish. PLoS One 4, e6596, https://doi.org/10.1371/journal.pone.0006596.

Takano, H., Komuro, I., Zou, Y., Kudoh, S., Yamazaki, T., Yazaki, Y., 1996. Activation of p70 S6 protein kinase is necessary for angiotensin II-induced hypertrophy in neonatal rat cardiac myocytes. FEBS Lett. 379, 255–259.

Talman, V., Ruskoaho, H., 2016. Cardiac fibrosis in myocardial infarction-from repair and remodeling to regeneration. Cell Tissue Res. 365 (3), 563–581.

Thorarensen, H., Young, C., Davie, P.S., 1996. 11-Ketotestosterone stimulates growth of heart and red muscle in rainbow trout. Can. J. Zool. 74, 912–917.

Todgham, A.E., Schulte, P.M., Iwama, G.K., 2005. Cross-tolerance in the tidepool sculpin: the role of heat shock proteins. Physiol. Biochem. Zool. 78, 133–144.

Todgham, A.E., Iwama, G.K., Schulte, P.M., 2006. Effects of the natural tidal cycle and artificial temperature cycling on Hsp levels in the tidepool sculpin *Oligocottus maculosus*. Physiol. Biochem. Zool. 79, 1033–1045.

Tschantz, D.R., Crockett, E.L., Niewiarowski, P.H., Londraville, R.L., 2002. Cold acclimation strategy is highly variable among the sunfishes (Centrarchidae). Physiol. Biochem. Zool. 75, 544–556.

van den Borne, S.W., Diez, J., Blankesteijn, W.M., Verjans, J., Hofstra, L., Narula, J., 2010. Myocardial remodeling after infarction: the role of myofibroblasts. Nat. Rev. Cardiol. 7, 30–37.

Visse, R., Nagase, H., 2003. Matrix metalloproteinases and tissue inhibitors of metalloproteinases: structure, function, and biochemistry. Circ. Res. 92, 827–839.

Vornanen, M., Hassinen, M., Koskinen, H., Krasnov, A., 2005. Steady-state effects of temperature acclimation on the transcriptome of the rainbow trout heart. Am. J. Physiol. Regul. Integr. Comp. Physiol. 289, R1177–1184.

Wang, J., Karra, R., Dickson, A.L., Poss, K.D., 2013. Fibronectin is deposited by injury-activated epicardial cells and is necessary for zebrafish heart regeneration. Dev. Biol. 382, 427–435.

Waxman, J.S., Keegan, B.R., Roberts, R.W., Poss, K.D., Yelon, D., 2008. Hoxb5b acts downstream of retinoic acid signaling in the forelimb field to restrict heart field potential in zebrafish. Dev. Cell 15, 923–934.

Weber, K.T., Sun, Y., Bhattacharya, S.K., Ahokas, R.A., Gerling, I.C., 2013. Myofibroblast-mediated mechanisms of pathological remodeling of the heart. Nat. Rev. Cardiol. 10, 15–26.

West, J.L., Driedzic, W.R., 1999. Mitochondrial protein synthesis in rainbow trout (*Oncorhynchus mykiss*) heart is enhanced in sexually mature males but impaired by low temperature. J. Exp. Biol. 202 (Pt. 17), 2359–2369.

Whittaker, P., Boughner, D.R., Kloner, R.A., 1989. Analysis of healing after myocardial infarction using polarized light microscopy. Am. J. Pathol. 134, 879–893.

Willenbrock, F., Crabbe, T., Slocombe, P.M., Sutton, C.W., Docherty, A.J., Cockett, M.I., O'Shea, M., Brocklehurst, K., Phillips, I.R., Murphy, G., 1993. The activity of the tissue inhibitors of metalloproteinases is regulated by C-terminal domain interactions: a kinetic analysis of the inhibition of gelatinase A. Biochemistry 32, 4330–4337.

Witman, N., Murtuza, B., Davis, B., Arner, A., Morrison, J.I., 2011. Recapitulation of developmental cardiogenesis governs the morphological and functional regeneration of adult newt hearts following injury. Dev. Biol. 354, 67–76.

Wright, J.K., Cawston, T.E., Hazleman, B.L., 1991. Transforming growth factor beta stimulates the production of the tissue inhibitor of metalloproteinases (TIMP) by human synovial and skin fibroblasts. Biochim. Biophys. Acta 1094, 207–210.

Yamazaki, T., Komuro, I., Kudoh, S., Zou, Y., Shiojima, I., Mizuno, T., Takano, H., Hiroi, Y., Ueki, K., Tobe, K., et al., 1995. Mechanical stress activates protein kinase cascade of phosphorylation in neonatal rat cardiac myocytes. J. Clin. Invest. 96, 438–446.

Yin, V., 2013. Cardiac regeneration. In: Evans, D., Claiborne, J., Currie, S. (Eds.), The Physiology of Fishes, fourth ed. CRC Press, Ottawa.

Yin, V.P., Lepilina, A., Smith, A., Poss, K.D., 2012. Regulation of zebrafish heart regeneration by miR-133. Dev. Biol. 365, 319–327.

Zhang, P., Lu, L., Yao, Q., Li, Y., Zhou, J., Liu, Y., Duan, C., 2012. Molecular, functional, and gene expression analysis of zebrafish hypoxia-inducible factor-3alpha. Am. J. Physiol. Regul. Integr. Comp. Physiol. 303, R1165–1174.

Zhang, R., Han, P., Yang, H., Ouyang, K., Lee, D., Lin, Y.F., Ocorr, K., Kang, G., Chen, J., Stainier, D.Y., Yelon, D., Chi, N.C., 2013. In vivo cardiac reprogramming contributes to zebrafish heart regeneration. Nature 498, 497–501.

Zhao, L., Borikova, A.L., Ben-Yair, R., Guner-Ataman, B., MacRae, C.A., Lee, R.T., Burns, C.G., Burns, C.E., 2014. Notch signaling regulates cardiomyocyte protection during zebrafish heart regeneration. Proc. Natl. Acad. Sci. U.S.A. 111, 1403–1408.

Zhou, X.L., Liu, J.C., 2014. Role of Notch signaling in the mammalian heart. Braz. J. Med. Biol. Res. 47, 1–10.

4

TEMPERATURE AND THE CARDIOVASCULAR SYSTEM

ERIKA J. ELIASON[*,1]
KATJA ANTTILA[†]

[*]University of California, Santa Barbara, Santa Barbara, CA, United States
[†]University of Turku, Turku, Finland
[1]Corresponding author: erika.eliason@lifesci.ucsb.edu

This chapter reviews the influence of environmental temperature on the cardiovascular system. Specifically, we examine how the cardiovascular system responds to both acute and chronic changes in temperature, and how these responses differ across fish species. Cardiovascular responses from the whole organ down to the molecular level are considered. Given that mitochondrial (tissue) oxygen demand increases as temperatures warm, we consider how the cardiovascular system's critical role in supporting aerobic metabolism is influenced by temperature change. Furthermore, we assess the mechanistic

The Cardiovascular System: Development, Plasticity and Physiological Responses, Volume 36B
FISH PHYSIOLOGY

DOI: https://doi.org/10.1016/bs.fp.2017.09.003

basis for cardiovascular collapse at high environmental temperatures. At the end of this chapter, we discuss the ecological implications of changes in environmental temperature, by specifically investigating the suggestion that cardiac function could be a key factor determining the distribution of fishes.

1. TEMPERATURE EFFECTS ON CARDIOVASCULAR FUNCTION

Temperature is a master environmental factor that has profound effects on all biological levels of organization, from molecules to ecosystems. This is especially true for ectothermic animals, such as fish, since their body temperature varies with the temperature of the surrounding water (with the rare exception of partly endothermic species like tunas and sharks) (Dickson and Graham, 2004). Fish can encounter substantial changes in environmental temperature (e.g., via diurnal cycles, seasonal variation, moving through a stratified water column, migrating to new areas). Since temperature affects the rates of many biological reactions (the movement of molecules, enzyme activities, and the function of ion channels to name a few) fish may employ a variety of strategies to compensate for these effects. These compensatory strategies include behavioral responses (e.g., avoiding harsh temperatures), physiological responses (e.g., changes in heart rate, f_H), and molecular responses (e.g., changes in the structure and expression of proteins and in lipid composition). In addition, these responses can occur over several distinct time scales including acute (short-term) reactions, phenotypic plasticity associated with chronic exposure (i.e., acclimation and acclimatization effects), and adaptation over generations. In this chapter, we specifically consider how the cardiovascular system is influenced by environmental temperature and the implications for whole animal function and thermal tolerance.

1.1. Morphology

One key strategy employed by fishes to compensate for changes in environmental temperature is to reversibly adjust the morphology of the heart (recently reviewed in Keen et al., 2017). Several species that remain active during cold winter periods increase relative ventricle mass (RVM) in response to a decrease in environmental temperature. This change has often been observed in rainbow trout (*Oncorhynchus mykiss*) during cold acclimation in the laboratory, but also when temperature has dropped naturally during seasonal change. The increase in RVM varies between 10% and 50% across studies, depending on the acclimation/environmental temperature (Graham and Farrell, 1989, 1990; Keen and Farrell, 1994; McClelland et al., 2005).

Similarly, cold acclimation has been observed to increase RVM in Atlantic salmon (*Salmo salar*) (~10%), European sea bass (*Dicentrarchus labrax*) (~28%), Atlantic cod (*Gadus morhua*) (~34%), channel catfish (*Ictalurus punctatus*) (~35%), green sunfish (*Lepomis cyanellus*) (~15%), and goldfish (*Carassius auratus*) (~26%) (Anttila et al., 2015; Farrell et al., 2007; Foster et al., 1993; Kent et al., 1988; Tsukuda et al., 1985). However, cardiac growth in response to temperature is not always observed (see below) and has been reported to vary between sexes. In this regard, male rainbow trout appear to have greater cardiac plasticity in RVM compared to females. For example, cold acclimation to 4°C increased the ventricle mass of male rainbow trout by ~62% when compared to male trout acclimated to 17°C (Klaiman et al., 2011), while no change in RVM was observed in female fish between acclimation temperatures (Keen et al., 2016; Klaiman et al., 2011). Thus, in studies where the sex of fish has not been determined, the percent changes in RVM could depend on the proportion of female and male fish. Furthermore, not all species respond equally. For example, wild European perch (*Perca fluviatilis*) have similar sized ventricles when living at different temperatures (i.e., 17 vs 23°C) (Ekström et al., 2016a) (also see discussion at the end of the section about species differences).

The fine-scale morphology of the cardiac tissue also shows remarkable thermal plasticity. Chronic cold acclimation in hatchery-reared rainbow trout induced hypertrophy of the spongy myocardium, with significant increases in the cross-sectional area of the bundles that make up the spongy trabeculae (Keen et al., 2016; Klaiman et al., 2011). The proportion of compact myocardium also decreased (by 13%–37%) with chronic cold acclimation (Keen et al., 2016; Klaiman et al., 2011), whereas collagen deposition increased several fold in both spongy and compact myocardium inducing fibrosis (Keen et al., 2016; Klaiman et al., 2011). This latter change has been associated with increased stiffness of the ventricular muscle during cold acclimation in rainbow trout (Keen et al., 2016). Interestingly, acclimation to 19°C, a temperature above the optimum growth temperature for Atlantic salmon, also resulted in increased collagen I protein levels in the compact myocardium (Jørgensen et al., 2014). The composition of the myofilaments has also been observed to change during cold acclimation (see review by Keen et al., 2017). A shift in troponin C (Ca^{2+} binds to troponin C and initiates contraction) isoforms with cold acclimation has been reported in zebrafish (*Danio rerio*) and rainbow trout (Genge et al., 2013), and the composition of troponin I isoforms has been described in rainbow trout (Alderman et al., 2012). These changes in filament composition could be connected with alterations in the contractile properties of the heart since different isoforms have different Ca^{2+} binding properties, and changes in the composition of the myocardial filaments can be linked to changes in contractility (Genge et al., 2013; Gillis et al., 2005; Keen et al., 2017).

Actually, most of the morphological changes can be related to functional changes of the heart at cold temperatures. The increase in ventricular mass, and especially in the proportion of spongy myocardium and in bundle cross-sectional area, would allow for greater force generation and maximum power output due to an increased proportion of contractile machinery. Furthermore, cold acclimation changes the composition and phosphorylation of the myofilaments, and this enhances the contractile performance of the myocytes (Keen et al., 2017). Indeed, cold acclimation has been shown to increase the maximum power output of Atlantic cod hearts by 80% (Lurman et al., 2012). The increase in ventricular size, and in the proportion of spongy myocardium that contains "lacunae" that can fill with blood during diastole, could also enhance diastolic filling volume. Furthermore, the trabecular nature of spongy myocardium reduces wall tension (LaPlace's law), i.e., the workload for cardiac contraction is lowered, and this is speculated to allow for high ejection fractions (Franklin and Davie, 1992, see also review by Keen et al., 2017). These structural changes are probably at least partly behind the observations that maximum stroke volume (V_{Smax}) increases with cold acclimation (Farrell, 1991; Graham and Farrell, 1989, 1990). For example, mathematical models have revealed that an increase in spongy myocardial area leads to increased stroke volume (V_{S}), stroke work, and ejection fraction (Kochová et al., 2015), and it has been suggested that increased amounts of collagen and fibrosis strengthen the myocardium and could protect the heart from overstretching at cold temperatures. Notably, however, there is a trade-off as greater amounts of collagen could also lead to myocardial stiffness, and this could reduce diastolic filling and lead to diastolic dysfunction (Collier et al., 2012). Collectively, the described changes in cardiac morphology enable fishes to at least partially compensate for the decrease in the heart's contractility at cold temperatures (see review by Keen et al., 2017), and allow them to maintain cardiac performance (see Sections 1.2.2, 1.3.2, and 1.4.2).

Whether the cold-induced growth of the ventricle and spongy myocardium occurs via cellular hypertrophy or hyperplasia is still under debate. While both appear to be involved, cellular hypertrophy likely plays a bigger role (Keen et al., 2016). McClelland et al. (2005) found that the ventricle DNA content remained unchanged during cold acclimation, though the mass of the ventricle increased. Furthermore, although the expression of proliferating cell nuclear antigen (PCNA, which is involved with DNA replication) increased with cold acclimation in rainbow trout (Keen et al., 2016), the increase in expression of genes related to hypertrophy was 3–5-times higher than that of PCNA. Several different markers of hypertrophy, protein turnover, transcription, and translation efficiency also increase with cold acclimation in rainbow trout, sockeye salmon (*Oncorhynchus nerka*), bluefin tuna (*Thunnus orientalis*), and common carp (*Cyprinus carpio*) (Anttila et al., 2014b; Gracey et al., 2004; Jayasundara et al., 2013; Vornanen et al., 2005), and these could be associated with the cold-induced hypertrophy of the heart through increased cell volume. Similarly, chronic warm acclimation reduced the expression of structural myocardial proteins such as different isoforms of myosin, actin and troponin in Atlantic salmon (Jørgensen et al., 2014).

On the other hand, cold acclimation-induced atrophy of the compact myocardium in hatchery-reared rainbow trout and Arctic char (*Salvelinus alpinus*) (Anttila et al., 2015; Keen et al., 2016; Klaiman et al., 2011). The reduced thickness of the compact layer would decrease this tissue's reliance on oxygen supplied via the coronary circulation at cold temperatures. This suggestion is supported by Egginton and Cordiner (1997) and Anttila et al. (2015) who both showed that capillary density in the compact layer is reduced during cold acclimation in these species (Fig. 1). Indeed, the heart's oxygen consumption is reduced at cold temperatures (Graham and Farrell, 1990), while the oxygen carrying capacity of both water and blood increase (Farrell and Clutterham, 2003). The atrophy of the compact myocardium could also be a compensatory strategy to reduce the stiffness of the ventricle associated with the increased amount of collagen (Keen et al., 2016; Klaiman et al., 2014).

Conversely, compact myocardium thickness has been observed to increase with warm acclimation in rainbow trout and Arctic char (Anttila et al., 2015; Keen et al., 2016; Klaiman et al., 2011). The increase in the proportion of compact myocardium may be a response to the atrophy of the spongy myocardium (Keen et al., 2016; Klaiman et al., 2011), which results in an overall decrease in RVM during warm acclimation (e.g., Anttila et al., 2015; Farrell et al., 2007; Foster et al., 1993; Keen et al., 2016; Kent et al., 1988; Klaiman et al., 2014; Tsukuda et al., 1985). The force of contraction is also lower in warm-acclimated fish as compared to cold-acclimated conspecifics (Bailey and Driedzic, 1990; Lurman et al., 2012). This is likely because of a reduction

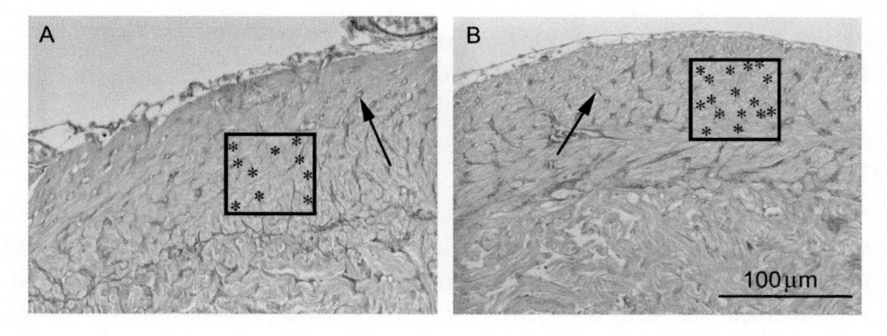

Fig. 1. Histological sections of amylase–periodic acid–Schiff stained ventricles of (A) cold-acclimated (8°C) and (B) warm-acclimated (15°C) Arctic char. *Arrows* point to capillaries. Inside the 5000 μm² *squares*, capillaries are marked with an *asterisk*.

in the amount of contracting muscle, but also because of the negative force–velocity relationship; i.e., the velocity and frequency of contraction both increase with temperature (Badr et al., 2016, 2017; Ballesta et al., 2012; Galli et al., 2009; Hassinen et al., 2014; Haverinen and Vornanen, 2009; Shiels et al., 2015; Vornanen et al., 2014: also see Sections 1.3 and 1.4). Cardiac oxygen consumption increases concurrently with the increase in heart rate (f_H) at warm temperatures (e.g., Graham and Farrell, 1990). In order to keep the heart supplied with oxygenated blood, the density of capillaries supplied by the coronary artery increases in the compact myocardium with warm acclimation in rainbow trout and Arctic char (Anttila et al., 2015; Egginton and Cordiner, 1997; Keen et al., 2016; Klaiman et al., 2011) (Fig. 1). Furthermore, protein levels of vascular endothelial growth factor (VEGF), which promotes proliferation and migration of endothelial cells and participates in the formation of new capillaries, increase with warm acclimation in hatchery-reared Atlantic salmon (Jørgensen et al., 2014); although a reduction in VEGF mRNA levels has been observed in domesticated rainbow trout (Keen et al., 2016).

Interestingly, when individuals acclimated to the same temperature are compared, individuals with bigger ventricles have a higher upper critical thermal maximum (CT_{max}—the temperature where fish lose equilibrium during acute heating). This correlation has been observed with domesticated Atlantic salmon and European sea bass (Anttila et al., 2013a; Ozolina et al., 2016). As mentioned above, a larger ventricle is associated with a larger V_S and a greater cardiac output ($\dot{Q}$) (Farrell, 1991; Hillman and Hedrick, 2015). Thus, individuals with a larger RVM could have enhanced blood supply to the tissues (including heart and brain). Owing to this increased blood supply capacity, brain and nervous function may be maintained longer at warm temperatures in these individuals (see Ozolina et al., 2016). However, the association between temperature tolerance and cardiac morphology (especially fine-scale morphology) and performance needs further study, especially since it has been shown that CT_{max}: only increases slightly with environmental hyperoxia (Ekström et al., 2016a); only decreases slightly with anemia (Wang et al., 2014); and is not impacted until water O_2 levels and aerobic scope are reduced by >50% and >70%, respectively (Ern et al., 2016). These data suggest that additional factors other than oxygen supply capacity are also related to CT_{max}.

While cold acclimation is most often associated with ventricular enlargement, there are intra- and inter-specific differences in the heart's response to temperature. For example, cold acclimation (8 vs 15°C) increased RVM by 10% in Atlantic salmon, while the same study observed no change in RVM of Arctic char (Anttila et al., 2015). Ruiz and Thorarensen (2001) actually found that Arctic char reared at 15°C had somewhat larger ventricles than individuals reared at 5°C. Similarly, the RVM did not change in European perch, white perch (*Morone americanus*) and yellow perch (*Perca flavescens*) with cold acclimation (17 vs 23°C for European perch and 5 vs 20°C for white and yellow perch) (Ekström et al., 2016a; Sephton and Driedzic, 1991). The RVM of crucian carp (*Carassius carassius*) was also unchanged when environmental temperature was decreased from 23 to 4°C (Tiitu and Vornanen, 2001). Thus, there must be other factors that play a role in the ventricular morphological plasticity besides temperature. One factor could be the activity of the species during the cold winter period. For example, crucian carp inhabit small lakes with ice coverage during the winter period, and these conditions are often concomitant with anoxia. Owing to these combined environmental factors, the crucian carp are dormant during the winter period (e.g., Holopainen et al., 1997), and thus, have a reduced f_H compared to during summer months (Matikainen and Vornanen, 1992; Tiitu and Vornanen, 2001). As such, there may be no need for cardiac growth to sustain the cardiorespiratory physiology of this species. Remarkably, crucian carp can maintain cardiac function during anoxia at cold temperatures (Stecyk et al., 2004), and thus, they are probably using some other mechanism beyond cardiac remodeling to achieve this unusual feat. One of them is clearly the capacity to produce ethanol using the enzyme pyruvate decarboxylase (Fagernes et al., 2017).

In general, species living in cold waters (Antarctic/Arctic areas) have larger ventricles when compared to species from temperate areas. For example, domesticated Arctic char have a higher RVM than Atlantic salmon (Anttila et al., 2015; Penney et al., 2014), and the thickness of the compact myocardium is higher in char than in salmon (Anttila et al., 2015). In wild Antarctic fishes, which lack hemoglobin, like *Chionodraco hamatus* and *C. aceratus*, RVM is 0.3%–0.4% (Harrison et al., 1991; Holeton, 1970; Tota et al., 1991). This is an impressive 3 × larger than in Atlantic salmon (around 0.1%; Anttila et al., 2015). These morphological differences enable these genera to have a higher V_S and $\dot{Q}$ than red-blooded fishes (Hemmingsen et al., 1972). Thus, even though f_H at the temperatures where these fish live is extremely low (around 14 beats per minute at 0.5°C; Holeton, 1970), the increased V_S and blood volume possessed by these species (e.g., Fitch et al., 1984) ensures adequate oxygen delivery to the tissues at low temperatures (Hemmingsen, 1991, see also review by Sidell and O'Brien, 2006).

In conclusion, cardiac morphology in fishes has evolved to deal with different kinds of environments and is also plastic so that individual fish can adjust, at least partly, to variations in environmental temperature (e.g., Anttila et al., 2015; Egginton and Cordiner, 1997; Farrell et al., 2007; Foster et al., 1993; Keen et al., 2016; Kent et al., 1988; Klaiman et al., 2011; Tsukuda et al., 1985).

1.2. Cardiac Output

Cardiac output ($\dot{Q}$) is the volume of blood pumped by the heart per unit time, is the product of heart rate (f_H) and stroke volume (V_S) ($\dot{Q}=f_H \times V_S$), and directly influences blood pressure (blood pressure $= \dot{Q} \times$ peripheral vascular resistance) (see Chapter 4, Volume 36A: Farrell and Smith, 2017). Several physiological factors (e.g., venous return and pressure, nervous tone, total blood volume, blood vessel diameter and their elasticity) have significant effects on these parameters and equations. In the following sections, we review how changes in temperature influence these parameters and their regulation. The effect of temperature on $\dot{Q}$, as well as the other rate variables, can be expressed using Q_{10} (temperature coefficient) values. Q_{10} values indicate the factorial change in the rate of a function when temperature is changed by 10°C. However, the rate of temperature change and the duration of exposure at the new (novel) temperature vary widely across studies, and this has critical implications for interpreting the results. For example, some studies employ a classic critical thermal maximum (CT_{max}) protocol, in which water temperature is increased rapidly (usually by 0.3°C min^{-1}) until the fish lose equilibrium (e.g., Anttila et al., 2013a), while the other studies acutely increase water temperature at a slower rate [e.g., 2°C h^{-1} (Mendonça and Gamperl, 2010; Steinhausen et al., 2008) or 4°C h^{-1} (Clark et al., 2008; Eliason et al., 2013)]. Furthermore, some studies hold fish at their test temperature for 24–48 h before conducting experiments (e.g., Eliason et al., 2013), while others acclimate fish for weeks (e.g., Franklin et al., 2007). The physiological, ecological, and evolutionary responses to temperature are all strongly influenced by the duration of the temperature exposure.

The influence of temperature on $\dot{Q}$ has been examined using several different methodologies (e.g., the indicator dye dilution method, Transonic® flow probes, Doppler flow probes, or indirectly calculated via the Fick equation (see Chapter 4, Volume 36A: Farrell and Smith, 2017)) in a wide variety of species. The early pioneering studies that used the Fick equation [Oxygen consumption ($\dot{M}O_2$) $= \dot{Q} \times$ the arterio-venous difference in blood oxygen content (A-VO_2)] to estimate $\dot{Q}$ (e.g., Brett, 1971; Butler and Taylor, 1975; Cech et al., 1976) are not included due to numerous issues associated with this method (i.e., overestimates of $\dot{Q}$, see Farrell et al., 2014). Similarly, studies that used the indicator dye dilution or microsphere methods to calculate $\dot{Q}$ (e.g., Barron et al., 1987; Taylor et al., 1996) are excluded.

1.2.1. Acute Responses

Acute increases in temperature result in an increase in resting $\dot{Q}$ in most fish species. However, the rate of the increase in $\dot{Q}$ with temperature varies greatly among species, as indicated by large differences in Q_{10} values: Arctic staghorn sculpin, *Gymnocanthus tricuspis* Q_{10}: 3.35 (Franklin et al., 2013); Arctic sculpin, *Myoxocephalus scorpioides* Q_{10}: 1.9–2.33 (Franklin et al., 2013; Gräns et al., 2013); Arctic shorthorn sculpin, *M. scorpius* Q_{10}: 1.75–2.1 (Franklin et al., 2013; Gräns et al., 2013); the bald notothen, *Pagothenia borchgrevinki* Q_{10}: 1.62 (Franklin et al., 2007); the emerald rockcod, *Pagothenia bernacchii* Q_{10}: 1.2 (Axelsson et al., 1992); rainbow trout Q_{10}: 1.2–2.0 (Brodeur et al., 2001; Ekström et al., 2014; Gamperl et al., 2011; Keen and Gamperl, 2012; Petersen et al., 2011; Sandblom and Axelsson, 2007); sockeye salmon Q_{10}: 1.7 (Eliason et al., 2013; Steinhausen et al., 2008); Chinook salmon (*Oncorhynchus tshawytscha*) Q_{10}: 1.8 (Clark et al., 2008); Atlantic salmon Q_{10}: ~2 (Penney et al., 2014); Arctic char Q_{10}: ~2 (Penney et al., 2014); Atlantic cod Q_{10}: 2.48 (Gollock et al., 2006); lingcod (*Ophiodon elongates*) Q_{10}: 2–3 (Stevens et al., 1972); spiny dogfish (*Squalus acanthias*) Q_{10}: 1.6 (Sandblom et al., 2009); winter flounder (*Pleuronectes americanus*) Q_{10}: 2.0 (Mendonça and Gamperl, 2010), European sea bass Q_{10}: 1.6 (Wang et al., 2014); and European perch Q_{10}: 2.0–2.3 (Ekström et al., 2016a; Sandblom et al., 2016a). Furthermore, resting $\dot{Q}$ has been observed to plateau and then decline in many species at temperatures approaching upper critical temperatures (e.g., Ekström et al., 2014, 2016a; Eliason et al., 2013; Gamperl et al., 2011; Gollock et al., 2006; Mendonça and Gamperl, 2010; Penney et al., 2014; Steinhausen et al., 2008; Wang et al., 2014), and to continue to increase up to critical temperatures in a few studies (e.g., Brodeur et al., 2001; Clark et al., 2008). In contrast to these numerous studies on water breathing fish, one study on the air-breathing Asian swamp eel (*Monopterus albus*) found that resting $\dot{Q}$ was insensitive to acute temperature increases (Lefevre et al., 2016).

The influence of acute temperature changes on dorsal aorta pressure (P_{DA}) is quite variable between species. P_{DA} increased with acute warming in rainbow trout (Heath and Hughes, 1973; Sandblom and Axelsson, 2007) and Japanese eel (*Anguilla japonica*) (Takei and Tsukada, 2001), whereas P_{DA} decreased with warming in winter flounder (Mendonça and Gamperl, 2010). Other studies report that P_{DA} remains constant during acute warming [Chinook salmon (Clark et al., 2008) and rainbow trout (Gamperl et al., 2011)]. P_{DA} may not directly follow changes in $\dot{Q}$ for all species due to differential changes in vascular resistance with temperature; e.g., in the gills (see Section 1.8 for further information on how regional blood flow is influenced by temperature).

Far fewer studies have examined the more complicated relationship between maximum $\dot{Q}$ and temperature. Studies have typically evaluated "maximum" $\dot{Q}$ either by progressively increasing the water velocity in a swim tunnel until fatigue is reached (e.g., constant acceleration test, Clark et al., 2011; critical swimming test, Eliason et al., 2013), swimming fish at a fixed velocity (e.g., Korsmeyer et al., 1997; Steinhausen et al., 2008) or by chasing fish to exhaustion for 3–10 min (e.g., Franklin et al., 2013; Sandblom et al., 2016a). These different methods should be taken into consideration when evaluating how maximum $\dot{Q}$ responds to temperature. When fish are swum across a range of temperatures well below their upper critical thermal limit, there is an increase in maximum $\dot{Q}$ with increasing temperature [European perch (Sandblom et al., 2016a); yellowfin tuna (*Thunnus albacares*) (Korsmeyer et al., 1997); and *M. scorpius* (Franklin et al., 2013)]. In contrast, when fish are swum to temperatures approaching their upper critical thermal limit, maximum $\dot{Q}$ reaches a plateau [sockeye salmon (Steinhausen et al., 2008); *M. scorpioides* (Franklin et al., 2013)] or declines [pink salmon (*Oncorhynchus gorbuscha*) (Clark et al., 2011); sockeye salmon (Eliason et al., 2013)]. Notably, maximum $\dot{Q}$ reaches a plateau phase and/or starts to decline at much lower temperatures than resting $\dot{Q}$ in the studies conducted to date (Eliason et al., 2013; Steinhausen et al., 2008) (Fig. 2). As such, cardiac scope (cardiac scope = maximum $\dot{Q}$ − resting $\dot{Q}$) declines at warm temperatures (Eliason et al., 2013; Steinhausen et al., 2008) (Figs. 2 and 3).

1.2.2. Acclimation Responses

Species have evolved different strategies to deal with their specific environmental conditions. Fishes continually exposed to a consistent temperature may optimize performance at that particular temperature, and may have limited capacity for phenotypic plasticity to alter performance in response to a changing environment (but see Seebacher et al., 2005). In contrast, fishes exposed to varying environmental temperatures may possess sufficient phenotypic plasticity to be able to largely compensate for the thermal effects on cardiovascular traits and/or performance. As such, the capacity for cardiac plasticity is predicted to differ across species, and also be connected to other environmental factors beyond temperature. Therefore, it is not surprisingly that the effect of thermal acclimation on $\dot{Q}$ varies across fish species.

Acclimation studies conducted on both resting and exercising fish show varying degrees of cardiac thermal compensation. Common carp display thermal compensation in resting $\dot{Q}$ at some acclimation temperatures, but not others. Specifically, two studies with common carp report a Q_{10} of 1.7–3.2 between fish seasonally acclimated to 5–6 and 10°C, but no differences in routine $\dot{Q}$ between 10 and 15°C (Stecyk and Farrell, 2002, 2006). Similarly, resting $\dot{Q}$ did not differ in large-scale sucker (*Catostomus macrocheilus*) seasonally acclimated at 5 and 10°C, but increased significantly at 16°C (Q_{10} = 2.1

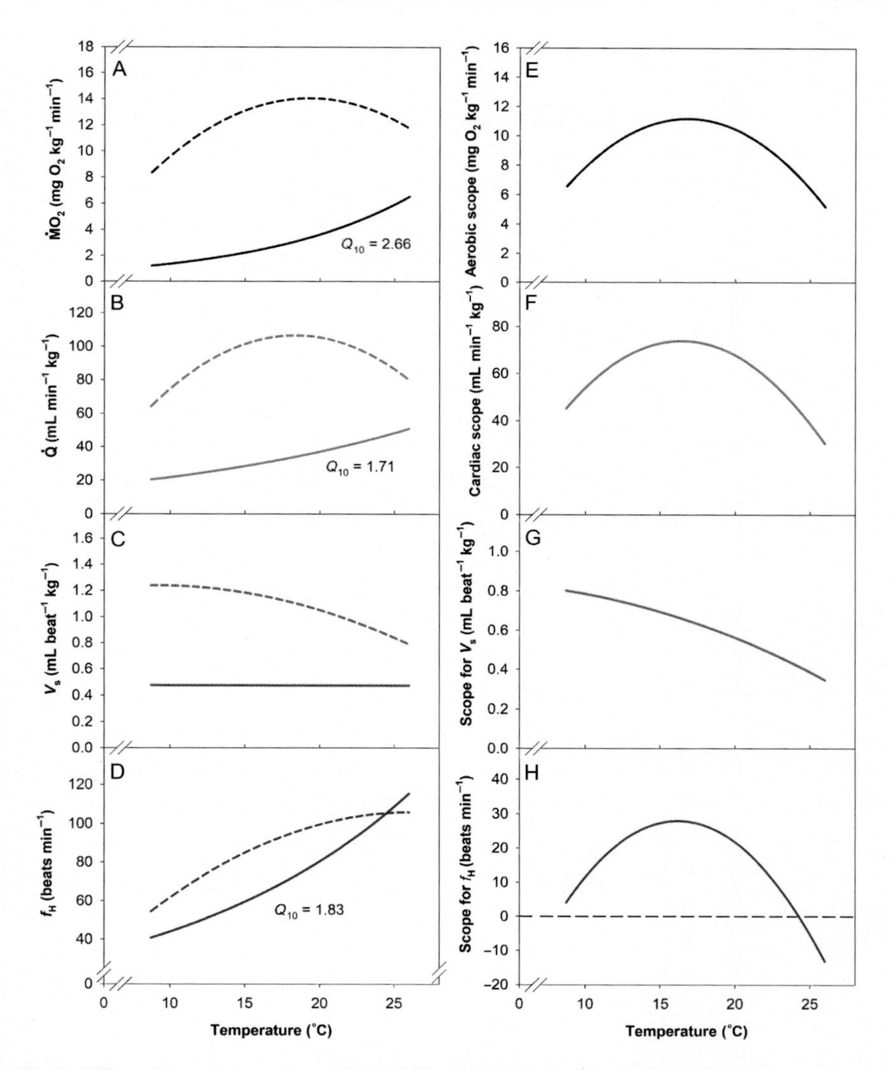

Fig. 2. Effect of temperature on resting (*solid lines*) and maximum (*dashed lines*) (A) oxygen consumption ($\dot{M}O_2$); (B) cardiac output ($\dot{Q}$); (C) cardiac stroke volume (V_S); and (D) heart rate (f_H) in Chilko sockeye salmon. The corresponding scope values (maximum – resting) are shown in panels (E)–(H). Data have been modified from Eliason, E.J., Clark, T.D., Hague, M.J., Hanson, L.M., Gallagher, Z.S., Jeffries, K.M., Gale, M.K., Patterson, D.A., Hinch, S.G., Farrell, A.P., 2011. Differences in thermal tolerance among sockeye salmon populations. Science 332, 109–112; Eliason, E.J., Clark, T.D., Hinch, S.G., Farrell, A.P., 2013. Cardiorespiratory collapse at high temperature in swimming adult sockeye salmon. Conserv. Physiol. 1, cot008.

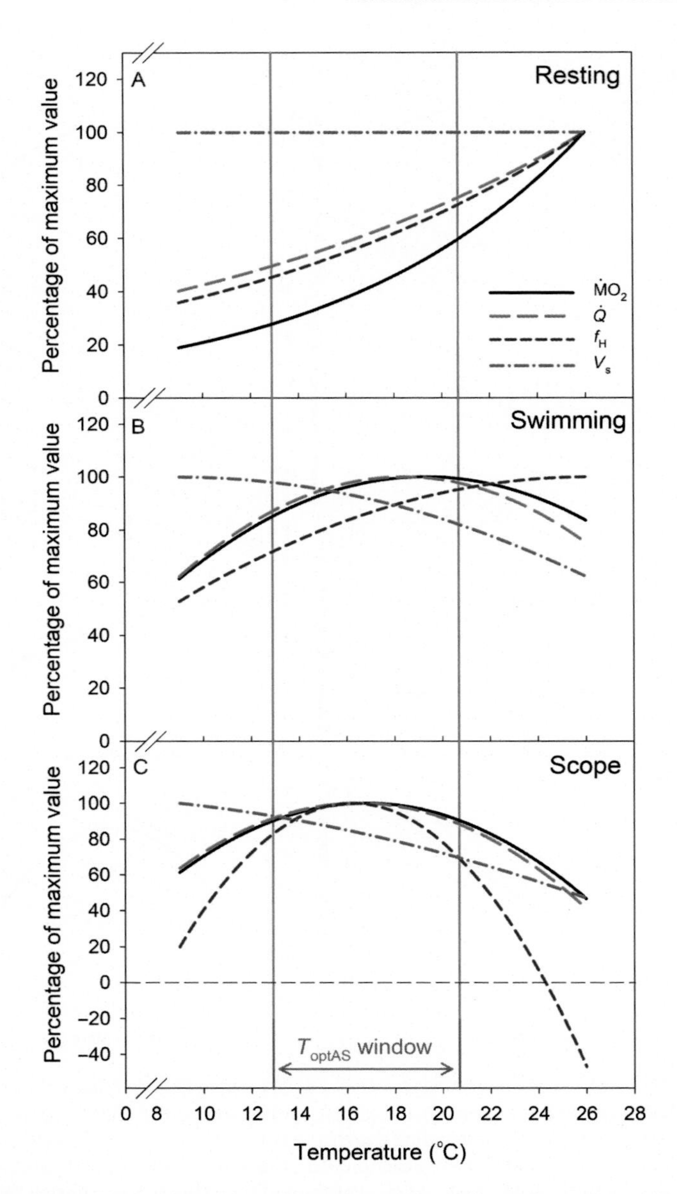

Fig. 3. Effect of temperature on percentage of maximum values for oxygen consumption ($\dot{M}O_2$), cardiac output ($\dot{Q}$), heart rate (f_H), and cardiac stroke volume (V_S) in Chilko sockeye salmon: (A) at rest; (B) when maximally swimming; and (C) scope (maximum – resting). *Lines* were calculated from the data in Fig. 2. The optimal thermal window for aerobic scope (T_{optAS} window = the range of temperatures where aerobic scope is maintained at 90% of maximum) is indicated by *vertical gray lines*.

between 10 and 16°C) (Kolok et al., 1993). Whereas, maximum $\dot{Q}$ increased significantly between 5 and 10°C (Q_{10}=2.96), but did not differ between 10 and 16°C, and this resulted in a large increase in absolute cardiac scope between 5 and 10°C and no difference between 10 and 16°C (Kolok et al., 1993). In contrast, seasonally acclimated northern squawfish (*Ptychocheilus oregonensis*) had higher resting and maximum $\dot{Q}$ at 16°C as compared to 5°C (Q_{10} values were 1.39 and 1.47, respectively) though cardiac scope did not differ between temperatures (Kolok and Farrell, 1994). A population of European perch field acclimated to 22°C had a higher maximum $\dot{Q}$, but no difference in resting $\dot{Q}$ or cardiac scope compared to a population field acclimated to 17°C (Sandblom et al., 2016a). Finally, resting $\dot{Q}$, maximum $\dot{Q}$, and cardiac factorial scope did not differ in *P. borchgrevinki* acclimated to −1 and 4°C, demonstrating complete thermal compensation of cardiac function with acclimation (Franklin et al., 2007; Seebacher et al., 2005).

The influence of thermal acclimation on P_{DA} also varies across species. The P_{DA} of rainbow trout was significantly higher at an intermediate acclimation temperature (11°C) compared to 4 or 18°C (Taylor et al., 1996). Yet, an earlier study with rainbow trout found that P_{DA} was unaffected by temperature acclimation to 5, 12, and 20°C (Wood et al., 1979). In the dogfish (*Scyliorhinus canicula*), P_{DA} increased by 72% across a 10°C acclimation range (Butler and Taylor, 1975). This species-specific variability in the P_{DA} response may be linked to differences in the effects of temperature acclimation on gill vascular resistance (R_{sys}), f_H, and/or V_S (see Sections 1.3.1 and 1.3.2).

1.3. Heart Rate

1.3.1. Acute Responses

Heart rate is one of the easiest variables to measure when assessing the influence of environmental temperature on cardiac function, and there are numerous studies that have examined both acute and long-term effects of temperature on f_H in different fish species. We have summarized these responses in Table 1. Similar to other rate functions, resting heart rate (f_H) rises with acute increases in environmental temperature (e.g., Brett, 1971; Brodeur et al., 2001; Farrell, 1991; Fry, 1947; Gamperl et al., 2011; Penney et al., 2014: see Table 1 and Figs. 2–5). This increase is not usually linear, and the Q_{10} value varies between 0.8 and 9.0 (see Table 1 and Fig. 5). In some species, resting f_H continues to increase until the maximum temperature tolerated by the fish has been reached (e.g., Clark et al., 2008) while in others, f_H plateaus and then decreases before CT_{max} is reached (e.g., Ekström et al., 2016a; Gollock et al., 2006; Mendonça and Gamperl, 2010; Penney et al., 2014). Furthermore,

Table 1

Effect of temperature on the heart rate (f_H) of different fish species

Region/family	Species	Acclimation temperature (°C)	Measurement temperature (°C)	Heart rate (bpm)	Q_{10} (acute)	Reference
Tropical/Serrasalmidae	Pacu, *Piaractus mesopotamicus*, 373 ± 14 g	15	15	~25 r	2.10	Aguiar et al. (2002)
		15	35	~110 r		
		35	15	~23 r	2.09	
		35	35	~100 r		
Tropical/Cichlidae	Nile tilapia, *Oreochromis niloticus*, 254 ± 17.6 g	25	25	~35 r	1.97	Costa et al. (2000)
		25	40	~97 r		
Tropical/Synbranchidae	Asian swamp eel, *Monopterus albus*, 188 ± 9 g	27	27	~49 r	2.60	Lefevre et al. (2016)
		27	32	~79 r		
		32	32	~75 r	1.78	
		32	37	~100 r		
Tropical/Ictaluridae	Channel catfish, *Ictalurus punctatus*, 228–733 g	22	26	~80 r	2.00	Burleson and Silva (2011)
		22	36.0	~160 r		
Tropical/Scombridae	Yellowfin tuna, *Thunnus albacares*, 3.16 ± 0.38 kg	20	10	~19.6 r	3.08	Blank et al. (2002)
		20	20	~70.8 r		
		20	25	~105.8 r		
Tropical/Scombridae	Bluefin tuna, *Thunnus orientalis*, 9.7–13.3 kg	20	10	~35 f	1.71	Clark et al. (2013)
		20	15	~40 f		
		20	20	~60 f		
Temperate/Carangidae	Jack mackerel, *Trachurus japonicus*, 18.5 ± 0.8 cm	10	10	~30 r		Nofrizal et al. (2009)
		15	15	~40 r		
		22	22	~70 r		
		10	10	~60 m		
		15	15	~120 m		
		22	22	~190 m		

Temperate/Squalidae	Pacific spiny dogfish, *Squalus acanthias*,	10	10	~19.0 r	2.19	Sandblom et al. (2009)
	1.89 ± 0.25 kg	10	16	~30.3 r		
Temperate/Cyprinidae	Goldfish, *Carassius auratus*,	12	20	~95 a	1.26	Ferreira et al. (2014)
	1.4–3.8 g	12	30	~120 a		
		28	20	~80 a	2.25	
		28	30	~180 a		
Temperate/Cyprinidae	Crucian carp, *Carassius carassius*,	5	5	~10 r	2.26	Matikainen and Vornanen (1992)
	13–32 g	5	27	~60 r		
		15	5	~10 r	2.57	
		15	27	~80 r		
Temperate/Cyprinidae	Common carp, *Cyprinus carpio*,	6	6	~7 r		Stecyk and Farrell (2002)
		10	10	~12 r		
	1.1 ± 0.05 kg	15	15	~17 r		
Tropical/Cyprinidae	Zebrafish, *Danio rerio*,	18	18	~80 r	2.10	Lee et al. (2016)
	~3.8 cm	18	28	~169 r		
		28	18	~78 r	2.08	
		28	28	~162 r		
	3.0–3.5 cm	26	15	~80 a	2.31	Sidhu et al. (2014)
		26	30	~280 a		
Temperate/Centrarchidae	Largemouth bass, *Micropterus salmoides*,	13	13	~32 r		Cooke et al. (2003)
		25	25	~49 r		
	592–953 g	13	13	~58 m		
		25	25	~108 m		
Temperate/Centrarchidae	Smallmouth bass, *Micropterus dolomieu*,	16	16	~35 r		Cooke et al. (2010)
		20	20	~45 r		
	477 g	16	16	~70 m		
		20	20	~105 m		

(*Continued*)

Table 1 (Continued)

Region/family	Species	Acclimation temperature (°C)	Measurement temperature (°C)	Heart rate (bpm)	Q_{10} (acute)	Reference
Temperate/Centrarchidae	Rock bass, *Ambloplites rupestris*, 228 g	16	16	~30 r		Cooke et al. (2010)
		23	23	~60 r		
		16	16	~80 m		
		23	23	~100 m		
Temperate/Centrarchidae	Pumpkinseed, *Lepomis gibbosus*, 151 g	16	16	~43 r		Cooke et al. (2010)
		24	24	~80 r		
		16	16	~90 m		
		24	24	~120 m		
Temperate/Centrarchidae	Bluegill, *Lepomis macrochirus*, 130 g	16	16	~50 r		Cooke et al. (2010)
		24	24	~80 r		
		16	16	~95 m		
		24	24	~130 m		
Temperate/Centrarchidae	Black crappie, *Pomoxis nigromaculatus*, 191 g	16	16	~22 r		Cooke et al. (2010)
		24	24	~60 r		
		16	16	~60 m		
		24	24	~90 m		
Temperate/Percichthyidae	Murray cod, *Maccullochella peelii* 1.81 ± 0.14 kg	~23	14	~18.2 r	1.70	Clark et al. (2005)
		~23	29	~40.4 r		
		~23	14	~35 m	2.01	
		~23	29	~100 m		
Temperate/Gadidae	Atlantic cod, *Gadus morhua*, 1.2–1.5 kg	5	5	~33 r	2.88	Claireaux et al. (1995)
		5	7.5	~43 r		
		5	5	~41 m	2.04	
		5	7.5	~49 m		
		10	11	~37 r	2.10	Gollock et al. (2006)
		10	20	~72 r		

Temperate/Percidae	European perch, *Perca fluviatilis*, 327–364 g	16	19	~50 r	2.40	Sandblom et al. (2016a)
		16	28	~110 r		
		16	16	~65 m	1.56	
		16	22	~85 m		
		22	22	~87 m		
		22	19	~35 r	3.21	
		22	28	~100 r		
Temperate/Gobiidae	Longjaw mudsucker, *Gillichthys mirabilis*, 27.31 g	9	9	~30 r	2.08	Jayasundara and Somero (2013)
		9	24	~90 r		
		19	19	~60 r	1.59	
		19	34	~120 r		
		26	26	~90 r	1.57	
		26	35	~135 r		
Temperate/Pleuronectidae	Winter flounder, *Pleuronectes americanus*, 453–843 g	4	4	~21 r		Joaquim et al. (2004)
		4	4	~32 m		
		10	10	~34 r		
		10	10	~52 m		
		8	8	~36 r	2.03	Mendonça and Gamperl (2010)
		8	18	~73 r		
Temperate/Salmonidae	Rainbow trout, *Oncorhynchus mykiss*, ~66–79 g	4	4	~31 r	1.97	Aho and Vornanen (2001)
		4	17	~75 r		
		4	20	~47 r		
		17	4	~25 r	2.28	
		17	17	~73 r		
		17	20	~84 r		
Temperate/Salmonidae	Chinook salmon, *Oncorhynchus tshawytscha*, 2.1–5.4 kg	13	13	~30 f	1.58	Clark et al. (2008)
		13	21	~45 f		
		13	25	~52 f		

(*Continued*)

Table 1 (Continued)

Region/family	Species	Acclimation temperature (°C)	Measurement temperature (°C)	Heart rate (bpm)	Q_{10} (acute)	Reference
Temperate/Salmonidae	Pink salmon,	14	8	~55 m	1.66	Clark et al. (2011)
	Oncorhynchus gorbuscha,	14	14	~90 m		
	1.6 ± 0.1 kg	14	25	~130 m		
Temperate/Salmonidae	Sockeye salmon,	13	15	~65 r	1.63	Steinhausen et al.
	Oncorhynchus nerka,	13	21	~87 r		(2008)
	2.2–2.9 kg	13	15	~82 m	1.51	
		13	21	~105 m		
Temperate/Salmonidae	Coho salmon,	10	13	~55 r	2.35	Casselman et al.
	Oncorhynchus kisutch,	10	21	~109 r		(2012)
	16.9–18.1 g	10	13	~97 m	1.37	
		10	21	~125 m		
Temperate/Salmonidae	Atlantic salmon,	12	12	~85 a	1.87	Anttila et al. (2014a)
	Salmo salar,	12	20	~140 a		
	5.2–7.0 g	20	12	~78 a	1.99	
		20	20	~135 a		
Temperate/Salmonidae	Brown trout, *Salmo trutta*	12	12	~35 r	2.51	Vornanen et al.
	fario, 113.3 ± 10.8 g	12	22	~88 r		(2014)
Temperate/Salmonidae	Brook charr, *Salvelinus*	6	6	~50.1 r	1.89	Benfey and Bennett
	fontinalis, larvae	6	12	~73.3 r		(2009)
Temperate/Lotidae	Burbot, *Lota lota*, 278 ± 39 g	1	1	~25 r	1.86	Tiitu and Vornanen
		1	18	~72 r		(2002)
Arctic/Gadidae	Arctic cod, *Boreogadus*	0	0	~23 a	2.17	Drost et al. (2014)
	saida, 31.8–39.7 g	0	10	~50 a		

Arctic/Gadidae	Navaga cod, *Eleginus navaga*, 70.3–94 g	−1	4	~18 r	1.73	Hassinen et al. (2014)
		−1	10	~25 r		
		8	4	~18 r	2.75	
		8	10	~33 r		
Antarctic/Nototheniidae	Emerald rockcod, *Pagothenia bernacchii*,	0	0	~10.5 r	9.00	Axelsson et al. (1992)
	50.5 ± 5.1 g	0	5	~31.5 r		
Antarctic/Nototheniidae	Bald notothen, *Pagothenia borchgrevinki*, 88.9–91.1 g	−1	−1	~20 r	2.16	Franklin et al. (2007)
		−1	4	~33 r		
		−1	8	~40 r		
		4	−1	~20 r	0.83	
		4	4	~20 r		
		4	8	~17 r		
		−1	−1	~27 m	1.72	
		−1	4	~36 m		
		−1	8	~44 m		
		4	−1	~30 m	1.57	
		4	4	~41 m		
		4	8	~45 m		
Antarctic/Channichthyidae	Blackfin icefish, *Chaenocephalus aceratus*,	1	1	~17.5 r	2.17	Hemmingsen and Douglas (1972)
	347–1917 g	1	10	~35 r		

Only studies where f_H (in beats per minute, bpm) was measured at different environmental temperatures are included. Furthermore, although there are multiple studies for most of the fish species, only one or two references per species are given in order to keep the table readable. When available, we have provided both routine and maximum values of f_H. Q_{10} is calculated for acute temperature effects on heart rate. r = routine, m = maximum, f = freely swimming, a = atropine and isoproterenol treatment to maximally stimulate f_H.

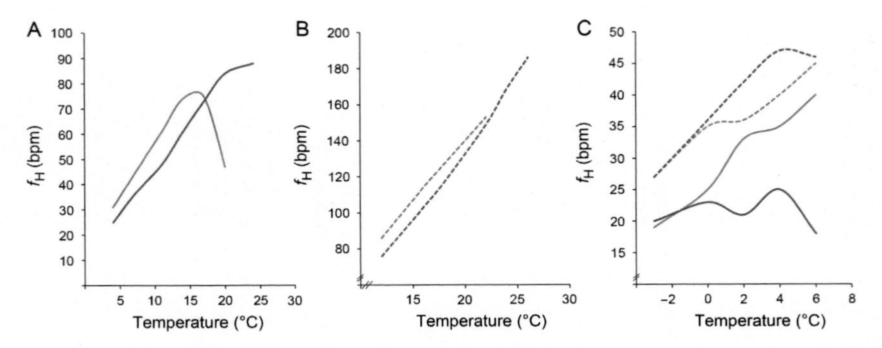

Fig. 4. Effects of acute changes in temperature on the resting (*solid lines*) and maximum (*dashed lines*) heart rate (f_H) of cold (*blue lines*) and warm (*red lines*) acclimated: (A) rainbow trout; (B) Atlantic salmon; and (C) Arctic bald notothen. Acclimation temperatures for rainbow trout were 4 and 17°C, for Atlantic salmon 12 and 20°C, and for bald notothen −1 and 4°C. Data for rainbow trout from Aho, E., Vornanen, M., 2001. Cold acclimation increases basal heart rate but decreases its thermal tolerance in rainbow trout (Oncorhynchus mykiss). J. Comp. Physiol. B 171, 173–179, for Atlantic salmon from Anttila, K., Couturier, C.S., Øverli, Ø., Johnsen, A., Marthinsen, G., Nilsson, G.E., Farrell, A.P., 2014a. Atlantic salmon show capability for cardiac acclimation to warm temperatures. Nat. Comm. 5, 4252 and for bald notothen from Franklin, C.E., Davison, W. Seebacher, F., 2007. Antarctic fish can compensate for rising temperatures: thermal acclimation of cardiac performance in Pagothenia borchgrevinki. J. Exp. Biol. 210, 3068–3074. bpm = beats per minute.

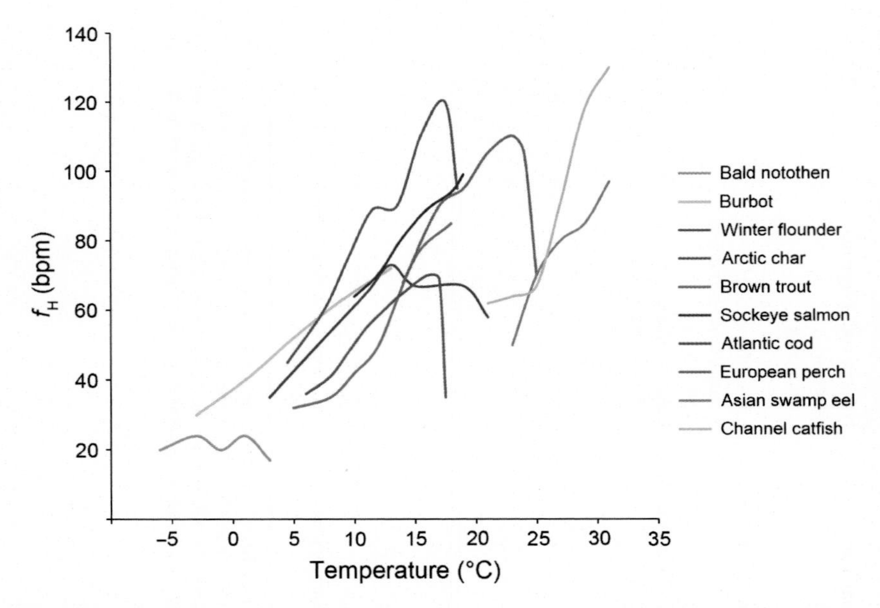

Fig. 5. The effect of an acute temperature increase on resting heart rate (f_H) of different fish species. The data for f_H curves were obtained from Franklin et al. (2007) for bald notothen, Tiitu and Vornanen (2002) for burbot, Mendonça and Gamperl (2010) for winter flounder, Penney et al. (2014) for Arctic char, Vornanen et al. (2014) for brown trout, Steinhausen et al. (2008) for sockeye salmon, Gollock et al. (2006) for Atlantic cod, Ekström et al. (2016a) for European perch, Lefevre et al. (2016) for Asian swamp eel, and Burleson and Silva (2011) for channel catfish. bpm = beats per minute.

cardiac arrhythmias are commonly seen near upper critical temperatures (Clark et al., 2008; Gollock et al., 2006; Heath and Hughes, 1973). Interspecific patterns of the effect of acute temperature increases on resting f_H are shown in Fig. 5.

While many studies have assessed how resting heart rate varies with acute temperature changes, far fewer studies have measured maximum heart rate (f_{Hmax}) in swimming fishes. In sockeye salmon, resting f_H continues to increase until the maximum upper temperature tolerance is reached, but f_{Hmax} reached a plateau at much lower temperatures (Eliason et al., 2013; Steinhausen et al., 2008; Figs. 2 and 3). In fact, f_{Hmax} actually decreased below resting levels in very warm sockeye salmon and arrhythmias were observed shortly before the fish quit swimming; the latter indicating extremely limited cardiac capacity at high temperatures in this species (Eliason et al., 2013 and see Fig. 2). As a result, scope for f_H (the difference between resting and maximum heart rate) was reduced at temperatures beyond the optimum temperature for aerobic scope (~17°C) (Eliason et al., 2013; Steinhausen et al., 2008; Figs. 2 and 3). A similar plateau in maximum f_H at high environmental temperatures was observed in pink salmon (Clark et al., 2011) and *P. borchgrevinki* (Franklin et al., 2007). In contrast, both resting f_H and f_{Hmax} increased linearly with temperature in Murray cod (*Maccullochella peelii peelii*) up to 29°C (Clark et al., 2005). In anesthetized, atropine (cholinergic blocker) and isoproterenol (β-adrenergic receptor agonist) treated fishes, the temperature where peak f_{Hmax} is reached is 2.5–8.3°C lower than their CT_{max}. This difference has been observed with Atlantic salmon (Anttila et al., 2014a), coho salmon (*Oncorhynchus kisutch*) (Casselman et al., 2012), rainbow trout (Anttila et al., 2013b), Chinook salmon (Muñoz et al., 2015), Arctic cod (*Boreogadus saida*) (Drost et al., 2014), goldfish (*C. auratus*) (Ferreira et al., 2014) and zebrafish (Sidhu et al., 2014). Furthermore, the Arrhenius break point temperature (temperature when f_{Hmax} first fails to continue increasing with the same logarithmic exponent during acute warming) of f_{Hmax} in fish treated with atropine and isoproterenol is similar to the optimum temperature for aerobic scope and growth in these species (Anttila et al., 2013b, 2014a; Casselman et al., 2012; Drost et al., 2014; Ferreira et al., 2014; Muñoz et al., 2015; Sidhu et al., 2014).

The increase in f_H with acute increases in temperature partly occurs through the influence of temperature on the membrane permeability of the heart's pacemaker cells and myocytes, and their ion currents. In-depth details about this regulation are provided by Vornanen (2017; Chapter 3, Volume 36A). In brief, the action potential of atrial and ventricular myocytes starts with the stable resting membrane potential (Phase 4) which is due to a small efflux of K^+ ions called the inward rectifier K^+ current (I_{K1}). The next phase (Phase 0) is the fast depolarization of the myocyte which occurs via the opening of Na^+-channels and the influx of Na^+-ions (inward Na^+ current, I_{Na}). This change in membrane potential also opens Ca^{2+}-channels leading to an influx of Ca^{2+}-ions (I_{Ca}) which initiates muscle contraction. At the same time,

K^+-channels open leading to an outflow of K^+-ions. During the long plateau phase of the action potential (Phase 2) both Ca^{2+}- and K^+-channels are open at the same time. During repolarization (Phase 3) only K^+-channels are open, and repolarization begins via both the rapid component of the delayed rectifier current (I_{Kr}) and I_{K1} (see Chapter 3, Volume 36A: Vornanen, 2016, 2017). Acute increases in temperature increase the density of all the currents until specific upper thermal limits are reached, as shown for example, with brown trout (*Salmo trutta fario*) (Vornanen et al., 2014). Importantly, the thermal dependencies of the various currents are different. In brown trout, the thermal dependencies as indicated by Q_{10} values are I_{Na}: 2.3, I_{K1}: 1.3–1.5, I_{Kr}: 2.3–3.2, and I_{Ca}: 1.7 (Vornanen et al., 2014). As a result, the depolarization begins faster, and the durations of Phases 2 and 3 are shorter, and this reduces the overall duration of the action potential during acute heating (see review by Vornanen, 2016). The decrease in action potential duration and, therefore, f_H has been observed with several different fish species (Badr et al., 2016; Ballesta et al., 2012; Galli et al., 2009; Haverinen and Vornanen, 2009; Shiels et al., 2015; Vornanen et al., 2002).

In addition to the direct temperature effects on ion channel function, temperature influences the autonomic control of f_H. The teleost heart is under both adrenergic (which causes positive chronotropic and inotropic effects) and cholinergic (which causes negative chronotropic and inotropic effects) control. Changes in the tone of either system can have direct effects on f_H (see review by Sandblom and Axelsson, 2011; Chapter 4, Volume 36A: Farrell and Smith, 2017); also note that autonomic control influences V_S, vascular resistance, and blood pressure, see Sections 1.4 and 1.8. Multispecies comparisons reveal that the inhibitory cholinergic tone is generally more active at both low and high environmental temperatures, while adrenergic tone predominates at moderate temperatures (Sandblom and Axelsson, 2011). In some species, acute heating reduces the cholinergic tone (Axelsson et al., 1992) and increases adrenergic tone and the concentration of circulating catecholamines (e.g., Currie et al., 2013; LeBlanc et al., 2012), and thus, increases f_H. However, the influence of temperature on autonomic control is species specific. For example, cholinergic tone is enhanced during acute temperature increases in European eel (*Anguilla anguilla*) and *P. borchgrevinki* (Franklin et al., 2001; Seibert, 1979). At high temperatures, cholinergic tone was also seen to increase in rainbow trout (Ekström et al., 2014). In fact, these differential patterns of autonomic control could be partly responsible for the different species-specific patterns in the response of f_H to acute warming (e.g., the reduction of heart rate before CT_{max} in some species, and the lack of an increase in f_H with acute temperature increases in *P. borchgrevinki*) (Franklin et al., 2001, 2007) (Figs. 4 and 5).

There is no clear explanation as to why f_{Hmax} in most fishes does not continue to increase until close to the upper temperature tolerance limit, or why routine f_H reaches a plateau or even decreases at high temperatures in many species. An increase in cholinergic tone at high temperatures could be a cardioprotective mechanism that allows enough time for cardiac filling, and helps maintain $\dot{Q}$ at high temperatures (Ekström et al., 2014). In addition, there could be limitations in energy metabolism of the heart, in oxygen delivery to the myocytes, and/or in electrical excitation of the myocytes (i.e., in I_{Na}, I_{K1}, I_{Kr} and I_{Ca}). For example, the activity of AMP-activated protein kinase (AMPK), which regulates the energy metabolism of cells (AMPK activates glycolytic ATP production pathways in situations when oxygen is limited; Hardie, 2011), increased significantly at temperatures beyond the optimum temperature for aerobic scope (rainbow trout >19°C, coho salmon >17°C) (Anttila et al., 2013b). This could be an indication that there are limitations in the energy metabolism of cardiac tissue at high temperatures. Indeed, it has been observed in Atlantic cod and *Notolabrus celidotus* that the efficiency of oxidative phosphorylation of cardiac mitochondria fails at temperatures approaching these species' upper thermal tolerance limits, even at saturating oxygen levels (e.g., see Iftikar and Hickey, 2013; Rodnick et al., 2014, also see Section 1.7 on energy metabolism). A recent study with European perch suggests that the plateau and reduction in routine f_H at high temperature was not due to a cardiac oxygen limitation since environmental hyperoxia did not affect routine f_H (Ekström et al., 2016a). Similarly, anemic sea bass at rest were able to compensate for reduced blood oxygen carrying capacity (anemia) by increasing $\dot{Q}$ via elevated f_H during an acute temperature increase, suggesting that the plateau in f_H at temperatures approaching CT_{max} was not due to an oxygen limitation (Wang et al., 2014). However, the possibility that cardiac oxygen delivery becomes limiting in maximally exercising fish as they approach their upper critical temperatures remains to be tested. Furthermore, this also implies that the limitation in f_H at high temperatures could be due to limitations in mitochondrial function/respiration *per se*, and independent of oxygen delivery to the tissue (see Section 1.7). Another possibility is that I_{Na} of cardiac (and possibly also neuronal) tissues starts to decline at high temperatures and eventually cannot induce action potentials in cardiomyocytes anymore. This appears as a variability in f_H, and finally as complete cessation of the heart beat as seen in brown trout (Vornanen et al., 2014: see further details in Chapter 3, Volume 36A: Vornanen, 2017). In addition, the deleterious venous blood environment associated with exercise and high temperature (low pH, high K^+) could impair cardiac contractility (see Sections 1.9.1 and 1.9.3). Thus, it seems that several interrelated cellular processes could be failing simultaneously at high temperatures (see Section 2.2).

1.3.2. Acclimation Responses

Some fish species have the ability to alter f_H in response to thermal acclimation. In salmonids, both resting and maximum f_H are higher at a given test temperature in cold-acclimated fish as compared to warm-acclimated conspecifics (e.g., Aho and Vornanen, 2001; Anttila et al., 2014a; Ekström et al., 2016b; see Table 1 and Fig. 4). Similar changes in resting f_H have been observed in other species as well [perch, sole (*Solea vulgaris*) and European eel (Ekström et al., 2016b; Sandblom et al., 2016b; Seibert, 1979; Sureau et al., 1989)]. However, cold acclimation reduces the absolute peak f_H (both resting and maximum) that fish can achieve during acute warming, and the temperature at which peak f_H is achieved is lower in cold-acclimated fish (e.g., Aho and Vornanen, 2001; Anttila et al., 2014a; Tsukuda et al., 1985; see Table 1 and Fig. 4). The former changes (i.e., higher heart rates at a given test temperature for cold-acclimated fish) are probably compensatory strategies that allow the fish to be active and sustain a high $\dot{Q}$, and thus, oxygen delivery to tissues at cold temperatures. The trade-off, however, is that the cold-acclimated fish cannot tolerate high temperatures and cardiac function fails at lower temperatures as compared to warm-acclimated fish.

The changes in heart rate during acclimation are again connected to changes at the cellular level. Cold acclimation, for example, increased I_{Kr} in several different fish species including rainbow trout, navaga cod (*Eleginus navaga*) and Pacific bluefin tuna (Abramochkin and Vornanen, 2015; Galli et al., 2009; Hassinen et al., 2008; Vornanen et al., 2002). Since I_{Kr} is the major repolarizing current in fish hearts, increasing the density of this current could reduce the duration of the action potential and permit a higher f_H (see Vornanen, 2016). Furthermore, cold acclimation increased sarco–endoplasmic reticulum Ca^{2+}-ATPase (SERCA) activity in the atrium of rainbow trout (Aho and Vornanen, 1998; Korajoki and Vornanen, 2012, 2013) and mRNA levels of SERCA in the atrium and ventricle of bluefin tuna and zebrafish (Jayasundara et al., 2013; Little and Seebacher, 2014). Increases in SERCA activity enhance Ca^{2+} resequestration back into the sarcoplasmic reticulum (SR) at the end of myocyte contraction, and thus, enables faster contraction rates (Aho and Vornanen, 1998) (for more details, see Chapter 3, Volume 36A: Vornanen, 2017). Cold acclimation also increased the protein and mRNA levels of 12-kDa FK506-binding protein (FKBP12) in rainbow trout. This protein increases the Ca^{2+} sensitivity of ryanodine receptors, and thus, accelerates Ca^{2+} release from the SR during contraction, and the rate of contraction (Korajoki and Vornanen, 2014).

In addition to these cellular changes, cold acclimation increases the sensitivity of the heart to adrenergic stimulation and alters the parasympathetic and sympathetic control on the heart (Wood et al., 1979, also see Section 1.6 on β-adrenoceptors). Cold-acclimated sole, European eel, rainbow trout, and European perch all have a lower cholinergic tone when compared to warm-acclimated individuals, which could be related to the higher f_H of cold-acclimated fish (Ekström et al., 2016b; Sandblom et al., 2016b; Seibert, 1979; Sureau et al., 1989). Interestingly, cold-acclimated European perch have a less pronounced tachycardia response to hypotension as compared to warm-acclimated individuals (Sandblom et al., 2016b). The functional significance of this response is currently unknown, but Sandblom et al. (2016b) have speculated that warm-acclimated fish require a more pronounced response so that they can ensure tissue blood flow at high temperatures when oxygen demand is high.

The capacity for f_H to respond (adjust) to thermal acclimation varies across species. Moreover, although resting f_H in European perch and rainbow trout has been shown to display positive thermal compensation with temperature acclimation, thermal compensation of f_{Hmax} is not as complete in these species (e.g., Ekström et al., 2016b; Sandblom et al., 2016a, 2016b). The reduced flexibility in f_{Hmax} could be detrimental for these fish as they could have a reduced cardiac scope with which to respond to warming events (see Sandblom et al., 2016a). In contrast, the winter dormant crucian carp, decreased routine f_H with cold acclimation (Matikainen and Vornanen, 1992; Tiitu and Vornanen, 2001). At the cellular level, cold acclimation reduced the protein and mRNA levels of FKBP12, and changed the isoforms of I_{K1} so that the slope of conductance increased with cold acclimation (this leads to a reduction in f_H) (Hassinen et al., 2008; Korajoki and Vornanen, 2014). Vornanen et al. (2002) speculated that one reason for this disparate response is the complete absence of oxygen during winter, and the need for crucian carp to enter to dormancy to survive. At the cardiac level, this is seen as a reduction in f_H and an increase in action potential duration (Tiitu and Vornanen, 2001; Vornanen et al., 2002). Ventricle size also remains unchanged, as discussed above. Similarly, the air-breathing Alaska blackfish (*Dallia pectoralis*) shows reduced ion channel densities with cold acclimation (Kubly and Stecyk, 2015). With regards to species living in consistently warm habitats, they also show a relatively low plasticity in routine f_H. For example, thermal plasticity of routine f_H is almost absent in zebrafish and Asian swamp eel (Lee et al., 2016; Lefevre et al., 2016). Even within salmonid species, thermal plasticity differs. For example, Atlantic salmon show significant plasticity in f_{Hmax} and the temperatures where arrhythmias are observed (Anttila et al., 2014a), while Chinook salmon lack this phenotypic plasticity (Muñoz et al., 2015). In Section 2.3, we discuss how cardiac capacity and its plasticity are related to thermal tolerance and the capacity of fish to respond to climate change in more detail.

1.4. Stroke Volume

1.4.1. Acute Responses

Cardiac stroke volume (V_S, volume of blood pumped per heart beat) is much less responsive to changes in temperature in comparison with f_H. Resting V_S has been reported to be insensitive to acute temperature increases in several fish species (Fig. 2): lingcod (Stevens et al., 1972); rainbow trout (Gamperl et al., 2011; Petersen et al., 2011); Chinook salmon (Clark et al., 2008); sockeye salmon (Steinhausen et al., 2008); Atlantic salmon and Arctic char (Penney et al., 2014); spiny dogfish (Sandblom et al., 2009); European sea bass (Wang et al., 2014); and European perch (Ekström et al., 2016a). However, studies have also reported moderate decreases in resting V_S with warming temperatures: *P. bernacchii* (Axelsson et al., 1992); *P. borchgrevinki* (Franklin et al., 2007); *M. scorpioides* and *M. scorpius* (Gräns et al., 2013); Asian swamp eel (Lefevre et al., 2016); rainbow trout (Brodeur et al., 2001; Sandblom and Axelsson, 2007); sockeye salmon (Eliason et al., 2013); and European perch (Sandblom et al., 2016a). Finally, one study with rainbow trout found a minor, but significant, increase in resting V_S (by 14%) during an acute temperature increase (Keen and Gamperl, 2012), and both Gollock et al. (2006) and Ekström et al. (2014) found no change in resting V_S during acute warming until fish approached their CT_{max}; at which time V_S increased significantly, concomitant with a reduction in f_H. Collectively, these studies show that that V_S does not normally increase greatly with an acute increase in temperature.

Fewer studies have assessed the effects of acute increases in temperature on V_S in swimming fish. Steinhausen et al. (2008) found that V_S was maintained during an acute temperature increase in sockeye salmon swum at 75% of their maximum swimming speed. Similarly, V_S was maintained after exhaustive exercise (i.e., a chase) across a 10°C temperature range in three species of Arctic fish (*G. tricuspis*, *M. scorpioides*, and *M. scorpius*) (Franklin et al., 2013). In contrast, V_S decreased at high temperatures in continuously swimming yellowfin tuna, maximally swum pink salmon and sockeye salmon, and following exhaustive exercise in *P. borchgrevinki* and European perch (Clark et al., 2011; Eliason et al., 2013; Franklin et al., 2007; Korsmeyer et al., 1997; Sandblom et al., 2016a) (Figs. 2 and 3). Scope for V_S was also unaffected by an acute temperature increase in sockeye salmon swum at 75% of their maximum (Steinhausen et al., 2008), but clearly declined at elevated temperatures when this species was swum maximally (Eliason et al., 2013 and Figs. 2 and 3). Acute temperature effects on the scope for V_S also differed with acclimation temperature in *P. borchgrevinki* (Franklin et al., 2007). The scope for V_S decreased with acute temperature increases in fish acclimated to −1°C, while it remained unchanged in fish acclimated to 4°C (note: the scope for V_S was zero at all test temperatures for 4°C acclimated fish) (Franklin et al., 2007).

Given that V_S can increase more than two-fold during aerobic exercise (e.g., Eliason et al., 2013; Fig. 2), it is curious why V_S does not generally increase with warming. This could be due to limitations in filling time, filling pressure, and/or contractility (Sandblom and Axelsson, 2007). Studies have repeatedly shown that as contraction frequency increases, the force of contraction decreases (termed the negative force–frequency relationship), and this is exacerbated with acute temperature increases (Shiels et al., 2002). As a result, the heart may not empty as much during contraction (Gamperl, 2011). Another strong possibility is that the heart is unable to increase its blood volume during diastole (i.e., end-diastolic volume), which would limit V_S. As f_H increases, filling time decreases (e.g., Sandblom and Axelsson, 2007), and this would limit the capacity of the heart to fill. In addition, V_S is inextricably linked with cardiac filling pressure in fishes. A mere 0.02 kPa increase in filling pressure can double V_S (Farrell et al., 2009). If cardiac filling pressure becomes compromised at warm temperatures, or cannot increase to compensate for the decrease in filling time, V_S would suffer. However, this does not appear to be the case for most species studied to date. Concurrent with a decrease in V_S, central venous blood pressure (P_{CV}) was stable, venous capacitance decreased, and mean circulatory filling pressure increased during an acute moderate temperature increase from 10 to 16°C in resting rainbow trout (Sandblom and Axelsson, 2007). Similarly, central venous blood pressure was maintained and the mean circulatory filling pressure increased (but V_S did not change) in resting dogfish acutely warmed from 10 to 16°C (Sandblom et al., 2009). Furthermore, P_{CV} actually increased, while V_S was maintained, as temperatures approached upper critical temperatures in resting Chinook salmon (Clark et al., 2008). These studies suggest that cardiac filling pressure is not compromised, or increases, when fishes are exposed to acute increases in temperature.

Systemic vascular resistance (R_{sys}) has been shown to decrease when fish are exposed to acute increases in temperature (Clark et al., 2008; Gamperl et al., 2011; Mendonça and Gamperl, 2010; Sandblom et al., 2009). In contrast, Sandblom and Axelsson (2007) reported no change in R_{sys} during acute warming in rainbow trout, though they only tested the fish across a moderate temperature range close to the species' optimum (10–16°C). Similarly, vascular resistance did not change in the Antarctic fish *P. bernacchii* during acute warming from 0 to 5°C (Axelsson et al., 1992). A reduction in R_{sys} with acute warming has been speculated to be associated with increased tissue perfusion to reduce diffusion distances, and thus, enhance tissue oxygen delivery (Clark et al., 2008). However, since the increase in $\dot{Q}$ typically outpaces the reduction in R_{sys} with warming, the heart must generate greater blood pressures to distribute blood to the tissues (Gamperl, 2011). This could further hinder the ability of the heart to increase or maintain V_S at high frequencies, especially given the well-documented negative force–frequency relationship (Shiels et al., 2002).

Numerous extrinsic factors (e.g., hormones, paracrine factors) also control the strength and rate of cardiac contraction (e.g., see Chapter 4, Volume 36A: Farrell and Smith, 2017; Chapter 5, Volume 36A: Imbrogno and Cerra, 2017), and autonomic nervous control is influenced by temperature (see Sections 1.3.1 and 1.6). Alterations in both these influences on cardiac function could lead to negative ionotropic effects, and thus, prevent V_S from increasing. On the other hand, cardiac contractility could become compromised at high temperature for many of the same reasons outlined earlier (e.g., the noxious hyperkalemic, acidotic, and/or hypoxic venous blood environment, or a limitation in energy metabolism in the heart (see Sections 1.3.1, 1.7, and 1.9.1; and Chapter 6, Volume 36A: Rodnick and Gesser, 2017).

Notably, elevated temperature, *per se*, does not limit the ability of the heart to increase V_S at rest (Gamperl et al., 2011). When resting rainbow trout were acutely warmed to 24°C after zatebradine administration to pharmacologically reduce f_H by 50%, routine $\dot{Q}$ was maintained by a doubling of V_S (Gamperl et al., 2011). Similarly, zatebradine injection in resting rainbow trout prior to a CT_{max} test significantly reduced f_H but $\dot{Q}$ remained comparable to control levels via an increase in V_S (Keen and Gamperl, 2012). As such, it is possible that central control mechanisms favor an increase in f_H over an increase in V_S with acute temperature increases (Gamperl, 2011). While the mechanisms preventing V_S from increasing during warming may not be resolved (see Keen and Gamperl, 2012), the outcome is that V_S does not compensate for the commonly observed plateau in maximum f_H at high temperature, and thus, cardiac performance has been observed to deteriorate in fish swimming at temperatures approaching upper critical limits (see Section 2.2 and Fig. 3).

1.4.2. Acclimation Responses

Thermal acclimation has different influences on V_S depending on the species studied. Both resting and maximum V_S were significantly greater in northern squawfish seasonally acclimated to 5 as compared to 16°C, which could at least partially be attributed to a larger RVM (Kolok and Farrell, 1994). In contrast, resting and maximum V_S were unaffected by seasonal acclimation temperature in large-scale sucker (Kolok et al., 1993). Two studies with common carp displayed a different pattern entirely, with resting V_S highest at an intermediate temperature (10°C) compared to low (5–6°C) and high (20°C) acclimation temperatures (Stecyk and Farrell, 2002, 2006). While resting V_S did not differ between −1 and 4°C acclimated *P. borchgrevinki*, maximum V_S was higher in −1°C acclimated fish (Franklin et al., 2007).

Taken all together, it is clear that thermal acclimation has a varying influence on the principal components of cardiac performance (i.e., see Section 1.3

on f_H and Section 1.4 on V_S). For example, the relative contributions of f_H and V_S to support the increased $\dot{Q}$ associated with swimming has been demonstrated to vary with acclimation temperature. *P. borchgrevinki* acclimated to −1°C increased $\dot{Q}$ with exercise via increases in both f_H and V_S. In contrast, exercise-induced increases in $\dot{Q}$ in 4°C acclimated fish were achieved exclusively by an increase in f_H (Franklin et al., 2007). In large-scale sucker, V_S was responsible for 54% of the exercise-induced increase in $\dot{Q}$ in fish seasonally acclimated to 5°C, 61% at 10°C and 30% at 16°C (Kolok et al., 1993). The ability of cardiac function to respond to thermal acclimation has important implications for the ability of a fish to cope with elevated temperatures, and for our understanding of how global climate change will impact fish species.

1.5. *In Vitro* and *In Situ* Cardiac Performance

One approach to investigating the influence of temperature on cardiac function is to perform experiments on *in situ* preparations, on isolated hearts, or on myocardial strips and/or cells. There are advantages and disadvantages with these techniques. By isolating the heart/cells, you can evaluate the direct influence of temperature on the specific parameter under investigation (e.g., rate and force of contraction) without confounding factors (such as circulating hormones—though these could be added to the perfusate to study the specific effects of various hormones at different temperatures). This, on the other hand, means that modulating factors that occur naturally are missing, and that the response could be somewhat different *in vivo* because many of these factors are interconnected. Nevertheless, studies using *in vitro* and *in situ* hearts have helped elucidate how temperature affects cardiac function and its regulation.

Temperature effects on f_H have been investigated using isolated hearts. For example, contraction rate and oxygen consumption increased in isolated, non-working, goldfish hearts when temperature was acutely increased (Tsukuda et al., 1985). Furthermore, f_H was higher at a given test temperature, and cardiac function failed at a lower test temperature, in cold- compared to warm-acclimated goldfish during an acute increase in temperature. Similar changes have been observed in a variety of other fish species in isolated systems and *in situ* [rainbow trout (Farrell et al., 1996; Graham and Farrell, 1989; Keen and Farrell, 1994; Overgaard et al., 2004), European sea bass (Farrell et al., 2007; Imbert-Auvray et al., 2013), common sole (*Solea solea*) (Imbert-Auvray et al., 2013), and Pacific bluefin and yellowfin tuna (Blank et al., 2002, 2004)], and are consistent with those seen *in vivo* (see Section 1.3). The influence of temperature on V_S has also been studied using isolated hearts/*in situ* heart preparations. Graham and Farrell (1989), Keen and Farrell (1994) and Farrell et al. (1996) all showed that the *in situ* V_S of hearts from cold-acclimated rainbow trout was greater when compared to that from warm-acclimated fish. This difference between acclimation temperatures has also been observed with European sea bass (Farrell et al., 2007). Acute warming decreased both the resting and maximum V_S of *in situ* yellowfin tuna hearts (Blank et al., 2002), which has also been observed *in vivo* in many species of fish (see Section 1.4). Maximum cardiac power output [CPO_{max}; i.e., the maximum pumping ability of heart which takes into account both $\dot{Q}$ and blood pressure: $CPO_{max} = \dot{Q} \times$ (output pressure − filling pressure)], on the other hand, appears to be the greatest near the species' optimum temperature for aerobic scope. This parameter increased in *in situ* rainbow trout hearts when acclimation temperature was increased from 8 to 18°C (Keen and Farrell, 1994), but decreased at temperatures beyond 18°C (Farrell et al., 1996). However, in *in situ* yellowfin tuna hearts, CPO_{max} appears to increase linearly with acute increases of temperature from 10 to 25°C (Blank et al., 2002). Interestingly, CPO_{max} in *in situ* European sea bass hearts increased with acclimation temperature (18–22°C) (Farrell et al., 2007).

Isolated myocardial cells and strips have also been used to study the influences of temperature at lower levels of biological organization. For example, the Ca^{2+} handling capacities and effects of temperature on contraction force have been studied with isolated systems such as ventricular strips. In rainbow trout and Pacific mackerel (*Scomber japonicus*), the importance of the release of Ca^{2+} from the SR for contraction increases with acute increases in temperature (Keen et al., 1994; Shiels and Farrell, 1997, 2000). Nonetheless, the acclimation responses again vary among species. In European sea bass, SR Ca^{2+} cycling is important for force generation at low acclimation temperatures, but in common sole, the importance of SR Ca^{2+} cycling increases at high acclimation temperatures (Imbert-Auvray et al., 2013). In yellow perch and smallmouth bass (*Micropterus dolomieu*), the SR's Ca^{2+} handling capacity increases with cold acclimation temperatures. This reduces the time for relaxation as a compensatory strategy to maintain cardiac function in the cold (Bailey and Driedzic, 1990). Furthermore, cold acclimation of rainbow trout increases: (i) the maximal rate of cardiac actomyosin Mg^{2+}-ATPase activity; (ii) the Ca^{2+} sensitivity of force generation by isolated myofilaments; and (iii) the rate and magnitude of pressure developed by the ventricle (Klaiman et al., 2011, 2014). These changes in Ca^{2+} cycling and myofilament function (via enhanced Ca^{2+} sensitivity and ATPase activity) at cold temperatures are linked with increased cardiac contractility—an effect that could only be revealed using isolated hearts.

1.6. Cardiac β-Adrenoceptors

Adrenergic stimulation is essential for maintaining cardiac performance during exercise and at temperature extremes in many fish species [though adrenergic stimulation appears to be less important for some species such as Atlantic cod (Petersen and Gamperl, 2010) and tilapia (Lague et al., 2012)]. Cardiac adrenergic stimulation is thought to be critical to both maximally stimulate $\dot{Q}$, and also to protect the heart against harmful conditions such as acidosis and hypoxia (Farrell and Milligan, 1986; Hanson et al., 2006; Hanson and Farrell, 2007). Adrenergic stimulation has inotropic (force) and chronotropic (rate) effects on the teleost heart, which are mediated through β-adrenoceptor signaling pathways (Ask, 1983; Ask et al., 1981). Specifically, catecholamines (mainly adrenaline and noradrenaline) are released via sympathetic nervous and/or humoral routes, bind to cardiac cell surface β-adrenoceptors, and ultimately increase intracellular Ca^{2+} delivery to the myocytes (see Chapter 2, Volume 36A: Shiels, 2017; Chapter 5, Volume 36A: Imbrogno and Cerra, 2017 for additional details). Heat stress is known to induce a substantial increase in circulating catecholamine levels in *P. borchgrevinki*, *P. bernacchii*, polar cod and rainbow trout (Currie et al., 2008, 2013; Forster et al., 1998; LeBlanc et al., 2011, 2012; Whiteley et al., 2006). For example, adrenaline and noradrenaline increased 100- and 50-fold in rainbow trout, respectively, following a 1-h heat shock to 25°C (Currie et al., 2008).

Temperature acclimation is known to alter the heart's sensitivity to adrenaline (Aho and Vornanen, 2001; Ask et al., 1981; Farrell et al., 1996; Graham and Farrell, 1989; Keen et al., 1993), which has been attributed to changes in ventricular β-adrenoceptor density (B_{max}) (Keen et al., 1993). At cold temperatures, adrenergic stimulation has been demonstrated to be critical for the maintenance of cardiac function in rainbow trout (Graham and Farrell, 1989; Keen et al., 1993). Specifically, cold-acclimated (8°C) rainbow trout hearts had an almost three-fold higher B_{max}, which corresponded to an increased sensitivity to adrenaline, as compared to warm-acclimated (18°C) fish (Keen et al., 1993). Similarly, B_{max} was almost twice as high in rainbow trout acclimated to 8°C as compared to 14°C (Gamperl et al., 1998).

However, enhanced cardiac function at warm temperatures has also been attributed to an elevated cardiac B_{max} in a population of sockeye salmon (Chilko) which has a remarkably high and broad thermal tolerance for this species. These fish can maintain maximum $\dot{Q}$ up to 22°C (Eliason et al., 2011), and ventricular B_{max} increases significantly when they are acclimated to both 19 and 22°C when compared to 13°C (Eliason et al., 2011). Notably, a co-migrating population (Nechako) with a narrow and lower thermal tolerance displayed no differences in B_{max} when acclimated to 13, 19 and 22°C (Eliason et al., 2011). Thus, sockeye salmon display intraspecific differences in the thermal plasticity of ventricular B_{max}, and these differences may be related to functional thermal tolerance. In contrast, pharmacological blockade of β-adrenoreceptors did not affect CT_{max} in hatchery-reared rainbow trout (Ekström et al., 2014), which suggests that CT_{max} is unaffected by adrenergic stimulation in resting rainbow trout. Notably, β-adrenergic blockade reduced f_H but $\dot{Q}$ was maintained via a compensatory increase in V_S. Thus, the importance of β-adrenergic stimulation for supporting maximum cardiac performance at high temperatures warrants further investigation.

Thermal plasticity of cardiac B_{max} also differs across species. In contrast to temperate salmonids, the African catfish (*Claris gariepinus*, a tropical fish species) demonstrated no thermal plasticity in B_{max} (Hanson et al., 2005). These species-specific differences in cardiac β-adrenoreceptor thermal plasticity may be due to differences in the temperature range that each species experiences in their local habitat. Cardiac β-adrenoceptor plasticity may be essential to maintain cardiac function across the broad range of seasonal temperatures experienced by temperate species such as rainbow trout, while such plasticity may be superfluous in tropical, stenothermal, species. This reduced plasticity in tropical species is evident when studying other aspects of cardiac function as well (e.g., f_H, see Section 1.3.2).

When cardiac B_{max} and β-adrenoceptor-binding affinity (K_d) are compared across fish species from tropical, temperate and Antarctic regions, differences cannot be attributed strictly to temperature (Table 2; Olsson et al., 2000). We see a 24-fold variation in B_{max} and an almost eightfold variation in K_d across the 10 species examined to date (Table 2). This more comprehensive species comparison refutes the previous suggestion that there is a positive relationship between B_{max} and temperature acclimation across species (Olsson et al., 2000). Instead, there is a general trend for more active fish to have a higher B_{max} (Table 2). A notable exception is the winter flounder, which has the highest B_{max} measured to date, but β-adrenoceptors with an exceptionally low binding affinity (Mendonça and Gamperl, 2009). This may be related to their reduced cardiac adrenergic sensitivity, benthic and inactive lifestyle, hypoxia tolerance, and perhaps an increased dependence on β_3-adrenoreceptors (Mendonça and Gamperl, 2009).

Several important research questions remain related to the role of the cardiac β-adrenoceptor signaling pathway in mediating temperature-dependent cardiac function. For example, how quickly can B_{max} change in response to temperature? Related to this, it is unknown if β-adrenoreceptors are sequestered and cycled to the membrane or if they are synthesized *de novo* in response to temperature change. The thermal plasticity of β-adrenoreceptors has only been examined in three species to date (Table 2; Eliason et al., 2011; Hanson et al., 2005), so it is difficult to make inferences about the remaining 30,000+ fish species. Given the broad range in circulating resting catecholamine concentrations across fish species, and between resting fish and those under conditions of severe stress (see Gamperl et al., 1994; Randall and Perry, 1992), the

Table 2

Ventricular β-adrenoceptor density (B_{max}) and binding affinity (K_d) as determined using the radiolabelled β_2-adrenoceptor ligand [^{3}H]CGP-12177

Region	Species	Population	T (°C)	Acclimation period	B_{max} (fmol mg protein^{-1})	K_d (nM)	Reference
Antarctic	*T. bernacchii*	McMurdo Station (otc)	1	2 weeks	10.5	0.18	Olsson et al. (2000)
Temperate	Rainbow trout	Hatchery (BC)	6	Months	36.3	0.23	Eliason et al. (2011)
Temperate	Rainbow trout	Hatchery (NS)	8	Several months	40	0.25	Gamperl et al. (1994)
Temperate	Rainbow trout	Hatchery (BC)	12	3 weeks	22.6	0.13	Olsson et al. (2000)
Temperate	Rainbow trout	Hatchery (BC)	14	Unknown	25	0.21	Gamperl et al. (1998)
Temperate	Rainbow trout	Hatchery (BC)	14	3 weeks	26.4	0.19	Hanson et al. (2005)
Temperate	Rainbow trout	Hatchery (BC)	14	Months	18.5	0.45	Goulding and Farrell (2016)
Temperate	Sockeye salmon	Chilko smolts (BC)	8	4 days	54.2	0.43	Goulding and Farrell (2016)
Temperate	Sockeye salmon	Chilko (BC)	13	4 days	78.5	0.18	Eliason et al. (2011)
Temperate	Sockeye salmon	Chilko (BC)	19	4 days	123.5	0.2	Eliason et al. (2011)
Temperate	Sockeye salmon	Chilko (BC)	21	4 days	128.2	0.22	Eliason et al. (2011)
Temperate	Sockeye salmon	Nechako (BC)	13	4 days	46.8	0.27	Eliason et al. (2011)
Temperate	Sockeye salmon	Nechako (BC)	19	4 days	66.2	0.21	Eliason et al. (2011)
Temperate	Sockeye salmon	Nechako (BC)	21	4 days	57.7	0.23	Eliason et al. (2011)
Temperate	Sockeye salmon	Stamp river (BC)—lab	10–13; then 16–17	3 weeks; 24–36 h	47.5	0.2	Olsson et al. (2000)
Temperate	Sockeye salmon	Stamp river (BC)—wild	21	None	45.4	0.36	Olsson et al. (2000)
Temperate	Chinook salmon	Robertson Creek (BC)	13	None	58	0.26	Gamperl et al. (1998)
Temperate	Pacific mackerel	British Columbia (otc)	13	1 week	27.2	0.25	Olsson et al. (2000)
Temperate	Winter flounder	Conception Bay (NL)	8	4 weeks	252.8	1.02	Mendonça and Gamperl (2009)

(Continued)

Table 2 (Continued)

Region	Species	Population	T (°C)	Acclimation period	B_{max} (fmol mg protein^{-1})	K_d (nM)	Reference
Tropical	African catfish	Singapore	15	4 weeks	17.8	0.48	Hanson et al. (2005)
Tropical	African catfish	Singapore	22	4 weeks	15.4	0.46	Hanson et al. (2005)
Tropical	African catfish	Singapore	32	4 weeks	14.3	0.88	Hanson et al. (2005)
Tropical	Yellowfin tuna	California (otc)	20	1 year	25.7	0.2	Olsson et al. (2000)
Tropical	Mahi	Hawaii (otc)	25	None	46.9	0.2	Olsson et al. (2000)
Tropical	Skipjack tuna	Hawaii (otc)	25	5 days	41.3	0.21	Olsson et al. (2000)

otc: off the coast; BC: British Columbia; NL: Newfoundland; NS: Nova Scotia.

relative importance of the β-adrenoceptor pathway for maintaining cardiac performance across temperatures likely varies considerably across species, and warrants further study.

1.7. Energy Metabolism and Mitochondrial Capacity

Acute alterations in environmental temperature lead to significant changes in the energy metabolism of cardiomyocytes and $\dot{Q}$. From a metabolic point of view, ATP consumption increases significantly with increasing temperature, which must be supported by enhanced ATP production via oxidative phosphorylation. Indeed, acute elevations in temperature increase oxidative phosphorylation in many fish species (e.g., Ekström et al., 2017b; Hilton et al., 2010; Iftikar et al., 2014, 2015; Iftikar and Hickey, 2013; Lemieux et al., 2010; Penney et al., 2014; Rodnick et al., 2014). However, there is an upper temperature limit beyond which the capacity for oxidative phosphorylation cannot increase further, or be maintained. Several studies with New Zealand wrasses (*N. celidotus*, *Notolabrus fucicola*, and *Thalassoma lunare*), Atlantic wolfish (*Anarhichas lupus*), Atlantic cod, Atlantic salmon, Arctic char, and European perch show that the efficiency of oxidative phosphorylation, and the absolute production of ATP, decrease at temperatures below that at which cardiac function collapses. This suggests that compromised mitochondrial performance at high temperatures contributes to the failure in cardiac performance (Ekström et al., 2017b; Hilton et al., 2010; Iftikar et al., 2014, 2015; Iftikar and Hickey, 2013; Lemieux et al., 2010; Penney et al., 2014; Rodnick et al., 2014). This reduction in efficiency occurs even at saturating oxygen levels (Iftikar and Hickey, 2013). In European perch, citrate synthase activity also decreases at temperatures beyond 30°C, but use of the malate–aspartate shuttle may be possible for energy production at high temperatures (Ekström et al., 2017b). In rainbow trout and coho salmon, AMPK activity increases when temperature rises beyond the species' optimum temperature for aerobic scope during acute heating, and this suggests that there is already a limitation in ATP production at temperatures well below the critical thermal maximum (a reduction in ATP levels and an increase in ADP/AMP levels activate AMPK) (Anttila et al., 2013b).

The thermal responses of oxidative phosphorylation (and its efficiency) vary between species, and this variation could be related to the stability of the environment, the geographical range of species, and/or other environmental factors like exposure to hypoxia. For example, when comparing wrasses living in different thermal environments (cold temperate vs warm tropical), species that reside at high environmental temperatures (either naturally or during acclimation) had more thermally robust mitochondria (Iftikar et al., 2014,

2015). Similarly, the triplefin *Bellapiscis medius*, which lives in the intertidal zone, has a more efficient oxidative phosphorylation at high temperatures in comparison with subtidal triplefin species (*Forsterygion varium* and *Forsterygion malcolmi*). This could be due to the fact that they routinely encounter fluctuating environmental temperatures and/or hypoxia (Hilton et al., 2010). However, as mentioned above, there is an upper limit to how much the efficiency of oxidative phosphorylation can increase, and species living in tropical regions are already near this upper limit (Iftikar et al., 2014).

There could be several mechanisms behind this decrease in the efficiency of oxidative phosphorylation at high temperatures. First, when the demand for ATP increases substantially, proton leak increases and this reduces the coupling of ATP production and oxygen consumption (Iftikar and Hickey, 2013; Rodnick et al., 2014). On the other hand, increases in temperature could negatively influence the tricarboxylic acid cycle, the electron transport chain, and metabolite transporters (see Rodnick et al., 2014). Furthermore, the affinity of ADP decreases at high temperatures, which could lead to an imbalance between ATP supply and demand, and result in an increase in reactive oxygen species (ROS) production (Rodnick et al., 2014). Increases in temperature also induce phase transition changes in membrane lipids, which can compromise membrane integrity, reduce membrane potential, and as a result impair ATP production (e.g., see Iftikar et al., 2014, 2015; Iftikar and Hickey, 2013; Rodnick et al., 2014). Increased permeability of membranes can promote cytochrome *c* release from the mitochondria (Borutaite and Brown, 2003; Dahlhoff and Somero, 1993; Hand and Menze, 2008; Iftikar and Hickey, 2013). This further depresses oxidative phosphorylation, and can affect the scavenging of ROS (i.e., resulting in elevated levels of ROS) (Mailer, 1990). Increased amounts of ROS could further damage both the inner and outer membrane of the mitochondria, and influence its ATP production efficiency (e.g., see review by Szewczyk et al., 2015). Thus, acute increases in temperature can have serious effects on mitochondrial efficiency, which in turn, will affect the energy metabolism of cardiomyocytes and could contribute to cardiac failure at high temperature (see Section 2.2).

Although an acute increase in environmental temperature leads to mitochondrial dysfunction, fish have different compensatory strategies to cope with long-term changes in environmental temperatures. Furthermore, a recent study with hatchery-reared rainbow trout demonstrated that it takes merely 2 days to achieve acclimation effects on mitochondrial function (Pichaud et al., 2017). This suggests that cardiac metabolic responses to thermal acclimation may be extremely rapid. Since environmental cooling leads to a reduction in all kinetic reactions, including enzyme activities, many fish species increase the amount and/or activity of enzymes in response to cold acclimation. For example, hexokinase activity increased with cold acclimation in the hearts of chain pickerel (*Esox niger*) (Kleckner and Sidell, 1985) and the oxidative metabolism of fatty acids increased in the hearts of white perch (*Morone americana*) and gilt-head bream (*Sparus aurata*) (Kyprianou et al., 2010; Sephton and Driedzic, 1991). Thus, cold acclimation appears to increase the use of both glycogen and fatty acid stores by fish hearts. In addition, cold acclimation, in general, increases mRNA levels for enzymes involved with lipid metabolism as well as enzymes related to the use of glycogen and glucose (Anttila et al., 2014b; Gracey et al., 2004; Jayasundara et al., 2013; Vornanen et al., 2005). In some species, mitochondrial properties also increase in response to cold acclimation or are reduced in response to warm acclimation. For example, cold acclimation increased myocardial cytochrome oxidase (COX) activity in common carp (Cai and Adelman, 1990), increased the thermal stability of mitochondrial function in New Zealand wrasse (*N. celidotus*) (Iftikar et al., 2015), and increased the expression of enzymes related to oxidative phosphorylation and the citric acid cycle in the hearts of common carp and sockeye salmon (Anttila et al., 2014b; Gracey et al., 2004). However, not all species respond similarly. Cold acclimation reduced COX activity in the Atlantic cod heart, and the expression of COX was reduced in rainbow trout hearts after cold acclimation (Foster et al., 1993; Vornanen et al., 2005). In Atlantic killifish (*Fundulus heteroclitus*), it was actually shown that mitochondrial function was reduced both at extremely cold (5°C) and extremely warm (33°C) temperatures as compared to 15°C (Chung et al., 2017). Finally, in yellow perch, European perch and striped bass, heart mitochondrial volume density remains unchanged following cold acclimation (Rodnick and Sidell, 1994; Sephton and Driedzic, 1991); although ventricular mass increased in striped bass with cold acclimation (see Section 1.1 on cardiac morphology), and thus, the total volume of mitochondria increased (Rodnick and Sidell, 1994; Sephton and Driedzic, 1991). Notably, the changes induced by cold acclimation such as increased oxidative metabolism, and especially the increased ability to oxidize fatty acids, are also apparent as adaptational changes. Antarctic species, for example, possess significantly higher enzyme activities related to fatty acid metabolism when compared to temperate species (Crockett and Sidell, 1990).

1.8. Regional Blood Flow

Blood flow must be partitioned among competing organ systems since the capacity of the heart is insufficient to simultaneously perfuse all capillary beds. Optimal physiological function depends upon sufficient oxygen delivery to meet the demands of the various organs. If blood flow is compromised, organ function may be impacted. Tissue perfusion is controlled by both local and remote mechanisms (Olson, 2011; also see Chapter 7, Volume 36A:

Sandblom and Gräns, 2017). Local control occurs primarily via vasodilators (e.g., metabolites such as ATP and adenosine, and nitric oxide), and remote control via neural (e.g., vasoconstriction via the sympathetic nervous system which is mediated by catecholamines) and endocrine routes (e.g., hormones that act as vasoconstrictors such as angiotensin II or vasodilators such as natriuretic peptides) (Olson, 2011). Given that tissue oxygen demand is strongly influenced by temperature and $\dot{Q}$ is temperature dependent (see Section 1.2), understanding the influence of temperature on blood flow to the tissues is critical to discern temperature effects on whole animal performance (e.g., growth, reproduction, swimming and digestion) and fitness.

The effect of temperature on regional blood flow has been examined in very few fish species to date. Although radiolabelled microspheres have been used to assess how regional blood flow changes with temperature in three studies (Barron et al., 1987; Egginton, 1997; Taylor et al., 1996), these studies must be interpreted with caution given that this method has been found to be unreliable when compared against direct flow measurements (Crocker et al., 2000). Two of the studies used radiolabelled microspheres to examine regional blood flow in rainbow trout acclimated between 4 and 18°C (Barron et al., 1987; Taylor et al., 1996). Barron et al. (1987) determined that warm-acclimated rainbow trout directed a greater percentage of blood flow to the white muscle compared to cold-acclimated fish. Similarly, Taylor et al. (1996) found that blood flow to the skin and white muscle was significantly higher in warm-acclimated trout at rest. When the fish were swimming aerobically, blood flow to the red muscle increased significantly at all temperatures (Taylor et al., 1996). Similar increases in blood flow to the muscle during swimming were observed in Antarctic black rockcod (*Notothenia coriiceps*) held at 0°C, although the hyperemia was less profound in the Antarctic fish compared to trout (Egginton, 1997).

Of all the organ systems, the effect of temperature on gut blood flow has received the most attention. Feeding increased gut blood flow equally in 10 and 16°C acclimated rainbow trout (by 236%–256%), which was accompanied by a 123%–142% increase in $\dot{Q}$ (Gräns et al., 2009a). Acute temperature effects on gut blood flow were also examined in two Arctic fish species [a thermal generalist with a broad distribution (*M. scorpius*) and a strictly Arctic species (*M. scorpioides*)], both acclimated to 9°C and exposed to an acute temperature increase (to 14°C) or decrease (to 4°C) (Gräns et al., 2013). There were no differences between the species. Gut blood flow increased with increasing water temperature ($Q_{10}=1.7$–2.0) and was associated with a comparable increase in $\dot{Q}$ ($Q_{10}=1.9$–2.1), as opposed to a redistribution of blood flow (Gräns et al., 2013). In contrast, gut blood flow increased by more than 150%, whereas $\dot{Q}$ only increased slightly when green sturgeon (*Acipenser medirostris*) were acutely heated from 19 to 24°C (Gräns et al., 2009b). When white sturgeon (*Acipenser transmontanus*) were allowed to freely choose a temperature between 13 and 24°C, gut blood flow was found to be insensitive to temperature before a meal, but tightly linked to temperature after ingesting a meal; the increase, on average, was 15% per °C (Gräns et al., 2010). Finally, one study has assessed the effect of an acute temperature change on hepatic venous blood flow. Hepatic venous blood flow remained constant in both well-fed and food-deprived rainbow trout when temperature decreased from 12 to 5°C (Petersen et al., 2011).

Temperature effects on the coronary vasculature have been examined in rainbow trout, steelhead trout (i.e., seawater acclimated rainbow trout), and leopard sharks (*Triakis semifasciata*). Acclimation temperature (1, 5, and 10°C) markedly affected how steelhead trout coronary arterioles responded to various vasoactive agents (Costa et al., 2015). The vasodilation induced by adrenaline, adenosine, serotonin and sodium nitroprusside were all reduced in cold-acclimated fish, but the vasoconstriction induced by acetylcholine was higher in cold-acclimated fish relative to those held at 10°C (Costa et al., 2015). In contrast, Farrell (1987) reported that the rainbow trout's main coronary artery was more sensitive to adrenaline-induced vasoconstriction when acclimated to 5°C as compared to 15°C. Both rainbow trout and leopard sharks exposed to an acute temperature challenge increased coronary blood flow (CBF) (Cox, 2015; Ekström et al., 2017a). Furthermore, chronic coronary ligation in juvenile rainbow trout was associated with an elevated f_H and reduced CT_{max} (Ekström et al., 2017a). These findings suggest that increased CBF at high temperature may permit myocardial oxygenation to be maintained even though the venous partial pressure of oxygen (P_vO_2) likely decreases, which could compromise oxygen delivery to the spongy myocardium. Accordingly, increased CBF appears to play a role in thermal tolerance for these species. Interestingly, female rainbow trout displayed a higher routine CBF and reduced scope for increasing CBF during warming compared to male rainbow trout (Ekström et al., 2017a), which could have negative ecological and evolutionary consequences for wild fish populations as climate change progresses.

Blood flow through the secondary circulatory system is also subject to temperature effects. Skin and gill epithelial tissues are able to directly take up oxygen from the water, bypassing the primary circulation as a source of oxygen. Oxygen consumption by these tissues has been reported to represent 12%–48% of $\dot{M}O_2$ in resting and swimming sockeye salmon and Chinook salmon at both optimal and supraoptimal temperatures for aerobic scope (Farrell et al., 2014). Absolute skin and gill tissue oxygen consumption increased with swimming in sockeye salmon swum at temperatures within the optimal range of aerobic scope (15–20°C), but not in fish swum above this temperature range (22–24°C). In addition, skin and gill tissue oxygen consumption tended to decrease when resting Chinook salmon were acutely heated to 26°C (Farrell et al., 2014). Thus, heat stress may negatively influence the ability of the skin and gills to extract oxygen directly from the environment, and this has been hypothesized to be associated with alterations in the secondary circulation flow (Farrell et al., 2014).

1.9. Blood Parameters

1.9.1. Hematological Variables

As discussed at the beginning of the chapter, when water temperature increases, oxygen consumption of the heart and other organs rises (e.g., Graham and Farrell, 1990). These increases in oxygen consumption (and thermal effects in general) lead to various compensatory responses at different levels of cardiovascular physiology, including in the blood. For example, hemoglobin oxygen affinity decreases significantly with acute increases in temperature, and thus, could have a significant influence on oxygen delivery to the tissues at high temperatures (see Chapter 1, Volume 36B: Harter and Brauner, 2017). However, there is no clear pattern with respect to how temperature affects blood hemoglobin concentration ([Hb]) or hematocrit (Hct), even though higher [Hb] and/or Hct could enable higher oxygen transport in situations when oxygen consumption is elevated. Acute increases in temperature have been shown to minimally effect Hct and/or [Hb] in largemouth bass (*Micropterus salmoides*) (Mulhollem et al., 2015), sockeye salmon (Eliason et al., 2013) and Chinook salmon (Clark et al., 2008). However, in rainbow trout, Hct increased by 15%–29% after environmental temperature was increased from 10 to 16°C and following an acute heat shock of 25°C (Sandblom and Axelsson, 2007; Templeman et al., 2014). Sandblom and Axelsson (2007) concluded that since thermal stress did not increase Hct in splenectomized rainbow trout, the increase in Hct was most likely due to the release of erythrocytes from the spleen. Thermal stress could also induce swelling of erythrocytes which would increase the Hct without any change in [Hb], as observed with European sea bass (Roche and Bogé, 1996) and rainbow trout (Templeman et al., 2014). One reason for this could be increased circulating adrenaline levels which induce erythrocyte swelling (e.g., Salama and Nikinmaa, 1988) and are known to increase with acute heat shock (e.g., Currie et al., 2013; LeBlanc et al., 2011; Templeman et al., 2014).

Thermal acclimation effects on Hct and [Hb] are also divergent across species. Acclimation to different temperatures did not affect Hct in Alaska blackfish (Lefevre et al., 2014), European perch (Ekström et al., 2016c), or largemouth bass (Mulhollem et al., 2015). However, rainbow trout acclimated to 18°C had higher Hct and [Hb] than trout acclimated to 9°C (Valenzuela et al., 2008). One factor mediating these disparate results could be that Hct and [Hb] are not solely influenced by temperature. Environmental hypoxia has significant effects on blood variables (see Chapter 1, Volume 36B: Harter and Brauner, 2017), and Hct and [Hb] could be influenced by circadian/seasonal rhythms (Morgan et al., 2008; Pascoli et al., 2011).

Many other hematological variables are also influenced by temperature. For example, blood lactate increased at supraoptimal temperatures in both resting Chinook salmon (>25°C) and sockeye salmon (>23°C) (Clark et al., 2008; Eliason et al., 2013; Steinhausen et al., 2008), and in swimming (>17°C) sockeye salmon (Eliason et al., 2013; Steinhausen et al., 2008), indicating that a mismatch between oxygen supply and demand had occurred, which increased the reliance upon anaerobic metabolism. It is also possible that some of this elevated lactate was associated with increased agitation, which is commonly observed in resting fishes at high temperatures. This increase in anaerobic metabolism is concomitant with a decrease in blood pH at high temperatures (Cech et al., 1976; Clark et al., 2008; Steinhausen et al., 2008), and with a decrease in blood glucose at temperatures above those where aerobic scope is maximal (17°C for Chinook salmon and 15–17°C for sockeye salmon) (Clark et al., 2008; Eliason et al., 2013).

The response of plasma ions to changes in temperature is more variable. Minimal temperature effects were observed in plasma Na^+ and Cl^- in resting Chinook salmon (Clark et al., 2008), while Na^+ and Cl^- decreased at high temperatures in resting and swimming sockeye salmon (Eliason et al., 2013). In these two studies, plasma K^+ was insensitive to temperature (Clark et al., 2008; Eliason et al., 2013), while others report an increase in plasma K^+ at elevated temperatures (Steinhausen et al., 2008). These hematological changes can have important implications for oxygen delivery and cardiac contractility. Specifically, decreases in blood pH facilitate oxygen delivery to the tissues via Bohr and Root effects (see Chapter 1, Volume 36B: Harter and Brauner, 2017), whereas low pH, high K^+, and low partial pressure of oxygen in the blood returning to the heart have deleterious effects on cardiac contractility (Farrell et al., 1986; Hanson et al., 2006; Hanson and Farrell, 2007; Kalinin and Gesser, 2002; Chapter 4, Volume 36A: Farrell and Smith, 2017). These latter changes could all contribute to cardiac collapse at high temperatures (see Section 2.2).

1.9.2. Arterial Blood Oxygenation

Given that the oxygen content of water decreases as water temperature increases, it has been proposed that fish may have a reduced capacity to extract oxygen from their environment at warm temperatures, and that this is a possible mechanism for the reduction in aerobic scope above optimal temperatures that has been reported for some species (Brett, 1971). This hypothesis can be directly assessed by measuring the partial pressure of oxygen (P_aO_2) and oxygen content (C_aO_2) in the arterial blood. P_aO_2 and C_aO_2 can change independently, and C_aO_2 is largely dependent on the oxygen carrying capacity of the blood. Therefore, studies with repeat blood sampling could reduce Hct and [Hb], and thus, cause C_aO_2 to decrease independent of any temperature effects. As such, caution must be used when interpreting studies that perform repeat blood sampling, but do not report Hct or if decreases in Hct are observed concurrently with warming.

Studies that have measured P_aO_2 and C_aO_2 in fish exposed to acute temperature increases provide conflicting results. Several studies report a decrease in P_aO_2 with warming temperatures, suggesting that there is a limitation in oxygen uptake at the gill at high temperatures. For example, P_aO_2 and C_aO_2 decreased steadily (and by ~50% and ~25% in total, respectively) in resting steelhead trout warmed from 12°C to their CT_{max} (Keen and Gamperl, 2012). Similarly, (1) Heath and Hughes (1973) reported small decreases in P_aO_2 and C_aO_2 in resting rainbow trout during an acute thermal challenge ($1.5°C\,h^{-1}$) and (2) although winter flounder displayed no change in P_aO_2 or C_aO_2 when acutely warmed from 5 to 10°C or from 10 to 15°C, P_aO_2 decreased in fish warmed from 15 to 20°C (Cech et al., 1976). Notably, all of these studies were conducted on resting fish held in fairly constrained spaces or with low water flow; conditions in which ventilation was very likely limited to buccal–opercular pumping (see Keen and Gamperl, 2012). In contrast, Chinook salmon maintained P_aO_2 during acute increases in water temperature at rest, except in the largest individuals which displayed a decrease in P_aO_2 at the highest test temperatures (Clark et al., 2008). Notably, water flow was maintained at $20–30\,cm\,s^{-1}$ during the experiment, which may have permitted ram ventilation to contribute to oxygen uptake (see below). The authors suggest that a diffusion limitation at the gills may have occurred in the largest fish, perhaps related to increased gill diffusion distance, though this idea remains to be investigated (see also Clark et al., 2012). Finally, C_aO_2 decreased with acute increases in temperature in air-breathing swamp eels acclimated to 27°C, though the vast majority of oxygen was derived from air breathing (Lefevre et al., 2016).

In contrast, three other studies found that P_aO_2 and C_aO_2 were maintained independent of temperature, suggesting that oxygen uptake at the gill is not compromised at high temperatures. Atlantic cod did not change P_aO_2 during an acute temperature increase from 10 to 19°C (Sartoris et al., 2003). Likewise, studies with sockeye salmon showed that P_aO_2 and C_aO_2 did not change (Eliason et al., 2013) and that P_aO_2 actually increased (Steinhausen et al., 2008) with increases in temperature in resting and swimming fish. In fact, even at exceptionally warm temperatures when swimming sockeye salmon displayed cardiac arrhythmias that led to post-exercise delayed mortality, C_aO_2 did not decline (Eliason et al., 2013). Notably, flow rates were set at ~20 cm s^{-1} for resting fish in the sockeye salmon experiments (see below).

The effect of acclimation temperature on P_aO_2 and C_aO_2 has been investigated in a limited number of studies. Rainbow trout acclimated to 4, 11, and 18°C had decreased values for P_aO_2 and C_aO_2 at the higher acclimation temperature; however, this change was associated with a ~50% decrease in Hct (Taylor et al., 1993) which would have dramatically affected the oxygen carrying capacity of the blood. Seasonal acclimation temperature (5, 10, and 15°C) had no effect on P_aO_2 in winter flounder, though C_aO_2 decreased at higher acclimation temperatures and was associated with a corresponding decrease in [Hb] (Cech et al., 1976). Small-spotted catshark displayed no differences in P_aO_2 and C_aO_2 across a 10°C acclimation range (7, 12, and 17°C), though Hct did increase significantly at 17°C (Butler and Taylor, 1975). Similarly, the air-breathing swamp eel displayed no difference in C_aO_2 with acclimation, though Hct and [Hb] were not reported (Lefevre et al., 2016).

Collectively, these data suggest that the ability of fish to maintain oxygen extraction from the environment when faced with warming temperatures varies across species and conditions. For some species that have a decrease in aerobic scope at high temperatures, the data suggest that this is not due to a gill limitation (sockeye salmon), while it remains a possibility for other species (winter flounder, other salmonids). However, future research should examine the role of the type of ventilation employed (e.g., winter flounder exclusively use buccal–opercular pumping at rest, salmon may be able to take advantage of ram ventilation at moderate to high water flows, and the swamp eel primarily relies on air breathing), the experimental conditions (low flow experimental chambers compared with a swim tunnel), the size of the fish (larger fish may have a diffusion distance constraint), and the potential for thermal plasticity. It also needs to be taken into account that other environmental factors (like hypoxia) can affect oxygen extraction, and that the activity of the species could also play a part in explaining interspecific differences.

1.9.3. Venous Blood Oxygenation

The increased tissue oxygen demand with increasing temperatures is met not only by increased $\dot{Q}$ (see Section 1.2), but also enhanced oxygen extraction from the blood. Thus, many studies have observed a decrease in the partial pressure of oxygen (P_vO_2) and oxygen content (C_vO_2) in the venous system at elevated temperatures (Clark et al., 2008; Ekström et al., 2016a; Eliason et al., 2013; Heath and Hughes, 1973; Lannig et al., 2004; Sartoris et al., 2003); though one study reported no change in P_vO_2 or C_vO_2 with warming (Steinhausen et al., 2008). A species-specific threshold in P_vO_2 has been suggested, beyond which oxygen diffusion to the spongy myocardium is insufficient to meet demand leading to impaired cardiac performance (Farrell and Clutterham, 2003; Steffensen and Farrell, 1998). As such, a cardiac limitation due to insufficient myocardial oxygen supply (i.e., a reduced P_vO_2) has been proposed as a possible mechanism limiting whole animal thermal tolerance (see Section 2.2).

Warm acclimation has been demonstrated to increase P_vO_2 in small-spotted catshark, rainbow trout, and winter flounder (Butler and Taylor, 1975; Cech et al., 1976; Farrell and Clutterham, 2003). Similarly, European perch chronically exposed to elevated temperatures had higher resting P_vO_2 compared to conspecifics living at cooler temperatures (Ekström et al., 2016a). However, when both groups were exposed to an acute thermal challenge to CT_{max}, cardiac output, and heart rate reached peak values at a similar P_vO_2 level (2.3–4.0 kPa), suggesting that cardiac failure occurred at a common species-specific P_vO_2 threshold. The compensatory changes that elevate P_vO_2 when acclimated to warm temperatures may serve to protect oxygen delivery to the spongy myocardium (Farrell and Clutterham, 2003). Warm acclimation resulted in an increase in C_vO_2 and Hct in small-spotted catshark (Butler and Taylor, 1975), but a decrease in C_vO_2 associated with a corresponding decrease in [Hb] in winter flounder (Cech et al., 1976).

2. ECOLOGICAL IMPLICATIONS

2.1. Role of the Cardiovascular System in Supporting Whole Animal Performance

Given the key role of the heart in supplying oxygen to working tissues, $\dot{Q}$ is predicted to mirror the relationship between oxygen uptake ($\dot{M}O_2$) and temperature (Farrell, 2009). Brett (1971) was the first to report such a relationship between $\dot{M}O_2$ and $\dot{Q}$ in both resting and swimming sockeye salmon, though $\dot{Q}$ was not directly measured; i.e., it was calculated via the Fick equation [$\dot{M}O_2 = \dot{Q} \times (C_aO_2 - C_vO_2)$]. Similarly, Butler and Taylor (1975) found that the resting $\dot{M}O_2$ and resting $\dot{Q}$ of dogfish had an equivalent Q_{10} (2.1) across a seasonal acclimation range (7, 10, and 17°C), and Cech et al. (1976) reported that increases in temperature resulted in increases in both resting $\dot{M}O_2$ ($Q_{10}=2.8$–3.4) and $\dot{Q}$ ($Q_{10}=1.7$–2.8) in winter flounder, again both using the Fick equation. However, there are several problems with estimating $\dot{Q}$ via the Fick equation (see Farrell et al., 2014), and more modern studies have evaluated the relationship between $\dot{M}O_2$ and $\dot{Q}$ using direct measurements.

Several studies have directly measured both $\dot{M}O_2$ and $\dot{Q}$ in resting fish exposed to an acute thermal challenge to temperatures approaching their CT_{max}. Resting Chinook salmon and rainbow trout increased both $\dot{M}O_2$ ($Q_{10}=2.2$, Chinook salmon; 2.42, rainbow trout) and $\dot{Q}$ ($Q_{10}=1.8$, Chinook salmon; 1.85, rainbow trout) when water temperature was increased (Clark et al., 2008; Keen and Gamperl, 2012). Resting $\dot{M}O_2$ and $\dot{Q}$ exhibited a similar pattern during an acute temperature increase in Atlantic cod, reaching maximum values 2.6- and 2.5-fold above baseline levels, respectively (Gollock et al., 2006). In contrast, a slightly different pattern was observed in Atlantic salmon and Arctic char. In these species, resting $\dot{M}O_2$ increased linearly when exposed to an acute increase in temperature, while resting $\dot{Q}$ tended to plateau or decline as the fish approached CT_{max} (Penney et al., 2014). While most studies have examined how cardiorespiratory variables change during an acute temperature increase, Brodeur et al. (2001) compared how variables change with acute warming vs acute cooling. Interestingly, when resting rainbow trout were exposed to an acute decrease in water temperature from 20 to 12°C, $\dot{M}O_2$ and $\dot{Q}$ were tightly linked, and there was only a minor decrease in blood oxygen extraction (Brodeur et al., 2001). However, when fish in this same study were acutely warmed from 12 to 28°C, the increase in $\dot{M}O_2$ was strongly supported by blood oxygen extraction, and the increase in $\dot{Q}$ was only weakly correlated with $\dot{M}O_2$.

The strongest evidence for the dependence of aerobic capacity on cardiovascular transport capacity during changes in temperature comes from studies with swimming salmon. When sockeye salmon were exposed to acute temperature increases while swimming at 75% of maximum, increases in $\dot{Q}$ paralleled increases in $\dot{M}O_2$ (Steinhausen et al., 2008). The changes (swimming values − resting values) for $\dot{M}O_2$ and $\dot{Q}$ were similarly maintained close to maximum between 17 and 21°C; however, at 23–24°C both $\dot{Q}$ and $\dot{M}O_2$ started to decline (Steinhausen et al., 2008). Similarly, $\dot{M}O_2$ closely matched $\dot{Q}$ in maximally swimming pink and sockeye salmon across a broad range of temperatures (Clark et al., 2011; Eliason et al., 2013 and Figs. 2 and 3), and aerobic scope and cardiac scope varied in parallel with temperature in four populations of sockeye salmon (Eliason et al., 2011). Notably, the collapse in aerobic scope above optimal temperatures in sockeye salmon has been associated with a cardiac limitation, which was attributed to reduced scope for f_H (Eliason et al., 2011, 2013; Steinhausen et al., 2008; and Figs. 2 and 3). Furthermore, pharmacological studies on anesthetized fish treated with atropine and isoproterenol have shown that the Arrhenius break point temperature for f_{Hmax} occurs at the optimum temperature for aerobic scope in several fish species, and that cardiac collapse occurs around the upper pejus temperature (Anttila et al., 2013b, 2014a; Casselman et al., 2012; Chen et al., 2015; Drost et al., 2014; Ferreira et al., 2014; Muñoz et al., 2015; Sidhu et al., 2014). However, more studies with different species are needed (including those on unanesthetized fish) to evaluate the use/validity of this method as a screening tool. In conclusion, it appears that the heart is particularly sensitive to acute temperature changes (see Section 2.2), and that its performance mirrors, and may even determine, the aerobic performance of the whole animal (see Section 2.3).

2.2. Cardiac Collapse at High Temperatures

It is interesting to consider why hearts fail at high temperature. Although many hypotheses have been put forward, no single "smoking gun" has been identified thus far. In all likelihood, a number of interacting factors likely underlie the collapse of cardiac function at high temperature. Furthermore, the mechanisms that cause cardiac failure in resting fish may be different from those mediating a similar effect in swimming fish. As discussed in the cardiac output (Section 1.2) and heart rate (Section 1.3) sections, resting/routine $\dot{Q}$ and f_H continue to increase until 1–2°C before the CT_{max} is reached (e.g., Eliason et al., 2013; Gollock et al., 2006; Mendonça and Gamperl, 2010; Penney et al., 2014; and see Figs. 3 and 5). However, in many fish species, swimming-induced maximum $\dot{Q}$ and maximum f_H reach their peak values at much lower temperatures, and either plateau or decline when temperatures increase beyond the species optimum for aerobic scope (Brett, 1971; Eliason et al., 2013; Franklin et al., 2013; Steinhausen et al., 2008; and see Fig. 3). Nevertheless, at least in one species, the Murray cod, both resting and maximum f_H and resting and maximum $\dot{M}O_2$ continue to increase linearly until very high temperatures (Clark et al., 2005).

One of the factors that could influence cardiac function at elevated temperatures is membrane integrity. Moyes and Ballantyne (2011) have reviewed these effects in fish. Temperature has significant effects on the fluidity of membranes, and increases in membrane fluidity are concomitant with increases in temperature. Membrane function is tightly linked with fluidity. When membranes become too fluid, their ability to act as a barrier is reduced. On the other hand, if a membrane is too rigid, membrane proteins cannot move easily, and their function is compromised (Moyes and Ballantyne, 2011). From a functional point of view, a compromised membrane structure at elevated temperatures would negatively influence membrane protein function (such as channels and pumps required in excitation–contraction coupling) (Moyes and Ballantyne, 2011) and also increases the leakiness of membranes; the latter having a deleterious impact on the function of mitochondria (e.g., Rodnick et al., 2014).

Temperature has significant influences on the function of membrane proteins like those involved in excitation–contraction (E-C) coupling and cardiomyocyte excitability (see Chapter 3, Volume 36A: Vornanen, 2017; and Section 1.3 on f_H), and since these proteins govern the beating of the heart, negative temperature-related impacts on their function could lead to cardiac collapse. Vornanen et al. (2014) examined which of the E-C coupling proteins (Na^+, Ca^{2+}, and K^+ channels) is the most sensitive to acute increases in temperature by measuring ion currents in brown trout. They discovered that Na^+ channels were the most sensitive, and that their function actually became compromised at 20.9°C; a temperature lower than where resting f_H began to decline (23.5°C). Unfortunately, in that study, f_{Hmax} was not measured, so it is not yet resolved if this reduced function of Na^+ channels corresponds to temperatures where f_{Hmax} fails. Nevertheless, Vornanen et al. (2014) suggested failure in the function of Na^+ channels at elevated temperatures influences the spread of the action potential across the heart, and that this causes a reduction in f_H (incl. missed ventricular beats), and finally, leads to total cardiac arrest (for further details see Chapter 3, Volume 36A: Vornanen, 2017).

On the other hand, it has also been observed that mitochondrial function fails before the CT_{max} of fish. Thus, there could be a limitation in ATP supply to the heart. This could be especially important in swimming fish as the cardiac ATP requirement increases because of both exercise- and temperature-dependent effects (see Sections 1.2 and 1.3). The same applies to E-C coupling, since the f_H of swimming fish is higher than that of resting fish (Figs. 2 and 3). Temperature effects on mitochondrial function are discussed in detail in Section 1.7. Briefly, an increase in temperature induces a phase transition in membrane lipids which increases the permeability of biological membranes (Moyes and Ballantyne, 2011). The increased leakiness of mitochondrial membranes reduces the membrane potential, which also increases the release of cytochrome *c* from the mitochondria into the cytoplasm, and both of these effects, depress ATP synthesis (Dahlhoff and Somero, 1993; Iftikar and Hickey, 2013; Rodnick et al., 2014). For example, the permeability of the inner membrane of mitochondria is 75% higher at 25°C than at 15°C in *N. celidotus*, and the capacity for oxidative phosphorylation is significantly reduced at 25°C (2.5°C before cardiac failure) (Iftikar and Hickey, 2013). In addition, the efficiency of ATP production has been shown to decrease at temperature near CT_{max} in many fish species (e.g., Hilton et al., 2010; Iftikar et al., 2014, 2015; Iftikar and Hickey, 2013; Penney et al., 2014; Rodnick et al., 2014).

Another factor leading to cardiac failure at high temperatures could be insufficient oxygen supply to the myocardium. As discussed in Section 1.9.3, acute increases in temperature often (but see Steinhausen et al., 2008) lead to a decrease in P_vO_2 and C_vO_2 (Clark et al., 2008; Ekström et al., 2016a; Eliason et al., 2013; Heath and Hughes, 1973; Lannig et al., 2004; Sartoris et al., 2003). Since the spongy myocardium receives oxygen exclusively via diffusion from venous blood in the majority of fish species, this reduction in P_vO_2 and C_vO_2 could compromise oxygen supply to the spongy myocardium and contribute to a reduction in cardiac ATP production. However, when the oxygen content of the blood is experimentally manipulated there is either no change in CT_{max} (Brijs et al., 2015; Wang et al., 2014) or a minor increase with hyperoxia (Ekström et al., 2016a). It should also be noted that mitochondria fail at temperatures near CT_{max}, even at saturating oxygen levels (Iftikar and Hickey, 2013; Rodnick et al., 2014). Nevertheless, none of these studies measured whether there was a change in maximum cardiac performance in swimming fish and, depending on the species, CT_{max} can be several °C higher than peak $\dot{Q}$. Thus, the question remains, is maximum cardiac performance compromised when P_vO_2 and C_vO_2 are reduced? Indirectly, there is some evidence that insufficient myocardial oxygen delivery could play a role in cardiac collapse at high temperatures, since AMPK activity (which is activated by reduced levels of ATP; Hardie, 2011) has been observed to increase in ventricles when environmental temperature is increased beyond the species' optimum temperature for aerobic scope (Anttila et al., 2013b). Also, heart myoglobin levels are positively correlated with the upper thermal tolerance of Atlantic salmon (Anttila et al., 2013a), and this suggests that oxygen supply could be one factor influencing the thermal tolerance of fishes.

Beyond P_vO_2 and C_vO_2, other factors in the blood are affected by temperature increases and could have detrimental effects on cardiac performance. As discussed earlier (see Section 1.9.1), the pH of blood drops (acidosis) and the K^+ concentration often increases (hyperkalemia) with warming and exercise in fish. Acidosis and hyperkalemia have negative inotropic and chronotropic effects on cardiac performance (e.g., Farrell et al., 1986; Hanson et al., 2006; Kalinin and Gesser, 2002). Thus, especially in maximally swimming fish, the deleterious venous blood environment could impair cardiac performance at high temperatures. Moreover, the sensitivity of the heart to adrenaline decreases with warming in some species (see Section 1.6; Aho and Vornanen, 2001; Ask et al., 1981; Farrell et al., 1996; Graham and Farrell, 1989; Keen et al., 1993), even though adrenergic stimulation is critical to maintaining maximum performance when venous blood is hypoxic, acidotic, and hyperkalemic (Hanson et al., 2006). Collectively, these effects could have downstream negative influences on cardiac function via changes in E-C coupling (Shiels et al., 2003).

2.3. Ecological, Evolutionary, and Conservation Implications

An important goal for conservation biologists is to determine the species most at risk of extinction, extirpation, and range shifts due to climate change. Some have suggested that fish at the poles are most at risk since these regions are expected to experience the largest temperature changes (IPCC, 2014), while others suggest that tropical species are the most susceptible since they are adapted to stable and warm environmental temperatures and may have reduced capacity for phenotypic plasticity, and thus, less able to respond to increases in temperature (Munday et al., 2008). Still others suggest that individual populations that are already experiencing temperatures near their critical thermal maximum are in the most danger (i.e., have a reduced warming tolerance, such as salmon and some equatorial species; Farrell et al., 2008; Rummer et al., 2014). In any case, a response to climate change will largely depend on: (1) the ability to alter phenotype in response to temperature change (thermal plasticity); (2) evolutionary processes that could lead to local adaptation to new temperature regimes; and (3) the ability to migrate to cooler habitats.

Phenotypic plasticity of cardiac function may be a critical avenue that allows some species to cope with climate change. For example, northern and southern wild populations of Atlantic salmon did not differ in their cardiac response to acute temperature increases, suggesting that there is limited capacity for local adaptation of cardiac performance to future temperatures in this species (Anttila et al., 2014a). However, both populations of Atlantic salmon showed remarkable cardiac plasticity, increasing both maximum heart rate and the temperature of cardiac collapse in response to warm acclimation (Anttila et al., 2014a). Similarly, the Antarctic fish *P. borchgrevinki* exhibited impressive cardiac plasticity in response to temperature acclimation, displaying complete cardiac thermal compensation, and similar resting $\dot{Q}$, maximum $\dot{Q}$, and cardiac scope across acclimation temperatures (Franklin et al., 2007; Seebacher et al., 2005).

Nonetheless, not all fishes have the capacity to improve their cardiovascular performance through phenotypic plasticity. For example, the air-breathing Asian swamp eel did not improve cardiac performance in response to warm acclimation. Even so, thermal tolerance was superior in the high temperature-acclimated group of fish (Lefevre et al., 2016). This suggests that the cardiovascular system may not be the key determinant of thermal tolerance in this species of fish, perhaps because they breathe air, and thus, the air-breathing organ can provide an oxygen-rich supply of blood to the heart.

Cardiovascular performance as a function of temperature has been associated with the evolutionary fitness of fishes. Aerobic scope and cardiac capacity have been suggested to be locally adapted to the prevailing environmental conditions in populations of migrating, adult sockeye salmon (Eliason et al., 2011). Indeed, the optimal thermal range for aerobic scope, cardiac scope and scope for f_H occurred at the typical temperatures encountered during upriver migration in four populations of sockeye salmon (Eliason et al., 2011). Moreover, cardiac adaptations (specifically, increased β-adrenoceptor density and thermal plasticity) were associated with enhanced thermal tolerance in one population of sockeye salmon (Eliason et al., 2011). From a conservation perspective, exposure to temperatures above those normally encountered during migration (i.e., outside the optimal thermal range for aerobic scope and cardiac performance) has been repeatedly associated with high mortality in sockeye salmon (Farrell et al., 2008; Martins et al., 2011, 2012). Given that sockeye salmon are semelparous and have a single opportunity to spawn, fish that perish *en route* to the spawning ground will have no lifetime reproductive success, raising clear conservation concerns.

As another example, Sandblom et al. (2016a) compared European perch living in water that had been warmed by 5–10°C by an adjacent nuclear power plant for over 3 decades with conspecifics from a cooler, natural environment. This model system allowed the researchers to examine how fish responded to environmental warming on a timescale relevant to climate change (i.e., decades). Chronically warmed fish adjusted basal energy requirements and resting cardiac performance (physiological "floors") as compared to the cooler conspecifics. In contrast, maximum cardiorespiratory performance and CT_{max} (physiological "ceilings") displayed limited thermal compensation. Thus, the adaptive capacity of fishes may be limited by physiological "ceilings" since they display limited thermal plasticity (Sandblom et al., 2016a). This study reiterates a point made throughout this chapter, namely, that studies examining resting fish in isolation may not be ideal to examine the role of cardiac function in thermal tolerance or to predict a species' response to climate change.

As many chapters of the two volumes (36A and 36B) on the cardiovascular system in fishes have demonstrated, there are several environmental variables beyond just temperature that have significant influences on cardiac performance. These environmental factors are also interacting (e.g., hypoxia and temperature), and stressors can have additive, synergistic, or antagonistic effects on fish performance (see McBryan et al., 2013). Therefore, from a conservation point of view, more research should examine how combined environmental stressors influence cardiac performance and fish fitness.

REFERENCES

Abramochkin, D.V., Vornanen, M., 2015. Seasonal acclimatization of the cardiac potassium currents (IK1 and IKr) in an arctic marine teleost, the navaga cod (*Eleginus navaga*). J. Comp. Physiol. B 185, 883–890.

Aguiar, L.H., Kalinin, A.L., Rantin, F.T., 2002. The effects of temperature on the cardio-respiratory function of the neotropical fish *Piaractus mesopotamicus*. J. Therm. Biol. 27, 299–308.

Aho, E., Vornanen, M., 1998. Ca^{2+}-ATPase activity and Ca^{2+}-uptake by sarcoplasmic reticulum in fish heart: effects of thermal acclimation. J. Exp. Biol. 201, 525–532.

Aho, E., Vornanen, M., 2001. Cold acclimation increases basal heart rate but decreases its thermal tolerance in rainbow trout (*Oncorhynchus mykiss*). J. Comp. Physiol. B 171, 173–179.

Alderman, S.L., Klaiman, J.M., Deck, C.A., Gillis, T.E., 2012. Effect of cold acclimation on troponin I isoform expression in striated muscle of rainbow trout. Am. J. Physiol. 303, R168–76.

Anttila, K., Casselman, M.T., Schulte, P.M., Farrell, A.P., 2013a. Optimum temperature in juvenile salmonids: connecting subcellular indicators to tissue function and whole-organism thermal optimum. Physiol. Biochem. Zool. 86, 245–256.

Anttila, K., Dhillon, R.S., Boulding, E.G., Farrell, A.P., Glebe, B.D., Elliott, J.A.K., Wolters, W.R., Schulte, P.M., 2013b. Variation in temperature tolerance among families of Atlantic salmon (*Salmo salar*) is associated with hypoxia tolerance, ventricle size and myoglobin level. J. Exp. Biol. 216, 1183–1190.

Anttila, K., Couturier, C.S., Øverli, Ø., Johnsen, A., Marthinsen, G., Nilsson, G.E., Farrell, A.P., 2014a. Atlantic salmon show capability for cardiac acclimation to warm temperatures. Nat. Commun. 5, 4252.

Anttila, K., Eliason, E.J., Kaukinen, K.H., Miller, K.M., Farrell, A.P., 2014b. Facing warm temperatures during migration—cardiac mRNA responses of two adult sockeye salmon *Oncorhynchus nerka* populations to warming and swimming challenges. J. Fish Biol. 84, 1439–1456.

Anttila, K., Lewis, M., Prokkola, J.M., Kanerva, M., Seppänen, E., Kolari, I., Nikinmaa, M., 2015. Warm acclimation and oxygen depletion induce species-specific responses in salmonids. J. Exp. Biol. 218, 1471–1477.

Ask, J.A., 1983. Comparative aspects of adrenergic receptors in the hearts of lower vertebrates. Comp. Biochem. Physiol. A Comp. Physiol. 76, 543–552.

Ask, J.A., Stene-Larsen, G., Helle, K.B., 1981. Temperature effects on the β_2-adrenoreceptors of the trout atrium. J. Comp. Physiol. 143, 161–168.

Axelsson, M., Davison, W., Forster, M.E., Farrell, A.P., 1992. Cardiovascular responses of the red blooded Antarctic fishes, *Pagothenia bernacchii* and *P. borchgrevinki*. J. Exp. Biol. 167, 179–201.

Badr, A., El-Sayed, M.F., Vornanen, M., 2016. Effects of seasonal acclimatization on temperature dependence of cardiac excitability in the roach, *Rutilus rutilus*. J. Exp. Biol. 219, 1495–1504.

Badr, A., Hassinen, M., El-Sayed, M.F., Vornanen, M., 2017. Effects of seasonal acclimatization on action potentials and sarcolemmal K^+ currents in roach (*Rutilus rutilus*) cardiac myocytes. Comp. Biochem. Physiol. A Mol. Integr. Physiol. 205, 15–27.

Bailey, J.R., Driedzic, W.R., 1990. Enhanced maximum frequency and force development of fish hearts following temperature acclimation. J. Exp. Biol. 149, 239–254.

Ballesta, S., Hanson, L.M., Farrell, A.P., 2012. The effect of adrenaline on the temperature dependency of cardiac action potentials in pink salmon *Oncorhynchus gorbuscha*. J. Fish Biol. 80, 876–885.

Barron, M.G., Tarr, B.D., Hayton, W.L., 1987. Temperature-dependence of cardiac output and regional blood flow in rainbow trout, *Salmo gairdneri* Richardson. J. Fish Biol. 31, 735–744.

Benfey, T.J., Bennett, L.E., 2009. Effect of temperature on heart rate in diploid and triploid brook charr, *Salvelinus fontinalis*, embryos and larvae. Comp. Biochem. Physiol. A Mol. Integr. Physiol. 152, 203–206.

Blank, J.M., Morrissette, J.M., Davie, P.S., Block, B.A., 2002. Effects of temperature, epinephrine and Ca(2+) on the hearts of yellowfin tuna (*Thunnus albacares*). J. Exp. Biol. 205, 1881–1888.

Blank, J.M., Morrissette, J.M., Landeira-Fernandez, A.M., Blackwell, S.B., Williams, T.D., Block, B.A., 2004. In situ cardiac performance of Pacific bluefin tuna hearts in response to acute temperature change. J. Exp. Biol. 207, 881–890.

Borutaite, V., Brown, G.C., 2003. Mitochondria in apoptosis of ischemic heart. FEBS Lett. 541, 1–5.

Brett, J.R., 1971. Energetic responses of salmon to temperature. A study of some thermal relations in the physiology and freshwater ecology of sockeye salmon (*Oncorhynchus nerka*). Am. Zool. 11, 99–113.

Brijs, J., Jutfelt, F., Clark, T.D., Gräns, A., Ekström, A., Sandblom, E., 2015. Experimental manipulations of tissue oxygen supply do not affect warming tolerance of European perch. J. Exp. Biol. 218, 2448–2454.

Brodeur, J.C., Dixon, D.G., McKinley, R.S., 2001. Assessment of cardiac output as a predictor of metabolic rate in rainbow trout. J. Fish Biol. 58, 439–452.

Burleson, M.L., Silva, P.E., 2011. Cross tolerance to environmental stressors: effects of hypoxic acclimation on cardiovascular responses of channel catfish (*Ictalurus punctatus*) to a thermal challenge. J. Therm. Biol. 36, 250–254.

Butler, P.J., Taylor, E.W., 1975. The effect of progressive hypoxia on respiration in the dogfish (*Scyliorhinus canicula*) at different seasonal temperatures. J. Exp. Biol. 63, 117–130.

Cai, Y.J., Adelman, I.R., 1990. Temperature acclimation in respiratory and cytochrome *c* oxidase activity in common carp (*Cyprinus carpio*). Comp. Biochem. Physiol. 95A, 139–144.

Casselman, M.T., Anttila, K., Farrell, A.P., 2012. Using maximum heart rate as a rapid screening tool to determine optimum temperature for aerobic scope in Pacific salmon *Oncorhynchus* spp. J. Fish Biol. 80, 358–377.

Cech Jr., J.J., Bridges, D.W., Rowell, D.M., Balzer, P.J., 1976. Cardiovascular responses of winter flounder *Pseudopleuronectes americanus* (Walbaum), to acute temperature increase. Can. J. Zool. 54, 1383–1388.

Chen, Z., Snow, M., Lawrence, C.S., Church, A.R., Narum, S.R., Devlin, R.H., Farrell, A.P., 2015. Selection for upper thermal tolerance in rainbow trout (*Oncorhynchus mykiss* Walbaum). J. Exp. Biol. 218, 803–812.

Chung, D.J., Bryant, H.J., Schulte, P.M., 2017. Thermal acclimation and subspecies-specific effects on heart and brain mitochondrial performance in a eurythermal teleost (*Fundulus heteroclitus*). J. Exp. Biol. 220, 1459–1471.

Claireaux, G., Webber, D.M., Kerr, S.R., Boutilier, R.G., 1995. Physiology and behaviour of free-swimming atlantic cod (*Gadus morhua*) facing fluctuating temperature conditions. J. Exp. Biol. 198, 49–60.

Clark, T.D., Ryan, T., Ingram, B.A., Woakes, A.J., Butler, P.J., Frappell, P.B., 2005. Factorial aerobic scope is independent of temperature and primarily modulated by heart rate in exercising murray cod (*Maccullochella peelii peelii*). Physiol. Biochem. Zool. 78, 347–355.

Clark, T.D., Sandblom, E., Cox, G.K., Hinch, S.G., Farrell, A.P., 2008. Circulatory limits to oxygen supply during an acute temperature increase in the Chinook salmon (*Oncorhynchus tshawytscha*). Am. J. Physiol. Regul. Integr. Comp. Physiol. 295, R1631–R1639.

Clark, T.D., Jeffries, K.M., Hinch, S.G., Farrell, A.P., 2011. Exceptional aerobic scope and cardiovascular performance of pink salmon (*Oncorhynchus gorbuscha*) may underlie resilience in a warming climate. J. Exp. Biol. 214, 3074–3081.

Clark, T.D., Donaldson, M.R., Pieperhoff, S., Drenner, S.M., Lotto, A., Cooke, S.J., Hinch, S.G., Patterson, D.A., Farrell, A.P., 2012. Physiological benefits of being small in a changing world: responses of coho salmon (*Oncorhynchus kisutch*) to an acute thermal challenge and a simulated capture event. PLoS One 7, e39079.

Clark, T.D., Farwell, C.J., Rodriguez, L.E., Brandt, W.T., Block, B.A., 2013. Heart rate responses to temperature in free-swimming Pacific bluefin tuna (*Thunnus orientalis*). J. Exp. Biol. 216, 3208–3214.

Collier, P., Watson, C.J., van Es, M.H., Phelan, D., McGorrian, C., Tolan, M., et al., 2012. Getting to the heart of cardiac remodeling; how collagen subtypes may contribute to phenotype. J. Mol. Cell. Cardiol. 52, 148–153.

Cooke, S.J., Ostrand, K.G., Bunt, C.M., Schreer, J.F., Wahl, D.H., Philipp, D.P., 2003. Cardiovascular responses of largemouth bass to exhaustive exercise and brief air exposure over a range of water temperatures. Trans. Am. Fish. Soc. 132, 1154–1165.

Cooke, S.J., Schreer, J.F., Wahl, D.H., Philipp, D.P., 2010. Cardiovascular performance of six species of field-acclimatized centrarchid sunfish during the parental care period. J. Exp. Biol. 213, 2332–2342.

Costa, M.J., Rivaroli, L., Rantin, F.T., Kalinin, A.L., 2000. Cardiac tissue function of the teleost fish *Oreochromis niloticus* under different thermal conditions. J. Therm. Biol. 25, 373–379.

Costa, I.A., Hein, T.W., Gamperl, A.K., 2015. Cold-acclimation leads to differential regulation of the steelhead trout (*Oncorhynchus mykiss*) coronary microcirculation. Am. J. Physiol. Regul. Integr. Comp. Physiol. 308, R743–R754.

Cox, G.K., 2015. The functional significance and evolution of the coronary circulation in sharks. PhD Thesis, University of British Columbia, Vancouver, Canada, p. 150.

Crocker, C.E., Farrell, A.P., Gamperl, A.K., Cech, J.J., 2000. Cardiorespiratory responses of white sturgeon to environmental hypercapnia. Am. J. Physiol. Regul. Integr. Comp. Physiol. 279, R617–R628.

Crockett, E.L., Sidell, B.D., 1990. Some pathways of energy metabolism are cold adapted in Antarctic fishes. Physiol. Zool. 63, 472–488.

Currie, S., Reddin, K., McGinn, P., McConnell, T., Perry, S.F., 2008. β-Adrenergic stimulation enhances the heat-shock response in fish. Physiol. Biochem. Zool. 81, 414–425.

Currie, S., Ahmady, E., Watters, M.A., Perry, S.F., Gilmour, K.M., 2013. Fish in hot water: hypoxaemia does not trigger catecholamine mobilization during heat shock in rainbow trout (*Oncorhynchus mykiss*). Comp. Biochem. Physiol. A Mol. Integr. Physiol. 165, 281–287.

Dahlhoff, E., Somero, G.N., 1993. Effect of temperature on mitochondria from a balone (Genus *haliotis*) adaptive plasticity and its limits. J. Exp. Biol. 185, 151–168.

Dickson, K.A., Graham, J.B., 2004. Evolution and consequences of endothermy in fishes. Physiol. Biochem. Zool. 77, 998–1018.

Drost, H., Carmack, E.C., Farrell, A.P., 2014. Upper thermal limits of cardiac function for Arctic cod *Boreogadus saida*, a key food web fish species in the Arctic Ocean. J. Fish Biol. 84, 1781–1792.

Egginton, S., 1997. Control of tissue blood flow at very low temperatures. J. Therm. Biol. 22, 403–407.

Egginton, S., Cordiner, S., 1997. Cold-induced angiogenesis in seasonally acclimatized rainbow trout (*Oncorhynchus mykiss*). J. Exp. Biol. 200, 2263–2268.

Ekström, A., Jutfelt, F., Sandblom, E., 2014. Effects of autonomic blockade on acute thermal tolerance and cardioventilatory performance in rainbow trout, *Oncorhynchus mykiss*. J. Therm. Biol. 44, 47–54.

Ekström, A., Brijs, J., Clark, T.D., Gräns, A., Jutfelt, F., Sandblom, E., 2016a. Cardiac oxygen limitation during an acute thermal challenge in the European perch: effects of chronic environmental warming and experimental hyperoxia. Am. J. Physiol. Regul. Integr. Comp. Physiol. 311, R440–449.

Ekström, A., Hellgren, K., Gräns, A., Pichaud, N., Sandblom, E., 2016b. Dynamic changes in scope for heart rate and cardiac autonomic control during warm acclimation in rainbow trout. J. Exp. Biol. 219, 1106–1109.

Ekström, A., Jutfelt, F., Sundström, F., Adill, A., Aho, T., Sandblom, E., 2016c. Chronic environmental warming alters cardiovascular and haematological stress responses in European perch (*Perca fluviatilis*). J. Comp. Physiol. B 186, 1023–1031.

Ekström, A., Axelsson, M., Gräns, A., Brijs, J., Sandblom, E., 2017a. Influence of the coronary circulation on thermal tolerance and cardiac performance during warming in rainbow trout. Am. J. Physiol. Regul. Integr. Comp. Physiol. 312, R549–R558.

Ekström, A., Sandblom, E., Blier, P.U., Dupont Cyr, B.A., Brijs, J., Pichaud, N., 2017b. Thermal sensitivity and phenotypic plasticity of cardiac mitochondrial metabolism in European perch, *Perca fluviatilis*. J. Exp. Biol. 220, 386–396.

Eliason, E.J., Clark, T.D., Hague, M.J., Hanson, L.M., Gallagher, Z.S., Jeffries, K.M., Gale, M.K., Patterson, D.A., Hinch, S.G., Farrell, A.P., 2011. Differences in thermal tolerance among sockeye salmon populations. Science 332, 109–112.

Eliason, E.J., Clark, T.D., Hinch, S.G., Farrell, A.P., 2013. Cardiorespiratory collapse at high temperature in swimming adult sockeye salmon. Conserv. Physiol. 1, cot008.

Ern, R., Norin, T., Gamperl, A.K., Esbaugh, A.J., 2016. Oxygen dependence of upper thermal limits in fishes. J. Exp. Biol. 219, 3376–3383.

Fagernes, C.E., Stenløkken, K.-O., Røhr, Å.K., Berenbrink, M., Ellefsen, S., Nilsson, G.E., 2017. Extreme anoxia tolerance in crucian carp and goldfish through neofunctionalization of dublicated genes creating a new ethanol-producing puruvate decarboxylase pathway. Sci. Rep. 7, 7884.

Farrell, A.P., 1987. Coronary flow in a perfused rainbow trout heart. J. Exp. Biol. 129, 107–123.

Farrell, A.P., 1991. From hagfish to tuna—a perspective on cardiac function in fish. Physiol. Zool. 64, 1137–1164.

Farrell, A.P., 2009. Environment, antecedents and climate change: lessons from the study of temperature physiology and river migration of salmonids. J. Exp. Biol. 212, 3771–3780.

Farrell, A.P., Clutterham, S.M., 2003. On-line venous oxygen tensions in rainbow trout during graded exercise at two acclimation temperatures. J. Exp. Biol. 206, 487–496.

Farrell, A.P., Milligan, C.L., 1986. Myocardial intracellular pH in a perfused rainbow trout heart during extracellular acidosis in the presence and absence of adrenaline. J. Exp. Biol. 125, 347–359.

Farrell, A.P., Smith, F., 2017. Cardiac form, function and physiology. In: Gamperl, A.K., Gillis, T.E., Farrell, A.P., Brauner, C.J. (Eds.), Fish Physiology. In: The Cardiovascular System: Morphology, Control and Function, vol. 36A. Academic Press, San Diego, pp. 155–264.

Farrell, A.P., MacLeod, K.R., Chancey, B., 1986. Intrinsic mechanical properties of the perfused rainbow trout heart and the effects of catecholamines and extracellular calcium under control and acidotic conditions. J. Exp. Biol. 125, 319–345.

Farrell, A.P., Gamperl, A.K., Hicks, J.M.T., Shiels, H.A., Jain, K.E., 1996. Maximum cardiac performance of rainbow trout (*Oncorhynchus mykiss*) at temperatures approaching their upper lethal limit. J. Exp. Biol. 199, 663–672.

Farrell, A.P., Axelsson, M., Altimiras, J., Sandblom, E., Claireaux, G., 2007. Maximum cardiac performance and adrenergic sensitivity of the sea bass *Dicentrarchus labrax* at high temperatures. J. Exp. Biol. 210, 1216–1224.

Farrell, A.P., Hinch, S.G., Cooke, S.J., Patterson, D.A., Crossin, G.T., Lapointe, M., Mathes, M.T., 2008. Pacific salmon in hot water: applying aerobic scope models and biotelemetry to predict the success of spawning migrations. Physiol. Biochem. Zool. 81, 697–709.

Farrell, A.P., Eliason, E.J., Sandblom, E., Clark, T.D., 2009. Fish cardiorespiratory physiology in an era of climate change. Can. J. Zool. 87, 835–851.

Farrell, A.P., Eliason, E., Clark, T., Steinhausen, M., 2014. Oxygen removal from water versus arterial oxygen delivery: calibrating the Fick equation in Pacific salmon. J. Comp. Physiol. B 184, 855–864.

Ferreira, E.O., Anttila, K., Farrell, A.P., 2014. Thermal optima and tolerance in the eurythermal goldfish (*Carassius auratus*): relationship between whole animal aerobic capacity and maximum heart rate. Physiol. Biochem. Zool. 87, 599–611.

Fitch, N.A., Johnston, I.A., Wood, R.E., 1984. Skeletal muscle capillary supply in a fish that lacks respiratory pigments. Respir. Physiol. 57, 201–211.

Forster, M.E., Davison, W., Axelsson, M., Sundin, L., Franklin, C.E., Gieseg, S., 1998. Catecholamine release in heat-stressed Antarctic fish causes proton extrusion by the red cells. J. Comp. Physiol. B 168, 345–352.

Foster, A.R., Hall, S.J., Houlihan, D.F., 1993. The effects of temperature acclimation on organ/tissue mass and cytochrome c oxidase activity in juvenile cod (*Gadus morhua*). J. Fish Biol. 42, 947–957.

Franklin, C.E., Davie, P.S., 1992. Sexual maturity can double heart mass and cardiac power output in male rainbow trout. J. Exp. Biol. 171, 139–148.

Franklin, C.E., Axelsson, M., Davison, W., 2001. Constancy and control of heart rate during an increase in temperature in the Antarctic fish *Pagothenia borchgrevinki*. Exp. Biol. Online 6, 1–8.

Franklin, C.E., Davison, W., Seebacher, F., 2007. Antarctic fish can compensate for rising temperatures: thermal acclimation of cardiac performance in *Pagothenia borchgrevinki*. J. Exp. Biol. 210, 3068–3074.

Franklin, C.E., Farrell, A.P., Altimiras, J., Axelsson, M., 2013. Thermal dependence of cardiac function in arctic fish: implications of a warming world. J. Exp. Biol. 216, 4251–4255.

Fry, F.E.J., 1947. Effects of the environment on animal activity. Pub. Ont. Fish. Res. Lab. 68, 1–62.

Galli, G.L.J., Lipnick, M.S., Block, B.A., 2009. Effect of thermal acclimation on action potentials and sarcolemmal K+ channels from Pacific bluefin tuna cardiomyocytes. Am. J. Physiol. 297, R502–R509.

Gamperl, A.K., 2011. Integrated responses of the circulatory system to temperature. In: Farrell, A.P. (Ed.), Encyclopedia of Fish Physiology: From Genome to Environment. Elsevier, Amsterdam, pp. 1197–1205.

Gamperl, A.K., Wilkinson, M., Boutilier, R.G., 1994. β-Adrenoceptors in the trout (*Oncorhynchus mykiss*) heart—characterization, quantification, and effects of repeated catecholamine exposure. Gen. Comp. Endocrinol. 95, 259–272.

Gamperl, A.K., Vijayan, M.M., Pereira, C., Farrell, A.P., 1998. β-receptors and stress protein 70 expression in hypoxic myocardium of rainbow trout and chinook salmon. Am. J. Physiol. Regul. Integr. Comp. Physiol. 43, R428–R436.

Gamperl, A.K., Swafford, B.L., Rodnick, K.J., 2011. Elevated temperature, per se, does not limit the ability of rainbow trout to increase stroke volume. J. Therm. Biol. 36, 7–14.

Genge, C.E., Davidson, W.S., Tibbits, G.F., 2013. Adult teleost heart expresses two distinct troponin C paralogs: cardiac TnC and a novel and teleost-specific ssTnC in a chamber- and temperature-dependent manner. Physiol. Genomics 45, 866–875.

Gillis, T.E., Liang, B., Chung, F., Tibbits, G.F., 2005. Increasing mammalian cardiomyocyte contractility with residues identified in trout troponin C. Physiol. Genomics 22, 1–7.

Gollock, M.J., Currie, S., Petersen, L.H., Gamperl, A.K., 2006. Cardiovascular and haematological responses of Atlantic cod (*Gadus morhua*) to acute temperature increase. J. Exp. Biol. 209, 2961–2970.

Goulding, A.T., Farrell, A.P., 2016. Quantification of ventricular β2-adrenoceptor density and ligand binding affinity in wild sockeye salmon *Oncorhynchus nerka* smolts using a novel modification to the tritiated ligand technique. J. Fish Biol. 88, 2081–2087.

Gracey, A.Y., Fraser, E.J., Li, W., Fang, Y., Taylor, R.R., Rogers, J., Brass, A., Cossins, A.R., 2004. Coping with cold: an integrative, multitissue analysis of the transcriptome of a poikilothermic vertebrate. Proc. Natl. Acad. Sci. U.S.A. 101, 16970–16975.

Graham, M.S., Farrell, A.P., 1989. The effect of temperature acclimation and adrenaline on the performance of a perfused trout heart. Physiol. Zool. 62, 38–61.

Graham, M.S., Farrell, A.P., 1990. Myocardial oxygen consumption in trout acclimated to 5°C and 15°C. Physiol. Zool. 63, 536–554.

Gräns, A., Albertsson, F., Axelsson, M., Olsson, C., 2009a. Postprandial changes in enteric electrical activity and gut blood flow in rainbow trout (*Oncorhynchus mykiss*) acclimated to different temperatures. J. Exp. Biol. 212, 2550–2557.

Gräns, A., Axelsson, M., Pitsillides, K., Olsson, C., Höjesjö, J., Kaufman, R., Cech, J., 2009b. A fully implantable multi-channel biotelemetry system for measurement of blood flow and temperature: a first evaluation in the green sturgeon. Hydrobiologia 619, 11–25.

Gräns, A., Olsson, C., Pitsillides, K., Nelson, H.E., Cech, J.J., Axelsson, M., 2010. Effects of feeding on thermoregulatory behaviours and gut blood flow in white sturgeon (*Acipenser transmontanus*) using biotelemetry in combination with standard techniques. J. Exp. Biol. 213, 3198–3206.

Gräns, A., Seth, H., Axelsson, M., Sandblom, E., Albertsson, F., Wiklander, K., Olsson, C., 2013. Effects of acute temperature changes on gut physiology in two species of sculpin from the west coast of Greenland. Polar Biol. 36, 775–785.

Hand, S.C., Menze, M.A., 2008. Mitochondria in energy-limited states: mechanisms that blunt the signaling of cell death. J. Exp. Biol. 211, 1829–1840.

Hanson, L.M., Farrell, A.P., 2007. The hypoxic threshold for maximum cardiac performance in rainbow trout *Oncorhynchus mykiss* (Walbaum) during simulated exercise conditions at 18° C. J. Fish Biol. 71, 926–932.

Hanson, L.M., Ip, Y.K., Farrell, A.P., 2005. The effect of temperature acclimation on myocardial β-adrenoceptor density and ligand binding affinity in African catfish (*Claris gariepinus*). Comp. Biochem. Physiol. A Mol. Integr. Physiol. 141, 164–168.

Hanson, L.M., Obradovich, S., Mouniargi, J., Farrell, A.P., 2006. The role of adrenergic stimulation in maintaining maximum cardiac performance in rainbow trout (*Oncorhynchus mykiss*) during hypoxia, hyperkalemia and acidosis at 10 °C. J. Exp. Biol. 209, 2442–2451.

Hardie, D.G., 2011. Energy sensing by the AMP-activated protein kinase and its effects on muscle metabolism. Proc. Nutr. Soc. 70, 92–99.

Harrison, P., Zummo, G., Farina, F., Tota, B., Johnston, I., 1991. Gross anatomy, myoarchitecture, and ultrastructure of the heart ventricle in the hemoglobinless icefish, *Chaenocephalus aceratus*. Can. J. Zool. 69, 1339–1347.

Harter, T.S., Brauner, C.J., 2017. The O_2 and CO_2 transport system in teleosts and the specialized mechanisms that enhance HB–O_2 unloading to tissues. In: Gamperl, K.A., Gillis, T.E., Farrell, A.P., Brauner, C.J. (Eds.), Fish Physiology. In: The Cardiovascular System: Development, Plasticity and Physiological Responses, vol. 36B. Academic Press, San Diego. pp. 1–106. Chapter 1.

Hassinen, M., Haverinen, J., Vornanen, M., 2008. Electrophysiological properties and expression of the delayed rectifier potassium (ERG) channels in the heart of thermally acclimated rainbow trout. Am. J. Physiol. 295, R297–R308.

Hassinen, M., Abramochkin, D.V., Vornanen, M., 2014. Seasonal acclimatization of the cardiac action potential in the Arctic navaga cod (*Eleginus navaga, Gadidae*). J. Comp. Physiol. B 184, 319–327.

Haverinen, J., Vornanen, M., 2009. Responses of action potential and K^+ currents to temperature acclimation in fish hearts: phylogeny or thermal preferences? Physiol. Biochem. Zool. 82, 468–482.

Heath, A.G., Hughes, G.M., 1973. Cardiovascular and respiratory changes during heat stress in rainbow trout (*Salmo gairdneri*). J. Exp. Biol. 59, 323–338.

Hemmingsen, E.A., 1991. Respiratory and cardiovascular adaptation in hemoglobin-free fish: resolved and unresolved problems. In: di Prisco, G., Maresca, B., Tota, B. (Eds.), Biology of Antarctic Fish. Springer-Verlag, New York, pp. 191–203.

Hemmingsen, E.A., Douglas, E.L., 1972. Respiratory, circulatory responses in a hemoglobin-free fish, *Chaenocephalus aceratus*, to changes in temperature and oxygen tension. Comp. Biochem. Physiol. A Comp. Physiol. 43, 1031–1043.

Hemmingsen, E.A., Douglas, E.L., Johansen, K., Millard, R.W., 1972. Aortic blood flow and cardiac output in the hemoglobin-free fish *Chaenocephalus aceratus*. Comp. Biochem. Physiol. 43A, 1045–1051.

Hillman, S.S., Hedrick, M.S., 2015. A meta-analysis of *in vivo* vertebrate cardiac performance: implications for cardiovascular support in the evolution of endothermy. J. Exp. Biol. 218, 1143–1150.

Hilton, Z., Clements, K.D., Hickey, A.J., 2010. Temperature sensitivity of cardiac mitochondria in intertidal and subtidal triplefin fishes. J. Comp. Physiol. B 180, 979–990.

Holeton, G.F., 1970. Oxygen uptake and circulation by a hemoglobinless Antarctic fish (*Chaenocephalus aceratus* Lonnberg) compared with three red-blooded Antarctic fish. Comp. Biochem. Physiol. 34, 457–471.

Holopainen, I.J., Tonn, W.M., Paszkowski, C.A., 1997. Tales of two fish: the dichotomous biology of crucian carp (*Carassius carassius* (L.)) in northern Europe. Ann. Zool. Fenn. 28, 1–22.

Iftikar, F.I., Hickey, A.J.R., 2013. Do mitochondria limit hot fish hearts? Understanding the role of mitochondrial function with heat stress in *Notolabrus celidotus*. PLoS One 8, e64120.

Iftikar, F.I., MacDonald, J.R., Baker, D.W., Renshaw, G.M.C., Hickey, A.J.R., 2014. Could thermal sensitivity of mitochondria determine species distributions in a changing climate? J. Exp. Biol. 217, 2348–2357.

Iftikar, F.I., Morash, A.J., Cook, D.G., Herbert, N.A., Hickey, A.J.R., 2015. Temperature acclimation of mitochondria function from the hearts of a temperate wrasse (*Notolabrus celidotus*). Comp. Biochem. Physiol. A Mol. Integr. Physiol. 184, 46–55.

Imbert-Auvray, N., Mercier, C., Huet, V., Bois, P., 2013. Sarcoplasmic reticulum: a key factor in cardiac contractility of sea bass *Dicentrarchus labrax* and common sole *Solea solea* during thermal acclimations. J. Comp. Physiol. B 183, 477–489.

Imbrogno, S., Cerra, M.C., 2017. Hormonal and autacoid control of cardiac function. In: Gamperl, A.K., Gillis, T.E., Farrell, A.P., Brauner, C.J. (Eds.), Fish Physiology. In: The Cardiovascular System: Morphology, Control and Function, vol. 36A. Academic Press, San Diego, pp. 265–315.

IPCC, Core Writing Team, 2014. In: Pachauri, R.K., Meyer, L.A. (Eds.), Climate Change 2014: Synthesis Report. Contribution of Working Groups I, II and III to the Fifth Assessment Report of the Intergovernmental Panel on Climate Change. IPCC, Geneva, Switzerland, p. 151.

Jayasundara, N., Somero, G.N., 2013. Physiological plasticity of cardiorespiratory function in a eurythermal marine teleost, the longjaw mudsucker, *Gillichthys mirabilis*. J. Exp. Biol. 216, 2111–2121.

Jayasundara, N., Gardner, L.D., Block, B.A., 2013. Effects of temperature acclimation on Pacific bluefin tuna (*Thunnus orientalis*) cardiac transcriptome. Am. J. Physiol. Regul. Integr. Comp. Physiol. 305, R1010–R1020.

Joaquim, N., Wagner, G.N., Gamperl, A.K., 2004. Cardiac function and critical swimming speed of the winter flounder (*Pleuronectes americanus*) at two temperatures. Comp. Biochem. Physiol. A Mol. Integr. Physiol. 138, 277–285.

Jørgensen, S.M., Castro, V., Krasnov, A., Torgersen, J., Timmerhaus, G., Hevrøy, E.M., Hansen, T.J., Susort, S., Breck, O., Takle, H., 2014. Cardiac responses to elevated seawater temperature in Atlantic salmon. BMC Physiol. 14, 2.

Kalinin, A., Gesser, H., 2002. Oxygen consumption and force development in turtle and trout cardiac muscle during acidosis and high extracellular potassium. J. Comp. Physiol. B 172, 145–151.

Keen, J.E., Farrell, A.P., 1994. Maximum prolonged swimming speed and maximum cardiac performance of rainbow trout acclimated to two different water temperatures. Comp. Biochem. Physiol. 108A, 287–295.

Keen, A.N., Gamperl, A.K., 2012. Blood oxygenation and cardiorespiratory function in steelhead trout (*Oncorhynchus mykiss*) challenged with an acute temperature increase and zatebradine-induced bradycardia. J. Therm. Biol. 37, 201–210.

Keen, J.E., Vianzon, D.M., Farrell, A.P., Tibbits, G.F., 1993. Thermal acclimation alters both adrenergic sensitivity and adrenoreceptor density in cardiac tissue of rainbow trout. J. Exp. Biol. 181, 27–48.

Keen, J.E., Vianzon, D.-M., Farrell, A.P., Tibbits, G.F., 1994. Effect of acute temperature change and temperature acclimation on excitation–contraction coupling in trout myocardium. J. Comp. Physiol. 164B, 438–443.

Keen, A.N., Fenna, A.J., McConnell, J.C., Sherratt, M.J., Gardner, P., Shiels, H.A., 2016. The dynamic nature of hypertrophic and fibrotic remodeling of the fish ventricle. Front. Physiol. 6, 427.

Keen, A.N., Klaiman, J.M., Shiels, H.A., Gillis, T.E., 2017. Temperature-induced cardiac remodelling in fish. J. Exp. Biol. 220, 147–160.

Kent, J., Koban, M., Prosser, C.L., 1988. Cold-acclimation-induced protein hypertrophy in channel catfish and green sunfish. J. Comp. Physiol. B 158, 185–198.

Klaiman, J.M., Fenna, A.J., Shiels, H.A., Macri, J., Gillis, T.E., 2011. Cardiac remodeling in fish: strategies to maintain heart function during temperature change. PLoS One 6, e24464.

Klaiman, J.M., Pyle, W.G., Gillis, T.E., 2014. Cold acclimation increases cardiac myofilament function and ventricular pressure generation in trout. J. Exp. Biol. 217, 4132–4140.

Kleckner, N.W., Sidell, B.D., 1985. Comparison of maximal activities of enzymes from tissues of thermally acclimated and naturally acclimatized chain pickerel (*Esox niger*). Physiol. Zool. 58, 18–28.

Kochová, P., Cimrman, R., Štengl, M., Ošťádal, B., Tonar, Z., 2015. A mathematical model of the carp heart ventricle during the cardiac cycle. J. Theor. Biol. 373, 12–25.

Kolok, A.S., Farrell, A.P., 1994. Individual Variation in the Swimming Performance and Cardiac Performance of Northern Squawfish, *Ptychocheilus oregonensis*. Physiol. Zool. 67, 706–722.

Kolok, A.S., Spooner, R.M., Farrell, A.P., 1993. The effect of exercise on the cardiac output and blood flow distribution of the largescale sucker *Catostomus macrocheilus*. J. Exp. Biol. 183, 301–322.

Korajoki, H., Vornanen, M., 2012. Expression of SERCA and phospholamban in rainbow trout (*Oncorhynchus mykiss*) heart: comparison of atrial and ventricular tissue and effects of thermal acclimation. J. Exp. Biol. 215, 1162–1169.

Korajoki, H., Vornanen, M., 2013. Temperature dependence of sarco(endo)plasmic reticulum Ca^{2+} ATPase expression in fish hearts. J. Comp. Physiol. B 183, 467–476.

Korajoki, H., Vornanen, M., 2014. Species- and chamber-specific responses of 12 kDa FK506-binding protein to temperature in fish heart. Fish Physiol. Biochem. 40, 539–549.

Korsmeyer, K.E., Lai, N.C., Shadwick, R.E., Graham, J.B., 1997. Heart rate and stroke volume contribution to cardiac output in swimming yellowfin tuna: response to exercise and temperature. J. Exp. Biol. 200, 1975–1986.

Kubly, K.L., Stecyk, J.A., 2015. Temperature-dependence of L-type Ca2+ current in ventricular cardiomyocytes of the Alaska blackfish (*Dallia pectoralis*). J. Comp. Physiol. B 185, 845–858.

Kyprianou, T.D., Pörtner, H.O., Anestis, A., Kostoglou, B., Feidantsis, K., Michaelidis, B., 2010. Metabolic and molecular stress responses of gilthead seam bream *Sparus aurata* during exposure to low ambient temperature: an analysis of mechanisms underlying the winter syndrome. J. Comp. Physiol. B 180, 1005–1018.

Lague, S.L., Speers-Roesch, B., Richards, J.G., Farrell, A.P., 2012. Exceptional cardiac anoxia tolerance in tilapia (*Oreochromis* hybrid). J. Exp. Biol. 215, 1354–1365.

Lannig, G., Bock, C., Sartoris, F.J., Portner, H.O., 2004. Oxygen limitation of thermal tolerance in cod, *Gadus morhua* L., studied by magnetic resonance imaging and on-line venous oxygen monitoring. Am. J. Physiol. Regul. Integr. Comp. Physiol. 287, R902–R910.

LeBlanc, S., Middleton, S., Gilmour, K.M., Currie, S., 2011. Chronic social stress impairs thermal tolerance in rainbow trout (*Oncorhynchus mykiss*). J. Exp. Biol. 214, 1721–1731.

LeBlanc, S., Hoglund, E., Gilmour, K.M., Currie, S., 2012. Hormonal modulation of the heat shock response: insights from fish with divergent cortisol stress responses. Am. J. Physiol. Regul. Integr. Comp. Physiol. 302, R184–R192.

Lee, L., Genge, C.E., Cua, M., Sheng, X., Rayani, K., et al., 2016. Correction: functional assessment of cardiac responses of adult zebrafish (*Danio rerio*) to acute and chronic temperature change using high-resolution echocardiography. PLoS One 11, e0145163.

Lefevre, S., Damsgaard, C., Pascale, D.R., Nilsson, G.E., Stecyk, J.A.W., 2014. Air breathing in the Arctic: influence of temperature, hypoxia, activity and restricted air access on respiratory physiology of the Alaska blackfish *Dallia pectoralis*. J. Exp. Biol. 217, 4387–4398.

Lefevre, S., Findorf, I., Bayley, M., Huong, D.T., Wang, T., 2016. Increased temperature tolerance of the air-breathing Asian swamp eel *Monopterus albus* after high-temperature acclimation is not explained by improved cardiorespiratory performance. J. Fish Biol. 88, 418–432.

Lemieux, H., Tardif, J.-C., Dutil, J.-D., Blier, P.U., 2010. Thermal sensitivity of cardiac mitochondrial metabolism in an ectothermic species from a cold environment Atlantic wolffish (*Anarhichas lupus*). J. Exp. Mar. Biol. Ecol. 384, 113–118.

Little, A.G., Seebacher, F., 2014. Thyroid hormone regulates cardiac performance during cold acclimation in zebrafish (*Danio rerio*). J. Exp. Biol. 217, 718–725.

Lurman, G.J., Petersen, L.H., Gamperl, A.K., 2012. In situ cardiac performance of Atlantic cod (*Gadus morhua*) at cold temperatures: long-term acclimation, acute thermal challenge and the role of adrenaline. J. Exp. Biol. 215, 4006–4014.

Mailer, K., 1990. Superoxide radical as electron donor for oxidative phosphorylation of ADP. Biochem. Biophys. Res. Commun. 170, 59–64.

Martins, E.G., Hinch, S.G., Patterson, D.A., Hague, M.J., Cooke, S.J., Miller, K.M., Lapointe, M.F., English, K.K., Farrell, A.P., 2011. Effects of river temperature and climate warming on stock-specific survival of adult migrating Fraser River sockeye salmon (*Oncorhynchus nerka*). Glob. Chang. Biol. 17, 99–114.

Martins, E.G., Hinch, S.G., Patterson, D.A., Hague, M.J., Cooke, S.J., Miller, K.M., Robichaud, D., English, K.K., Farrell, A.P., 2012. High river temperature reduces survival of sockeye salmon (*Oncorhynchus nerka*) approaching spawning grounds and exacerbates female mortality. Can. J. Fish. Aquat. Sci. 69, 330–342.

Matikainen, N., Vornanen, M., 1992. Effect of season and temperature acclimation on the function of crucian carp (*Carassius carassius*) heart. J. Exp. Biol. 167, 203–220.

McBryan, T.L., Anttila, K., Healy, T.M., Schulte, P.M., 2013. Responses to temperature and hypoxia as interacting stressors in fish: implications for adaptation to environmental change. Integr. Comp. Biol. 53, 648–659.

McClelland, G.B., Dalziel, A.C., Fragoso, N.M., Moyes, C.D., 2005. Muscle remodeling in relation to blood supply: implications for seasonal changes in mitochondrial enzymes. J. Exp. Biol. 208, 515–522.

Mendonça, P.C., Gamperl, A.K., 2009. Nervous and humoral control of cardiac performance in the winter flounder (*Pleuronectes americanus*). J. Exp. Biol. 212, 934–944.

Mendonça, P.C., Gamperl, A.K., 2010. The effects of acute changes in temperature and oxygen availability on cardiac performance in winter flounder (*Pseudopleuronectes americanus*). Comp. Biochem. Physiol. A Mol. Integr. Physiol. 155, 245–252.

Morgan, A.L., Thompson, K.D., Auchinachie, N.A., Migaud, H., 2008. The effect of seasonality on normal haematological and innate immune parameters of rainbow trout *Oncorhynchus mykiss* L. Fish Shellfish Immunol. 25, 791–799.

Moyes, C.D., Ballantyne, J.S., 2011. Membranes and temperature: homoviscous adaptation. In: Farrell, A.P. (Ed.), Encyclopedia of Fish Physiology: From Genome to Environment. Elsevier, Amsterdam, pp. 1725–1731.

Mulhollem, J.J., Suski, C.D., Wahl, D.H., 2015. Response of largemouth bass (*Micropterus salmoides*) from different thermal environments to increased water temperature. Fish Physiol. Biochem. 41, 833–842.

Munday, P.L., Jones, G.P., Pratchett, M.S., Williams, A.J., 2008. Climate change and the future for coral reef fishes. Fish Fish. 9, 261–285.

Muñoz, N.J., Farrell, A.P., Heath, J.W., Neff, B.D., 2015. The adaptive capacity of a Pacific salmon challenged by climate change. Nat. Clim. Chang. 5, 163–166.

Nofrizal, Yanase, K., Arimoto, T., 2009. Effect of temperature on the swimming endurance and post-exercise recovery of jack mackerel *Trachurus japonicus* as determined by ECG monitoring. Fish. Sci. 75, 1369–1375.

Olson, K.R., 2011. Integrated control and response of the circulatory system. In: Farrell, A.P. (Ed.), Encyclopedia of Fish Physiology: From Genome to Environment. Elsevier, Amsterdam, pp. 1169–1177.

Olsson, H.I., Yee, N., Shiels, H.A., Brauner, C., Farrell, A.P., 2000. A comparison of myocardial β-adrenoreceptor density and ligand binding affinity among selected teleost fishes. J. Comp. Physiol. B 170, 545.

Overgaard, J., Stecyk, J.A., Gesser, H., Wang, T., Farrell, A.P., 2004. Effects of temperature and anoxia upon the performance of in situ perfused trout hearts. J. Exp. Biol. 207, 655–665.

Ozolina, K., Shiels, H.A., Ollivier, H., Claireaux, G., 2016. Intraspecific individual variation of temperature tolerance associated with oxygen demand in the European sea bass (*Dicentrarchus labrax*). Conserv. Physiol. 4, cov060.

Pascoli, F., Lanzanoa, G.S., Negratoa, E., Poltronieria, C., Trocinob, A., Radaellia, G., Bertottoa, D., 2011. Seasonal effects on hematological and innate immune parameters in sea bass *Dicentrarchus labrax*. Fish Shellfish Immunol. 31, 1081e1087.

Penney, C.M., Nash, G.W., Gamperl, A.K., 2014. Cardiorespiratory responses of seawater-acclimated adult Arctic char (*Salvelinus alpinus*) and Atlantic salmon (*Salmo salar*) to an acute temperature increase. Can. J. Fish. Aquat. Sci. 71, 1096–1105.

Petersen, L.H., Gamperl, A.K., 2010. *In situ* cardiac function in Atlantic cod (*Gadus morhua*): effects of acute and chronic hypoxia. J. Exp. Biol. 213, 820–830.

Petersen, L.H., Dzialowski, E., Huggett, D.B., 2011. The interactive effects of a gradual temperature decrease and long-term food deprivation on cardiac and hepatic blood flows in rainbow trout (*Oncorhynchus mykiss*). Comp. Biochem. Physiol. A Mol. Integr. Physiol. 160, 311–319.

Pichaud, N., Ekström, A., Hellgren, K., Sandblom, E., 2017. Dynamic changes in cardiac mitochondrial metabolism during warm acclimation in rainbow trout. J. Exp. Biol. 220, 1674–1683.

Randall, D.J., Perry, S.F., 1992. Catecholamines. In: Hoar, W.S., Randall, D.J., Farrell, A.P. (Eds.), The Cardiovascular System. In: Fish Physiology, vol. XII, Part B. Academic Press, New York, pp. 255–300.

Roche, H., Bogé, G., 1996. Fish blood parameters as a potential tool for identification of stress caused by environmental factors and chemical intoxication. Mar. Environ. Res. 41, 27–43.

Rodnick, K.J., Gesser, H., 2017. Cardiac energy metabolism. In: Gamperl, A.K., Gillis, T.E., Farrell, A.P., Brauner, C.J. (Eds.), Fish Physiology. In: The Cardiovascular System: Morphology, Control and Function, vol. 36A. Academic Press, San Diego, pp. 317–367.

Rodnick, K.J., Sidell, B.D., 1994. Cold acclimation increases carnitine palmitoyltransferase I activity in oxidative muscle of striped bass. Am. J. Physiol. Regul. Integr. Comp. Physiol. 266, R405–R412.

Rodnick, K.J., Gamperl, A.K., Nash, G.W., Syme, D.A., 2014. Temperature and sex dependent effects on cardiac mitochondrial metabolism in Atlantic cod (*Gadus morhua* L.). J. Therm. Biol. 44, 110–118.

Ruiz, M.A.M., Thorarensen, H., 2001. Genetic and environmental effects on the size of the cardio-respiratory organs in Arctic charr (*Salvelinus alpinus*). In: Abstract, Voluntary Food Intake in Fish, COST 827 Workshop. Reykjavik, Iceland.

Rummer, J.L., Couturier, C.S., Stecyk, J.A.W., Gardiner, N.M., Kinch, J.P., Nilsson, G.E., Munday, P.L., 2014. Life on the edge: thermal optima for aerobic scope of equatorial reef fishes are close to current day temperatures. Glob. Chang. Biol. 20, 1055–1066.

Salama, A., Nikinmaa, M., 1988. The adrenergic responses of carp (*Cyprinus carpio*) red cells: effects of PO2 and pH. J. Exp. Biol. 136, 405–416.

Sandblom, E., Axelsson, M., 2007. Venous hemodynamic responses to acute temperature increase in the rainbow trout (*Oncorhynchus mykiss*). Am. J. Physiol. Regul. Integr. Comp. Physiol. 292, R2292–R2298.

Sandblom, E., Axelsson, M., 2011. Autonomic control of circulation in fish: a comparative view. Auton. Neurosci. 165, 127–139.

Sandblom, E., Gräns, A., 2017. Form, function, and control of the vasculature. In: Gamperl, A.K., Gillis, T.E., Farrell, A.P., Brauner, C.J. (Eds.), Fish Physiology. In: The Cardiovascular System: Morphology, Control and Function, vol. 36A. Academic Press, San Diego, pp. 369–433.

Sandblom, E., Cox, G.K., Perry, S.F., Farrell, A.P., 2009. The role of venous capacitance, circulating catecholamines, and heart rate in the hemodynamic response to increased temperature and hypoxia in the dogfish. Am. J. Physiol. Regul. Integr. Comp. Physiol. 296, R1547–1556.

Sandblom, E., Clark, T.D., Gräns, A., Ekström, A., Brijs, J., Sundström, L.F., Odelström, A., Adill, A., Aho, T., Jutfelt, F., 2016a. Physiological constraints to climate warming in fish follow principles of plastic floors and concrete ceilings. Nat. Commun. 7, 11447.

Sandblom, E., Ekström, A., Brijs, J., Sundström, L.F., Jutfelt, F., Clark, T.D., Adill, A., Aho, T., Gräns, A., 2016b. Cardiac reflexes in awarming world: thermal plasticity of barostatic control and autonomic tones in a temperate fish. J. Exp. Biol. 219, 2880–2887.

Sartoris, F.J., Bock, C., Serendero, I., Lannig, G., Portner, H.O., 2003. Temperature-dependent changes in energy metabolism, intracellular pH and blood oxygen tension in the Atlantic cod. J. Fish Biol. 62, 1239–1253.

Seebacher, F., Davison, W., Lowe, C.J., Franklin, C.E., 2005. A falsification of the thermal specialization paradigm: compensation for elevated temperatures in Antarctic fishes. Biol. Lett. 1, 151–154.

Seibert, H., 1979. Thermal adaptation of heart rate and its parasympathetic control in the European eel *Anguilla anguilla* (L.). Comp. Biochem. Physiol. 64C, 275–278.

Sephton, D.H., Driedzic, W.R., 1991. Effect of acute and chronic temperature transition on enzymes of cardiac metabolism in white perch (*Morone americanus*), yellow perch (*Perca jlavescens*), and smallmouth bass (*Micropterus dolomieui*). Can. J. Zool. 69, 258–262.

Shiels, H.A., 2017. Cardiomyocyte morphology and physiology. In: Gamperl, A.K., Gillis, T.E., Farrell, A.P., Brauner, C.J. (Eds.), Fish Physiology. In: The Cardiovascular System: Morphology, Control and Function, vol. 36A. Academic Press, San Diego, pp. 55–98.

Shiels, H.A., Farrell, A.P., 1997. The effect of temperature and adrenaline on the relative importance of the sarcoplasmic reticulum in contributing Ca^{2+} to force development in isolated ventricular trabeculae from rainbow trout. J. Exp. Biol. 200, 1607–1621.

Shiels, H.A., Farrell, A.P., 2000. The effect of ryanodine on isometric tension development in isolated ventricular trabeculae from Pacific mackerel (*Scomber japonicus*). Comp. Biochem. Physiol. A Mol. Integr. Physiol. 125, 331–341.

Shiels, H.A., Vornanen, M., Farrell, A.P., 2002. The force–frequency relationship in fish hearts—a review. Comp. Biochem. Physiol. A Mol. Integr. Physiol. 132, 811–826.

Shiels, H.A., Vornanen, M., Farrell, A.P., 2003. Acute temperature change modulates the response of ICa to adrenergic stimulation in fish cardiomyocytes. Physiol. Biochem. Zool. 76, 816–824.

Shiels, H.A., Galli, G.L., Block, B.A., 2015. Cardiac function in an endothermic fish: cellular mechanisms for overcoming acute thermal challenges during diving. Proc. Biol. Sci. 282, 20141989.

Sidell, B.D., O'Brien, K.M., 2006. When bad things happen to good fish: the loss of hemoglobin and myoglobin expression in Antarctic icefishes. J. Exp. Biol. 209, 1791–1802.

Sidhu, R., Anttila, K., Farrell, A.P., 2014. Upper thermal tolerance of closely related Danio species. J. Fish Biol. 84, 982–995.

Stecyk, J.A.W., Farrell, A.P., 2002. Cardiorespiratory responses of the common carp (*Cyprinus carpio*) to severe hypoxia at three acclimation temperatures. J. Exp. Biol. 205, 759–768.

Stecyk, J.A.W., Farrell, A.P., 2006. Regulation of the cardiorespiratory system of common carp (*Cyprinus carpio*) during severe hypoxia at three seasonal acclimation temperatures. Physiol. Biochem. Zool. 79, 614–627.

Stecyk, J.A.W., Stensløkken, K.O., Farrell, A.P., Nilsson, G.E., 2004. Maintained cardiac pumping in anoxic crucian carp. Science 306, 77.

Steffensen, J.F., Farrell, A.P., 1998. Swimming performance, venous oxygen tension and cardiac performance of coronary-ligated rainbow trout, *Oncorhynchus mykiss*, exposed to progressive hypoxia. Comp. Biochem. Physiol. 119A, 585–592.

Steinhausen, M.F., Sandblom, E., Eliason, E.J., Verhille, C., Farrell, A.P., 2008. The effect of acute temperature increases on the cardiorespiratory performance of resting and swimming sockeye salmon (*Oncorhynchus nerka*). J. Exp. Biol. 211, 3915–3926.

Stevens, E.D., Bennion, G.R., Randall, D.J., Shelton, G., 1972. Factors affecting arterial pressures and blood flow from the heart in intact, unrestrained lingcod, *Ophiodon elongatus*. Comp. Biochem. Physiol. A Physiol. 43, 681–695.

Sureau, D., Lagardere, J.P., Pennect, J.P., 1989. Heart rate and its cholinergic control in the Sole (*Solea vulgaris*) acclimatized to different temperatures. Comp. Biochem. Physiol. A Physiol. 92, 49–51.

Szewczyk, A., Jarmuszkiewicz, W., Koziel, A., Sobieraj, I., Nobik, W., Lukasiak, A., Skup, A., Bednarczyk, P., Drabarek, B., Dymkowska, D., Wrzosek, A., Zablocki, K., 2015. Mitochondrial mechanisms of endothelial dysfunction. Pharmacol. Rep. 67, 704–710.

Takei, Y., Tsukada, T., 2001. Ambient temperature regulates drinking and arterial pressure in eels. Zoolog. Sci. 18, 963–967.

Taylor, S.E., Egginton, S., Taylor, E.W., 1993. Respiratory and cardiovascular responses in rainbow trout (*Oncorhynchus mykiss*) to aerobic exercise over a range of acclimation temperatures. J. Physiol. 459, 19.

Taylor, S., Egginton, S., Taylor, E., 1996. Seasonal temperature acclimatisation of rainbow trout: cardiovascular and morphometric influences on maximal sustainable exercise level. J. Exp. Biol. 199, 835–845.

Templeman, N.M., LeBlanc, S., Perry, S.F., Currie, S., 2014. Linking physiological and cellular responses to thermal stress: β-adrenergic blockade reduces the heat shock response in fish. J. Comp. Physiol. B 184, 719–728.

Tiitu, V., Vornanen, M., 2001. Cold adaptation suppresses the contractility of both atrial and ventricular muscle of the crucian carp (*Carassius carassius* L.) heart. J. Fish Biol. 59, 141–156.

Tiitu, V., Vornanen, M., 2002. Regulation of cardiac contractility in a cold stenothermal fish, the burbot *Lota lota* L. J. Exp. Biol. 205, 1597–1606.

Tota, B., Acierno, R., Agnisola, C., 1991. Mechanical performance of the isolated and perfused heart of the haemoglobinless Antarctic icefish *Chionodraco hamatus* (Lonnberg): effects of loading conditions and temperature. Philos. Trans. R. Soc. B 332, 191–198.

Tsukuda, H., Liu, B., Fujii, K.-I., 1985. Pulsation rate and oxygen consumption of isolated hearts of the goldfish, *Carassius auratus*, acclimated to different temperatures. Comp. Biochem. Physiol. A Physiol. 82, 281–283.

Valenzuela, A.E., Silva, V.M., Klempau, A.E., 2008. Effects of different artificial photoperiods and temperatures on haematological parameters of rainbow trout (*Oncorhynchus mykiss*). Fish Physiol. Biochem. 34, 159–167.

Vornanen, M., 2016. The temperature dependence of electrical excitability in fish hearts. J. Exp. Biol. 219, 1941–1952.

Vornanen, M., 2017. Electrical excitability of the fish heart and its autonomic regulation. In: Gamperl, A.K., Gillis, T.E., Farrell, A.P., Brauner, C.J. (Eds.), Fish Physiology. In: The Cardiovascular System: Morphology, Control and Function, vol. 36A. Academic Press, San Diego, pp. 99–153.

Vornanen, M., Shiels, H.A., Farrell, A.P., 2002. Plasticity of excitation–contraction coupling in fish cardiac myocytes. Comp. Biochem. Physiol. A Mol. Integr. Physiol. 132, 827–846.

Vornanen, M., Hassinen, M., Koskinen, H., Krasnov, A., 2005. Steady-state effects of temperature acclimation on the transcriptome of the rainbow trout heart. Am. J. Physiol. Regul. Integr. Comp. Physiol. 289, R1177–R1184.

Vornanen, M., Haverinen, J., Egginton, S., 2014. Acute heat tolerance of cardiac excitation in the brown trout (*Salmo trutta fario*). J. Exp. Biol. 217, 299–309.

Wang, T., Lefevre, S., Iversen, N.K., Findorf, I., Buchanan, R., McKenzie, D.J., 2014. Anaemia only causes a small reduction in the upper critical temperature of sea bass: is oxygen delivery the limiting factor for tolerance of acute warming in fishes? J. Exp. Biol. 217, 4275–4278.

Whiteley, N.M., Christiansen, J.S., Egginton, S., 2006. Polar cod, *Boreogadus saida* (Gadidae), show an intermediate stress response between Antarctic and temperate fishes. Comp. Biochem. Physiol. A Mol. Integr. Physiol. 145, 493–501.

Wood, C.M., Pieprzak, P., Trott, J.N., 1979. Influence of temperature and anemia on the adrenergic and cholinergic mechanisms controlling heart rate in the rainbow trout. Can. J. Zool. 57, 2440–2447.

5

CARDIOVASCULAR RESPONSES TO LIMITING OXYGEN LEVELS

JONATHAN A.W. STECYK[1]

Department of Biological Sciences, University of Alaska Anchorage, Anchorage, AK, United States
[1]Corresponding author: jstecyk@alaska.edu

The capacity for metabolic energy (adenosine triphosphate, ATP) production via oxidative phosphorylation is much greater (~15×) than by anaerobic glycolysis, and vertebrates have, therefore, evolved efficient respiratory and

The Cardiovascular System: Development, Plasticity and Physiological Responses, Volume 36B
FISH PHYSIOLOGY

DOI: https://doi.org/10.1016/bs.fp.2017.09.005

circulatory systems to take oxygen up from the environment and transport it to the tissues. Consequently, reduced oxygen availability (hypoxia) or no oxygen (anoxia) in the environment poses a significant challenge to most vertebrates. For example, the vertebrate heart itself, being an aerobic organ with a high ATP demand, is inherently sensitive to oxygen limitation. The freshwater, estuarine and marine aquatic ecosystems in which fish reside are prone to hypoxia and low dissolved oxygen conditions of varying severity, periodicity and duration. Thus, hypoxia is a common environmental challenge faced by fishes. The focus of this chapter is two-fold. A significant proportion of the chapter summarizes the cardiovascular responses to oxygen limitation and their regulation in water-breathing and air-breathing fishes intolerant of surviving prolonged periods of severe hypoxia or anoxia, as well as those species that can survive prolonged periods of very little to no oxygen. Second, the survival strategies that the severe hypoxia/anoxia-tolerant species employ to endure prolonged periods of oxygen deprivation are described. Here, the differing anoxia survival strategies utilized by severe hypoxia/anoxia-tolerant species to balance cardiac ATP supply and demand, and to cope with metabolic wastes, are highlighted. In this section, there is special emphasis on the anoxic survival strategies used by one of the champions of vertebrate anoxia survival, the crucian carp (*Carassius carassius*). This species spends up to half of its life deprived of oxygen.

1. INTRODUCTION

1.1. Oxygen: The Molecule of Life

Oxygen is of utmost importance for the survival of most multicellular life forms on Earth, and its presence in cells is intricately coupled to the efficient production of the metabolic energy (adenosine triphosphate, ATP) that is required to power endergonic biochemical cellular processes. In fact, ATP production is most efficient when carbohydrate, lipid or amino acid fuels are completely catabolized to CO_2 and H_2O in the presence of oxygen (Hochachka and Somero, 2002). For example, the complete oxidation of 1 mol of glucose yields 29 mol of ATP (Brand, 2003). By comparison, the only major biochemical pathway for ATP production in the absence of oxygen, the oxygen-independent enzyme reactions of glycolysis, yield only 2 mol of ATP per 1 mol of glucose. This equates to approximately 6%–10% of the ATP yield of oxidative phosphorylation, depending on how well the mitochondria are coupled under normoxic conditions (Brand, 2003; Hochachka and Somero, 2002). It is not surprising, then, that: (1) the vast majority of vertebrate species are highly dependent upon molecular oxygen for survival, a phenomenon reflected in the evolution of efficient respiratory and circulatory systems to ensure adequate oxygen delivery to the tissues; and (2) most vertebrate species are unable to successfully cope with prolonged periods of reduced oxygen availability (hypoxia) or no oxygen (anoxia).

Death from oxygen deficiency is primarily attributed to a mismatch between anaerobic ATP production and ATP demand, the disruption of cellular processes associated with this energetic shortfall, and the waste products of anaerobic glycolysis, especially in critical tissues such as the brain and heart that have a high ATP demand. With oxygen deprivation, anoxia-intolerant species attempt to maintain normoxic cellular ATP levels through the rapid depletion of phosphogen reserves (Lutz and Storey, 1997) and the upregulation of anaerobic ATP supply through glycolysis; a phenomenon termed the Pasteur effect (Storey, 1985). However, phosphogens are almost instantaneously depleted (Lutz and Storey, 1997) and finite glycogen stores are also rapidly diminished with augmented glycolytic flux, which to match aerobic ATP production rates would need to be upregulated ~14-fold (Storey, 1985). Therefore, ATP production soon falls passively with oxygen deprivation. This has several consequences. For example, ATP-dependent ion-motive pumps such as the Na^+/K^+-ATPase fail, and this results in the disruption of cellular resting membrane potentials and loss of the ionic integrity of cellular membranes (Boutilier, 2001). The subsequent drift of intracellular and extracellular ions toward their thermodynamic equilibrium eventually depolarizes the cell membrane and leads to the uncontrolled influx of Ca^{2+} through voltage-gated Ca^{2+} channels (Hochachka, 1986). This increase in intracellular Ca^{2+} concentration activates Ca^{2+}-dependent phospholipases and proteases that perpetuate the rate of membrane depolarization, leading to cellular swelling and, ultimately, cell death (i.e., necrosis) (Boutilier, 2001).

The vertebrate heart has an ATP requirement greatly exceeding that of most other tissues. For the heart to function as a muscular pump, a continual ATP supply is needed to power myosin-ATPase. ATP is also essential to power the various ATP-dependent ion-motive pumps (i.e., Na^+/K^+-ATPase and Ca^{2+}-ATPases) that are requisite for repeated action potential (AP) generation, intracellular Ca^{2+} homeostasis, and membrane ion transport (Aho and Vornanen, 1997; Huss and Kelly, 2005; Rolfe and Brown, 1997; Taha and Lopaschuk, 2007). For mammalian hearts, mechanical activity has been estimated to account for 75%–85% of cardiac ATP demand, whereas the Na^+/K^+-ATPase and Ca^{2+}-ATPases account for approximately 15%–25% (Rolfe and Brown, 1997; Schramm et al., 1994), and it is assumed that ~2% of the cellular ATP pool is consumed in each heartbeat (Balaban, 2002). In the fish heart, the cardiac ATP budget may differ from mammals due to differences in cellular structure and Ca^{2+} management (Aho and Vornanen, 1997; Santer, 1985). However, the energetic cost of mechanical work is still likely to be the greater fraction of total cardiac ATP expenditure because the myofibrillar volume density of cardiomyocytes is similar in both fish and mammals (Aho and Vornanen, 1997; Santer, 1985). Thus, if ATP supply is not matched to ATP demand, the heart's ATP pool will quickly be depleted, and this will lead to cardiomyocyte death by necrosis, cardiac failure, and ultimately the death of the organism.

An additional contributing factor to anoxic cardiac failure is the accumulation of protons from anaerobic metabolism (acidosis). Acidosis dramatically decreases the ability of the heart to pump blood by reducing contractile force and promoting fatal ventricular arrhythmias (Gesser and Jørgensen, 1982; Orchard and Kentish, 1990; Williamson et al., 1976). Moreover, acidosis combined with oxygen deprivation can trigger apoptotic cell death (Graham et al., 2004; Hochachka et al., 1996). Briefly, in mammalian cardiac myocytes, the accumulation of hypoxia-inducible factor-1 during hypoxia induces the transcription of the death-promoting *BNIP3* gene. At neutral pH, *BNIP3* is inactive. However, with acidosis, *BNIP3* is translocated from the cytosolic compartment into the mitochondrial membrane where it stimulates opening of the mitochondrial permeability transition pore. This leads to the release of the apoptosis-inducing factors, cytochrome *c*, and Ca^{2+}, which stimulate proteases and DNases involved in cell death.

1.2. Hypoxia and Anoxia in the Aquatic Environment

While hypoxia, and even more so anoxia, are rare conditions in terrestrial ecosystems (apart from high-altitude environments, which are always hypoxic), hypoxia and anoxia are prevalent in freshwater and marine aquatic ecosystems (Diaz and Breitburg, 2009). The high incidence of hypoxia in aquatic ecosystems is due to several factors. Foremost, the physicochemical properties of water and air differ tremendously. In general, water holds 20–40 times less oxygen than air, depending on temperature and salinity (colder water holds more oxygen than warmer water, and the solubility of oxygen decreases with increased salinity), and the diffusion of oxygen is about 10,000 times slower in water as compared to air. The relatively small amount of oxygen in water, coupled with its slow movement, contributes to the development of aquatic hypoxia, especially in waters that circulate poorly or are stagnant. In addition, temperature- and density-dependent stratification of the water column can limit oxygen exchange between the surface water that is in contact with the atmosphere and deeper water, leading to the stratification of oxygen levels. The presence of hypoxia at the bottom of freshwater lakes and ponds is further facilitated by the decomposition of organic matter and/or ice and snow cover. Not only does ice and snow cover prevent diffusion of atmospheric oxygen into the water column, it also restricts the photosynthetic production of oxygen. At the extreme, small lakes and ponds in northern temperate environments can become completely anoxic for weeks to months in the winter because of these "winter-kill" conditions. In tropical climates, reduced precipitation in the dry season curtails river water flow, thereby isolating bodies of water. At night, dissolved oxygen levels in these isolated bodies of water decrease because photosynthesis by aquatic plants and algae is curtailed while respiration is increased (Val and Almeida-Val, 1995).

In the marine environment, hypoxia can be localized or widespread. On a local scale, intertidal pools of the nearshore marine environment become isolated from the ocean for periods of time during each tidal cycle. Consequently, they can exhibit large fluctuations in temperature, pH and oxygen (Richards, 2011). Similarly, waters of tropical coral reefs can be periodically cut off from the surrounding ocean during low tide (Nilsson and Ostlund-Nilsson, 2004). When this occurs, respiration by coral makes the enclosed water hypoxic. In addition, the microhabitat within branching coral that many fish species utilize to feed or escape from predators has been shown to be a low-oxygen environment (Nilsson and Ostlund-Nilsson, 2004). On a broad scale, low-oxygen conditions are associated with naturally occurring oceanic oxygen minimum zones and coastal upwelling zones. Oceanic oxygen minimum zones are extensive and stable. They persist for decades, have oxygen levels below normal levels (i.e., from 4 to 6 mg O_2 L^{-1} to as low as 0.1 mg O_2 L^{-1}), and occur at depths between 400 and 1000 m (Diaz and Breitburg, 2009; Helly and Levin, 2004). These zones are distributed worldwide along continental margins where surface productivity is high and circulation is slow. Hypoxia develops because of high primary production in surface water that sinks, and is subsequently degraded by microbial processes that consume the available oxygen. Coastal upwelling zones are associated with the eastern boundaries of the Pacific and Atlantic oceans. These areas are highly productive due to natural forces (i.e., offshore winds and the earth's rotation) that move surface water offshore necessitating its replacement by nutrient-rich deeper waters (i.e., the upwelling) (Diaz and Rosenberg, 2008). Much like in oxygen minimization zones, hypoxia (<0.5 mL O_2 L^{-1}) develops in coastal upwelling zones because the high productivity that ensues in the nutrient-rich surface water eventually sinks and decomposes (Diaz and Breitburg, 2009). However, hypoxia in coastal upwelling zones is not as persistent as the hypoxia in oxygen minimization zones.

Finally, anthropogenic influences contribute to the prevalence of hypoxia in both freshwater and marine environments. Excessive nutrient release into the environment through human-related activities leads to rapid eutrophication. Levels of dissolved oxygen in the water are depleted when the explosive production of organic matter is decomposed by microbial metabolism. Excess eutrophication has led to the formation of "dead zones" in coastal environments, as well as large lakes, around the world (Diaz and Rosenberg, 2008; Dybas, 2005).

Overall, aquatic hypoxia and anoxia occur in a wide range of aquatic systems, both naturally and due to anthropogenic causes. Moreover, the frequency and duration of oxygen deprivation is variable. Oxygen depletion can be aperiodic (<1 event per year), periodic (>1 event per year), diel (1 event per day), or seasonal (1 event per year), and can last for a few hours or even persist for multiple years (Diaz and Breitburg, 2009; Diaz and Rosenberg, 2008).

1.3. Focus of the Chapter

Regardless of why, how, or when hypoxia or anoxia occur, oxygen-limiting conditions present a stressor to fish to which they must respond. Given the variability in the severity, periodicity, and duration of environmental oxygen limitation, as well as the great diversity of fishes [at more than 32,000 species (Nelson, 2016)], it is not surprising that differing tolerances to oxygen deprivation are exhibited (e.g., survival times range from a few minutes to many weeks) and that a wide array of physiological, anatomical and behavioral adjustments and adaptations to environmental hypoxia are displayed. In addition, the responses to oxygen deprivation are life-stage specific and can be further modulated by other confounding environmental factors. For example, the amount of time that low-oxygen conditions can be tolerated by fish increases as ambient temperature, and therefore body temperature, decreases (Blazka, 1958, 1960; Piironen and Holopainen, 1986). The primary goal of this chapter is to present an overview of the cardiovascular responses that fish display when faced with oxygen limitation (for comprehensive reviews of the behavioral, morphological, physiological, metabolic, molecular and biochemical responses, and the adaptive strategies used by fishes to survive low-oxygen conditions, the reader is referred to Volume 27 of *Fish Physiology*; Chapman and McKenzie, 2009; Farrell and Richards, 2009; Perry et al., 2009a; Pörtner and Lannig, 2009; Richards, 2009; Wang et al., 2009; Wells, 2009; Wu, 2009). Volume 27 of *Fish Physiology* also includes an exceptional synopsis of cardiovascular function and cardiac energetics in hypoxic fish that is broadly organized by the duration of hypoxia exposure (i.e., acute vs chronic; Gamperl and Driedzic, 2009). To avoid undue redundancy with this earlier review, and to emphasize recent advancements that have been made in the study of fish cardiovascular responses to limiting oxygen levels, this chapter is organized so that it compares and contrasts the responses of severe hypoxia/anoxia-intolerant and severe hypoxia/anoxia-tolerant species to oxygen limitation. In this regard, emphasis is placed on the cardiovascular responses of adult fish to oxygen deprivation. Severe hypoxia/anoxia-intolerant species are defined as those that cannot survive severe hypoxia (water oxygen level of less than 1 kPa = 7.5 mmHg, 0.44 mg O_2 L^{-1}, or 4.82% of air saturation in freshwater at 20°C; = 7.5 mmHg, 0.61 mg O_2 L^{-1}, or 4.75% of air saturation in freshwater at 5°C) for more than a few minutes even at low acclimation temperatures, whereas severe hypoxia/anoxia-tolerant species are defined as those that can survive severe hypoxia/anoxia exposure for hours, days or weeks (depending on acclimation temperature). Moreover, severe hypoxia/anoxia-tolerant species are defined as those whose cardiac activity reaches a new steady state during prolonged severe hypoxia/anoxia exposure.

The species highlighted in this chapter as the principle examples of tolerance to severe oxygen deprivation include the Pacific hagfish (*Eptatretus stoutii*), common carp (*Cyprinus carpio*) and crucian carp (*Carassius carassius*). Hagfish encounter hypoxic and anoxic environments at the great depths where they reside, as well as while feeding on, or from within, decomposing carcasses (Hansen and Sidell, 1983; Martini, 1998; Perry et al., 1993, 2009b). In the laboratory, Pacific hagfish can survive and recover from an impressive 36 h of anoxia at 10°C (Cox et al., 2010, 2011). Common carp inhabit shallow ponds that are highly eutrophic and become diurnally hypoxic, especially in the summer months when aquatic plant respiration is high and the oxygen content of water is already low (Beamish, 1964; Garey and Rahn, 1970; Scott and Crossman, 1973). The distribution of common carp also extends into northern latitudes where ice and snow cover of lakes may occur for prolonged periods during the winter months (McCrimmon, 1968). In the laboratory, common carp tolerate hours of severe hypoxia, but remain motionless on the bottom of aquaria and ventilate very shallowly at a normal rate (Stecyk and Farrell, 2002, 2006). Temperature plays a central role in the survival time of common carp in severely hypoxic water. It is ~3 h at 15°C, ~ 6 h at 10°C and ~24 h at 5°C (Stecyk and Farrell, 2002, 2006). The crucian carp is undoubtedly one of the champions of vertebrate anoxia survival (Vornanen et al., 2009). These fish overwinter under ice in ponds that become progressively hypoxic, and ultimately anoxic, for many months (Vornanen and Paajanen, 2004). Consequently, this fish exhibits the remarkable ability to survive days of anoxia at room temperature, and several months of anoxia at temperatures near 0°C (Blazka, 1958, 1960; Piironen and Holopainen, 1986).

Further, there are additional species that are tolerant of severely low-oxygen conditions, and available data from these species will also be presented throughout the chapter, when relevant. In particular, the goldfish (*Carassius auratus*) exhibits a profound, but lesser, anoxia tolerance than its close relative the crucian carp (Van Waversveld et al., 1989b). The tilapia (*Oreochromis* hybrid: *Oreochromis niloticus* × *mossambicus hornorum*) recovers from 8 h of severe hypoxia (2.5% of air saturation = 3.9 mmHg, 0.5 kPa) at 22°C

(Speers-Roesch et al., 2010). The Atlantic hagfish (*Myxine glutinosa*) (Axelsson et al., 1990; Hansen and Sidell, 1983; Perry et al., 1993) and New Zealand hagfish (*Eptatretus cirrhatus*) (Forster et al., 1992) are also recognized for their tolerance of low-oxygen conditions. The coral reef inhabiting epaulette shark (*Hemiscyllium ocellatum*) is probably the most hypoxia-tolerant elasmobranch (Nilsson and Renshaw, 2004; Routley et al., 2002; Stensløkken et al., 2004; Wise et al., 1998). It can survive 3.5 h of severe hypoxia (Wise et al., 1998) and at least 1 h of anoxia at temperatures near 25°C (Renshaw et al., 2002). Other warmer climate fish that have developed considerable hypoxia tolerance include the oscar (*Astronotus ocellatus*), which are reported to tolerate 5%–20% air saturation for 20–50 h and to survive up to 6 h of anoxia at 28°C (Almeida-Val and Hochachka, 1995; Almeida-Val et al., 2000; Muusze et al., 1998; Val and Almeida-Val, 1995), and the toadfish (*Opsanus tau*) (McDonald et al., 2010; Panlilio et al., 2016; Ultsch et al., 1981).

An additional focus of the chapter is on how the heart of severe hypoxia/anoxia-tolerant species can continue to function under extremely low-oxygen conditions. In this regard, when relevant, similarities and differences are drawn between the mechanisms of cardiac anoxia tolerance in the crucian carp and freshwater turtles of the genera *Chrysemys* and *Trachemys*, the other champions of vertebrate anoxia survival. These turtles, like the crucian carp, overwinter in ice-covered ponds, which become progressively hypoxic, and ultimately anoxic, as thick ice coverage inhibits both photosynthesis and oxygen diffusion from the air (Herbert and Jackson, 1985a,b; Jackson and Ultsch, 1982; Reese et al., 2002; Ultsch and Jackson, 1982; Warren et al., 2006). Ice cover also prevents the turtles from reaching the surface to breathe, and the low oxygen levels are insufficient to support aerobic respiration via extrapulmonary O_2 uptake (Jackson and Ultsch, 1982, 2010).

2. *IN VIVO* CARDIOVASCULAR RESPONSES TO OXYGEN LIMITATION IN SEVERE HYPOXIA/ANOXIA-INTOLERANT SPECIES

2.1. Brief Overview of Fish Cardiovascular Physiology

Before delving into the cardiovascular responses of severe hypoxia/anoxia-intolerant and severe hypoxia/anoxia-tolerant fish to oxygen deprivation, and the cardiac survival strategies of species tolerant to oxygen deprivation, brief overviews of the fish cardiovascular system, cardiac physiology and cardiovascular regulation are warranted. For detailed descriptions of fish cardiac anatomy, cardiac excitation–contraction coupling, and cardiovascular control, the reader is directed to Chapters 1–5 and 7 of Volume 36A: Icardo (2017), Shiels

(2017), Vornanen (2017), Farrell and Smith (2017), Imbrogno and Cerra (2017) and Sandblom and Gräns (2017).

The cardiorespiratory system of water-breathing fishes, like other vertebrates, is designed to efficiently transport oxygen and nutrients to, and remove waste products from, the tissues. In the typical piscine circulation, the heart, which is an aerobic organ and the muscular pump that circulates oxygen in the blood from the respiratory organ to the tissues, is composed of six morphologically distinct regions (segments) arranged in series. These are the *sinus venosus*, atrium, atrioventricular (AV) segment, ventricle, and outflow tract (OFT, containing components of both the *conus arteriosus* and *bulbus arteriosus* in different proportions) (see Chapter 1, Volume 36A: Icardo, 2017). The single atrium receives blood from the sinus venosus (a reservoir and receiving chamber for venous blood) and atrial contraction expels blood into the more muscular (single) ventricle. Ventricular contraction propels blood successively through the OFT, into the ventral aorta, and then onto the gills (branchial circulation) where gas and ion transfer occurs. Blood flow then continues via the dorsal aorta to the peripheral tissues (systemic circulation) before returning to the heart. Thus, the branchial and systemic circulations are in series. In air-breathing fishes, the branchial and systemic circulations are still in series, but the circulatory system is modified to accommodate blood flow to and from the air-breathing organ, the location of which varies depending on species (Graham, 1997; Ishimatsu, 2012; Lefevre et al., 2014d; Olson, 1994).

With each heartbeat, the ventricle ejects a volume of blood termed cardiac stroke volume (V_S) at a specific blood pressure. The number of heartbeats in one minute is termed heart rate (f_H), and cardiac output ($\dot{Q}$), the product of V_S and f_H, is the amount of blood pumped per unit time. Cardiac power output (CPO) is calculated as the product of $\dot{Q}$ and ventral aortic pressure (P_{VA}), and is an indirect measure of cardiac ATP demand. Vascular resistance is the resistance that must be overcome by the heart to propel blood through the circulatory system. Vasoconstriction leads to an increase in vascular resistance, whereas vasodilation decreases vascular resistance. In fishes, vascular resistance of the branchial circulation (R_{gill}) is estimated by dividing the difference between P_{VA} and dorsal aortic (P_{DA}) blood pressures by $\dot{Q}$, and systemic vascular resistance (R_{sys}) is calculated by dividing P_{DA} by $\dot{Q}$ (these formulas assume that central venous pressure (P_{CV}) is negligible: which is not entirely accurate—see Chapter 7, Volume 36A: Sandblom and Gräns, 2017). Total peripheral resistance (R_{tot}) is the sum of the branchial and systemic resistances and can also be calculated by dividing P_{VA} by $\dot{Q}$.

The coordinated pumping action of the fish heart is initiated by action potentials (APs) generated by the pacemaker cells located in a morphologically distinct ring of tissue located at the base of the sinoatrial valve

(Haverinen and Vornanen, 2007). The pacemaker cells set the spontaneous rhythm of cardiac contraction, and this results in a synchronized propagation of excitation throughout the atrium and ventricle. The generation of the AP, and its spread throughout the heart, requires the integrated activities of several sarcolemma (SL) ionic currents that pass through pore-forming ion channel proteins, namely, voltage-gated Na^+ channels (I_{Na}), L-type Ca^{2+} channels (I_{Ca}), delayed-rectifier K^+ channels (I_{Kr}), and inward-rectifier K^+ channels (I_{K1}) (Roden et al., 2002). The linkage of excitation of the cardiac myocyte SL to the contraction of the myofilaments is termed excitation–contraction (E–C) coupling. In mammalian cardiomyocytes, when an AP excites the SL, Ca^{2+} entry via L-type Ca^{2+} channels is initiated. Ca^{2+} entry into the cell triggers the release of a greater store of Ca^{2+} from the sarcoplasmic reticulum (SR) through Ca^{2+} release channels termed ryanodine receptors (i.e., Ca^{2+}-induced Ca^{2+} release) (Bers and Despa, 2006). The intracellular Ca^{2+} transient ($[Ca^{2+}]_i$), which controls myofilament contraction by binding to regulatory sites on troponin C, is the sum of SL Ca^{2+} influx and SR Ca^{2+} release. Relaxation occurs when Ca^{2+} is returned to diastolic levels by reuptake into the SR via the SR Ca^{2+}-ATPase (SERCA). Also, Ca^{2+} is extruded across the SL via forward-mode Na^+/Ca^{2+} exchanger (NCX) activity, which exchanges one Ca^{2+} ion for three Na^+ ions, and to a much lesser extent, via the sarcolemmal Ca^{2+}-ATPase. The amplitude and rate of change of $[Ca^{2+}]_i$ largely determines the strength and rate of myocyte contraction. In fish cardiomyocytes, the contribution of SL Ca^{2+} to cardiac contraction is variable among species (Shiels, 2011; Shiels and Galli, 2014). In general, the contribution of SL Ca^{2+} to cardiac contraction is minimal to none in slow and sluggish fish such as carp, whereas it plays a more important role in athletic fish such as salmon and tuna, which have comparatively high f_H and P_{VA}.

Regulation of the piscine cardiovascular system occurs primarily through the action of the autonomic nervous system, with the exception of hagfish, in which nervous control of the circulatory system appears to be absent (Sandblom and Axelsson, 2011). In addition, a plethora of circulating (humoral) and local (paracrine) biochemical factors, hormones, and gasotransmitters play important roles in cardiovascular control (see Chapters 5 and 7, Volume 36A: Imbrogno and Cerra, 2017; Sandblom and Gräns, 2017). The autonomic nervous system consists of two opposing pathways, a parasympathetic vagal inhibitory component and a sympathetic (adrenergic) excitatory component. Cardioinhibitory parasympathetic control via the cardiac branch of the vagus nerve is the predominant form of f_H regulation in teleosts (see Chapter 4, Volume 36A: Farrell and Smith, 2017), whereas the sympathetic component is of paramount importance for the regulation of blood pressure, cardiac contractility and the distribution of blood flow among vascular beds (Farrell and Jones, 1992; Nilsson and Axelsson, 1987; Sandblom and Axelsson, 2011; Satchell, 1991; Taylor, 1992). Adrenergic stimulation of the cardiovascular system in teleosts occurs through both neuronal and humoral means (Nilsson and Axelsson, 1987; Satchell, 1991). Catecholamines, adrenaline and noradrenaline, liberated from the postganglionic terminals of sympathetic nerves near their site of action, evoke rapid responses in effector tissues. A more widely distributed adrenergic response occurs through catecholamine release directly into the blood stream from chromaffin cells. These cells are localized within the walls of the posterior cardinal vein and in close proximity to the lymphoid tissue in the region of the anterior (head) kidney (Perry and Bernier, 1999; Randall and Perry, 1992). Adrenaline is the primary neurotransmitter and circulating catecholamine in teleost fish. However, in the common carp, noradrenaline is the predominant circulating catecholamine (Farrell and Jones, 1992).

Since catecholamines cannot penetrate the cell membrane, a transduction pathway is needed for transmembrane signaling. Transmembrane signaling commences with the binding of catecholamines to cell-surface adrenoreceptors, of which two main types, α and β, have been identified in teleosts. However, the location of each type varies among species (Sandblom and Axelsson, 2011). Usually, α-adrenoceptors are predominantly found in the vascular smooth muscle of arterioles, and β-adrenoceptors are found in both the heart and vascular smooth muscle. Adrenergic stimulation typically results in an α-adrenergically mediated systemic vasoconstriction and β-adrenergically mediated increases in cardiac inotropy and chronotropy (Farrell, 1984). A β-adrenergic vasodilatory effect can also occur in the arterioles of the systemic circulation, but is often masked by the more potent α-vasoconstriction (Wood and Shelton, 1975). In addition, in some fish such as the common carp (Temma et al., 1989), European perch (*Perca fluviatilis*) (Tirri and Ripatti, 1982), Japanese eel (*Anguilla japonica*) (Chan and Chow, 1976) and European eel (*Anguilla anguilla*) (Forster, 1976; Peyraud-Waitzenegger et al., 1980), stimulation of cardiac α-adrenoceptors mediates negative chronotropy and inotropy. Adrenergic stimulation of the branchial vasculature also exists in teleosts (Nilsson and Axelsson, 1987; Randall and Perry, 1992; Sandblom and Axelsson, 2011). Pharmacological adrenergic stimulation in perfused rainbow trout gills results in a transient α-adrenergic vasoconstriction followed by a sustained β-adrenergically mediated vasodilation (Wood, 1974, 1975; Wood and Shelton, 1975). Similar results occur with sympathetic stimulation in Atlantic cod (*Gadus morhua*) gills (Nilsson and Pettersson, 1981; Pettersson and Johansen, 1982; Pettersson and Nilsson, 1979), and with adrenergic drug infusion in lingcod (*Ophiodon elongatus*) *in vivo* (Farrell, 1981). Thus, in contrast to the systemic circulation, the β-adrenergic-mediated vasodilatory response dominates in the gill arterio-arterial vasculature (for more details on control of the gill vasculature, see Chapter 7, Volume 36A: Sandblom and Gräns, 2017).

2.2. Water-Breathers

Anoxia-intolerant vertebrate species exhibit numerous physiological responses designed to facilitate oxygen delivery to the tissues during periods of moderate oxygen deprivation. In mammals and birds exposed to hypoxic conditions, oxygen delivery to cells is sustained through increased respiration frequency (f_R), f_H and $\dot{Q}$ (Lutz and Storey, 1997; Wood, 1991). Additionally, hyperventilation decreases the partial pressure of CO_2 in the blood, which partially offsets the decrease in blood pH resulting from anaerobic lactic acid accumulation, and thus, limits the decrease in hemoglobin–oxygen affinity (Hurtado, 1964). The cardiorespiratory responses to hypoxia are mediated through increases in sympathetic neural tone and circulating catecholamines (Padbury et al., 1987; Rose et al., 1983), which are also important in stimulating erythrocyte production (Nikinmaa, 1990), hepatic glycogenolysis and the release of glucose (the fuel for anaerobic glycolysis) into the blood in mammals (Hems and Whitton, 1980; Jones et al., 1988).

Some of the physiological responses of severe hypoxia/anoxia-intolerant water-breathing fishes to moderate hypoxic exposure reflect those in mammals. Increased f_R and ventilatory stroke volume during hypoxic exposure in fish are well documented, and catecholamine release from the chromaffin tissue during hypoxia aids in oxygen transport (Burleson and Smatresk, 1990b; Cech and Wohlschlag, 1973; Holeton and Randall, 1967b; Hughes, 1973; Hughes and Shelton, 1962; Kinkead and Perry, 1990, 1991; Nonnotte et al., 1993; Perry and Bernier, 1999; Perry et al., 1999; Peyraud-Waitzenegger and Soulier, 1989; Randall and Jones, 1973; Randall and Shelton, 1963; Saunders, 1962; Saunders and Sutterlin, 1971; Smith and Jones, 1978). Specifically, catecholamines trigger the release of erythrocytes from the spleen (Pearson et al., 1992) and stimulate the activation of Na^+/H^+ exchange across the erythrocyte membrane. The latter response increases red blood cell intracellular pH relative to the plasma, and facilitates oxygen binding (Boutilier et al., 1986; Nikinmaa, 1986). However, the f_H response of water-breathing fishes to oxygen limitation, which commonly consists of a slowing of f_H, differs from the increased f_H observed in mammals. The f_H responses, as well as other important patterns in the cardiovascular responses of hypoxia/anoxia-intolerant fishes to oxygen deprivation, are detailed below and summarized in Table 1 (which compiles the findings from studies published over the last half century).

In contrast to the heightened f_H displayed by hypoxic mammals and birds, for many water-breathing fish, the response to aquatic hypoxia is a cholinergically mediated reflex slowing of f_H (termed hypoxic bradycardia; see Table 1). The response is mediated through the activation of O_2 chemoreceptors located in the gills; in most species these chemoreceptors sense and respond to changes in oxygen in the inspired water (Milsom, 2012). However,

Table 1

Summary of *in vivo* cardiovascular responses to oxygen limitation

Species	Common name	Temperature (°C)	PO_2 (kPa)	Exposure time	f_H (min^{-1}) Normoxia	f_H (min^{-1}) Hyp. or Anx.	f_H (% change)	Bradycardia (kPa)	$\dot{Q}$	V_S	P_{VA}	CPO	P_{DA}	R_{gill}	R_{sys}	R_{tot}	Reference
Severe hypoxia/anoxia-intolerant water-breathers																	
Elasmobranchs																	
Aptychotrema rostrata	Shovelnose ray	28	3.1	4 h	56	22	60 ↓	7.7	↓	↑	—	↓	↓	—	=	—	Speers-Roesch et al. (2012)
Scyliorhinus canicula	Dogfish	12; anesthetized	0.0	2 min	28	5	82 ↓	—	—	—	—	—	↓	—	—	—	Satchell (1961)
		12	4.0	31 min	25	19	24 ↓	9.3	—	—	=	—	=	—	—	—	Butler and Taylor (1971)
		12	2.3	5 min	25	19	24 ↓	—	—	—	↓	—	↓	—	—	—	
		7	5.7	30 min	20	19	=	—	↑	↑	=	—	=	=	↓	—	Butler and Taylor (1975)
		12	5.6	30 min	29	20	31 ↓	8.0	=	↑	=	—	↓	=	↓	—	
		17	5.2	30 min	41	20	51 ↓	12.0	=	↑	↓	—	↓	=	↓	—	
		13.5	4.0	10 min	31	18	41 ↓	—	—	↑	—	—	—	—	—	—	Short et al. (1977)
		15	10.7	1 h	32	17	46 ↓	—	=	↑	=	=	=	=	=	=	Short et al. (1979)
		15	6.7	—	31	17	45 ↓	—	—	—	—	—	—	—	—	—	Taylor and Barrett (1985)
		7	4.7	1 h	20	18	10 ↓	—	↓	↑	—	=	—	—	—	—	Taylor et al. (1977)
		12	4.7	1 h	25	19	24 ↓	—	↓	↑	—	=	—	—	—	—	
		17	4.7	1 h	37	18	51 ↓	—	↓	↑	—	↓	—	—	—	—	
		15	4.7	1 h	35	16	54 ↓	—	↓	↑	↓	—	↓	—	—	—	
		15	4.0	3 min	29	20	31 ↓	—	—	—	↓	—	—	—	—	—	Butler et al. (1977)
Scyliorhinus stellaris	Larger spotted dogfish	17	7.2	1 h	39	7	82 ↓	—	—	—	↓	—	↓*	—	—	—	Piiper et al. (1970)
Chondrosteans																	
Acipenser baeri	Siberian sturgeon	18	1.3	30 min	52	17	67 ↓	—	—	—	—	—	↓	—	—	—	Maxime et al. (1995)
Acipenser naccarii	Adriatic sturgeon	23	4.6	30 min	60	41	31 ↓	—	=	—	—	—	↓	—	—	—	Agnisola et al. (1999)
		23	2.5	20 min	60	52	13 ↓	—	—	—	—	—	=	—	—	—	McKenzie et al. (1995)
Teleosts																	
Anguilla anguilla	European eel	15	5.3	1 h	37	23	38 ↓	—	↓	↑	↓	—	↓	↑	=	—	Peyraud-Waitzenegger and Soulier (1989)
Anguilla japonica	Japanese eel	22	5.3	2 h	56	26	53 ↓	9.3–10.7	—	—	↓	—	↓	—	—	—	Chan (1986)

(*Continued*)

Table 1 (Continued)

Species	Common name	Temperature (°C)	PO_2 (kPa)	Exposure time	f_H (min^{-1}) Normoxia	Hyp. or Anx.	f_H (% change)	Bradycardia (kPa)	$\dot{Q}$	V_S	P_{VA}	CPO	P_{DA}	R_{gill}	R_{sys}	R_{tot}	Reference
Chaenocephalus aceratus	Blackfin icefish	1–2	6.7–9.3	—	17–18	17–18	=	—	↑	↑	—	—	↓	—	—	—	Hemmingsen et al. (1972)
		1–2	6.7–9.3	—	18	16	11 ↓	—	—	—	—	—	—	—	—	—	Hemmingsen and Douglas (1972)
		0.5	5.3–8	10–20 min; 3–4 h	—	—	60–80 ↓	6.7–9.3	—	—	↑	—	↑	—	—	—	Holeton (1972)
Ciliata mustela	Five-bearded rockling	9–11	4.0–5.3	5 min	79	74	=	—	—	—	↑	—	—	—	—	—	Fritsche (1990)
Colossoma macropomum	Tambaqui	25	1.3	10 min	48	18	63 ↓	16.0	—	—	↓	—	—	—	—	—	Sundin et al. (2000)
Gadus morhua	Atlantic cod	10	4.0–5.3	6 min	41	22	46 ↓	—	=	↑	↑	—	↑	=	↑	—	Fritsche and Nilsson (1989)
		10–12	4.0–5.3	8 min	40	33	18 ↓	—	—	—	↑	—	↑	—	—	—	Fritsche and Nilsson (1990)
		10	3.0	2 h	40	27	33 ↓	6.1	—	—	—	—	—	—	—	—	McKenzie et al. (2009)
		10	8–9	3 h	29	34	−17	—	↑	=	—	—	—	—	—	—	Petersen and Gamperl (2010)
		10; hypoxia -acclimated	8–9	3 h	30	36	−20	—	=	=	—	—	—	—	—	—	
		10	2.7	—	30	20	33 ↓	—	—	—	—	—	—	—	—	—	
		10–11	4.0–5.3	8 min	38	38	=	—	↑	↑	↑	—	↑	—	=	—	Axelsson and Fritsche (1991)
		10–12	5.3–6.5	12 min	30	30	=	—	↑	↑	↑	—	↑	—	↑	—	Sundin (1995)
Gobius cobitis	Giant goby	12.5	2.0	—	26	20	23 ↓	2.1	—	—	—	—	—	—	—	—	Berschick et al. (1987)
Hemitripterus americanus	Sea raven	2–4; anesthetized	1.7	10 min	—	—	↓	4.0	—	—	↑	—	↑	—	—	—	Saunders and Sutterlin (1971)
Hoplias lacerdae	Trairão	25	0.7	1 h	44	24	45 ↓	4.7	—	—	—	—	—	—	—	—	Rantin et al. (1993)
		25	1.3	4 min	59	18	69 ↓	—	—	—	↓	—	—	—	—	—	Micheli-Campbell et al. (2009)
		25	0.7	30 min	45	25	44 ↓	2.7	—	—	—	—	—	—	—	—	Rantin et al. (1995)
Hoplias malabaricus	Traíra	25	1.5	10 min	50	27	46 ↓	—	—	—	↓	—	—	—	—	—	Sundin et al. (1999)
		25	0.7	1 h	61	23	62 ↓	2.7	—	—	—	—	—	—	—	—	Rantin et al. (1993)
		25	0.7	30 min	60	36	40 ↓	1.3	—	—	—	—	—	—	—	—	Rantin et al. (1995)

Ictalurus nebulosu	Brown bullhead catfish	25	1.0	—	75	21	72 ↓	18.5	—	—	—	—	—	—	—	—	Marvin and Heath (1968)
		25	3.2	60 min	67	32	52 ↓	—	—	—	—	—	—	—	—	—	Marvin and Burton (1973)
Ictalurus punctatus	Channel catfish	20–25; anesthetized	6.1	10 min	96	83	13 ↓	—	—	—	—	—	↑,↓	—	—	—	Burleson and Smatresk (1990a)
		22; anesthetized	5.5	1 h	53	32	40 ↓	—	—	—	—	—	↑	—	—	—	Burleson and Smatresk (1989)
Katsuwonus pelamis	Skipjack tuna	25	6.7	2 min	126	78	38 ↓	17.3	↓	↓	=	↓	=	↑	=	↑	Bushnell and Brill (1992)
		25	6.7	3–4 min	117	80	32 ↓	11.3–13.9	↓	↑	—	—	—	—	—	—	Bushnell et al. (1990)
Leiopotherapon unicolor	Spangled perch	23; measured at 5	1.1	—	12	6	50 ↓	4.2	—	—	—	—	—	—	—	—	Gehrke and Fielder (1988)
		23; measured at 5	1.1	—	40	17	58 ↓	4.2–8.4	—	—	—	—	—	—	—	—	
		23; measured at 10	1.1	—	27	13	52 ↓	4.2–8.4	—	—	—	—	—	—	—	—	
		23; measured at 20	1.0	—	64	17	73 ↓	8.3	—	—	—	—	—	—	—	—	
		23; measured at 25	1.0	—	73	30	59 ↓	8.2	—	—	—	—	—	—	—	—	
		23; measured at 30	1.0	—	102	40	60 ↓	12.2	—	—	—	—	—	—	—	—	
		23; measured at 35	1.0	—	105	55	48 ↓	12.0	—	—	—	—	—	—	—	—	
Lepomis macrochiru	Bluegill sunfish	25	3.1	—	88	17	81 ↓	10.3	—	—	—	—	—	—	—	—	Marvin and Heath (1968)
		25	4.0	60 min	104	36	65 ↓	—	—	—	—	—	—	—	—	—	Marvin and Burton (1973)
Micropterus dolomieu	Smallmouth bass	22	6.0	1 h	38	24	37 ↓	—	↓	↑	—	—	—	—	—	—	Furimsky et al. (2003)
Micropterus salmoides	Largemouth bass	22	6.0	1 h	42	37	12 ↓	—	=	↑	—	—	—	—	—	—	Furimsky et al. (2003)
Mugil cephalus	Gray mullet	24.5	2.1	15 min	100	18	82 ↓	8.4	—	—	↓	—	—	—	—	—	Shingles et al. (2005)
Myoxocephalus scorpius	Short-horned sculpin	8	3.5	6 h	20	8	60 ↓	—	↓	=	—	—	—	—	—	—	MacCormack and Driedzic (2004)
		9–11	4.0–5.3	5 min	40	23	43 ↓	—	—	—	=	—	—	—	—	—	Fritsche (1990)
Oncorhynchus kisutch	Coho salmon	8–10	10.0	10 min	45	36	20 ↓	—	↑	—	—	—	↑	—	↑	—	Axelsson and Farrell (1993)
Oncorhynchus mykiss	Rainbow trout	9–19	2.6	—	70	21.5	69 ↓	15.2	—	—	↑	—	↑	—	—	—	Holeton and Randall (1967a)
		12	5.1	—	68	11	84 ↓	16.7	—	—	—	—	—	—	—	—	Marvin and Heath (1968)
		12	7.5	60–70 min	71	24	66 ↓	—	—	—	—	—	—	—	—	—	Marvin and Burton (1973)

(*Continued*)

Table 1 (Continued)

Species	Common name	Temperature (°C)	PO_2 (kPa)	Exposure time	f_H (min^{-1}) Normoxia	Hyp. or Anx.	f_H (% change)	Bradycardia (kPa)	$\dot{Q}$	V_S	P_{VA}	CPO	P_{DA}	R_{gill}	R_{sys}	R_{tot}	Reference
		8.5–15	5.3	—	67	35	48 ↓	8.4	—	—	—	—	—	—	—	—	Randall and Smith (1967)
		7	4.0	—	45	28	38 ↓	12.0	—	—	—	—	—	—	—	—	Smith and Jones (1978)
		16	4.0	—	57	33	42 ↓	9.0	—	—	—	—	—	—	—	—	
		14.5	6.7–8.7	20–45 min	86	53	38 ↓	—	↑	↑	—	—	↑	—	—	—	Wood and Shelton (1980)
		10–12	12.0	30 min	64	59	82 ↓	13.0–14.5	↑	↑	—	—	↑	—	—	—	Gamperl et al. (1994)
		15	11.5	8 min	48	50	=	—	↑	↑	—	—	=	—	=	—	Sandblom and Axelsson (2005b)
		15	7.3	8 min	68	52	24 ↓	—	=	↑	—	—	↓	—	↓	—	
		10; anesthetized	1.1	6 min	47	24	49 ↓	—	↓	—	↑	—	↓	↑	=	—	Sundin and Nilsson (1997)
		13	5.3–6.0	40 min	66	31	53 ↓	—	=	↑	—	—	↑	—	—	—	Perry and Desforges (2006)
		12–14	8.7	15–20 min	78	53	32 ↓	8.7	↓	=	=	—	=	↑	↑	—	Perry et al. (1999)
		15	9.0		66	52	21 ↓	—	=	↑	↑	=	=	—	=	—	Sandblom and Axelsson (2006)
Ophiodon elongatus	Lingcod	9–11	3.3–6.0	20–35 min	29	12	58 ↓	7.9	↓	↑	↓	↓	↓	↑	↑	—	Farrell (1982)
Pagothenia bernacchii	Emerald rockcod	0	6.7	40 min	10	12	20 ↑	—	=	—	↑	—	—	—	—	↑	Axelsson et al. (1992)
Pagothenia borchgrevinki	Bald notothen	0	6.7	40 min	15	15	=	—	↑	=	↑	—	—	—	—	=	
Pagrus auratus	Silver seabream	10; anesthetized	5.1	60 min	40	27	33 ↓	11.5	—	—	—	—	—	—	—	—	Janssen et al. (2010)
Piaractus mesopotamicus	Pacu	25; with air access	1.3	1 h 15 min	54	37	31 ↓	2.7	—	—	—	—	—	—	—	—	Rantin et al. (1998)
		25; without air access	1.3	1 h 15 min	47	15	68 ↓	4.0	—	—	—	—	—	—	—	—	
		25	1.3	10 min	63	30	52 ↓	4.0	—	—	—	—	—	—	—	—	Leite et al. (2007)
		25	1.3	1 h	52	32	38 ↓	2.7	—	—	—	—	—	—	—	—	Rantin et al. (1995)
Pseudopleuronectes americanus	Winter flounder	8	4.0	3.5 h	31	18	41 ↓	13.0	↓	↑	—	—	=	—	=	—	Mendonça and Gamperl (2010)
		15	4.0	3.5 h	52	53	=	—	=	=	—	—	↑,=	—	↑,=	—	
		10	8.5	1–2 h	35	39	=	—	↑	=	—	=	↓*	—	—	—	Cech et al. (1977)

Salminus maxillosus	Dourado	25	2.7	30 min	95	22	77 ↓	9.3	↓	↑	—	—	—	—	—	—	de Salvo Souza et al. (2001)
Thunnus albacares	Yellowfin tuna	25	6.7	2 min	97	70	28 ↓	6.7	↓	=	=	↓	=	↑	↑	↑	Bushnell and Brill (1992)
		25	6.7	3–4 min	99	85	14 ↓	11.3–13.9	↓	=	—	—	—	—	—	—	Bushnell et al. (1990)
Thunnus obesus	Bigeye tuna	25	6.7	3–4 min	99	67	32 ↓	8.7–11.2	↓	=	—	—	—	—	—	—	Bushnell et al. (1990)
Tinca tinca	Tench	12–17; anesthetized	1.5	30 min	40	19	53 ↓	5.3	—	—	—	—	—	—	—	—	Randall and Shelton (1963)
Zoarces viviparous	Eelpout	9–11	4.0–5.3	5 min	49	25	49 ↓	—	—	—	↓	—	—	—	—	—	Fritsche (1990)
Air-breathers																	
Obligate air-breathers																	
Lepidosiren paradoxa	South American lungfish	25	4.7	1 h	38	40	=	—	—	—	—	—	—	—	—	—	Sanchez et al. (2001)
Protopterus aethiopicus	Marbled lungfish	25	3.5	30 min	48	46	=	—	—	—	—	—	=	—	—	—	Perry et al. (2005)
Facultative air-breathers																	
Hoplerythrinus unitaeniatus	Jeju	18–26	1.0	1 h	—	—	=; ↑↓ with AB	—	—	—	—	—	—	—	—	—	McKenzie et al. (2007)
		25; without air access	2.7	1 h	39	22	44 ↓	8.5	—	—	—	—	—	—	—	—	Oliveira et al. (2004)
		25; without air access	2.3	10 min	43	21	51 ↓	5.3	—	—	—	—	↑*	—	—	—	Lopes et al. (2010)
Clarias gariepinus	African sharptooth catfish	25	1.3	1 h	35	28	=; ↑↓ with AB	—	—	—	—	—	—	—	—	—	Belão et al. (2011)
		25; without air access	1.3	1 h	43	19	56 ↓	8.0	—	—	—	—	—	—	—	—	
		25	1.3	1 h	—	—	=; ↑↓ with AB	—	—	—	—	—	—	—	—	—	Belão et al. (2015)
		25; without air access	1.3	1 h	40	19	52 ↓	8.0	—	—	—	—	—	—	—	—	
Amia calva	Bowfin	20	6.3	15 min	30	25	17 ↓	—	—	—	—	—	↑	—	—	—	McKenzie et al. (1991b)
		14–16	4.0–5.3	1 h	25–27	27–29	=	—	—	—	—	—	↑	—	—	—	Hedrick et al. (1991)
Lepisosteus osseus	Longnose gar	19–22	3.5	10–15 min	57	65	14 ↑	—	—	—	=	—	—	—	—	—	Smatresk et al. (1986)
Lepisosteus oculatus	Spotted gar	18–21	1.6	1 h	33	36	9 ↑	—	↑	—	↑	—	↑	=	—	—	Smatresk and Cameron (1982)
Glyptoperichthys gibbiceps	Armored catfish	28; without air access	1.5	50–60 min	98	98	=	—	—	—	—	—	—	—	—	—	MacCormack et al. (2003)
Liposarcus pardalis	Armored catfish	28; without air access	1.5	50–60 min	75	54	=	—	—	—	—	—	—	—	—	—	MacCormack et al. (2003)
Synbranchus marmoratus	Marbled swamp eel	25	6.7	5 h	39	52	=	—	↑	↑	—	—	↓	—	—	—	Skals et al. (2006)

(*Continued*)

Table 1 (Continued)

Species	Common name	Temperature (°C)	PO_2 (kPa)	Exposure time	f_H (min^{-1}) Normoxia	Hyp. or Anx.	f_H (% change)	Bradycardia (kPa)	$\dot{Q}$	V_S	P_{VA}	CPO	P_{DA}	R_{gill}	R_{sys}	R_{tot}	Reference
Monopterus albus	Asian swamp eel	30; gill ventilating	0.3	30 min	15	13	=	—	↓	↓	—	↓	=	—	—	—	Iversen et al. (2013)
		30; N_2-breathing	0.3	30 min	62	52	=	—	=	=	—	=	=	—	—	—	
		30; combined response	0.3	30 min	71	48	=	—	=	=	—	=	=	—	—	—	
Neoceratodus forsteri	Australian lungfish	18–22	3.0	60 min	26	27	=	—	—	—	—	—	=	—	—	—	Fritsche et al. (1993)
Megalops cyprinoides	Pacific tarpon	27; juveniles	2.0	20 min	35	58	65 ↑	8.0	↑	↑	—	—	—	—	—	—	Clark et al. (2007)
		27; adults	2.0	20 min	63	61	=	—	=	=	—	—	—	—	—	—	
Anabas testudineus	Climbing perch	25	6.7	1 h	—	—	=; ↑↓ with AB	—	—	—	—	—	—	—	—	—	Singh and Hughes (1973)
		25; without air access	6.7	1 h	42	15	64 ↓	16.0	—	—	—	—	—	—	—	—	
Severe hypoxia/anoxia-tolerant water-breathers																	
Agnathans																	
Eptatretus cirrhatus	New Zealand hagfish	17	5.3	20–30 min	25	24	=	—	↑	↑	↑	—	↑	↑	=	—	Forster et al. (1992)
Myxine glutinosa	Atlantic hagfish	10–11	1.5–2.2	15–35 min	22	20	=	—	=	=	↑	=	↑	=	=	—	Axelsson et al. (1990)
Eptatretus stoutii	Pacific hagfish	10	0.0	36 h	10	4	60 ↓	—	↓	↑	↑,=	↑,=	—	—	—	—	Cox et al. (2010)
Elasmobranchs																	
Hemiscyllium ocellatum	Epaulette shark	28	2.0	4 h	60	21	65 ↓	3.9	↓	↑	—	↓	↓		↑		Speers-Roesch et al. (2012)
		25; anesthetized	<0.8	20 min	58	39	33 ↓	—	—	—	↓	—	↓	—	—	—	Stensløkken et al. (2004)
Teleosts																	
Carassius carassius	Crucian carp	2	<0.2	6 weeks	10	4	60 ↓	—	—	—	—	—	—	—	—	—	Tikkanen et al. (2017)
		8	<0.02	5 days	17	14	↓,=	—	=	↑,=	↓	=	—	—	—	↓	Stecyk et al. (2004b)
		15	1.6	up to 16 h	78	30	62 ↓	—	—	—	—	—	—	—	—	—	Vornanen (1994a)
		22	1.9	14 min	64	53	17 ↓	—	—	—	—	—	—	—	—	—	Vornanen and Tuomennoro (1999)

Cyprinus carpio	Common carp	25	0.7	15 min	37	18	51 ↓	5.3	—	—	—	—	—	—	—	—	Glass et al. (1991)
		25		30 min	36	19	47 ↓	1.3	—	—	—	—	—	—	—	—	Rantin et al. (1995)
		6	<0.5	24 h	7	2	71 ↓	—	↓	↓	—	—	—	—	—	—	Stecyk and Farrell (2002)
		10	<0.6	6 h	12	5	58 ↓	—	↓	=	—	—	—	—	—	—	
		15	<0.6	2.5 h	17	3	82 ↓	—	↓	↑	—	—	—	—	—	—	
		5	<0.2	12.5 h	9	3	66 ↓	—	↓	↑,=	↓	↓	—	—	—	↑	Stecyk and Farrell (2006)
		10	<0.2	3.5 h	16	5	69 ↓	—	↓	=	↓	↓	—	—	—	↑	
		15	<0.2	1.17 h	24	6	75 ↓	—	↓	↑,=	↓	↓	—	—	—	↑	
Oreochromis niloticus	Nile tilapia	25	2.7	40 min	45	42	=	—	—	—	—	—	—	—	—	—	Zeraik et al. (2013)
Oreochromis hybrid sp.	Tilapia	22	1.0	8 h	36	20	44 ↓	4.1	↓	=	=	↓	—	—	—	↑	Speers-Roesch et al. (2010)

The "Bradycardia" column indicates the PO_2 at which bradycardia occurs.

mmHg in the literature was converted to kPa (1 mmHg=0.133 kPa).

mg O_2 L^{-1} in the literature was converted to kPa accounting for temperature and salinity.

% air saturation in the literature was converted to kPa assuming standard barometric pressure and accounting for temperature.

In many instances f_H data were obtained from textual descriptions or estimated from the mean values in figures. Therefore, all f_H data are presented to the nearest whole value.

For studies in which the hypoxia exposure was graded, the lowest PO_2 and exposure time at that PO_2 are listed.

Air-breathers had access to atmospheric air unless otherwise noted in the "Temperature" column.

An equality sign (=) indicates no statistically significant change.

An up arrow (↑) indicates a statistically significant increase.

A down arrow (↓) indicates a statistically significant decrease.

Two symbols indicate a transient or multidirectional change (e.g., ↑,= signifies a transient increase that returned to the control normoxic level).

An asterisk (*) in the P_{DA} column indicates systemic pressure measured in the caudal or other systemic artery.

A dash (—) indicates that a variable was not measured or described.

AB, air breath; Anx., anoxia; CPO, cardiac power output; f_H, heart rate; Hyp., hypoxia; P_{DA}, blood pressure in the dorsal aorta; P_{VA}, blood pressure in the ventral aorta; $\dot{Q}$, cardiac output; R_{gill}, vascular resistance of the gill circulation; R_{sys}, vascular resistance of the systemic circulation; R_{tot}, sum of the branchial and systemic resistances; V_S, cardiac stroke volume.

variation exists among species with regard to the water oxygen level at which hypoxic bradycardia is initiated, as well is in the magnitude of f_H depression. The variability relates to differences in ecology, lifestyle and hypoxia tolerance (Gamperl and Driedzic, 2009). In general, for fish that are active and have high energetic demands, and/or that normally inhabit well-oxygenated waters, such as the rainbow trout, largemouth bass, Japanese eel and dourado, the PO_2 threshold for the induction of bradycardia is higher (Table 1). Conversely, fish that are sluggish and/or normally inhabit waters that are regularly hypoxic, such as the tench, five-bearded rockling, smallmouth bass, traíra and trairão, exhibit a lower threshold for the induction of bradycardia and a less pronounced bradycardia (for details, see Fritsche, 1990; Furimsky et al., 2003; Rantin et al., 1993) (Table 1). Another important trend that emerges from the f_H data in Table 1 is that the oxygen tension at which bradycardia commences increases with temperature (and thus metabolic rate). This phenomenon is nicely demonstrated in 23°C-acclimated spangled perch exposed to hypoxia at temperatures ranging between 5°C and 35°C, and in dogfish seasonally acclimated to 7°C, 12°C and 17°C (Table 1; Butler and Taylor, 1975; Gehrke and Fielder, 1988).

It is evident from Table 1 that some species do not always exhibit hypoxic bradycardia, whereas others do not exhibit hypoxic bradycardia at all. For example, rainbow trout exposed to mild hypoxia do not exhibit bradycardia, whereas f_H slows by 24% in animals exposed to severe hypoxia (Sandblom and Axelsson, 2005b). Similarly, while 8°C-acclimated winter flounder exhibit hypoxic bradycardia (Mendonça and Gamperl, 2010), winter flounder acclimated to 10°C and 15°C do not (Cech et al., 1977; Mendonça and Gamperl, 2010). In Atlantic cod, no clear pattern emerges between the level of hypoxia, or temperature, and the initiation of bradycardia (Axelsson and Fritsche, 1991; Fritsche and Nilsson, 1989, 1990; McKenzie et al., 2009; Petersen and Gamperl, 2010; Sundin, 1995). The differing responses among studies could reflect differences in acclimation/test temperature, the instrumentation of the animals and/or the level of hypoxia, or the rate at which water oxygen levels were reduced. However, in sea raven, intraindividual variation existed under identical experimental conditions (Saunders and Sutterlin, 1971). Further, red-blooded (i.e., with hemoglobin; *Pagothenia bernacchii* and *Pagothenia borchgrevinki*) and white-blooded (i.e., without hemoglobin; *Chaenocephalus aceratus*) Antarctic fish do not consistently display hypoxic bradycardia. *P. bernacchii* exhibited tachycardia when exposed to hypoxia, and of three individual *P. borchgrevinki* exposed to hypoxia, only one showed bradycardia, whereas the others displayed tachycardia (Axelsson et al., 1992). Overall, the mean response for *P. borchgrevinki*, as calculated from the control normoxic and final hypoxic values read from the figure in the paper, was no change in f_H. For *C. aceratus*, one study reports the maintenance of f_H with hypoxia exposure (Hemmingsen et al., 1972), another reports a "slight decrease" in heart rate (Hemmingsen and Douglas, 1972), whereas a third reports a considerable bradycardia to 20%–40% of normoxic values (Holeton, 1972). The profound bradycardia in the latter study could reflect an extended period of hypoxic exposure (3–4 h) compared to the other studies.

It has been hypothesized that the bradycardia displayed by fishes in response to oxygen limitation may have several direct benefits to the heart (reviewed by Farrell, 2007). Specifically, Farrell (2007) proposed that a slower f_H during hypoxia exposure would: (1) allow the heart to beat more forcefully due to the negative force–frequency relationship typically displayed by the fish myocardium, and thus, offset the negative effects of decreased oxygen availability on cardiac contractility; (2) enhance myocardial oxygen extraction/delivery due to an increase in blood diastolic residence time; (3) enhance myocardial oxygen delivery due to an increased end-diastolic volume, which would stretch the myocardium and reduce the effective oxygen diffusion distance between the blood and trabeculae; (4) augment coronary blood flow (for those species that have coronary arteries); and (5) reduce myocardial oxygen demand by lowering the rate of ventricular pressure development and CPO. In addition, the occurrence of hypoxic bradycardia has been suggested to improve gas exchange at the gill surface through the slowing of blood flow through the branchial circulation (Randall and Shelton, 1963) and the recruitment of the gill secondary lamellae, and thus, the functional surface area of the gills available for gas exchange, the latter effect due to a bradycardia-mediated increase in cardiac filling pressure, and thus, V_S and arterial pulse pressure (Farrell, 1980; Soivio and Tuurala, 1981). Likewise, a decrease in $\dot{Q}$ would increase capillary transit time in the peripheral tissues, and diffusional gas exchange (Farrell, 1982).

Taken together, it appears that the acute cardiorespiratory responses to hypoxia function as mechanisms to maintain oxygen transfer at the gills and to various tissues (including the heart) in the face of declining oxygen availability. However, empirical evidence in support of the potential benefits of hypoxic bradycardia on oxygen uptake is equivocal. Studies conducted with dogfish, rainbow trout and Atlantic cod have assessed the effect of hypoxic bradycardia on branchial gas exchange by preventing (or suppressing the extent of) hypoxic bradycardia through vagotomy or the application of atropine (an antagonist of cardiac muscarinic acetylcholine receptors). While some studies conclude that hypoxia improves oxygen uptake at the gills (Taylor and Barrett, 1985; Taylor et al., 1977), others report no beneficial effects of hypoxic bradycardia on branchial gas exchange (McKenzie et al., 2009; Perry and Desforges, 2006; Short et al., 1979).

The number of water-breathing fish species in which the responses of $\dot{Q}$, V_S, blood pressures (P_{VA} and P_{DA}), CPO and vascular resistances (R_{gill}, R_{sys}, and R_{tot}) during hypoxia exposure have been documented is considerably less than those in which f_H has been reported (Table 1). Yet, some general trends (with exceptions) in the response of these variables to oxygen deprivation emerge. The response of $\dot{Q}$ to hypoxia is dependent on the V_S response and the magnitude of the bradycardia. $\dot{Q}$ is maintained or increased via a compensatory increase in V_S in rainbow trout (all but one study), dogfish (some studies), Atlantic cod (some studies), largemouth bass, and *C. aceratus* (Axelsson and Fritsche, 1991; Butler and Taylor, 1975; Fritsche and Nilsson, 1989; Furimsky et al., 2003; Gamperl et al., 1994; Hemmingsen et al., 1972; Perry and Desforges, 2006; Sandblom and Axelsson, 2005b, 2006; Short et al., 1977, 1979; Sundin, 1995; Wood and Shelton, 1980). In rainbow trout, the increased V_S arises from the increase in cardiac filling time that accompanies bradycardia, as well as from an increase in P_{CV} and decreased venous capacitance mediated by humoral and neural α-adrenergic control systems (Perry et al., 1999; Sandblom and Axelsson, 2005a,b, 2006). On the opposite end of the response spectrum, V_S and $\dot{Q}$ decrease concomitantly in skipjack tuna (Bushnell and Brill, 1992). A range of intermediate responses is also displayed, including: unchanged V_S and unchanged $\dot{Q}$ in 10°C hypoxia-acclimated Atlantic cod and 15°C-acclimated winter flounder (Mendonça and Gamperl, 2010; Petersen and Gamperl, 2010); unchanged V_S, but increased,$\dot{Q}$ due to hypoxic tachycardia in 10°C-acclimated winter flounder, Atlantic cod (some studies) and *P. borchgrevinki* (Axelsson and Fritsche, 1991; Axelsson et al., 1992; Cech et al., 1977; Sundin, 1995); unchanged V_S and decreased $\dot{Q}$ in short-horned sculpin, rainbow trout (one study), yellowfin tuna and bigeye tuna (Bushnell and Brill, 1992; Bushnell et al., 1990; MacCormack and Driedzic, 2004; Perry et al., 1999); decreased V_S, but unchanged $\dot{Q}$ in *P. bernacchii* (Axelsson et al., 1992); and increased V_S, but decreased $\dot{Q}$ in lingcod, European eel, smallmouth bass, dogfish (some studies), 8°C-acclimated winter flounder, shovelnose ray, dourado and skipjack tuna (Bushnell et al., 1990; de Salvo Souza et al., 2001; Farrell, 1982; Furimsky et al., 2003; Mendonça and Gamperl, 2010; Peyraud-Waitzenegger and Soulier, 1989; Speers-Roesch et al., 2012; Taylor et al., 1977).

It is regularly repeated in the literature that the typical response of hypoxia/anoxia-intolerant teleosts to hypoxia exposure includes an increased R_{gill}, as well as an increased P_{DA} that arises from an increase in R_{sys}. Indeed, an increased R_{gill} has been reported for rainbow trout, lingcod, European eel, yellowfin tuna and bluefin tuna (Bushnell and Brill, 1992; Farrell, 1982; Perry et al., 1999; Peyraud-Waitzenegger and Soulier, 1989; Sundin and Nilsson, 1997); an increased R_{sys} is displayed by rainbow trout, lingcod, Atlantic cod, 15°C-acclimated winter flounder, coho salmon and yellowfin tuna (Axelsson and Farrell, 1993; Bushnell and Brill, 1992; Farrell, 1982; Fritsche and Nilsson, 1989; Mendonça and Gamperl, 2010; Perry et al., 1999; Sundin, 1995); an increased P_{VA} is exhibited by rainbow trout, five-bearded rockling, Atlantic cod, sea raven, *P. bernacchii, P. borchgrevinki* and *C. aceratus* (Axelsson and Fritsche, 1991; Axelsson et al., 1992; Fritsche, 1990; Fritsche and Nilsson, 1989, 1990; Holeton, 1972; Holeton and Randall, 1967a; Sandblom and Axelsson, 2006; Saunders and Sutterlin, 1971; Sundin, 1995; Sundin and Nilsson, 1997); and an increased P_{DA} is shown by rainbow trout, Atlantic cod, channel catfish, sea raven, 15°C-acclimated winter flounder, coho salmon and *C. aceratus* (Axelsson and Farrell, 1993; Axelsson and Fritsche, 1991; Burleson and Smatresk, 1989, 1990a; Cech et al., 1977; Fritsche and Nilsson, 1989, 1990; Gamperl et al., 1994; Holeton, 1972; Holeton and Randall, 1967a; Mendonça and Gamperl, 2010; Perry and Desforges, 2006; Saunders and Sutterlin, 1971; Smith, 1978; Sundin, 1995) (see Table 1). Moreover, an increased R_{tot} has been reported for *P. bernacchii*, yellowfin tuna and skipjack tuna (Axelsson et al., 1992; Bushnell and Brill, 1992), and isolated perfused gill preparations from rainbow trout and Atlantic cod exhibit an increased R_{gill} when the perfusate and/or water is made hypoxic (Pettersson and Johansen, 1982; Ristori and Laurent, 1977; Smith et al., 2001). The change in vasomotor tone of the branchial vasculature in rainbow trout and Atlantic cod with hypoxia is ascribed to cholinergic vasoconstriction of the efferent filamental arteries, presumably at the efferent filamental artery sphincter (Sundin, 1995; Sundin and Nilsson, 1997). This vasoconstriction contributes to an increased intralamellar transmural pressure, and leads to the recruitment of additional gill lamellae and an increased functional respiratory surface area available for gas exchange. Studies on Atlantic cod have also provided insight into the regulation of the systemic vasculature during hypoxia. Under normoxic conditions, a tonic α-adrenergic tone controls resting P_{DA} and an increased α-adrenergic tone mediates the hypoxic systemic vasoconstriction that serves to regulate blood flow distribution, namely, a reduced gastrointestinal blood flow (Axelsson and Fritsche, 1991; Fritsche and Nilsson, 1989). The α-adrenergic tonus is neurally mediated, as treatment of fish with bretylium, which blocks the release of transmitters from adrenergic nerves, decreased P_{DA}. Further, subsequent injection of the α-adrenergic receptor phentolamine had no effect on P_{DA}, indicating that circulating catecholamines were not important in mediating the increased systemic vasoconstriction during hypoxia.

Nonetheless, the responses of the branchial and systemic circulations to hypoxia that are described above are far from universal among fishes (Table 1). Foremost, in contrast to most teleost species, P_{DA} and R_{sys} are decreased or unchanged in elasmobranchs exposed to hypoxia (Butler and

Taylor, 1971, 1975; Piiper et al., 1970; Satchell, 1961; Short et al., 1979; Speers-Roesch et al., 2012; Taylor et al., 1977). Moreover, the response of R_{sys}, P_{DA}, and P_{VA} is highly variable among species, even within species, including the rainbow trout and Atlantic cod, in which the "typical" piscine responses have been most intensively studied. Specifically, decreased R_{sys} has been reported for rainbow trout (Sandblom and Axelsson, 2005b), whereas unchanged R_{sys} has been reported for rainbow trout, Atlantic cod, 8°C-acclimated winter flounder and skipjack tuna (Axelsson and Fritsche, 1991; Bushnell and Brill, 1992; Mendonça and Gamperl, 2010; Sandblom and Axelsson, 2005b, 2006), and unchanged R_{tot} has been reported for *P. borchgrevinki* (Axelsson et al., 1992). An arterial hypotension accompanying hypoxic bradycardia has also been reported for rainbow trout, lingcod, channel catfish, sea raven, European eel, Japanese eel, 10°C-acclimated winter flounder, Siberian sturgeon, Adriatic sturgeon and *C. aceratus* (Agnisola et al., 1999; Burleson and Smatresk, 1990a; Cech et al., 1977; Chan, 1986; Farrell, 1982; Hemmingsen et al., 1972; Maxime et al., 1995; Peyraud-Waitzenegger and Soulier, 1989; Sandblom and Axelsson, 2005b; Saunders and Sutterlin, 1971; Sundin and Nilsson, 1997), and no change in P_{DA} has been reported for rainbow trout, 8°C winter flounder, Adriatic sturgeon, yellowfin tuna and skipjack tuna (Bushnell and Brill, 1992; McKenzie et al., 1995; Mendonça and Gamperl, 2010; Perry et al., 1999; Sandblom and Axelsson, 2005b, 2006). P_{VA} has been reported to decrease in lingcod, eel pout, dogfish, spotted dogfish, European eel, Japanese eel, traíra, trairão, tambaqui, and gray mullet (Butler and Taylor, 1971, 1975; Butler et al., 1977; Chan, 1986; Farrell, 1982; Fritsche, 1990; Micheli-Campbell et al., 2009; Peyraud-Waitzenegger and Soulier, 1989; Piiper et al., 1970; Shingles et al., 2005; Sundin et al., 1999, 2000; Taylor et al., 1977), and remain unchanged in rainbow trout, short-horned sculpin, dogfish, yellowfin tuna and skipjack tuna (Bushnell and Brill, 1992; Butler and Taylor, 1971, 1975; Fritsche, 1990; Perry et al., 1999; Short et al., 1979). Like the variable f_H response to hypoxia observed within species, some of the intraspecific variation in blood pressures and vascular resistances may be attributed to differences in the duration of hypoxia exposure. For example, in channel catfish, P_{DA} is decreased with short-term (10 min) hypoxia exposure, but increases with longer (1 h) hypoxia exposure (Burleson and Smatresk, 1989, 1990a). Similarly, the response of dogfish to hypoxia varies with the rate at which hypoxia is induced (Butler and Taylor, 1971). With a rapid drop in water oxygen levels, a pronounced bradycardia was accompanied by hypotension in the dorsal and ventral aortae, whereas during slowly induced hypoxia, P_{DA} and P_{VA} were unaltered. In other instances, intraspecific variation can be attributed to different acclimation temperatures. Cold-acclimated (8°C) winter flounder do not exhibit a change in arterial blood pressure, whereas 10°C-acclimated animals exhibit a decrease and 15°C-acclimated fish display a transient increase (Cech et al., 1977; Mendonça and Gamperl, 2010). Nevertheless, the regulatory mechanism(s) that underlie the time- and temperature-dependent responses remain incompletely understood. Overall, while important insights have been made into the regulation of R_{gill}, R_{sys}, R_{tot}, P_{VA} and P_{DA} responses to oxygen limitation, it is clear from the abundance of inter- and intraspecific differences that additional studies are required to develop a more comprehensive understanding of the control of blood pressure and vascular resistances in hypoxic water-breathing fishes.

2.3. Air-Breathers

The ability to breathe air may have evolved independently numerous times in fishes. Within the bony fishes (Class Osteichthyes) 49 air-breathing fish families exist, and within the largest group of bony fishes, the teleosts, 450 species can air-breathe (Graham, 1997). Air-breathing has long been hypothesized to have evolved to benefit oxygen delivery to the tissues, especially the myocardium, in response to regular occurrences of aquatic hypoxia in the freshwater environment and periodic emersion (Barrell, 1916; Carter and Beadle, 1931; Clark et al., 2007; Graham, 1997; Gunther, 1871; Lefevre et al., 2013, 2014a). Indeed, the vast majority of extant air-breathing fishes reside in tropical freshwater environments, where drought and hypoxia are regular occurrences. Moreover, although considerable variation exists among air-breathing fish in terms of the type of air-breathing organ and the location of the air-breathing organ within the circulation, in most species, the location of the air-breathing organ within the circulation results in a mixing of oxygenated blood from the air-breathing organ with systemic venous return just before the heart (Graham, 1997; Ishimatsu, 2012; Lefevre et al., 2014d; Olson, 1994). While this design may not necessarily provide an increased oxygen supply to the systemic tissues during exposure to aquatic hypoxia, due to the potential for oxygen loss from the blood to the water in the branchial circulation if a branchial shunt is absent or inefficient, it serves to increase the partial pressure of oxygen in the blood supplying the heart.

Nonetheless, among the air-breathing fishes, there exists a wide diversity in the habitat in which they reside, the environmental and intrinsic physiological conditions under which air-breathing is employed, and in the need to breathe air. In addition to being found in tropical environments, air-breathing fish also inhabit freshwater and marine habitats at temperature latitudes. One species, the Alaska blackfish (*Dallia pectoralis*), even lives above the Arctic Circle, where it overwinters in hypoxic waters under ice and snow cover that precludes air-breathing (Campbell and Lopéz, 2014; Kubly and Stecyk, 2015; Lefevre et al., 2014b; Ostdiek and Roland, 1959). In addition to being a response to exposure to aquatic hypoxia, several species also employ air-breathing to enhance oxygen delivery to the myocardium, and thereby cardiac performance. This occurs under conditions of increased oxygen demand when venous oxygen return to the heart may otherwise become limiting, including during swimming (Farmer and Jackson, 1998; reviewed by Lefevre et al., 2014c), digestion (Iftikar et al., 2008; Lefevre et al., 2012) and elevated temperature (Johansen, 1970). Finally, some species are facultative air-breathers, meaning that they only breathe air in response to falling water oxygen levels, whereas some species are obligate air-breathers, meaning that they must periodically breathe air in order not to suffocate, even when in normoxic water.

Despite the great diversity of air-breathing fishes, every study conducted to date has demonstrated that hypoxia (aquatic or aerial depending on the species) stimulates an increase in air-breathing frequency. Concomitantly, air-breathing is associated with a suite of cardiorespiratory adjustments, including changes in the activity of the heart, vascular tone and branchial ventilation. These responses are integrated through vagal and other sensory and motor neurons that interconnect the medulla, the gills, the heart, the air-breathing organ, and other structures affected by aerial respiration (Graham, 1997). However, given the multiplicity of air-breathing fishes, it is not surprising that the cardiovascular changes that accompany aquatic hypoxia and air-breathing are variable among species, and that there is no fixed species-specific relationship between air-breathing and cardiorespiratory parameters (Graham, 1997). Yet, some predominant trends among species emerge. These are highlighted below.

The typical cardiac response of air-breathing fish with bimodal respiration when exposed to aquatic hypoxia is bradycardia prior to an air breath and a subsequent transient tachycardia that commences with or just after each air breath (Graham, 1997). A prebreath bradycardia and postbreath tachycardia have been reported for marbled swamp eel (*Synbranchus marmoratus*) (Johansen, 1966; Johansen et al., 1970; Roberts and Graham, 1985; Skals et al., 2006), African sharptooth catfish (*Clarias gariepinus*) (Belão et al., 2011, 2015), the electric eel (*Electrophorus electricus*) (Johansen et al., 1968, 1970), the climbing perch (*Anabas testudineus*) (Singh and Hughes, 1973), the walking catfish (*Clarias batrachus*) (Jordan, 1976), jeju (*Hoplerythrinus unitaeniatus*) (McKenzie et al., 2007), *Hypostomus regani* (Nelson et al., 2007), Asian swamp eel (*Monopterus albus*) (Iversen et al., 2013) and adult and juvenile Pacific tarpon (*Megalops cyprinoides*) (Clark et al., 2007) exposed to aquatic hypoxia. These phasic changes in f_H and branchial ventilatory activity with air-breathing are hypothesized to maximize oxygen uptake from the air-breathing organ via augmented blood flow to the air-breathing organ when the gradient of oxygen to the blood is highest (i.e., just following air inspiration) (Graham et al., 1995; Johansen, 1966). An increased f_H also rapidly distributes the oxygen to tissues to preserve aerobic metabolism (Graham et al., 1995; Johansen, 1966). Concurrently, a diminished branchial ventilation serves to limit blood oxygen loss to the hypoxic water. Indeed, in most air-breathing fishes, deflation of the air-breathing organ alone elicits bradycardia and reduces air-breathing organ perfusion, while spontaneous or experimental air-breathing organ inflation initiates tachycardia, increases air-breathing organ perfusion, and reduces aquatic ventilation (Graham, 1997). In hypoxia-exposed *S. marmoratus*, the tachycardia associated with air-breathing was accompanied by an almost doubling of $\dot{Q}$ and a 34% increase in V_S, which was facilitated by marked increases in P_{CV} and mean circulatory filling pressure (Skals et al., 2006). Thus, like for rainbow trout, where an increase in P_{CV} plays an important role in setting V_S and $\dot{Q}$ during hypoxia exposure (Perry et al., 1999; Sandblom and Axelsson, 2005a,b), the venous system plays an active role in regulating venous return, cardiac filling, V_S and $\dot{Q}$ during air-breathing in *S. marmoratus*. Nevertheless, despite the proposed benefits of the phasic response of f_H with air-breathing, in jeju, the cyclical changes in f_H associated with air-breathing do not function to facilitate oxygen uptake from the air-breathing organ (McKenzie et al., 2007).

Much like water-breathing fishes, air-breathing fish exhibit a spectrum of cardiovascular responses to aquatic hypoxia. For the f_H response, given that air-breathing provides a more secure supply of oxygen to the myocardium as compared to water-breathers, it has been suggested that hypoxic bradycardia is redundant in air-breathing fishes (Farrell, 2007). Indeed, the Dipnoans, namely, the Australian lungfish (*Neoceratodus forsteri*), South American lungfish (*Lepidosiren paradoxa*) and marbled lungfish (*Protopterus dolloi*), which have true lungs and separate efferent vessels leading from the lungs to the right side of the atrium (Ishimatsu, 2012), maintain f_H during exposure to aquatic hypoxia (Fritsche et al., 1993; Perry et al., 2005; Sanchez et al., 2001) (Table 1). Similarly, the mean f_H of spotted gar (*Lepisosteus oculatus*) (Smatresk and Cameron, 1982), jeju (McKenzie et al., 2007), and African sharptooth catfish (Belão et al., 2011, 2015) was unaltered from normoxic levels during exposure to aquatic hypoxia when air-breathing was possible (Table 1), although f_H declined prior to each air breath and tachycardia occurred immediately after inhalation, as detailed above. However, when jeju and African sharptooth catfish were exposed to aquatic hypoxic, and denied air access, they displayed an approximate 50% reduction in f_H (Belão et al., 2011, 2015; Lopes et al., 2010; Oliveira et al., 2004) (Table 1). Walking perch also exhibit a pronounced bradycardia when submerged and prevented from surfacing in hypoxic water (Singh and Hughes, 1973). Further, the longnose gar (*Lepisosteus osseus*), which exhibited tachycardia upon aquatic hypoxia exposure with access to air, exhibited bradycardia when the air phase, in addition to the water within the experimental chamber, was made hypoxic (Smatresk et al., 1986). The differential responses of f_H to aquatic hypoxia, depending on whether oxygen can be obtained by aerial respiration, indicate that hypoxic bradycardia is not lost in these species despite their ability to breathe air. In stark contrast, armored catfish (*Glyptoperichthys gibbiceps* and *Liposarcus pardalis*) did not exhibit hypoxic bradycardia when exposed to hypoxia and denied the opportunity for aerial respiration in the laboratory or in a simulated pond environment (MacCormack et al., 2003). Likewise, the Asian swamp eel (*Monopterus albus*) did not exhibit the typical piscine hypoxic bradycardia when exposed to 30 min of combined water and air anoxia (Iversen et al., 2013). The overall f_H response of *M. albus* to anoxia exposure (calculated during periods of gill ventilation and air-breathing) was unaltered from normoxia. However, reductions in $\dot{Q}$ and CPO, driven by decreased V_S, occurred during periods of gill ventilation, presumably to limit branchial loss of oxygen to the anoxic water. In other species, the f_H response to aquatic hypoxia varies with temperature and developmental stage (Table 1). Bowfin (*Amia calva*) exposed to hypoxia with air access at low temperature (14–16°C) do not exhibit bradycardia (Hedrick et al., 1991), whereas bradycardia is reported for bowfin exposed to hypoxia with air access at high temperature (20°C) (McKenzie et al., 1991b). Juvenile Pacific tarpon exhibited a pronounced tachycardia during hypoxia exposure, whereas adult tarpon did not exhibit a difference in f_H in hypoxic compared to normoxic water (Clark et al., 2007).

As for water-breathing fish, the number of air-breathing fish species in which the overall responses of $\dot{Q}$, V_S, blood pressures (P_{VA} and P_{DA}), CPO, and vascular resistances (R_{gill}, R_{sys}, and R_{tot}) to aquatic hypoxia have been directly measured is considerably less than those in which the overall f_H response has been documented (Table 1). Moreover, unlike for f_H, which as detailed in the preceding paragraph has been measured during aquatic hypoxia exposure with and without air access in some air-breathers, none of these cardiovascular variables have been measured under both conditions to explore if the responses to aquatic hypoxia differ depending on whether air-breathing is possible. A prevailing response to aquatic hypoxia exposure with air access is an increase in $\dot{Q}$. $\dot{Q}$ increased by ~50% in spotted gar exposed to 1 h of hypoxia at 18–21°C (Smatresk and Cameron, 1982), approximately doubled in marbled swamp eel exposed to 5 h of hypoxia at 25°C (Skals et al., 2006), and roughly tripled in juvenile Pacific tarpon exposed to graded hypoxia at 27°C (Clark et al., 2007). This elevated $\dot{Q}$ was accompanied by increases in P_{VA} and P_{DA} in spotted gar, a decreased P_{DA} in marbled swamp eel, and increased V_S in juvenile pacific tarpon. Moreover, in spotted gar, the fraction of $\dot{Q}$ directed toward the air-breathing organ (i.e., lungs) increased from 9% in normoxia to 30% in hypoxia without any change in systemic perfusion (Smatresk and Cameron, 1982). Similarly, Randall et al. (1981) calculated the distribution of $\dot{Q}$ between the air-breathing organ and the systemic circulation in hypoxia-exposed bowfin using the Fick principle (Randall et al., 1981). These calculations revealed that $\dot{Q}$ increased from 69 to 96 mL min^{-1} - kg^{-1} during hypoxia, and that blood flow distribution to the air-breathing organ increased by 17%. Like for the spotted gar, total systemic blood flow remained constant. Taken together, these findings indicate that adjustments in the relative vascular resistance of the air-breathing organ and the systemic capillary beds during aquatic hypoxia exposure can augment blood flow to the air-breathing organ while minimizing O_2 loss across the gills; the latter ensuring that there is an adequate supply of oxygen to the tissues. Interestingly, while juvenile Pacific tarpon exposed to hypoxia increased $\dot{Q}$ through increases in both f_H and V_S, adult Pacific tarpon exposed to hypoxia maintained $\dot{Q}$, f_H and V_S at normoxic levels (Clark et al., 2007) (Table 1). The underlying reason for the differential response between life stages may be the greater proportion of compact myocardium in adult tarpon, which is more susceptible to low-oxygen conditions than the trabecular myocardium found in juveniles. The predominance of compact myocardium may have also reduced the compliance of the heart.

In water-breathers, cardiorespiratory adjustments to hypoxia depend upon externally oriented O_2-sensitive chemoreceptors that monitor PO_2 oscillations in the inspired water and/or internally oriented chemoreceptors that monitor PO_2 oscillations in the arterial blood (reviewed by Milsom, 2012). However, the distribution and orientation of the receptors is not uniform among species. Similarly, the integrated cardiorespiratory response of air-breathing fish exposed to aquatic hypoxia is also mediated through central regulation and O_2-sensitive chemoreceptors in the gills, the location, orientation and innervation of which vary by species (Milsom, 2012). In bowfin, the O_2-sensitive chemoreceptors responsible for f_H responses are located on all four gill arches, are externally oriented, and innervated by the glossopharyngeal and vagus nerves; whereas the O_2-sensitive chemoreceptors involved in gill amplitude and frequency responses are located on all four gill arches as well as extra-branchially, are internally and externally oriented, and are innervated by the glossopharyngeal and vagus nerves; and the O_2-sensitive chemoreceptors involved in air-breathing are located on all four gill arches as well as the pseudobranch, are externally oriented, and innervated by the glossopharyngeal, vagus and facial nerves (McKenzie et al., 1991a,b). In jeju, hypoxic bradycardia, gill amplitude and gill frequency responses are elicited by externally and internally oriented O_2-sensitive chemoreceptors that are located on all four gill arches and are innervated by the glossopharyngeal and vagus nerves, whereas the

O_2-sensitive chemoreceptors involved in air-breathing are also located on all four gill arches and innervated by the glossopharyngeal and vagus nerves, but are internally oriented (Lopes et al., 2010). In African sharptooth catfish, the O_2-sensitive chemoreceptors eliciting cardiac responses are both internally and externally oriented and distributed on all gill arches or extra-branchially, whereas air-breathing responses are predominantly mediated by receptors in the first pair of gill arches (Belão et al., 2015). In longnose gar, the O_2-sensitive chemoreceptors involved in gill frequency responses are located on all four gill arches, are externally and internally oriented, and innervated by the glossopharyngeal and vagus nerves, whereas the receptors mediating air-breathing responses are located on all four gill arches, are internally oriented, and innervated by the glossopharyngeal and vagus nerves (Smatresk, 1988, 1989; Smatresk et al., 1986). Finally, in African lungfish, the response to air-breathing is mediated by internally oriented O_2-sensitive chemoreceptors located on all four gill arches and innervated by the glossopharyngeal and vagus nerves (Lahiri et al., 1970). In addition to O_2-sensitive chemoreceptors, stimulation of mechanoreceptors within the buccal–pharyngeal chamber of synbranchid fish (by air inflation) has been shown to be involved in mediating the cardiac response to surface air-breathing (Graham et al., 1995).

Evidence exists that the cardiovascular changes with air-breathing are under autonomic control. In resting South American lungfish, the heart is under a tonic cholinergic vagal inhibitory tone and a stable stimulatory adrenergic tone, and the cyclic changes in f_H associated with air-breathing are due to modulation of the cholinergic tonus (Axelsson et al., 1989; Sandblom et al., 2010). Similarly, in the Asian swamp eel, pharmacological blockade of muscarinic and adrenergic receptors revealed that the low f_H during water-breathing stems from a high cholinergic tone that inhibits the cardiac pacemaker (Iversen et al., 2011), whereas the tachycardia associated with air-breathing and inflation of the buccal cavity is primarily a result of the withdrawal of this inhibitory tone (Iversen et al., 2011), mediated by the stimulation of mechanoreceptors within the buccal cavity (Graham et al., 1995). Likewise, in jeju exposed to deep aquatic hypoxia, pharmacological blockade revealed that both the adrenergic and cholinergic systems contribute to the cycling of f_H with each air breath, but that modulation of inhibitory cholinergic tone was responsible for the major proportion of the variability in f_H (McKenzie et al., 2007). In hypoxia-exposed *S. marmoratus*, in which the venous system plays an important regulatory role in cardiac filling and V_S during air-breathing, bolus infusions of α- and β-adrenergic agonists (adrenaline, phenylephrine, and isoproterenol) revealed that venous tone is regulated by the sympathetic nervous system, with vasoconstriction mediated by α-adrenergic receptors and vasodilation mediated by β-receptors (Skals et al., 2006).

3. STRATEGIES OF SURVIVAL, CARDIOVASCULAR RESPONSES AND CARDIOVASCULAR CONTROL IN SEVERE HYPOXIA/ANOXIA-TOLERANT SPECIES DURING OXYGEN DEPRIVATION

3.1. Overview of Strategies to Balance ATP Supply and Demand During Oxygen Deprivation

As detailed above (see Section 1.1), death from oxygen limitation is primarily ascribed to the disruption of cellular processes in critical tissues with a high ATP demand like the brain and heart, which arises from a mismatching of anaerobic ATP production to ATP demand. It is not surprising, then, that vertebrate species tolerant of surviving prolonged periods with little to no oxygen employ a plethora of physiological strategies and biochemical mechanisms that presumably allow for a matching of ATP synthesis and ATP demand such that the ionic integrity of cellular membranes is preserved beyond the minutes of survival of anoxia-intolerant vertebrates (Boutilier, 2001). Theoretically, severe hypoxia/anoxia-tolerant vertebrates could balance ATP supply and demand during prolonged anoxia exposure by upregulating glycolytic ATP production to match an unchanged demand (termed the Pasteur effect), reducing energy demand to a level that can be supported by the reduced ATP available from anaerobic metabolism (hypometabolism), or through a combination of both responses (Lutz and Nilsson, 1997). There are clear trade-offs with both strategies. Upregulating glycolysis allows for continued activity during anoxia. However, like for anoxia-intolerant vertebrates, fermentable fuel stores become limiting. Therefore, these stores need to be substantial for long-term anoxia survival. Also, the upregulation of glycolysis will lead to a rapid accumulation of potentially harmful anaerobic end-products and acidosis (Lutz, 1989). In contrast, a benefit of the hypometabolic strategy is that it greatly slows fermentable fuel depletion and waste accumulation, both of which can be critical for long-term anoxia survival. However, hypometabolism may cause the brain to shut down to a comatose-like state and, consequently, impair the animal's ability to respond to external stimuli (Lutz and Nilsson, 1997).

The strategy of upregulating glycolytic ATP production to meet somewhat reduced ATP demands is displayed by the anoxia-tolerant crucian carp. Crucian carp remain active during 5.5 h of anoxia exposure at 9°C and continue to swim, albeit at a reduced level compared to normoxia (Nilsson et al., 1993). Consequently, their brain must remain functional (active) if it is to coordinate locomotion (Lutz and Nilsson, 1997; Nilsson, 2001). Accordingly, there is no evidence of reduced protein synthesis in the brain of crucian carp during anoxia exposure (Smith et al., 1996). Similarly, there is no evidence of "channel arrest"; a theoretical construct that predicts that the long-term survival of hypothermic and/or anoxic conditions is facilitated by the coordinated suppression of transmembrane ion flow through ion channels and active ion transport by ion pumps such as the Na^{+}-K^{+}-ATPase (Hochachka, 1986; Lutz et al., 1985) in the brain crucian carp during anoxia (Johansson and Nilsson, 1995). However, some aspects of central nervous system function are reversibly reduced during anoxia in *Carassius*. For example, auditory nerve activity is suppressed during anoxia in the closely related goldfish (Suzue et al., 1987) and anoxic crucian carp become blind (Johansson et al., 1997). Further, microcalorimetric studies of crucian carp brain slices show that there is at least a 30%–40% reduction in ATP turnover during anoxia, a finding indicative of metabolic depression (Johansson et al., 1995). However, despite this reduction, an approximate 8- to 10-fold increase in glycolytic rate is still needed to maintain ATP supply. Similarly, the reduction of body heat production to one-third of the normoxic levels in *Carassius* during anoxia (Van Waversveld et al., 1989a,b), which is mediated in part by the metabolic sensor AMP-activated kinase (AMPK; Stensløkken et al., 2008), is not large enough to avoid activation of the Pasteur effect. Indeed, ATP supply in the brain of crucian carp is maintained through a sustained increase in glycolytic rate (Johansson et al., 1995). Specifically, the increase in glycolysis arises from an upregulation of key glycolytic enzymes and an increase in fructose 2,6-bisphosphate activity, a potent activator of glycolysis (Storey, 1987). Simultaneously, brain blood flow is increased and sustained at an elevated level to deliver fermentable fuel and remove wastes (Nilsson et al., 1994).

In contrast to the crucian carp, freshwater turtles of the genera *Chrysemys* and *Trachemys* exhibit the anoxia survival strategy of downregulating ATP demand and ATP production in concert, such that ATP supply and demand are matched. With the onset of anoxia exposure, the turtle first attempts to compensate for the reduction in ATP generation by increasing brain blood flow by 171%–242% (Hylland et al., 1994, 1996) and upregulating glycolysis, as evidenced by the quick depletion of brain, heart and skeletal muscle glycogen stores (Daw et al., 1967; Lutz and Nilsson, 1993; Wasser and Jackson, 1991). However, with protracted periods of anoxia (i.e., hours to days), a biochemical reorganization occurs that leads to a coordinated metabolic depression and a suppression of the Pasteur effect. For example, turtle brain blood flow returns to preanoxic levels within 1–2 h of hypoxic exposure (Hylland et al., 1994, 1996), and the initially upregulated pyruvate kinase activity in turtle muscle is inhibited (Kelly and Storey, 1988). Ultimately, whole-body metabolic rate of warm-acclimated turtles (20–24°C) is depressed to 15%–18% of the normoxic metabolic rate during prolonged anoxia exposure (Herbert and Jackson, 1985b; Jackson, 1968). For cold-acclimated turtles (3°C), the decreases in metabolic rate are even greater, reaching values less than 10%

of the normoxic metabolic rate at 12 weeks of anoxia exposure (Herbert and Jackson, 1985b). The decrease in metabolic rate in the anoxia-tolerant turtle during prolonged anoxia exposure occurs through the suppression of ion pumping and protein turnover, the key ATP-consuming processes of a cell (Hochachka, 1986; Hochachka and Somero, 2002; Hochachka et al., 1996). Moreover, several studies have directly or indirectly demonstrated the occurrence of "channel arrest" in the anoxic turtle brain at high temperature. In particular, voltage-gated Na^+ channel density, K^+ channel leakage rate, Ca^{2+} channel activity, Ca^{2+}-activated K^+ channel open probability, *N*-methyl-D-aspartate (NMDA) receptor open probability, and whole-cell current and alpha-amino-3-hydroxy-5-methylisoxazole-4-propionic acid (AMPA) receptor whole-cell peak current all decrease with anoxia exposure at room temperature (Bickler, 1992; Bickler and Buck, 1998; Buck and Bickler, 1998; Chih et al., 1989; Pamenter et al., 2008; Perez-Pinzon et al., 1992; Rodgers-Garlick et al., 2013; Shin and Buck, 2003). These changes, combined with a reduction of brain Na^+-K^+-ATPase activity by 30% and 55% at high and low acclimation temperature, respectively (Hylland et al., 1997; Stecyk et al., 2017), result in the anoxic brain [ATP] being either stable (Buck and Bickler, 1995; Doll et al., 1994; Kelly and Storey, 1988; Lutz et al., 1984) or only moderately lower (Buck et al., 1998). Suppressed protein synthesis, termed translational arrest, also occurs in turtle heart (Bailey and Driedzic, 1996), hepatocytes (Land et al., 1993) and brain (Fraser et al., 2001) during anoxia, although some stress proteins (Chang et al., 2000; Stecyk et al., 2012) and glycolytic enzymes (Hochachka et al., 1996) can be upregulated during anoxia. Additionally, increased levels of inhibitory neurotransmitters (Nilsson and Lutz, 1991) contribute to energy conservation in the anoxic turtle by reducing the electrical activity and firing frequency of brain cells (Chih et al., 1989; Sick et al., 1993); a phenomenon termed spike arrest. The Pasteur effect, like metabolism, is inhibited by a variety of means during prolonged anoxia in the turtle. The activities of phosphofructokinase-1, glycogen phosphorylase and pyruvate kinase are reduced through posttranslational phosphorylation (i.e., covalent modification) in the anoxic turtle (Brooks and Storey, 1988, 1989, 1990; Kelly and Storey, 1988; Mehrani and Storey, 1995). Further, since the association of enzymes with the particulate fraction of the cell (i.e., membrane fractions, glycogen particles, or structural portions of the cell) increases the efficiency of enzymatic pathways (Storey, 1985), the proportion of glycolytic enzymes (e.g., pyruvate kinase and lactate dehydrogenase) associating with the particulate vs soluble fractions of the cell is significantly reduced following a long-term (20 h) anoxic exposure compared to a short (5 h) anoxic exposure in the turtle as a means to limit the Pasteur effect (Duncan and Storey, 1992). Finally, the activity of glycolytic enzymes can also be depressed by the fall in pH that accompanies prolonged anoxia (Storey, 1996).

3.2. Cardiovascular Responses of Severe Hypoxia/Anoxia-Tolerant Species to Oxygen Deprivation

The heart is a unique tissue because its metabolic activity and performance (i.e., $\dot{Q}$) are dependent on the whole-body demand for blood flow, which reflects whole-animal metabolic rate (Jackson, 2000b). Thus, examination of the working heart provides a perspective on whole-animal metabolic rate. For instance, under resting conditions, cardiac performance will be at some basal level to support the routine metabolic needs of the organism. However, if metabolic needs increase, for example during exercise, cardiac performance will also increase to maintain sufficient oxygen delivery to metabolically active tissues. Similarly, if metabolic rate is reduced, cardiac performance will decrease accordingly. In severe hypoxia/anoxia-tolerant species during prolonged anoxia exposure, the heart continues its role in internal convection despite the absence of oxygen. Specifically, continued cardiac activity is needed to transport metabolites and anaerobic waste products between tissues. Thus, like in normoxia, the level of cardiac performance during severe hypoxia/anoxia exposure will be set by whole-body demand for blood flow, and species with differing anoxia survival strategies would be expected to exhibit different cardiovascular responses to prolonged anoxia exposure.

Studies of the working heart provide an excellent solution to the challenge of studying the simple principle that anoxic cardiac failure can be traced to an inadequate matching of ATP supply to ATP demand. CPO, estimated from the product of $\dot{Q}$ and P_{VA} (an indirect measure of cardiac ATP demand), is directly related to cardiac ATP supply up to some maximal level during both normoxia (ATP supply estimated from myocardial O_2 consumption) and anoxia (ATP supply estimated from lactate production rates) (Arthur et al., 1992; Farrell et al., 1985; Graham and Farrell, 1989; Reeves, 1963). Therefore, under both aerobic and anaerobic conditions, a stable CPO indicates a matching of ATP supply and demand. Conversely, a decreasing CPO suggests that ATP supply and demand are not balanced.

Consistent with the general discussion on potential anoxia survival strategies (Section 3.1), it has been proposed that severe hypoxia/anoxia-tolerant vertebrates utilize one of two potential strategies to match CPO to the less efficient generation of ATP from anaerobic glycolysis that occurs during prolonged anoxia exposure (Farrell and Stecyk, 2007). Based on available measurements of cardiac activity during oxygen deprivation in hagfish, common carp, crucian carp and the red-eared slider turtle (*Trachemys scripta*), and a predicted value of the maximum level of cardiac work (i.e., CPO) that could be supported by anaerobic glycolysis (termed the maximum cardiac glycolytic potential and estimated to be ~0.7 mW g^{-1} for hearts of ectothermic vertebrates at 15°C), Farrell and Stecyk (2007) proposed that cardiac anoxia survival is possible by one of two strategies: (1) having a routine CPO that falls within the maximum cardiac glycolytic potential; or (2) temporarily and substantially decreasing cardiac metabolism so that PO is below maximum cardiac glycolytic potential during prolonged periods of oxygen deprivation (Farrell and Stecyk, 2007). Hagfish and crucian carp have been predicted to utilize the former strategy, whereas common carp and the red-eared slider turtle use the latter. The reasoning is as follows. The routine CPO of Atlantic hagfish (*Myxine glutinosa*) and New Zealand hagfish (*Eptatretus cirrhatus*) measured under normoxia (0.1 and 0.4 mW g^{-1}, respectively) was below that found in elasmobranches and teleosts, as well as below the estimated maximum cardiac glycolytic potential (i.e., 0.7 mW g^{-1}). Further, hagfish cardiac performance during brief severe hypoxia exposures (15 min at 10°C and 30 min at 17°C), as assessed by perfused heart preparations, was comparable to that measured during normoxia (Axelsson et al., 1990; Farrell, 1991; Forster et al., 1991) (Table 1). Similarly, crucian carp, with an estimated CPO of 0.46 mW g^{-1} under normoxia at 8°C, retained normal CPO for at least 5 days of anoxia (Stecyk et al., 2004b) (Fig. 1; Table 1). Although, the immediate and acute cardiovascular response to anoxia in the crucian carp is a substantial bradycardia (Stecyk et al., 2004b; Vornanen, 1994a; Vornanen and Tuomennoro, 1999), by 48 h of anoxia at 8°C, CPO, $\dot{Q}$, f_H and V_S all returned to preanoxic levels, while P_{VA} and R_{tot} decreased significantly (by 30% and 40%, respectively); the latter signifying vasodilation in peripheral tissues (Stecyk et al., 2004b) (Fig. 1). Of importance to recall here is that a decrease in R_{tot} has never been documented to occur in any severe hypoxia/anoxia-intolerant or any other severe/hypoxia-tolerant species exposed to oxygen deprivation (Table 1). In contrast, in common carp, f_H and $\dot{Q}$ are quickly and substantially reduced (by approximately 80% of normoxic levels) throughout severe hypoxia exposure (Stecyk and Farrell, 2002, 2006) (Fig. 2; Table 1). This large depression in cardiac function results in a significant arterial hypotension (Stecyk and Farrell, 2006), where P_{VA} is maximally reduced by as much as 38%, 38% and 29% at 5°C, 10°C and 15°C, respectively (Fig. 2). However, the extent of the hypotension is attenuated by a parallel increase in R_{tot} of 2.0-fold at 5°C, 3.4-fold at 10°C and 2.9-fold at 15°C (Fig. 2). Combined, the marked decreases in $\dot{Q}$ and P_{VA} result in maximal reductions in CPO of 76% at 15°C, 87% at 10°C and 72% at 5°C that result in minimum CPO values of 0.21 mW g^{-1} at 15°C, 0.09 mW g^{-1} at 10°C and 0.08 mW g^{-1} at 5°C; all these are well below the theoretical anoxic capability of 0.7 mW g^{-1} (Stecyk and Farrell, 2006) (Fig. 2). Nonetheless, the 3.2- to 7.7-fold reductions in CPO fall short of the 14-fold reduction needed to preclude an activation of a Pasteur effect. This suggests that the severely hypoxic common carp heart depletes metabolic fuel at a faster rate than in normoxia,

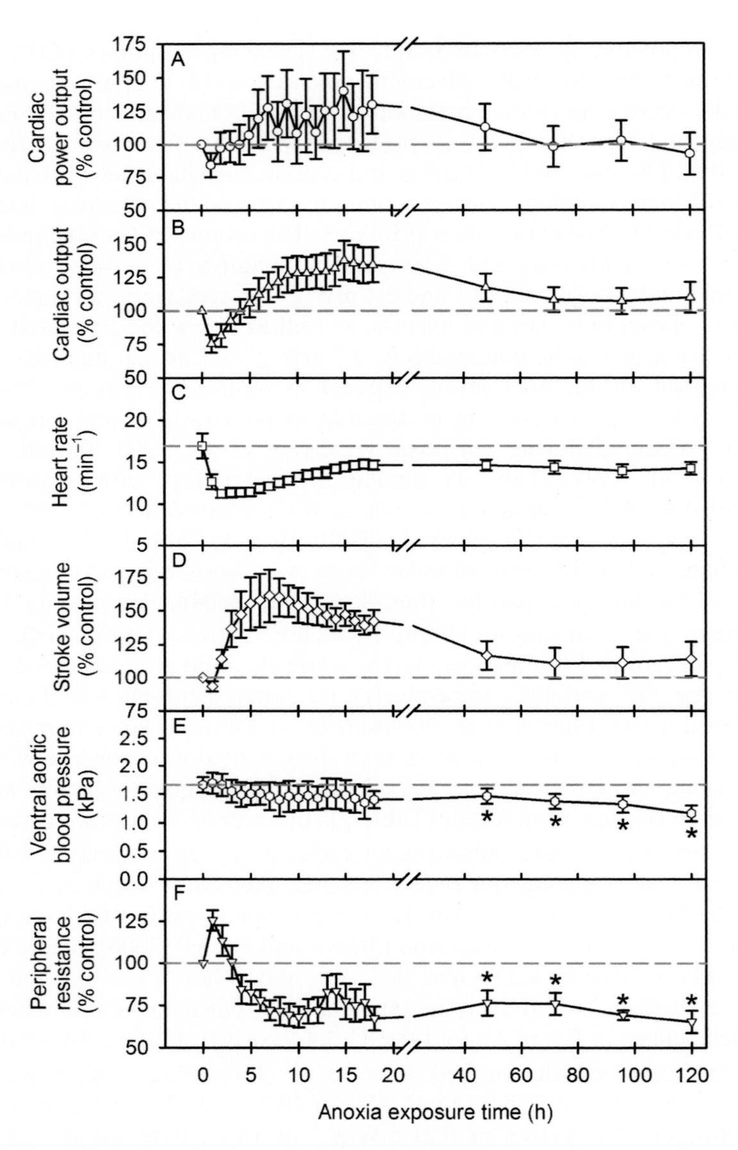

Fig. 1. Changes in the cardiovascular status of 8°C-acclimated crucian carp (*Carassius carassius*) exposed to 5 days of anoxia. Significant differences ($P<0.05$) between normoxic control (time zero) and hours of anoxic exposure (48, 72, 96, and 120) are indicated by *asterisks*. The *dashed line* indicates normoxic control values. Values are means ± S.E.M.; $N=6$–18. Data adapted from Stecyk, J.A., Stensløkken, K.-O., Farrell, A.P., Nilsson, G.E., 2004. Maintained cardiac pumping in anoxic crucian carp. Science 306, 77.

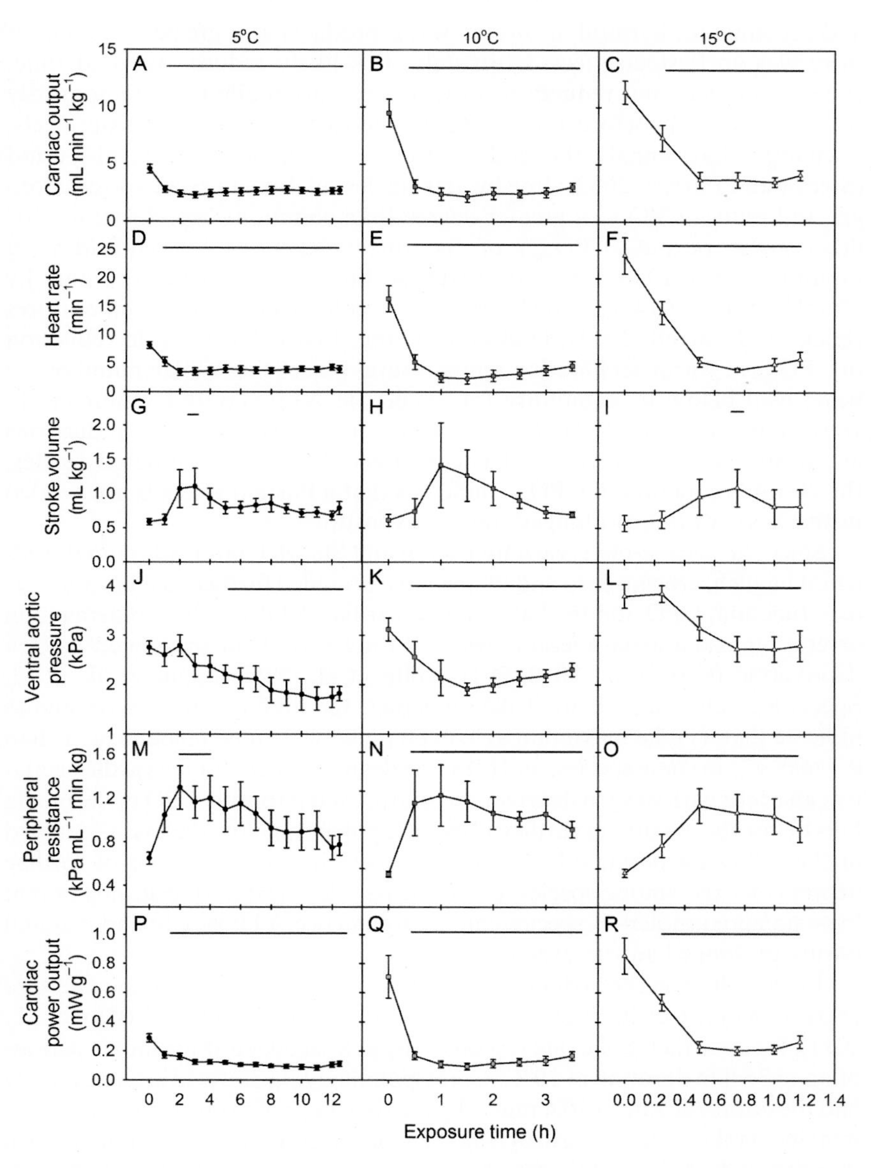

Fig. 2. Changes in the cardiovascular status of 5°C-, 10°C-, and 15°C-acclimated common carp (*Cyprinus carpio*) when exposed to prolonged severe hypoxia (0.3 mg O_2 L^{-1}). Significant differences ($P < 0.05$) from normoxic control (time zero) values are indicated by *solid lines* above the traces. Values are means ± S.E.M.; $N = 8$–14 at 5°C, 6 at 10°C, and 6–8 at 15°C. Data adapted from Stecyk, J.A., Farrell, A.P., 2006. Regulation of the cardiorespiratory system of common carp (*Cyprinus carpio*) during severe hypoxia at three seasonal acclimation temperatures. Physiol. Biochem. Zool. 79, 614–627.

and accumulates harmful anaerobic waste products at a greater rate than if there was no Pasteur effect. Ultimately, this likely reduces survival time. Indeed, the maximal reductions in CPO are only maintained temporarily by common carp before they begin to increase toward normoxic levels, a change that signals the end of the severe hypoxia survival period (Stecyk and Farrell, 2002). Freshwater turtles tolerate anoxia exposure longer and reduce CPO to a greater degree than common carp. Systemic cardiac power output (CPO_{sys}) of the turtle decreases by 6.6-fold to a minimum of 0.10 mW g^{-1} at warm acclimation temperatures and by 20-fold from 0.044 to 0.002 mW g^{-1} at cold acclimation temperatures (Hicks and Farrell, 2000a; Hicks and Wang, 1998). The 6.6-fold reduction in CPO_{sys} at warm acclimation temperatures places the ATP demand of the heart well below its capability for anaerobic ATP supply (Arthur et al., 1997; Farrell et al., 1994; Reeves, 1963). Nevertheless, these reductions are insufficient to prevent a Pasteur effect. For 5°C-acclimated turtles, the 20-fold reduction of CPO_{sys} indicates that a Pasteur effect is not needed in the heart during prolonged anoxia exposure.

Since the decade-old review by Farrell and Stecyk (2007), additional studies on hagfish, crucian carp and tilapia have provided further insights into cardiac function, CPO and the balancing of cardiac ATP supply and demand in severe hypoxia/anoxia-tolerant species during periods of prolonged oxygen deprivation (Cox et al., 2010, 2011; Gillis et al., 2015; Lague et al., 2012; Speers-Roesch et al., 2010). Taken together, the findings from these studies indicate that: (1) the maximum glycolytic potential can be much greater than 0.7 mW g^{-1} in some species; and (2) some degree of reduction in cardiac activity can occur with oxygen deprivation *in vivo*, even if routine CPO in normoxia falls below the maximum cardiac glycolytic potential. Nonetheless, as detailed in the following paragraphs, large differences in the magnitude of cardiac depression exist among species, and these parallel the differing strategies that hypoxia/anoxia-tolerant species employ to balance ATP supply and demand during prolonged anoxia exposure.

In hagfish, it has been discovered that *E. stoutii* utilizes metabolic rate suppression as part of its strategy for longer-term anoxia survival (Cox et al., 2011). Routine metabolic rate is reduced by at least 50% during anoxic periods of up to 36 h in duration at 10°C. Correspondingly, f_H slows to approximately half the normoxic rate of 10.4 min^{-1} by the sixth hour of exposure, where it then remains stable (Cox et al., 2010) (Fig. 3). Concurrently, V_S nearly doubles such that $\dot{Q}$ is only decreased by 33% from the routine normoxic value of 12.3 mL min^{-1} kg^{-1} (Fig. 3). P_{VA}, which initially increases by 60% upon exposure to anoxia, is restored to and maintained at the normoxic level after 3 h of exposure (Fig. 3). Consequently, CPO increases with the commencement of anoxia

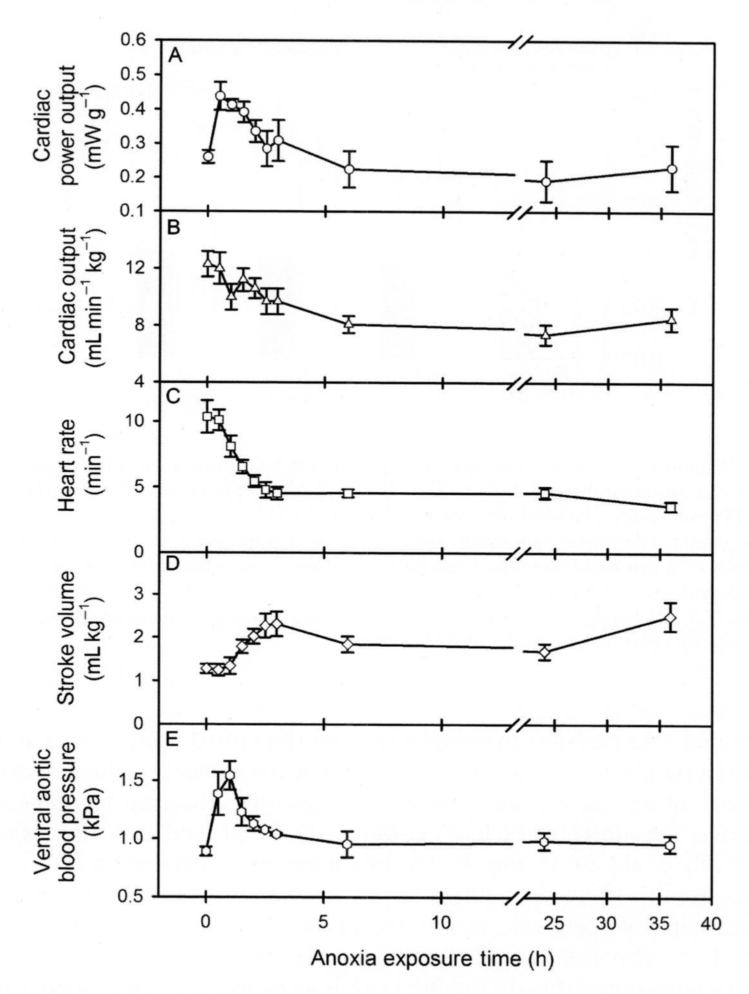

Fig. 3. Changes in cardiovascular status of Pacific hagfish (*Eptatretus stoutii*) exposed to 36 h of anoxia at 10°C. Values are means ± S.E.M.; $N=2$–10. Data adapted from Cox, G.K., Sandblom, E., Farrell, A.P., 2010. Cardiac responses to anoxia in the Pacific hagfish, *Eptatretus stoutii*. J. Exp. Biol. 213, 3692–3698.

exposure, but then stabilizes at a level just 25% below the routine value of 0.26 mW g^{-1} (Fig. 3). Upon reoxygenation, f_H and $\dot{Q}$ rapidly increase to levels approximately double that of routine normoxia before returning to normoxic levels within 3–6 h. In agreement with the *in vivo* measurements of $\dot{Q}$ during prolonged anoxia exposure in Pacific hagfish, the metabolic response of the excised,

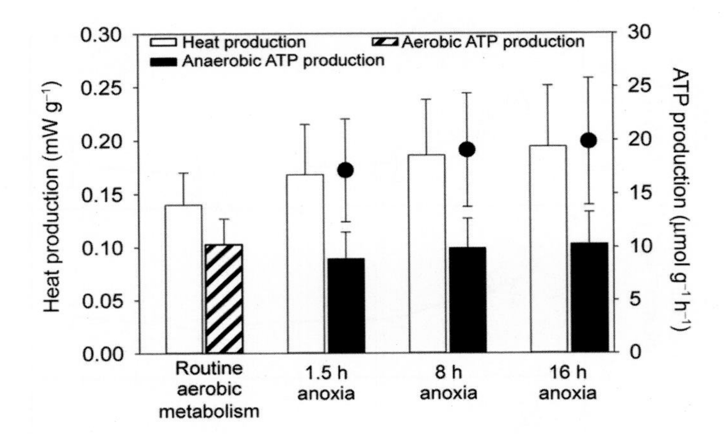

Fig. 4. Metabolic heat (white bars) and ATP production by excised hagfish hearts during normoxia (i.e., routine aerobic metabolism) and after 1.5, 8 and 16 h of anoxia. The *striped bar* is aerobic ATP production. The *black bars* are glycolytic ATP yields calculated assuming that 2 mol of ATP is generated per mole of substrate. The *black circles* are glycolytic ATP yields assuming that 3 mol of ATP is generated per mole of substrate. There are no statistical differences between any of the measurements. $N=5$ hearts. Data are from Gillis, T.E., Regan, M.D., Cox, G.K., Harter, T.S., Brauner, C.J., Richards, J.G., Farrell, A.P., 2015. Characterizing the metabolic capacity of the anoxic hagfish heart. J. Exp. Biol. 218, 3754–3761.

cannulated, hagfish heart to complete anoxia (measured using direct calorimetry) was a significant reduction in the rate of metabolic heat production over the first hour of anoxia exposure, which subsequently recovered to near-routine levels over the subsequent 6 h of exposure when expressed as ATP production rates (Gillis et al., 2015) (Fig. 4). The initial metabolic depression is thought to be the result of an immediate arrest of aerobic respiration, while the reestablishing of metabolic rate is the result of the activation of anaerobic metabolism supported by glycogen. Combined, the results of the *in vivo* and *in vitro* studies reveal that the hagfish heart is an incredibly robust organ capable of maintaining metabolic activity during periods of prolonged anoxia. Even though CPO decreases slightly with anoxia, such a small decrease in cardiac performance during prolonged anoxia exposure is impressive and is surpassed only by 8°C-acclimated anoxic crucian carp.

In crucian carp, it was recently reported that winter-acclimated fish exposed to 6 weeks of anoxia at 2°C displayed a sustained bradycardia from 10.3 to 4.1 min^{-1} (Tikkanen et al., 2017). Clearly, the ~60% reduction in f_H at 2°C contrasts with the maintained f_H displayed by 8°C-acclimated fish (Stecyk et al., 2004b). Indeed, from an energetic perspective, the anoxic bradycardia should conserve metabolic energy. Fewer heart beats translate into less myofilament ATPase activity and fewer APs, which would require less restoration of Na^+, Ca^{2+} and K^+ gradients. However, without knowledge of the $\dot{Q}$ and P_{VA} responses to prolonged anoxia at 2°C, it is impossible to discern CPO (i.e., ATP demand). If it is assumed that V_S, $\dot{Q}$ and P_{VA} exhibit similar responses as during anoxia exposure at 8°C, a moderate fall in CPO in concert with the anoxic bradycardia would be expected. Nevertheless, compared to 5°C-acclimated freshwater turtles, which display a greater than five-fold decrease in f_H from ~5 min^{-1} to less than 1 min^{-1} within 48 h of the commencement of anoxia exposure (Hicks and Farrell, 2000a; Stecyk et al., 2004a, 2008, 2009), the magnitude of bradycardia displayed by 2°C-acclimated anoxic crucian carp is considerably less (Fig. 5). Thus, the relatively moderate bradycardia displayed by anoxic crucian carp at 2°C is in line with the contrasting anoxia survival strategies displayed by crucian carp and freshwater turtles.

Investigation into the tolerance of the tilapia heart to anoxia and acidosis using an *in situ* perfused heart preparation revealed that the maximum glycolytic potential of the tilapia heart is exceptional (Lague et al., 2012). Under normoxic conditions at 22°C, maximum CPO was nearly 5 mW g^{-1}, and with anoxic perfusion, maximum CPO and $\dot{Q}$ only decreased by 30% and 22%, respectively (Fig. 6). These findings imply that no major depression of cardiac CPO is required by the tilapia heart to balance ATP supply and demand during anoxia exposure. However, tilapia exposed to 8 h of hypoxia (water PO_2=1 kPa) at 22°C substantially downregulated cardiovascular status (Speers-Roesch et al., 2010) (Fig. 7; Table 1). *In vivo*, CPO was decreased by 50%–60%, driven largely by a fall in f_H that reduced $\dot{Q}$. V_S and P_{VA} remained stable throughout hypoxia exposure, whereas R_{tot} increased when the water oxygen fell below 15% of air saturation (Fig. 7). The disparate results between the *in situ* and *in vivo* experiments signify that other factors also play a role in setting cardiac CPO *in vivo*. For tilapia, it is hypothesized that a reduction in cardiac contractile activity may be more important for minimizing fuel use and waste production than matching ATP supply and demand. Hypoxia exposure *in vivo* was associated with an increase in heart lactate content and a decrease in cardiac intracellular pH (Speers-Roesch et al., 2010), and the additive effect of acidosis and anoxia *in situ* significantly decreased maximum cardiac performance (Lague et al., 2012).

3.3. Cardiovascular Control in Severe Hypoxia/Anoxia-Tolerant Species During Prolonged Oxygen Deprivation

The brain of anoxic crucian carp remains functional (active) during hypoxia exposure (Lutz and Nilsson, 1997; Nilsson, 2001). Therefore, it should not be surprising that autonomic control of the heart and peripheral circulation also remains intact during anoxia exposure in crucian carp (Stecyk et al.,

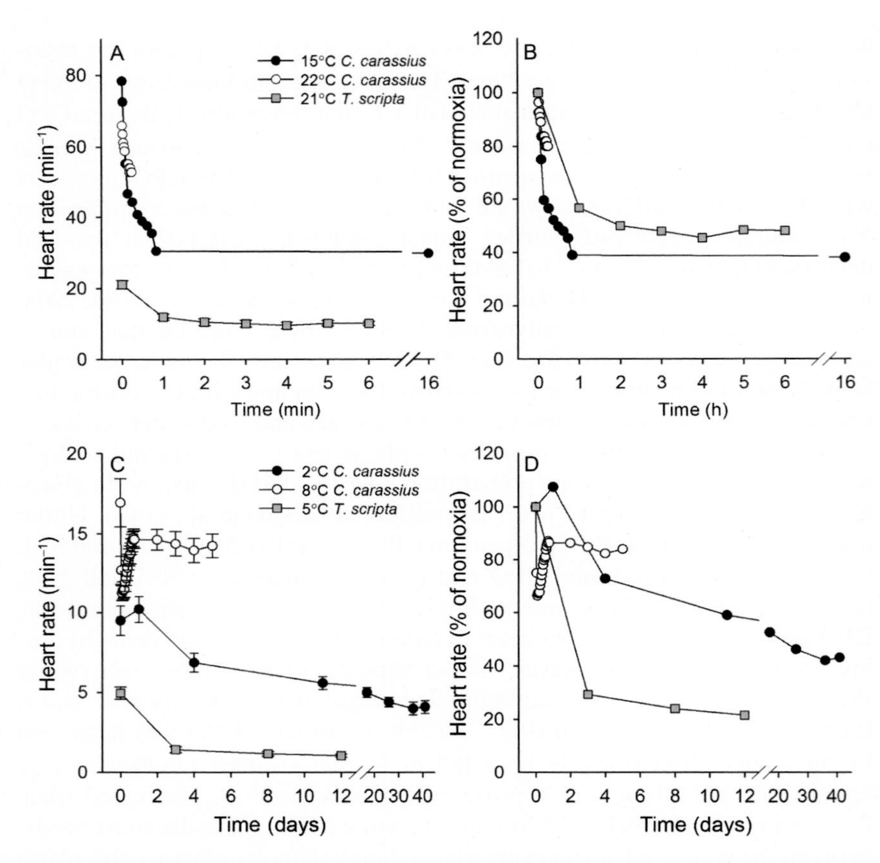

Fig. 5. Comparison of the heart rate (f_H) response of crucian carp (*Carassius carassius*) and red-eared slider (freshwater turtle, *Trachemys scripta*) to anoxia exposure at high (A and B) and low (C and D) acclimation temperatures. Absolute values are presented in panels A and C, whereas in panels B and D, f_H is normalized to the rate during normoxia. Note the different scaling of the *y*-axis between panels A and C. In panel A, the break in the *x*-axis indicates a variable period of 30 min to 16 h for the 15°C fish. $N=6$ for *C. carassius* at 15°C and 22°C, 16–18 for *C. carassius* at 8°C, 10 for *C. carassius* at 2°C, 8 for *T. scripta* at 21°C, and 6 for *T. scripta* at 5°C. Data adapted from various sources: Stecyk, J.A., Overgaard, J., Farrell, A.P., Wang, T., 2004. α-adrenergic regulation of systemic peripheral resistance and blood flow distribution in the turtle *Trachemys scripta* during anoxic submergence at 5°C and 21°C. J. Exp. Biol. 207, 269–283. Stecyk, J.A., Stensløkken, K.-O., Farrell, A.P., Nilsson, G.E., 2004. Maintained cardiac pumping in anoxic crucian carp. Science 306, 77. Tikkanen, E., Haverinen, J., Egginton, S., Hassinen, M., Vornanen, M., 2017. Effects of prolonged anoxia on electrical activity of the heart in Crucian carp (*Carassius carassius*). J. Exp. Biol. 220, 445–454. Vornanen, M., 1994. Seasonal adaptation of crucian carp (*Carassius carassius* L.) heart: glycogen stores and lactate dehydrogenase activity. Can. J. Zool. 72, 433–442. Vornanen, M., Tuomennoro, J., 1999. Effects of acute anoxia on heart function in crucian carp: importance of cholinergic and purinergic control. Am. J. Physiol. 277, R465–R475.

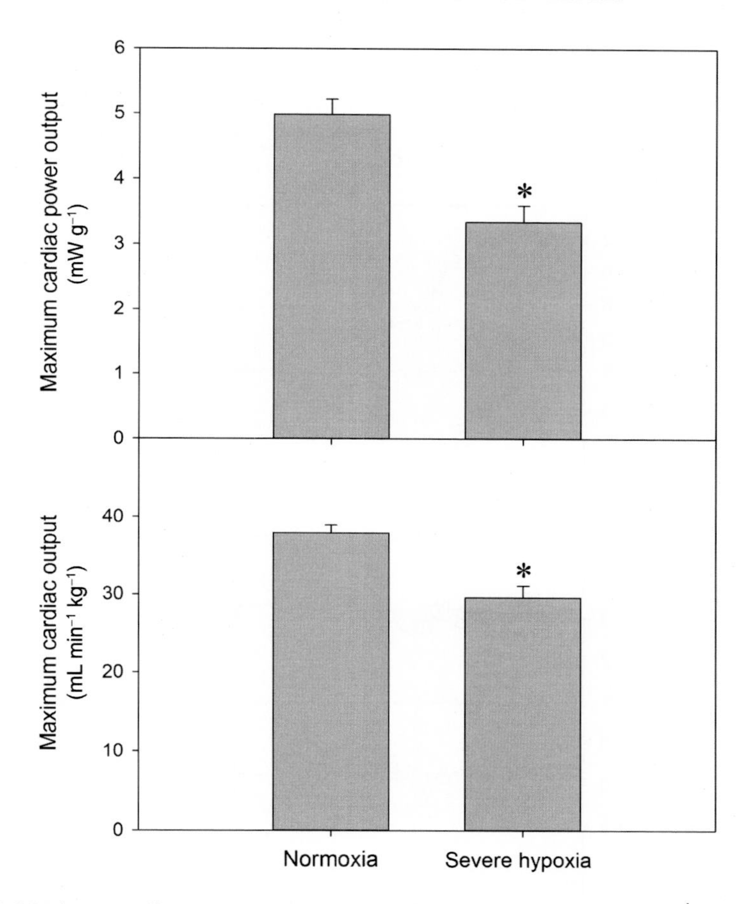

Fig. 6. Maximum cardiac power output (CPO; A) and maximum cardiac output ($\dot{Q}$; B) of *in situ* perfused tilapia (*Oreochromis niloticus* × *mossambicus* × *hornorum*; strain origin: Ace Developments, Bruneau, ID, USA) hearts during normoxia and exposure to severe hypoxia (<0.20 kPa O_2) at 22°C. An *asterisk* indicates a significant difference ($P < 0.05$) between normoxia and severe hypoxia. $N = 6$. Data adapted from Lague, S.L., Speers-Roesch, B., Richards, J.G., Farrell, A.P., 2012. Exceptional cardiac anoxia tolerance in tilapia (*Oreochromis hybrid*). J. Exp. Biol. 215, 1354–1365.

2004b; Vornanen and Tuomennoro, 1999), unlike in cold-acclimated anoxic turtles (Hicks and Farrell, 2000b; Stecyk et al., 2004a). Injections of atropine (cholinergic antagonist) increase, and propranolol (β-adrenergic antagonist) decrease, f_H during anoxia at 8°C, as also occurs during normoxia (Stecyk et al., 2004b). Further, tonic α-adrenergic vasoconstriction is present in crucian carp under normoxia and with prolonged anoxia exposure (Stecyk

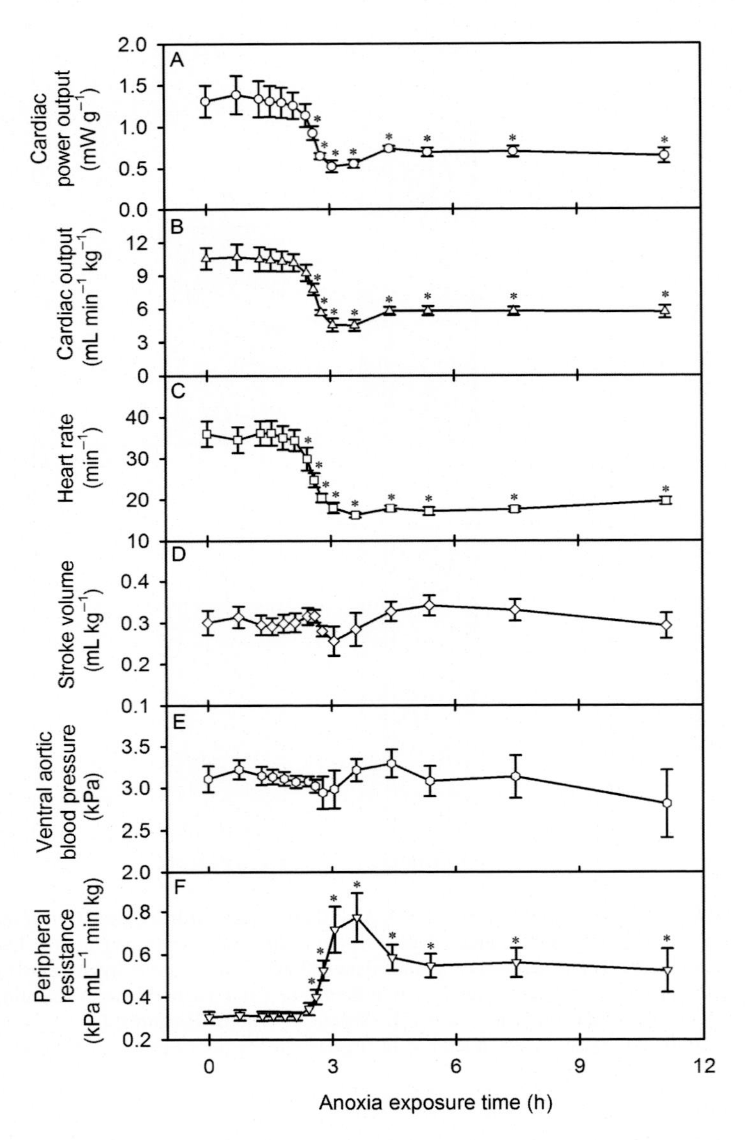

Fig. 7. Changes in the cardiovascular status of 22°C-acclimated tilapia (*Oreochromis niloticus* × *mossambicus* × *hornorum*; strain origin: Ace Developments, Bruneau, ID, USA) when exposed to a progressive decrease in water oxygen content from 92% air saturation to 2.5% air saturation over 3.6 h, followed by 8 h at 5% air saturation. Significant differences ($P < 0.05$) as compared to normoxic control (time zero) values are indicated by *asterisks*. Values are means ± S.E.M.; $N = 5$–6. Data adapted from Speers-Roesch, B., Sandblom, E., Lau, G.Y., Farrell, A.P., Richards, J.G., 2010. Effects of environmental hypoxia on cardiac energy metabolism and performance in tilapia. Am. J. Physiol. Regul. Integr. Comp. Physiol. 298, R104–R119.

et al., 2004b). Pharmacological α-adrenergic blockade with phentolamine decreases P_{VA} and R_{tot} under both normoxia and anoxia. Thus, crucian carp retain cholinergic inhibitory and adrenergic excitatory cardiac control, as well as excitatory adrenergic vascular control, during 5 days of anoxia at 8°C. It appears likely that autonomic cardiovascular control, at least a vagal cholinergic inhibitory tone, persists beyond 5 days of anoxia and at colder acclimation temperatures. The bradycardia displayed by winter-acclimated crucian carp during 6 weeks of anoxia exposure at 2°C was associated with a marked increase in heart rate variability, which is indicative of increased vagal tone in fish and mammals (Tikkanen et al., 2017). There is no adenosinergic inhibition of the anoxic crucian carp heart. Despite adenosine having weak negative chronotropic and inotropic effects on normoxic carp cardiac tissue *in vitro* (Vornanen and Tuomennoro, 1999), pharmacological blockade of adenosine receptors had no significant effect on cardiac activity during either short-term anoxia at 22°C (Vornanen and Tuomennoro, 1999) or 5 days of anoxia at 8°C (Stecyk et al., 2007).

In common carp, the cardiovascular responses to prolonged severe hypoxia are centrally regulated, independent of acclimation temperature (Stecyk and Farrell, 2006). Intraarterial injections of α- and β-adrenergic and cholinergic antagonists clearly reveal that an inhibitory cholinergic cardiac tonus, an α-adrenergic-mediated peripheral vasoconstriction, and a large stimulatory cardiac β-adrenergic tone are present. Specifically, a vagal-mediated bradycardia reduces $\dot{Q}$ and creates an arterial hypotension, which is partially attenuated by an α-adrenergic-mediated increase in R_{tot} (Stecyk and Farrell, 2006). The underlying stimulatory cardiac β-adrenergic tone may possibly protect the heart from attendant acidotic conditions during prolonged severe hypoxia as it does in rainbow trout (*Oncorhynchus mykiss*) (Farrell and Milligan, 1986; Farrell et al., 1983, 1986; Hanson et al., 2006).

Nevertheless, the bradycardia displayed by severe hypoxia/anoxia-tolerant fishes during oxygen deprivation is not always the result of cholinergic inhibition. With brief bouts (20–30 min) of severe hypoxia (0.3–0.8 mg O_2 L^{-1}) at 25°C, the anesthetized epaulette shark exhibited a noncholinergically mediated 33% reduction in f_H (Stensløkken et al., 2004) (Table 1). Similarly, a direct hypoxemic effect on f_H is hypothesized to contribute to the bradycardia displayed by the unanesthetized epaulette shark during progressive hypoxia at 28°C (Speers-Roesch et al., 2012) (Table 1). Moreover, the hagfish heart is aneural and, thus, lacks cardiac vagal innervation (Nilsson and Axelsson, 1987). However, as recent work has demonstrated, despite a lack of autonomic innervation, f_H of Pacific hagfish slows by half during prolonged anoxia exposure, and then increases by four-fold upon reoxygenation (Cox et al., 2010). This begs the question: how are the drastic changes in cardiac activity in hagfish regulated? Several possibilities exist. First, the bradycardia displayed by anoxic hagfish during prolonged anoxia could involve negative chronotropic effects of extrinsic factors associated with anoxia, such as extracellular acidosis and hyperkalemia, both of which reduce f_H and cardiac contractile force in the turtle (Overgaard et al., 2005, 2007; Stecyk and Farrell, 2007). However, the ability of the hagfish heart to tightly regulate intracellular pH during anoxia exposure (Gillis et al., 2015) argues against extracellular acidosis contributing significantly to the bradycardia. Moreover, the dissipation of the extrinsic factors upon reoxygenation does not explain the tachycardia. Second, the bradycardia during anoxia could result from a remodeling of ion channels in the cardiac pacemaker itself. Pharmacological characterization of cardiac pacemaking in the hagfish has revealed that hyperpolarization-activated cyclic nucleotide-gated (HCN) channels, which are implicated in the control of f_H in the vertebrate heart by spontaneously depolarizing pacemaker cells via a mixed inward Na^+ and K^+ current, play a predominant role in setting intrinsic f_H in hagfish during normoxia, whereas the SR plays a very minor role (Wilson and Farrell, 2012). However, although a high abundance of HCN mRNA expression was found in the hagfish heart, HCN mRNA expression was largely unaffected by anoxia exposure and reoxygenation (Wilson et al., 2013). Nevertheless, changes in HCN protein expression have not been measured, and it is possible that changes in HCN channel density could play an important role in controlling f_H of anoxic hagfish. Finally, the bradycardia during anoxia exposure, as well as the tachycardia displayed upon reoxygenation, could result from regulation of the inward HCN current by altered levels of cyclic nucleotides, which, as the name of the channel indicates, activate the channel and result in an elevation of f_H. Indeed, cardiac β-adrenergic stimulation, which stimulates the production of the secondary messenger cAMP by transmembrane adenylyl cyclase, appears to be important for maintaining cardiac performance in normoxia (Axelsson et al., 1990; Johnsson and Axelsson, 1996). Moreover, the hagfish heart has its own endogenous store of catecholamines. Thus, the slow development of bradycardia could result from a depletion of endogenous catecholamine stores, which likely cannot be replenished during anoxia as the rate-limiting step in catecholamine production, the conversion of tyrosine to 3,4-dihydroxyphenylalanine, is oxygen-dependent. Indeed, Wilson et al. (2016) reported that intracellular cAMP levels fall by approximately 30% in the anoxic hagfish heart, and that the bradycardia associated with anoxia is associated with withdrawal of an autocrine adrenergic tonus mediated by the transmembrane adenylyl cyclase pathway (Wilson et al., 2016). Moreover, Wilson et al. (2016) provide evidence that upon reoxygenation, an additional bicarbonate-activated soluble adenylyl cyclase pathway is activated and functions in tandem with the transmembrane adenylyl cyclase pathway to elevate intracellular cAMP and stimulate f_H (Wilson et al., 2016).

4. ADDITIONAL INSIGHTS INTO THE MECHANISMS OF ANOXIC CARDIAC SURVIVAL IN CRUCIAN CARP

4.1. Cold Temperature Prepares the Heart for Anoxia Exposure

Regardless of the survival strategy employed to balance ATP supply and demand during oxygen deprivation, exposure or acclimation of severe hypoxia/anoxia-tolerant vertebrates to low temperature serves to enhance anoxia survival time. Decreased temperature reduces the kinetic energy of molecules. Therefore, in the absence of compensatory measures, exposure of ectothermic vertebrates to decreased temperature will result in decreased rates of chemical reactions, physiological processes, and ultimately whole-body metabolic rate and overall ATP demand (the effects of low temperature on the cardiac metabolism of fish are detailed in Chapter 6, Volume 36A: Rodnick and Gesser, 2017). If the upregulation of glycolysis is central to a species' anoxia survival strategy, cold temperature exposure extends the availability of fermentable fuel stores, whereas for the hypometabolic anoxia survival strategy, cold temperature exposure aids in reducing metabolic rate. In addition, for both strategies, cold temperature will slow the rate of accumulation of harmful metabolic waste products.

The significant role of cold temperature in facilitating anoxia survival is exemplified by the crucian carp, a fish that can survive days of anoxia at room temperature, but several months at temperatures near 0°C (Blazka, 1958, 1960; Piironen and Holopainen, 1986). However, the protracted anoxia survival time of crucian carp at low temperature is not simply due to the passive depressive effect of low temperature on physiological rate processes. Laboratory-based cold acclimation and acute cold exposure studies, as well as those with seasonally acclimatized fish captured from their natural ponds, have revealed that the fish utilize decreased temperature as a cue for the induction of numerous physiological changes that serve to prepare its tissues, including its heart, for winter anoxia; a phenomenon termed inverse thermal acclimation. Most prominently, prior to the winter months, the crucian carp increases its tissue glycogen stores to sustain prolonged glycolysis during anoxia exposure. The fractional weight of the liver increases from 2% to 14%, and 30% of the liver mass is glycogen (Holopainen and Hyvärinen, 1985; Hyvärinen et al., 1985). Large glycogen reserves are also accumulated in the brain and heart (Schmidt and Wegener, 1988; Vornanen, 1994a; Vornanen and Paajanen, 2004, 2006; Vornanen et al., 2011). However, it is the liver glycogen stores that are used to fuel anaerobic metabolism during prolonged anoxia. In fact, it has been shown that at 8°C, a temperature at which crucian carp survives approximately 2 weeks of anoxia, the only factor that eventually limits anoxic survival of the crucian carp is the total exhaustion of liver glycogen stores (Nilsson, 1990).

Modifications of cardiac morphology, physiology, *in vivo* f_H, contraction kinetics and electrophysiology by cold temperature are also involved in the conditioning of the crucian carp heart for anoxia exposure. A cold-induced enlargement of the ventricle, which is characteristic for many cold-active fish species, does not occur in seasonally acclimatized crucian carp. Instead, there is a marked decrease in cardiac water content in late autumn, probably due to the accumulation of glycogen in the heart, which is only depleted once anoxia ensues (Aho and Vornanen, 1997; Vornanen, 1994a; Vornanen and Paajanen, 2004; Vornanen et al., 2011). Also, in the summer months, fast and slow myosin heavy chain isoforms are expressed in the crucian carp heart. By comparison, in the winter months, only the slow myosin heavy chain isoform is present (Vornanen, 1994b). The slow myosin heavy chain isoform is believed to produce force more efficiently than its fast summer counterpart (Alpert and Mulieri, 1982), and due to it having a lower myosin-ATPase activity, it may improve energetic economy of contraction and contribute to anoxia tolerance. Lower total ATPase activity of the carp heart in winter than summer may similarly benefit winter anoxia survival (Aho and Vornanen, 1997). The *in vivo* f_H of the crucian carp is also temperature-dependent. It is ~65 min^{-1} at 22°C, ~35 min^{-1} at 15°C, ~17 min^{-1} at 8°C, ~15 min^{-1} at 5°C, and ~10 min^{-1} at 2°C (Matikainen and Vornanen, 1992; Stecyk et al., 2004b; Tikkanen et al., 2017; Vornanen, 1994b, 1999). Part of the cold-induced reduction in f_H is intrinsic. At temperatures above 10°C, the spontaneous f_H of 5°C-acclimated heart preparations is significantly lower than that of 15°C-acclimated heart preparations, indicating a modification of electrophysiological processes with cold acclimation, but no difference in f_H exists at 4°C (Matikainen and Vornanen, 1992). Consistent with a seasonal reduction in myofibrillar ATPase activity and myosin heavy chain isoform swapping, the velocity of cardiac contraction slows and contraction duration increases, especially the relaxation phase, suggesting differences in cardiac contractility (Tiitu and Vornanen, 2001). Also, the refractory period of atrial and ventricular muscle is lengthened (Matikainen and Vornanen, 1992; Tiitu and Vornanen, 2001), and the rate of SR Ca^{2+} uptake is decreased (Aho and Vornanen, 1998) without an alteration in the number of SR Ca^{2+} release channels in the ventricle (Tiitu and Vornanen, 2003). Further, the minor contribution of SR Ca^{2+} to atrial contraction at warm temperature is decreased in the cold (Tiitu and Vornanen, 2001). At the electrophysiological level, cold acclimation from 18°C to 4°C prolongs AP duration in crucian carp ventricular muscle by more than two-fold (i.e., from 1.3 to 2.8 s) (Paajanen and Vornanen, 2004), which may explain the increase in refractory period (Vornanen, 1996). However, the density of the two major K^+ currents of the fish heart, the inward-rectifier K^+ current (I_{K1}) which maintains the negative resting membrane potential and contributes to the final rate of AP repolarization, and the rapid component of the delayed-rectifier K^+ current (I_{Kr}) which is important in the regulation of AP plateau duration, are increased to partially compensate for the depressive effects of low temperature and limit the prolongation of AP duration (Haverinen and Vornanen, 2009). Acclimation to 4°C also reduces the density of I_{Na} (which determines the rate of impulse propagation in the heart) in ventricular myocytes to one-fifth that in 18°C-acclimated fish (Haverinen and Vornanen, 2004). Further, the recovery of I_{Na} from inactivation is 44% slower after cold acclimation. Similarly, peak I_{Ca} density at 4°C is one-sixth of that at 18°C, with no resulting changes in L-type Ca^{2+} channel kinetics, whereas the number of pore-forming alpha subunits of the L-type Ca^{2+} channels is halved in winter as compared to mid-summer (Vornanen, 1998; Vornanen and Paajanen, 2004). Finally, acclimatization to winter temperatures induces extensive changes in gene expression in the heart of crucian carp that preconditions the organ for the approaching anoxic winter period. Normoxic crucian carp acclimated to 8°C exhibit an 11-fold higher mRNA expression of heat-shock protein 70a, a protein which plays an important role in limiting cellular damage during times of stress (Stensløkken et al., 2010). Winter-acclimatized crucian carp also exhibited extensive changes in the expression of 21 genes involved in E–C coupling, including suppression of the expression of genes related to the SR Ca^{2+} pump (Tikkanen et al., 2017). Overall, cold acclimation plays a central role in preparing cardiac tissues for winter anoxia by reducing their energy requirement as well as provisioning them with glycogen stores. Nevertheless, sarcolemmal Na^+/K^+-ATPase activity does not differ between warm- and cold-acclimated or acclimatized carp, but instead decreases with the onset of hypoxia/anoxia (Aho and Vornanen, 1997; Paajanen and Vornanen, 2003).

4.2. Absence of "Channel Arrest" in the Anoxic Crucian Carp Heart

The "channel arrest" hypothesis predicts that the long-term survival of hypothermic and/or anoxic conditions is facilitated by the coordinated suppression of transmembrane ion flow through ion channels and active ion transport by ion pumps (Hochachka, 1986; Lutz et al., 1985). Indeed, in line with the contrasting strategies for anoxia survival displayed by crucian carp and freshwater turtles (see Section 3.1), evidence for channel arrest exists in the brain of the anoxic turtle, but not in the brain of the anoxic crucian carp. Similarly, cardiac electrophysiology of the crucian carp is largely unaffected by severe hypoxia and anoxia. AP shape for cardiomyocytes isolated from warm-acclimated crucian carp is only marginally affected by cyanide, an inhibitor of mitochondrial cytochrome *c* oxidase and aerobic ATP production (Vornanen and Tuomennoro, 1999). In atrial cardiomyocytes, cyanide poisoning depolarizes resting membrane potential (V_m), but does not affect AP shape, whereas in ventricular cardiomyocytes, AP duration is lengthened by ~15% in addition to V_m depolarization. In the ventricle, AP duration doubles after 6 weeks of anoxia exposure at 2°C (Tikkanen et al., 2017). However, the response appears to be directly induced by oxygen shortage, not active downregulation of ion channels as anoxia only caused minor changes in the expression of 21 genes involved in E–C coupling. Moreover, whole-cell conductance, single-channel conductance, and the open probability of ventricular I_{K1} are unaffected after 4 weeks of severe hypoxia exposure (0.4 mg O_2 L^{-1}) at 4°C (Paajanen and Vornanen, 2003). Thus, channel arrest does not appear to apply to crucian carp I_{K1}. Even so, SL Na^+/K^+-ATPase activity is reduced by one-third within 4 days of anoxia exposure at 4°C, as well as with the onset of hypoxic conditions in the natural environment (Aho and Vornanen, 1997). This change likely conserves ATP, but is at odds with the channel arrest hypothesis because it is not accompanied by a concomitant reduction in the major K^+ current. Seasonal studies of cardiac L-type Ca^{2+} channel abundance and I_{Ca} density provide further evidence against anoxic channel arrest as an ATP-conserving mechanism in the anoxic crucian carp heart. The number of ventricular dihydropyridine receptors (the subunit of the L-type Ca^{2+} cardiac channel that triggers channel opening) and the density of I_{Ca} do not change with the seasonal decrease in water oxygen content in the natural environment (Vornanen and Paajanen, 2004). Therefore, cardiac downregulation of L-type Ca^{2+} channels is not triggered by seasonal anoxia.

4.3. Management of Metabolic Wastes

In addition to being able to balance ATP supply and demand during anoxia, anoxia-tolerant vertebrates have developed creative waste management strategies to successfully cope with the production of metabolic waste products, namely, lactate and H^+ ions that accompany anaerobic metabolism. Indeed, the negative effects of extracellular acidosis on the fish heart are well reported, and include a decrease in the force development of ventricular muscle strips in a variety of teleost species (Gesser and Jørgensen, 1982; Gesser et al., 1982; reviewed by Gesser and Overgaard, 2009; Gesser and Poupa, 1983) and a reduction of $\dot{Q}$ in *in situ* perfused hearts (Farrell, 1985; Farrell and Milligan, 1986; Farrell et al., 1983, 1986; Hanson et al., 2006; Turner and Driedzic, 1980). Hagfish have a high blood volume (~15% of body mass) (Forster et al., 1989), which is five times higher than teleosts (Olson, 1992), and may be important in buffering metabolic wastes. In addition, hagfish cardiac cells have a peculiarly thick glycocalyx, which may be important in protecting the extracellular Ca^{2+} supply to cardiac myocytes from the effects of extracellular acidosis (Poupa et al., 1984). Indeed, as noted above, intracellular pH during anoxia exposure is tightly regulated in the anoxic hagfish heart (Gillis et al., 2015). Freshwater turtles utilize their bones and shell to reduce the harmful effects of H^+ accumulation during prolonged anoxia exposure (Jackson, 2000a, 2002, 2004). Specifically, Ca^{2+}, Mg^{2+} and Na^+ carbonates are released from the shell into the extracellular fluid in exchange for lactate to supplement extracellular buffering of H^+. Fish of the genus *Carassius* combat the decreased pH accompanying anaerobic metabolism by converting the lactate produced by glycolysis into ethanol and CO_2, which are excreted into the ambient water (Fagernes et al., 2017; Nilsson, 2001; Shoubridge and Hochachka, 1980). The conversion of lactate into ethanol and CO_2 restores the NAD^+/NADH ratio, but a disadvantage is the loss of chemical potential energy to the environment (van Waarde, 1991). Nonetheless, the benefit gained (i.e., survival) from the conversion of lactate to ethanol and CO_2 during periods of prolonged anoxia outweighs the associated cost.

The conversion of lactate to ethanol and CO_2 has two important implications for cardiovascular activity in anoxic crucian carp. Primarily, this waste management strategy has been proposed to require normal levels of $\dot{Q}$ in 8°C anoxic crucian carp for several reasons (Farrell and Stecyk, 2007; Stecyk et al., 2004b). First, a sustained $\dot{Q}$ may be needed to rapidly distribute glucose from the crucian carp's large liver glycogen store to metabolically active tissues, especially given that they remain active during anoxia. Second, a sustained $\dot{Q}$, coupled with a decrease in R_{tot}, which as noted above is the opposite of the anoxic response seen in common carp (Stecyk and Farrell, 2006), anoxia-tolerant turtles (Hicks and Farrell, 2000a; Stecyk et al., 2004a, 2008) and severe hypoxia/anoxia-intolerant fishes (Table 1), may reflect an increase in perfusion of skeletal muscle, the sole site of the lactate fermenting enzymes (Fagernes et al., 2017; Nilsson, 2001). Third, a sustained $\dot{Q}$ may be essential for shuttling ethanol to the gills for excretion. In addition, it appears that the prevention of severe acidosis through ethanol production is a key process that enables the maintenance of a high level of anoxic cardiac activity. Using spontaneously contracting crucian carp hearts exposed to graded acidosis under normoxic or anoxic conditions, Stecyk et al. (2011) demonstrated that the seemingly unique ability of 8°C-acclimated crucian carp heart to perform at normoxic levels during prolonged anoxia exposure was not due to the myocardium displaying an extraordinary tolerance of acidosis, but rather appeared to be permitted, in part, by avoidance of the severe extracellular acidosis afforded by ethanol production; the latter shelters the heart from the otherwise debilitating effects of severe acidosis on its intrinsic contractile properties. The heart preparations were tolerant of acidosis if oxygen was available or if extracellular pH remained at or above pH 7.4 (Fig. 8). However,

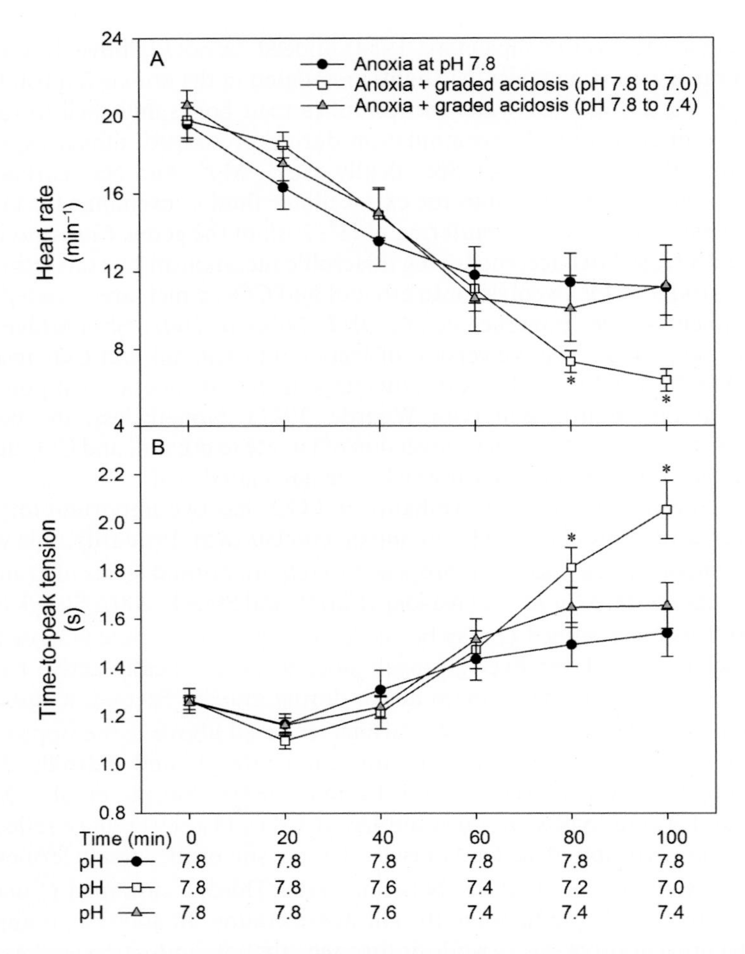

Fig. 8. Comparison of the heart rate (A) and time-to-peak tension (B) of 6.5°C spontaneously contracting crucian carp (*Carassius carassius*) heart preparations exposed to anoxia at pH 7.8 ($N=8$), graded acidosis from pH 7.8 to 7.0 in anoxia ($N=9$), and graded acidosis limited to pH 7.4 in anoxia ($N=6$). *Asterisks* indicate a significant difference ($P<0.05$) from hearts exposed to anoxia at pH 7.8 at a specific time point. Values are means ± S.E.M. Data adapted from Stecyk, J.A.W., Larsen, B.C., Nilsson, G.E., 2011. Intrinsic contractile properties of the crucian carp (*Carassius carassius*) heart during anoxic and acidotic stress. Am. J. Physiol. 301, R1132–R1142.

when extracellular pH was decreased below 7.4 in anoxia, f_H was impaired and the time-to-peak force was increased (Fig. 8). The extracellular pH level of 7.4 is similar to the plasma pH measured in a single crucian carp exposed to 26 h of anoxia at 15°C (van Waarde, 1991). In this animal, blood pH progressively decreased from the normoxic level of ~7.7 during the initial 1.5 h of anoxia exposure. However, corresponding with the commencement of ethanol production and stabilization of plasma lactate concentration, plasma pH stabilized at 7.4. Similarly, we have measured intracellular pH (using the ratiometric pH indicator BCECF) in isolated quiescent ventricular myocytes from 21°C-acclimated goldfish that were exposed to graded acidosis from 7.9 to 6.9. Intracellular pH was defended down to an extracellular pH of 7.2, below which intracellular pH acidified (Carlson and Stecyk, unpublished data) (Fig. 9). Combined, these findings suggest that for anoxic *Carassius*, preventing severe acidosis through ethanol production could be the real key for maintaining cardiac activity, and thus survival, provided that the glycolytic flux rate can be maintained to keep up with ATP demands. However, additional studies on the mechanisms of pH regulation in the anoxic *Carassius* heart are required to confirm this hypothesis.

5. CONCLUDING REMARKS

Whether natural or anthropogenically produced, low aquatic oxygen levels are a challenge faced by many fish species. Overall, the cardiovascular responses to oxygen limitation among severe hypoxia/anoxia-intolerant water-breathers, severe hypoxia/anoxia-intolerant air-breathers, and severe hypoxia/anoxia-tolerant water-breathing species reveal that many of the responses, most notably hypoxic bradycardia, are shared among diverse species. However, it is important to recognize that exceptions to the predominant "normal" responses exist, even within species. An important goal for the field moving forward will be to understand why these differences exist, how they are related to the physiological status of other organ systems and tissues, and the regulatory mechanisms through which the inter- and intraspecific differences are manifested. Beyond gaining an understanding of basic vertebrate cardiovascular physiology, the examination of how fishes respond to, and cope with, low oxygen levels has important ramifications for other areas of science. In an era of climate change (i.e., higher temperatures, lower water pH, and the increasing prevalence of aquatic hypoxia due to anthropogenic influences), it is important to understand species-specific responses to hypoxia so that informed conservation and management decisions can be made. Indeed, an efficiently functioning cardiovascular system is paramount to a range of activities in fish, including locomotion, digestion, and reproduction. Moreover, the potential for the study of fish cardiovascular physiology as a tool to investigate questions relevant to human health and various oxygen deprivation-related diseases has not been exploited. Indeed, the zebrafish (*Danio rerio*) has become a popular model for human cardiac diseases and pharmacology including

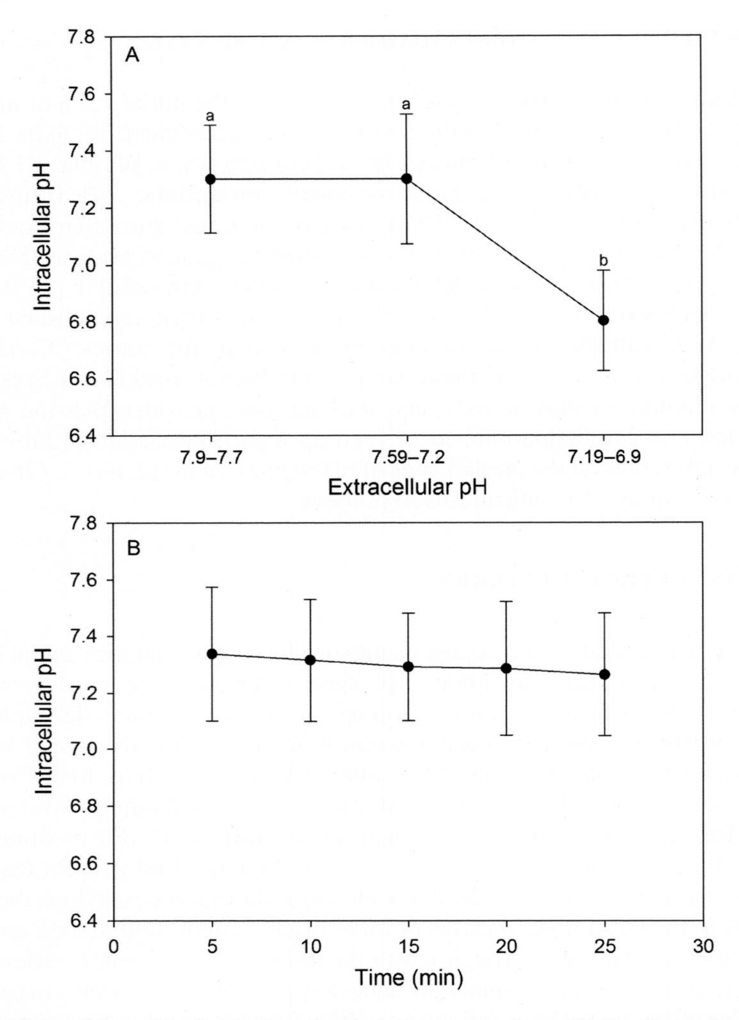

Fig. 9. Intracellular pH of quiescent ventricular myocytes from 21°C-acclimated goldfish. (A) Experimental cells ($N=9$) were exposed to graded extracellular acidosis from pH 7.9 to 6.9 over a 25 min period. (B) Control cells ($N=4$–6) were exposed to an extracellular pH of 7.8 for a 25 min period. Ventricular cells were enzymatically isolated (0.7 mg mL^{-1} collagenase Type-1; 0.5 mg mL^{-1} trypsin Type IX-S; 0.5 mg mL^{-1} fatty acid-free BSA) using standard solutions and perfusion times for teleost fish (Kubly and Stecyk, 2015). Intracellular pH was measured in cells loaded with the fluorescent ratiometric pH indicator BCECF, AM (2 μM; 1 h at 4°C). Fluorescent measurements (excitation at 490 and 440 nm with emission at 530 nm) were made with an Optosource Xe lamphouse, Optoscan monochromator (Cairn Research) and Nikon Ti-S Eclipse microscope. Signals were digitized and analyzed with ClampEx v10.4 (Molecular Devices). Cells were superfused with a 5 mM HCO_3^--buffered saline solution relevant for teleost fish that was continually gassed with 90% oxygen and CO_2 (ranging between 0.1% and 1.6%, balance N_2) to manipulate pH. All measurements were calibrated using the nigericin/high K^+ technique with pH standards of 6.5, 7.0, 7.5 and 8.0. Dissimilar letters indicate statistical significance ($P<0.05$; one-way repeated-measures analysis of variance; Holm–Sidak post hoc test). Values are means ± S.E.M. (Carlson and Stecyk, unpublished data).

cardiac arrhythmias and their electrophysiological basis (Vornanen and Hassinen, 2016). However, one of the most anoxia-tolerant vertebrates is a fish whose heart can continue to beat for weeks without oxygen. This remarkable ability contrasts starkly with the mere seconds that the human heart can beat without oxygen, despite both hearts utilizing the same basic "molecular toolbox" and underlying physiological basis of cardiac contraction. Probing the strategies that permit the hearts of anoxia-tolerant fishes to continue to beat in anoxia has the potential to provide a much deeper understanding of the connections between oxygen, metabolism, and electrical excitation, and novel treatments for pathologies related to hypoxia and rhythm disturbances in the human heart.

REFERENCES

Agnisola, C., McKenzie, D.J., Pellegrino, D., Bronzi, P., Tota, B., Taylor, E.W., 1999. Cardiovascular responses to hypoxia in the Adriatic sturgeon (*Acipenser naccarii*). J. Appl. Ichthyol. 15, 67–72.

Aho, E., Vornanen, M., 1997. Seasonality of ATPase activities in crucian carp (*Carassius carassius* L.) heart. Fish Physiol. Biochem. 16, 355–364.

Aho, E., Vornanen, M., 1998. Ca^{2+}-ATPase activity and Ca^{2+} uptake by sarcoplasmic reticulum in fish heart: effects of thermal acclimation. J. Exp. Biol. 201, 525–532.

Almeida-Val, V.M.F., Hochachka, P.W., 1995. Air-breathing fishes: metabolic biochemistry of the first diving vertebrates. In: Hochachka, P.W., Mommsen, T.P. (Eds.), Biochemistry and Molecular Biology of Fishes, vol. 5. Elsevier, Amsterdam, NY, pp. 45–55.

Almeida-Val, V.M.F., Val, A.L., Duncan, W.P., Souza, F.C.A., Paula-Silva, M.N., Land, S., 2000. Scaling effects on hypoxia tolerance in the Amazon fish *Astronotus ocellatus* (Perciformes: Cichlidae): contribution of tissue enzyme levels. Comp. Biochem. Physiol. B 125, 219–226.

Alpert, N.R., Mulieri, L.A., 1982. Increased myothermal economy of isometric force generation in compensated cardiac hypertrophy induced by pulmonary artery constriction in the rabbit. A characterization of heat liberation in normal and hypertrophied right ventricular papillary muscles. Circ. Res. 50, 491.

Arthur, P.G., Keen, J.E., Hochachka, P.W., Farrell, A.P., 1992. Metabolic state of the in situ perfused trout heart during severe hypoxia. Am. J. Physiol. 263, R798–804.

Arthur, P.G., Franklin, C.E., Cousins, K.L., Thorarensen, H., Hochachka, P.W., Farrell, A.P., 1997. Energy turnover in the normoxic and anoxic turtle heart. Comp. Biochem. Physiol. A Physiol. 117, 121–126.

Axelsson, M., Farrell, A.P., 1993. Coronary blood flow in vivo in the coho salmon (*Oncorhynchus kisutch*). Am. J. Physiol. 264, R963.

Axelsson, M., Fritsche, R., 1991. Effects of exercise, hypoxia and feeding on the gastrointestinal blood flow in the Atlantic cod Gadus morhua. J. Exp. Biol. 158, 181–198.

Axelsson, M., Abe, A.S., Eduardo, J., Bicudo, P.W., Nilsson, S., 1989. On the cardiac control in the South American lungfish, *Lepidosiren paradoxa*. Comp. Biochem. Physiol. A Physiol. 93, 561–565.

Axelsson, M., Farrell, A.P., Nilsson, S., 1990. Effects of hypoxia an drugs on the cardiovascular dynamics of the Atlantic hagfish *Myxine glutinosa*. J. Exp. Biol. 151, 297–316.

Axelsson, M., Davison, W., Forster, M.E., Farrell, A.P., 1992. Cardiovascular responses of the red-blooded antarctic fishes *Pagothenia bernacchii* and *P. borchgrevinki*. J. Exp. Biol. 167, 179–201.

Bailey, J.R., Driedzic, W.R., 1996. Decreased total ventricular and mitochondrial protein synthesis during extended anoxia in turtle heart. Am. J. Physiol. 271, R1660–1667.

Balaban, R.S., 2002. Cardiac energy metabolism homeostasis: role of cytosolic calcium. J. Mol. Cell. Cardiol. 34, 1259–1271.

Barrell, J., 1916. The influence of Silurian-Devonian climates on the rise of air-breathing vertebrates. Proc. Natl. Acad. Sci. U. S. A. 2, 499–504.

Beamish, F.W.H., 1964. Respiration of fishes with special emphasis on standard oxygen consumption: II: influence of weight and temperature on several species. Can. J. Zool. 42, 177–188.

Belão, T.C., Leite, C.A.C., Florindo, L.H., Kalinin, A.L., Rantin, F.T., 2011. Cardiorespiratory responses to hypoxia in the African catfish, *Clarias gariepinus* (Burchell 1822), an air-breathing fish. J. Comp. Physiol. B 181, 905.

Belão, T.C., Zeraik, V.M., Florindo, L.H., Kalinin, A.L., Leite, C.A.C., Rantin, F.T., 2015. Control of cardiorespiratory function in response to hypoxia in an air-breathing fish, the African sharptooth catfish, *Clarias gariepinus*. Comp. Biochem. Physiol. A Mol. Integr. Physiol. 187, 130–140.

Bers, D.M., Despa, S., 2006. Cardiac myocytes Ca^{2+} and Na^{+} regulation in normal and failing hearts. J. Pharmacol. Sci. 100, 315–322.

Berschick, P., Bridges, C.R., Grieshaber, M.K., 1987. The influence of hyperoxia, hypoxia and temperature on the respiratory physiology of the intertidal rockpool fish *Gobius cobitis* Pallas. J. Exp. Biol. 130, 368.

Bickler, P.E., 1992. Cerebral anoxia tolerance in turtles: regulation of intracellular calcium and pH. Am. J. Physiol. 263, R1298–302.

Bickler, P.E., Buck, L.T., 1998. Adaptations of vertebrate neurons to hypoxia and anoxia: maintaining critical Ca^{2+} concentrations. J. Exp. Biol. 201, 1141–1152.

Blazka, P., 1958. The anaerobic metabolism of fish. Physiol. Zool. 31, 117–128.

Blazka, B., 1960. On the biology of crucian carp (*Carassius carassius* L. morpha humilis Heckel). Zool. Zhurnal 39, 1384–1389.

Boutilier, R.G., 2001. Mechanisms of cell survival in hypoxia and hypothermia. J. Exp. Biol. 204, 3171–3181.

Boutilier, R.G., Iwama, G.K., Randall, D.J., 1986. The promotion of catecholamine release in rainbow trout, *Salmo gairdneri*, by acute acidosis: interactions between red cell pH and haemoglobin oxygen-carrying capacity. J. Exp. Biol. 123, 145–157.

Brand, M., 2003. Approximate yield of ATP from glucose, designed by Donald Nicholson: commentary. Biochem. Mol. Biol. Educ. 31, 2–4.

Brooks, S.P., Storey, K.B., 1988. Anoxic brain function: molecular mechanisms of metabolic depression. FEBS Lett. 232, 214–216.

Brooks, S.P., Storey, K.B., 1989. Regulation of glycolytic enzymes during anoxia in the turtle *Pseudemys scripta*. Am. J. Physiol. 257, R278–283.

Brooks, S.P., Storey, K.B., 1990. Phosphofructokinase from a vertebrate facultative anaerobe: effects of temperature and anoxia on the kinetic parameters of the purified enzyme from turtle white muscle. Biochim. Biophys. Acta 1037, 161–164.

Buck, L.T., Bickler, P.E., 1995. Role of adenosine in NMDA receptor modulation in the cerebral cortex of an anoxia-tolerant turtle (*Chrysemys picta belli*). J. Exp. Biol. 198, 1621–1628.

Buck, L.T., Bickler, P.E., 1998. Adenosine and anoxia reduce N-methyl-D-aspartate receptor open probability in turtle cerebrocortex. J. Exp. Biol. 201, 289–297.

Buck, L., Espanol, M., Litt, L., Bickler, P., 1998. Reversible decreases in ATP and PCr concentrations in anoxic turtle brain. Comp. Biochem. Physiol. A Mol. Integr. Physiol. 120, 633–639.

Burleson, M.L., Smatresk, N.J., 1989. The effect of decerebration and anesthesia on the reflex responses to hypoxia in catfish. Can. J. Zool. 67, 630–635.

Burleson, M.L., Smatresk, N.J., 1990a. Effects of sectioning cranial nerves IX and X on cardiovascular and ventilatory reflex responses to hypoxia and NaCN in channel catfish. J. Exp. Biol. 154, 407–420.

Burleson, M.L., Smatresk, N.J., 1990b. Evidence for two oxygen-sensitive chemoreceptor loci in channel catfish, *Ictalurus punctatus*. Physiol. Zool. 63, 208–221.

Bushnell, P.G., Brill, R.W., 1992. Oxygen transport and cardiovascular responses in skipjack tuna (*Katsuwonus pelamis*) and yellowfin tuna (*Thunnus albacares*) exposed to acute hypoxia. J. Comp. Physiol. B 162, 131–143.

Bushnell, P.G., Brill, R.W., Bourke, R.E., 1990. Cardiorespiratory responses of skipjack tuna (*Katsuwonus pelamis*), yellowfin tuna (*Thunnus albacares*), and bigeye tuna (*Thunnus obesus*) to acute reductions of ambient oxygen. Can. J. Zool. 68, 1857–1865.

Butler, P.J., Taylor, E.W., 1971. Response of the dogfish (*Scyliorhinus canicula* L.) to slowly induced and rapidly induced hypoxia. Comp. Biochem. Physiol. A Physiol. 39, 307–323.

Butler, P.J., Taylor, E.W., 1975. The effect of progressive hypoxia on respiration in the dogfish (Scyliorhinus canicula) at different seasonal temperatures. J. Exp. Biol. 63, 117–130.

Butler, P.J., Taylor, E.W., Short, S., 1977. The effect of sectioning cranial nerves V, VII, IX and X on the cardiac response of the dogfish *Scyliorhinus canicula* to environmental hypoxia. J. Exp. Biol. 69, 233.

Campbell, M.A., Lopéz, J.A., 2014. Mitochondrial phylogeography of a Beringian relict: the endemic freshwater genus of blackfish Dallia (Esociformes). J. Fish Biol. 84, 523–538.

Carter, G.S., Beadle, L.C., 1931. Reports of an Expedition to Brazil and Paraguay in 1926–7, supported by the Trustees of the Percy Sladen Memorial Fund and by the Executive Committee of the Carnegie Trust for the Universities of Scotland. The Fauna of the Swamps of the Paraguayan Chaco in relation to its Environment.—II. Respiratory Adaptations in the Fishes. Zool. J. Linn. Soc. 37, 327–368.

Cech, J.J., Wohlschlag, D.E., 1973. Respiratory responses of the striped mullet, *Mugil cephalus* (L.) to hypoxic conditions. J. Fish Biol. 5, 421–428.

Cech, J.J., Rowell, D.M., Glasgow, J.S., 1977. Cardiovascular responses of the winter flounder *Pseudopleuronectes americanus* to hypoxia. Comp. Biochem. Physiol. 57A, 123–125.

Chan, D.K.O., 1986. Cardiovascular, respiratory, and blood adjustments to hypoxia in the Japenese eel, *Anguilla japonica*. Fish Physiol. Biochem. 2, 179–193.

Chan, D.K.O., Chow, P.H., 1976. The effects of acetylcholine, biogenic amines and other vasoactive agents on the cardiovascular functions of the eel, *Anguilla japonica*. J. Exp. Zool. 196, 13–26.

Chang, J., Knowlton, A.A., Wasser, J.S., 2000. Expression of heat shock proteins in turtle and mammal hearts: relationship to anoxia tolerance. Am. J. Physiol. Regul. Integr. Comp. Physiol. 278, R209–214.

Chapman, L.J., McKenzie, D.J., 2009. Behavioral responses and ecological consequences. In: Jeffrey, A.P.F., Richards, G., Colin, J.B. (Eds.), Fish Physiology, vol. 27. Academic Press, San Diego, pp. 25–77.

Chih, C.P., Rosenthal, M., Sick, T.J., 1989. Ion leakage is reduced during anoxia in turtle brain: a potential survival strategy. Am. J. Physiol. 257, R1562–1564.

Clark, T.D., Seymour, R.S., Christian, K., Wells, R.M.G., Baldwin, J., Farrell, A.P., 2007. Changes in cardiac output during swimming and aquatic hypoxia in the air-breathing Pacific tarpon. Comp. Biochem. Physiol. A Mol. Integr. Physiol. 148, 562–571.

Cox, G.K., Sandblom, E., Farrell, A.P., 2010. Cardiac responses to anoxia in the Pacific hagfish, *Eptatretus stoutii*. J. Exp. Biol. 213, 3692–3698.
Cox, G., Sandblom, E., Richards, J., Farrell, A., 2011. Anoxic survival of the Pacific hagfish (*Eptatretus stoutii*). J. Comp. Physiol. B 181, 361–371.
Daw, J.C., Wenger, D.P., Berne, R.B., 1967. Relationship between cardiac glycogen and tolerance to anoxia in the western painted turtle, *Chrysemys picta bellii*. Comp. Biochem. Physiol. 22, 69–73.
de Salvo Souza, R., Soncini, R., Glass, M., Sanches, J., Rantin, F., 2001. Ventilation, gill perfusion and blood gases in dourado, *Salminus maxillosus* Valenciennes (Teleostei, Characidae), exposed to graded hypoxia. J. Comp. Physiol. B 171, 483–489.
Diaz, R.J., Breitburg, D.L., 2009. The hypoxic environment. In: Jeffrey, A.P.F., Richards, G., Colin, J.B. (Eds.), Fish Physiology, vol. 27. Academic Press, San Diego, pp. 1–23.
Diaz, R.J., Rosenberg, R., 2008. Spreading dead zones and consequences for marine ecosystems. Science 321, 926.
Doll, C., Hochachka, P., Hand, S., 1994. A microcalorimetric study of turtle cortical slices: insights into brain metabolic depression. J. Exp. Biol. 191, 141–153.
Duncan, J.A., Storey, K.B., 1992. Subcellular enzyme binding and the regulation of glycolysis in anoxic turtle brain. Am. J. Physiol. 262, R517–523.
Dybas, C.L., 2005. Dead zones spreading in world oceans. BioScience 55, 552–557.
Fagernes, C.E., Stensløkken, K.-O., Røhr, Å.K., Berenbrink, M., Ellefsen, S., Nilsson, G.E., 2017. Extreme anoxia tolerance in crucian carp and goldfish through neofunctionalization of duplicated genes creating a new ethanol-producing pyruvate decarboxylase pathway. Sci. Rep. 7, 7884.
Farmer, C.G., Jackson, D.C., 1998. Air-breathing during activity in the fishes *Amia calva* and *Lepisosteus oculatus*. J. Exp. Biol. 201, 943.
Farrell, A.P., 1980. Gill morphometrics, vessel dimensions, and vascular resistance in ling cod, *Ophiodon elongatus*. Can. J. Zool. 58, 807–818.
Farrell, A.P., 1981. Cardiovascular changes in the lingcod (*Ophidon elongatus*) following adrenergic and cholinergic drug infusions. J. Exp. Biol. 91, 203–305.
Farrell, A.P., 1982. Cardiovascular changes in the unanaesthetized lingcod (*Ophidon elongatus*) during short-term, progressive hypoxia and spontaneous activity. Can. J. Zool. 60, 933–941.
Farrell, A.P., 1984. A review of cardiac performance in the teleost heart: intrinsic and humoral regulation. Can. J. Zool. 62, 523–536.
Farrell, A.P., 1985. A protective effect of adrenaline on the acidotic teleost heart. J. Exp. Biol. 116, 503–508.
Farrell, A.P., 1991. Cardiac scope in lower vertebrates. Can. J. Zool. 69, 1981–1984.
Farrell, A.P., 2007. Tribute to P. L. Lutz: a message from the heart—why hypoxic bradycardia in fishes? J. Exp. Biol. 210, 1715–1725.
Farrell, A.P., Jones, D.R., 1992. The heart. In: Hoar, W.S., Randall, D.J., Farrell, A.P. (Eds.), Fish Physiology, vol. 12A. Academic Press, San Diego, pp. 1–88.
Farrell, A.P., Milligan, C.L., 1986. Myocardial intracellular pH in a perfused rainbow trout heart during extracellular acidosis in the presence and absence of adrenaline. J. Exp. Biol. 125, 347–359.
Farrell, A.P., Richards, J.G., 2009. Defining hypoxia: an integrative synthesis of the responses of fish to hypoxia. In: Jeffrey, A.P.F., Richards, G., Colin, J.B. (Eds.), Fish Physiology, vol. 27. Academic Press, San Diego, pp. 487–503.
Farrell, A.P., Smith, F., 2017. Cardiac form, function and physiology. In: Gamperl, A.K., Gillis, T.E., Farrell, A.P., Brauner, C.J. (Eds.), The Cardiovascular System: Morphology, Control and Function. Fish Physiology, vol. 36A. Academic Press, San Diego, pp. 155–264.
Farrell, A.P., Stecyk, J.A., 2007. The heart as a working model to explore themes and strategies for anoxic survival in ectothermic vertebrates. Comp. Biochem. Physiol. A Mol. Integr. Physiol. 147, 300–312.

Farrell, A.P., MacLeod, K.R., Driedzic, W.R., Wood, S., 1983. Cardiac performance in the in situ perfused fish heart during extracellular acidosis: interactive effects of adrenaline. J. Exp. Biol. 107, 415–429.

Farrell, A.P., Wood, S., Hart, T., Driedzic, W.R., 1985. Myocardial oxygen consumption in the sea raven, *Hemitripterus americanus*: the effects of volume loading, pressure loading and progressive hypoxia. J. Exp. Biol. 117, 237–250.

Farrell, A.P., MacLeod, K.R., Chancey, B., 1986. Intrinsic mechanical properties of the perfused rainbow trout heart and the effects of catecholamines and extracellular calcium under control and acidotic conditions. J. Exp. Biol. 125, 319–345.

Farrell, A.P., Franklin, C., Arthur, P., Thorarensen, H., Cousins, K., 1994. Mechanical performance of an *in situ* perfused heart from the turtle *Chrysemys scripta* during normoxia and anoxia at 5°C and 15°C. J. Exp. Biol. 191, 207–229.

Forster, M.E., 1976. Effects of adrenergic blocking drugs on the cardiovascular system of the eel, *Anguilla anguilla* (L.). Comp. Biochem. Physiol. C 55, 33–36.

Forster, M.E., Davison, W., Satchell, G.H., Taylor, H.H., 1989. The subcutaneous sinus of the hagfish, *Eptatretus cirrhatus* and its relation to the central circulating blood volume. Comp. Biochem. Physiol. A Physiol. 93, 607–612.

Forster, M.E., Axelsson, M., Farrell, A.P., Nilsson, S., 1991. Cardiac function and circulation in hagfishes. Can. J. Zool. 69, 1985–1992.

Forster, M.E., Davison, W., Axelsson, M., Farrell, A.P., 1992. Cardiovascular responses to hypoxia in the hagfish, Eptatretus cirrhatus. Respir. Physiol. 88, 373–386.

Fraser, K.P., Houlihan, D.F., Lutz, P.L., Leone-Kabler, S., Manuel, L., Brechin, J.G., 2001. Complete suppression of protein synthesis during anoxia with no post-anoxia protein synthesis debt in the red-eared slider turtle *Trachemys scripta elegans*. J. Exp. Biol. 204, 4353–4360.

Fritsche, R., 1990. Effects of hypoxia on blood pressure and heart rate in three marine teleosts. Fish Physiol. Biochem. 8, 85–92.

Fritsche, R., Nilsson, S., 1989. Cardiovascular responses to hypoxia in the Atlantic cod, *Gadus morhua*. Exp. Biol. 48, 153–160.

Fritsche, R., Nilsson, S., 1990. Autonomic nervous control of blood pressure and heart rate during hypoxia in the cod, *Gadus morhua*. J. Comp. Physiol. B 160, 287–292.

Fritsche, R., Axelsson, M., Franklin, C.E., Grigg, G.G., Holmgren, S., Nilsson, S., 1993. Respiratory and cardiovascular responses to hypoxia in the Australian lungfish. Respir. Physiol. 94, 173–187.

Furimsky, M., Cooke, S.J., Suski, C.D., Wang, Y., Tufts, B.L., 2003. Respiratory and circulatory responses to hypoxia in largemouth bass and smallmouth bass: Implications for "live-release" angling tournaments. Trans. Am. Fish. Soc. 132, 1065–1075.

Gamperl, A.K., Driedzic, W.R., 2009. Cardiovascular function and cardiac metabolism. In: Jeffrey, A.P.F., Richards, G., Colin, J.B. (Eds.), Fish Physiology, vol. 27. Academic Press, San Diego, pp. 301–360.

Gamperl, A.K., Pinder, A.W., Grant, R.R., 1994. Influence of hypoxia and adrenaline administration on coronary blood flow and cardiac performance in seawater rainbow trout (*Oncorhynchus mykiss*). J. Exp. Biol. 193, 209–232.

Garey, W.F., Rahn, H., 1970. Gas tensions in tissues of trout and carp exposed to diurnal changes in oxygen tension of the water. J. Exp. Biol. 52, 575–582.

Gehrke, P.C., Fielder, D.R., 1988. Effects of temperature and dissolved oxygen on heart rate, ventilation rate and oxygen consumption of spangled perch, *Leiopotherapon unicolor* (Gunther 1859), (Percoidei, Teraponidae). J. Comp. Physiol. B 157, 771–782.

Gesser, H., Jørgensen, E., 1982. pH_i, contractility and Ca-balance under hypercapnic acidosis in the myocardium of different vertebrate species. J. Exp. Biol. 96, 405–412.

Gesser, H., Overgaard, J., 2009. Comparative aspects of hypoxia tolerance of the ectothermic vertebrate heart. In: Glass, M.L., Wood, S.C. (Eds.), Cardio-Respiratory Control in Vertebrates: Comparative and Evolutionary Aspects. Springer Berlin Heidelberg, Berlin, Heidelberg, pp. 263–284.

Gesser, H., Poupa, O., 1983. Acidosis and cardiac muscle contractility: comparative aspects. Comp. Biochem. Physiol. A Comp. Physiol. 76, 559–566.

Gesser, H., Andersen, P., Brams, P., Sund-Larsen, J., 1982. Inotropic effects of adrenaline on the anoxic or hypercapnic myocardium of rainbow trout and eel. J. Comp. Physiol. 147, 123–138.

Gillis, T.E., Regan, M.D., Cox, G.K., Harter, T.S., Brauner, C.J., Richards, J.G., Farrell, A.P., 2015. Characterizing the metabolic capacity of the anoxic hagfish heart. J. Exp. Biol. 218, 3754–3761.

Glass, M.L., Rantin, F.T., Verzola, R.M.M., Fernandes, M.N., Kalinin, A.L., 1991. Cardiorespiratory synchronization and myocardial function in hypoxic carp, *Cyprinus carpio* L. J. Fish Biol. 39, 143–149.

Graham, J.B., 1997. Air-Breathing Fishes: Evolution, Diversity and Adaptation. Academic Press, San Diego.

Graham, M.S., Farrell, A.P., 1989. The effect of temperature acclimation and adrenaline on the performance of a perfused trout heart. Physiol. Zool. 62, 38–61.

Graham, J., Lai, N., Chiller, D., Roberts, J., 1995. The transition to air breathing in fishes. V. Comparative aspects of cardiorespiratory regulation in *Synbranchus marmoratus* and *Monopterus albus* (Synbranchidae). J. Exp. Biol. 198, 1455.

Graham, R.M., Frazier, D.P., Thompson, J.W., Haliko, S., Li, H., Wasserlauf, B.J., Spiga, M.-G., Bishopric, N.H., Webster, K.A., 2004. A unique pathway of cardiac myocyte death caused by hypoxia–acidosis. J. Exp. Biol. 207, 3189.

Gunther, A., 1871. The new Ganoid fish (Ceratodus) recently discovered in Queensland. Nature, 406–408.

Hansen, C.A., Sidell, B.D., 1983. Atlantic hagfish cardiac muscle: metabolic basis of tolerance to anoxia. Am. J. Physiol. 244, R356–362.

Hanson, L.M., Obradovich, S., Mouniargi, J., Farrell, A.P., 2006. The role of adrenergic stimulation in maintaining maximum cardiac performance in rainbow trout (*Oncorhynchus mykiss*) during hypoxia, hyperkalemia and acidosis at 10 degrees C. J. Exp. Biol. 209, 2442–2451.

Haverinen, J., Vornanen, M., 2004. Temperature acclimation modifies Na^+ current in fish cardiac myocytes. J. Exp. Biol. 207, 2823–2833.

Haverinen, J., Vornanen, M., 2007. Temperature acclimation modifies sinoatrial pacemaker mechanism of the rainbow trout heart. Am. J. Physiol. Regul. Integr. Comp. Physiol. 292, R1023–1032.

Haverinen, J., Vornanen, M., 2009. Responses of action potential and K + currents to temperature acclimation in fish hearts: phylogeny or thermal preferences? Physiol. Biochem. Zool. 82, 468–482.

Hedrick, M.S., Burleson, M.L., Jones, D.R., Milsom, W.K., 1991. An examination of central chemosensitivity in an air-breathing fish (*Amia calva*). J. Exp. Biol. 155, 165.

Helly, J.J., Levin, L.A., 2004. Global distribution of naturally occurring marine hypoxia on continental margins. Deep-Sea Res. I Oceanogr. Res. Pap. 51, 1159–1168.

Hemmingsen, E.A., Douglas, E.L., 1972. Respiratory and circulatory responses in a hemoglobin-free fish, *Chaenocephalus aceratus*, to changes in temperature and oxygen tension. Comp. Biochem. Physiol. 43, 1031–1043.

Hemmingsen, E.A., Douglas, E.L., Johansen, K., Millard, R.W., 1972. Aortic blood flow and cardiac output in the hemoglobin-free fish *Chaenocephalus aceratus*. Comp. Biochem. Physiol. A Comp. Physiol. 43, 1045–1051.

Hems, D.A., Whitton, P.D., 1980. Control of hepatic glycogenolysis. Physiol. Rev. 60, 1–50.

Herbert, C.V., Jackson, D.C., 1985a. Temperature effects on the responses to prolonged submergence in the turtle *Chrysemys bellii*. I. Blood acid-base and ionic changes during and following anoxic submergence. Physiol. Zool. 58, 655–669.

Herbert, C.V., Jackson, D.C., 1985b. Temperature effects on the responses to prolonged submergence in the turtle *Chrysemys picta bellii*. II. Metabolic rate, blood acid-base and ionic changes, and cardiovascular function in aerated and anoxic water. Physiol. Zool. 58, 670–681.

Hicks, J.M., Farrell, A.P., 2000a. The cardiovascular responses of the red-eared slider (*Trachemys scripta*) acclimated to either 22 or 5°C. I. Effects of anoxic exposure on *in vivo* cardiac performance. J. Exp. Biol. 203, 3765–3774.

Hicks, J.M., Farrell, A.P., 2000b. The cardiovascular responses of the red-eared slider (*Trachemys scripta*) acclimated to either 22 or 5°C. II. Effects of anoxia on adrenergic and cholinergic control. J. Exp. Biol. 203, 3775–3784.

Hicks, J.W., Wang, T., 1998. Cardiovascular regulation during anoxia in the turtle: an *in vivo* study. Physiol. Zool. 71, 1–14.

Hochachka, P.W., 1986. Defense strategies against hypoxia and hypothermia. Science 231, 234–241.

Hochachka, P.W., Somero, G.N., 2002. Biochemical Adaptation: Mechanisms and Process in Physiological Evolution. Oxford University Press, Oxford.

Hochachka, P.W., Buck, L.T., Doll, C.J., Land, S.C., 1996. Unifying theory of hypoxia tolerance: molecular/metabolic defense and rescue mechanisms for surviving oxygen lack. Proc. Natl. Acad. Sci. U. S. A. 93, 9493–9498.

Holeton, G.F., 1972. Gas exchange in fish with and without hemoglobin. Respir. Physiol. 14, 142–150.

Holeton, G.F., Randall, D.J., 1967a. Changes in blood pressure in the rainbow trout during hypoxia. J. Exp. Biol. 46, 297–305.

Holeton, G.F., Randall, D.J., 1967b. The effect of hypoxia upon the partial pressure of gases in the blood and water afferent and efferent to the gills of rainbow trout. J. Exp. Biol. 46, 317–327.

Holopainen, I.J., Hyvärinen, H., 1985. Ecology and physiology of crucian carp (*Carassius carassius* L.) in small Finnish ponds with anoxic conditions in winter. Verh. Int. Ver. Theor. Angew. Limnol. 22, 2566–2570.

Hughes, G.M., 1973. Respiratory responses to hypoxia in fish. Am. Zool. 13, 475–489.

Hughes, G.M., Shelton, G., 1962. Respiratory mechanisms and their nervous control in fish. Adv. Comp. Physiol. Biochem. 1, 275–364.

Hurtado, A., 1964. Animals in high altitudes; resident man. In: Dill, D.B., Adolph, E.F. (Eds.), Handbook of Physiology. Adaptation to the Environment, Section 4. American Physiology Society, Washington, DC, pp. 843–860.

Huss, J.M., Kelly, D.P., 2005. Mitochondrial energy metabolism in heart failure: a question of balance. J. Clin. Invest. 115, 547–555.

Hylland, P., Nilsson, G.E., Lutz, P.L., 1994. Time course of anoxia-induced increase in cerebral blood flow rate in turtles: evidence for a role of adenosine. J. Cereb. Blood Flow Metab. 14, 877–881.

Hylland, P., Nilsson, G.E., Lutz, P.L., 1996. Role of nitric oxide in the elevation of cerebral blood flow induced by acetylcholine and anoxia in the turtle. J. Cereb. Blood Flow Metab. 16, 290–295.

Hylland, P., Milton, S., Pek, M., Nilsson, G.E., Lutz, P.L., 1997. Brain Na^+/K^+-ATPase activity in two anoxia tolerant vertebrates: crucian carp and freshwater turtle. Neurosci. Lett. 235, 89–92.

Hyvärinen, H., Holopainen, I.J., Piironen, J., 1985. Anaerobic wintering of crucian carp (*Carassius carassius* L). I. Annual dynamics of glycogen reserves in nature. Comp. Biochem. Physiol. A Physiol. 82, 797–803.

Icardo, J.M., 2017. Heart morphology and anatomy. In: Gamperl, A.K., Gillis, T.E., Farrell, A.P., Brauner, C.J. (Eds.), The Cardiovascular System: Morphology, Control and Function. Fish Physiology, vol. 36A. Academic Press, San Diego, pp. 1–54.

Iftikar, F.I., Patel, M., Ip, Y.K., Wood, C.M., 2008. The influence of feeding on aerial and aquatic oxygen consumption, nitrogenous waste excretion, and metabolic fuel usage in the African lungfish, *Protopterus annectens*. Can. J. Zool. 86, 790–800.

Imbrogno, S., Cerra, M.C., 2017. Hormonal and autacoid control of cardiac function. In: Gamperl, A.K., Gillis, T.E., Farrell, A.P., Brauner, C.J. (Eds.), The Cardiovascular System: Morphology, Control and Function. Fish Physiology, vol. 36A. Academic Press, San Diego, pp. 265–315.

Ishimatsu, A., 2012. Evolution of the cardiorespiratory system in air-breathing fishes. Aqua BioSci. Monogr. 5, 1–28.

Iversen, N.K., Huong, D.T.T., Bayley, M., Wang, T., 2011. Autonomic control of the heart in the Asian swamp eel (Monopterus albus). Comp. Biochem. Physiol. A Mol. Integr. Physiol. 158, 485–489.

Iversen, N.K., Lauridsen, H., Huong, D.T.T., Van Cong, N., Gesser, H., Buchanan, R., Bayley, M., Pedersen, M., Wang, T., 2013. Cardiovascular anatomy and cardiac function in the air-breathing swamp eel (*Monopterus albus*). Comp. Biochem. Physiol. A Mol. Integr. Physiol. 164, 171–180.

Jackson, D.C., 1968. Metabolic depression and oxygen depletion in the diving turtle. J. Appl. Physiol. 24, 503–509.

Jackson, D.C., 2000a. How a turtle's shell helps it survive prolonged anoxic acidosis. News Physiol. Sci. 15, 181–185.

Jackson, D.C., 2000b. Living without oxygen: lessons from the freshwater turtle. Comp. Biochem. Physiol. A Mol. Integr. Physiol. 125, 299–315.

Jackson, D.C., 2002. Hibernating without oxygen: physiological adaptations of the painted turtle. J. Physiol. 543, 731–737.

Jackson, D.C., 2004. Acid-base balance during hypoxic hypometabolism: selected vertebrate strategies. Respir. Physiol. Neurobiol. 141, 273–283.

Jackson, D.C., Ultsch, G.R., 1982. Long-term submergence at 3°C of the turtle, *Chrysemys picta bellii*, in normoxic and severely hypoxic water: II. Extracellular ionic responses to extreme lactic acidosis. J. Exp. Biol. 96, 29–43.

Jackson, D.C., Ultsch, G.R., 2010. Physiology of hibernation under the ice by turtles and frogs. J. Exp. Zool. A Ecol. Genet. Physiol. 313A, 311–327.

Janssen, G.J.A., Jerrett, A.R., Black, S.E., Forster, M.E., 2010. The effects of progressive hypoxia and re-oxygenation on cardiac function, white muscle perfusion and haemoglobin saturation in anaesthetised snapper (*Pagrus auratus*). J. Comp. Physiol. B 180, 503–510.

Johansen, K., 1966. Air breathing in the teleost *Symbranchus marmoratus*. Comp. Biochem. Physiol. 18, 383–395.

Johansen, K., 1970. Air breathing in fishes. In: Hoar, W.S., Randall, D.J. (Eds.), Fish Physiology, vol. 4. Academic Press, San Diego, pp. 361–411.

Johansen, K., Lenfant, C., Schmidt-Nielsen, K., Petersen, J.A., 1968. Gas exchange and control of breathing in the electric eel, *Electrophorus electricus*. Z. Vgl. Physiol. 61, 137–163.

Johansen, K., Lenfant, C., Hanson, D., 1970. Phylogenetic development of pulmonary circulation. Fed. Proc. 29, 1135–1140.

Johansson, D., Nilsson, G., 1995. Roles of energy status, K_{ATP} channels and channel arrest in fish brain K^+ gradient dissipation during anoxia. J. Exp. Biol. 198, 2575–2580.

Johansson, D., Nilsson, G., Tornblom, E., 1995. Effects of anoxia on energy metabolism in crucian carp brain slices studied with microcalorimetry. J. Exp. Biol. 198, 853–859.

Johansson, D., Nilsson, G.E., Døving, K.B., 1997. Anoxic depression of light-evoked potentials in retina and optic tectum of crucian carp. Neurosci. Lett. 237, 73–76.
Johnsson, M., Axelsson, M., 1996. Control of the systemic heart and the portal heart of *Myxine glutinosa*. J. Exp. Biol. 199, 1429.
Jones, C.T., Roebuck, M.M., Walker, D.W., Johnston, B.M., 1988. The role of the adrenal medulla and peripheral sympathetic nerves in the physiological responses of the fetal sheep to hypoxia. J. Dev. Physiol. 10, 17–36.
Jordan, J., 1976. The influence of body weight on gas exchange in the air-breathing fish, *Clarias batrachus*. Comp. Biochem. Physiol. A Physiol. 53, 305–310.
Kelly, D.A., Storey, K.B., 1988. Organ-specific control of glycolysis in anoxic turtles. Am. J. Physiol. 255, R774–9.
Kinkead, R., Perry, S.F., 1990. An investigation of the role of circulation catecholamines in the control of ventilation during acute moderate hypoxia in rainbow trout (*Oncorhynchus mykiss*). J. Comp. Physiol. B 160, 441–448.
Kinkead, R., Perry, S.F., 1991. The effects of catecholamines on ventilation in rainbow trout during hypoxia or hypercapnia. Respir. Physiol. 84, 77–92.
Kubly, K.L., Stecyk, J.A.W., 2015. Temperature-dependence of L-type Ca^{2+} current in ventricular cardiomyocytes of the Alaska blackfish (*Dallia pectoralis*). J. Comp. Physiol. B 185, 845–858.
Lague, S.L., Speers-Roesch, B., Richards, J.G., Farrell, A.P., 2012. Exceptional cardiac anoxia tolerance in tilapia (*Oreochromis hybrid*). J. Exp. Biol. 215, 1354–1365.
Lahiri, S., Szidon, J.P., Fishman, A.P., 1970. Potential respiratory and circulatory adjustments to hypoxia in the African lungfish. Fed. Proc. 29, 1141–1148.
Land, S.C., Buck, L.T., Hochachka, P.W., 1993. Response of protein synthesis to anoxia and recovery in anoxia-tolerant hepatocytes. Am. J. Physiol. 265, R41–48.
Lefevre, S., Huong, D.T.T., Phuong, N.T., Wang, T., Bayley, M., 2012. Effects of hypoxia on the partitioning of oxygen uptake and the rise in metabolism during digestion in the air-breathing fish *Channa striata*. Aquaculture 364–365, 137–142.
Lefevre, S., Wang, T., Huong, D., Phuong, N., Bayley, M., 2013. Partitioning of oxygen uptake and cost of surfacing during swimming in the air-breathing catfish Pangasianodon hypophthalmus. J. Comp. Physiol. B 183, 215–221.
Lefevre, S., Bayley, M., McKenzie, D.J., Craig, J.F., 2014a. Air-breathing fishes. J. Fish Biol. 84, 547–553.
Lefevre, S., Damsgaard, C., Pascale, D.R., Nilsson, G.E., Stecyk, J.A.W., 2014b. Air breathing in the Arctic: influence of temperature, hypoxia, activity and restricted air access on respiratory physiology of the Alaska blackfish (*Dallia pectoralis*). J. Exp. Biol. 217, 4387–4398.
Lefevre, S., Domenici, P., McKenzie, D.J., 2014c. Swimming in air-breathing fishes. J. Fish Biol. 84, 661–681.
Lefevre, S., Wang, T., Jensen, A., Cong, N.V., Huong, D.T.T., Phuong, N.T., Bayley, M., 2014d. Air-breathing fishes in aquaculture. What can we learn from physiology? J. Fish Biol. 84, 705–731.
Leite, C.A.C., Florindo, L.H., Kalinin, A.L., Milsom, W.K., Rantin, F.T., 2007. Gill chemoreceptors and cardio-respiratory reflexes in the neotropical teleost pacu, *Piaractus mesopotamicus*. J. Comp. Physiol. A 193, 1001–1011.
Lopes, J., de Lima Boijink, C., Florindo, L., Leite, C., Kalinin, A., Milsom, W., Rantin, F., 2010. Hypoxic cardiorespiratory reflexes in the facultative air-breathing fish jeju (*Hoplerythrinus unitaeniatus*): role of branchial O2 chemoreceptors. J. Comp. Physiol. B 180, 797–811.
Lutz, P.L., 1989. Interaction between hypometabolism and acid-base balance. Can. J. Zool. 67, 3018–3023.

Lutz, P.L., Nilsson, G.E., 1993. Metabolic transitions to anoxia in the turtle brain: the role of neurotransmitters. In: Bicudo, E., Glass, M. (Eds.), The Vertebrate Gas Transport Cascade. CRC Press, Boca Raton, pp. 323–329.
Lutz, P.L., Nilsson, G.E., 1997. Contrasting strategies for anoxic brain survival—glycolysis up or down. J. Exp. Biol. 200, 411–419.
Lutz, P.L., Storey, K.B., 1997. Adaptations to variations in oxygen tension by vertebrates and invertebrates. In: Dantzler, W.H. (Ed.), Handbook of Physiology, vol. II. Oxford University Press, New York, pp. 1472–1522. Section 13.
Lutz, P.L., McMahon, P., Rosenthal, M., Sick, T.J., 1984. Relationships between aerobic and anaerobic energy production in turtle brain in situ. Am. J. Physiol. 247, R740–4.
Lutz, P.L., Rosenthal, M., Sick, T.J., 1985. Living without oxygen: turtle brain as a model of anaerobic metabolism. Mol. Physiol. 8, 411–425.
MacCormack, T.J., Driedzic, W.R., 2004. Cardiorespiratory and tissue adenosine responses to hypoxia and reoxygenation in the short-horned sculpin *Myoxocephalus scorpius*. J. Exp. Biol. 207, 4157.
MacCormack, T.J., McKinley, R.S., Roubach, R., Almeida-Val, V.M.F., Val, A.L., Driedzic, W.R., 2003. Changes in ventilation, metabolism, and behaviour, but not bradycardia, contribute to hypoxia survival in two species of Amazonian armoured catfish. Can. J. Zool. 81, 272–280.
Martini, F.H., 1998. The ecology of hagfishes. In: Jørgensen, J.M., Lomholt, J.P., Weber, R.E., Malte, H. (Eds.), The Biology of Hagfishes. Springer Netherlands, Dordrecht, pp. 57–77.
Marvin Jr., D.E., Burton, D.T., 1973. Cardiac and respiratory responses of rainbow trout, bluegills and brown bullhead catfish during rapid hypoxia and recovery under normoxic conditions. Comp. Biochem. Physiol. A Physiol. 46, 755–765.
Marvin, D.E., Heath, A.G., 1968. Cardiac and respiratory responses to gradual hypoxia in three ecologically distinct species of fresh-water fish. Comp. Biochem. Physiol. 27, 349–355.
Matikainen, N., Vornanen, M., 1992. Effect of season and temperature acclimation on the function of crucian carp (*Carassius carassius*) heart. J. Exp. Biol. 167, 203–220.
Maxime, V., Nonnotte, G., Peyraud, C., Williot, P., Truchot, J.P., 1995. Circulatory and respiratory effects of an hypoxic stress in the Siberian sturgeon. Respir. Physiol. 100, 203–212.
McCrimmon, H.R., 1968. Carp in Canada. Bull. Fish. Res. Board Can. 165, 1–93.
McDonald, M.D., Gilmour, K.M., Walsh, P.J., Perry, S.F., 2010. Cardiovascular and respiratory reflexes of the gulf toadfish (*Opsanus beta*) during acute hypoxia. Respir. Physiol. Neurobiol. 170, 59–66.
McKenzie, D.J., Aota, S., Randall, D.J., 1991a. Ventilatory and cardiovascular responses to blood pH, plasma PCO2, blood O2 content, and catecholamines in an air-breathing fish, the bowfin (*Amia calva*). Physiol. Zool. 64, 432–450.
McKenzie, D.J., Burleson, M.L., Randall, D.J., 1991b. The effects of branchial denervation and pseudobranch ablation on cardioventilatory control in an air-breathing fish. J. Exp. Biol. 161, 347.
McKenzie, D.J., Taylor, E.W., Bronzi, P., Bolis, C.L., 1995. Aspects of cardioventilatory control in the adriatic sturgeon (*Acipenser naccarii*). Respir. Physiol. 100, 45–53.
McKenzie, D.J., Campbell, H.A., Taylor, E.W., Micheli, M., Rantin, F.T., Abe, A.S., 2007. The autonomic control and functional significance of the changes in heart rate associated with air breathing in the jeju, *Hoplerythrinus unitaeniatus*. J. Exp. Biol. 210, 4224–4232.
McKenzie, D.J., Skov, P.V., Taylor, E.W.T., Wang, T., Steffensen, J.F., 2009. Abolition of reflex bradycardia by cardiac vagotomy has no effect on the regulation of oxygen uptake by Atlantic cod in progressive hypoxia. Comp. Biochem. Physiol. A Mol. Integr. Physiol. 153, 332–338.
Mehrani, H., Storey, K.B., 1995. Enzymatic control of glycogenolysis during anoxic submergence in the fresh-water turtle Trachemys scripta. Int. J. Biochem. Cell Biol. 27, 821–830.

Mendonça, P.C., Gamperl, A.K., 2010. The effects of acute changes in temperature and oxygen availability on cardiac performance in winter flounder (*Pseudopleuronectes americanus*). Comp. Biochem. Physiol. A Mol. Integr. Physiol. 155, 245–252.

Micheli-Campbell, M.A., Campbell, H.A., Kalinin, A.L., Rantin, F.T., 2009. The relationship between O_2 chemoreceptors, cardio-respiratory reflex and hypoxia tolerance in the neotropical fish *Hoplias lacerdae*. Comp. Biochem. Physiol. A Mol. Integr. Physiol. 154, 224–232.

Milsom, W.K., 2012. New insights into gill chemoreception: receptor distribution and roles in water and air breathing fish. Respir. Physiol. Neurobiol. 184, 326–339.

Muusze, B., Marcon, J., van den Thillart, G., Almeida-Val, V., 1998. Hypoxia tolerance of Amazon fish: respirometry and energy metabolism of the cichlid *Astronotus ocellatus*. Comp. Biochem. Physiol. A Mol. Integr. Physiol. 120, 151–156.

Nelson, J.S., 2016. Fishes of the World. John Wiley & Sons, Hoboken, NJ.

Nelson, J.A., Rios, F.S., Sanches, J.R., Fernandes, M.N., Rantin, F.T., 2007. Environmental influences on the respiratory physiology and gut chemistry of a facultative air-breathing, tropical herbivorous fish *Hypostomus regani* (Ihering, 1905). In: Fernandes, M.N., Rantin, F.T., Glass, M.L., Kapoor, B.G. (Eds.), Fish Respiration and Environment. Sciences Publishers, Enfield, pp. 191–217.

Nikinmaa, M., 1986. Control of red cell pH in teleost fishes. Ann. Zool. Fenn. 23, 223–235.

Nikinmaa, M., 1990. Vertebrate Red Blood Cells Adaptations of Function to Respiratory Requirements. Springer, Berlin, Heidelberg.

Nilsson, G.E., 1990. Long-term anoxia in crucian carp: changes in the levels of amino acid and monoamine neurotransmitters in the brain, catecholamines in chromaffin tissue, and liver glycogen. J. Exp. Biol. 150, 295–320.

Nilsson, G.E., 2001. Surviving anoxia with the brain turned on. News Physiol. Sci. 16, 217–221.

Nilsson, S., Axelsson, M., 1987. Cardiovascular control systems in fish. In: Taylor, E.W. (Ed.), Neurobiology of the Cardiorespiratory System. Manchester University Press, Manchester, pp. 115–133.

Nilsson, G.E., Lutz, P.L., 1991. Release of inhibitory neurotransmitters in response to anoxia in turtle brain. Am. J. Physiol. 261, R32–7.

Nilsson, G.E., Ostlund-Nilsson, S., 2004. Hypoxia in paradise: widespread hypoxia tolerance in coral reef fishes. Proc. Biol. Sci. 271 (Suppl. 3), S30–3.

Nilsson, S., Pettersson, K., 1981. Sympathetic nervous control of blood flow in the gills of the Atlantic cod, *Gadus morhua*. J. Comp. Physiol. 144, 157–163.

Nilsson, G.E., Renshaw, G.M.C., 2004. Hypoxic survival strategies in two fishes: extreme anoxia tolerance in the North European crucian carp and natural hypoxic preconditioning in a coral-reef shark. J. Exp. Biol. 207, 3131.

Nilsson, G.E., Rosen, P.R., Johansson, D., 1993. Anoxic depression of spontaneous locomotor activity in crucian carp quantified by a computerized imaging technique. J. Exp. Biol. 180, 153–162.

Nilsson, G.E., Hylland, P., Lofman, C.O., 1994. Anoxia and adenosine induce increased cerebral blood flow rate in crucian carp. Am. J. Physiol. 267, R590–5.

Nonnotte, G., Maxime, V., Truchot, J.P., Williot, P., Peyraud, C., 1993. Respiratory responses to progressive ambient hypoxia in the sturgeon, *Acipenser baeri*. Respir. Physiol. 91, 71–82.

Oliveira, R.D., Lopes, J.M., Sanches, J.R., Kalinin, A.L., Glass, M.L., Rantin, F.T., 2004. Cardiorespiratory responses of the facultative air-breathing fish jeju, *Hoplerythrinus unitaeniatus* (Teleostei, Erythrinidae), exposed to graded ambient hypoxia. Comp. Biochem. Physiol. A Mol. Integr. Physiol. 139, 479–485.

Olson, K.R., 1992. Blood and extracellular fluid volume regulation: role of the renin-angiotensin system, kallikrein-kinin system, and atrial natriuretic peptides. In: Randall, D.J., Hoar, W.S., Farrell, A.P. (Eds.), Fish Physiology, vol. 12B. Academic Press, San Diego, pp. 135–254.

Olson, K.R., 1994. Circulatory anatomy in bimodally breathing fish. Am. Zool. 34, 280–288.
Orchard, C.H., Kentish, J.C., 1990. Effects of changes of pH on the contractile function of cardiac muscle. Am. J. Physiol. Cell Physiol. 258, C967–981.
Ostdiek, J.L., Roland, M.N., 1959. Studies on the Alaskan blackfish *Dallia pectoralis* I. Habitat, size and stomach analyses. Am. Midl. Nat. 61, 218–229.
Overgaard, J., Wang, T., Nielsen, O.B., Gesser, H., 2005. Extracellular determinants of cardiac contractility in the cold anoxic turtle. Physiol. Biochem. Zool. 78, 976–995.
Overgaard, J., Gesser, H., Wang, T., 2007. Tribute to P. L. Lutz: cardiac performance and cardiovascular regulation during anoxia/hypoxia in freshwater turtles. J. Exp. Biol. 210, 1687–1699.
Paajanen, V., Vornanen, M., 2003. Effects of chronic hypoxia on inward rectifier K + current (IKI) in ventricular myocytes of crucian carp (Carrassius carassius) heart. J. Membrane Biol. 194, 119–127.
Paajanen, V., Vornanen, M., 2004. Regulation of action potential duration under acute heat stress by $I_{K,ATP}$ and I_{K1} in fish cardiac myocytes. Am. J. Physiol. Regul. Integr. Comp. Physiol. 286, R405–15.
Padbury, J.F., Ludlow, J.K., Ervin, M.G., Jacobs, H.C., Humme, J.A., 1987. Thresholds for physiological effects of plasma catecholamines in fetal sheep. Am. J. Physiol. 252, E530–537.
Pamenter, M.E., Shin, D.S., Buck, L.T., 2008. AMPA receptors undergo channel arrest in the anoxic turtle cortex. Am. J. Physiol. 294, R606–13.
Panlilio, J.M., Marin, S., Lobl, M.B., McDonald, M.D., 2016. Treatment with the selective serotonin reuptake inhibitor, fluoxetine, attenuates the fish hypoxia response. Sci. Rep. 6, 31148.
Pearson, M., van der Kraak, G., Stevens, D., 1992. *In vivo* pharmacology of spleen contraction in rainbow trout. Can. J. Zool. 70, 625–627.
Perez-Pinzon, M.A., Rosenthal, M., Sick, T.J., Lutz, P.L., Pablo, J., Mash, D., 1992. Down-regulation of sodium channels during anoxia: a putative survival strategy of turtle brain. Am. J. Physiol. 262, R712–5.
Perry, S.F., Bernier, N.J., 1999. The acute humoral adrenergic stress response in fish: facts and fiction. Aquaculture 177, 285–295.
Perry, S.F., Desforges, P.R., 2006. Does bradycardia or hypertension enhance gas transfer in rainbow trout (*Oncorhynchus mykiss*)? Comp. Biochem. Physiol. A Mol. Integr. Physiol. 144, 163–172.
Perry, S.F., Fritsche, R., Thomas, S., 1993. Storage and release of catecholamines from the chromaffin tissue of the Atlantic hagfish *Myxine glutinosa*. J. Exp. Biol. 183, 165.
Perry, S.F., Fritsche, R., Hoagland, T.M., Duff, D.W., Olson, K.R., 1999. The control of blood pressure during external hypercapnia in the rainbow trout (*Oncorhynchus mykiss*). J. Exp. Biol. 202, 2177–2190.
Perry, S.F., Gilmour, K.M., McNeill, B., Chew, S., Ip, Y.K., 2005. Circulating catecholamines and cardiorespiratory responses in hypoxic lungfish (*Protopterus dolloi*): a comparison of aquatic and aerial hypoxia. Physiol. Biochem. Zool. 78, 325–334.
Perry, S.F., Jonz, M.G., Gilmour, K.M., 2009a. Oxygen sensing and the hypoxic ventilatory response. In: Jeffrey, A.P.F., Richards, G., Colin, J.B. (Eds.), Fish Physiology, vol. 27. Academic Press, San Diego, pp. 193–253.
Perry, S.F., Vulesevic, B., Braun, M., Gilmour, K.M., 2009b. Ventilation in Pacific hagfish (*Eptatretus stoutii*) during exposure to acute hypoxia or hypercapnia. Respir. Physiol. Neurobiol. 167, 227–234.
Petersen, L.H., Gamperl, A.K., 2010. Effect of acute and chronic hypoxia on the swimming performance, metabolic capacity and cardiac function of Atlantic cod (*Gadus morhua*). J. Exp. Biol. 213, 808.

Pettersson, K., Johansen, K., 1982. Hypoxic vasoconstriction and the effects of adrenaline on gas exchange efficiency in fish gills. J. Exp. Biol. 97, 263–272.

Pettersson, K., Nilsson, S., 1979. Nervous control of the branchial vasculature resistance of the Atlantic cod, *Gadus morhua*. J. Comp. Physiol. 129, 179–183.

Peyraud-Waitzenegger, M., Soulier, P., 1989. Ventilatory and circulatory adjustments in the European eel (*Anguilla anguilla* L.) exposed to short term hypoxia. Exp. Biol. 48, 107–122.

Peyraud-Waitzenegger, M., Barthelemy, L., Peyraud, C., 1980. Cardiovascular and ventilatory effects of catecholamines in unrestrained eels (*Anguilla anguilla* L.). J. Comp. Physiol. 138, 367–375.

Piiper, J., Baumgarten, D., Meyer, M., 1970. Effects of hypoxia upon respiration and circulation in the dogfish *Scyliorhinus stellaris*. Comp. Biochem. Physiol. 36, 513–520.

Piironen, J., Holopainen, I.J., 1986. A note on seasonality in anoxia tolerance of crucian carp (*Carassius carassius* (L.)) in the laboratory. Ann. Zool. Fenn. 23, 335–338.

Pörtner, H.O., Lannig, G., 2009. Oxygen and capacity limited thermal tolerance. In: Jeffrey, A.P.F., Richards, G., Colin, J.B. (Eds.), Fish Physiology, vol. 27. Academic Press, San Diego, pp. 143–191.

Poupa, O., Ask, J.A., Helle, K.B., 1984. Absence of calcium paradox in the cardiac ventricle of the Atlantic hagfish (*Myxine glutinosa*). Comp. Biochem. Physiol. A Physiol. 78, 181–183.

Randall, D.J., Jones, D.R., 1973. The effect of deafferentation of the pseudobranch on the respiratory response to hypoxia and hyperoxia in the trout (*Salmo gairdneri*). Respir. Physiol. 17, 291–301.

Randall, D.J., Perry, S.F., 1992. Catecholamines. In: Randall, D.J., Hoar, W.S., Farrell, A.P. (Eds.), Fish Physiology, vol. 12B. Academic Press, San Diego, pp. 255–300.

Randall, D.J., Shelton, G., 1963. The effects of changes in environmental gas concentrations of the breathing and heart rate of a teleost fish. Comp. Biochem. Physiol. 9, 229–239.

Randall, D.J., Smith, J.C., 1967. The regulation of cardiac activity in fish in a hypoxic environment. Physiol. Zool. 40, 104–113.

Randall, D.J., Cameron, J.N., Daxboeck, C., Smatresk, N., 1981. Aspects of bimodal gas exchange in the bowfin, *Amia calva* L. (Actinopterygii: Amiiformes). Respir. Physiol. 43, 339–348.

Rantin, F.T., Glass, M.L., Kalinin, A.L., Verzola, R.M.M., Fernandes, M.N., 1993. Cardiorespiratory responses in two ecologically distinct erythrinids (*Hoplias malabaricus* and *Hoplias lacerdae*) exposed to graded environmental hypoxia. Environ. Biol. Fishes 36, 93–97.

Rantin, F.T., Kalinin, A.L., Guerra, C.D., Maricondi-Massari, M., 1995. Electrocardiographic characterization of myocardial function in normoxic and hypoxic teleosts. Braz. J. Med. Biol. Res. 28, 1277–1289.

Rantin, F.T., Del Rosario Guerra, C., Kalinin, A.L., Lesner Glass, M., 1998. The influence of aquatic surface respiration (ASR) on cardio-respiratory function of the serrasalmid fish *Piaractus mesopotamicus*. Comp. Biochem. Physiol. A Mol. Integr. Physiol. 119, 991–997.

Reese, S.A., Jackson, D.C., Ultsch, G.R., 2002. The physiology of overwintering in a turtle that occupies multiple habitats, the common snapping turtle (Chelydra serpentina). Physiol. Biochem. Zool. 75, 432–438.

Reeves, R.B., 1963. Energy cost of work in aerobic and anaerobic turtle heart muscle. Am. J. Physiol. 205, 17–22.

Renshaw, G.M.C., Kerrisk, C.B., Nilsson, G.E., 2002. The role of adenosine in the anoxic survival of the epaulette shark, Hemiscyllium ocellatum. Comp. Biochem. Physiol. B 131, 133–141.

Richards, J.G., 2009. Metabolic and molecular responses of fish to hypoxia. In: Jeffrey, A.P.F., Richards, G., Colin, J.B. (Eds.), Fish Physiology, vol. 27. Academic Press, San Diego, pp. 443–485.

Richards, J.G., 2011. Physiological, behavioral and biochemical adaptations of intertidal fishes to hypoxia. J. Exp. Biol. 214, 191–199.
Ristori, M.T., Laurent, P., 1977. Action de l'hypoxie sur le systeme vasculaire branchial de la tete perfusee de truite. C. R. Seances Soc. Biol. Fil. 171, 809–813.
Roberts, J.L., Graham, J.B., 1985. Adjustments of cardiac rate to changes in respiratory gases by a bimodal breather, the Panamanian swamp eel, *Synbranchus marmoratus*. Am. Zool. 25, 51A.
Roden, D.M., Balser, J.R., George Jr., A.L., Anderson, M.E., 2002. Cardiac ion channels. Annu. Rev. Physiol. 64, 431–475.
Rodgers-Garlick, C.I., Hogg, D.W., Buck, L.T., 2013. Oxygen-sensitive reduction in Ca^{2+}-activated K^{+} channel open probability in turtle cerebrocortex. Neuroscience 237, 243–254.
Rodnick, K.J., Gesser, H., 2017. Cardiac energy metabolism. In: Gamperl, A.K., Gillis, T.E., Farrell, A.P., Brauner, C.J. (Eds.), The Cardiovascular System: Morphology, Control and Function. Fish Physiology, vol. 36A. Academic Press, San Diego, pp. 317–367.
Rolfe, D.F., Brown, G.C., 1997. Cellular energy utilization and molecular origin of standard metabolic rate in mammals. Physiol. Rev. 77, 731–758.
Rose Jr., C.E., Althaus, J.A., Kaiser, D.L., Miller, E.D., Carey, R.M., 1983. Acute hypoxemia and hypercapnia: increase in plasma catecholamines in conscious dogs. Am. J. Physiol. 245, H924–929.
Routley, M.H., Nilsson, G.E., Renshaw, G.M.C., 2002. Exposure to hypoxia primes the respiratory and metabolic responses of the epaulette shark to progressive hypoxia. Comp. Biochem. Physiol. A Mol. Integr. Physiol. 131, 313–321.
Sanchez, A., Soncini, R., Wang, T., Koldkjaer, P., Taylor, E.W., Glass, M.L., 2001. The differential cardio-respiratory responses to ambient hypoxia and systemic hypoxaemia in the South American lungfish, *Lepidosiren paradoxa*. Comp. Biochem. Physiol. A Mol. Integr. Physiol. 130, 677–687.
Sandblom, E., Axelsson, M., 2005a. Baroreflex mediated control of heart rate and vascular capacitance in trout. J. Exp. Biol. 208, 821.
Sandblom, E., Axelsson, M., 2005b. Effects of hypoxia on the venous circulation in rainbow trout (*Oncorhynchus mykiss*). Comp. Biochem. Physiol. A Mol. Integr. Physiol. 140, 233–239.
Sandblom, E., Axelsson, M., 2006. Adrenergic control of venous capacitance during moderate hypoxia in the rainbow trout (*Oncorhynchus mykiss*): role of neural and circulating catecholamines. Am. J. Physiol. 291, R711–R718.
Sandblom, E., Axelsson, M., 2011. Autonomic control of circulation in fish: a comparative view. Auton. Neurosci. 165, 127–139.
Sandblom, E., Gräns, A., 2017. Form, function and control of the vasculature. In: Gamperl, A.K., Gillis, T.E., Farrell, A.P., Brauner, C.J. (Eds.), The Cardiovascular System: Morphology, Control and Function. Fish Physiology, vol. 36A. Academic Press, San Diego, pp. 369–433.
Sandblom, E., Gräns, A., Seth, H., Axelsson, M., 2010. Cholinergic and adrenergic influences on the heart of the African lungfish *Protopterus annectens*. J. Fish Biol. 76, 1046–1054.
Santer, R.M., 1985. Morphology and innervation of the fish heart. Adv. Anat. Embryol. Cell Biol. 89, 1–102.
Satchell, G.H., 1961. The response of the dogfish to anoxia. J. Exp. Biol. 38, 531.
Satchell, G.H., 1991. Physiology and Form of Fish Circulation. Cambridge University Press, Cambridge.
Saunders, R.L., 1962. The irrigation of the gills in fishes: II. Efficiency of oxygen uptake in relation to respiratory flow activity and concentrations of oxygen and carbon dioxide. Can. J. Zool. 40, 817–862.
Saunders, R.L., Sutterlin, A.M., 1971. Cardiac and respiratory responses to hypoxia in the sea raven *Hemipterus americanus* and investigation of possible control mechanisms. J. Fish. Res. Board Can. 28, 491–503.

Schmidt, H., Wegener, G., 1988. Glycogen phosphorylase in fish brain (*Carassius carassius*) during hypoxia. Biochem. Soc. Trans. 16, 621.

Schramm, M., Klieber, H.G., Daut, J., 1994. The energy-expenditure of actomyosin-ATPase, Ca^{2+}-ATPase and Na^+,K^+-ATPase in guinea-pig cardiac ventricular muscle. J. Physiol. 481, 647–662.

Scott, W.B., Crossman, E.J., 1973. Freshwater fishes of Canada. Bull. Fish. Res. Board Can. 184, 1–966.

Shiels, H.A., 2011. Cardiac excitation-contraction coupling: routes of cellular calcium flux. In: Farrell, A.P. (Ed.), Design and Physiology of the Heart. In: Encyclopedia of Fish Physiology, vol. 2. Academic Press, San Diego, pp. 1045–1053.

Shiels, H.A., 2017. Cardiomyocyte morphology and physiology. In: Gamperl, A.K., Gillis, T.E., Farrell, A.P., Brauner, C.J. (Eds.), The Cardiovascular System: Morphology, Control and Function. Fish Physiology, 36A. Academic Press, San Diego, pp. 55–98.

Shiels, H.A., Galli, G.L.J., 2014. The sarcoplasmic reticulum and the evolution of the vertebrate heart. Physiology 29, 456–469.

Shin, D.S.-H., Buck, L.T., 2003. Effect of anoxia and pharmacological anoxia on whole-cell NMDA receptor currents in cortical neurons from the Western painted turtle. Physiol. Biochem. Zool. 76, 41.

Shingles, A., McKenzie, D.J., Claireaux, G., Domenici, P., 2005. Reflex cardioventilatory responses to hypoxia in the flathead gray mullet (*Mugil cephalus*) and their behavioral modulation by perceived threat of predation and water turbidity. Physiol. Biochem. Zool. 78, 744–755.

Short, S., Butler, P.J., Taylor, E.W., 1977. The relative importance of nervous, humoral and intrinsic mechanisms in the regulation of heart rate and stroke volume in the dogfish *Scyliorhinus canicula*. J. Exp. Biol. 70, 77.

Short, S., Taylor, E.W., Butler, P.J., 1979. The effectiveness of oxygen transfer during normoxia and hypoxia in the dogfish (*Scyliorhinus canicula* L.) before and after cardiac vagotomy. J. Comp. Physiol. 132, 289–295.

Shoubridge, E.A., Hochachka, P.W., 1980. Ethanol: novel end product of vertebrate anaerobic metabolism. Science 209, 308–309.

Sick, T.J., Pérez-Pinón, M.A., Rosenthal, M., 1993. Maintaining coupled metabolism and membrane function in anoxic brain: a comparison between the turtle and the rat. In: Hochachka, P.W., Lutz, P.L., Sick, T.J., Rosenthal, M., van den Thillart, G. (Eds.), Surviving Hypoxia: Mechanisms of Control and Adaptation. CRC Press, Boca Raton, pp. 351–364.

Singh, B.N., Hughes, G.M., 1973. Cardiac and respiratory responses in the climbing perch *Anabas testudineus*. J. Comp. Physiol. 84, 205–226.

Skals, M., Skovgaard, N., Taylor, E.W., Leite, C.A.C., Abe, A.S., Wang, T., 2006. Cardiovascular changes under normoxic and hypoxic conditions in the air-breathing teleost *Synbranchus marmoratus*: importance of the venous system. J. Exp. Biol. 209, 4167.

Smatresk, N.J., 1988. Control of the respiratory mode in air-breathing fishes. Can. J. Zool. 66, 144–151.

Smatresk, N., 1989. Chemoreflex control of respiration in an air-breathing fish. In: Lahiri, S., Foster II, R.E., Davies, R.O., Pack, A.I. (Eds.), Chemoreceptors and Chemoreflexes in Breathing: Cellular and Molecular Aspects. Oxford University Press, London, pp. 29–52.

Smatresk, N.J., Cameron, J.N., 1982. Respiration and acid-base physiology of the spotted gar, a bimodal breather: I. Normal values, and the response to severe hypoxia. J. Exp. Biol. 96, 263.

Smatresk, N.J., Burleson, M.L., Azizi, S.Q., 1986. Chemoreflexive responses to hypoxia and NaCN in longnose gar: evidence for two chemoreceptor loci. Am. J. Physiol. 251, R116.

Smith, D.G., 1978. Neural control of blood pressure in rainbow trout (*Salmo gairdneri*). Can. J. Zool. 56, 1678–1683.

Smith, F.M., Jones, D.R., 1978. Localization of receptors causing hypoxic bradycardia in trout (*Salmo gairdneri*). Can. J. Zool. 56, 1260–1265.

Smith, R.W., Houlihan, D.F., Nilsson, G.E., Brechin, J.G., 1996. Tissue-specific changes in protein synthesis rates *in vivo* during anoxia in crucian carp. Am. J. Physiol. 271, R897–904.

Smith, M.P., Russell, M.J., Wincko, J.T., Olson, K.R., 2001. Effects of hypoxia on isolated vessels and perfused gills of rainbow trout. Comp. Biochem. Physiol. A Mol. Integr. Physiol. 130, 171–181.

Soivio, A., Tuurala, H., 1981. Structural and circulatory responses to hypoxia in the secondary lamellae of *Salmo gairdneri* gills at two temperatures. J. Comp. Physiol. 145, 37–43.

Speers-Roesch, B., Sandblom, E., Lau, G.Y., Farrell, A.P., Richards, J.G., 2010. Effects of environmental hypoxia on cardiac energy metabolism and performance in tilapia. Am. J. Physiol. Regul. Integr. Comp. Physiol. 298, R104–119.

Speers-Roesch, B., Brauner, C.J., Farrell, A.P., Hickey, A.J.R., Renshaw, G.M.C., Wang, Y.S., Richards, J.G., 2012. Hypoxia tolerance in elasmobranchs. II. Cardiovascular function and tissue metabolic responses during progressive and relative hypoxia exposures. J. Exp. Biol. 215, 103–114.

Stecyk, J.A., Farrell, A.P., 2002. Cardiorespiratory responses of the common carp (*Cyprinus carpio*) to severe hypoxia at three acclimation temperatures. J. Exp. Biol. 205, 759–768.

Stecyk, J.A., Farrell, A.P., 2006. Regulation of the cardiorespiratory system of common carp (*Cyprinus carpio*) during severe hypoxia at three seasonal acclimation temperatures. Physiol. Biochem. Zool. 79, 614–627.

Stecyk, J.A., Farrell, A.P., 2007. Effects of extracellular changes on spontaneous heart rate of normoxia- and anoxia-acclimated turtles (*Trachemys scripta*). J. Exp. Biol. 210, 421–431.

Stecyk, J.A., Overgaard, J., Farrell, A.P., Wang, T., 2004a. α-adrenergic regulation of systemic peripheral resistance and blood flow distribution in the turtle Trachemys scripta during anoxic submergence at 5°C and 21°C. J. Exp. Biol. 207, 269–283.

Stecyk, J.A., Stensløkken, K.-O., Farrell, A.P., Nilsson, G.E., 2004b. Maintained cardiac pumping in anoxic crucian carp. Science 306, 77.

Stecyk, J.A., Stensløkken, K.-O., Nilsson, G.E., Farrell, A.P., 2007. Adenosine does not save the heart of anoxia-tolerant vertebrates during prolonged oxygen deprivation. Comp. Biochem. Physiol. A Mol. Integr. Physiol. 147, 961–973.

Stecyk, J.A., Galli, G.L., Shiels, H.A., Farrell, A.P., 2008. Cardiac survival in anoxia-tolerant vertebrates: an electrophysiological perspective. Comp. Biochem. Physiol. C 148, 339–354.

Stecyk, J.A.W., Bock, C., Overgaard, J., Wang, T., Farrell, A.P., Pörtner, H.-O., 2009. Correlation of cardiac performance with cellular energetic components in the oxygen-deprived turtle heart. Am. J. Physiol. 297, R756–768.

Stecyk, J.A.W., Larsen, B.C., Nilsson, G.E., 2011. Intrinsic contractile properties of the crucian carp (*Carassius carassius*) heart during anoxic and acidotic stress. Am. J. Physiol. 301, R1132–R1142.

Stecyk, J.A.W., Couturier, C.S., Fagernes, C.E., Ellefsen, S., Nilsson, G.E., 2012. Quantification of heat shock protein mRNA expression in warm and cold anoxic turtles (*Trachemys scripta*) using an external RNA control for normalization. Comp. Biochem. Physiol. Part D Genomics Proteomics 7, 59–72.

Stecyk, J.A.W., Farrell, A.P., Vornanen, M., 2017. Na^+/K^+-ATPase activity in the anoxic turtle (*Trachemys scripta*) brain at different acclimation temperature. Comp. Biochem. Physiol. A Mol. Integr. Physiol. 206, 11–16.

Stensløkken, K.O., Sundin, L., Renshaw, G.M., Nilsson, G.E., 2004. Adenosinergic and cholinergic control mechanisms during hypoxia in the epaulette shark (*Hemiscyllium ocellatum*), with emphasis on branchial circulation. J. Exp. Biol. 207, 4451–4461.

Stensløkken, K.O., Ellefsen, S., Stecyk, J.A., Dahl, M.B., Nilsson, G.E., Vaage, J., 2008. Differential regulation of AMP-activated kinase and AKT kinase in response to oxygen availability in crucian carp (Carassius carassius). Am. J. Physiol. Regul. Integr. Comp. Physiol. 295, R1803–R1814.

Stensløkken, K.-O., Ellefsen, S., Larsen, H.K., Vaage, J., Nilsson, G.E., 2010. Expression of heat shock proteins in anoxic crucian carp (*Carassius carassius*): support for cold as a preparatory cue for anoxia. Am. J. Physiol. 298, R1499–1508.

Storey, K.B., 1985. A re-evaluation of the Pasteur effect: new mechanisms in anaerobic metabolism. Mol. Physiol. 8, 439–461.

Storey, K.B., 1987. Tissue-specific controls on carbohydrate catabolism during anoxia in goldfish. Physiol. Zool. 60, 601–607.

Storey, K.B., 1996. Metabolic adaptations supporting anoxia tolerance in reptiles: recent advances. Comp. Biochem. Physiol. B 113, 23–35.

Sundin, L.I., 1995. Responses of the branchial circulation to hypoxia in the Atlantic cod, *Gadus morhua*. Am. J. Physiol. 268, R771.

Sundin, L., Nilsson, G.E., 1997. Neurochemical mechanisms behind gill microcirculatory responses to hypoxia in trout: in vivo microscopy study. Am. J. Physiol. 272, R576.

Sundin, L.I., Reid, S.G., Kalinin, A.L., Rantin, F.T., Milsom, W.K., 1999. Cardiovascular and respiratory reflexes: the tropical fish, traira (*Hoplias malabaricus*) O_2 chemoresponses. Respir. Physiol. 116, 181–199.

Sundin, L., Reid, S.G., Rantin, F.T., Milsom, W.K., 2000. Branchial receptors and cardiorespiratory reflexes in a neotropical fish, the tambaqui (*Colossoma macropomum*). J. Exp. Biol. 203, 1225.

Suzue, T., Wu, G.B., Furukawa, T., 1987. High susceptibility to hypoxia of afferent synaptic transmission in the goldfish sacculus. J. Neurophysiol. 58, 1066–1079.

Taha, M., Lopaschuk, G.D., 2007. Alterations in energy metabolism in cardiomyopathies. Ann. Med. 39, 594–607.

Taylor, E.W., 1992. Nervous control of the heart and cardiorespiratory interactions. In: Randall, D.J., Hoar, W.S., Farrell, A.P. (Eds.), Fish Physiology, vol. 12B. Academic Press, San Diego, pp. 343–387.

Taylor, E.W., Barrett, D.J., 1985. Evidence of a respiratory role for the hypoxic bradycardia in the dogfish *Scyliorhinus canicula* L. Comp. Biochem. Physiol. A Physiol. 80, 99–102.

Taylor, E.W., Short, S., Butler, P.J., 1977. The role of the cardiac vagus in the response of the dogfish *Scyliorhinus canicula* to hypoxia. J. Exp. Biol. 70, 57.

Temma, K., Komazu, Y., Shiraki, Y., Kitazawa, T., Kondo, H., 1989. The roles of alpha- and beta-adrenoceptors in the chronotropic responses to norepinephrine in carp heart (*Cyprinus carpio*). Comp. Biochem. Physiol. C 92, 149–153.

Tiitu, V., Vornanen, M., 2001. Cold adaptation suppresses the contractility of both atrial and ventricular muscle of the crucian carp heart. J. Fish Biol. 59, 141–156.

Tiitu, V., Vornanen, M., 2003. Ryanodine and dihydropyridine receptor binding in ventricular cardiac muscle of fish with different temperature preferences. J. Comp. Physiol. B 173, 285–291.

Tikkanen, E., Haverinen, J., Egginton, S., Hassinen, M., Vornanen, M., 2017. Effects of prolonged anoxia on electrical activity of the heart in crucian carp (*Carassius carassius*). J. Exp. Biol. 220, 445–454.

Tirri, R., Ripatti, P., 1982. Inhibitory adrenergic control of heart rate of perch (*Perca fluviatilis*) *in vitro*. Comp. Biochem. Physiol. 73C, 399–401.

Turner, J.D., Driedzic, W.R., 1980. Mechanical and metabolic response of the perfused isolated fish heart to anoxia and acidosis. Can. J. Zool. 58, 886–889.

Ultsch, G.R., Jackson, D.C., 1982. Long-term submergence at 3C of the turtle, *Chrysemys picta bellii*, in normoxic and severely hypoxic water: I. Survival, gas exchange and acid-base balance. J. Exp. Biol. 96, 11–28.

Ultsch, G.R., Jackson, D.C., Moalli, R., 1981. Metabolic oxygen conformity among lower vertebrates: the toadfish revisited. J. Comp. Physiol. 142, 439–443.

Val, A.L., Almeida-Val, V.M.F., 1995. Fishes of the Amazon and Their Environments. Physiological and Biochemical Features. Springer Verlag, Heidelberg.

van Waarde, A., 1991. Alcoholic fermentation in multicellular organisms. Physiol. Zool. 64, 895–920.

Van Waversveld, J., Addink, A.D.F., Van den Thillart, G., 1989a. The anaerobic energy metabolism of goldfish determined by simultaneous direct and indirect calorimetry during anoxia and hypoxia. J. Comp. Physiol. B 159, 263–268.

Van Waversveld, J., Addink, A.D.F., Van Den Thillart, G., 1989b. Simultaneous direct and indirect calorimetry on normoxic and anoxic goldfish. J. Exp. Biol. 142, 325–335.

Vornanen, M., 1994a. Seasonal adaptation of crucian carp (*Carassius carassius* L.) heart: glycogen stores and lactate dehydrogenase activity. Can. J. Zool. 72, 433–442.

Vornanen, M., 1994b. Seasonal and temperature-induced changes in myosin heavy chain composition of crucian carp hearts. Am. J. Physiol. 267, R1567–73.

Vornanen, M., 1996. Effect of extracellular calcium on the contractility of warm-and cold-acclimated crucian carp heart. J. Comp. Physiol. B 165, 507–517.

Vornanen, M., 1998. L-type Ca^{2+} current in fish cardiac myocytes: effects of thermal acclimation and beta-adrenergic stimulation. J. Exp. Biol. 201, 533–547.

Vornanen, M., 1999. Na^{+}/Ca^{2+} exchange current in ventricular myocytes of fish heart: contribution to sarcolemmal Ca^{2+} influx. J. Exp. Biol. 202, 1763–1775.

Vornanen, M., 2017. Electrical excitability of the fish heart and its autonomic regulation. In: Gamperl, A.K., Gillis, T.E., Farrell, A.P., Brauner, C.J. (Eds.), The Cardiovascular System: Morphology, Control and Function. Fish Physiology, 36A. Academic Press, San Diego, pp. 99–153.

Vornanen, M., Hassinen, M., 2016. Zebrafish heart as a model for human cardiac electrophysiology. Channels 10, 101–110.

Vornanen, M., Paajanen, V., 2004. Seasonality of dihydropyridine receptor binding in the heart of an anoxia-tolerant vertebrate, the crucian carp (*Carassius carassius* L.). Am. J. Physiol. 287, R1263–R1269.

Vornanen, M., Paajanen, V., 2006. Seasonal changes in glycogen content and Na^{+}-K^{+}-ATPase activity in the brain of crucian carp. Am. J. Physiol. 291, R1482–1489.

Vornanen, M., Tuomennoro, J., 1999. Effects of acute anoxia on heart function in crucian carp: importance of cholinergic and purinergic control. Am. J. Physiol. 277, R465–475.

Vornanen, M., Stecyk, J.A.W., Nilsson, G.E., 2009. The anoxia-tolerant crucian carp (*Carassius carassius* L.). In: Richards, J.G., Farrell, A.P., Brauner, C.J. (Eds.), Fish Physiology, vol. 27. Academic Press, London, pp. 397–441.

Vornanen, M., Asikainen, J., Haverinen, J., 2011. Body mass dependence of glycogen stores in the anoxia-tolerant crucian carp (*Carassius carassius* L.). Naturwissenschaften 98, 225–232.

Wang, T., Lefevre, S., Thanh Huong, D.T., Cong, N.v., Bayley, M., 2009. The effects of hypoxia on growth and digestion. In: Jeffrey, A.P.F., Richards, G., Colin, J.B. (Eds.), Fish Physiology, vol. 27. Academic Press, San Diego, pp. 361–396.

Warren, D.E., Reese, S.A., Jackson, D.C., 2006. Tissue glycogen and extracellular buffering limit the survival of red-eared slider turtles during anoxic submergence at 3°C. Physiol. Biochem. Zool. 79, 736–744.

Wasser, J.S., Jackson, D.C., 1991. Effects of anoxia and graded acidosis on the levels of circulating catecholamines in turtles. Respir. Physiol. 84, 363–377.

Wells, R.M.G., 2009. Blood-gas transport and hemoglobin function: adaptations for functional and environmental hypoxia. In: Jeffrey, A.P.F., Richards, G., Colin, J.B. (Eds.), Fish Physiology, vol. 27. Academic Press, San Diego, pp. 255–299.

Williamson, J.R., Safer, B., Rich, T., Schaffer, S., Kobayashi, K., 1976. Effects of acidosis on myocardial contractility and metabolism. Acta Med. Scand. Suppl. 587, 95–112.

Wilson, C.M., Farrell, A.P., 2012. Pharmacological characterization of the heartbeat in an extant vertebrate ancestor, the Pacific hagfish, *Eptatretus stoutii*. Comp. Biochem. Physiol. A Mol. Integr. Physiol. 164, 258–263.

Wilson, C.M., Stecyk, J.A.W., Couturier, C.S., Nilsson, G.E., Farrell, A.P., 2013. Phylogeny and effects of anoxia on hyperpolarization-activated cyclic nucleotide-gated channel gene expression in the heart of a primitive chordate, the Pacific hagfish (Eptatretus stoutii). J. Exp. Biol. 216, 4462–4472.

Wilson, C.M., Roa, J.N., Cox, G.K., Tresguerres, M., Farrell, A.P., 2016. Introducing a novel mechanism to control heart rate in the ancestral Pacific hagfish. J. Exp. Biol. 219, 3227–3236.

Wise, G., Mulvey, J.M., Renshaw, G.M.C., 1998. Hypoxia tolerance in the epaulette shark (*Hemiscyllium ocellatum*). J. Exp. Zool. 281, 1–5.

Wood, C.M., 1974. A critical examination of the physical and adrenergic factors affecting blood flow through the gills of the rainbow trout. J. Exp. Biol. 60, 241.

Wood, C.M., 1975. A pharmacological analysis of the adrenergic and cholinergic mechanisms regulating branchial vascular resistance in the rainbow trout (*Salmo gairdneri*). Can. J. Zool. 53, 1569–1577.

Wood, S.C., 1991. Interactions between hypoxia and hypothermia. Annu. Rev. Physiol. 53, 71–85.

Wood, C.M., Shelton, G., 1975. Physical and adrenergic factors affecting systemic vascular resistance in the rainbow trout: a comparison with branchial vascular resistance. J. Exp. Biol. 63, 505–523.

Wood, C.M., Shelton, G., 1980. The reflex control of heart rate and cardiac output in the rainbow trout: interactive influences of hypoxia, haemorrhage and systemic vasomotor tone. J. Exp. Biol. 87, 271–284.

Wu, R.S.S., 2009. Effects of hypoxia on fish reproduction and development. In: Jeffrey, A.P.F., Richards, G., Colin, J.B. (Eds.), Fish Physiology, vol. 27. Academic Press, San Diego, pp. 79–141.

Zeraik, V.M., Belão, T.C., Florindo, L.H., Kalinin, A.L., Rantin, F.T., 2013. Branchial O_2 chemoreceptors in Nile tilapia *Oreochromis niloticus*: control of cardiorespiratory function in response to hypoxia. Comp. Biochem. Physiol. A Mol. Integr. Physiol. 166, 17–25.

6

ENVIRONMENTAL POLLUTION AND THE FISH HEART

JOHN P. INCARDONA[1]
NATHANIEL L. SCHOLZ

Northwest Fisheries Science Center, National Marine Fisheries Service, National Oceanic and Atmospheric Administration, Seattle, WA, United States
[1]Corresponding author: john.incardona@noaa.gov

As a nonstop pump driven by physiologically complex excitable cells, the heart is especially susceptible to a variety of insults, in particular exogenous chemicals. In humans, for example, heart-related toxicities are the most

The Cardiovascular System: Development, Plasticity and Physiological Responses, Volume 36B
FISH PHYSIOLOGY

DOI: https://doi.org/10.1016/bs.fp.2017.09.006

common adverse drug reaction. Fish living in polluted environments may or may not be able to avoid exposure to cardiotoxic chemicals. Nevertheless, juvenile and adult fish have a variety of robust mechanisms for protecting the heart from toxic chemicals, such as hepatic metabolism. The situation is very different for fish early life history stages, in particular very small embryos and larvae. The developing heart is the first organ to become functional in embryos, at a point very early in its morphogenesis. Owing to the intimate and interacting links between form and function in the developing heart, very subtle cardiotoxicity during embryogenesis can cause serious and long-lasting outcomes. Fish embryos have minimal capacity for metabolic detoxification, and therefore, developmental cardiotoxicity is a real-world environmental health concern for wild populations. This chapter focuses on the exquisite sensitivity of the developing fish heart to chemical contaminants, as demonstrated through aquatic pollution case studies focused on legacy organochlorine compounds and recent oil spills. These studies have revealed a diversity of mechanisms linking cardiomyocyte physiology to heart development, and abnormal development in turn to latent impacts on physiology at later life stages.

1. INTRODUCTION

1.1. The Significance of the Fish Heart as a Target Organ for Widespread Environmental Pollution

Cardiovascular disease is the largest source of human morbidity and mortality globally. While diet and coronary vascular disease are the largest contributors, lower cardiorespiratory fitness is strongly associated with chronic disease and increased risk of death (Despres, 2016; Lee et al., 2010). In addition, more recent epidemiological studies have also identified other major environmental etiologies, in particular ambient air pollution (Cosselman et al., 2015). At the same time, human pharmacology and toxicology provide ample evidence for the chemical sensitivity of cardiac function. Adverse reactions involving the heart are the most common reason drugs are pulled from the market, and cardiotoxicity screening is now one of the most important facets of drug development (MacDonald and Robertson, 2009; Wilke et al., 2007). Moreover, congenital heart diseases make up the most common class of birth defects in human newborns (van der Bom et al., 2011). Yet, heart disease is rarely considered a factor in the health of wild animal populations. However, molecular processes governing heart development and function are highly conserved across all vertebrates, so the same factors that render the human heart susceptible to chemical injury would seem to apply to other terrestrial and aquatic vertebrates, including fish. For example, due to idiosyncrasies of its structure, a key potassium channel regulating heart rate and rhythm is particularly susceptible to blockade by a very large, structurally diverse group of drugs, leading to severe consequences such as sudden death (Knape et al., 2011; Mitcheson, 2008). The same amino acids that confer this susceptibility are conserved in the orthologous genes encoding this protein in fish (Langheinrich et al., 2003).

Despite decades-long efforts to decrease industrial pollution, expansion of human populations along coastal areas and inland waterways has led to increased aquatic loadings of a broad array of chemically diverse and pervasive chemical pollutants from land-based sources. Given the chemical susceptibility of cardiac function known for our own species, it is not surprising that in recent decades the fish heart has been identified as a key target of aquatic pollutants. Similarly, recent findings also indicate a broader role for cardiorespiratory performance in fish health, for example, as a contributor to infectious disease resistance (Castro et al., 2013b). In the modern era, few chemicals enter the environment at concentrations high enough to produce outright fish kills. However, the ecotoxicological studies described here highlight the importance of sublethal impacts on cardiac function to individual physiological performance and survival, population recruitment and abundance, species conservation, and ecosystem-based natural resource management. Finally, the striking overlap between some of the most abundant classes of pollutants in both water and air suggest how studies of pollution effects in fish may inform research on human cardiovascular health.

1.2. Major Sources of Toxic Chemicals in Fish Habitats: Past, Present and Future

Fish are exposed to many sources of contaminants, and these can be broadly divided into a variety of categories. These include a distinction between the >80,000 chemicals that are in modern societal use vs so-called legacy contaminants that were produced by industry during the 20th century, many of which were later banned in response to environmental concerns (Scholz and McIntyre, 2015). Examples of the former include current-use pesticides, industrial solvents, pharmaceuticals, plasticizers, petroleum compounds, flame retardants, and many more. Examples of the latter include polychlorinated biphenyls (PCBs) and numerous synthetic chlorinated hydrocarbons, including dichlorodiphenyltrichloroethane (DDT) and related compounds. A distinction is also drawn between chemicals that enter aquatic habitats from a point source (e.g., a discharge pipe) and those that are transported by stormwater, atmospheric deposition and other diffuse nonpoint sources. Lastly, whereas most chemicals are intentionally released by human activities (or their release is known), others are a consequence of unforeseen accidents. Major oil spills, including the 2010 Deepwater Horizon incident in the northern Gulf of Mexico, have been among the most catastrophic pollution events in recent history (Scholz and McIntyre, 2015).

Historically, the cardiorespiratory responses of fish have been a common focus in toxicological studies (Hughes and Adeney, 1977; Lunn et al., 1976). Given the overwhelming number of potentially toxic chemicals entering aquatic systems, a classification of behavioral and cardiorespiratory responses in small fish (i.e., the fathead minnow) was established in an attempt to develop a toxicity screening tool (McKim et al., 1987a). However, these early studies almost always focused on lethal concentrations of chemicals and cardiorespiratory responses leading up to death (Bradbury et al., 1989; McKim et al., 1987b, c). In terms of understanding cardiotoxicity, these studies are often difficult to interpret. First, as detailed throughout the rest of this chapter, in the vast majority of cases, chemical contamination in fish habitats does not occur at concentrations that are lethal to juvenile or adult fish. Second, determining direct vs indirect effects on cardiac function is often extremely difficult in intact juvenile or adult fish. In this chapter, we consider primarily chemicals that have direct effects on the heart, rather than chemicals that impact the cardiovascular system indirectly, for example, by influencing the autonomic nervous system (e.g., anticholinergic pesticides). In contrast, more recent studies have focused on the developing heart in fish embryos for two key reasons: (1) major aquatic pollution situations or incidents have revealed the vulnerability of the developing fish heart to low-level chemical exposure; and (2) because the embryonic heart acquires intrinsic function prior to development of external influences such as the autonomic nervous system, it actually serves as a simple and sensitive indicator of chemicals with direct cardiotoxic activity. In essence, the developing hearts of very small fish embryos are virtually equivalent to isolated cardiomyocytes *in vitro*. From a practical standpoint for resource management, this means that understanding developmental cardiotoxicity is critical for assessing the impacts of aquatic pollution on fish populations. At the same time, this has allowed model fish embryos (i.e., zebrafish) to become exceptional systems for the basic study of cardiac pharmacology and toxicology (Langheinrich, 2003; Rennekamp and Peterson, 2015).

The differences between embryonic and postlarval (i.e., juvenile and adult) fish are highlighted by oil spill studies. In response to development of the Trans-Alaska pipeline and shipping of Prudhoe Bay crude oil, toxicity tests were performed on a broad selection of Alaskan marine species, including nine fish species representing pelagic, benthic and intertidal habitats (Rice et al., 1979). Lethality tests were performed using juvenile and adult fish, generating 96-h tolerance limits. Although measurements of toxic hydrocarbons in water were somewhat different than current methods, lethal concentrations for aromatic hydrocarbons for all species were all higher than 1 mg L^{-1} (parts per million). In contrast, water samples from areas heavily contaminated by the 1989 Exxon Valdez oil spill in Alaska's Prince William Sound, collected 1 week after the spill, showed concentrations of aromatic hydrocarbons in the range of 1–6 μg L^{-1} (parts per billion) (Short and Harris, 1996). The nearly 1000 times higher lethal concentrations for local fish species, determined in juveniles and adults, were no longer relevant. As detailed in this chapter, however, concentrations of aromatic hydrocarbons in the parts per billion range have profound consequences for the hearts of developing fish embryos.

Urban stormwater entering fish habitats has recently received intensive focus, in particular runoff from roadways with high traffic density (Scholz and McIntyre, 2015). Urbanization produces a distinctive suite of contaminants (Gobel et al., 2007; Paul and Meyer, 2001; Van Bohemen and Van de Laak, 2003), including: a variety of metals from motor vehicle brake pads and tires; petroleum-derived compounds from oil, grease, and vehicle exhaust; surfactants; and pesticides. Although none of these chemicals have been found to be present at lethal concentrations (when known), field studies in the United States Pacific Northwest identified a species-specific lethal response to urban stormwater in adult and juvenile coho salmon (*Oncorhynchus kisutch*) (Feist et al., 2011; McIntyre et al., 2015; Scholz et al., 2011). As mentioned above, a variety of chemicals can affect the cardiorespiratory responses of fish and produce lethality, including some metals and pesticides (Hughes and Adeney, 1977; Lunn et al., 1976; Majewski and Giles, 1981; Sargent et al., 1980). However, none of these have been shown to specifically produce cardiotoxicity at environmental concentrations. Indeed, even an artificial stormwater mimic containing petroleum, combustion-derived hydrocarbons, and common metals failed to reproduce the observed coho salmon lethality (Spromberg et al., 2016). While the pathophysiology of this syndrome is still unknown, and is more likely derived from respiratory toxicity, developing fish embryos exposed to stormwater clearly display cardiac defects that overlap significantly with what arises when they are exposed to petroleum hydrocarbons alone (McIntyre et al., 2016b).

In this context, the classes of cardiotoxic contaminants that are most intensively studied in fish include petroleum-derived polycyclic aromatic hydrocarbons (PAHs), as well as persistent organochlorine pollutants including dioxins and PCBs (Fig. 1). Both classes of compounds are associated with developmental cardiotoxicity at environmental concentrations in wild fish. The PAH family of chemicals also encompasses related nitrogen, oxygen and sulfur-containing heterocyclic compounds. Because PAHs and related

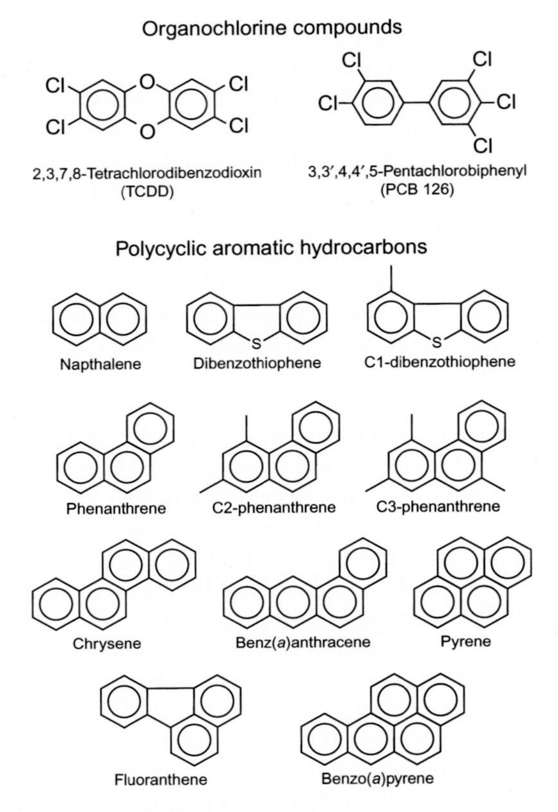

Fig. 1. Major cardiotoxic pollutants in fish. Only representative compounds are shown, as both organochlorine compounds and polycyclic aromatic hydrocarbons (PAHs)/polycyclic aromatic compounds (PACs) make up very large families. PAHs are largely derived from fossil fuels, where they typically occur in a homologous series of increasing alkylation (C1-, C2-, etc.) with methyl and larger side groups.

heterocycles cooccur in environmental mixtures, and they share similar pathways of toxicity and metabolism, they are often collectively referred to as polycyclic aromatic compounds (PACs; used interchangeably with PAHs). Dioxins, PCBs and PACs all share a planar multiring structure and are consequently agonists for the aryl hydrocarbon receptor (AHR), the ligand-activated transcription factor that controls expression of a battery of genes involved in metabolic detoxification (Fig. 2). This battery includes so-called Phase I enzymes such as cytochrome P450 1 family members (e.g., CYP1A, CYP1B, and CYP1C) and Phase II enzymes such as glutathione *S*-transferases and UDP-glucuronosyltransferases. Phase I enzymes oxidize hydrophobic

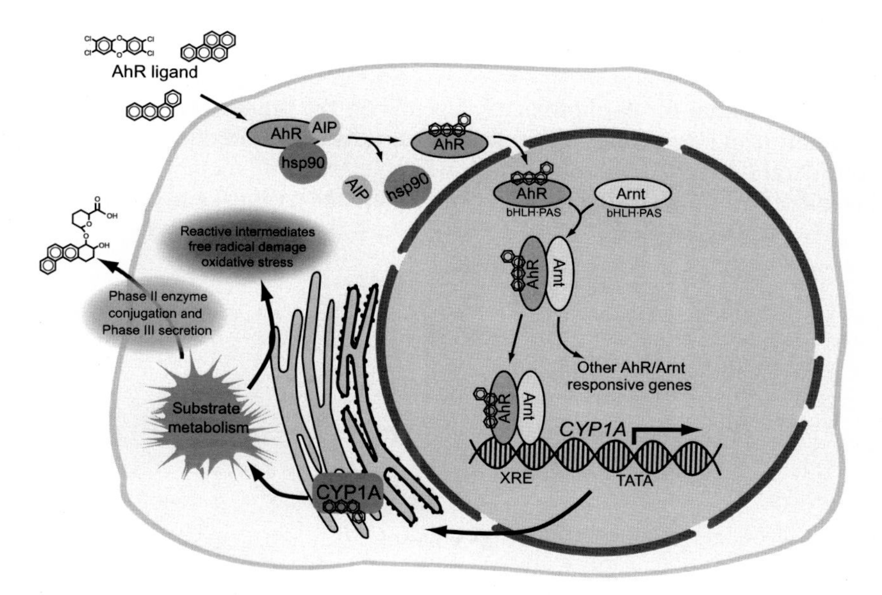

Fig. 2. Schematic of the basic xenobiotic metabolism pathway controlled by the aryl hydrocarbon receptor (AHR). The AHR resides in a cytoplasmic complex with hsp90 (heat shock protein 90) and AIP (AHR interacting protein), which dissociates upon ligand binding, allowing the ligand-bound AHR to translocate into the nucleus. After dimerizing with the AHR nuclear translocator (ARNT), the AHR–ARNT dimer binds to xenobiotic response elements (XRE) in the promoter region of genes involved in detoxification, with *cyp1a* being the most prominent. CYP1A enzyme accumulates in the endoplasmic reticulum and oxidizes planar aromatic compounds to produce reactive intermediates that are modified by Phase II enzymes for secretion. Uncoupling of CYP1A metabolism from Phase II enzymes can lead to accumulation of these reactive intermediates, which can cause cellular damage by creating DNA adducts, for example. This is the basis of carcinogenesis by PAHs such as benzo(*a*)pyrene.

aromatic compounds, thereby allowing Phase II enzymes to add a hydrophilic group to the metabolite. This conversion eventually facilitates excretion (at the cellular level involving Phase III transporters and pumps) and clearance from fish, typically via the bile. As discussed in more detail below, a key distinction between PACs and the organochlorine compounds is their relative metabolism. PACs are good substrates for Phase I enzymes and are thus detoxified and eliminated. Conversely, DDTs, PCBs and other polychlorinated compounds are not metabolized (or metabolized extremely slowly), and therefore, accumulate over both the lifetime of an individual fish and at higher trophic levels in aquatic communities and ecosystems. These chemicals are therefore termed persistent and bioaccumulative. Differences in metabolism determine the overall toxicity of a chemical to a given species of fish and, more specifically, underpin precise mechanisms of cardiotoxicity.

1.3. Routes of Chemical Uptake and Transport to the Fish Cardiovascular System

The potential for cardiotoxicity depends on the uptake, distribution and elimination of chemicals in fish tissues (toxicokinetics), and this process varies considerably across different life stages (Fig. 3). Scales protect the epidermis of juvenile and adult fish, and thus the primary uptake pathways are by ingestion

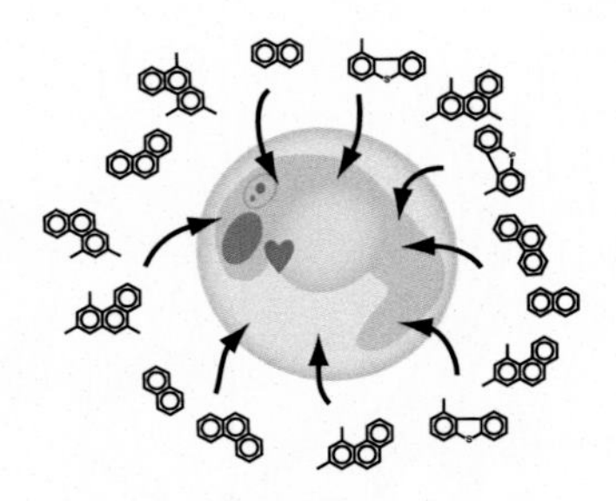

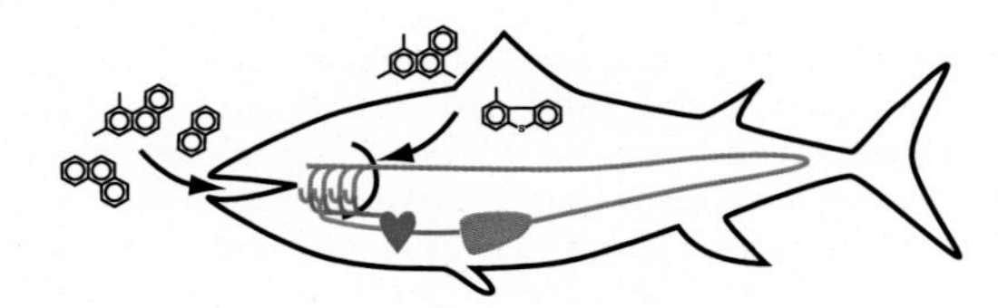

Fig. 3. Routes of pollutant uptake in relation to the heart in fish early life history stages relative to juveniles and adults. At top, fish embryos and larvae take up hydrophobic contaminants such as polycyclic aromatic hydrocarbons (PAHs) across the epidermis (and chorion) directly from surrounding water. The heart is relatively unprotected and in very close proximity to the very thin outer epidermis/yolk-sac membrane, and also obtains oxygen by simple diffusion. At bottom, the heart in juvenile and adult fish is encased in the pericardial cavity and not directly exposed. Pollutant uptake occurs across via the gills or the gastrointestinal tract. Although some fish have a coronary circulation, the majority of blood-borne contaminants pass through the liver before reaching the heart, providing a robust first-pass metabolic protection.

of contaminants in prey or sediments (bioaccumulation), or across the gills for chemicals in the dissolved phase (bioconcentration). The induction of detoxifying enzymes in the digestive tract and the gills (as well as other tissues) can often provide a first line of defense against chemical toxicity in older fish. As evidence of this, many studies have used CYP1A induction as a marker for a protective metabolic response in the tissues of fish exposed to AHR agonists (Costa et al., 2011; Ortiz-Delgado et al., 2005; Van Veld et al., 1997). For example, dietary exposures to the 5-ring PAH benzo(*a*)pyrene in many species results in a robust increase in CYP1A in the intestinal epithelium, with little induction in the liver or elsewhere. This indicates an effective first-pass defense, wherein proximal metabolism in the intestine prevents a cardiotoxic agent from reaching the vasculature and the heart. Conversely, waterborne exposures generally lead to more widespread CYP1A induction beyond the gill, including the vascular endothelium, kidney and liver. For both bioaccumulation and bioconcentration uptake pathways, there are multiple sites for detoxification before a chemical reaches cardiac tissue in postmetamorphic fish. Contaminants that are, in general, rapidly metabolized and excreted (e.g., PAHs) do not accumulate in juvenile and adult fish, and thus have much lower cardiotoxicity at these older life stages as compared to embryos and larvae.

Exposure dynamics are very different for embryos and early larvae. In the absence of scales, the primary route of uptake in early life stages is bioconcentration across the outer epithelium or epidermis (Fig. 3). Moreover, during early embryonic development, protective organs such as the liver are just beginning to form. The surface epithelium is a single layer of cells during organogenesis and remains thin (e.g., two cell layers at $\sim 10\,\mu m$) in hatched larvae (Kim et al., 2008; Le Guellec et al., 2004; Nakamura et al., 2002; Ottesen and Olafsen, 1997; Saadatfar et al., 2010; Sawada et al., 1999). Embryos and larvae exposed to environmentally relevant concentrations of PAHs (i.e., low $\mu g\,L^{-1}$) typically show robust CYP1A induction throughout the surface epithelium, indicating broad uptake over the entire surface (Incardona et al., 2005, 2009; Jung et al., 2013). Nevertheless, PACs can accumulate to very high tissue concentrations in developing fish (e.g., Carls et al., 1999; Heintz et al., 1999; Incardona et al., 2009; Jung et al., 2015; Sørhus et al., 2016), a clear indication that the first-pass detoxification response in the epidermis is not sufficiently protective.

Overall, the potential for contaminant uptake and transport to cardiovascular tissues is much greater in fish early life stages. Also, the developing heart is exquisitely sensitive to a wide range of toxic pollutants that are common in fish habitats worldwide. Consequently, the remainder of this chapter focuses on developmental cardiotoxicity and uses several case studies to explore the environmental dysregulation of heart form and function.

2. DEVELOPMENTAL CARDIOTOXICITY

2.1. Factors Contributing to the Particular Vulnerability of Fish Early Life Stages to Cardiotoxic Agents

It is commonly assumed that developing organisms are more sensitive to toxic contaminants because of the high degree of cell–cell signaling and determination that occurs during embryogenesis. However, modern research suggests that developmental sensitivity may have more mundane causes. Very small size, a lack of maternal protection (for most fish embryos), and the absence of mature detoxification pathways (e.g., liver development) are key determinants of vulnerability. For these reasons, fish eggs and embryos can rapidly achieve high tissue concentrations of hydrophobic chemicals that are present at very low concentrations in the surrounding water. The ability of fish embryos to bioconcentrate trace contaminants is evidenced by studies on persistent organochlorine compounds such as dioxins and PCBs. The embryos and larvae of several species were shown to accumulate PCBs at subnanomolar water concentrations to high micromolar tissue levels in a span of 3–6 days, representing several 100,000-fold bioconcentration factors (Petersen and Kristensen, 1998).

In many ways, the interconnectedness of heart form and function during cardiogenesis determines how the heart and vasculature will respond to a cardiotoxic agent. Although the fish heart continues to grow throughout life, the juvenile heart is basically a smaller version of the adult heart, nearly identical in form. However, the embryonic heart becomes functional at the onset of organogenesis, while it still has a very primitive form and must undergo dramatic morphogenetic changes (see Chapter 2, Volume 36B: Burggren et al., 2017). Disruption of cardiac function by any means (i.e., either chemical or genetic) during early organogenesis subsequently alters cardiac morphology, which in turn will further alter function, and initiate a cascade of adverse effects on cardiorespiratory performance later in life. For these and other reasons, fish early life stages are often orders of magnitude more sensitive to chemical contaminants relative to older life stages.

2.2. Brief Summary of Heart Development in Embryos and Subsequent Growth in Larvae and Juveniles (Also See Chapter 2, Volume 36B: Burggren et al., 2017)

During cardiogenesis and heart growth, multiple and often interrelated processes are vulnerable to chemical disruption. The final form of the heart, and in particular the ventricular chamber, differs markedly among species based on general physiological traits relating to swimming capacity. Generally, cryptic species that do not swim continuously (e.g., flatfish, sculpins) have hearts with saccular ventricles without a coronary circulation, while fast, continuous swimmers (e.g., salmonids, scombrids) have hearts with more muscular pyramidal ventricles and coronary vessels (Santer et al., 1983). By far, heart development has been most intensively studied in zebrafish, which is a fast-swimming species with a pyramidal ventricle. Most of what follows, therefore, applies most directly to species with pyramidal ventricles. After initial formation, the fish heart grows continuously from the juvenile stage through adulthood by both the proliferative addition and enlargement (hypertrophy) of cardiomyocytes, not by hypertrophy of a fixed number of cells alone (the latter typical of mammals) (see Chapter 3, Volume 36B: Gillis and Johnson, 2017). The general cardiac form for fish that are continuous swimmers is established in the juvenile period. It consists of a thin walled sinus venosus, a relatively saccular atrium with a wall only a few cardiomyocytes thick, a ventricle comprised of an outer compact myocardial layer and a highly trabeculated spongy internal myocardium, and an outflow tract (OFT) comprised of both the conus arteriosus and bulbus arteriosus (in different proportions depending on species) (see Chapter 1, Volume 36A: Icardo, 2017). This latter chamber/segment attenuates the pressure developed by the ventricle during ejection and delivers blood into the ventral aorta and onto the gills. All stages of heart development, from early cardiogenesis to adoption of the juvenile form, have been described in considerable detail in zebrafish (Hu et al., 2000; Singleman and Holtzman, 2012; Staudt and Stainier, 2012; Chapter 2, Volume 36B: Burggren et al., 2017).

In all fish species, the heart becomes functional, or begins pumping, at a very early stage of development. At this point, the heart is a simple tube with a wall thickness of only a single cardiomyocyte. After establishment of a regular heartbeat, valves and specialized cells for atrioventricular conduction begin to form, and ventricular cardiomyocytes proliferate to build a wall multiple cells thick. Other morphological changes that occur subsequently include cardiac looping, which brings the atrial and ventricular chambers from a linear to a side-by-side orientation, and regional specialization of the inner and outer curvatures of the ventricle. The outer curvature of the ventricle balloons out as cardiomyocytes both elongate and proliferate via specifically oriented cell divisions (Auman et al., 2007), adding mass to the ventricle as it grows toward its ultimate form. Trabeculae are formed as cardiomyocytes migrate interiorly, delaminate from the initial ventricular wall, and then proliferate (Liu et al., 2010; Staudt and Stainier, 2012). As demonstrated largely through genetic analysis in zebrafish, all of these processes depend on normal electrical activity and contractility.

Cardiomyocyte contractility, intracardiac hemodynamic forces, and cardiac conduction all influence morphogenesis (Berdougo et al., 2003; Chi et al., 2010; Lin et al., 2012; Samsa et al., 2015; Chapter 2, Volume 36B: Burggren et al., 2017). For example, mutations that disrupt the function of ion channels that regulate cardiomyocyte physiology typically lead to defects in chamber morphology (Arnaout et al., 2007; Berdougo et al., 2003; Chi et al., 2010; Ebert et al., 2005; Hove et al., 2003; Rottbauer et al., 2001). Ventricular cardiomyocyte proliferation is also dependent on normal cardiac function, as mutations disrupting either L-type Ca^{2+} channels (LTCC) or repolarizing potassium channels (e.g., ERG/*kcnh2*) lead to smaller ventricles (Arnaout et al., 2007; Rottbauer et al., 2001).

While the vast majority of zebrafish studies have focused on early heart development, there have been important advances in terms of understanding key aspects of postembryonic organ growth. After initial rotation of the cardiac cone and looping of the chambers, the heart rotates again into its final position with the atrium lying dorsal to the ventricle (Singleman and Holtzman, 2012). From this point, the heart possesses the general morphology it will retain through adulthood, with additional growth paralleling that of the whole fish. The ventricle lengthens in proportion to the overall length of the fish, thereby changing the length-to-width ratio of the chamber. Sophisticated clonal analyses in zebrafish have revealed the relationship between the spongy and compact myocardial cells of the juvenile heart and the pool of embryonic ventricular cardiomyocytes (Gupta and Poss, 2012). These authors identified three adult ventricular cardiomyocyte lineages that were created in sequence from each other, a single-cell thick "primordial" layer that lies at the base of the compact myocardium, trabecular cells (spongy myocardium), and the outer or "cortical" layer of compact myocardium. The primordial monolayer, evident in the juvenile compact myocardium, is derived from the cardiomyocytes that originally make up the single-cell thick embryonic ventricular wall. The trabecular cardiomyocyte lineage is then derived from this layer during the larval phase by delamination and subsequent migration, intermixing, and proliferation. During juvenile to adult growth, the primordial layer retains its single-cell thickness, but becomes covered by the outer cortical layer of compact myocardial cells. The origin of the cortical cells appears to be clones of cells that had emerged through the primordial layer from the trabeculae. While the primordial layer descends from roughly several dozen embryonic ventricular cardiomyocytes, the cortical layer arises from fewer than 10 clonally dominant cells derived from the embryonic ventricle. Given how few embryonic cells ultimately contribute to the adult compact myocardium, there is considerable potential for embryonic cardiotoxicity to disrupt heart morphogenesis, thereby causing lasting impacts on heart form.

2.3. Brief Summary Relating Cardiac Anatomy to Swimming Performance (Also See Chapter 4, Volume 36A: Farrell and Smith, 2017)

There is increasing evidence that transient and sublethal exposures to low levels of contaminants can alter the form–function relationship during heart development. As a consequence, surviving fish at later life stages may have permanent changes in heart structure, with associated negative consequences for a diversity of swimming-dependent life history traits. Important anatomical features include, for example, ventricular shape and the thickness of the compact myocardium. As mentioned above, both overall shape and the cellular structure of the ventricle is related to swimming performance/activity (Santer et al., 1983; Santer and Greer Walker, 1980). At a histological level, the majority of teleost fish (~70%–80%) have ventricles composed entirely of spongy trabecular myocardium. The rest have a "mixed" myocardium with an additional outer compact layer, as described above for developing zebrafish. All species with a mixed myocardium either swim continuously (e.g., pelagic planktivores or ram ventilating apex predators) or swim intensively for extended periods (e.g., migratory salmonids). Overall ventricular shapes occur in three forms (Santer et al., 1983), with energetic swimmers all possessing roughly pyramidal ventricles. The most common form is a sac- or purse-like ventricle that has a rounded apex, and generally shows more morphological variation. For both forms, the bulbus arteriosus forms the majority of the OFT and this structure projects dorsally from the ventricle. The third and least common form in fishes has a tube-like and almost cylindrical ventricle, with the OFT (primarily composed of bulbus arteriosus) that is directed anteriorly along the same axis as the ventricle. While the shape, and presence of compact myocardium and a coronary circulation optimize the output of hearts with pyramidal ventricles, the functional consequences/benefits of possessing tubular and saccular ventricles are poorly understood. Both forms are found among Gadidae, for example, that occupy roughly similar habitats (Santer et al., 1983). Zebrafish are a very small, but unusually fast swimming species (up to 13 body lengths s^{-1}) (Plaut and Gordon, 1994), with an adult ventricular shape consistent with that found in continuous swimmers. However, studies in zebrafish report that ventricular dimensions change considerably though early larval to juvenile development (Singleman and Holtzman, 2012). While the relationship between ventricular shape and cardiac output ($\dot{Q}$) has not been established for larval or juvenile fish, several lines of evidence link thickness of the compact myocardium to pumping ability and swimming speed in both juvenile and adult fish (Anttila et al., 2014; Eliason et al., 2011; Poupa et al., 1974). Notably, differences in life history (e.g., hatchery vs wild fish) are associated with altered ventricular form, reduced swimming performance, and stress-related mortality

(Poppe et al., 2003). Accordingly, environmental contaminants that produce similar deviations in heart shape might be expected to influence individual survival and population recruitment.

2.4. Case Example: Dioxins and PCBs

2.4.1. The Collapse of Lake Trout Populations in the Great Lakes

The potential role of dioxins in the precipitous decline of the Great Lake's lake trout fishery is one of the most extensively studied examples of how early life stage cardiotoxicity in fish can translate into population-level consequences. Increasing industrialization in the first half of the 20th century led to the discharge of many persistent and bioaccumulative organic pollutants, including 2,3,7,8-tetrachlorodibenzo-*p*-dioxin (TCDD) and related chemicals. Although sources of dioxins varied, human activities involving combustion were a significant contributor, leading to the atmospheric deposition of dioxins and other legacy contaminants into the Great Lake's ecosystems. As top predators, lake trout (*Salvelinus namaycush*) are more vulnerable to contaminants that bioconcentrate through aquatic food webs. Adult female trout maternally transfer TCDD to eggs, producing a cardiovascular injury phenotype in sac fry larvae (Walker et al., 1994). The etiology of this syndrome (Fig. 4) includes circulatory failure, edema accumulation and vascular defects (e.g., hemorrhaging), and is phenocopied by TCDD exposures via water (Spitsbergen et al., 1991) or injection (Guiney et al., 1997). This dioxin-induced syndrome is often termed "blue sac disease," due to the opalescent bluish color of the yolk sac in affected alevins. Historically, blue sac disease is a nonspecific term applied to syndromes of multiple etiologies, all of which lead to yolk-sac edema (Wolf, 1954, 1957). With TCDD exposure, cardiac abnormalities appear shortly before hatching and become progressively more severe until affected fish die at the fry stage from circulatory dysregulation, anemia, hypoxia, and secondary lesions in the brain, retina, liver and other organs (Spitsbergen et al., 1991). Among fish species, lake trout are exceptionally sensitive to TCDD cardiotoxicity (Elonen et al., 1998), with mortality occurring at tissue concentrations in the low parts-per-trillion range (Walker et al., 1991). While less potent, other polychlorinated dibenzo-*p*-dioxins, dibenzofurans, and biphenyls (PCBs) can also contribute to the sac fry syndrome, illustrating the importance of contaminant mixtures in the Great Lakes food webs (Jayasundara et al., 2015; Walker and Peterson, 1991; Wright and Tillitt, 1999). Using this and other information, Cook et al. (2003) retrospectively calculated historical exposures to TCDD-like chemicals from lakebed sediment cores (1937–80), and concluded that

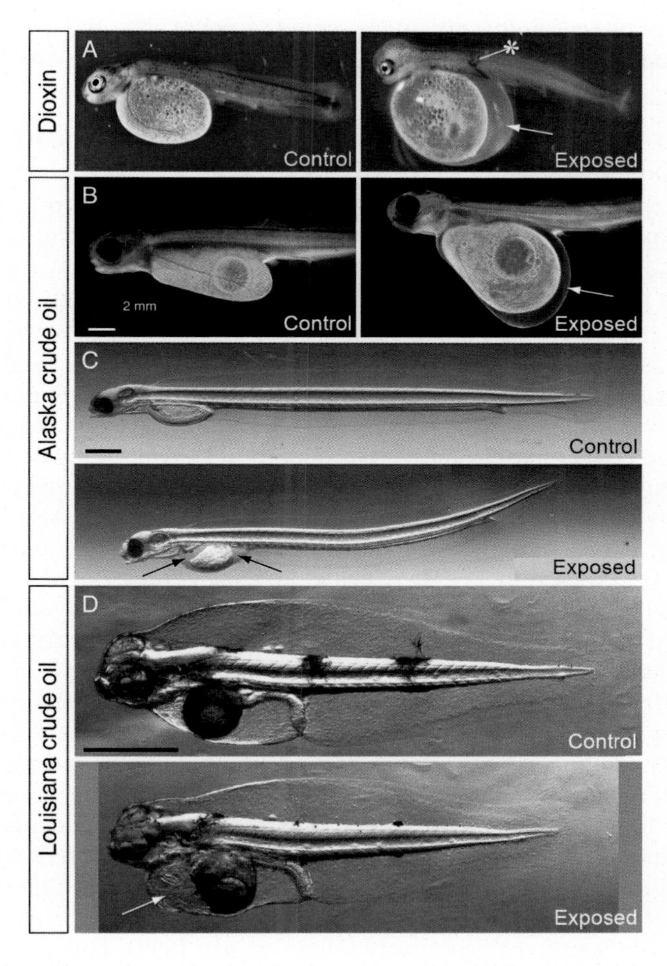

Fig. 4. Gross phenotypes associated with exposure to organochlorine compounds or crude oil. Each pair represents a control and an exposed animal, as indicated in the lower right corner. (A) Dioxin (TCDD) exposure syndrome in lake trout alevins. (B–D) Crude oil exposure syndrome resulting from Alaska North Slope crude oil (B and C) and Louisiana crude oil from the Deepwater Horizon–MC252 well (D). (B) Pink salmon. (C) Pacific herring. (D) Red drum. *Arrows* indicate yolk-sac and/or pericardial edema. Asterisk in (A) indicates peripheral vascular hemorrhage. Scale bars are 2 mm (B) and 1 mm (C and D). Images in (A) courtesy of Richard Peterson (Spitsbergen, J.M., Walker, M.K., Olson, J.R., Peterson, R.E., 1991. Pathologic alterations in early life stages of lake trout Salvelinus namaycush exposed to 2,3,7,8-tetrachlorodibenzo-p-dioxin as fertilized eggs. Aquat. Toxicol. 19, 41–72).

developmental cardiotoxicity alone could account for the reproductive failure and consequent collapse of the Lake Ontario Lake trout fishery by 1960.

2.4.2. Cellular Pathways and Mechanisms of Organochlorine Cardiotoxicity

While dioxin toxicity in mammals had been linked to the AHR by the early 1980s (Poland and Knutson, 1982), the higher sensitivity of fish early life history stages led to intensive mechanistic studies. The known affinity of TCDD and other organochlorine compounds for the AHR (Safe, 1990), the consistency of the syndrome across multiple fish species in laboratory studies (Helder, 1980, 1981; Prince and Cooper, 1989, 1990; Walker et al., 1991), and its association with observations from the Great Lakes (Symula et al., 1990) provided the basis for developing TCDD toxicity equivalence factors (TEFs) for TCDD-like chemicals (Hornung et al., 1996). Many early studies focused on the distribution and potential role for CYP1A, as it was the most extensively characterized downstream target gene for the AHR/ARNT dimer. Moreover, it was strongly upregulated in the vascular endothelium of lake trout sac fry exposed to TCDD (Chen et al., 2008; Guiney et al., 1997) and thus hypothesized to be involved in edema formation. These studies were challenged by the difficulties in distinguishing between primary effects on the peripheral vasculature and direct effects on the developing heart.

The 1990s also saw an increasing use of zebrafish in environmental toxicology. An extensive understanding of developmental genetics, together with an expanding set of experimental tools for studying specific mechanisms underlying developmental disorders, has since made the zebrafish system a foremost experimental model, particularly for toxicants such as TCDD that target the developing heart (Heideman et al., 2005). As in lake trout and other wild fish species, waterborne exposure to TCDD produces pericardial and yolk-sac edema late in zebrafish development, at hatching and shortly thereafter, respectively (Henry et al., 1997). Numerous hemodynamic alterations were also evident, along with extensive lesions, leading to larval mortality (Henry et al., 1997). These findings set the stage for the use of zebrafish to study dioxin toxicity in fish, propelled in particular by the ability to knock down gene function with antisense oligonucleotides (Carney et al., 2006b).

While mammals possess a single AHR gene, AHRs are encoded by a multigene family in teleosts, due to both gene and whole genome duplications (Hahn et al., 2006). Generally two clades, AHR1 and AHR2, comprise the teleost AHR family, with individual species showing duplication or loss within each clade. The zebrafish possesses two AHR1 genes (*ahr1a* and *ahr1b*) and a single *ahr2*. In most species, AHR2 isoforms appear to be the predominant receptor with the broadest expression pattern. Experiments using antisense morpholino injections provided the first definitive evidence that TCDD developmental cardiotoxicity is mediated by AHR2 (Prasch et al., 2003). A similar approach also revealed a key role for zebrafish ARNT1 (Antkiewicz et al., 2006). Notably, although TCDD exposures upregulate CYP1A throughout the heart and vasculature of zebrafish embryos (Andreasen et al., 2002; Teraoka et al., 2003), gene knockdown experiments with antisense oligonucleotides to CYP1A do not prevent canonical developmental toxicity, including pericardial edema, reduced circulation, craniofacial defects, and abnormal erythropoiesis (Antkiewicz et al., 2006; Carney et al., 2004). The AHR results were confirmed more recently via the construction of a constitutively active AHR2, targeted specifically to developing cardiomyocyctes. Overexpression of AHR phenocopies TCDD-like heart failure, including defects in cardiac form and function, as well as downstream abnormalities involving the swim bladder and craniofacial development (Lanham et al., 2014). This finding unequivocally links the major extracardiac features of the syndrome to secondary consequences of cardiotoxicity.

Downstream targets for the AHR2/ARNT1 dimer are still under investigation. Detailed examination of the cardiac phenotype in zebrafish embryos and its progression from the earliest onset showed that AHR activation in the developing heart primarily impacts the early ventricular cardiomyocyte proliferation step (Antkiewicz et al., 2005). This leads to poor looping of the chambers, and ultimately loss of contractility. Similar effects were observed in embryos exposed to PCB126 (Grimes et al., 2008). Presumably, impacts on cardiac morphogenesis result from excess or inappropriate transcriptional activity of the AHR (Fig. 5). However, it is unknown whether direct targets of AHR drive abnormal morphogenesis, or whether more distal, indirect targets are responsible. Several studies have characterized both dioxin-induced transcriptional changes in whole hearts isolated from embryos and genetic interactions modifying toxicity (Carney et al., 2006a; Chen et al., 2008; Waits and Nebert, 2011). TCDD exposure led to downregulation of several genes with known roles in heart development, including two transcription factors (forkhead box M1, *foxm1*; hairy/enhancer-of-split-related YRPM motif 2, *hey2*) and a morphogen receptor (bone morphogenetic protein receptor 1a, *bmpr1a*) (Carney et al., 2006a). These authors also observed a more general downregulation of cell cycle and proliferation genes (Carney et al., 2006a), that correlated with the phenotype of reduced cardiomyocyte proliferation. How AHR-mediated transcriptional activation ultimately leads to downregulation of these genes remains to be determined, but could involve induction of the transcriptional repressor, BCL6 corepressor (*bcor*) (Carney et al., 2006a; Waits and Nebert, 2011).

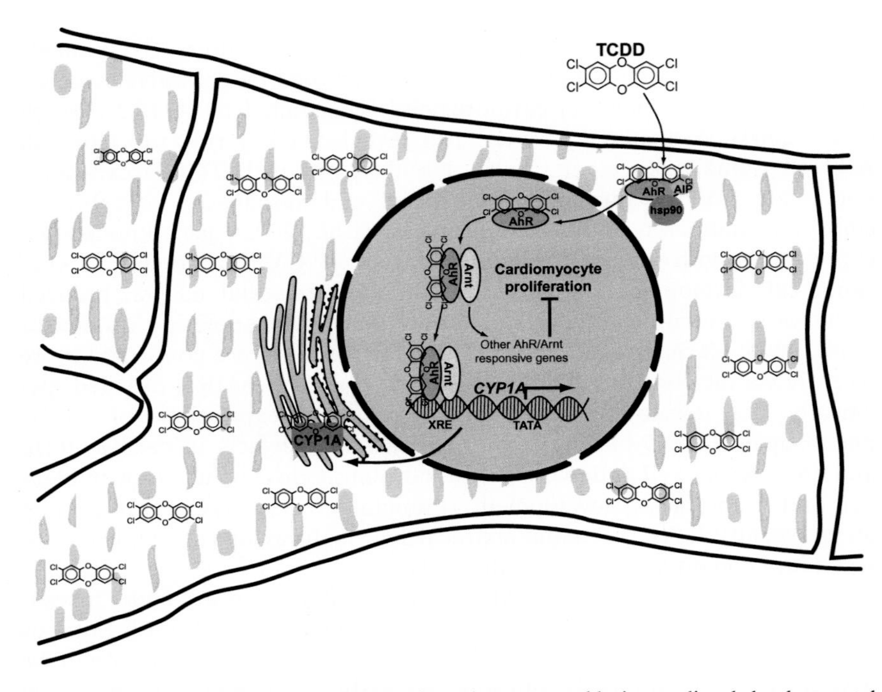

Fig. 5. Aryl hydrocarbon receptor (AHR)-dependent organochlorine-mediated developmental cardiotoxicity. Accumulation of compounds such as dioxin (TCDD) in developing cardiomyocytes leads to persistent activation of AHR and transcription of *cyp1a* and other xenobiotic response elements (XRE)-containing AHR target genes. Although persistently elevated cytochrome P450 1A (CYP1A) could lead to oxidative stress in cardiomyocytes, knockdown studies indicate that CYP1A is not required for abnormal heart development. Inappropriate AHR activation ultimately blocks cardiomyocyte proliferation, presumably through the action of other AHR target genes.

2.4.3. Legacy Pollution and the Evolution of Chemical Resistance

Several working waterways along the Atlantic coast of the United States have a legacy of industrial pollution. As a consequence, sediments in fish spawning habitats are extensively contaminated with a wide array of dioxin-like chemicals that are known to disrupt cardiovascular development and function in fish. These include polychlorinated dibenzo-*p*-dioxins (PCDDs), PCBs, and PAHs. For example, in the mid-20th century, two General Electric facilities discharged >500,000 kg of PCBs into the Hudson River in upstate New York (Wirgin et al., 2011). The resulting selective anthropogenic pressure (early life-stage mortality) led to a rapid evolutionary adaptation among Atlantic tomcod (*Microgadus tomcod*), a common and anadromous resident fish species. Relative to clean reference locations, tomcod collected from polluted sites showed resistance to AHR2 activation by a coplanar PCB (PCB-77), as evidenced by a significantly reduced upregulation of CYP1A mRNA expression (Yuan et al., 2006). This adaptive response was attributable to a single structural change in the coding region of AHR2 (Wirgin et al., 2011). It is likely that resistance to halogenated aromatic hydrocarbons evolved rapidly (over 50–100 years), driven by high tissue levels of contaminants, panmixia within the Hudson River tomcod population, minimal gene flow between neighboring populations, and extreme selective pressure (Wirgin et al., 2011).

Resistance to TCDD, PCBs and PAHs has also been studied extensively in the Atlantic killifish (*Fundulus heteroclitus*), particularly in populations from New Bedford Harbor in Massachusetts (PCB resistant) and the Elizabeth River in southern Virginia (PAH resistant). Estuarine killifish populations vary in their sensitivities to PCBs—e.g., as indicated by substantially reduced embryo and larval cardiotoxicity in response to PCB-126 exposure (Nacci et al., 2010). As in tomcod, PCB resistance in killifish from New Bedford Harbor appears to have evolved rapidly, with genetic differentiation in PCB-tolerant fish attributable to selection at specific nucleotides in AHR1 and AHR2 (Reitzel et al., 2014). This is consistent with earlier evidence for a broad, genome-wide disruption of AHR-dependent signaling (Oleksiak et al., 2011). It suggests that ancestral killifish populations had genotypes that enabled rapid, repeated, heritable, and convergent evolution in response to 20th century legacy pollution (Whitehead et al., 2011).

Killifish resistant to the toxic effects of PAHs were first discovered in and around the Atlantic Wood Industries Superfund Site on the Elizabeth River. Unlike fish from reference sites, adult killifish from this highly polluted location did not show an inducible CYP1A response to a model PAH (3-methylcholanthrene; Van Veld and Westbrook, 1995). Moreover, whereas embryos from reference fish exposed to field-collected sediment samples from the Atlantic Woods site suffered a variety of cardiac abnormalities (e.g., tube hearts, pericardial edema, poor circulation), embryos from Atlantic Wood parents were nearly entirely resistant to this developmental toxicity (reviewed by Di Giulio and Clark, 2015). Crosses between sensitive and resistant families showed variable heritability of the trait, despite suggestions of AHR2 modification and reduced responsiveness of AHR-dependent pathways (Di Giulio and Clark, 2015).

A complex basis for tolerance to organochlorine compounds and PAHs in killifish was more recently revealed through high-throughput sequencing of genomes in multiple sensitive and resistant populations (Reid et al., 2016). Using a strategy previously unavailable prior to rapid advancements in sequencing technology, these authors compared the entire genomic sequences of up to 50 individual fish from resistant populations at four distinct polluted sites to those in sensitive populations from nearby clean sites. In addition, they exposed embryos of each population to PCB-126 and compared their transcriptomic responses. With this combined approach, they were able to identify alterations in functional pathways and adaptive signatures of selection for pollutant tolerance across the populations. Tolerance to these compounds in each population was associated with deletions affecting *AHR* loci. At the same time, some populations showed duplications of *CYP1A* genes, potentially as a compensatory mechanism for detoxification of nonorganochlorine compounds. Importantly, there were distinct genomic signatures associated with sites that were more heavily contaminated with PAHs relative to organochlorines. At these sites, additional selection was observed at loci encoding voltage-gated potassium channels and sarcoplasmic reticulum (SR) calcium handling proteins, both targets of AHR-independent toxicity from PAHs (detailed in Section 2.5). This complexity of adaptive genotypes reflects the complexity of the mixtures at these polluted sites and the multiple mechanisms driving toxicity, and provides the basis for the lack of simple single-gene heritability.

2.5. Case Example: Oil Spill-Derived PAHs

2.5.1. Broad Insights Gained From the 1989 Exxon Valdez Disaster

The Exxon Valdez oil spill occurred in the spring of 1989, coincident with the initiation of the spawning season for the large herring population in Prince William Sound. Storms and tidal exchanges deposited large quantities of oil along extensive stretches of the Sound's rocky shorelines. Anadromous pink salmon subsequently spawned in the summer following the spill. Whereas herring eggs are adherent, sticking to nearshore macroalgae and seagrass, pink salmon eggs develop in gravel nests (redds) at the termini of streams and rivers, near the entry point to the nearshore environment. At the time of the spill, the two species comprised the most commercially important fisheries in Prince William Sound. This, and an inopportune life history (i.e., spawning in proximity to oiled coastlines) made herring and pink salmon the focus for field studies assessing the impacts of the spill (reviewed by Peterson et al., 2003; Rice et al., 2001). The life history of pink salmon, a top-value fishery, had been studied intensively prior to the spill, including field evaluations of embryo survival as a tool for estimating future adult returns. Redd sampling was expanded by the Alaska Department of Fish and Game in 1989, comparing oiled and nonoiled streams. Elevated mortality was observed at oiled streams (Bue et al., 1996, 1998), and correlated with shoreline oiling in the vicinity of streambeds (Murphy et al., 1999). Similarly, herring larvae collected in plankton samples showed higher mortality along oiled shorelines

(McGurk and Brown, 1996), and elevated rates of pathological abnormalities, in particular the accumulation of edema or ascites (Marty et al., 1997a).

These field studies set a new precedent for oil spill impact science by providing evidence of toxicity in the absence of direct physical contact between spilled oil and fish embryos. The science suggested that residual oil coating rock or intertidal sediments was a source of water-soluble compounds that were transported to and taken up by nearby embryos—i.e., that crude oil-derived contaminants in the dissolved phase were bioavailable. This discovery represented a paradigm shift in oil spill natural resource injury assessment (Peterson et al., 2003).

Despite observations of elevated mortality of pink salmon and herring embryos and larvae, population-level impacts to these species from the spill were most likely underestimated due to an inability to account for longer-term effects arising from sublethal toxicity (Geiger et al., 1996; Rice et al., 2001). There was, however, some evidence for oil impacts that were delayed in time. For example, fry from the 1988 brood that emerged around the time of the spill showed reduced growth in coastal waters near oiled sites (Wertheimer and Celewycz, 1996; Willette, 1996). Repeated surveys of redds in oiled streams documented mortality persisting several years after the spill, suggesting long-term adult reproductive failure. Poor embryo survival was also observed in the progeny of pink salmon that were exposed to oil during early development, as evidence of transgenerational impacts on fitness (Bue et al., 1996).

Scientists with the National Oceanic and Atmospheric Administration's (NOAA) Auke Bay Laboratories designed experiments to address the potential long-term impacts of sublethal crude oil exposure on pink salmon. These studies did not address reproductive or transgenerational toxicity, but they provided the first hints of delayed cardiotoxicity in embryos exposed transiently to very low concentrations of oil. At hatching, these fish had very low frequencies of malformations (edema; e.g., 1.5%), and normal-appearing fry were tagged by the thousands with coded wires and released to the ocean. To complete this mark-and-recapture approach, adults were recovered when they returned to spawn a year later. This effort was repeated across three pink salmon cohorts, over the course of a decade. Exposure to $20\,\mu g\,L^{-1}$, total PAHs consistently led to ~40% reduction in marine survival (Heintz, 2007; Heintz et al., 2000). As discussed in more detail below, these important findings spurred mechanistic studies using the zebrafish model that eventually implicated developmental cardiotoxicity as the cause of delayed mortality in the landmark pink salmon field studies.

2.5.2. Linking Field and Laboratory Findings

Prior to the Exxon Valdez spill, it was generally assumed that direct physical contact between oil and an organism was required for toxicity.

However, pioneering studies carried out at the NOAA Auke Bay Laboratories showed that direct contact was unnecessary. These investigators used a generator column approach to mimic the translocation of dissolved-phase petroleum compounds from oiled gravel beaches to nearby surface waters and pore waters where fish spawn (Short and Heintz, 1997). Crude oil is a highly complex mixture of hundreds of thousands of chemicals, but most have very limited water solubility. This line of research focused on PAHs because of their known toxicity at the time (primarily carcinogenicity), their relative water solubility, and observations of CYP1A induction in pink salmon embryos from oiled streams (Wiedmer et al., 1996). The oiled gravel column studies demonstrated that as water washed, or weathered, oiled substrates PAHs were gradually lost from the oil, dissolved in water (Fig. 6), and then bioconcentrated by lipid-rich fish embryos exposed to the effluent (Carls et al., 1999; Heintz et al., 1999; Marty et al., 1997b; Short and Heintz, 1997). Moreover, as oiled gravel weathered over time, the composition of PAHs in the water and taken up into tissues shifted from being dominated by two-ringed compounds (e.g., naphthalenes) to being dominated by three-ringed compounds (fluorenes, dibenzothiophenes, and phenanthrenes), as shown in Fig. 6. This shift corresponded to a marked increase in toxicity (Carls et al., 1999; Heintz et al., 1999). Importantly, the morphological abnormalities observed in experimentally exposed herring and salmon embryos were virtually indistinguishable from those evident in field-collected specimens (Fig. 4). For herring larvae from the lab and the field, the most sensitive visible endpoint was an accumulation of ascites or yolk-sac edema (Carls et al., 1999; Marty et al., 1997a).

In recent decades, these and other methods have been used to mix crude oil or oil sands from distinct geological sources worldwide into water. Irrespective of exposure protocol, virtually all teleosts from both marine and fresh water show a consistent syndrome of edema affecting the pericardial space, yolk sac, or both. These species include Atlantic herring (*Clupea harengus*) (McIntosh et al., 2010), Atlantic killifish (*F. heteroclitus*) (Couillard, 2002), Gulf killifish (*Fundulus grandis*) (Dubansky et al., 2013), crimson-spotted rainbowfish (*Melanotaenia fluviatilis*) (Pollino and Holdway, 2002), white sucker (*Catostomus commersoni*) (Colavecchia et al., 2006), fathead minnow (*Pimephales promelas*) (Colavecchia et al., 2004), Japanese medaka (*Oryzias latipes*) (González-Doncel et al., 2008; Madison et al., 2015), marine medaka (*Oryzias melastigma*) (Mu et al., 2014), inland silverside (*Menidia beryllina*) (Chi et al., 2010), olive flounder (*Paralichthys olivaceus*), and Japanese seabass (*Lateolabrax japonicus*) (Jung et al., 2015); bluefin tuna (*Thunnus thynnus*), yellowfin tuna (*Thunnus albacares*), and yellowtail amberjack (*Seriola lalandi*) (Incardona et al., 2014); and mahi mahi (*Coryphaena hippurus*) (Edmunds et al., 2015), red drum (*Sciaenops ocellatus*) (Morris et al., 2017), Atlantic

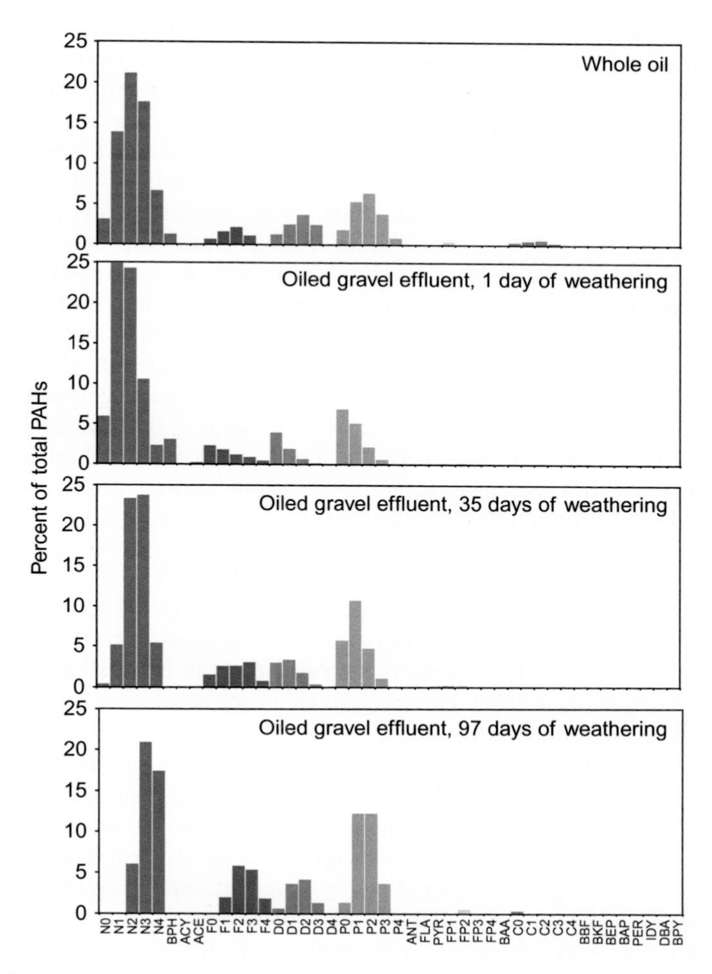

Fig. 6. Changes in polycyclic aromatic hydrocarbon (PAH) mixture composition over time with crude oil weathering. PAHs measured in whole crude oil (Alaska North Slope) are shown at top; *y*-axes are percent of total summed PAHs. The three plots below show PAHs measured in water (effluent from an oiled gravel column) after 1, 35 and 97 days of weathering (i.e., continuous flow through the column). The plots highlight the shift from dominance by two-ringed (naphthalenes) and parent nonalkylated three-ringed compounds (at the left end of the *x*-axes and left end of common shaded groups, respectively) to alkylated three-ringed compounds. Abbreviations: *ACE*, acenaphthene; *ACY*, acenaphthylene; *ANT*, anthracene; *BAA*, benz[*a*]anthracene; *BAP*, benzo[*a*]-pyrene; *BBF*, benzo[*b*]fluoranthene; *BEP*, benzo[*e*]pyrene; *BKF*, benzo[*j*]fluoranthene/benzo[*k*] fluoranthene; *BPH*, biphenyl; *BPY*, benzo[*ghi*]perylene; *C*, chrysene; *D*, dibenzothiophene; *DBA*, dibenz[*a*,*h*]anthracene/dibenz[*a*,*c*]anthracene; *F*, fluorene; *FLA*, fluoranthene; *FP*, fluoranthenes/pyrenes; *IDY*, indeno[1,2,3-*cd*]pyrene, *N*, naphthalenes; *P*, phenanthrene; *PER*, perylene; *PYR*, pyrene. Parent compound is indicated by a 0 (e.g., N0), while numbers of additional carbons (e.g., methyl groups) for alkylated homologs are indicated as N1, N2, etc.

haddock (*Melanogrammus aeglefinus*) (Sørhus et al., 2015, 2016) and Atlantic cod (*Gadus morhua*) (Sørensen et al., 2017). The lessons learned from the Exxon Valdez disaster have been applied widely to subsequent spills and were the genesis for most of the research that has shaped our current understanding of future oil spill threats to fish spawning habitats. This is true for both marine spills and freshwater areas potentially impacted by inland activities such as oil sands extraction (Colavecchia et al., 2004, 2006; Madison et al., 2015) or, as mentioned in Section 1.2, stormwater runoff from impervious surfaces (McIntyre et al., 2014, 2016a). Derived from both motor vehicles (burned and unburned fuel) and paving sources, PAHs are abundant and ubiquitous contaminants in stormwater runoff (Mahler et al., 2005; Van Metre et al., 2000; Van Metre and Mahler, 2003). Although stormwater also contains myriad other chemicals, the effects of embryonic exposure in zebrafish clearly indicate considerable overlap with isolated oil spills (McIntyre et al., 2014, 2016a). PAH contamination, is therefore, as relevant an issue for inland waterways impacted by urbanization as for pristine marine areas potentially impacted by oil production activities and shipping accidents.

2.5.3. Sorting Out the Causes of a Complex Heart Failure Syndrome

The Exxon Valdez studies described above identified fluid accumulation or edema as a major outcome of embryonic oil exposure in both herring and salmon, and also noted abnormally shaped hearts in exposed pink salmon alevins (Marty et al., 1997b). However, at that time there were no direct links between PAHs (or other crude oil compounds) and cardiotoxicity. The overall gross similarity of the visible phenotype to dioxin- or PCB-exposed fish embryos suggested that the crude oil syndrome was "dioxin-like." The etiology of the syndrome was subsequently clarified using zebrafish, as the strengths of this model for genetic screens apply also to relatively high-throughput screening for chemical toxicity. The first step was to identify which specific chemical components of crude oil were causing developmental defects. A parallel goal was to use the advanced experimental tools available for zebrafish, a major model system for developmental biology, to decipher underlying molecular and cellular pathways for PAH-induced injury.

A series of studies on zebrafish embryos showed that (1) a subset of three-ring PAHs are cardiotoxic, (2) PAHs disrupted cardiac function prior to accumulation of edema or other visible morphological defects, and (3) many of the gross morphological defects not involving the heart (e.g., jaw defects, small eyes) are secondary to loss of circulation, or heart failure (Fig. 7) (Incardona et al., 2004, 2005). To test the hypothesis that crude oil toxicity is mediated by a "dioxin-like" mechanism, the antisense morpholino knockdown technique was used to determine the roles of the AHR pathway and CYP1A-mediated metabolism. The tricyclic PAH compounds fluorene, dibenzothiophene and phenanthrene were shown to cause defects in heart rate

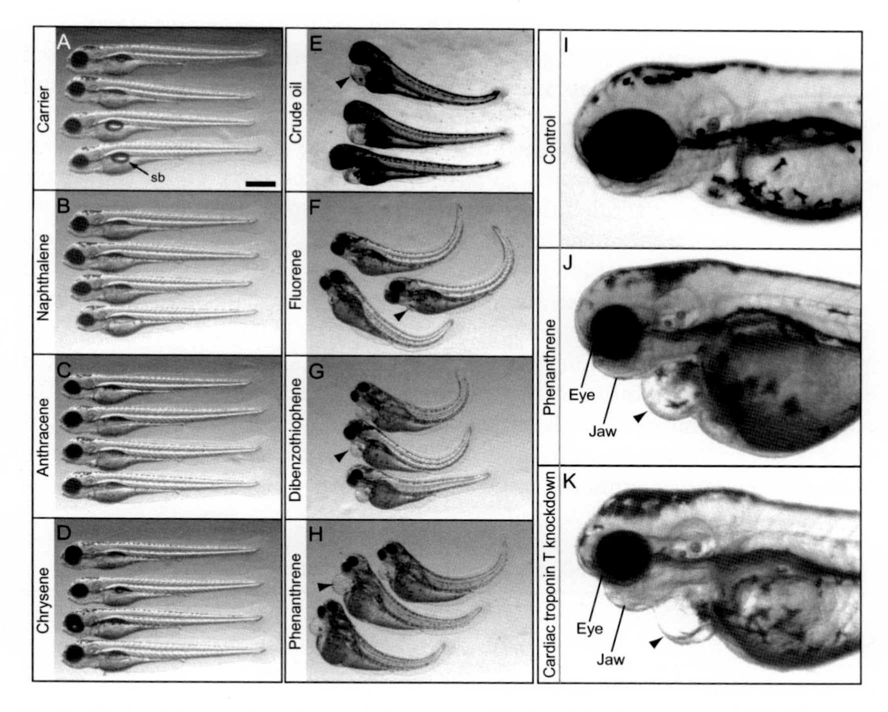

Fig. 7. Effects of single polycyclic aromatic hydrocarbons (PAHs) on zebrafish development compared to crude oil and genetic loss of cardiac function. (A–H) Zebrafish larvae 4 days postfertilization (dpf) that were exposed beginning shortly after fertilization to a nontoxic carrier solvent (A; dimethylsulfoxide) or the indicated PAHs naphthalene (two rings; B), anthracene (three rings, C), chrysene (four rings; D), fluorene (three rings, F), dibenzothiophene (three rings; G), and phenanthrene (three rings; H). For reference, crude oil-exposed larvae are shown in (E). Naphthalene and chrysene are abundant in crude oil but noncardiotoxic, while anthracene is typically present in very low concentrations relative to the other three-ringed compounds. (I)–(K) Comparison of the phenanthrene exposure phenotype (J) to the phenotype following loss of cardiac function by knockdown of troponin T (K), a structural component of cardiomyocyte sarcomeres, which leads to a noncontractile heart. *Arrowheads* indicate pericardial edema. Derived from Incardona, J.P., Collier, T.K., Scholz, N.L., 2004. Defects in cardiac function precede morphological abnormalities in fish embryos exposed to polycyclic aromatic hydrocarbons. Toxicol. Appl. Pharmacol. 196, 191–205.

(f_H) and rhythm through an AHR-independent pathway, as detailed below (Incardona et al., 2005). Moreover, genetic knockdown of CYP1A significantly reduced the threshold for crude oil cardiotoxicity (Hicken et al., 2011), demonstrating that CYP1A-mediated metabolism (detoxification) is protective. Therefore, cardiotoxic PAHs in crude oil are very likely to be CYP1A substrates. Of the tricyclic compounds, phenanthrene in particular caused a dose-dependent reduction in f_H (bradycardia) and arrhythmias characteristic of atrioventricular conduction blockade. This effect was reversible, as

f_H and rhythm returned to normal after transfer to clean water (Incardona et al., 2004). However, prolonged exposure and continuous arrhythmia caused secondary defects in cardiac morphogenesis, culminating in heart failure and edema accumulation.

The major findings from the zebrafish model were then validated in a diversity of marine species prompted, in part, by additional large oil spills over the past decade. These included the 2007 Hebei Spirit spill in the Yellow Sea off the Taean Region of Korea, the 2007 Cosco Busan spill in San Francisco Bay, and the 2010 Deepwater Horizon spill in the Gulf of Mexico. The embryos for the focal fish species for these three spills responded to crude oil exposure with two overlapping classes of cardiac function defects: (1) reduced chamber contractility; and (2) abnormalities in f_H and rhythm. As discussed in more detail below, the relative severity of either injury category appears to depend on the ecological physiology of a given species (see Table 1). There are also differences between the single compound studies and the effects of crude oil-derived mixtures. Zebrafish embryos exposed to relatively high concentrations of

Table 1
Summary of crude oil-induced cardiac function defects across species

Species	Habitat	Bradycardia	Arrhythmia	Contractility Defect
Zebrafish (*Danio rerio*)	Freshwater, tropical	+/–	—	+ (V)
Pacific herring (*Clupea pallasi*)	Boreal, pelagic	++	++	+ (A)
Atlantic haddock (*Melanogrammus aeglefinus*)	Boreal-Arctic, benthic	+	++	+ (A, V)
Southern bluefin tuna (*Thunnus maccoyii*)	Temperate–tropical, epipelagic–bathypelagic	++	ND	ND
Yellowfin tuna (*Thunnus albacares*)	Temperate–tropical, epipelagic–mesopelagic	++	++	
Yellowtail amberjack (*Seriola lalandi*)	Temperate–tropical, epipelagic–mesopelagic	++	++	ND
Mahi mahi (*Coryphaena hippurus*)	Temperate–tropical, epipelagic–mesopelagic	+/–	—	++ (A, V?)
Red drum (*Sciaenops ocellatus*)	Temperate–tropical, nearshore benthic	+/–	—	++ (A, V)

ND, not determined; A, atrial; V, ventricular; –, absent; +/–, present or absent only at high doses; +, present, dose-dependency not determined; and ++, present and strongly dose dependent.

phenanthrene respond with increasingly severe bradycardia, advancing to partial then complete AV block in a concentration-dependent manner (Incardona et al., 2004, 2005). In contrast, the phenotype of zebrafish exposed to complex mixtures from crude oils, at lower PAH concentrations, is characterized primarily by reduced ventricular contractility (Incardona et al., 2013; Jung et al., 2013). Similarly, the marine perciforms mahi mahi (*C. hippurus*) and red drum (*S. ocellatus*) responded to Deepwater Horizon—MC252 crude oil exposure with only modest effects on f_H (sinus bradycardia), but stronger reductions in contractility of both the atrium and ventricle (Edmunds et al., 2015; Morris et al., 2017). On the other hand, Pacific herring (Incardona et al., 2009), bluefin tuna (*T. maccoyii*), yellowfin tuna (*T. albacares*), yellowtail amberjack (*S. lalandi*) (Incardona et al., 2014) and Atlantic haddock (*M. aeglefinus*) (Sørhus et al., 2016), showed reduced contractility, but also more profound bradycardia, irregular arrhythmia, delayed diastole, and partial to complete AV block (silent ventricle). In all cases, prolonged exposure (i.e., through the first ~2/3 of embryonic development or longer) led to secondary defects in cardiac morphogenesis, including poor chamber looping and reduced ventricle size.

Differential effects on heart function across species correspond to differences in specific downstream morphological defects. In some cases, this may be attributable to interspecific variation in toxicokinetics and internal dose, for which data are limited. Nevertheless, virtually all species show defects in looping of the cardiac chambers, and most show reductions in ventricle size at the early cardiomyocyte proliferation stages (Fig. 8). For example, red drum, a species with that has a sharp ventrally oriented loop at the AV junction (Fig. 8A), show a concentration-dependent failure of looping measured as an increase in the angle between the chambers from about 75 degrees in controls to nearly 180 degrees in oil-exposed fish (Fig. 8B and C). Similarly, Atlantic haddock with severe functional defects (e.g., silent ventricle) showed near complete failure of cardiac looping (Fig. 8D–G). In newly hatched haddock larvae, there is a nearly 180 degree fold between the atrium and ventricle that is oriented anteriorly (as seen from a ventral view), which failed to form in embryos that accumulated $>1000\,ng\,g^{-1}$ PAHs (Sørhus et al., 2016) (Fig. 8D and F). At these concentrations, the ventricle also showed a near complete lack of the cell proliferation that normally occurs during the first several days after hatch (Fig. 8D and E compared to F and G). Although Pacific herring embryos that accumulated similarly high tissue levels of PAHs also showed severe cardiac defects (Incardona et al., 2009), exposure to much lower concentrations also caused a marked reduction in ventricular growth in hatching stage larvae (Fig. 8H and I). Although the tissue concentrations have not been measured for mahi mahi, this species shows comparatively milder

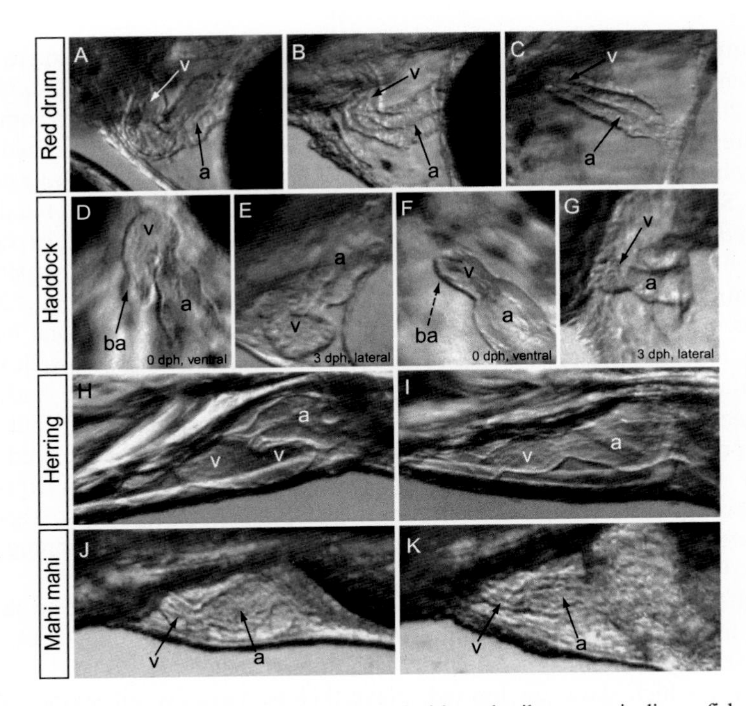

Fig. 8. Abnormal cardiac morphogenesis associated with crude oil exposure in diverse fish species. All views are lateral, with anterior to the left except where indicated. Chambers are indicated atrium (a), ventricle (v), and bulbus arteriosus (ba). (A–C) Red drum exposed to Deepwater Horizon–MC252 crude oil, ~12 h posthatch (36 hours postfertilization, hpf). (A) Control and high-energy water-accommodated fractions (WAF) with $\sum$PAH 5.9 μg L^{-1} (B) and 31.5 μg L^{-1} (C). (D)–(G) Atlantic haddock exposed to mechanically dispersed Norwegian Sea crude oil ($\sum$PAH 7 μg L^{-1}), day of hatch/12 dpf (D, control; F, exposed), and 3 days posthatch (E, control; G, exposed). Ventral views are shown in (D) and (F). *Arrow with dashed line* indicates bulbus arteriosus out of the focal plane behind the ventricle. Pacific herring control (H) and (I) exposed to Alaska North Slope crude-oiled gravel effluent ($\sum$PAH ~2 μg L^{-1}) at 3 days posthatch (14 dpf). Mahis mahi control (J) and (K) exposed to high-energy WAF of Deepwater Horizon–MC252 crude oil ($\sum$PAH 20 μg L^{-1}), ~12 h posthatch (48 hpf).

morphological defects at waterborne PAH concentrations that lead to larval mortality (Fig. 8J and K). However, compared to other species, the heart in hatching stage mahi mahi is relatively underdeveloped, with a very small ventricle and shallow chamber looping. Although larval heart development has not been described in detail in mahi mahi, this is consistent with a significant heterochrony wherein major ventricular morphogenesis occurs later into the larval period.

The whole heart impacts observed in all of these species were strongly suggestive of several distinct cardiotoxic mechanisms. First, they highlight an important distinction between cardiac phenotypes arising from dioxin/PCB exposure and exposure to crude oil or tricyclic PAHs. Embryos or larvae exposed to dioxins or PCBs show minimal effects on cardiac function at early time points where there are already visible defects in morphology. Exposure of zebrafish embryos or larvae to TCDD or PCB-126 leads to poor looping and reduced ventricular cardiomyocyte proliferation without major effects on f_H or rhythm (Antkiewicz et al., 2005; Grimes et al., 2008). Instead, the crude oil cardiotoxicity phenotypes are highly reminiscent of mutations in zebrafish that effect ion channels, transporters, and pumps involved in regulating cardiomyocyte excitability and excitation–contraction (E–C) coupling (for details on cardiomyocyte physiology, see Chapters 2 and 3, Volume 36A: Shiels, 2017; Vornanen, 2017, respectively). For example, like phenanthrene exposure, knockdown of or null mutations in the zebrafish *kcnh2* potassium channel gene leads to bradycardia, and partial to complete AV block (Arnaout et al., 2007; Langheinrich et al., 2003; Milan et al., 2003). Indeed, the effects of phenanthrene mimic a large suite of human pharmaceuticals known to block the repolarizing potassium channel (ERG/I_{Kr}) encoded by *kcnh2* (Langheinrich et al., 2003; Milan et al., 2003). Exposure of adult animals to these compounds would simply lead to arrhythmia (and possibly sudden death) without affecting cardiac structure. However, as has been established for cardiac function mutants in zebrafish (Andrés-Delgado and Mercader, 2016; Glickman and Yelon, 2002; Staudt and Stainier, 2012), the inseparable link between function and morphogenesis in the developing heart is the basis for secondary cardiac malformation, and ultimately heart failure, in oil-exposed embryos. For example, genetic disruption of either potassium or calcium channels leads to ventricular malformation (Arnaout et al., 2007; Rottbauer et al., 2001). Remarkably, both the functional and morphological phenotypes of oil-exposed haddock embryos closely mimic partial and complete loss-of-function mutants for the ERG/*kcnh2* gene (Arnaout et al., 2007; Langheinrich et al., 2003; Rottbauer et al., 2001; Sørhus et al., 2016). As detailed in Section 2.5.5, even exposure to lower concentrations of crude oil that do not cause overt embryonic heart failure can lead to significant impacts on cardiac morphogenesis, and ultimately cardiac function in fish that survive.

2.5.4. Mechanisms of Acute Toxicity to the Embryonic Fish Heart From Crude Oil Exposure

As introduced above, exposure to crude oil or tricyclic PAHs leads to cardiac dysfunction followed by abnormal morphogenesis, whereas exposure to organochlorine compounds leads to malformation first followed by declining function. The comparison of the specific developmental cardiac phenotypes between embryos exposed to crude oil or organochlorine AHR agonists suggests two distinct pathways to heart malformation, and thus, distinct underlying mechanisms. These are detailed in Sections 2.5.4.1–2.5.4.3.

2.5.4.1. Roles of the AHR. Although the PAH fractions of crude oil are dominated by two- and three-ring PAHs that are weak AHR agonists, there are clearly more potent agonists in crude oil-derived mixtures. Virtually all studies of early life history stage exposure demonstrate robust induction of CYP1A. However, assessment of tissue-specific patterns of CYP1A induction following exposure to crude oil or single compounds provides additional insight.

As discussed in Section 2.4.2, cardiac malformation following organochlorine exposure requires activation of the AHR in cardiomyocytes. Some individual higher molecular weight PAH compounds were shown to cause similar AHR-dependent cardiotoxicity. For example, benz(*a*)anthracene (four rings), benzo(*a*)pyrene and benzo(*k*)fluoranthene (five rings) caused a true "dioxin-like" cardiotoxicity in model species (zebrafish and *F. heteroclitus*), in that the functional and morphological consequences of exposure are entirely depending on AHR signaling, as shown by morpholino knockdown (Clark et al., 2010; Incardona et al., 2006, 2011; Van Tiem and Di Giulio, 2011). Benz(*a*)anthracene and benzo(*a*)pyrene activated the AHR in cardiomyocytes in addition to the endocardium and vascular endothelial cells (measured by CYP1A induction, Fig. 9), but loss of myocardial but not endocardial CYP1A induction by AHR knockdown correlated with prevention of toxicity (Incardona et al., 2006, 2011). Similar results were shown for retene, a C4-alkylated phenanthrene (7-isopropyl-1-methyl phenanthrene). Retene activated the AHR in outer curvature cardiomyocytes in zebrafish embryos, leading to a poorly looped heart with a smaller ventricle. Normal heart development was restored with loss of myocardial CYP1A induction in retene-exposed embryos by AHR knockdown (Scott et al., 2011).

In contrast, the cardiotoxicity of the tricyclic compounds dibenzothiophene and phenanthrene was exacerbated by AHR knockdown (Incardona et al., 2005). While these compounds are poor AHR agonists, they are CYP1A substrates, so these results indicated that AHR-mediated CYP1A metabolism detoxifies these lower molecular weight PAHs. Similarly, knockdown of both AHR and CYP1A demonstrated that AHR activity was not required for crude oil cardiotoxicity in zebrafish (Incardona et al., 2005), and that metabolism by CYP1A was protective (Hicken et al., 2011). In these experiments, embryos were exposed to oiled gravel effluent at a relatively early stage of weathering when the PAH compositions were dominated by two-ring compounds and parent (nonalkylated) tricyclic compounds. Oil-exposed embryos with poor cardiac function showed CYP1A induction only in endocardial and vascular endothelial cells, but not myocardial cells (Fig. 9). Cardiotoxicity in the absence of myocardial CYP1A induction was shown for three different crude oils: Alaska North Slope (representative of the Exxon Valdez spill) (Incardona et al., 2005); an Iranian heavy crude oil (Jung et al., 2013); and MC252 oil

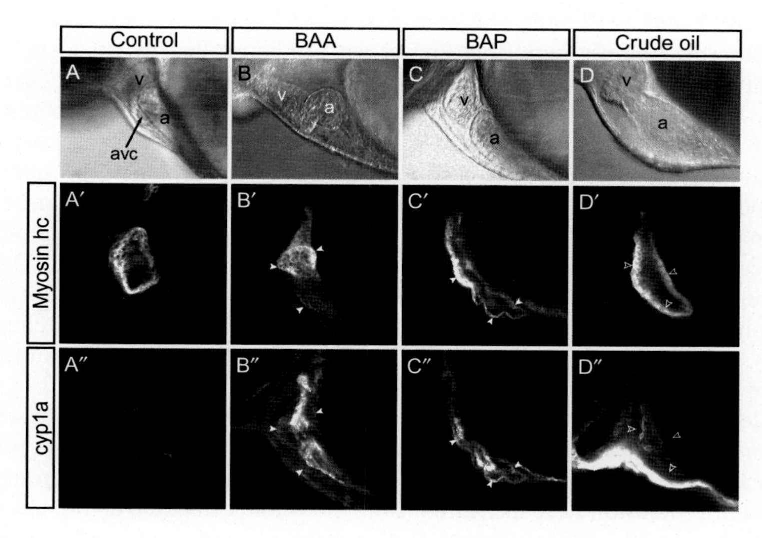

Fig. 9. Differential activation of the aryl hydrocarbon receptor (AHR) pathway by cardiotoxic four- and five-ring polycyclic aromatic hydrocarbons (PAHs) and crude oil. All images are lateral views with anterior to the left. *Top panels* show light micrographs of hearts in live embryos, *bottom panels* show confocal images of immunofluorescence for myosin heavy chain (*middle panels*, marking cardiomyocytes), and cytochrome P450 1A (CYP1A) (*bottom panels*). *Filled arrowheads* indicate CYP1A-positive myocardium, *unfilled arrowheads* indicate myocardium without CYP1A labeling. Control embryos (A) have a normally looped heart with the ventricle to the right of the atrium, out of focus and obscured, and no CYP1A labeling in the heart. The atrioventricular canal (avc) is visible as a distinct ring parallel to the focal plane. Embryos exposed to high concentrations of (B) benz(*a*)anthracene (four rings) or (C) benzo(*a*)pyrene have poorly looped hearts (AV canal not visible) with CYP1A-positive myocardium. Embryos exposed to crude oil (Iranian heavy crude oil high-energy water-accommodated fractions) have unlooped hearts, but no myocardial CYP1A induction, consistent with AHR-independent toxicity.

from the Deepwater Horizon spill (Incardona et al., 2013). Although AHR knockdown has not been carried out in marine species, severe cardiotoxicity following exposure to Norwegian Sea crude oil in Atlantic haddock embryos also occurred in the absence of myocardial AHR activation (Sørhus et al., 2016). Combined, these studies indicate that AHR activation is not required for cardiotoxicity from crude oil or individual tricyclic PAHs, and support the existence of AHR-independent mechanisms.

2.5.4.2. AHR-Independent Effects on Intracellular Ion Handling and Cardiomyocyte Physiology. As described in Section 2.5.3, the whole-heart functional phenotypes observed in multiple fish species suggested that specific ion channels or transporters which regulate membrane excitability and E–C coupling were potential direct targets of cardiotoxic PAHs. As detailed in

Chapters 2 and 3, Volume 36A (Shiels, 2017; Vornanen, 2017, respectively), and described briefly here, these processes are controlled by fluxes of Na^+, K^+, and Ca^{2+} ions. Action potentials are generated by inward Na^+ currents carried by voltage-gated Na^+ channels, which is followed by the opening of LTCC. This leads to extracellular Ca^{2+} entry and the subsequent release of additional internal Ca^{2+} from the SR into the cytoplasm through the ryanodine receptor (RyR; RyR2 isoform in cardiomyocytes). Elevated cytoplasmic Ca^{2+} then initiates myofiber contraction by regulating the interaction between actin and myosin filaments. Baseline levels of cytoplasmic Ca^{2+} ions are restored by pumps and exchangers such as sarcoplasmic–endoplasmic reticulum calcium ATPase 2 (SERCA2) and the sodium/calcium exchanger 1 (NCX1). Finally, membrane potentials are reestablished by the outward efflux of K^+ through voltage-gated channels such as ERG.

Electrophysiological evidence for direct effects of crude oil exposure on action potentials and E–C coupling was obtained in isolated tuna and mackerel cardiomyocytes exposed to DWH-MC252 crude oil. Effects on action potential generation were measured using the whole-cell patch clamp method. Intracellular Ca^{2+} cycling during E–C coupling was assessed using live-cell confocal microscopy with Ca^{2+}-sensitive dyes. Water-accommodated fractions of MC252 crude oil blocked the ERG-encoded I_{Kr} current, leading to action potential prolongation, and also disrupted intracellular Ca^{2+} cycling through blockade of either SERCA2 or RyR (Brette et al., 2014). As in the oiled gravel column studies following the Exxon Valdez spill, weathering studies with MC252 crude oil linked these cellular effects to tricyclic PAHs. The potency for action potential prolongation increased with more weathered oil samples that were relatively enriched in phenanthrenes, and dose–response modeling showed a strong correlation with the concentration of tricyclic compounds rather than total PAHs (Brette et al., 2014). Moreover, these authors tested a number of single compounds, including naphthalene, fluorene, dibenzothiophene, carbazole, phenanthrene and pyrene. Phenanthrene ($5\,\mu M$) alone had complex effects on Ca^{2+} dynamics in both atrial and ventricular cardiomyocytes, and also rapidly blocked the I_{Kr} potassium current, prolonging action potentials. These effects of phenanthrene were, overall, very similar to the effects of the complex mixture from MC252 crude oil, strongly supporting the idea that phenanthrene and related homologs are the primary drivers of crude oil cardiotoxicity. Both crude oil and phenanthrene reduced the peak intracellular $[Ca^{2+}]$, impaired its return to baseline levels, and depleted SR Ca^{2+}. The precise mechanisms remain to be determined, but phenanthrene (and related homologs) could block the entry of extracellular Ca^{2+} through the LTCC, limit the reuptake of Ca^{2+} into the SR through SERCA2, or reduce SR Ca^{2+} release through the RyR (Fig. 10).

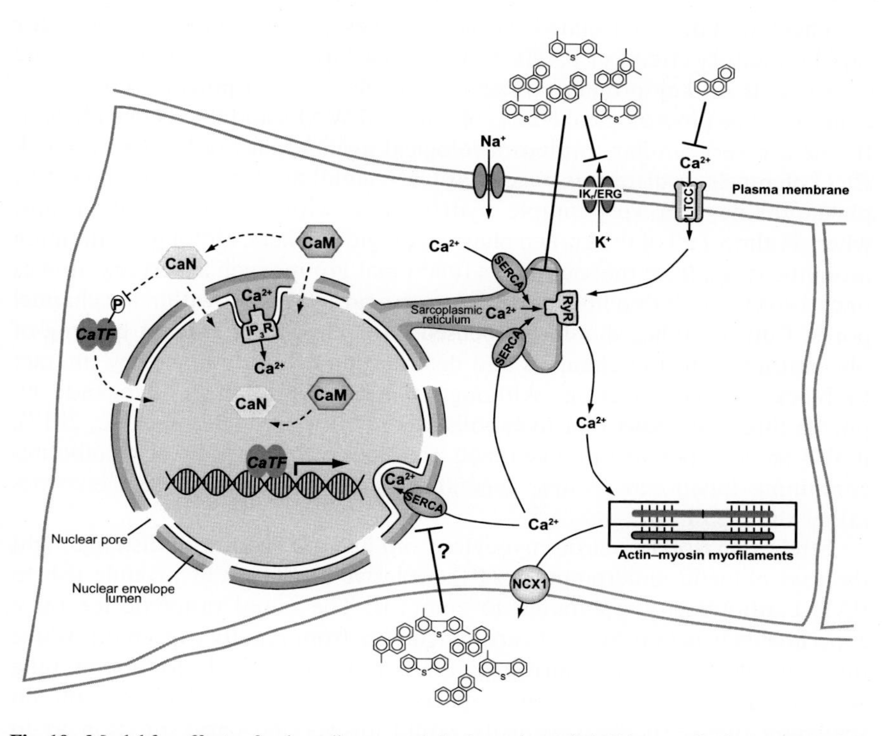

Fig. 10. Model for effects of polycyclic aromatic hydrocarbons (PAHs) on excitation–contraction and excitation–transcription coupling in cardiomyocytes. This simplified schematic of E–C (excitation–contraction) and E–T (excitation–transcription) coupling shows the interconnectedness of sarcoplasmic reticulum (SR) and nuclear Ca^{2+} pools. Generally, depolarization by action potentials causes entry of extracellular Ca^{2+} through L-type Ca^{2+} channels (LTCC) in the plasma membrane, in turn triggering Ca^{2+}-induced calcium release from the SR via the ryanodine receptor (RyR). Elevated cytoplasmic Ca^{2+} binds actin–myosin myofilaments leading to contraction. Resting Ca^{2+} levels are restored by sarcoendoplasmic reticulum Ca^{2+}-ATPase (SERCA2)-mediated pumping back into the SR and extracellular transport by the Na^{+}–Ca^{2+} exchanger NCX1. SERCA2 also restores nuclear envelope Ca^{2+} levels. Resting membrane potentials are restored for the next contraction cycle in part by the activity of rectifying potassium channels such as ERG (ether-à-go-go-related gene K^{+} channel). Elevated cytoplasmic and nuclear Ca^{2+} (also derived from release by inositol triphosphate receptors; IP_3R) leads to activation of calmodulin (CaM) and calcineurin (CaN), the latter a phosphatase that modifies and activates calcium-sensitive transcription factors (CaTF) such as NFATc and myocardin. Crude oil water-accommodated fractions (WAFs) and three-ring PAHs have been shown to block the I_{kr} current/ERG channel, leading to action potential prolongation and arrhythmia, as well as SR Ca^{2+} depletion either through effects on SERCA2 or RyR. Phenanthrene may also block LTCC. The potential action of PAHs on SERCA2 or other proteins controlling nuclear Ca^{2+} suggest a mechanism for direct effects on cardiomyocyte gene expression. Model derived from Brette, F., Machado, B., Cros, C., Incardona, J.P., Scholz, N.L., Block, B.A., 2014. Crude oil impairs cardiac excitation–contraction coupling in fish. Science 343, 772–776; Brette, F., Shiels, H.A., Galli, G.L.J., Cros, C., Incardona, J.P., Scholz, N.L., Block, B.A., 2017. A novel cardiotoxic mechanism for a pervasive global pollutant. Sci. Rep. 7, 41476.

These findings in isolated cardiomyocytes provide direct evidence for rapid inhibitory effects of PAHs or other cooccurring water-soluble aromatic compounds on key protein targets that regulate action potentials and E–C coupling. The physiological effects of crude oil WAFs and PAHs like phenanthrene are very similar to pharmacological agents that are known to block ERG channels or disrupt intracellular Ca^{2+} handling. Structurally, however, phenanthrene is a very simple hydrocarbon with no functional groups, whereas almost all of the known pharmacologic blockers, although containing aromatic rings, have more complex functional groups such as tertiary amines or carboxylic acids that interact with amino acid side chains within ion channel pores. Future studies should be focused on demonstrating direct binding of phenanthrenes to ion channels and determining precisely how they interact to block channel function. Although the rapid rate of I_{Kr} blockade by phenanthrene suggests that metabolism is not required (Brette et al., 2017), it also remains possible that channels are blocked by toxic PAH metabolites containing functional groups, generated by constitutively present enzymes rather than CYP1A.

Studies on isolated cardiomyocytes from juvenile scombrids also highlight the toxicokinetic underpinnings of the relative resistance of juvenile fish to PAH cardiotoxicity, compared to embryos. The initial rationale for these experiments was that isolated cardiomyocytes from readily captured juvenile tunas would serve as a surrogate for extremely difficult to obtain tuna embryos. For assessment of the impacts of the Deepwater Horizon spill on spawning tunas, such experiments would answer the basic question as to whether tunas were susceptible to the same form of toxicity that had been primarily documented in cold-water species with prolonged development times. The suspect tricyclic PAHs typically accumulate to the equivalent of low micromolar concentrations in exposed embryos (e.g., Incardona et al., 2009). Application of single compounds or WAF-derived PAH mixtures at these concentrations to isolated cardiomyocytes *in vitro* thus closely mimics tissue concentrations of an intact embryonic heart. Therefore, I_{Kr} blockade and disruption of SR Ca^{2+} handling in isolated tuna cardiomyocytes at these physiologically relevant concentrations (Brette et al., 2014, 2017) are consistent with the whole-heart effects observed in oil-exposed tuna embryos (e.g., Incardona et al., 2014). However, because of the much larger metabolic detoxification of the intact ~10 kg juvenile tuna, and much larger volume of distribution, the intact juvenile heart would never see PAHs at micromolar concentrations from a parts per billion exposure.

2.5.4.3. Potential for Interactions and Synergistic Cardiotoxicity. Given the complexity of mixtures derived from crude oil and the minimal number of single compounds that have been tested, there are still open questions regarding the mechanisms of crude oil/PAH cardiotoxicity. First, although phenanthrene can accumulate to very high tissue concentrations (micromolar range) with relatively unweathered oil (e.g., Incardona et al., 2009; Jung et al., 2015), with increased weathering, mixtures are dominated by alkyl-phenanthrenes, and the parent phenanthrene may be nearly absent. Although retene is variably present at low concentrations of crude oil, there is no information on cardiotoxic activities of the other more abundant C1-, C2-, and C3-phenanthrenes that are at much higher concentrations. Thus, the structural requirements that convert phenanthrene from a weak AHR agonist with ion channel-blocking activity to a more potent alkylated AHR agonist with potential "dioxin-like" activity are unknown. Nevertheless, there are some indications (i.e., observed myocardial CYP1A induction) that AHR agonism could be involved with crude oil cardiotoxicity under some circumstances, particularly with prolonged weathering (Incardona et al., 2009; Jung et al., 2013). However, even the more potent four- and five-ring PAHs that do act through an AHR-dependent "dioxin-like" mechanism require very high concentrations to cause toxicity. Pore water concentrations of benzo(*a*)pyrene in highly contaminated sediments are typically below 5 $\mu g\,L^{-1}$ (Lu et al., 2004, 2006; Yu et al., 2009), whereas AHR-dependent cardiotoxicity was elicited at concentrations $\geq$100 $\mu g\,L^{-1}$ (Incardona et al., 2011; Jayasundara et al., 2015). Yet no single compound has shown the same potency as mixtures of WAFs, which have effects with total PAH concentrations below 1 $\mu g\,L^{-1}$ (e.g., Carls et al., 1999; Incardona et al., 2015). Thus, it is highly likely that the potency of crude oil-derived mixtures arises from interactions between multiple mechanisms. As described below, both ion-channel blockade and AHR agonism converge on genetic pathways of heart malfunction and malformation, so these two mechanisms alone would likely synergize.

2.5.5. Long-Term Impacts to Cardiac Anatomy and Function in Fish That Survive Embryonic Oil Exposures

A major lesson from genetic and functional analyses of heart development in zebrafish is the recognition of the interplay between cardiac function and morphogenesis (Andrés-Delgado and Mercader, 2016; Glickman and Yelon, 2002; Staudt and Stainier, 2012). This occurs on multiple levels, from simple mechanical forces derived from cardiomyocyte contractility and fluid flow within the chambers, and how this relates to directional cell growth and division (e.g., Auman et al., 2007; Hove et al., 2003), to the myriads role that intracellular Ca^{2+} has as a second messenger for both contractility and gene expression (Ljubojevic and Bers, 2015; Winslow et al., 2016). Disruption of function in the atrium alone can influence the morphology of the ventricle in developing zebrafish embryos (Berdougo et al., 2003) and lead to abnormal ventricular form persisting into adulthood (Singleman and Holtzman, 2012).

Based on these analyses alone, transient disruptions of cardiac function during development by crude oil exposure would be expected to have lasting impacts on cardiac form, and in turn, function.

An initial indication that transient exposure to PAHs could lead to lasting changes in heart morphology came from studies in phenanthrene-exposed zebrafish embryos. As discussed in Section 2.5.3, exposure to phenanthrene produced dose-dependent bradycardia with progression through partial to complete AV block. Although exposure started shortly after fertilization, these effects were not observed until 36 h postfertilization; about 12 h after the regular heart beat is established, and at the beginning of looping and formation of the cardiac conduction system. Embryos with either partial or complete AV block were transferred to clean water, then reassessed for cardiac function and morphology 14 h later, after looping was complete (Incardona et al., 2004). Although normal rhythm was restored after phenanthrene washout, increasingly poor looping was correlated with the severity of prior AV block (Fig. 11). These embryos did not have edema or other external morphological defects, raising the possibility that long-term impacts on growth and survival observed in sublethal pink salmon exposures had a cardiac origin. Thus, dose-dependent cardiotoxicity from oil-derived PAHs would be reflected in a gradient of defects, ranging from embryonic heart failure and larval lethality, to subtle effects on heart development that allow survival well into the juvenile period.

This hypothesis was first tested in zebrafish, using critical swimming speed (U_{crit}) as a measure of cardiac function (Hicken et al., 2011). Because of (1) the established relationship between heart shape, cardiac output and U_{crit} and (2) the established relationship between disruption of cardiac function and altered morphogenesis in embryos, these authors predicted that sublethal embryonic exposure to crude oil would lead to altered ventricular shape and reduced aerobic capacity in adults. With the same Alaskan crude oil used in the pink salmon studies, zebrafish embryos were exposed to oiled gravel effluent (~30 μg L^{-1} total PAHs) through the cardiac looping stage (48 h postfertilization, hpf), then transferred to clean water for hatching and growth. This exposure level led to low levels of edema in the larvae, and increased larval mortality slightly (3.4-fold higher than controls). Externally normal survivors were then reared in a standard zebrafish colony until they were 11 months old, well past reproductive maturity (3 months). Oil-exposed fish had reduced U_{crit}, both absolute and relative to body length, that was on average 18% lower than in controls. Ventricular shape was assessed in sagittal sections, and oil-exposed fish had reduced length–width ratios, indicating rounder ventricles. These findings were, therefore, consistent with observations made in salmonids that correlated rounder ventricles (from natural variation) with reduced cardiac output and lower U_{crit} (Claireaux et al., 2005).

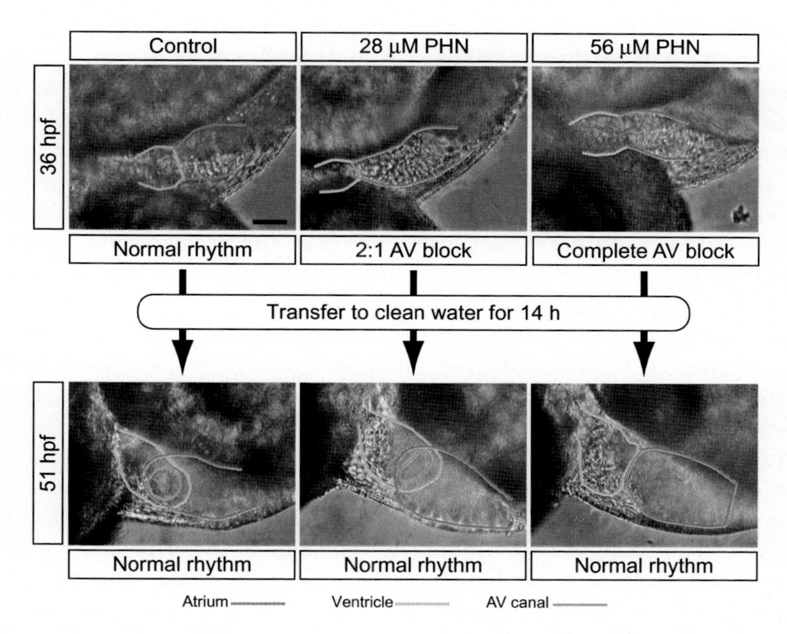

Fig. 11. Altered cardiac morphogenesis after transient exposure to a cardiotoxic PAH compound. Zebrafish embryos were exposed transiently to the indicated concentrations of phenanthrene (PHN) from shortly after fertilization to 36 h postfertilization (hpf), prior to cardiac looping. The indicated functional defects were observed at 36 hpf, partial or complete atrioventricular block (*2:1 AV block* and *complete AV block*, respectively). After transfer to clean water, normal rhythm was restored within 1 h, and heart morphology was assessed in the same individual 14 h later. Defective looping [indicated by altered orientation of the AV (atrioventricular) canal] and altered ventricular shape correlated with the severity of the antecedent arrhythmia. Figure derived from supplemental data in Incardona, J.P., Collier, T.K., Scholz, N.L., 2004. Defects in cardiac function precede morphological abnormalities in fish embryos exposed to polycyclic aromatic hydrocarbons. Toxicol. Appl. Pharmacol. 196, 191–205.

Based on these findings from a laboratory model species, these authors returned to pink salmon and Pacific herring to ground truth the hypothesis in species impacted by the Exxon Valdez spill (Incardona et al., 2015). Again, using the same oil and exposure methods, pink salmon embryos were exposed to a series of doses that overlapped with the pink salmon mark–recapture studies (10–45 $\mu g L^{-1}$ vs 5–20 $\mu g L^{-1}$ total PAHs), while herring embryos were exposed to much lower concentrations of more highly weathered oil (0.23 $\mu g L^{-1}$ total PAHs). Both species were exposed roughly two-thirds of the way through embryonic development (relative to hatching time), which is just past the critical period of early heart development. Because of the much smaller size and greater surface-to-volume ratio of herring eggs relative to those of salmon, herring embryos accumulated tissue PAHs from this low dose exposure at a concentration (1800 $ng\,g^{-1}$ lipid) similar to the 10 $\mu g\,L^{-1}$ salmon exposure (2000 $ng\,g^{-1}$ lipid). Similar to the previous experiments in pink salmon, these exposure levels resulted in elevated, but very low levels of visible edema in hatched larvae. The pink salmon fry were also monitored for growth and showed the same dose-dependent reduction throughout the juvenile period as observed previously (Heintz et al., 2000). Both herring and salmon were reared for 7–9 months past embryonic exposure in clean water, at which point their U_{crit} was measured and their hearts dissected for anatomical and histological characterization.

Consistent with the findings in zebrafish, transient oil exposure during embryogenesis led to 15%–20% reductions in U_{crit} in both salmon and herring juveniles. Both species also had very similar changes in cardiac anatomy. However, rather than reduced length-to-width ratios as observed in fully mature zebrafish, the ventricles of both salmon and herring juveniles were more elongated. In addition, there were impacts on the OFT (bulbus arteriosus). Ventricles from the juvenile salmon were examined in more detail by histology. The spongy myocardium showed signs of hypertrophy, including a significant increase in cardiomyocyte nuclei and a trend to increase cardiomyocyte diameter. Remarkably, instead of a hypertrophic response, the compact myocardium was reduced in thickness. While the relationship between ventricular shape (and dimensions) and pumping activity has not been established in juvenile fish, as it has for adults, the thickness of the compact myocardium has been correlated with higher swimming speed in juvenile Atlantic salmon (*Salmo salar*) (Anttila et al., 2014; Wang et al., 2006). At the same time, physiological hypertrophy in fish involves coordinated responses in both the spongy and compact myocardium (Castro et al., 2013a), indicating that oil-exposed fish most likely showed pathological hypertrophy. How changes in ventricular shape seen in juveniles following exposure to oil influence the cardiac morphology of mature or senescent adults has not been detailed, but these finding suggest that ongoing cardiac stress resulting from an initial developmental insult ultimately leads to a more rounded, less functional ventricle as fish approach senescence. Moreover, these findings suggest that oil exposure during early heart development affects the process that leads to clonal population of the compact myocardium from trabecular precursors (Gupta and Poss, 2012). Because in these studies oil exposure ended before the onset of trabeculation, further examination of this effect could shed light on molecular mechanisms underlying the trabecular–compact myocardial lineages. In addition, the pathophysiology of abnormal hypertrophy in fish is likely to involve unique molecules, as genes encoding atrial and B-type natriuretic peptides (markers of hypertrophy in mammalian hearts) were not significantly upregulated in juvenile pink salmon hearts.

A key question from this line of research is what are the dose and timing thresholds for irreversible effects on heart development? For Pacific herring, lasting impacts on ventricular shape occurred with extremely low PAH concentrations (hundreds of parts per trillion) (Incardona et al., 2015), near background levels for urbanized waterways such as Washington's Puget Sound or San Francisco Bay (West et al., 2014). So far, a no-effects concentration has not been reached for this species. Presumably, there is likely to be some threshold where functional defects in the embryonic phase do not lead to permanent structural changes. While this can readily be determined in model species such as zebrafish, this is much more challenging in wild fish in which embryos are not widely available, and methods for rearing through metamorphosis are not well established.

2.5.6. Adverse Impacts on Cardiac Function and Development Through Effects on Gene Expression

2.5.6.1. Effects on Genes Involved in Action Potential Generation and E–C Coupling. While changes in cardiac morphology clearly must be associated with changes in gene expression, a key question that needs to be answered is whether crude oil also impacts cardiac function through alterations in gene expression. Given that AHR activation (a bona fide transcriptional mechanism) appears to only play a minor role, how crude oil and its PAHs might affect specific cardiac genes is difficult to predict. Moreover, a role for transcription in homeostatic regulation of ion channels is not well defined (Rosati and McKinnon, 2004); although some studies using pharmacologic blockers have indicated that there is a compensatory upregulation of the blocked channel's gene (Duff et al., 1992; Leoni et al., 2005). Therefore, characterizing the impacts of oil exposure on ion channel/E–C coupling genes may also provide novel insight(s) into normal developmental regulation of cardiomyocyte physiology. Understanding changes in cardiac gene expression in response to oil exposure would further clarify mechanistic links between the effects of PAHs on cardiac function and the morphological defects that occur secondarily. Finally, from a standpoint of developing relatively simple assays or biomarkers for assessing oil-induced injury in the field, measurements of mRNA levels by quantitative reverse transcriptase polymerase chain reaction (qPCR) assays are easier to achieve than measures of protein function, or physiological and morphological endpoints. Here, we first consider evidence for alterations in cardiac gene expression that relates to functional cardiotoxicity, followed by studies that have identified changes in gene expression related to the longer-term sequelae of embryonic exposure.

As described in Section 2.5.3, exposure of Atlantic haddock embryos and larvae to dispersed crude oil led to several phenotypes that exhibited defects in cardiomyocyte physiology (Sørhus et al., 2016). Embryos exposed to only

6.7 μg L^{-1} of total PAHs showed complete AV block, with a noncontractile ventricle, that in many cases was smaller than normal. In larvae exposed to this PAH level after cardiac looping was complete and ventricular trabeculation had started, cardiac morphology was less impacted, consistent with a phase of primarily cardiac growth. However, animals displayed various degrees of partial AV block. Several candidate target E–C coupling genes were measured by qPCR, including *cacna1c* (LTCC alpha 1C subunit), *ncx1* (Na^+/Ca^{2+} exchanger 1), *atp2a2* (SERCA2), and *kcnh2* (ERG). Although *kcnh2* and *ncx1* expression levels were significantly reduced (by >6- and ~3-fold, respectively), this reduction was not apparent until well after the onset of cardiac dysfunction, suggesting a secondary mechanism. Moreover, analysis in individual animals showed that *kcnh2* expression was only reduced in those that had ventricles that were both silent and severely malformed. No significant effects were observed on any of these genes in the larval exposure, despite the presence of AV block.

High-throughput RNA sequencing (RNA–Seq) was subsequently used to analyze the entire transcriptomes of the same oil-exposed embryos and larvae (Sørhus et al., 2017). This highly sensitive method (Goodale et al., 2013) confirmed the late downregulation of *kcnh2*, but detected some additional changes. Key genes involved in SR Ca^{2+} cycling were downregulated at an early developmental stage (cardiac cone), prior to the onset of a regular heart beat and visible cardiac dysfunction, including *ryr2*, one of two *ncx1* paralogs, and two of three *atp2a2* paralogs. A third *atp2a2* paralog was downregulated at the hatching stage, when arrhythmia was severe and *kcnh2* was also reduced. A very different picture emerged for fish exposed during the larval period. At the point during exposure when AV block was apparent, there were no changes in E–C coupling genes. However, several days later, there was significant downregulation of *ncx1*, one of the same *atp2a2* paralogs downregulated in embryos, and the *kcnj12* gene. The latter encodes the IRK2 subunit that contributes to the I_{K1} inward rectifying potassium current (Tamargo et al., 2004). Remarkably, there was also strong upregulation of two *kcnj2* paralogs, which encode the IRK1 subunit of I_{K1}. The I_{K1} controls the cell's resting membrane potential, but also functions in repolarization, and *kcnj12* and *kcnj2* mutations are associated with long QT syndrome in humans. The findings from the embryonic exposure studies suggest that crude oil compounds could also have an inhibitory effect on the expression of genes related to E–C coupling, and a dual impact on Ca^{2+} handling (Fig. 10). In contrast, the responses in the larval exposure are more consistent with both longer-term pathological (*kcnj12* downregulation) and compensatory responses (*kcnj2* upregulation) to protein-level impacts of PAHs.

Similar results were obtained in a very different species, the mahi mahi, in which RNA–Seq was applied to early and late stage yolk-sac larvae following embryonic exposure (Xu et al., 2016). Cardiac defects that occurred in mahi mahi following embryonic exposure to DWH–MC252 crude oil were characterized primarily by reduced contractility without arrhythmia or serious bradycardia, leading to poor looping (Edmunds et al., 2015). In close correlation with this phenotype, there was overlap in altered Ca^{2+} handling genes compared to haddock, but no effect on K^+ channel genes. At the earlier time point, the genes for calcium–calmodulin kinase II gamma (*cam2kg*) and calsequestrin 2 (*casq2*) were upregulated. Calsequestrin 2 is a major SR Ca^{2+}-buffering protein in cardiomyocytes and also forms a complex with Ryr2 to regulate its activity (Faggioni and Knollmann, 2012). Subsequently, 2 days later *casq2* remained upregulated while *ryr2* was downregulated, along with *cacna1c* encoding the alpha 1C unit of the major cardiac LTCC. In addition, the Ca^{2+}-sensitive endocardial transcription factor *nfatc1* was downregulated. Although this study did not analyze time points that preceded the onset of cardiac dysfunction, these results are also consistent with adverse impacts on cardiomyocyte Ca^{2+} handling at both the protein and gene levels.

As also supported by the analysis of genes functioning directly in heart development (described in the next section), it is possible that PAHs in crude oil affect gene expression through impacts on nuclear Ca^{2+}, in parallel to their impacts on SR Ca^{2+} handling (Fig. 10). Recent studies have characterized the process of excitation–transcription (E–T) coupling in cardiomyocytes and other excitable cells (Bers, 2008; Winslow et al., 2016). Because the SR pool of Ca^{2+} is contiguous with the nuclear envelope pool (Wu and Bers, 2006), and both utilize SERCA2 (Ljubojevic and Bers, 2015), altered Ca^{2+} cycling as a consequence of PAH exposure may have more direct effects on gene expression than previously appreciated.

2.5.6.2. Effects on Genes Related to Terminal Differentiation of Cardiomyocytes and Cardiac Form. While the effects of PAHs on gene expression may not be important drivers of acute toxicity, relative to protein-level effects, there are very clear linkages of altered gene expression to the structural developmental defects that follow acute toxicity. In many ways, this is not surprising as the phenotypic effects are similar to loss-of-function mutants in zebrafish genes that control cardiac function and morphogenesis. A combination of hypothesis-driven gene selection based on zebrafish genetics and RNA–Seq studies in both mahi mahi and Atlantic haddock have provided major insights into cardiac differentiation and morphogenesis pathways that are disrupted by crude oil exposure. As might be expected based on both the different functional and morphological responses to crude oil exposure in different species described above (Section 2.5.3), species-specific alterations in gene expression have been observed. Despite these differences, however, there is evidence for disruption of some common underlying pathways.

Using qPCR, Edmunds et al. (2015) examined a set of 10 genes in mahi mahi embryos exposed to MC-252 crude oil that was linked to cardiac function, morphogenesis and pathological hypertrophy in zebrafish and humans. In this case, altered gene expression (downregulation) was observed only after the detection of morphological changes, linking these genes to secondary developmental consequences rather than initiating events. Consistent with the phenotype of reduced contractility, MC-252 exposure resulted in downregulation of genes for myofibrillar proteins, including atrial and ventricular myosin heavy chain isoforms (*amhc* and *vhmc*, respectively), and the cardiac myosin light chain 2 isoform (*cmlc2*). Also downregulated was the four-and-a-half LIM domain family protein 2 (*fhl2*), which is a protein that is relatively restricted to the developing heart and plays multiple roles through a myriad protein–protein interactions. However, many of the interacting partners are clearly related to the abnormal cardiac function phenotypes observed in oil-exposed embryos, such as contractility (maintenance of myofiber elasticity) and the regulation of repolarizing K^+ currents (Tran et al., 2016). Two major transcription factors involved in cardiomyocyte differentiation, NK2 homeobox 5 (*nkx2.5*) and T-box 5 (*tbx5*), were also downregulated, along with the NKX2.5 target gene encoding atrial natriuretic peptide (*nppa*). As a homeostatic regulator of contractility, reduced *nppa* expression was also consistent with the functional phenotype.

Consistent additional findings were obtained from the previously mentioned RNA–Seq study on oil-exposed mahi mahi larvae (Xu et al., 2016). In this case, developmental time points flanked the stages assessed by Edmunds et al. (2015), earlier hatching stage larvae and 2 days later feeding stage larvae. Members of the bone morphogenetic protein (BMP) family play multiple roles in heart development from the very earliest stages (van Wijk et al., 2007). In zebrafish embryos, bone morphogenetic protein 4 (*bmp4*) undergoes dynamic expression initially in both atrial–ventricular chambers and then later becoming restricted to the atrial–ventricular boundary (through the activity of *nkx2.5*) (Tu et al., 2009). Oil exposure altered the dynamic pattern of *bmp4* expression in mahi mahi larvae in a complex manner, but either excess or reduced BMP4 activity leads to looping defects in zebrafish (Chen et al., 1997). Also consistent with the effects on intracellular Ca^{2+} handling, levels of several key genes involved in Ca^{2+} homeostasis and E–C coupling were altered. In hatching stage larvae, the gene encoding the Ca^{2+}-buffering protein calsequestrin 2 (*casq2*) was upregulated by oil exposure, while both *casq2* and calsequestrin 1 (*casq1*) were upregulated by the feeding stage. Importantly, CASQ2 plays a key role in buffering Ca^{2+} in the SR, but also interacts with (and regulates) the activity of the RyR (Ernst et al., 1977). It remains to be determined whether this is a compensatory or pathophysiological response. Finally, downregulation of both *cacna1c* and *ryr2* was observed at the later time point.

RNA–Seq in Atlantic haddock also revealed a key role for BMP signaling in oil-induced morphological defects affecting the ventricle. Because of the relatively protracted development of haddock, a cold water species (e.g., 12 days to hatch compared to ~36h for mahi mahi), a wider range of developmental stages were analyzed during and after exposure, in particular during organogenesis. In this case, changes in developmental signaling pathways were observed prior to visible morphological abnormalities (Sørhus et al., 2017). As in most species, heart development proceeded normally in oil-exposed haddock embryos through the cardiac progenitor, cardiac cone and initial looping stages. However, the *bmp10* gene was upregulated at the cardiac cone stage. This finding is significant for several reasons. First, as discussed above, BMPs are potent morphogens whose activities must be tightly regulated for normal development, and BMP10 is a more potent ligand than BMP4 (Lichtner et al., 2013). In zebrafish, chick and mouse embryos, *bmp10* is normally expressed at later stages of heart development; i.e., during trabeculation when it drives cardiomyocyte proliferation (Grego-Bessa et al., 2007). In contrast, *bmp4* is normally expressed at the cardiac cone stage, where it shifts from a radially symmetric pattern to being asymmetrically elevated on the left side. This lateralization of *bmp4* expression is necessary for the normal rightward looping of the ventricle, and misexpression of *bmp4* leads to failed looping (Chen et al., 1997). Importantly, upregulation of *bmp10* also links disruption of intracellular Ca^{2+} handling to abnormal morphogenesis, possibly through impacts on E–T coupling. Transcription of *bmp10* is mediated by myocardin, a Ca^{2+}-sensitive transcriptional activator (Huang et al., 2012). In mouse embryos, loss of the FK506 binding protein FKBP12, which interacts with RyR2 to prevent SR Ca^{2+} leakage, leads to SR Ca^{2+} depletion, an abnormal upregulation of *bmp10*, and consequently severe morphological defects of the ventricle (Chen et al., 2004). As with BMP4, loss of or excess BMP10 leads to severe defects in heart development, and the premature upregulation of a more potent family member at the cone stage is extremely likely to be the cause of the looping defects observed in oil-exposed embryos. These findings demonstrate a transcriptional cascade that tightly links defects in cardiac function (cardiomyocyte intracellular Ca^{2+} cycling) and form (heart chamber growth) through *bmp10*.

This also potentially explains why exposure at later developmental stages has much less severe impacts on morphogenesis, when BMP signaling primarily influences proliferation and growth of the now-formed ventricle, rather than key morphogenetic movements. Once the larval heart has passed through these dramatic morphogenetic movements, it primarily undergoes simple linear growth henceforth (Singleman and Holtzman, 2012). For example, while haddock larvae exposed during feeding stages showed oil-induced arrhythmia, heart morphology at this stage was not dramatically altered

(Sørhus et al., 2016). Correspondingly, there were alterations in expression of E–C coupling genes (both up- and downregulation) and downregulation of genes encoding myofibrillar components, but no changes in key cardiac developmental regulator genes (Sørhus et al., 2017). The embryo–larval comparison suggests that during early development, intracellular Ca^{2+} handling is linked to the expression of morphogenetic regulators, but at stages following this critical window, Ca^{2+} handling is linked to cardiomyocyte growth. Presumably, however, these later processes have some commonality with hypertrophic pathways underlying cardiac remodeling. Thus, it remains to be determined if chronic exposure during later larval and juvenile growth can lead to similar types of abnormal hypertrophy. However, although reduced cardiac function has been measured in larger fish exposed to crude oil (Claireaux and Davoodi, 2010), these later life stages exposures did not lead to lasting impacts (Mauduit et al., 2016). These latter studies thus provide further impetus to focus on exposure during early development.

3. FOCAL AREAS FOR ONGOING AND FUTURE RESEARCH

3.1. An Improved Mechanistic Understanding of Environmental Cardiotoxicity

The case example of crude oil toxicity (Section 2.5) highlights the discovery of several novel mechanisms for environmental impacts on heart development at multiple levels, from molecules to the whole organism. Improved mechanistic insights clarify the cause–effect relationships between exposure to aquatic pollutants such as PAHs and adverse impacts on organismal health, which in turn enhances our predictive capacity for environmental impacts. Detailed mechanistic information at each scale of biological organization is needed for the construction of adverse outcome pathways (AOPs). AOPs consist of detailed toxicological cause and effect links that span multiple levels of biological organization, ideally from molecular initiating events to species, community, or ecosystem scale responses of regulatory concern (e.g., reduction in a fisheries abundance target). There is an ongoing movement to use AOPs to improve risk assessments for both human and environmental health (Ankley et al., 2010; Garcia-Reyero, 2015; Kramer et al., 2011; Villeneuve et al., 2014). At the same time, improved mechanistic understanding at the molecular level will enhance the development of tools for monitoring the effects of human-induced environmental change, in this case oil spills, on key fish populations. An important consideration is that the science of oil spill ecotoxicology has been driven by damage assessments in the wake of large environmental disasters (e.g., the Exxon Valdez and Deepwater Horizon oil spills). Few other pollutants have received this level of effort, especially on a proactive basis. Compared to other types of pollution, such as stormwater runoff or municipal wastewater treatment discharges, PAH mixtures derived from oil spills are relatively simple. The two decades of research on crude oil and fish heart development can serve as an example both for studies of other potentially cardiotoxic pollutants, and for studies of impacts on other organ systems and processes.

3.2. Next-Generation Molecular Indicators to Assess Fish Health in Polluted Habitats

A long-term goal in aquatic ecotoxicology is the development of new and more sensitive indicators of contaminant exposure, as well as fish health, in habitats potentially degraded by pollution. The induction of detoxification pathways in the liver and other tissues is a classical example of a biomarker. Over the past several decades, investigators have monitored cytochrome P-450 monooxygenanse systems as indicators of exposure to AHR agonists and other chemicals, in both the field and the laboratory (Stegeman and Lech, 1991). Genes in this superfamily are responsive to a large number of environmental contaminants, and CYP1A in particular can be exquisitely sensitive. Moreover, in certain cases, CYP1A induction has been shown to be predictive of adverse health outcomes—e.g., in pink salmon exposed to Alaska North Slope crude oil (Carls et al., 2005).

Nevertheless, there remain important limitations in terms of using CYP1A and related detoxifying enzymes for environmental surveillance. First, many of the classical biomarkers are upregulated in response to multiple classes of contaminants. This makes them more useful for monitoring discrete pollution events (e.g., an oil spill) in relatively pristine habitats (e.g., Prince William Sound), and less useful when an oil spill occurs in a river or along a coastline with a history of industrial activity. In the latter case, background levels of PCBs in the tissues of a given fish species might confound the interpretation that an observed CYP1A response is attributable to PAHs from the spill. Second, evidence of detoxification is not necessarily evidence of injury. If a first-pass response is successful in peripheral tissues such as the skin or the gills, there may be little or no toxicity to the heart.

Ideally, the cardiovascular-specific biomarkers of the future will be diagnostic for both contaminant exposure as well as adverse health outcomes that are consequential for wild fish. This dovetails with the construction of AOPs, as initiating events or responses at the molecular or cellular level are generally easier to measure than responses and outcomes at the organismal or population scale. Ideally, the power of AOPs is realized when simple molecular measures can be used to robustly predict higher-level outcomes.

The aforementioned advances in mechanism-oriented research are creating new opportunities in this area, particularly at a molecular scale. As in humans, the path forward will not rely on a single indicator, but rather batteries of markers with known mechanistic interconnections and clear phenotypic anchors (Gerszten and Wang, 2008). Measuring responsive genes that are expressed exclusively in the heart will reduce exposure uncertainty—i.e., show that toxic chemicals are accumulating in target tissues at levels sufficient to cause cardiotoxicity. A more detailed understanding of how individual chemicals within mixtures impact heart development and function will make it easier to establish causation in fish habitats where complex mixtures of contaminants may be common. For example, recent studies have focused on a subset of PAHs—the phenanthrenes—as pharmacological blockers of delayed-rectifying potassium channels in cardiomyocytes, as a cellular basis for heart rhythm defects (Brette et al., 2017). New biomarkers specific to this molecular initiating event will help distinguish phenanthrene toxicity from toxicity caused by other classes of compounds with different mechanisms of action. Lastly, the functional association of new biomarkers with demonstrable injury will allow for more direct determinations of fish health in polluted habitats. Overall, these new methods may prove less expensive and labor-intensive than traditional assessment techniques based on histology, immunohistochemistry, enzyme assays, and live digital imaging. Oil-induced changes in gene expression (as described in Section 2.5.6) are a potential starting point for the development of new molecular injury indicators.

3.3. Biological Scaling: Application to Fisheries Management

Environmental pollution has been a longstanding challenge for the management of recreational and commercial fisheries, as well as the aquatic ecosystems upon which they depend (Scholz and McIntyre, 2015). In the developed world, pollution reduction legislation such as the 1977 Clean Water Act in the United States largely ended the mass mortality events that were common in earlier decades. However, the number of chemicals in societal use continues to grow, and many of these contaminants adversely affect the health of fish in ways that are much more nuanced. Developmental cardiotoxicity is a hallmark example of how chemicals can reduce the fitness (i.e., survival and lifetime reproductive success) of fish. How this loss of individual fitness affects the abundance of populations and species remains an active area of investigation (Scholz and Incardona, 2015). This is important because fisheries are generally managed at these higher biological scales, and species life history plays a central role in the propagation of environmental health impacts across generations.

Much of this chapter has focused on mechanisms of developmental cardiotoxicity, as molecular initiating events for AOPs (e.g., Incardona and

Scholz, 2016). With respect to individual fitness, adverse outcomes can be divided into two categories. The first encompasses severe developmental abnormalities that preclude larval survival. These can be caused by disruptions of cardiovascular function, malformations of the heart, or downstream impacts on other developing tissues. The second category includes more subtle impacts that reduce the physiological performance (Hicken et al., 2011; Incardona et al., 2015) and survival of fish later in life. Whereas acute embryo and larval toxicity have been studied relatively extensively, the processes that lead to delayed mortality are still poorly understood. From the standpoint of natural resource management, both types of adverse outcomes can potentially limit the abundance of wild fish populations. More importantly, the two categories of cardiotoxicity can influence the interpretation of environmental exposure information. Physiological perturbations that are delayed in time occur at much lower contaminant exposure concentrations than those that produce grossly visible developmental abnormalities (Heintz, 2007; Heintz et al., 2000; Incardona et al., 2015). This, in turn, determines the proportion of a managed fishery that may be impacted by an intentional or accidental chemical discharge.

For individual fish, biological scaling entails connecting a molecular event (e.g., AHR activation, cardiac K^+ channel blockade, etc.) to various endpoints that determine individual fitness. For example, a sublethal disruption of heart form and function during early development can lead to a reduced cardiorespiratory performance later in life. A reduced cardiorespiratory capability can, in turn, increase the likelihood of starvation, predation vulnerability, and disease susceptibility. These quantifiable metrics for survival can be used to parameterize individual-based models, which are then coupled to population models to estimate the potential for long-term impacts on the intrinsic growth rate (i.e., lambda) and abundance of fisheries (e.g., Baldwin et al., 2009). Practical applications include Natural Resource Damage Assessments following environmental disasters, Environmental Impact Assessments for future development projects, and recovery planning for threatened and endangered species. Notably, population modeling is inherently tied to the life history characteristics of a given species. Factors such as density dependence, stage-specific recruitment, iteroparity (vs semelparity), and baseline population abundance can determine to a large extent how losses of individuals will resonate at a population scale.

3.4. Fish Models for Genetics: Crossover Applications for Human and Fish Cardiovascular Health

The late George Streisinger was a bacteriophage geneticist who initially selected zebrafish to develop as a genetic model to study development of the vertebrate brain (Stahl, 1995). As a fish hobbyist, he was aware that small tropical aquarium fish had many of the advantages of the fruit fly *Drosophila* for genetics and embryology, such as rapid development of easily accessible embryos, short generation time, and ability to culture large numbers of families in a small space. While the zebrafish model was ultimately developed as a premier model for human disease through a concerted effort from the National Institutes of Health (Rasooly et al., 2003), Streisinger was probably less aware of the even greater similarity between fish and human hearts than humans have with small rodent models. An even greater potential for the use of zebrafish for genetic analysis of cardiac function and development is derived from the ability of small fish embryos and larvae to survive early development while completely lacking circulation, due to oxygen uptake by simple diffusion. Similar analyses in very early mammalian hearts are impossible, as embryos quickly succumb without adequate placental circulation.

Irrespective of size, almost all fish species have heart rates closer in range to humans (~60–120 beats min^{-1}) than small rodents with vastly higher heart rates (e.g., ~600 beats min^{-1} for mice). Consequently, fish and humans have a closer alignment with respect to ion channels that control excitability. For example, the repolarization phase (corrected QT interval on the electrocardiogram) in humans is in the range of 350–450 ms (Johnson and Ackerman, 2009), 440–480 ms in newly hatched to early feeding stage zebrafish larvae (Dhillon et al., 2013), but only about 50 ms in mice (Salama and London, 2007). Accordingly, humans and fish have much higher densities of I_{Kr}/ERG than mice (Arnaout et al., 2007; Salama and London, 2007), which use a different suite of K^+ channels for more rapid repolarization. ERG is the most frequent target of adverse drug reactions in humans (Wilke et al., 2007), largely because of its uniquely large pore volume and presence of drug-interacting aromatic amino acid side chains (Knape et al., 2011; Mitcheson, 2008). Moreover, consistent with similar sensitivity to drug-induced arrhythmia, the amino acid sequences of the drug-binding pore domains of human and zebrafish are 99% identical (Langheinrich et al., 2003; Milan et al., 2003). The nearly identical drug sensitivity of zebrafish and human ERG channels strongly suggests the converse is true: human ERG channels are likely to have the same sensitivity to environmental pollutants identified as cardiotoxic in zebrafish (or other fish species). Thus, tricyclic PAHs would be predicted to induce arrhythmia in human hearts, and, remarkably, drug studies have correlated the presence of three aromatic rings with higher potency of ERG blockade (Ritchie and Macdonald, 2009).

Perhaps not coincidentally, urban traffic-derived air pollution is strongly associated with serious cardiac arrhythmias, including atrial fibrillation (Cosselman et al., 2015; Link et al., 2013; Monrad et al., 2017), and tricyclic PAHs (phenanthrenes in particular) are the most abundant PAHs in both the particulate and vapor phases of urban air (Rogge et al., 2011). Although the heart in adult fish is relatively protected from PAH cardiotoxicity by first pass metabolism (Section 1.3), airborne PAHs are readily taken up by mammalian lungs (Gerde et al., 1993; Li et al., 2010), and would be delivered directly to the left atrium and coronary circulation of the heart. In this context, it is notable that atrial fibrillation originates primarily from regions adjacent to the junction of the pulmonary veins and left atrium (Haïssaguerre et al., 1998). PAHs still have not been considered in the etiology of arrhythmia and other acute cardiac events associated with urban air pollution (Cosselman et al., 2015), but the body of work derived from the effects of aquatic pollution on the fish heart makes this a very reasonable and easily testable hypothesis.

On a broader level, the similarities in ventricular anatomy between fast swimming fish and humans (i.e., presence of compact myocardium) will continue to make fish models like zebrafish an important contributor to understanding human cardiac development and function. For studies of the environmental health of wild fish species, the selection of a fast swimming species for development into arguably the most sophisticated genetic model was fortuitous. Zebrafish also provides an important laboratory model applicable to understanding the hearts of a majority of economically and ecologically important species, including pelagic planktivores comprising key forage fish species, salmonids, and large pelagic predators that are the focus of important commercial and recreational fisheries. In this regard, the potential for laboratory fish models is only beginning to be realized.

REFERENCES

Andreasen, E.A., Spitsbergen, J.M., Tanguay, R.L., Stegeman, J.J., Heideman, W., Peterson, R.E., 2002. Tissue-specific expression of AHR2, ARNT2, and CYP1A in zebrafish embryos and larvae: effects of developmental stage and 2,3,7,8-tetrachlorodibenzo-p-dioxin exposure. Toxicol. Sci. 68, 403–419.

Andrés-Delgado, L., Mercader, N., 2016. Interplay between cardiac function and heart development. Biochim. Biophys. Acta 1863, 1707–1716.

Ankley, G.T., Bennett, R.S., Erickson, R.J., Hoff, D.J., Hornung, M.W., Johnson, R.D., Mount, D.R., Nichols, J.W., Russom, C.L., Schmieder, P.K., Serrrano, J.A., Tietge, J.E., Villeneuve, D.L., 2010. Adverse outcome pathways: a conceptual framework to support ecotoxicology research and risk assessment. Environ. Toxicol. Chem. 29, 730–741.

Antkiewicz, D.S., Burns, C.G., Carney, S.A., Peterson, R.E., Heideman, W., 2005. Heart malformation is an early response to TCDD in embryonic zebrafish. Toxicol. Sci. 84, 368–377.

Antkiewicz, D.S., Peterson, R.E., Heideman, W., 2006. Blocking expression of AHR2 and ARNT1 in zebrafish larvae protects against cardiac toxicity of 2,3,7,8-tetrachlorodibenzo-p-dioxin. Toxicol. Sci. 94, 175–182.

Anttila, K., Jørgensen, S.M., Casselman, M.T., Timmerhaus, G., Farrell, A.P., Takle, H., 2014. Association between swimming performance, cardiorespiratory morphometry and thermal tolerance in Atlantic salmon (Salmo salar L.). Front. Mar. Sci. 1, 76.

Arnaout, R., Ferrer, T., Huisken, J., Spitzer, K., Stainier, D.Y., Tristani-Firouzi, M., Chi, N.C., 2007. Zebrafish model for human long QT syndrome. Proc. Natl. Acad. Sci. U.S.A. 104, 11316–11321.

Auman, H.J., Coleman, H., Riley, H.E., Olale, F., Tsai, H.J., Yelon, D., 2007. Functional modulation of cardiac form through regionally confined cell shape changes. PLoS Biol. 5, e53.

Baldwin, D.H., Spromberg, J.A., Collier, T.K., Scholz, N.L., 2009. A fish of many scales: extrapolating sublethal pesticide exposures to the productivity of wild salmon populations. Ecol. Appl. 19, 2004–2015.

Berdougo, E., Coleman, H., Lee, D.H., Stainier, D.Y., Yelon, D., 2003. Mutation of *weak atrium/atrial myosin heavy chain* disrupts atrial function and influences ventricular morphogenesis in zebrafish. Development 130, 6121–6129.

Bers, D.M., 2008. Calcium cycling and signaling in cardiac myocytes. Annu. Rev. Physiol. 70, 23–49.

Bradbury, S.P., Henry, T.R., Niemi, G.J., Carlson, R.W., Snarski, V.M., 1989. Use of respiratory-cardiovascular responses of rainbow trout (*Salmo gairdneri*) in identifying acute toxicity syndromes in fish: part 3. Polar narcotics. Environ. Toxicol. Chem. 8, 247–261.

Brette, F., Machado, B., Cros, C., Incardona, J.P., Scholz, N.L., Block, B.A., 2014. Crude oil impairs cardiac excitation–contraction coupling in fish. Science 343, 772–776.

Brette, F., Shiels, H.A., Galli, G.L.J., Cros, C., Incardona, J.P., Scholz, N.L., Block, B.A., 2017. A novel cardiotoxic mechanism for a pervasive global pollutant. Sci. Rep. 7, 41476

Burggren, W., Dubansky, B., Bautista, N.M., 2017. Cardiovascular development in embryonic and larval fishes. In: Gamperl, A.K., Gillis, T.E., Farrell, A.P., Brauner, C.J. (Eds.), The Cardiovascular System: Development, Plasticity and Physiological Responses. Fish Physiology, vol. 36B. Elsevier, Amsterdam. pp. 107–184. Chapter 2.

Bue, B.G., Sharr, S., Moffitt, S.D., Craig, A.K., 1996. Effects of the Exxon Valdez oil spill on pink salmon embryos and preemergent fry. Am. Fish. Soc. Symp. 18, 619–627.

Bue, B.G., Sharr, S., Seeb, J.E., 1998. Evidence of damage to pink salmon populations inhabiting Prince William Sound, Alaska, two generations after the Exxon Valdez oil spill. Trans. Am. Fish. Soc. 127, 35–43.

Carls, M.G., Rice, S.D., Hose, J.E., 1999. Sensitivity of fish embryos to weathered crude oil: part I. Low-level exposure during incubation causes malformations, genetic damage, and mortality in larval Pacific herring (*Clupea pallasi*). Environ. Toxicol. Chem. 18, 481–493.

Carls, M.G., Heintz, R.A., Marty, G.D., Rice, S.D., 2005. Cytochrome P4501A induction in oil-exposed pink salmon Oncorhynchus gorbuscha embryos predicts reduced survival potential. Mar. Ecol. Prog. Ser. 301, 253–265.

Carney, S.A., Peterson, R.E., Heideman, W., 2004. 2,3,7,8-Tetrachlorodibenzo-p-dioxin activation of the aryl hydrocarbon receptor/aryl hydrocarbon receptor nuclear translocator pathway causes developmental toxicity through a CYP1A-independent mechanism in zebrafish. Mol. Pharmacol. 66, 512–521.

Carney, S.A., Chen, J., Burns, C.G., Xiong, K.M., Peterson, R.E., Heideman, W., 2006a. Aryl hydrocarbon receptor activation produces heart-specific transcriptional and toxic responses in developing zebrafish. Mol. Pharmacol. 70, 549–561.

Carney, S.A., Prasch, A.L., Heideman, W., Peterson, R.E., 2006b. Understanding dioxin developmental toxicity using the zebrafish model. Birth Defects Res. A Clin. Mol. Teratol. 76, 7–18.

Castro, V., Grisdale-Helland, B., Helland, S.J., Torgersen, J., Kristensen, T., Claireaux, G., Farrell, A.P., Takle, H., 2013a. Cardiac molecular-acclimation mechanisms in response to swimming-induced exercise in Atlantic salmon. PLoS One 8, e55056

Castro, V., Grisdale-Helland, B., Jorgensen, S., Helgerud, J., Claireaux, G., Farrell, A.P., Krasnov, A., Helland, S., Takle, H., 2013b. Disease resistance is related to inherent swimming performance in Atlantic salmon. BMC Physiol. 13, 1.

Chen, J.N., van Eeden, F.J., Warren, K.S., Chin, A., Nusslein-Volhard, C., Haffter, P., Fishman, M.C., 1997. Left–right pattern of cardiac BMP4 may drive asymmetry of the heart in zebrafish. Development 124, 4373–4382.

Chen, H., Shi, S., Acosta, L., Li, W., Lu, J., Bao, S., Chen, Z., Yang, Z., Schneider, M.D., Chien, K.R., Conway, S.J., Yoder, M.C., Haneline, L.S., Franco, D., Shou, W., 2004. BMP10 is essential for maintaining cardiac growth during murine cardiogenesis. Development 131, 2219–2231.

Chen, J., Carney, S.A., Peterson, R.E., Heideman, W., 2008. Comparative genomics identifies genes mediating cardiotoxicity in the embryonic zebrafish heart. Physiol. Genomics 33, 148–158.

Chi, N.C., Bussen, M., Brand-Arzamendi, K., Ding, C., Olgin, J.E., Shaw, R.M., Martin, G.R., Stainier, D.Y., 2010. Cardiac conduction is required to preserve cardiac chamber morphology. Proc. Natl. Acad. Sci. U.S.A. 107, 14662–14667.

Claireaux, G., Davoodi, F., 2010. Effect of exposure to petroleum hydrocarbons upon cardio-respiratory function in the common sole (*Solea solea*). Aquat. Toxicol. 98, 113–119.

Claireaux, G., McKenzie, D.J., Genge, A.G., Chatelier, A., Aubin, J., Farrell, A.P., 2005. Linking swimming performance, cardiac pumping ability and cardiac anatomy in rainbow trout. J. Exp. Biol. 208, 1775–1784.

Clark, B.W., Matson, C.W., Jung, D., Di Giulio, R.T., 2010. AHR2 mediates cardiac teratogenesis of polycyclic aromatic hydrocarbons and PCB-126 in Atlantic killifish (*Fundulus heteroclitus*). Aquat. Toxicol. 99, 232–240.

Colavecchia, M.V., Backus, S.M., Hodson, P.V., Parrott, J.L., 2004. Toxicity of oil sands to early life stages of fathead minnows (*Pimephales promelas*). Environ. Toxicol. Chem. 23, 1709–1718.

Colavecchia, M.V., Hodson, P.V., Parrott, J.L., 2006. CYP1A induction and blue sac disease in early life stages of white suckers (*Catostomus commersoni*) exposed to oil sands. J. Toxicol. Environ. Health A 69, 967–994.

Cook, P.M., Robbins, J.A., Endicott, D.D., Lodge, K.B., Guiney, P.D., Walker, M.K., Zabel, E.W., Peterson, R.E., 2003. Effects of aryl hydrocarbon receptor-mediated early life stage toxicity on lake trout populations in Lake Ontario during the 20th century. Environ. Sci. Technol. 37, 3864–3877.

Cosselman, K.E., Navas-Acien, A., Kaufman, J.D., 2015. Environmental factors in cardiovascular disease. Nat. Rev. Cardiol. 12, 627–642.

Costa, J., Ferreira, M., Rey-Salgueiro, L., Reis-Henriques, M.A., 2011. Comparision of the waterborne and dietary routes of exposure on the effects of Benzo(a)pyrene on biotransformation pathways in Nile tilapia (*Oreochromis niloticus*). Chemosphere 84, 1452–1460.

Couillard, C.M., 2002. A microscale test to measure petroleum oil toxicity to mummichog embryos. Environ. Toxicol. 17, 195–202.

Despres, J.P., 2016. Physical activity, sedentary behaviours, and cardiovascular health: when will cardiorespiratory fitness become a vital sign? Can. J. Cardiol. 32, 505–513.

Dhillon, S.S., Doro, E., Magyary, I., Egginton, S., Sik, A., Muller, F., 2013. Optimisation of embryonic and larval ECG measurement in zebrafish for quantifying the effect of QT prolonging drugs. PLoS One 8, e60552

Di Giulio, R.T., Clark, B.W., 2015. The Elizabeth River story: a case study in evolutionary toxicology. J. Toxicol. Environ. Health B Crit. Rev. 18, 259–298.

Dubansky, B., Whitehead, A., Miller, J.T., Rice, C.D., Galvez, F., 2013. Multitissue molecular, genomic, and developmental effects of the Deepwater Horizon oil spill on resident Gulf killifish (*Fundulus grandis*). Environ. Sci. Technol. 47, 5074–5082.

Duff, H.J., Offord, J., West, J., Catterall, W.A., 1992. Class I and IV antiarrhythmic drugs and cytosolic calcium regulate mRNA encoding the sodium channel alpha subunit in rat cardiac muscle. Mol. Pharmacol. 42, 570–574.

Ebert, A.M., Hume, G.L., Warren, K.S., Cook, N.P., Burns, C.G., Mohideen, M.A., Siegal, G., Yelon, D., Fishman, M.C., Garrity, D.M., 2005. Calcium extrusion is critical for cardiac morphogenesis and rhythm in embryonic zebrafish hearts. Proc. Natl. Acad. Sci. U.S.A. 102, 17705–17710.

Edmunds, R.C., Gill, J.A., Baldwin, D.H., Linbo, T.L., French, B.L., Brown, T.L., Esbaugh, A.J., Mager, E.M., Stieglitz, J.D., Hoenig, R., Benetti, D.D., Grosell, M., Scholz, N.L., Incardona, J.P., 2015. Corresponding morphological and molecular indicators of crude oil toxicity to the developing hearts of mahi mahi. Sci. Rep. 5, 17326.

Eliason, E.J., Clark, T.D., Hague, M.J., Hanson, L.M., Gallagher, Z.S., Jeffries, K.M., Gale, M.K., Patterson, D.A., Hinch, S.G., Farrell, A.P., 2011. Differences in thermal tolerance among sockeye salmon populations. Science 332, 109–112.

Elonen, G.E., Spehar, R.L., Holcombe, G.W., Johnson, R.D., Fernandez, J.D., Erickson, R.J., Tietge, J.E., Cook, P.M., 1998. Comparative toxicity of 2,3,7,8-tetrachlorodibenzo-p-dioxin to seven freshwater fish species during early life-stage development. Environ. Toxicol. Chem. 17, 472–483.

Ernst, V.V., Neff, J.M., Anderson, J.W., 1977. The effects of the water-soluble fractions of no. 2 fuel oil on the early development of the estuarine fish, *Fundulus grandis* Baird and Girard. Environ. Pollut. 14, 25–35.

Faggioni, M., Knollmann, B.C., 2012. Calsequestrin 2 and arrhythmias. Am. J. Physiol. Heart Circ. Physiol. 302, H1250–1260.

Farrell, A.P., Smith, F., 2017. Cardiac form, function and physiology. In: Gamperl, A.K., Gillis, T.E., Farrell, A.P., Brauner, C.J. (Eds.), The Cardiovascular System: Morphology, Control and Function. Fish Physiology, vol. 36A. Elsevier, Amsterdam, pp. 155–264.

Feist, B.E., Buhle, E.R., Arnold, P., Davis, J.W., Scholz, N.L., 2011. Landscape ecotoxicology of coho salmon spawner mortality in urban streams. PLoS One 6, e23424.

Garcia-Reyero, N., 2015. Are adverse outcome pathways here to stay? Environ. Sci. Technol. 49, 3–9.

Geiger, H.J., Bue, B.G., Sharr, S., Wertheimer, A.C., Willette, T.M., 1996. A life history approach to estimating damage to Prince William Sound pink salmon caused ty the Exxon Valdez oil spill. Am. Fish. Soc. Symp. 18, 487–498.

Gerde, P., Muggenburg, B.A., Hoover, M.D., Henderson, R.F., 1993. Disposition of polycyclic aromatic hydrocarbons in the respiratory tract of the beagle dog. I. The alveolar region. Toxicol. Appl. Pharmacol. 121, 313–318.

Gerszten, R.E., Wang, T.J., 2008. The search for new cardiovascular biomarkers. Nature 451, 949–952.

Gillis, T.E., Johnson, E.F., 2017. Cardiac preconditioning, remodeling, and regeneration. In: Gamperl, A.K., Gillis, T.E., Farrell, A.P., Brauner, C.J. (Eds.), The Cardiovascular System: Development, Plasticity and Physiological Responses. Fish Physiology, vol. 36B. Elsevier, Amsterdam. pp. 185–233. Chapter 3.

Glickman, N.S., Yelon, D., 2002. Cardiac development in zebrafish: coordination of form and function. Semin. Cell Dev. Biol. 13, 507–513.

Gobel, P., Dierkes, C., Coldewey, W.C., 2007. Storm water runoff concentration matrix for urban areas. J. Contam. Hydrol. 91, 26–42.

González-Doncel, M., Gonzalez, L., Fernández-Torija, C., Navas, J.M., Tarazona, J.V., 2008. Toxic effects of an oil spill on fish early life stages may not be exclusively associated to PAHs: studies with prestige oil and medaka (*Oryzias latipes*). Aquat. Toxicol. 87, 280–288.

Goodale, B.C., Tilton, S.C., Corvi, M.M., Wilson, G.R., Janszen, D.B., Anderson, K.A., Waters, K.M., Tanguay, R.L., 2013. Structurally distinct polycyclic aromatic hydrocarbons induce differential transcriptional responses in developing zebrafish. Toxicol. Appl. Pharmacol. 272, 656–670.

Grego-Bessa, J., Luna-Zurita, L., del Monte, G., Bolos, V., Melgar, P., Arandilla, A., Garratt, A.N., Zang, H., Mukouyama, Y.S., Chen, H., Shou, W., Ballestar, E., Esteller, M., Rojas, A., Perez-Pomares, J.M., de la Pompa, J.L., 2007. Notch signaling is essential for ventricular chamber development. Dev. Cell 12, 415–429.

Grimes, A.C., Erwin, K.N., Stadt, H.A., Hunter, G.L., Gefroh, H.A., Tsai, H.J., Kirby, M.L., 2008. PCB126 exposure disrupts zebrafish ventricular and branchial but not early neural crest development. Toxicol. Sci. 106, 193–205.
Guiney, P.D., Smolowitz, R.M., Peterson, R.E., Stegeman, J.J., 1997. Correlation of 2,3,7,8-tetrachlorodibenzo-p-dioxin induction of cytochrome P4501A in vascular endothelium with toxicity in early life stages of lake trout. Toxicol. Appl. Pharmacol. 143, 256–273.
Gupta, V., Poss, K.D., 2012. Clonally dominant cardiomyocytes direct heart morphogenesis. Nature 484, 479–484.
Hahn, M.E., Karchner, S.I., Evans, B.R., Franks, D.G., Merson, R.R., Lapseritis, J.M., 2006. Unexpected diversity of aryl hydrocarbon receptors in non-mammalian vertebrates: insights from comparative genomics. J. Exp. Zool. A: Comp. Exp. Biol. 305A, 693–706.
Haïssaguerre, M., Jais, P., Shah, D.C., Takahashi, A., Hocini, M., Quiniou, G., Garrigue, S., Le Mouroux, A., Le Metayer, P., Clementy, J., 1998. Spontaneous initiation of atrial fibrillation by ectopic beats originating in the pulmonary veins. N. Engl. J. Med. 339, 659–666.
Heideman, W., Antkiewicz, D.S., Carney, S.A., Peterson, R.E., 2005. Zebrafish and cardiac toxicology. Cardiovasc. Toxicol. 5, 203–214.
Heintz, R.A., 2007. Chronic exposure to polynuclear aromatic hydrocarbons in natal habitats leads to decreased equilibrium size, growth, and stability of pink salmon populations. Integr. Environ. Assess. Manage. 3, 351–363.
Heintz, R.A., Short, J.W., Rice, S.D., 1999. Sensitivity of fish embryos to weathered crude oil: part II. Increased mortality of pink salmon (*Oncorhynchus gorbuscha*) embryos incubating downstream from weathered Exxon Valdez crude oil. Environ. Toxicol. Chem. 18, 494–503.
Heintz, R.A., Rice, S.D., Wertheimer, A.C., Bradshaw, R.F., Thrower, F.P., Joyce, J.E., Short, J.W., 2000. Delayed effects on growth and marine survival of pink salmon *Oncorhynchus gorbuscha* after exposure to crude oil during embryonic development. Mar. Ecol. Prog. Ser. 208, 205–216.
Helder, T., 1980. Effects of 2,3,7,8-tetrachlorodibenzo-p-dioxin (TCDD) on early life stages of the pike (*Esox lucius* L.). Sci. Total Environ. 14, 255–264.
Helder, T., 1981. Effects of 2,3,7,8-tetrachlorodibenzo-dioxin (TCDD) on early life stages of rainbow trout (*Salmo gairdneri*, Richardson). Toxicology 19, 101–112.
Henry, T.R., Spitsbergen, J.M., Hornung, M.W., Abnet, C.C., Peterson, R.E., 1997. Early life stage toxicity of 2,3,7,8-tetrachlorodibenzo-p-dioxin in zebrafish (*Danio rerio*). Toxicol. Appl. Pharmacol. 142, 56–68.
Hicken, C.E., Linbo, T.L., Baldwin, D.H., Willis, M.L., Myers, M.S., Holland, L., Larsen, M., Stekoll, M.S., Rice, G.S., Collier, T.K., Scholz, N.L., Incardona, J.P., 2011. Sub-lethal exposure to crude oil during embryonic development alters cardiac morphology and reduces aerobic capacity in adult fish. Proc. Natl. Acad. Sci. U.S.A. 108, 7086–7090.
Hornung, M.W., Zabel, E.W., Peterson, R.E., 1996. Toxic equivalency factors of polybrominated dibenzo-p-dioxin, dibenzofuran, biphenyl, and polyhalogenated diphenyl ether congeners based on rainbow trout early life stage mortality. Toxicol. Appl. Pharmacol. 140, 227–234.
Hove, J.R., Koster, R.W., Forouhar, A.S., Acevedo-Bolton, G., Fraser, S.E., Gharib, M., 2003. Intracardiac fluid forces are an essential epigenetic factor for embryonic cardiogenesis. Nature 421, 172–177.
Hu, N., Sedmera, D., Yost, H.J., Clark, E.B., 2000. Structure and function of the developing zebrafish heart. Anat. Rec. 260, 148–157.
Huang, J., Elicker, J., Bowens, N., Liu, X., Cheng, L., Cappola, T.P., Zhu, X., Parmacek, M.S., 2012. Myocardin regulates BMP10 expression and is required for heart development. J. Clin. Invest. 122, 3678–3691.

Hughes, G.M., Adeney, R.J., 1977. The effects of zinc on the cardiac and ventilatory rhythms of rainbow trout (*Salmo Gairdneri*, Richardson) and their responses to environmental hypoxia. Water Res. 11, 1069–1077.

Icardo, J.M., 2017. Heart morphology and anatomy. In: Gamperl, A.K., Gillis, T.E., Farrell, A.P., Brauner, C.J. (Eds.), The Cardiovascular System: Morphology, Control and Function. Fish Physiology, vol. 36A. Elsevier, Amsterdam, pp. 1–54.

Incardona, J.P., Scholz, N.L., 2016. The influence of heart developmental anatomy on cardiotoxicity-based adverse outcome pathways in fish. Aquat. Toxicol. 177, 515–525.

Incardona, J.P., Collier, T.K., Scholz, N.L., 2004. Defects in cardiac function precede morphological abnormalities in fish embryos exposed to polycyclic aromatic hydrocarbons. Toxicol. Appl. Pharmacol. 196, 191–205.

Incardona, J.P., Carls, M.G., Teraoka, H., Sloan, C.A., Collier, T.K., Scholz, N.L., 2005. Aryl hydrocarbon receptor-independent toxicity of weathered crude oil during fish development. Environ. Health Perspect. 113, 1755–1762.

Incardona, J.P., Day, H.L., Collier, T.K., Scholz, N.L., 2006. Developmental toxicity of 4-ring polycyclic aromatic hydrocarbons in zebrafish is differentially dependent on AH receptor isoforms and hepatic cytochrome P450 1A metabolism. Toxicol. Appl. Pharmacol. 217, 308–321.

Incardona, J.P., Carls, M.G., Day, H.L., Sloan, C.A., Bolton, J.L., Collier, T.K., Scholz, N.L., 2009. Cardiac arrhythmia is the primary response of embryonic Pacific herring (*Clupea pallasi*) exposed to crude oil during weathering. Environ. Sci. Technol. 43, 201–207.

Incardona, J.P., Linbo, T.L., Scholz, N.L., 2011. Cardiac toxicity of 5-ring polycyclic aromatic hydrocarbons is differentially dependent on the aryl hydrocarbon receptor 2 isoform during zebrafish development. Toxicol. Appl. Pharmacol. 257, 242–249.

Incardona, J.P., Swarts, T.L., Edmunds, R.C., Linbo, T.L., Edmunds, R.C., Aquilina-Beck, A., Sloan, C.A., Gardner, L.D., Block, B.A., Scholz, N.L., 2013. Exxon Valdez to Deepwater Horizon: comparable toxicity of both crude oils to fish early life stages. Aquat. Toxicol. 142–143, 303–316.

Incardona, J.P., Gardner, L.D., Linbo, T.L., Brown, T.L., Esbaugh, A.J., Mager, E.M., Stieglitz, J.D., French, B.L., Labenia, J.S., Laetz, C.A., Tagal, M., Sloan, C.A., Elizur, A., Benetti, D.D., Grosell, M., Block, B.A., Scholz, N.L., 2014. Deepwater Horizon crude oil impacts the developing hearts of large predatory pelagic fish. Proc. Natl. Acad. Sci. U.S.A. 111, E1510–E1518.

Incardona, J.P., Carls, M.G., Holland, L., Linbo, T.L., Baldwin, D.H., Myers, M.S., Peck, K.A., Rice, S.D., Scholz, N.L., 2015. Very low embryonic crude oil exposures cause lasting cardiac defects in salmon and herring. Sci. Rep. 5, 13499

Jayasundara, N., Van Tiem Garner, L., Meyer, J.N., Erwin, K.N., Di Giulio, R.T., 2015. AHR2-Mediated transcriptomic responses underlying the synergistic cardiac developmental toxicity of PAHs. Toxicol. Sci. 143, 469–481.

Johnson, J.N., Ackerman, M.J., 2009. QTc: how long is too long? Br. J. Sports Med. 43, 657–662.

Jung, J.-H., Hicken, C.E., Boyd, D., Anulacion, B.F., Carls, M.G., Shim, W.J., Incardona, J.P., 2013. Geologically distinct crude oils cause a common cardiotoxicity syndrome in developing zebrafish. Chemosphere 91, 1146–1155.

Jung, J.H., Kim, M., Yim, U.H., Ha, S.Y., Shim, W.J., Chae, Y.S., Kim, H., Incardona, J.P., Linbo, T.L., Kwon, J.H., 2015. Differential toxicokinetics determines the sensitivity of two marine embryonic fish exposed to Iranian heavy crude oil. Environ. Sci. Technol. 49, 13639–13648.

Kim, C.H., Park, J.Y., Park, M.K., Kang, E.J., Kim, J.H., 2008. Minute tubercles on the skin surface of larvae in the Korean endemic bitterling, *Rhodeus pseudosericeus* (Pisces, Cyprinidae). J. Appl. Ichthyol. 24, 269–275.

Knape, K., Linder, T., Wolschann, P., Beyer, A., Stary-Weinzinger, A., 2011. In silico analysis of conformational changes induced by mutation of aromatic binding residues: consequences for drug binding in the hERG K+ channel. PLoS One 6, e28778.

Kramer, V.J., Etterson, M.A., Hecker, M., Murphy, C.A., Roesijadi, G., Spade, D.J., Spromberg, J.A., Wang, M., Ankley, G.T., 2011. Adverse outcome pathways and ecological risk assessment: bridging to population-level effects. Environ. Toxicol. Chem. 30, 64–76.

Langheinrich, U., 2003. Zebrafish: a new model on the pharmaceutical catwalk. Bioessays 25, 904–912.

Langheinrich, U., Vacun, G., Wagner, T., 2003. Zebrafish embryos express an orthologue of HERG and are sensitive toward a range of QT-prolonging drugs inducing severe arrhythmia. Toxicol. Appl. Pharmacol. 193, 370–382.

Lanham, K.A., Plavicki, J., Peterson, R.E., Heideman, W., 2014. Cardiac myocyte-specific AHR activation phenocopies TCDD-induced toxicity in zebrafish. Toxicol. Sci. 141, 141–154.

Le Guellec, D., Morvan-Dubois, G., Sire, J.Y., 2004. Skin development in bony fish with particular emphasis on collagen deposition in the dermis of the zebrafish (*Danio rerio*). Int. J. Dev. Biol. 48, 217–231.

Lee, D.-C., Artero, E.G., Sui, X., Blair, S.N., 2010. Mortality trends in the general population: the importance of cardiorespiratory fitness. J. Psychopharmacol. 24, 27–35.

Leoni, A.L., Marionneau, C., Demolombe, S., Le Bouter, S., Mangoni, M.E., Escande, D., Charpentier, F., 2005. Chronic heart rate reduction remodels ion channel transcripts in the mouse sinoatrial node but not in the ventricle. Physiol. Genomics 24, 4–12.

Li, Z., Mulholland, J.A., Romanoff, L.C., Pittman, E.N., Trinidad, D.A., Lewin, M.D., Sjodin, A., 2010. Assessment of non-occupational exposure to polycyclic aromatic hydrocarbons through personal air sampling and urinary biomonitoring. J. Environ. Monit. 12, 1110–1118.

Lichtner, B., Knaus, P., Lehrach, H., Adjaye, J., 2013. BMP10 as a potent inducer of trophoblast differentiation in human embryonic and induced pluripotent stem cells. Biomaterials 34, 9789–9802.

Lin, Y.-F., Swinburne, I., Yelon, D., 2012. Multiple influences of blood flow on cardiomyocyte hypertrophy in the embryonic zebrafish heart. Dev. Biol. 362, 242–253.

Link, M.S., Luttmann-Gibson, H., Schwartz, J., Mittleman, M.A., Wessler, B., Gold, D.R., Dockery, D.W., Laden, F., 2013. Acute exposure to air pollution triggers atrial fibrillation. J. Am. Coll. Cardiol. 62, 816–825.

Liu, J.D., Bressan, M., Hassel, D., Huisken, J., Staudt, D., Kikuchi, K., Poss, K.D., Mikawa, T., Stainier, D.Y.R., 2010. A dual role for ErbB2 signaling in cardiac trabeculation. Development 137, 3867–3875.

Ljubojevic, S., Bers, D.M., 2015. Nuclear calcium in cardiac myocytes. J. Cardiovasc. Pharmacol. 65, 211–217.

Lu, X., Reible, D.D., Fleeger, J.W., 2004. Bioavailability and assimilation of sediment-associated benzo[a]pyrene by *Ilyodrilus templetoni* (oligochaeta). Environ. Toxicol. Chem. 23, 57–64.

Lu, X., Reible, D.D., Fleeger, J.W., 2006. Bioavailability of polycyclic aromatic hydrocarbons in field-contaminated Anacostia River (Washington, DC) sediment. Environ. Toxicol. Chem. 25, 2869–2874.

Lunn, C.R., Toews, D.P., Pree, D.J., 1976. Effects of three pesticides on respiration, coughing, and heart rates of rainbow trout (*Salmo gairdneri* Richardson). Can. J. Zool. 54, 214–219.

MacDonald, J.S., Robertson, R.T., 2009. Toxicity testing in the 21st century: a view from the pharmaceutical industry. Toxicol. Sci. 110, 40–46.

Madison, B.N., Hodson, P.V., Langlois, V.S., 2015. Diluted bitumen causes deformities and molecular responses indicative of oxidative stress in Japanese medaka embryos. Aquat. Toxicol. 165, 222–230.

Mahler, B.J., Van Metre, P.C., Bashara, T.J., Wilson, J.T., Johns, D.A., 2005. Parking lot sealcoat: an unrecognized source of urban polycyclic aromatic hydrocarbons. Environ. Sci. Technol. 39, 5560–5566.

Majewski, H.S., Giles, M.A., 1981. Cardiovascular-respiratory responses of rainbow trout (Salmo gairdneri) during chronic exposure to sublethal concentrations of cadmium. Water Res. 15, 1211–1217.

Marty, G.D., Hose, J.E., McGurk, M.D., Brown, E.D., Hinton, D.E., 1997a. Histopathology and cytogenetic evaluation of Pacific herring larvae exposed to petroleum hydrocarbons in the laboratory or in Prince William Sound, Alaska, after the Exxon Valdez oil spill. Can. J. Fish. Aquat. Sci. 54, 1846–1857.

Marty, G.D., Short, J.W., Dambach, D.M., Willits, N.H., Heintz, R.A., Rice, S.D., Stegeman, J.J., Hinton, D.E., 1997b. Ascites, premature emergence, increased gonadal cell apoptosis, and cytochrome P4501A induction in pink salmon larvae continuously exposed to oil-contaminated gravel during development. Can. J. Zool. 75, 989–1007.

Mauduit, F., Domenici, P., Farrell, A.P., Lacroix, C., Le Floch, S., Lemaire, P., Nicolas-Kopec, A., Whittington, M., Zambonino-Infante, J.L., Claireaux, G., 2016. Assessing chronic fish health: an application to a case of an acute exposure to chemically treated crude oil. Aquat. Toxicol. 178, 197–208.

McGurk, M.D., Brown, E.D., 1996. Egg-larval mortality of Pacific herring in Prince William Sound, Alaska, after the *Exxon Valdez* oil spill. Can. J. Fish. Aquat. Sci. 53, 2343–2354.

McIntosh, S., King, T., Wu, D., Hodson, P.V., 2010. Toxicity of dispersed weathered crude oil to early life stages of Atlantic herring (Clupea harengus). Environ. Toxicol. Chem. 29, 1160–1167.

McIntyre, J.K., Davis, J.W., Incardona, J.P., Stark, J.D., Anulacion, B.F., Scholz, N.L., 2014. Zebrafish and clean water technology: assessing soil bioretention as a protective treatment for toxic urban runoff. Sci. Total Environ. 500, 173–180.

McIntyre, J.K., Davis, J.W., Hinman, C., Macneale, K.H., Anulacion, B.F., Scholz, N.L., Stark, J.D., 2015. Soil bioretention protects juvenile salmon and their prey from the toxic impacts of urban stormwater runoff. Chemosphere 132, 213–219.

McIntyre, J.K., Edmunds, R.C., Anulacion, B.F., Davis, J.W., Incardona, J.P., Stark, J.D., Scholz, N.L., 2016a. Severe coal tar sealcoat runoff toxicity to fish is prevented by bioretention filtration. Environ. Sci. Technol. 50, 1570–1578.

McIntyre, J.K., Edmunds, R.C., Redig, M.G., Mudrock, E.M., Davis, J.W., Incardona, J.P., Stark, J.D., Scholz, N.L., 2016b. Confirmation of stormwater bioretention treatment effectiveness using molecular indicators of cardiovascular toxicity in developing fish. Environ. Sci. Technol. 50, 1561–1569.

McKim, J.M., Bradbury, S.P., Niemi, G.J., 1987a. Fish acute toxicity syndromes and their use in the QSAR approach to hazard assessment. Environ. Health Perspect. 71, 171–186.

McKim, J.M., Schmieder, P.K., Carlson, R.W., Hunt, E.P., Niemi, G.J., 1987b. Use of respiratory-cardiovascular responses of rainbow trout (*Salmo gairdneri*) in identifying acute toxicity syndromes in fish: part 1. Pentachlorophenol, 2,4-dinitrophenol, tricaine methanesulfonate and 1-octanol. Environ. Toxicol. Chem. 6, 295–312.

McKim, J.M., Schmieder, P.K., Niemi, G.J., Carlson, R.W., Henry, T.R., 1987c. Use of respiratory-cardiovascular responses of rainbow trout (*Salmo gairdneri*) in identifying acute toxicity syndromes in fish: part 2. malathion, carbaryl, acrolein and benzaldehyde. Environ. Toxicol. Chem. 6, 313–328.

Milan, D.J., Peterson, T.A., Ruskin, J.N., Peterson, R.T., MacRae, C.A., 2003. Drugs that induce repolarization abnormalities cause bradycardia in zebrafish. Circulation 107, 1355–1358.

Mitcheson, J.S., 2008. hERG potassium channels and the structural basis of drug-induced arrhythmias. Chem. Res. Toxicol. 21, 1005–1010.

Monrad, M., Sajadieh, A., Christensen, J.S., Ketzel, M., Raaschou-Nielsen, O., Tjonneland, A., Overvad, K., Loft, S., Sorensen, M., 2017. Long-term exposure to traffic-related air pollution and risk of incident atrial fibrillation: a cohort study. Environ. Health Perspect. 125, 422–427.

Morris, J., Baldwin, D. H., Edmunds, R. C., Forth, H., French, B. L., Gill, J. A., Krasnec, M., Labenia, J. S., Linbo, T. L., Takeshita, R., Scholz, N. L. and Incardona, J. P. (2017) Deepwater horizon crude oil toxicity to red drum early life stages is independent of dispersion energy. submitted.

Mu, J.L., Jin, F., Ma, X.D., Lin, Z.S., Wang, J.Y., 2014. Comparative effects of biological and chemical dispersants on the bioavailability and toxicity of crude oil to early life stages of marine medaka (*Oryzias melastigma*). Environ. Toxicol. Chem. 33, 2576–2583.

Murphy, M.L., Heintz, R.A., Short, J.W., Larsen, M.L., Rice, S.D., 1999. Recovery of pink salmon spawning areas after the Exxon Valdez oil spill. Trans. Am. Fish. Soc. 128, 909–918.

Nacci, D.E., Champlin, D., Jayaraman, S., 2010. Adaptation of the estuarine fish *Fundulus heteroclitus* (Atlantic killifish) to polychlorinated biphenyls (PCBs). Estuar. Coasts 33, 853–864.

Nakamura, O., Saeki, M., Kamiya, H., Muramoto, K., Watanabe, T., 2002. Development of epidermal and mucosal galectin containing cells in metamorphosing leptocephali of Japanese conger. J. Fish Biol. 61, 822–833.

Oleksiak, M.F., Karchner, S.I., Jenny, M.J., Franks, D.G., Welch, D.B., Hahn, M.E., 2011. Transcriptomic assessment of resistance to effects of an aryl hydrocarbon receptor (AHR) agonist in embryos of Atlantic killifish (*Fundulus heteroclitus*) from a marine superfund site. BMC Genomics 12, 263.

Ortiz-Delgado, J.B., Segner, H., Sarasquete, C., 2005. Cellular distribution and induction of CYP1A following exposure of gilthead seabream, *Sparus aurata*, to waterborne and dietary benzo(a)pyrene and 2,3,7,8-tetrachlorodibenzo-p-dioxin: an immunohistochemical approach. Aquat. Toxicol. 75, 144–161.

Ottesen, O.H., Olafsen, J.A., 1997. Ontogenetic development and composition of the mucous cells and the occurrence of saccular cells in the epidermis of Atlantic halibut. J. Fish Biol. 50, 620–633.

Paul, M.J., Meyer, J.L., 2001. Streams in the urban landscape. Annu. Rev. Ecol. Syst. 32, 333–365.

Petersen, G.I., Kristensen, P., 1998. Bioaccumulation of lipophilic substances in fish early life stages. Environ. Toxicol. Chem. 17, 1385–1395.

Peterson, C.H., Rice, S.D., Short, J.W., Esler, D., Bodkin, J.L., Ballachey, B.E., Irons, D.B., 2003. Long-term ecosystem response to the Exxon Valdez oil spill. Science 302, 2082–2086.

Plaut, I., Gordon, M.S., 1994. Swimming metabolism of wild-type and cloned zebrafish *Brachydanio rerio*. J. Exp. Biol. 194, 209–223.

Poland, A., Knutson, J.C., 1982. 2,3,7,8-Tetrachlorodibenzo-p-dioxin and related halogenated aromatic hydrocarbons: examination of the mechanism of toxicity. Annu. Rev. Cell Dev. Biol. 22, 517–554.

Pollino, C.A., Holdway, D.A., 2002. Toxicity testing of crude oil and related compounds using early life stages of the crimson-spotted rainbow fish (*Melanotaenia fluviatilis*). Ecotoxicol. Environ. Saf. 52, 180–189.

Poppe, T.T., Johansen, R., Gunnes, G., Tørud, B., 2003. Heart morphology in wild and farmed Atlantic salmon *Salmo salar* and rainbow trout *Oncorhynchus mykiss*. Dis. Aquat. Organ. 57, 103–108.

Poupa, O., Gesser, H., Jonsson, S., Sullivan, L., 1974. Coronary-supplied compact shell of ventricular myocardium in salmonids—growth and enzyme pattern. Comp. Biochem. Physiol. 48, 85–95.

Prasch, A.L., Teraoka, H., Carney, S.A., Dong, W., Hiraga, T., Stegeman, J.J., Heideman, W., Peterson, R.E., 2003. Aryl hydrocarbon receptor 2 mediates 2,3,7,8-tetrachlorodibenzo-p-dioxin developmental toxicity in zebrafish. Toxicol. Sci. 76, 138–150.

Prince, R., Cooper, K., 1989. Differential embryo sensitivity to 2, 3, 7, 8-tetrachlorodibenzo-p-dioxin (TCDD) in *Fundulus heteroclitus*. The Toxicologist: Supplement to Toxicol. Sci. 9, abstract 169.

Prince, R., Cooper, K., 1990. Biological effects in and deposition to eggs produced by female Japanese medaka (*Oryzias latipes*) exposed to 2, 3, 7, 8-tetrachlorodibenzo-p-dioxin (TCDD). The Toxicologist: Supplement to Toxicol. Sci. 10, abstract 1255.

Rasooly, R.S., Henken, D., Freeman, N., Tompkins, L., Badman, D., Briggs, J., Hewitt, A.T., 2003. Genetic and genomic tools for zebrafish research: the NIH zebrafish initiative. Dev. Dyn. 228, 490–496.

Reid, N.M., Proestou, D.A., Clark, B.W., Warren, W.C., Colbourne, J.K., Shaw, J.R., Karchner, S.I.I., Hahn, M.E., Nacci, D., Oleksiak, M.F., Crawford, D.L., Whitehead, A., 2016. The genomic landscape of rapid repeated evolutionary adaptation to toxic pollution in wild fish. Science 354, 1305–1308.

Reitzel, A.M., Karchner, S.I., Franks, D.G., Evans, B.R., Nacci, D., Champlin, D., Vieira, V.M., Hahn, M.E., 2014. Genetic variation at aryl hydrocarbon receptor (AHR) loci in populations of Atlantic killifish (*Fundulus heteroclitus*) inhabiting polluted and reference habitats. BMC Evol. Biol. 14, 6.

Rennekamp, A.J., Peterson, R.T., 2015. 15 years of zebrafish chemical screening. Curr. Opin. Chem. Biol. 24, 58–70.

Rice, S.D., Moles, A., Taylor, T.L., Karinen, J.F., 1979. In: Sensitivity of 39 Alaskan marine species to Cook Inlet crude oil and no. 2 fuel oil. 1979 Oil Spill Conference (Prevention, Behavior, Control, Cleanup). American Petroleum Institute, Los Angeles, CA, pp. 549–554.

Rice, S.D., Thomas, R.E., Carls, M.G., Heintz, R.A., Wertheimer, A.C., Murphy, M.L., Short, J.W., Moles, A., 2001. Impacts to pink salmon following the Exxon Valdez oil spill: persistence, toxicity, sensitivity, and controversy. Rev. Fish. Sci. 9, 165–211.

Ritchie, T.J., Macdonald, S.J., 2009. The impact of aromatic ring count on compound developability—are too many aromatic rings a liability in drug design? Drug Discov. Today 14, 1011–1020.

Rogge, W.F., Ondov, J.M., Bernardo-Bricker, A., Sevimoglu, O., 2011. Baltimore PM2.5 Supersite: highly time-resolved organic compounds—sampling duration and phase distribution—implications for health effects studies. Anal. Bioanal. Chem. 401, 3069–3082.

Rosati, B., McKinnon, D., 2004. Regulation of ion channel expression. Circ. Res. 94, 874–883.

Rottbauer, W., Baker, K., Wo, Z.G., Mohideen, M.A., Cantiello, H.F., Fishman, M.C., 2001. Growth and function of the embryonic heart depend upon the cardiac-specific L-type calcium channel alpha1 subunit. Dev. Cell 1, 265–275.

Saadatfar, Z., Shahsavani, D., Fatemi, F.S., 2010. Study of epidermis development in sturgeon (*Acipenser persicus*) larvae. Anat. Histol. Embryol. 39, 440–445.

Safe, S., 1990. Polychlorinated biphenyls (PCBs), dibenzo-p-dioxins (PCDDs), dibenzofurans (PCDFs), and related compounds: environmental and mechanistic considerations which support the development of toxic equivalency factors (TEFs). Crit. Rev. Toxicol. 21, 51–88.

Salama, G., London, B., 2007. Mouse models of long QT syndrome. J. Physiol. 578, 43–53.

Samsa, L.A., Givens, C., Tzima, E., Stainier, D.Y., Qian, L., Liu, J., 2015. Cardiac contraction activates endocardial notch signaling to modulate chamber maturation in zebrafish. Development 142, 4080–4091.

Santer, R.M., Greer Walker, M., 1980. Morphological studies on the ventricle of teleost and elasmobranch hearts. J. Zool. 190, 259–272.

Santer, R.M., Greer Walker, M., Emerson, L., Witthames, P.R., 1983. On the morphology of the heart ventricle in marine teleost fish (teleostei). Comp. Biochem. Physiol. A Physiol. 76, 453–457.

Sargent, J.R., Bell, M.V., Kelly, K.F., Johnstone, A., Hawkins, A., Hall, C.D., 1980. Sodium orthovanadate and cardiac activity in teleost fish. Comp. Biochem. Physiol. C 66, 111–114.

Sawada, Y., Kato, K., Okada, T., Kurata, M., Mukai, Y., Miyashita, S., Murata, O., Kumai, H., 1999. Growth and morphological development of larval and juvenile *Epinephelus bruneus* (Perciformes:Serranidae). Ichthyol. Res. 46, 245–257.

Scholz, N.L., Incardona, J.P., 2015. In response: scaling polycyclic aromatic hydrocarbon toxicity to fish early life stages: a governmental perspective. Environ. Toxicol. Chem. 34, 459–461.

Scholz, N.L., McIntyre, J.K., 2015. Chemical pollution. In: Closs, G.P., Krkosek, M., Olden, J.D. (Eds.), Conservation of Freshwater Fishes. Cambridge University Press, Cambridge, UK, pp. 149–178.

Scholz, N.L., Myers, M.S., McCarthy, S.G., Labenia, J.S., McIntyre, J.K., Ylitalo, G.M., Rhodes, L.D., Laetz, C.A., Stehr, C.M., French, B.L., McMillan, B., Wilson, D., Reed, L., Lynch, K.D., Damm, S., Davis, J.W., Collier, T.K., 2011. Recurrent die-offs of adult coho salmon returning to spawn in puget sound lowland urban streams. PLoS One 6, e28013.

Scott, J.A., Incardona, J.P., Pelkki, K., Shepardson, S., Hodson, P.V., 2011. AhR2-mediated; CYP1A-independent cardiovascular toxicity in zebrafish (*Danio rerio*) embryos exposed to retene. Aquat. Toxicol. 101, 165–174.

Shiels, H.A., 2017. Cardiomyocyte morphology and physiology. In: Gamperl, A.K., Gillis, T.E., Farrell, A.P., Brauner, C.J. (Eds.), The Cardiovascular System: Morphology, Control and Function. Fish Physiology, vol. 36A. Elsevier, Amsterdam, pp. 55–98.

Short, J.W., Harris, P.M., 1996. Chemical sampling and analysis of petroleum hydrocarbons in near-surface seawater of Prince William Sound after the Exxon Valdez oil spill. Am. Fish. Soc. Symp. 18, 17–28.

Short, J.W., Heintz, R.A., 1997. Identification of *Exxon Valdez* oil in sediments and tissues from Prince William Sound and the Northwestern Gulf of Alaska based on a PAH weathering model. Environ. Sci. Technol. 31, 2375–2384.

Singleman, C., Holtzman, N.G., 2012. Analysis of postembryonic heart development and maturation in the zebrafish, *Danio rerio*. Dev. Dyn. 241, 1993–2004.

Sørensen, L., Sørhus, E., Nordtug, T., Incardona, J.P., Linbo, T.L., Giovanetti, L., Karlsen, O., Meier, S., 2017. Oil droplet fouling and differential PAH toxicokinetics in embryos of Atlantic haddock and cod. PLoS One 1, e0180048.

Sørhus, E., Edvardsen, R.B., Karlsen, O., Nordtug, T., van der Meeren, T., Thorsen, A., Harman, C., Jentoft, S., Meier, S., 2015. Unexpected interaction with dispersed crude oil droplets drives severe toxicity in Atlantic haddock embryos. PLoS One 10, e0124376.

Sørhus, E., Incardona, J.P., Karlsen, Ø., Linbo, T.L., Sørensen, L., Nordtug, T., van der Meeren, T., Thorsen, A., Thorbjørnsen, M., Jentoft, S., Edvardsen, R.B., Meier, S., 2016. Effects of crude oil on haddock reveal roles for intracellular calcium in craniofacial and cardiac development. Sci. Rep. 6, 31058.

Sørhus, E., Incardona, J.P., Furmanek, T., Goetz, G.W., Scholz, N.L., Meier, S., Edvardsen, R.B., Jentoft, S., 2017. Novel adverse outcome pathways revealed by chemical genetics in a developing marine fish. eLife 6, e20707.

Spitsbergen, J.M., Walker, M.K., Olson, J.R., Peterson, R.E., 1991. Pathologic Alterations in Early Life Stages of lake trout *Salvelinus namaycush* exposed to 2,3,7,8-tetrachlorodibenzo-p-dioxin as fertilized eggs. Aquat. Toxicol. 19, 41–72.

Spromberg, J.A., Baldwin, D.H., Damm, S.E., McIntyre, J.K., Huff, M., Sloan, C.A., Anulacion, B.F., Davis, J.W., Scholz, N.L., 2016. Coho salmon spawner mortality in western US urban watersheds: bioinfiltration prevents lethal storm water impacts. J. Appl. Ecol. 53, 398–407.

Stahl, F.W., 1995. George Streisinger—December 27, 1927–September 5, 1984. Biogr. Mem. Natl. Acad. Sci. 68, 353–361.

Staudt, D., Stainier, D., 2012. Uncovering the molecular and cellular mechanisms of heart development using the zebrafish. Annu. Rev. Genet. 46, 397–418.

Stegeman, J.J., Lech, J.J., 1991. Cytochrome P-450 monooxygenase systems in aquatic species: carcinogen metabolism and biomarkers for carcinogen and pollutant exposure. Environ. Health Perspect. 90, 101–109.

Symula, J., Meade, J., Skea, J.C., Cummings, L., Colquhoun, J.R., Dean, H.J., Miccoli, J., 1990. Blue-sac disease in Lake Ontario lake trout. J. Great Lakes Res. 16, 41–52.

Tamargo, J., Caballero, R., Gomez, R., Valenzuela, C., Delpon, E., 2004. Pharmacology of cardiac potassium channels. Cardiovasc. Res. 62, 9–33.

Teraoka, H., Dong, W., Tsujimoto, Y., Iwasa, H., Endoh, D., Ueno, N., Stegeman, J.J., Peterson, R.E., Hiraga, T., 2003. Induction of cytochrome P450 1A is required for circulation failure and edema by 2,3,7,8-tetrachlorodibenzo-p-dioxin in zebrafish. Biochem. Biophys. Res. Commun. 304, 223–228.

Tran, M.K., Kurakula, K., Koenis, D.S., de Vries, C.J., 2016. Protein–protein interactions of the LIM-only protein FHL2 and functional implication of the interactions relevant in cardiovascular disease. Biochim. Biophys. Acta 1863, 219–228.

Tu, C.T., Yang, T.C., Tsai, H.J., 2009. Nkx2.7 and Nkx2.5 function redundantly and are required for cardiac morphogenesis of zebrafish embryos. PLoS One 4, e4249.

Van Bohemen, H.D., Van de Laak, W.H.J., 2003. The influence of road infrastructure and traffic on soil, water, and air quality. Environ. Manage. 31, 50–68.

van der Bom, T., Zomer, A.C., Zwinderman, A.H., Meijboom, F.J., Bouma, B.J., Mulder, B.J., 2011. The changing epidemiology of congenital heart disease. Nat. Rev. Cardiol. 8, 50–60.

Van Metre, P.C., Mahler, B.J., 2003. The contribution of particles washed from rooftops to contaminant loading to urban streams. Chemosphere 52, 1727–1741.

Van Metre, P.C., Mahler, B.J., Furlong, E.T., 2000. Urban sprawl leaves its PAH signature. Environ. Sci. Technol. 34, 4064–4070.

Van Tiem, L.A., Di Giulio, R.T., 2011. AHR2 knockdown prevents PAH-mediated cardiac toxicity and XRE- and ARE-associated gene induction in zebrafish (*Danio rerio*). Toxicol. Appl. Pharmacol. 254, 280–287.

Van Veld, P.A., Westbrook, D.J., 1995. Evidence for depression of cytochrome P4501A in a population of chemically resistant mummichog (*Fundulus heteroclitus*). Environ. Sci. 3, 221–234.

Van Veld, P.A., Vogelbein, W.K., Cochran, M.K., Goksoyr, A., Stegeman, J.J., 1997. Route-specific cellular expression of cytochrome P4501A (CYP1A) in fish (*Fundulus heteroclitus*) following exposure to aqueous and dietary benzo[a]pyrene. Toxicol. Appl. Pharmacol. 142, 348–359.

van Wijk, B., Moorman, A.F., van den Hoff, M.J., 2007. Role of bone morphogenetic proteins in cardiac differentiation. Cardiovasc. Res. 74, 244–255.

Villeneuve, D.L., Crump, D., Garcia-Reyero, N., Hecker, M., Hutchinson, T.H., LaLone, C.A., Landesmann, B., Lettieri, T., Munn, S., Nepelska, M., Ottinger, M.A., Vergauwen, L., Whelan, M., 2014. Adverse outcome pathway (AOP) development I: strategies and principles. Toxicol. Sci. 142, 312–320.

Vornanen, M., 2017. Electrical excitability of the fish heart and its autonomic regulation. In: Gamperl, A.K., Gillis, T.E., Farrell, A.P., Brauner, C.J. (Eds.), The Cardiovascular System: Morphology, Control and Function. Fish Physiology, vol. 36A. Elsevier, Amsterdam, pp. 99–153.

Waits, E.R., Nebert, D.W., 2011. Genetic architecture of susceptibility to PCB126-induced developmental cardiotoxicity in zebrafish. Toxicol. Sci. 122, 466–475.

Walker, M.K., Peterson, R.E., 1991. Potencies of polychlorinated dibenzo-p-dioxin, dibenzofuran, and biphenyl congeners, relative to 2,3,7,8-tetrachlorodibenzo-p-dioxin, for producing early life stage mortality in rainbow trout (*Oncorhynchus mykiss*). Aquat. Toxicol. 21, 219–237.

Walker, M.K., Spitsbergen, J.M., Olson, J.R., Peterson, R.E., 1991. 2,3,7,8-Tetrachlorodibenzo-p-dioxin (TCDD) toxicity during early life stage development of lake trout (*Salvelinus namaycush*). Can. J. Fish. Aquat. Sci. 48, 875–883.

Walker, M.K., Cook, P.M., Batterman, A.R., Butterworth, B.C., Berini, C., Libal, J.J., Hufnagle, L.C., Peterson, R.E., 1994. Translocation of 2,3,7,8-Tetrachlorodibenzo-p-dioxin from adult female lake trout (*Salvelinus namaycush*) to oocytes: effects on early life stage development and Sac fry survival. Can. J. Fish. Aquat. Sci. 51, 1410–1419.

Wang, Z., Stout, S.A., Fingas, M., 2006. Forensic fingerprinting of biomarkers for oil spill characterization and source identification. Environ. Forensic 7, 105–146.

Wertheimer, A.C., Celewycz, A.G., 1996. Abundance and growth of juvenile pink salmon in oiled and non-oiled locations of western Prince William Sound after the Exxon Valdez oil spill. Am. Fish. Soc. Symp. 18, 518–532.

West, J.E., O'Neill, S.M., Ylitalo, G.M., Incardona, J.P., Doty, D.C., Dutch, M.E., 2014. An evaluation of background levels and sources of polycyclic aromatic hydrocarbons in naturally spawned embryos of Pacific herring (*Clupea pallasii*) from Puget Sound, Washington, USA. Sci. Total Environ. 499, 114–124.

Whitehead, A., Pilcher, W., Champlin, D., Nacci, D., 2011. Common mechanism underlies repeated evolution of extreme pollution tolerance. Proc. Biol. Sci. 279, 427–433.

Wiedmer, M., Fink, M.J., Stegeman, J.J., Smolowitz, R., Marty, G.D., Hinton, D.E., 1996. Cytochrome P-450 induction and histopathology in preemergent pink salmon from oiled spawning sites in Prince William Sound. Am. Fish. Soc. Symp. 18, 509–517.

Wilke, R.A., Lin, D.W., Roden, D.M., Watkins, P.B., Flockhart, D., Zineh, I., Giacomini, K.M., Krauss, R.M., 2007. Identifying genetic risk factors for serious adverse drug reactions: current progress and challenges. Nat. Rev. Drug Discov. 6, 904–916.

Willette, M., 1996. Impacts of the Exxon Valdez oil spill on the migration, growth, and survival of juvenile pink salmon in Prince William Sound. Am. Fish. Soc. Symp. 18, 533–550.

Winslow, R.L., Walker, M.A., Greenstein, J.L., 2016. Modeling calcium regulation of contraction, energetics, signaling, and transcription in the cardiac myocyte. WIREs Syst. Biol. Med. 8, 37–67.

Wirgin, I., Roy, N.K., Loftus, M., Chambers, R.C., Franks, D.G., Hahn, M.E., 2011. Mechanistic basis of resistance to PCBs in Atlantic tomcod from the Hudson River. Science 331, 1322–1325.

Wolf, K., 1954. Progress report on blue-sac disease. Progress. Fish Cult. 16, 51–59.

Wolf, K., 1957. Experimental induction of Blue-Sac disease. Trans. Am. Fish. Soc. 86, 61–70.

Wright, P.J., Tillitt, D.E., 1999. Embryotoxicity of Great Lakes lake trout extracts to developing rainbow trout. Aquat. Toxicol. 47, 77–92.

Wu, X., Bers, D.M., 2006. Sarcoplasmic reticulum and nuclear envelope are one highly interconnected Ca^{2+} store throughout cardiac myocyte. Circ. Res. 99, 283–291.

Xu, E.G., Mager, E.M., Grosell, M., Pasparakis, C., Schlenker, L.S., Stieglitz, J.D., Benetti, D., Hazard, E.S., Courtney, S.M., Diamante, G., Freitas, J., Hardiman, G., Schlenk, D., 2016. Time- and oil-dependent transcriptomic and physiological responses to deepwater horizon oil in mahi-mahi (*Coryphaena hippurus*) embryos and larvae. Environ. Sci. Technol. 50, 7842–7851.

Yu, Y., Xu, J., Wang, P., Sun, H.W., Dai, S.G., 2009. Sediment-porewater partition of polycyclic aromatic hydrocarbons (PAHs) from Lanzhou Reach of Yellow River, China. J. Hazard. Mater. 165, 494–500.

Yuan, Z., Courtenay, S., Chambers, R.C., Wirgin, I., 2006. Evidence of spatially extensive resistance to PCBs in an anadromous fish of the Hudson River. Environ. Health Perspect. 114, 77–84.

7

CARDIOVASCULAR EFFECTS OF DISEASE: PARASITES AND PATHOGENS

MARK D. POWELL[*,†,1]
MUHAMMAD N. YOUSAF[‡]

[*]Institute of Marine Research, Bergen, Norway
[†]Institute for Biology, University of Bergen, Bergen, Norway
[‡]Norwegian Veterinary Institute, Harstad, Norway
[1]Corresponding author: mark.powell@niva.no

Cardiovascular diseases can be divided into those of non-infectious and infectious etiology. The non-infectious diseases include morphological anomalies, and diet/nutritional and toxicological responses. Infectious diseases include those caused by pathogens. Of particular interest are morphological anomalies or pathological conditions that result in cardiac dysfunction, although the impacts of diseases on cardiac physiology have rarely been examined. For example, there are several conditions that result in hyperplasia or hypertrophy of the myocardium (i.e., an increase in cardiac mass), but not

The Cardiovascular System: Development, Plasticity and Physiological Responses, Volume 36B
FISH PHYSIOLOGY

DOI: https://doi.org/10.1016/bs.fp.2017.09.007

always an increase in cardiac output ($\dot{Q}$). Inflammation of the heart, primarily within the myocardium, occurs in response to infectious agents. Viral diseases such as infectious salmon anemia (ISA), heart and skeletal muscle inflammation (HSMI), cardiomyopathy syndrome (CMS), and pancreas disease (PD) all lead to myocardial inflammation and/or necrosis. The effects of inflammation on cardiac function are unclear, although an increase in nitric oxide synthase (NOS) signaling appears to be involved. Parasitic diseases, where fibrosis or myocardial necrosis can interfere with myocardial mechanics, appear to reduce stroke volume (V_S) (perhaps by limiting end-diastolic volume), and thus, maximal cardiac output ($\dot{Q}_{max}$) and swimming stamina. Alternatively, partial occlusion of the afferent branchial arteries or ventral aorta can directly restrict blood flow. Other pathogens may exert indirect effects on cardiac function, for example, by increasing systemic vascular resistance; the result a compensatory increase in the amount of compact myocardium and remodeling of the ventricle into a more pyramidal shape. In general, there are four main categories of physiological response to pathogens/disease: (1) the induction of inflammatory responses; (2) the upregulation of iNOS and vasodilatory processes; (3) biomechanical alterations that lead to impairment of the myocardium or coronary endothelium; and (4) myocardial hyperplasia and hypertrophy that result from (3).

1. INTRODUCTION

Relating pathology to physiology is surprisingly challenging. Although there is extensive knowledge of physiological function and how various parameters impact it, our understanding of the consequences of physiological disturbances caused by pathology in fishes is limited. One reason for this is that the physiological disturbances one might predict based on pathological examinations are not always borne out. This raises the issue of how much pathology must there be to cause physiological dysfunction? The study of fish pathophysiology has been increasing in recent years and has primarily focused on fish species of commercial importance. This work has yielded some interesting insights with regard to the physiological disturbances associated with various pathogens and diseases. In particular, there has been considerable work on the effects of disease and disorders on fish cardiac physiology. This chapter summarizes recent findings in this area, and highlights some of the significant cardiac diseases in fish and how they may affect cardiovascular physiology.

2. NON-INFECTIOUS DISEASES AND DISORDERS

2.1. Morphological Anomalies

Structural and morphological cardiac anomalies are often reported as singular/curious findings, and it is often not clear if these conditions occur under both natural conditions and in intensive fish culture. Commonly, the shape of the ventricle ranges from pyramidal (e.g., in salmonids) to having a more rounded apex that creates a "sac-like" shape (e.g., in flounder). However, there are several examples of cardiac structural anomalies (Fig. 1) and evidence that these conditions impact cardiac function. For example, work by Poppe and Ferguson (2006) suggested that cardiac function may be impeded in fish with *situs invertus*, a condition where the position of the heart is reversed within the pericardial cavity. In addition, aplasia or hypoplasia (Poppe and Taksdal, 2000) of the *septum transversum* where the caudal constraining wall of the pericardium is absent or incomplete, and the ventricle protrudes into the peritoneum and is surrounded by the liver (Fig. 1D), may result from high egg

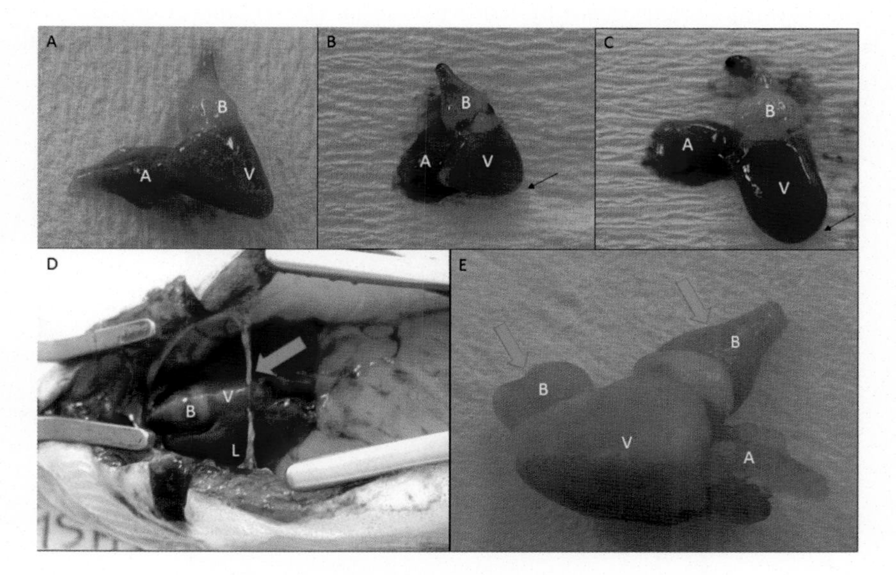

Fig. 1. Heart shape variation in Atlantic salmon: (A) normal, pyramidal, shaped ventricle; (B) ventricle with a rounded apex (see *arrow*); (C) sac-like heart with loss of pyramidal shape and rounded ventricle (see *arrow*); (D) *septum transversum* with the ventricle (V) of the heart protruding caudally and the liver (L) protruding cranially; and (E) heart from a triploid Atlantic salmon with two bulbi (*large arrows*). A, atrium; B, bulbus arteriosus; and V, ventricle.

incubation temperatures (>9°C) (Poppe and Ferguson, 2006). Dissecting hemorrhage (Poppe and Tørud, 2009) has also been described, where hemorrhage occurs within the compact myocardium and extends into the trabecular myocardial region of the heart. The epidemiology of such conditions has not been extensively investigated. Finally, one to two hyaline cartilage foci have recently been reported in the bulbus arteriosus of diseased and non-diseased Atlantic salmon (*Salmo salar*) hearts. The source of this cartilage tissue and its functional significance remain to be determined. This cartilage tissue in the bulbus arteriosus is not normal, and may have negative effects on Atlantic salmon heart function (Yousaf and Poppe, 2017) (Fig. 2). However, because of the lack of consistency by which many of these anomalies occur, empirical studies of abnormal morphologies and their specific functional consequences have not been undertaken (Gamperl and Farrell, 2004).

2.2. Coronary Arteriosclerosis

There have been considerable advances in our understanding of the process of arteriosclerosis in fish since its first observation in mature Chinook salmon (*Oncorhynchus tshawytscha*) by Robertson et al. (1961), and arteriosclerotic lesions have been extensively described in a variety of salmonids under several

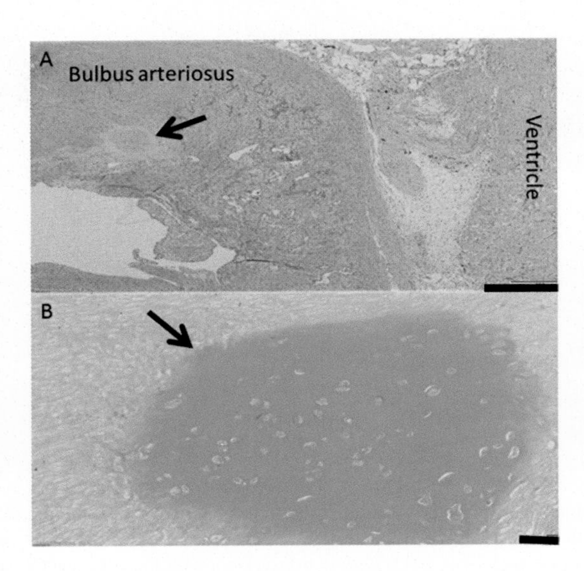

Fig. 2. Cartilagenous tissue (foci) in the Atlantic salmon heart. (A). Hyaline cartilage (*arrow*) identified in endocardial ridges of the distal bulbus arteriosus. (B) Alcian blue stain-identified acidic sulfated glycosaminoglycans (*arrow*) as extracellular matrix.

conditions (for a review, see Iwama and Farrell, 1998). Arteriosclerotic lesions form primarily in the coronary arteries of *Salmo* and *Oncorhynchus* species, with few reports in other fish taxa, and become progressively more extensive with age. These lesions consist of proliferations of the vascular smooth muscle within the coronary intima that result in a disrupted elastic lamina (Dalum et al., 2016; Iwama and Farrell, 1998; Seierstad et al., 2008). However, unlike in mammals, there is a marked absence of fat and calcium deposits in fish coronary lesions (Iwama and Farrell, 1998). Coronary arteriosclerosis has been reported to occupy 66%–80% of the length of the coronary artery and to occlude the arterial lumen by 10%–50% (Iwama and Farrell, 1998). It is believed that the etiology of the sclerotic lesion may be linked to natural stressors, including feeding and predator avoidance. These lead to a generalized hypertension, which causes coronary vascular injury as a consequence of over distension of the bulbus arteriosus. Similarly, factors such as sexual maturation, diet and growth rate, and the process of vascular repair itself, are all associated risk factors for lesion progression (Iwama and Farrell, 1998).

The consequences of this occlusion on the heart are unclear as the effects of prolonged ischemia on the compact myocardium are not fully understood. Although this muscle layer is only found in approx. 30% of fish species and is generally a relatively small proportion of the total ventricular mass when it is present (~5% to ~60%: Chapter 4, Volume 36A: Farrell and Smith, 2017), it appears to be very important for cardiac pressure generation, and thus, cardiac function in salmonids/fishes. Although ablation of the coronary artery does not result in mortality of rainbow trout (*Oncorhynchus mykiss*) or Chinook salmon (*O. tshawytscha*), it reduces their maximum swimming capacity during normoxia by 70%–80% (Farrell et al., 1990; Farrell and Steffensen, 1987). Coronary blood flow increases when salmonids are exposed to hypoxia and increases in current velocity (Gamperl et al., 1995), and Steffensen and Farrell (1998) showed that trout with ablated coronary arteries were unable to generate normal increases in ventral aortic pressure (P_{VA}) when swimming in hypoxic water. Finally, an increase in heart rate (f_H) was observed as a compensatory mechanism for maintaining cardiac contractility and $\dot{Q}$ when coronary-ablated rainbow trout were exposed to elevated temperatures (Ekström et al., 2017). Thus, coronary arteriosclerosis may represent a significant limit to cardiac and exercise performance, and potentially threaten the survival of highly aerobic species with compact myocardium.

2.3. Nutritional Effects

Cultured fish are generally at low at risk for nutritional deficiencies that may lead to cardiac pathology. However, where this does occur, the ramifications with regard to cardiac morphology and pathology are significant. Salmonids and turbot (*Scophthalmus maximus*) are at risk for vitamin E and selenium deficiencies (Fjølstad and Hyeraas, 1985; Poppe et al., 1986). These deficiencies have been associated with eosinophilic lesions, the loss of myocardial striation, the hypertrophy of existing myocytes and hyperplasia of the endocardium, and myopathy of the non-trabecular spongiosa following rectification of the deficiency (Ferguson, 1989).

Dietary polyunsaturated fatty acids (PUFA), in particular eicosapentanoic acid (EPA), arachidonic acid (AA) and eicosatetranoic acid (ETA), have mitogenic activities in fish cardiac tissue at concentrations greater than $80\,\mu M\,g^{-1}$ tissue. However, the efficacy- and dose-dependent effects of these fatty acids on cardiac growth are variable. For example, at low concentrations ($20\,\mu M\,g^{-1}$ tissue), AA can be a potent mitogen, whereas EPA is ineffective, and ETA can inhibit mitogenesis (Gong et al., 1997 as cited in Iwama and Farrell, 1998). Although the significance of this is unclear, deficiencies in ω-3 PUFA (this group includes EPA, ETA and AA) result in significantly reduced heart size in post-smolt Atlantic salmon, and atrophy of the compact and trabecular (spongy) myocardium that results in a thinning of the ventricular wall and an associated muscular necrosis (Bell et al., 1991). It is difficult to determine how these changes in cardiac morphology might affect heart function (i.e., V_S and $\dot{Q}$) due to possible effects on both cardiac filling (i.e., end-diastolic volume) and ejection (i.e., end-systolic volume). However, a negative impact on cardiac function is very likely since the fish in Bell et al. (1991) experienced high rates of mortality (30%) when subjected to transportation stress.

Modern commercial diets for salmonids, driven by the need for the aquaculture industry to decrease feed costs and its reliance on fish meal and oil, have seen an increasing inclusion of plant-derived lipids. The high proportions of plant lipids (e.g., up to 50% of total oil) have been associated with cardiac hypertrophy, but no apparent patterns with regard to inflammation or immunological responses (Seierstad et al., 2005). Furthermore, studies show that the incorporation of significant proportions of plant lipids into Atlantic salmon diets has little impact on heart morphology, or inflammatory or immunological gene expression, as compared with wild-caught fish (Dalum et al., 2016). Interestingly, although the type of lipid in the diet had no effect on the prevalence of coronary artery lesions, arteriosclerotic plaques did form at the major bifurcation of the coronary artery at the ventricular apex (Seierstad et al., 2008). These plaques were described as being of putative myointimal smooth muscle origin (using immunohistochemistry) (Seierstad et al., 2008). The presence of arteriosclerotic plaques is not uncommon in cultured salmonids (see above), and gene expression data suggest that cysteine-rich angiogenic inducer 61 (Cyr61) and connective tissue growth factor (CTGF) are upregulated in the plaques. Both of these genes code for stretch-induced growth factors indicating the potential for cardiac and cardiovascular remodeling in response to the occurrence of coronary arteriosclerosis (Dalum et al., 2016).

2.4. Anemia

Anemia is a common pathology in fishes and is associated with many different causes, including nutritional (Pillay and Kutty, 2005) and environmental factors (Scarano et al., 1984), as well as infection with pathogens (Woo et al., 2003). Depending upon the type and severity of anemia, external aspects of the fish (e.g., skin discoloration; Beutler, 2005) as well as physiological and clinical parameters (e.g., blood oxygen transport, differential white blood cell count, etc.) can be affected.

2.4.1. Types of Anemia

Anemia, whereby red blood cells and hemoglobin (Hb) are removed from the circulation at a greater rate than they can be replaced (Roberts and Rodger, 2001), can be classified as one of three types:

Hemorrhagic anemia is described as a sequential and progressive removal of erythrocytes from the circulation. This can occur through physical disruption of the blood vessels, through injury, trauma or damage to the vessel wall. This loss of erythrocytes from the circulation has been modeled by the repeated removal of blood via vessels such as the caudal vein in various fishes including rainbow trout (*O. mykiss*) and starry flounder (*Platichthys stellatus*) (Wood et al., 1979, 1982).

Hemolytic anemia is the result of hemolysis or a shortening of the life span of erythrocytes in the circulation, a phenomenon commonly caused by chemical intoxication (Beutler, 2005). The process of intoxication typically either interacts with Hb, erythrocytic enzymes, or elicits redox reactions resulting in erythrocytic membrane damage (Beutler, 2005). Hemolytic anemia has been induced in several fish using phenylhydrazine (PHZ) (e.g., in *Lagodon rhomoides*, Cameron and Wolschlag, 1969; in *O. tshawytscha*, Smith et al., 1971; in *O. mykiss*, Simonot and Farrell, 2007; in *Gadus morhua*, Powell et al., 2011; in *Hippoglossus hippoglossus*, Powell et al., 2012; in *S. salar*, Burke, 2009; and in the yellowtail kingfish, *Seriola lalandi*, M.D. Powell, unpublished). Phenylhyradzine is a redox reactive compound that reacts with Hb forming a number of bioreactive radicals, with one mole of unbound oxygen being consumed per reaction (Shetler and Hill, 1985). Ultimately, this process reduces the total functional hemoglobin available and results in cell lysis (Berger, 2007).

Functional anemia reflects the effects of direct oxidation of Fe II to Fe III and results in methemoglobinemia, which may form toward the end of the erythrocytic lifespan as suggested by Scarano et al. (1984). A frequent cause of functional anemia in fish is the presence of NO_2^- following the partial oxidation of ammonia as reported by Scarano et al. (1984). However, Scarano and coworkers also ascribed hemolytic anemia, to the nitrite-induced functional anemia in sea bass (*Dicentrarchus labrax*).

2.4.2. Functional Responses to Anemia

Anemia ultimately leads to a decrease in the blood's oxygen carrying capacity, and a functional hypoxia that has the potential to limit the fish's performance and/or increase the reliance of the animal on glycolysis for ATP generation (Prchal, 2005). To compensate, there is increased transcription of hypoxia inducible factor 1α (HIF-1α), among other genes (Terova et al., 2008), and this increases erythropoietin (EPO) transcription and erythrocyte production (Prchal, 2005). For example, Atlantic salmon injected with PHZ to induce an acute hemolytic anemia showed an induction and upregulation of an erythropoietin gene (*epo*) in the heart, the primary site for EPO production in fish (Krasnov et al., 2013; Lai et al., 2006). Indeed, a putative EPO has been identified in a number of model species including puffer fish (*Takifugu rubripes*), zebrafish (*Danio rerio*), and rainbow trout (Chou et al., 2004, 2007; Wickramasinghe, 1993). Moreover, the transcript for the EPO receptor (*epor*) was apparently upregulated in the spleen shortly after EPO expression was enhanced in the heart of anemic salmon, suggesting that EPO is transported from the heart to the spleen where it stimulates *epor* upregulation (Krasnov et al., 2013). Interestingly, acute PHZ-induced hemolytic anemia also stimulated the upregulation of a suite of genes associated with innate immune responses, but a downregulation of those related to the adaptive immune response (Krasnov et al., 2013). This concurs with studies that report a general decrease in circulating lymphocytes in Atlantic salmon (Burke, 2009) and Atlantic cod (Powell et al., 2011). However, platelet (thrombocyte) markers were upregulated in the Krasnov et al. (2013) study, and both Burke (2009) and Powell et al. (2011) also reported an increasing trend in the proportion of circulating monocytes and neutrophils following PHZ-induced hemolytic anemia. This trend, however, was not evident in Atlantic halibut (Burke, 2009). Thus, the response of the immune system to acute hemolytic anemia may vary significantly between species.

The functional responses to the hypoxemia associated with anemia (particularly when hematocrit becomes critically low; i.e., $<10\%$) include an increase in $\dot{Q}$, that is the result of enhanced V_S but minimal changes in f_H (Simonot and Farrell, 2007). This increase in V_S appears to be achieved through increases in end-diastolic volume rather than reducing end-systolic volume (Franklin and Davie, 1992). Ultimately, the increased demands on cardiac function caused by anemia lead to a structural remodeling of the heart and an increase in relative ventricular mass (RVM) (McClelland et al., 2005; Powell et al., 2011, 2012; Simonot and Farrell, 2007, 2009). This has also been observed when fish are challenged by a variety of other factors including low temperature, sexual maturation, and chronic stress (see Chapter 3, Volume 36B: Gillis and Johnson, 2017; Johansen et al., 2011, 2017). This anemia-induced cardiac remodeling, however, is temperature dependent. For example, Simonot and Farrell (2007) repeatedly injected warm- and cold-acclimated (~17.6 and 6.4°C, respectively) rainbow trout with PHZ for 8 weeks to reduce hematocrit. In anemic, warm-acclimated fish, RVM started to increase at 2 weeks post-injection and was 41% greater than control fish by 8 weeks. In contrast, anemic cold-acclimated rainbow trout only had an RVM that was 18% greater than controls, and this change was largely due to a 28% increase in the mass of the compact myocardium. Based on increased myocardial DNA levels, McClelland et al. (2005) suggested that anemia-induced cardiac hypertrophy in rainbow trout is due to cellular hyperplasia. However, the drivers for this cardiac remodeling are unclear. In contrast, cardiac remodeling in the form of cellular hypertrophy, rather than hyperplasia, has been demonstrated in response to elevated blood cortisol levels (chronic stress) (Johansen et al., 2017). In fish fed a cortisol-laced diet, there was an ~30% increase in RVM that was also associated with increased relative expression of slow myosin light chain 2 (SMLC2), muscle LIM protein (MLP), and regulator of calcineurin 1 (RCAN1). Whereas ventricular myosin heavy chain (VMHC) and proliferating cell nuclear antigen (PCNA) were not significantly upregulated; these latter results indicating that there was an increase in cellular myosin concentrations, but no increase in myocyte numbers (Johansen et al., 2017). Interestingly, this cardiac enlargement was associated with a decrease in $\dot{Q}_{max}$ and V_{Smax}, and scope for $\dot{Q}$ and f_H. These findings indicate that cortisol-induced cardiac remodeling results in a pathologic cardiac hypertrophy that causes significant cardiac dysfunction in rainbow trout (Johansen et al., 2017).

In some cases, as with juvenile Atlantic halibut, anemia-induced cardiac growth is also associated with changes in heart shape, with significant lengthening of the ventricle with respect to its height and width (Powell et al., 2012). This response, with the heart becoming more pyramidal in shape, is likely to increase the efficiency of contraction (Olsen, 1998). A pyramidal ventricular shape is a consistent feature of more athletic fish species such as salmonids, and in general, concomitant with higher relative proportions of compact myocardium that enhances the mechanical performance of the heart (Cerra et al., 2004).

3. INFECTIOUS DISEASES AND DISORDERS

3.1. Immune Cell Populations in the Heart and Immune Responses

The prevalence of viral cardiac diseases has increased in farmed Atlantic salmon in recent years (Bornø and Linaker, 2015). Viral diseases pose a serious threat to the aquaculture industry, and the specific viral cardiac diseases of farmed Atlantic salmon include cardiomyopathy syndrome (CMS), pancreas disease (PD), and heart and skeletal muscle inflammation (HSMI) (Poppe and Ferguson, 2006; Silva et al., 2008).

Inflammation has been suggested as a generalized protective mechanism following tissue damage, irrespective of the cause (Roberts and Rodger, 2001). The inflammatory cell populations that are associated with cardiac diseases in fish are largely comprised of mononuclear lymphocyte-like cells (Christie et al., 2007; Ferguson et al., 1990; Kongtorp et al., 2004a, b, 2006; Poppe and Ferguson, 2006; Taksdal et al., 2007). By routine histology, lymphocytes have been located in the trabecular (PD, HSMI, and CMS) or non-trabecular (compact) (HSMI and PD) layers of the heart (Kongtorp, 2009; Poppe and Ferguson, 2006; Yousaf, 2012). However, in the case of coinfections of CMS with HSMI, myocardial inflammation is also reported in the non-trabecular myocardium and epicardial layer of the heart. Recent advances in the development of salmonid-specific immune markers (antibodies) have enabled the inflammatory cellular response in diseased hearts to be characterized in detail. For example, antibodies to lymphocyte receptors such as CD3ε (Boardman et al., 2012; Koppang et al., 2010) which activates other T cells, and CD8α (Hetland et al., 2010, 2011; Olsen et al., 2011) that is specific for cytotoxic T cells and cytokines such as recombinant tumor necrosis factor-α (rTNF-α) (Zou et al., 2003), have been developed. An increased influx of CD3ε-positive cells (to moderate levels) has been reported in HSMI-affected fish compared with non-diseased fish, where only a few CD3ε-positive cells were identified (Koppang et al., 2010; Yousaf et al., 2012). This observation was supported by moderately severe lesions reported in HSMI-affected fish by histology (Kongtorp et al., 2006). However, different cardiac diseases show differences in CD3ε-positive cell immune expression: HSMI-affected hearts have high numbers of CD3ε-positive cells compared with moderate–low levels in CMS- and PD-affected hearts (Yousaf et al., 2013).

The CD8 protein is the main marker of cytotoxic T lymphocytes and forms part of the T-cell receptor (TCR). Cell-mediated cytotoxicity has been suggested by CD8α-positive lymphocytes in salmonids such as the rainbow trout (Takizawa et al., 2011). Although they have relatively low levels of CD8α-positive cells, CMS-infected hearts express more CD8α-positive cells as compared to other diseased (HSMI- and PD-affected hearts) and non-diseased hearts (Yousaf et al., 2013). This CD8α subunit immunostaining was confined to lymphocyte-like cells in diseased and non-diseased hearts (Yousaf et al., 2012, 2013). On the contrary, in fish infected with piscine reovirus (PRV), the putative causative agent of HSMI, CD3ε-positive T cells were found in the epicardium and non-trabecular layers of the ventricle in addition to the CD8-positive T-cell myocarditis at 9 weeks after viral challenge (Mikalsen et al., 2012). HSMI heart lesions have also been associated with CD3ε-positive lymphocytes in naturally infected fish, although the numbers of CD8α-positive T cells were fewer (Mikalsen et al., 2012; Yousaf et al., 2012). In addition, predominantly CD3ε-positive lymphocytes were immunolocalized in Atlantic salmon hearts collected from field outbreaks of CMS, HSMI and PD (Yousaf et al., 2012, 2013). Recent studies suggest that lymphocytic responses dominate over granulocytic responses in diseased hearts, and that the inflammatory cells associated with the cardiac pathology mainly consist of CD3ε-positive T lymphocytes in CMS-, PD- and HSMI-affected hearts (Yousaf et al., 2012, 2013).

Viral pathogens, such as those that cause cardiac diseases in Atlantic salmon (HSMI, CMS, and PD), induce pathogen-specific or non-specific changes in gene expression. As a result, molecular approaches including real-time PCR (qPCR) have become important tools to identify, and study, immune markers of these pathogens (Johansen et al., 2016; Mikalsen et al., 2012; Timmerhaus et al., 2011). For example, Mikalsen et al. (2012) showed significant upregulation of transcription for CD4 (in naïve T cells), CD8 cytotoxic T-cell markers, IFNγ, granzyme A and the cytokine IL-10 at 9–10 weeks post-PRV challenge. Timmerhaus et al., 2011 reported that increases in cardiac CD8-positive T-cell gene expression preceeded decreases in viral load and histopathology. This suggested the involvement of CD8-positive cytotoxic T lymphocytes in clearing the experimentally induced PMCV infection in Atlantic salmon (PMCV—piscine myocarditis virus; the causative agent of CMS—see below). Recently, microarray studies have been used to demonstrate that maximum levels of PRV-associated gene expression occur at 8 weeks or greater post-challenge in fresh (parr) and salt (smolt) water-acclimated Atlantic salmon. Innate antiviral response genes that were upregulated at 8 weeks post-PRV infection included viperin, interferon alpha (IFNα), interferon gamma (IFNγ), interferon-stimulated gene 15 (ISG15), CC chemokine with stalk CK2, C—C motif chemokine 19–2, IL-8 receptor and the IL-6 subfamily member M17. Similarly, transcription for adaptive immune markers such as CDα, CDβ cytotoxic T cells, IgM and granzyme A were upregulated shortly after 10 weeks post-challenge (Johansen et al., 2016).

In addition to changes in gene expression, moderate levels of rTNFα-positive cells (including macrophage-like and eosinophilic granular cells, and putative mast cell-like cells) were observed using a recombinant TNFα antiserum in the areas surrounding inflammatory lesions in all three of the above-mentioned heart diseases (HSMI, CMS, and PD) (Yousaf et al., 2012, 2013). Owing to the central role of TNFα in inflammation and immunity, this gene transcript has been suggested as an inflammatory biomarker in salmonids (Haugland, 2008). Macrophages are the main antigen-presenting cells (APC), although dendritic cells have also been identified in fish, and have shown to present antigen to T lymphocytes and to secrete cytokines (incl. TNFα) (Lovy et al., 2008; Magnasdottir, 2010). A comparative transcriptional study of Atlantic salmon hearts affected by HSMI and PD identified a stronger induction of the innate immune genes in HSMI-affected fish as compared to those infected with PD (Johansen et al., 2015). More cells presenting MHC class II were also found in HSMI-affected hearts as compared with PD-affected hearts. Finally, upregulation of B- and T-cell-specific genes such as CD3 (epsilon, gamma delta-A, zeta), CD8 (alpha, beta), CD2 T-cell surface antigen, CD28 T-cell-specific surface glycoprotein, CD40, granzyme and IL-16 was more pronounced in HSMI as compared to PD-affected hearts (Johansen et al., 2015). These data suggest that inflammatory responses (particularly T-cell responses) are more pronounced in response to HSMI as compared with PD.

3.2. Viral Diseases and Cardiac Inflammation

Viral fish diseases are among the most destructive diseases in fish aquaculture (Silva et al., 2008). Although there is limited information regarding the functional effects of viral inflammatory diseases of the heart, or where the heart is one of the primary targets, important viral diseases such as infectious salmon anemia (ISA) do affect cardiac performance (Gattuso et al., 2002). A unifying feature of cardiac viral diseases in salmonids is extensive endocardial, myocardial, and epicardial inflammation, and myocardial necrosis (see discussion above). For the purposes of this chapter, these diseases are discussed in terms of their pathology in relation to potential physiological dysfunction.

3.2.1. Infectious Salmon Anemia

In fish, ISA causes acute anemia and hemorrhaging in vital organs including the kidney and was first diagnosed in farmed salmon in Norway (Thorud and Djupvik, 1988). Since then, it has been reported in farmed Atlantic salmon in Scotland, Canada, the Faroe Islands and the United States. In particular, ISA epidemics in Chile in the mid-2000s decimated the burgeoning aquaculture industry. This disease is caused by an orthomyxovirus and is characterized by severe anemia, ascites, congestion, petechial hemorrhage and acute high mortalities (Koren and Nylund, 1997). The primary target of the virus appears to be the endothelium and endocardium (Nylund et al., 1995) with virus particles seen budding into the lumen of blood vessels and the heart (Nylund et al., 1995, 1996). ISA virus-infected fish show significant impediments in cardiac performance due to a weakened Frank–Starling response, which appears to be related to increased inducible nitric oxide synthase (iNOS) activity or proinflammatory cytokine production (Gattuso et al., 2002). For example, while the hearts from healthy salmon appeared to show no substantial iNOS activity, the $\dot{Q}$ and V_S of *ex vivo* isolated hearts from ISA-infected fish displayed a marked sensitivity to the specific iNOS inhibitor, L-NIL (Gattuso et al., 2002). These data suggest that NO signaling has an important role in modulating fish cardiac performance in disease, and may act by altering either systolic (involving intracellular calcium modulation) or diastolic (reducing diastolic stiffness) function (Gattuso et al., 2002). However, further studies are required to elucidate the mechanisms by which NO is involved in the pathogenesis of fish diseases.

3.2.2. Cardiomyopathy Syndrome

Cardiomyopathy syndrome (CMS) is a cardiac disease of Atlantic salmon that mainly affects the atrium and trabecular (spongy) ventricular myocardium without involvement of the non-trabecular (compact) ventricular myocardium. CMS shares similar features with HSMI, for example, both diseases cause myocarditis and pericarditis, and CMS has been proposed as a late stage of HSMI (Amin and Trasti, 1988; Ferguson et al., 1990; Kongtorp et al., 2006; Poppe and Ferguson, 2006); although this has not been substantiated. CMS was first reported in the late 1980s in farmed Atlantic salmon in Norway (Amin and Trasti, 1988; Ferguson et al., 1990) and has been subsequently reported in the Faroe Islands and Scotland (Bruno and Poppe, 1996; Rodger and Turnbull, 2000) with significant economic impact (Brun et al., 2010). More recently, this disease has been identified in wild Atlantic salmon and Chinook salmon (*O. tshawytscha*) in British Columbia, Canada (Brocklebank and Raverty, 2002; Poppe and Seierstad, 2003). Cardiomyopathy syndrome was originally considered a disease affecting adult Atlantic salmon. However, outbreaks have also been reported in juvenile postsmolt fish. These fish have the same type and severity of lesions as reported for adults, but they are present in lower numbers (Fritsvold et al., 2009).

Amin and Trasti (1988) first proposed a viral etiology for CMS based on the presence of intranuclear eosinophilic inclusion bodies in unaffected cardiomyocytes situated adjacent to degenerated myocardium. Later CMS challenge studies using the intraperitoneal injection of affected tissue homogenates demonstrated that CMS lesions could be reproduced in unvaccinated Atlantic salmon smolts (Fritsvold et al., 2009). Similarly, CMS has been successfully transmitted in adult salmon using Scottish and Norwegian tissue homogenates, leading to the conclusion that similar disease conditions occur in both countries (Bruno and Noguera, 2009). The above-mentioned studies also supported the viral etiology of CMS in Atlantic salmon (Bruno and Noguera, 2009; Fritsvold et al., 2009), and PMCV, a naked double-stranded RNA totivirus, has been associated with CMS in this species (Haugland et al., 2011; Løvoll et al., 2010). PMCV was identified from several natural CMS outbreaks, and from fish where CMS had been induced through experimental transmission. PMCV is not ubiquitous and appears to be more specific to CMS-affected fish (Løvoll et al., 2010). PMCV RNA has also been identified in healthy Atlantic salmon brood fish and fertilized eggs, and it has been suggested that PMCV is transferred vertically from parental fish to progeny (Wiik-Nielsen et al., 2012). In addition, a distinct strain of PMCV has been identified in healthy Atlantic argentine (*Argentina silus*) (Bockerman et al., 2011; Tengs and Bockerman, 2012), and more recently in corkwing (*Symphodus melops*) and ballan wrasse (*Labrus bergylta*) (Scholz et al., 2017). Recently, a challenge trial involving the intraperitoneal injection of PMCV induced CMS-specific lesions in Atlantic salmon, and viral loads were correlated with histopathological changes (Timmerhaus et al., 2011).

Macroscopically, skin hemorrhage, raised scales and edema can be seen in fish affected with CMS. In addition, rupture of the atrium or sinus venosus has been reported in the terminal stages of CMS (Ferguson et al., 1990; Poppe and Ferguson, 2006). Necropsy findings include atrial thrombosis, ascetic fluid, fibrinous peritonitis, and blood clots on the liver and heart (Ferguson et al., 1990; Poppe and Seierstad, 2003; Rodger and Turnbull, 2000). The cardiac lesions vary from being multifocal to diffuse, and include/result in necrosis and inflammation of ventricular and atrial trabecular myocardium, epicarditis, and the presence of a myocardial cellular infiltrate comprised mainly of lymphocytes and macrophages. The non-trabecular compact myocardial layer is usually not affected (Ferguson et al., 1990; Poppe and Ferguson, 2006). Protein casts in kidney collecting tubules and melanin deposits in CMS-affected hearts have also been identified in CMS fish (Fritsvold et al., 2009; Yousaf et al., 2012). The most significant physiological impact of CMS appears to be the weakening of the atrial wall that leads to an extensive stretch of the atrium and its potential rupture. This occurs particularly during periods of acute stress such as fish handling or ectoparasite treatments (such as those associated with removing the sea lice *Lepeophtheirus salmonis*). The more chronic (subacute) effects of this disease on cardiac performance have not been fully investigated.

3.2.3. Heart and Skeletal Muscle Inflammation

HSMI is a disease of farmed Atlantic salmon that mainly affects both trabecular and non-trabecular (compact) myocardium (Fig. 3) and the aerobic (red) skeletal muscle. HSMI was first reported in Norway in 1999 and has

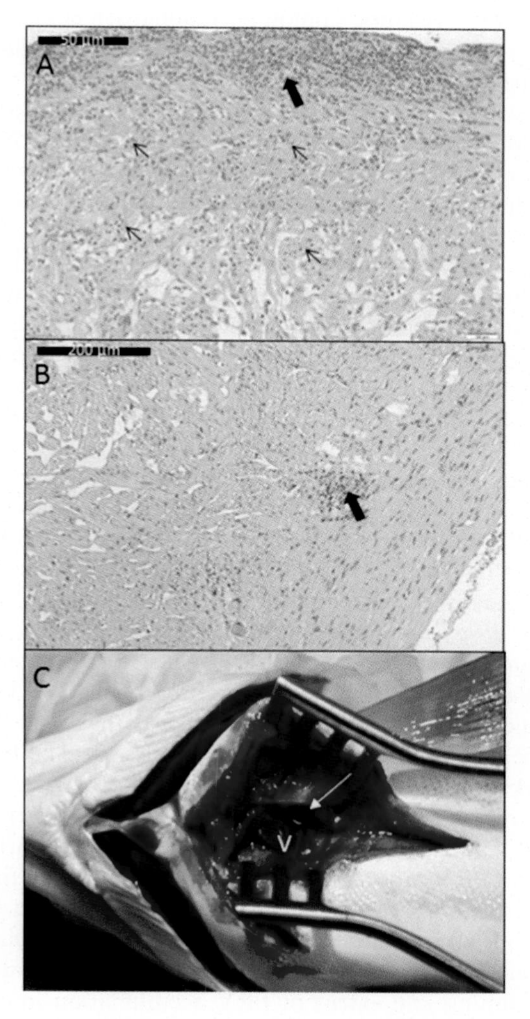

Fig. 3. The effect of heart and skeletal muscle inflammation (HMSI) on the heart of a freshwater salmon (parr). (A) Extensive inflammatory cell infiltration of the trabecular (*thin arrow*) and compact myocardium (*thick arrow*). (B) Focal inflammation in the trabecular myocardium of a heart (*thick arrow*) testing positive for piscine reovirus (PRV). (C) Fish from which the heart shown in (B) was collected showing extensive pericardial hemorrhage (*white arrow*). V, ventricle. Sections were stained with H&E.

since been reported in Scotland (Ferguson et al., 2005; Kongtorp et al., 2004b). HSMI is characterized as a disease of low mortality (~20%), but high morbidity (~100%), that commonly affects fish 5–9 months after transfer to sea (Kongtorp et al., 2004b). Recently, outbreaks of HSMI have also been reported in freshwater Atlantic salmon parr (Johansen et al., 2016). HSMI is currently one of the most significant issues associated with salmon farming in Northern Europe.

The double-stranded RNA orthoreovirus PRV is suggested as an etiological agent of HSMI (Markussen et al., 2013; Palacios et al., 2010). The PRV virus has been detected by qPCR in Norway, Chile, Scotland, Ireland and Canada in farmed salmon. It has also been detected in wild asymptomatic Atlantic salmon, but at a much lower level of infection (i.e., high C_t values) than in diseased fish (Løvoll et al., 2012; Palacios et al., 2010). A recent study showed susceptibility of Atlantic salmon and sockeye salmon to Western North American PRV without the development of clinical HSMI lesions (Graver et al., 2016). HSMI-like disease has also reported in rainbow trout, although the associated reovirus only has 85% sequence similarity to PRV (Olsen et al., 2015). Other PRV susceptible salmonids include brown trout (*Salmo trutta*) in Norway, and wild cutthroat trout (*Oncorhynchus clarkii*), wild chum salmon (*Oncorhynchus keta*), farmed steelhead trout (*O. mykiss*), and wild coho (*Oncorhynchus kisutch*) and Chinook salmon in Canada (Kibenge et al., 2013; Marty et al., 2015). Low titers of PRV mRNA have also been measured in marine fish along the Norwegian coastline by PCR screening, including the great silver smelt (*A. silus*), Atlantic horse mackerel (*Trachurus trachurus*), Atlantic herring (*Clupea harengus*) and capelin (*Mallotus villosus*) (Garseth et al., 2013; Wiik-Nielsen et al., 2012). PRV mRNA has been found in Atlantic salmon brood fish, with no clinical signs of HSMI, but not in fertilized eggs. This suggests that vertical transmission is unlikely to be a major route for PRV transmission. The fish that survive HSMI outbreaks are suggested to be lifelong PRV carriers (Wiik-Nielsen et al., 2012). A recent immunohistochemical study strengthened the association between PRV and HSMI, by staining HSMI-affected hearts using specific antibodies against PRV capsid proteins (Finstad et al., 2012). Erythrocytes are also a major target for PRV, and become infected early in the pathogenesis of the disease. The PRV virus subsequently accumulates in the spleen, at a similar viral load as in erythrocytes. Histopathological changes typical for HSMI correlate with the timing and extent of PRV accumulation in cardiomyocytes (Finstad et al., 2014). It is also worth noting that PRV in Atlantic salmon is associated with focal pathological changes in red and white skeletal muscle (Bjorgen et al., 2015; Ola, 2016). To date, PRV pathogenesis remains obscure and failure of *in vitro* cultivation of PRV hinders further investigations.

Moribund fish affected by PRV are anorexic, show abnormal swimming behaviors and have increased levels of mortality. In addition, although there is no change in hematocrit, the liver is pale. Macroscopically, hearts of infected fish appear pale with a “loose” texture, and they often exhibit pericardial hemorrhage and ascities. Microscopically, the cardiac and red skeletal muscle exhibit the most significant and diagnostic histopathological lesions, and these have extensive infiltration of inflammatory cells. The pathological findings first appear, and are more frequent, in the heart than in the red skeletal muscle (Kongtorp et al., 2004a, b). Since the discovery of PRV, HSMI has been diagnosed by histopathology of the cardiac and white/red muscle tissue, and by qPCR of PRV mRNA from the ventricle and blood. In addition to PRV, a new virus named Atlantic salmon calcivirus has been identified from HSMI diseased fish (Mikalsen et al., 2014). However, the involvement of this virus in the development of HSMI remains unknown. The HSMI diagnosis is based upon histopathological changes, and presents as: epi-, endo- and myocarditis; a pronounced mononuclear cellular infiltration of both the trabecular (spongy) and non-trabecular (compact) myocardium of the ventricle; myocytic necrosis; and myositis and necrosis of red skeletal muscle. Affected myocytes are characterized by degeneration, the loss of striation and eosinophilia, vacuolation, centralized nuclei, and karyorhexis. Inflammatory changes, comprised mainly of infiltrated mononuclear lymphocytes, are often more pronounced than necrotic changes in the heart and skeletal muscle (Kongtorp et al., 2004a, b, 2006; Poppe and Ferguson, 2006). Functional feeds, defined as "foods with dietary ingredients that provide healthy and economic benefits beyond basic nutrition" (Olmos et al., 2011) have been suggested as precautionary measures to alleviate HSMI pathology (Alne et al., 2009; Grammes et al., 2012; Martinez-Rubino et al., 2012). The specific actions of such functional ingredients appear to be in reducing the inflammatory responses in the heart (see detailed explanation above) so sustaining normal cardiac function.

Functionally, the impact of PRV infection and HSMI is likely to be extensive. However, this has not been shown by studies to date. For example, while it appears that the hypoxia tolerance of PRV-infected fish is reduced, there are limited effects on hemoglobin–oxygen binding or maximum f_H (Lund et al., 2017). Furthermore, the $\dot{Q}$, f_H, and V_S of PRV-infected Atlantic salmon showing only mild signs of HSMI were similar to those reported for uninfected fish under resting conditions, even though there was extensive pericardial hemorrhage (Fig. 3) (Powell, unpublished).

3.2.4. Pancreas Disease

The single-stranded RNA salmonid alphavirus (SAV), which belongs to the family *Togaviridea* (Weston et al., 2002), is recognized as the causative agent of PD in marine farmed Atlantic salmon (*S. salar*) and sleeping disease (SD) in freshwater rainbow trout (*O. mykiss*) (Cano et al., 2015; Herath et al., 2012; Kongtorp et al., 2010; McLoughlin and Graham, 2007; Weston et al., 1999). The term "pancreas disease" has been known in salmonid aquaculture since 1976 and was given its name based on the histopathological changes in the pancreas in addition to the skeletal and cardiac muscle (Ferguson et al., 1986; Munro et al., 1984). PD is confined mainly to Europe and has been reported in Ireland, Scotland, the United Kingdom, Spain, Italy and Norway (Christie et al., 1998; Ferguson et al., 1990; Graham et al., 2007, 2012; Herath et al., 2012; Kongtorp et al., 2010; McVicar, 1987; Poppe et al., 1989; Rowley et al., 1998; Taksdal et al., 2007). However, pathology consistent with PD has been reported in the United States; although the virus has not been isolated and there is no genetic data to confirm the presence of SAV (Kent and Elston, 1987).

To date, at least six subtypes of SAV exist (SAV 1–6), and all subtypes affect farmed Atlantic salmon held in seawater. SAV 2 is divided into two variants. The first is a fresh water variant that causes SD, which is an atypical alphavirus and exhibits similar histopathology to PD. The second is the marine variant of SAV 2 that causes PD in seawater-reared Atlantic salmon and rainbow trout (Boucher and Baudin-Laurencin, 1994, 1996; Fringuelli et al., 2008; Hjortaas et al., 2013; Hodneland et al., 2005; Jansen et al., 2016; Villoing et al., 2000; Weston et al., 1999, 2002). Recently, in Scottish waters, SAV mRNA has also been detected by qPCR in wild marine flatfish such as common dab (*Limanda limanda*), long rough dab (*Hippoglossoides platessoides*) and plaice (*Pleuronectes platessa*) (Snow et al., 2010). For all six SAV subtypes, cohabitant challenge models have been established and, in turn, produced long-term immunity in fish (Graham et al., 2011; Houghton, 1994; Lopez-Doriga et al., 2001). Both fecal and mucosal routes of transmission between fish, and the shedding of virions, have been suggested (Graham et al., 2011). As opposed to mammalian alphavirus, which requires an intermediate host, SAV in fish is transmitted horizontally, i.e., directly between individuals (Houghton and Ellis, 1996; McLoughlin et al., 1996). However, SAV is also capable of replicating in an arthropod cell line derived from the Asian tiger mosquito *Aedes albopictus*, which suggests that SAV is not restricted to vertebrate hosts (Hikke et al., 2014). Vertical transmission from parents to offspring is probably a minor route of transmission (Rimstad et al., 2011).

During the last decade, PD has re-emerged and has become a major economic and animal welfare issue for farmed Atlantic salmon in Europe (McLoughlin and Graham, 2007). Clinical signs of PD include loss of appetite, lethargy, abnormal swimming, yellow fecal casts, a cachectic appearance as well as increased mortality. PD-related mortality ranges from 10% to 50% in natural outbreaks, and an individual outbreak may last for 3–4 months (Christie et al., 2007; McLoughlin et al., 2002). The acute phase of the disease lasts for 10 days post-infection at 2–14°C, with lesions in the heart and pancreas being the most prevalent features. The acute phase is followed by a subacute phase that lasts between 10 and 21 days after the onset of clinical signs, and is characterized by lesions in the pancreas, heart and skeletal muscles. This phase is followed by a chronic phase (21–42 days post-infection) where lesions in the muscles are the predominant feature. Finally, there is a recovery phase, where fish growth is significantly stunted and the survivors become "runts" (McLoughlin et al., 2002; McLoughlin and Graham, 2007). The most important pathological changes include a severe loss of exocrine pancreas, cardiac and skeletal myopathies, epicarditis, focal gliosis of the brain stem, white skeletal muscle degeneration, and the presence of functionally unknown cells in kidney with cytoplasmic eosinophilic granules (Christie et al., 2007; Taksdal et al., 2007).

The effects of PD and SAV infection on the function of the cardiorespiratory system have not yet been investigated. However, due to the similarities between HSMI and PD in terms of cardiac inflammation and immune cell infiltration, it is likely that the responses would be similar to those of HSMI.

3.3. Parasitic Infection and Cardiovascular Performance

Recent examination of the impacts of parasitic diseases on fish have provided novel insights into how adaptive the cardiovascular system is, and in particular, into how the heart responds to a pathogenic insult. The four best studied models include the microsporidian parasite *Loma* sp. (Powell et al., 2005; Powell and Gamperl, 2016); the Ichthyosporean *Ichthyophonus* sp. (Kocan et al., 2006, 2009); the amoebaozoan, *Neoparamoeba perurans* (see review by Powell et al., 2008); and the copepod *Lernaeocera branchialis* (Baily et al., 2011). Although *N. perurans* is primarily a gill ectoparasite, the inflammatory response in Atlantic salmon induces cardiovascular-related pathology and physiological dysfunction, particularly at lower levels of infestation (Powell et al., 2008).

3.3.1. Cardiac Microsporidiosis in Atlantic Cod

Loma morhua is a microsporidian parasite that is transmitted by the ingestion of zoospores, and subsequently infects endothelial cells (primary target organs are the heart, liver, spleen, and gills) (Murchelano et al., 1986). The spores enter a merogeny-like phase in the heart following infection, and using macrophages, the parasite spreads to other organs and develops in a sporogonic stage that forms a xenoma (an asexual spore forming stage) containing numerous small spores (Powell et al., 2005, 2006; Rodriguez-Tovar et al., 2003; Sanchez et al., 2000, 2001). Cardiac xenoma formation, with the infection of endothelial cells, is a characteristic of *L. morhua* in Atlantic cod. Rupture of the spore-filled xenoma then releases spores into the surrounding tissues, circulation, or water depending upon the location of the xenoma. Autoinfection, is therefore, a common feature of the pathogenesis of *Loma* infections, and this leads to fish reinfecting themselves and the presence of xenomas at different stages of development within one individual (Rodriguez-Tovar et al., 2003). Infected fish show a range of pathological reactions to the presence of xenomas in the heart and gills, from minimal tissue reaction to extensive fibrosis (Fig. 4) (Powell and Gamperl, 2016). Although RVM and ventricular shape were not affected by infection, infected Atlantic cod showed a number of changes in cardiac function as compared with non-infected fish from other studies. For example, infected cod had significantly lower values for V_{Smax} (by 48%) and $\dot{Q}_{max}$ (by 41%), and a significantly higher resting f_H (by 29%) (Powell and Gamperl, 2016). Furthermore, while both V_S and $\dot{Q}_{max}$ were negatively correlated with the density of cardiac xenomas, f_H showed the opposite relationship. Ultimately, these changes limited the cod's capacity to increase oxygen consumption, and resulted in a reduced metabolic scope as measured in both critical thermal maximum (CT_{Max}) and critical swimming speed (U_{crit}) tests (Powell and Gamperl, 2016). As there was no change in the ventricle's shape (i.e., its dimensions or their ratios), as has been reported for a number of disease states including systemic hypertension (Powell et al., 2002b) and hemolytic anemia in Atlantic halibut *H. hippoglossus* (Powell et al., 2012), it is unlikely that remodeling of the ventricle affected systolic function (i.e., ejection of blood from the heart). However, it is possible that the presence of xenomas in the myocardium resulted in a change in the elastic properties of the heart chambers, and thus, limited myocardial stretch and end-diastolic volume. Indeed, such an impact on myocardial properties has been suggested for other parasites that infect the heart such as *Stephanostomum tenue* (McGladdery et al., 1990) and *Kudoa thyrsites* (Kabata and Whitaker, 1988).

3.3.2. Ichthyophoniasis

Ichthyophonus sp. is an important pathogen that, although it belongs to a unique clade of protistan parasites, superficially resembles a fungus and produces spores and hyphae (Kocan et al., 2009). Commonly, *Ichthyophonus* sp. infect a number of fish tissues and numerous epidemics have been reported. These include outbreaks in herring (*C. harengus*), farmed rainbow trout and Atlantic salmon, as well as Pacific salmon species such as coho, sockeye (*Oncorhynchus nerka*) and Chinook (as reviewed by Kocan et al., 2009). Although showing extensive tissue tropism, the heart is a primary target organ for infection which results in necrosis, separation of the muscle fibers by large multinucleate bodies, and an associated infiltration of inflammatory cells. In later stages, the spherical multinucleate bodies become encapsulated by fibrous tissue (Kocan et al., 2009; McVicar, 1999; McVicar and McLay, 1985).

In spawn-run (migrating mature adults returning upstream to the spawning grounds) wild sockeye salmon, naturally infected with *Ichthyophonus* sp. among other pathogens, there were extensive pathological signs. These signs of cardiac

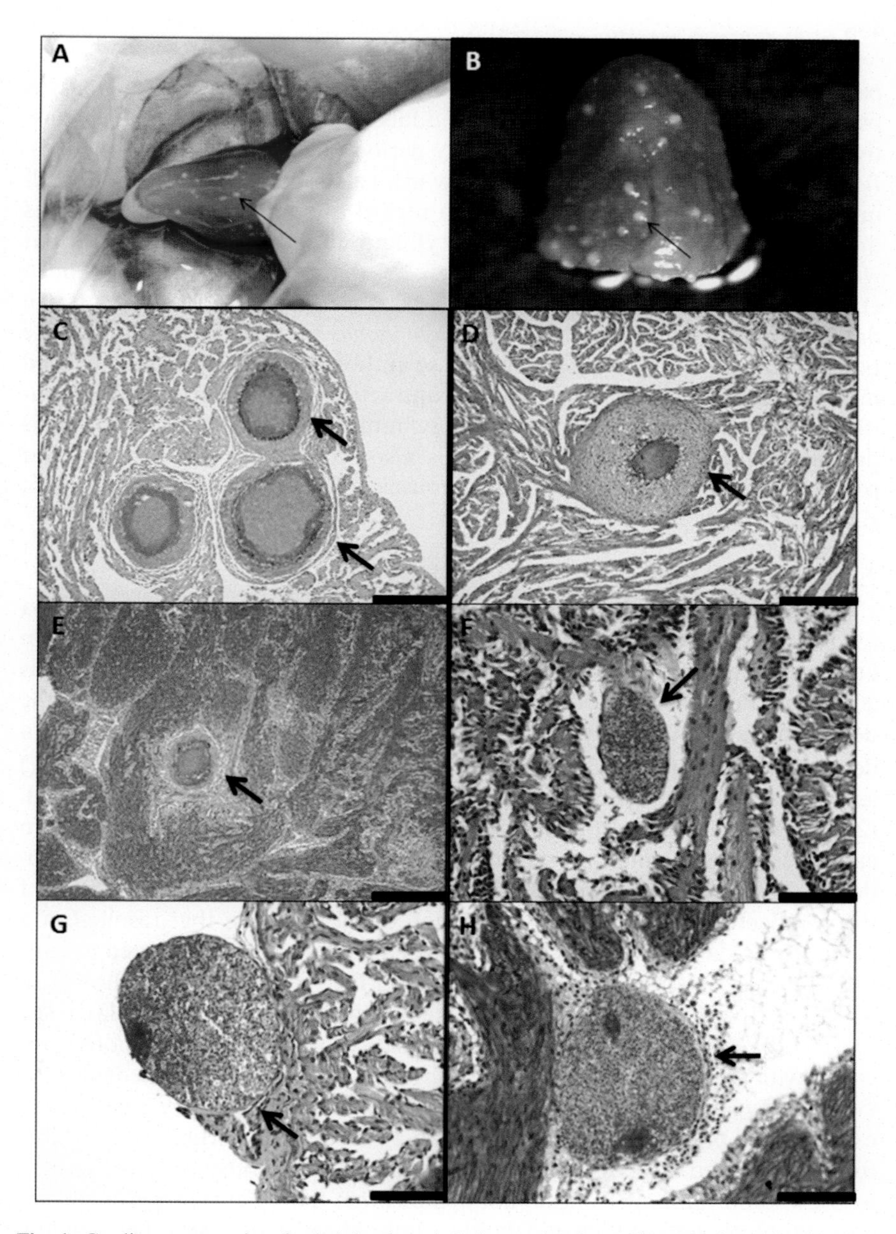

Fig. 4. Cardiac xenomas in a fresh Atlantic cod *Gadus morhua* heart (A) and a heart fixed in formalin showing *Loma morhua* xenomas in the atrium, ventricle, and bulbus arteriosus of the heart (C, D, and E, respectively) with an accompanied fibrocytic and granulomatous inflammatory reaction (bar = 200 μm). Endocardial (F), epicardial (G) and bulbus arteriosus (H) showing endothelial xenomas with limited inflammatory reaction (bar = 50 μm). *Arrows* indicate xenomas. Sections were stained with H&E. Panels (C)–(H) from Powell, M.D., Gamperl, A.K., 2016. Effects of *Loma morhua* (microsprodia) infection on the cardiorespiratory performance of Atlantic cod *Gadus morhua* (L.). J. Fish Dis. 39, 189–204.

pathology were also associated with impaired swimming performance. This was particularly noticeable in infected individuals during a second bout of exhaustive exercise following a short recovery period (Tierney and Farrell, 2004). *Ichthyophonus* sp.-only infected rainbow trout also showed significant impediments to swimming performance with regard to stamina (characterized as distance swum at a given water velocity), and no improved performance at warmer water temperatures as was observed for uninfected fish (Kocan et al., 2009). This reduced stamina has been attributed to induced cardiac damage due to the presence of the *Ichthyophonus* sp. spherical multinucleate bodies (Kocan et al., 2006). Indeed, these authors hypothesized that reduced myocardial blood supply and reduced contractility of the heart probably limited cardiac function during sustained swimming (Kocan et al., 2006). This parasite-induced cardiac dysfunction is also seen with *Trypanosoma crusi* pseudocysts in mammalian cardiac muscle (Roberts and Janovy, 2005; Zhang and Tarleton, 1999).

3.3.3. Amoebic Gill Disease

Amoebic gill disease (AGD), caused by the amphizoic amoeba (*N. perurans*) affects salmonids and other species, and cardiovascular function has been demonstrated to be impacted in Atlantic salmon suffering from this condition (Powell et al., 2008). For example, Leef et al. (2005) showed that the development of AGD after experimental infection in the lab resulted in a decrease in resting $\dot{Q}$ by 35%, but no change in dorsal aortic pressure (P_{DA}) as systemic vascular resistance (R_{sys}) was increased. However, Powell et al. (2002b) reported that Atlantic salmon with AGD that were sampled from cage-sites had systemic hypertension that was associated with cardiac remodeling (Powell et al., 2002a); specifically, a small increase in the thickness of the compact myocardium and elongation of the ventricle that resulted in a more pyramidal shape. This change in cardiac morphology is believed to have increased the pressure generating capacity of the ventricle, and to allow it to overcome the higher afterload associated with the increased vascular resistance and P_{DA} (Powell et al., 2002a). Dissipation of the branchial lesions associated with AGD by treatment with freshwater, and a subsequent reduction in systemic hypertension, suggests that hyperplastic inflammatory tissue or an extracellular product from the amoeba may be at least partially responsible for the observed hypertension (Powell et al., 2002a). Other salmonids such as brown and rainbow trout do not demonstrate the same changes in cardiovascular physiology as seen in Atlantic salmon (Leef et al., 2005). The Atlantic salmon appears to be the most susceptible (in terms of mortality) salmonid species to AGD (Munday et al., 2001), and there is also evidence to suggest that rainbow trout are less susceptible to AGD as compared to other salmonids (Munday et al., 2001; Roberts and Powell, 2005).

The involvement of prostaglandin E2 (PGE2) and/or nitric oxide (NO) is thought to be important in mediating the systemic hypertension downstream of the affected gill, despite several studies that have failed to show consistent changes in cyclooxygenase-2 (COX-2) or inducible nitric oxide synthase (iNOS) activity in AGD-infected Atlantic salmon (Nowak et al., 2013; Powell et al., 2008; Rosenlund, 2017). This is because both PGE2 and NO are powerful systemic vasodilators in fish (Stensløkken et al., 2002), including salmonids like rainbow trout and Atlantic salmon (M.D. Powell and J.A. Becker, unpublished). Furthermore, Leef et al. (2007) showed that hypertensive AGD-infected fish given an intraarterial injection of sodium nitroprusside (an NO donor) responded with a significant decrease in systemic vascular resistance (i.e., vasodilation), whereas, non-AGD infected (control) fish were only marginally affected. Interestingly, treatment of fish with other vasodilators such as the angiotensin-converting enzyme (ACE) inhibitor, captopril, resulted in a similar decrease in R_{sys} in both AGD-infected and control fish (Leef et al., 2007). Collectively, these data suggest that, a reduction (or absence) of NO production in the gills of AGD-affected fish was potentially involved in the observed hypertension.

3.3.4. *Lernaeocera branchialis*

The sanguiphageous (blood sucking) parasitic copepod *L. branchialis* is widely distributed around the North Atlantic and primarily infects gadoids such as the Atlantic cod. It is believed that flatfish species such as sole (*Solea solea*), plaice (*P. platessa*), dab (*L. limanda*) and flounder (*Platichthys platessa*) as well as the lumpfish (*Cyclopterus lumpus*), act as intermediate hosts (Baily et al., 2011; Smith et al., 2007). The adult females attach to the gill arch (Fig. 5A) and extend a proboscis into the afferent branchial arteries (Smith et al., 2007). The proboscis then extends caudally toward the heart, either within the lumen of the afferent branchial arteries and ventral aorta or along the vessel walls (Baily et al., 2011). The anterior of the proboscis develops into an antler-shaped ampulla that anchors itself into the vessel wall and/or bulbus arteriosus (Baily et al., 2011) (Fig. 5C and D). Associated with the intrusion of the ampulla and head of the copepod is an acute inflammatory reaction of the blood vessel intima, and this leads to extensive fibrosis and the infiltration of inflammatory cells. This infiltration results in a epithelioid macrophage layer of several cell thickness which forms a granuloma (Baily et al., 2011; Behrens et al., 2014).

The consequence of the invasion of the afferent branchial arteries and ventral aorta is the partial occlusion of the vascular lumen, either by the parasite or the luminal invasion of the associated granuloma (Behrens et al., 2014). This results in partly or substantially distorted ventral aortic blood flow profiles, a reduction in the amplitude of changes in blood flow during the cardiac

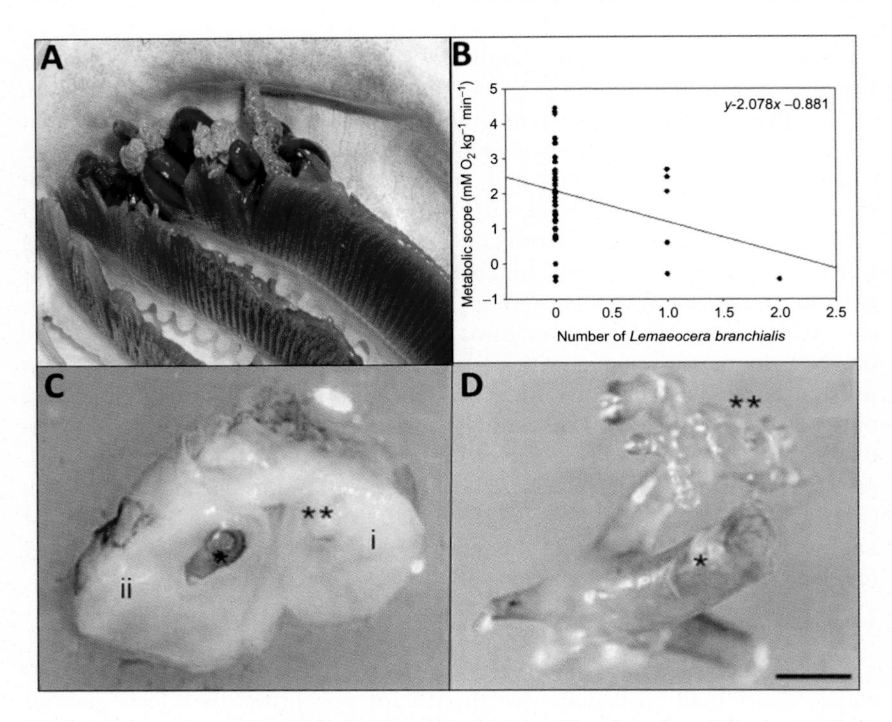

Fig. 5. (A) *Lernaeocera branchialis* female attached to the gill arches of an Atlantic cod *Gadus morhua* (Photo: Egil Karlsbakk). (B) Correlation between metabolic scope ($MO_{2active} - MO_{2rest}$) and the number of *L. branchialis* individuals on the gills of Atlantic cod (*Gadus morhua*), Pearson coefficient −0.290, $P = 0.032$ (Søreng and Powell, unpublished); (C) Cross-section through the bulbus arteriosus (i) and connective tissue capsule (ii) encapsulating the parasite, where * and ** indicate the location of the parasite's cephalic anchors. (D) Chitinous exoskeleton of the cephalic anchors (ampulla) that was embedded in the bulbus and capsule. The anchor in the bulbus (**) was occluding a large part of the lumen. Scale bar in D, 1 mm. Panels (C) and (D) from Gamperl and Farrell (2004).

cycle, and a limited ability to increase $\dot{Q}$ when fed (i.e., a reduced post-prandial increase in $\dot{Q}$). Interestingly, although the specific dynamic action of cod infected with *L. branchialis* was reduced by 30%, neither resting oxygen consumption (MO_2) nor maximum (post-feeding) MO_2 were affected by infection. This latter result contrasts with Reidy et al. (2000) who showed that *L. branchialis* infection reduced the critical swim speed (U_{crit}) of Atlantic cod and the oxygen available to support high swim speeds. Recent results show that the overall effect of this parasite is a limitation of $\dot{Q}$ (Behrens et al., 2014), that leads to a direct negative effect on metabolic scope (Fig. 5B and E. Søreng and M.D. Powell, unpublished).

4. CONCLUDING REMARKS

Four main themes emerge with regard to the functional effects of disease on fish cardiovascular performance. The first being that of inflammation. However, it still unclear whether inflammation, by itself, has a direct impact on cardiac or cardiovascular performance. Certainly, the development of sclerotic plaques or myocardial necrosis will lead to the occlusion of blood flow and a weakening of the contractile performance or structural integrity of the heart, respectively. However, the presence of inflammatory cells remains enigmatic. Furthermore, inflammatory cytokines such as IL1-β and TNFα have wide-ranging cellular effects, and besides mediating the infiltration of inflammatory cells, can mediate a range of physiological cascades. Associated with this theme, are the potential roles played by the induction of iNOS (and potentially prostaglandin-mediated effects due to the effects of COX-2 upregulation) in cardiac function and endothelial vasodilation. This seems be a common feature of disease-related alterations in fish cardiovascular physiology, particularly where inflammation is involved. However, the mechanism(s) by which these processes interact is still unknown, and information in this area may provide important insights toward understanding the pathophysiology of cardiovascular disease in fish.

Third is the physical effect of tissue repair or reaction to an insult, be it a pathogen (virus, bacteria, or parasite) or a non-pathogenic insult (e.g., physical trauma, etc.). It appears likely that the presence of an acute tissue reaction producing either intimal plaques or fibrosis has a direct effect on the biomechanics of the cardiovascular system and/or the endo-, myo- and epicardium. The latter most likely affects ventricular stiffness and reduces $\dot{Q}$ by limiting end-diastolic volume, although this can be partially or largely compensated for by an increase in cardiac frequency (f_H).

The fourth theme is the fish's response to reduced blood oxygen transport, be it due to pathogen insult, or other secondary responses to pathological conditions such as anemia or systemic hypertension/vascular resistance. When cardiac or cardiovascular pathology becomes limiting, a remodeling of the myocardium, primarily in the ventricle, appears to compensate. This takes place in two main ways, by an increase in myocardial mass (principally through cardiac hypertrophy and/or hyperplasia) and increases in the relative mass of the heart with respect to the fish. Alternatively, the ventricle can remodel and take on a more pyramidal shape, this cardiac morphology characteristic of more athletic fish species.

Importantly, the range of physiological responses to cardiovascular and cardiac diseases is varied. Despite advances in our understanding of how diseases impact physiology and give rise to physiological dysfunction, the story is often complex. As the toolbox for understanding the cellular mechanisms of pathology and physiology expands, and a better understanding of the physiology of diseased fish develops, the intricacies and interactions between pathogens and their host will be revealed. The interaction of multiple cellular and immunological factors, in concert with ongoing compensatory changes, makes the study of fish cardiovascular pathophysiology an exciting and challenging field.

ACKNOWLEDGMENTS

The authors would like to thank the contributions of graduate students and collaborators over many years including specifically Egil Karlsbakk, Espen Søreng, Matt Jones, Morten Lund, Gerrit Timmerhaus, Sven-Martin Jorgensen, Maria Dahle, and Joy Becker.

REFERENCES

Alne, H., Thomassen, M.S., Takle, H., Terjesen, B.F., Grammes, F., Oehme, M., Refstie, S., Sigholt, T., Berge, R.K., Rorvik, K.A., 2009. Increased survival by feeding tetradecylthioacetic acid during a natural outbreak of heart and skeletal muscle inflammation in S0 Atlantic salmon, *Salmo salar* L. J. Fish Dis. 32, 953–961.

Amin, A.B., Trasti, J., 1988. Endomyocarditis in Atlantic salmon in Norwegian seafarms. Bull. Eur. Assoc. Fish Pathol. 8, 70–71.

Baily, J.E., Smith, J.L., Wootten, R., Sommervill, C., 2011. The fate of *Lernaeocera branchialis* (L.) (Crustacea; copepod) in Atlantic cod, *Gadus morhua* L. J. Fish Dis. 34, 139–147.

Behrens, J.W., Seth, H., Axelsson, M., Buchmann, K., 2014. The parasitic copepod *Lernaeocera branchialis* negatively affects cardiorespiratory function in *Gadus morhua*. J. Fish Biol. 84, 1599–1606.

Bell, J.G., McVicar, A.H., Park, M.T., Sargent, J.R., 1991. High dietary linoleic acid affects the fatty acid compositions of individual phospholipids from tissues of Atlantic salmon (*Salmo salar*): association with stress susceptibility and cardiac lesion. J. Nutrition 121, 1163–1172.

Berger, J., 2007. Phenylhydrazine haemotoxicity. J. Appl. Biomed. 5, 125–130.

Beutler, E., 2005. Haemolytic anaemia resulting from chemical and physical agents. In: Lichtman, M.A., Williams, W.J., Beutler, E., Kaushansky, K., Kipps, T.J., Seligsohn, U., Prchal, J. (Eds.), Williams Haemtology, seventh ed. McGraw-Hill Professional, Chicago, IL, pp. 717–721.

Bjorgen, H., Koppang, E., Kaldhusdal, M., Moldal, T., Dale, O.B., 2015. Generalised inflammation with CD3E and MHC class II positive cells and dislocated epithelial cells in the mid-intestine in farmed Atlantic salmon (*Salmo salar* L.). J. Immunol. 194. VET2P.1132.

Boardman, T., Warner, C., Ramirez–Gomez, F., Matrisciano, J., Bromage, E., 2012. Characterization of an anti-rainbow trout (*Oncorhynchus mykiss*) CD3ε monoclonal antibody. Vet. Immunol. Immunopathol. 145, 511–515.

Bockerman, I., Wiik-Nielsen, C.R., Sindre, H., Johansen, R., Tengs, T., 2011. Prevalence of piscine myocarditis virus (PMCV) in marine fish species. J. Fish Dis. 34, 955–957.

Bornø, G., Linaker, M.L., 2015. Fiskehelserapporten 2014. Veterinærinstituttet, Harstad. ISSN 1893–1480, elektronisk utgave.

Boucher, P., Baudin-Laurencin, F., 1994. Sleeping disease of salmonids. Bull. Eur. Assoc. Fish Pathol. 4, 179–180.

Boucher, P., Baudin-Laurencin, F., 1996. Sleeping disease and pancreas disease comparative histopathology and acquired cross-protection. J. Fish Dis. 19, 303–310.

Brocklebank, J., Raverty, S., 2002. Sudden mortality caused by cardiac deformities following seining of preharvest farmed Atlantic salmon (*Salmo salar*) and by cardiomyopathy post intraperitoneally vaccinated Atlantic salmon parr in British Columbia. Can. Vet. J. 43, 129–130.

Brun, E., Poppe, T., Skrudland, A., Jarp, J., 2010. Cardiomyopathy syndrome in farmed Atlantic salmon *Salmo salar*: occurrence and direct financial losses for Norwegian aquaculture. Dis. Aquat. Organ. 56, 241–247.

Bruno, D.W., Noguera, P.A., 2009. Comparative experimental transmission of cardiomyopathy syndrome (CMS) in Atlantic salmon *Salmo salar*. Dis. Aquat. Organ. 87, 235–242.

Bruno, D.W., Poppe, T.T., 1996. Diseases of uncertain aetiology. Cardiac myopathy syndrome. In: Bruno, D.W., Poppe, T.T. (Eds.), A Color Atlas of Salmonid Diseases. Academic Press, London, pp. 140–141.

Burke, M., 2009. Effects of simulated anaemia on the blood chemistry of Atlantic cod, Atlantic halibut and, Atlantic salmon. MSc thesis, Bodø University College, Bodø, Norway, 66 p.

Cameron, J.N., Wolschlag, D.E., 1969. Respiratory response to experimentally induced anaemia in the pin fish (*Lagodon rhomboids*). J. Exp. Biol. 50, 307–317.

Cano, I., Joiner, C., Bayley, A., Rimmer, G., Bateman, K., Feist, S.W., Stone, D., Paley, R., 2015. An experimental means of transmitting pancreas disease in Atlantic salmon *Salmo salar* L. fry in fresh water. J. Fish Dis. 38, 271–281.

Cerra, M.C., Imbrogno, S., Amelio, D., Garofalo, F., Colvee, E., Tota, B., Icardo, J.M., 2004. Caradiac morphodenynamic remodeling in the growing eel (*Anguilla anguilla* L.). J. Exp. Biol. 207, 2867–2875.

Chou, C.F., Tohari, S., Brenner, S., Venkatesh, B., 2004. Erythropoetin gene from a telesot fish *Fugu rubripes*. Blood 104, 1498–1503.

Chou, C.Y., Cheng, C.H., Chen, G.D., Chen, Y.C., Hung, C.C., Huang, K.Y., et al., 2007. The zebrafish erythropoietin: functional identification and biochemical characterization. FEBS Lett. 581, 4265–4271.

Christie, K.E., Fyrand, K., Holtet, L., Rowley, H.M., 1998. Isolation of pancreas disease virus from farmed Atlantic salmon, *Salmo salar* L., in Norway. J. Fish Dis. 21, 391–394.

Christie, K.E., Graham, D.A., McLoughlin, M.F., Villoing, S., Todd, D., Knappskog, D., 2007. Experimental infection of Atlantic salmon *Salmo salar* pre-smolts by i.p. injection with new Irish and Norwegian salmonid alphavirus (SAV) isolates: a comparative study. Dis. Aquat. Organ. 75, 13–22.

Dalum, A., Tangen, R., Falk, K., Hordvik, I., Rosenlund, G., Torstensen, B., Koppang, E.O., 2016. Coronary changes in the Atlantic salmon *Salmo salar* L: characterization and impact of dietary fatty acid compositions. J. Fish Dis. 39, 41–45.

Ekström, A., Axelsson, M., Gräns, A., Brijs, J., Sandblom, E., 2017. Influence of the coronary circulation on thermal tolerance and cardiac performance during warming in rainbow trout. Am. J. Physiol. Regul. Integr. Comp. Physiol. 312, R549–R558.

Farrell, A.P., Smith, F.M., 2017. Cardiac form, function and physiology. In: Gamperl, A.K., Gillis, T.E., Farrell, A.P., Brauner, C.J. (Eds.), The Cardiovascular System: Morphology, Control and Function. Fish Physiology, vol. 36A. Elsevier, Amsterdam, pp. 155–264.

Farrell, A.P., Steffensen, J.F., 1987. Coronary ligation reduces the maximal sustained swimming speed in Chinook salmon, *Oncorhynchus tshwaytscha*. Comp. Biochem. Physiol. 87A, 35–37.

Farrell, A.P., Johansen, J.A., Moyes, C.D., Steffensen, J.F., West, T.G., Saurez, R.K., 1990. Effects of exercise training and coronary ablation on swimming performance heart size and cardiac enzymes in rainbow trout, *Oncorhynchus mykiss*. Can. J. Zool. 68, 1174–1179.

Ferguson, H.W., 1989. Systemic Pathology of Fish. Iowa State University Press, Ames IA.

Ferguson, H.W., Rice, D.A., Lynes, J.K., 1986. Clinical pathology of myodegeneration (pancreas disease) in Atlantic salmon (*Salmo salar*). Vet. Rec. 119, 297–299.

Ferguson, H.W., Poppe, T.T., Speare, D.J., 1990. Cardiomyopathy in farmed Norwegian salmon. Dis. Aquat. Organ. 8, 225–231.

Ferguson, H.W., Kongtorp, R.T., Taksdal, T., Graham, D., Falk, K., 2005. An outbreak of disease resembling heart and skeletal muscle inflammation in Scottish farmed salmon, *Salmo salar* L., with observations on myocardial regeneration. J. Fish Dis. 28, 119–123.

Finstad, O.W., Falk, K., Lovoll, M., Evensen, O., Rimstad, E., 2012. Immunohistochemical detection of piscine reovirus (PRV) in hearts of Atlantic salmon coincide with the course of heart and skeletal muscle inflammation (HSMI). Vet. Res. 43, 27.

Finstad, O.W., Dahle, M.K., Lindholm, T.H., Nyman, I.B., Lovoll, M., Wallace, C., Olsen, C.M., Storset, A.K., Rimstad, E., 2014. Piscine orthoreovirus (PRV) infects Atlantic salmon erythrocytes. Vet. Res. 45, 35.

Fjølstad, M., Hyeraas, A.L., 1985. Muscular and myocardial degeneration in cultured Atlantic salmon, *Salmo salar* L., suffering from 'Hitra disease'. J. Fish Dis. 8, 367–372.

Franklin, C.E., Davie, P.S., 1992. Myocardial power output of an isolated eel (*Anguilla dieffenbachii*) heart preparation in response to adrenaline. Comp. Biochem. Physiol. C 101, 293–298.

Fringuelli, E., Rowley, H.M., Wilson, J.C., Hunter, R., Rodger, H., Graham, D.A., 2008. Phylogenetic analyses and molecular epidemiology of European salmonid alphaviruses (SAV) based on partial E2 and nsP3 gene nucleotide sequences. J. Fish Dis. 31, 811–823.

Fritsvold, C., Kongtorp, R.T., Taksdal, T., Ørpetveit, I., Heum, M., Poppe, T.T., 2009. Experimental transmission of cardiomyopathy syndrome (CMS) in Atlantic salmon *Salmo salar*. Dis. Aquat. Organ. 87, 225–234.

Gamperl, A.K., Farrell, A.P., 2004. Cardiac plasticity in fishes: environmental influences and intraspecific differences. J. Exp. Biol. 207, 2539–2550.

Gamperl, A.K., Axelsson, M., Farrell, A.P., 1995. Effects of swimming and environmental hypoxia on coronary blood flow in rainbow trout. Am. J. Physiol. 269, R1258–1266.

Garseth, A.H., Fritsvold, C., Opheim, M., Skjerve, E., Biering, E., 2013. Piscine reovirus (PRV) in wild Atlantic salmon, *Salmo salar* L., and sea-trout, *Salmo trutta* L., in Norway. J. Fish Dis. 36, 483–493.

Gattuso, A., Mazza, R., Imbrogno, S., Sverdrup, A., Tota, B., Nylund, A., 2002. Cardiac performance in *Salmo salar* with infectious salmon anaemia (ISA): putative role of nitric oxide. Dis. Aquat. Organ. 52, 11–20.

Gillis, T.E., Johnson, E.F., 2017. Cardiac Preconditioning, Remodeling, and Regeneration. In: Gamperl, A.K., Gillis, T.E., Farrell, A.P., Brauner, C.J. (Eds.), The Cardiovascular System: Development, Plasticity and Physiological Responses. Fish Physiology, vol. 36B. Elsevier, Amsterdam. pp. 185–233. Chapter 3.

Gong, B.Q., Townley, R., Farrell, A.P., 1997. Effects of polyunstaurated fatty acids and some of their metabolites on mitotic activity of vascular smooth muscle explants from the coronary artery of rainbow trout (*Oncorhynchus mykiss*). Can. J. Zool. 75, 80–86.

Graham, D.A., Rowley, H.M., Fringuelli, E., Bovo, G., Amedeo, M., McLoughlin, M.F., et al., 2007. First laboratory confirmation of salmonid alphavirus infection in Italy and Spain. J. Fish Dis. 30, 569–572.

Graham, D.A., Frost, P., McLaughlin, K., Rowly, H.M., Gabestad, I., Gordon, A., McLoughlin, M.F., 2011. A comparative study of marine salmonid alphavirus subtypes 1–6 using an experimental cohabitation challenge model. J. Fish Dis. 34, 273–286.

Graham, D.A., Fringuelli, E., Rowley, H.M., Cockerill, D., Cox, D.I., Turnbull, T.T., Rodger, H., Morris, D., McLoughlin, M.F., 2012. Geographical distribution of salmonid alphavirus subtypes in marine farmed Atlantic salmon, *Salmo salar* L., in Scotland and Ireland. J. Fish Dis. 35, 755–765.

Grammes, F., Rorvik, K.–.A., Takle, H., 2012. Tetradecylthioacetic acid modulates cardiac transcription in Atlantic salmon, *Salmo salar* L., suffering from heart and skeletal muscle inflammation. J. Fish Dis. 35, 109–117.

Graver, K.A., Johnson, S.C., Polinski, M.P., Bradshaw, J.C., Marty, G.D., Snyman, H.N., et al., 2016. Piscine orthoreovirus from Western North America is transmissible to Atlantic salmon and sockeye salmon but fails to cause heart and skeletal muscle inflammation. PLoS One 11, e0146229.

Haugland, Ø., 2008. Studies of Inflammation and Immunity in Atlantic Salmon with Focus on TNFα Expression. Norwegian School of Veterinary Sciences, Oslo, Norway (PhD thesis).

Haugland, Ø., Mikalsen, A.B., Nilsen, P., Lindmo, K., Thu, B.J., Eliassen, T.M., et al., 2011. Cardiomyopathy syndrome of Atlantic salmon (*Salmo salar* L.) is caused by a double-stranded RNA virus of the Totiviridae family. J. Virol. 85, 5275–5286.

Herath, T.K., Bron, J.E., Thompson, K.D., Taggart, J.B., Adams, A., Ireland, J.H., Richards, R.H., 2012. Transcriptomic analysis of the host response to early stage salmonid alphavirus (SAV–1) infection in Atlantic salmon *Salmo salar* L. Fish Shellfish Immunol. 32, 796–807.

Hetland, D.L., Jørgensen, S.M., Skjødt, K., Dale, O.B., Falk, K., Xu, C., Mikalsen, A.S., et al., 2010. In situ localisation of major histocompatibility complex class I and class II and CD8 positive cells in infectious salmon anaemia virus (ISAV)-infected Atlantic salmon. Fish Shellfish Immunol. 28, 30–39.

Hetland, D.L., Dale, O.B., Skjødt, K., Press, C.M., Falk, K., 2011. Depletion of CD8 alpha cells from tissues of Atlantic salmon during the early stages of infection with high or low virulent strains of infectious salmon anaemia virus (ISAV). Dev. Comp. Immunol. 35, 817–826.

Hikke, M.C., Verest, M., Vlak, J.M., Pijlman, G.P., 2014. Salmonid alphavirus replication in mosquito cells: towards a novel vaccine production system. Microb Biotechnol. 7, 480–484.

Hjortaas, M.J., Skjelstad, H.R., Taksdal, T., Olsen, A.B., Johansen, R., Bang-Jensen, B., Ørpetveit, I., Sindre, H., 2013. The first detection of subtype 2-related salmonid alphavirus (SAV2) in Atlantic salmon, *Salmo salar* L., in Norway. J. Fish Dis. 36, 71–74.

Hodneland, K., Bratland, A., Christie, K.E., Endresen, C., Nylund, A., 2005. New subtype of salmonid alphavirus (SAV), *Togaviridae*, from Atlantic salmon *Salmo salar* and rainbow trout *Oncorhynchus mykiss* in Norway. Dis. Aquat. Organ. 66, 113–120.

Houghton, G., 1994. Acquired protection in Atlantic salmon *Salmo salar* parr and post-smolts against pancreas disease. Dis. Aquat. Organ. 18, 109–118.

Houghton, G.H., Ellis, A.E., 1996. Pancreas disease in Atlantic salmon: serum neutralisation and passive immunisation. Fish Shellfish Immunol. 6, 465–472.

Iwama, G.K., Farrell, A.P., 1998. Disorders of the cardiovascular and respiratory systems. In: Leatherland, J.F., Woo, P.T.K. (Eds.), Fish Diseases and Disorders. In: Non-Infectious Disorders, Vol. 2. CAB International, New York, pp. 245–278.

Jansen, M.D., Jensen, B.B., McLoughlin, M.F., Rodger, H.D., Taksdal, T., Sindre, H., Graham, D.A., Lillehaug, A., 2016. The epidemiology of pancreas disease in salmonid aquaculture: a summary of the current state of knowledge. J. Fish Dis. https://doi.org/10.1111/jfd.12478.

Johansen, I.B., Lunde, I.G., Røsjø, H., Christensen, G., Nilsson, G.E., Bakken, M., Øverli, Ø., 2011. Cortisol response to stress is associated with myocardial remodeling in salmonid fishes. J. Exp. Biol. 214, 1313–1321.

Johansen, L.H., Thim, H.L., Jørgensen, S.M., Afanasyev, S., Stranskog, G., Taksdal, T., Fremmerlid, K., McLoughlin, M., Jørgensen, J.B., Krasnov, A., 2015. Comparison of transcriptomic responses to pancreas disease (PD) and heart and skeletal muscle inflammation (HSMI) in heart of Atlantic salmon (*Salmo salar* L.). Fish Shellfish Immunol. 46, 612–623.

Johansen, L.H., Dahle, M.K., Wessel, Ø., Timmerhaus, G., Løvoll, M., Røsæg, M., Jørgensen, S.M., Rimstad, E., 2016. Differences in the gene expression in Atlantic salmon parr and smolt after challenge with piscine reovirus (PRV). Mol. Immunol. 73, 138–150.

Johansen, I.B., Sandblom, E., Skov, P.V., Gräns, A., Ekström, A., Lunde, I.G., Vindas, M.A., Zhang, L., Högland, E., Frisk, M., Sjaastad, I., Nilsson, G., Øverli, Ø., 2017. Bigger is not better: cortisol-induced cardiac growth and dysfunction in salmonids. J. Exp. Biol. 220, 2545–2553.

Kabata, Z., Whitaker, D.J., 1988. *Kudoa thyrsites* (Gilchrist 1924) (Myxozoa) in the cardiac muscle of Pacific salmon (*Oncorhynchus* spp.) and steelhead trout (*Salmo gairdneri*). Can. J. Zool. 67, 341–342.

Kent, M.L., Elston, R.A., 1987. Pancreas disease in pen-reared Atlantic salmon in North America. Bull. Eur. Assoc. Fish Pathol. 7, 29–31.

Kibenge, M.J.T., Iwamoto, T., Wang, Y., Morton, A., Godoy, M.G., Kibenge, F.S.B., 2013. Whole-genome analysis of piscine reovirus (PRV) shows PRV represents a new genus in family *Reoviridae* and its genome segment S1 sequences group it into two separate sub-genotypes. Virol. J. 10, 230–250.

Kocan, R., LaPatra, S., Gregg, J., Winton, J., Hershberger, P., 2006. *Ichthyophonus*-induced cardiac damage: a mechanism for reduced swimming stamina in salmonids. J. Fish Dis. 29, 521–527.

Kocan, R., Hershberger, P., Sanders, G., Winton, J., 2009. Effects of temperature on disease progression and swimming stamina in *Ichthyophonus*-infected rainbow trout, *Oncorhynchus mykiss* (Walbaum). J. Fish Dis. 32, 835–843.

Kongtorp, R.T., 2009. Heart and skeletal muscle inflammation (HSMI) in Atlantic salmon, *Salmo salar*: pathology, pathogenesis and experimental infection. Unipublished AS: PhD thesis, ISBN: 978-82-90550-66-5.

Kongtorp, R.T., Taksdal, T., Lyngøy, A., 2004a. Pathology of heart and skeletal muscle inflammation (HSMI) in farmed Atlantic salmon *Salmo salar*. Dis. Aquat. Organ. 59, 217–224.

Kongtorp, R.T., Kjerstad, A., Taksdal, T.T., Guttvik, A., Falk, K., 2004b. Heart and skeletal muscle inflammation in Atlantic salmon, *Salmo salar* L.: a new infectious disease. J. Fish Dis. 27, 351–358.

Kongtorp, R.T., Halse, M., Taksdal, T., Falk, K., 2006. Longitudinal study of a natural outbreak of heart and skeletal muscle inflammation in Atlantic salmon, *Salmo salar* L. J. Fish Dis. 29, 233–244.

Kongtorp, R.T., Stene, A., Andreassen, P.A., Aspehaug, V., Graham, D.A., Lyngstad, T.M., et al., 2010. Lack of evidence for vertical transmission of SAV 3 using gametes of Atlantic salmon, *Salmo salar* L., exposed by natural and experimental routes. J. Fish Dis. 33, 879–888.

Koppang, E.O., Fischer, U., Moore, L., Tranulis, M.A., Dijkstra, J.M., Kollner, B., et al., 2010. Salmonid T cells assemble in the thymus, spleen and in novel intrabranchial lymphoid tissue. J. Anat. 217, 728–739.

Koren, C.W.R., Nylund, A., 1997. Morphology and morphogenesis of infectious salmon anaemia virus replicating in the endothelium of Atlantic salmon *Salmo salar* L. Dis. Aquat. Organ. 29, 99–109.

Krasnov, A., Timmerhaus, G., Afanasyev, S., Takle, H., Jorgensen, S.-M., 2013. Induced erythropoiesis during acute anemia in Atlantic salmon: a transcriptomic survey. Gen. Comp. Endocrinol. 192, 181–190.

Lai, J.C., Kakuta, I., Mok, H.O., Rummer, J.L., Randall, D., 2006. Effects of moderate and substantial hypoxia on erythropoietin levels in rainbow trout kidney and spleen. J. Exp. Biol. 209, 2734–2738.

Leef, M.J., Harris, J.O., Hill, J., Powell, M.D., 2005. Cardiovascular responses of three salmonid species affected with amoebic gill disease (AGD). J. Comp. Physiol. B 175, 523–532.

Leef, M.J., Hill, J.V., Harris, J.O., Powell, M.D., 2007. Increased systemic vascular resistance in Atlantic salmon, *Salmo salar* L., affected with amoebic gill disease. J. Fish Dis. 30, 601–613.

Lopez-Doriga, M.V., Smail, D.A., Smith, R.J., Domenech, A., Castric, J., Smith, P.D., Ellis, A.E., 2001. Isolation of salmon pancreas disease virus (SPDV) in cell culture and its ability to protect against infection by the 'wild type' agent. Fish Shellfish Immunol. 11, 505–522.

Løvoll, M., Wiik-Nielsen, J., Grove, S., Wiik-Nielsen, C.R., Kristoffersen, A.B., Faller, R., et al., 2010. A novel totivirus and piscine reovirus (PRV) in Atlantic salmon (*Salmo salar*) with cardiomyopathy syndrome (CMS). Virol. J. 309, 1–7.

Løvoll, M., Alarcon, M., Jensen, B.B., Taksdal, T., Kristoffersen, A.B., Tengs, T., 2012. Quantification of piscine reovirus (PRV) at different stages of Atlantic salmon *Salmo salar* production. Dis. Aquat. Org. 99, 7–12.

Lovy, J., Wright, G.M., Speare, D.J., 2008. Comparative cellular morphology suggesting the existence of resident dendritic cells within immune organs of salmonids. Anat. Rec. 291, 456–462.

Lund, M., Dahle, M.K., Timmerhaus, G., Aspehaug, V., Rimstad, E., Powell, M.D., Jørgensen, S.-M., 2017. Hypoxia tolerance and responses to acute hypoxic stress during heart and skeletal muscle inflammation in Atlantic salmon (*Salmo salar*). PLoS One 12 (7), e0181109.

Magnasdottir, B., 2010. Immunological control of fish diseases. Marine Biotechnol. 12, 361–379.

Markussen, T., Dahle, M.K., Tengs, T., Løvoll, M., Finstad, Ø.W., et al., 2013. Correction: sequence analysis of the genome of piscine orthoreovirus (PRV) associated with heart and skeletal muscle inflammation (HSMI) in Atlantic Salmon (*Salmo salar*). PLoS One 8, e70075.

Martinez-Rubino, L., Morais, S., Evensen, Ø., Wadsworth, S., Ruohonen, K., Vecino, J.L., Bell, J.G., Tocher, D.R., 2012. Functional feeds reduce heart inflammation and pathology in Atlantic salmon (*Salmo salar* L.) following experimental challenge with Atlantic salmon reovirus (ASRV). PLoS One 7, e40266.

Marty, G.D., Morrison, D.B., Bidulka, J., Joeseph, T., Siah, A., 2015. Piscine reovirus in wild and farmed salmonids in British Columbia, Canada: 1974–2013. J. Fish Dis. 38, 713–728.

McClelland, G.B., Dalziel, A.C., Fragoso, N.M., Moyes, C.D., 2005. Muscle remodeling in relation to blood supply: implications for seasonal changes in mitochondrial enzymes. J. Exp. Biol. 203, 515–522.

McGladdery, S.E., Murphy, L., Hicks, B.D. & Wagner, S.K. (1990). The effects of Stephanstomum tenue (Digenea: acanthocolpidae) on marine aquaculture of rainbow trout, Salmo gairdneri. In: Pathology in Marine Science(Ed. by F.O. Perkins & T.C. Cheng). Academic Press Inc. San Diego pp. 305–315.

McLoughlin, M.F., Graham, D.A., 2007. Alphavirus infections in salmonids—a review. J. Fish Dis. 30, 511–531.

McLoughlin, M., Nelson, R.T., Rowley, H.M., Cox, D.I., Grant, A.N., 1996. Experimental pancreas disease in Atlantic salmon *Salmo salar* post–smolts induced by salmon pancreas disease virus (SPDV). Dis. Aquat. Organ. 26, 117–124.

McLoughlin, M.F., Nelson, R.T., McCormick, J.I., Rowley, H.M., Bryson, D.G., 2002. Clinical and histopathological features of naturally occurring pancreas disease in farmed Atlantic salmon, *Salmo salar* L. J. Fish Dis. 25, 33–43.

McVicar, A.H., 1987. Pancreas disease of farmed Atlantic salmon, *Salmo salar* in Scotland: epidemiology and early pathology. Aquaculture 67, 71–78.

McVicar, A.H., 1999. *Ichthyophonus* and related organisms. In: Woo, P.T.K., Bruno, D.W. (Eds.), Fish Diseases and Disorders. In: Viral, Bacterial and Fungal Infections, *Vol. 3*. CABI Publishing, New York, pp. 661–687.

McVicar, A.H., McLay, H.A., Ellis, A.E., 1985. Tissue responses of plaice, haddock and rainbow trout to the systemic fungus *Ichthyophonus*. In: Fish and Shellfish Pathology. Academic Press, London, pp. 329–346.

Mikalsen, A.B., Haugland, O., Rode, M., Solbakk, I.T., Evensen, O., 2012. Atlantic salmon reovirus infection causes a CD8 T cell myocarditis in Atlantic salmon (*Salmo salar* L.). PLoS One 7, e37269.

Mikalsen, A.B., Nilsen, P., Frøystad-Saugen, M., Lindmo, K., Eliassen, T.M., et al., 2014. Characterization of a novel calicivirus causing systemic infection in Atlantic salmon (*Salmo salar* L.): proposal for a new genus of caliciviridae. PLoS One 9, e107132.

Munday, B.L., Zilberg, D., Findlay, V., 2001. Gill disease of marine fish caused by infection with *Neoparamoeba pemaquidensis*. J. Fish Dis. 24, 497–507.

Munro, A.L.S., Ellis, A.E., McVicar, A.H., Mclay, H.A., Needham, E.A., 1984. An exocrine pancreas disease of farmed Atlantic salmon in Scotland. Helgolander Meeresuntersuchungen 37, 571–586.

Murchelano, R.A., Despres-Patanjo, L., Ziskowski, J., 1986. A histopathological evaluation of gross lesions excised from commercially important North Atlantic marine fishes. NOAA Tech. Rep. NMFS 37, 1–14.

Nowak, B.F., Valdenegro-Vega, V., Crosbie, P., Bridle, A., 2013. Immunity to amoeba. Devel. Comp. Immunol 43, 257–267.

Nylund, A., Hovland, T., Watanabe, K., Endresen, C., 1995. Presence of infectious salmon anaemia virus in different organs of *Salmo salar* L. collected from three fish farms. J. Fish Dis. 18, 135–145.

Nylund, A., Krossøy, B., Watanabe, K., Holm, J.A., 1996. Target cells for the ISA virus in Atlantic salmon (*Salmo salar* L). Bull. Eur. Assoc. Fish. Pathol. 16, 68–72.

Ola, K.B., 2016. Påvisning av melaninflekker I filet hos slakteklar Atlantisk laks (*Salmo salar* L.). Masteroppgave i Akvamedsine, Norges Arktiske Universitet, Tromsø, Norway.

Olmos, J., Ochoa, L., Panaiagua-Michel, J., Contreras, R., 2011. Functional feed assessment on *Litopenaeus vannamei* using 100% fish meal replacement by soybean meal, high levels of complex carbohydrates and Bacillus probiotic strains. Mar. Drugs 9, 1119–1132.

Olsen, K.R., 1998. The cardiovascular system. In: Evans, D.H. (Ed.), The Physiology of Fishes, second ed. CRC Press Inc, Boca Raton, LA, pp. 129–154.

Olsen, M.M., Kania, P.W., Heinecke, R.D., Skjoedt, K., Rasmussen, K.J., Buchmann, K., 2011. Cellular and humoral factors involved in the response of rainbow trout gills to *Ichthyophthirius multifillis* infections: molecular and immunohistochemical studies. Fish Shellfish Immunol. 30, 859–869.

Olsen, A.B., Hjortaas, M., Tengs, T., Hellberg, H., Johansen, R., 2015. First description of a new disease in rainbow trout (*Oncorhynchus mykiss* (Walbaum)) similar to heart and skeletal muscle inflammation (HSMI) and detection of a gene sequence related to piscine orthoreovirus (PRV). PLoS One 10, e0131638.

Palacios, G., Lovoll, M., Tengs, T., Hornig, M., Hutchison, S., et al., 2010. Heart and skeletal muscle inflammation of farmed salmon is associated with infection with a novel reovirus. PLoS One 5, e11487.

Pillay, T.V.R., Kutty, M.N., 2005. Nutrition and feeds. In: Aquaculture Principles and Practices, second ed. Wiley-Blackwell Ltd, Oxford, UK, pp. 117–165.

Poppe, T.T., Ferguson, H.W., 2006. Cardiovascular system. In: Ferguson, H.W. (Ed.), Systemic Pathology of Fish. A Text and Atlas of Normal Tissue Responses in Teleosts, and Their Responses in Disease. Scotian Press Ltd, London, UK, pp. 141–167.

Poppe, T.T., Seierstad, S.L., 2003. First description of cardiomyopathy syndrome (CMS)-related lesions in wild Atlantic salmon *Salmo salar* in Norway. Dis. Aquat. Organ. 56, 87–88.

Poppe, T.T., Taksdal, T., 2000. Ventricular hypoplasia in farmed Atlantic salmon *Salmo salar*. Dis. Aquat. Organ. 42, 35–40.

Poppe, T., Tørud, B., 2009. Intramyocardial dissecting haemorrhage in farmed rainbow trout *Oncorhynchus mykiss* (Walbaum). J. Fish Dis. 32, 1041–1043.

Poppe, T.T., Håstein, T., Frøslie, A., Koppang, N., Norheim, G., 1986. Nutritional aspects of Haemorrhagic Syndrome ('Hitra Disease') in farmed Atlantic salmon, *Salmo salar*. Dis. Aquat. Org. 1, 155–162.

Poppe, T.T., Rimstad, E., Hyllseth, B., 1989. Pancreas disease of Atlantic salmon (*Salmo salar*, L.) post-smolts infected with infectious pancreatic necrosis virus (IPNV). Bull. Eur. Assoc. Fish Pathol. 9, 83–85.

Powell, M.D., Gamperl, A.K., 2016. Effects of *Loma morhua* (Microsprodia) infection on the cardiorespiratory performance of Atlantic cod *Gadus morhua* (L.). J. Fish Dis. 39, 189–204.

Powell, M.D., Nowak, B.F., Adams, M.B., 2002a. Cardiac morphology in relation to amoebic gill disease history in Atlantic salmon, *Salmo salar* L. J. Fish Dis. 25, 209–215.

Powell, M.D., Forster, M.E., Nowak, B.F., 2002b. Apparent vascular hypertension associated with amoebic gill disease affected Atlantic salmon (*Salmo salar*) in Tasmania. Bull. Eur. Assoc. Fish Pathol. 22, 328–333.

Powell, M.D., Speare, D.J., Daley, J., Lovy, J., 2005. Differences in metabolic response to *Loma salmonae* infection in juvenile rainbow trout *Onchorhynchus mkykiss* and brook trout *Salvelinus fontinalis*. Dis. Aquat. Organ. 67, 233–237.

Powell, M.D., Speare, D.J., Becker, J.A., 2006. Whole body net ion fluxes, plasma electrolyte concentrations and haematology during *Loma salmonae* infection in juvenile rainbow trout *Oncorhynchus mykiss* (Walbaum). J. Fish Dis. 29, 727–735.

Powell, M.D., Leef, M.J., Roberts, S.D., Jones, M.A., 2008. Neoparamoebic gill infections: host response and physiology in salmonids. J. Fish Biol. 73, 2161–2183.

Powell, M.D., Bruke, M.S., Dahle, D., 2011. Cardiac remodeling, blood chemistry, haematology and oxygen consumption of Atlantic cod, *Gadus morhua* L., induced by experimental haemolytic anaemia with phenylhydrazine. Fish Physiol. Biochem. 37, 31–41.

Powell, M.D., Burke, M.S., Dahle, D., 2012. Cardiac remodeling of Atlantic halibut, *Hippoglossus hippoglossus* induced by experimental anaemia with phenylhydrazine. J. Fish Biol. 81, 335–344.

Prchal, J., 2005. Clinical manifestations and classification of erythrocyte disorders. In: Lichtman, M.A., Williams, W.J., Beutler, E., Kaushansky, K., Kipps, T.J., Seligsohn, U., Prchal, J. (Eds.), William's Haemtology, seventh ed. McGraw-Hill Professional, Chicago, IL, pp. 411–418.

Reidy, S.P., Kerr, S.R., Nelson, J.A., 2000. Aerobic and anaerobic swimming performance of individual Atlantic cod. J. Exp. Biol. 203, 347–357.

Rimstad, E., Dalsgaard, I., Hjeltnes, B., Håstein, T., 2011. Risikovurdering-stamfiskovervåkning og vertikal smitteoverføring. Vurdering av sannsynlighet for og risiko ved vertikal overføring av smitte hos oppdrettsfisk. Vitenskapeligkomiteen for mattrygghet (VKM). ISBN:978-82-8082-384-7.

Roberts, L.S., Janovy Jr., J., 2005. Foundations of Parasitology, seventh ed. McGraw-Hill, New York, pp. 61–88.

Roberts, S.D., Powell, M.D., 2005. The viscosity and glycoprotein biochemistry of salmonid mucus varies with species, salinity and presence of amoebic gill disease. J. Comp. Physiol. B 175, 1–11.

Roberts, R.J., Rodger, H.D., 2001. The pathophysiological and asystemic pathology of teleosts. In: Roberts, R.J. (Ed.), Fish Pathology, third ed. Elsevier Health Services, Amsterdam, Netherlands, pp. 55–132.

Robertson, O.H., Wexler, B.C., Miller, B.F., 1961. Degenerative changes in the cardiovascular system of the spawning Pacific salmon (*Oncorhynchus tshawytscha*). Circ. Res. 9, 826–834.

Rodger, H., Turnbull, T., 2000. Cardiomyopathy syndrome in farmed Scottish salmon. Vet. Rec. 146, 500–501.

Rodriguez-Tovar, L.E., Wadowska, D.W., Wright, G.W., Groman, D.B., Speare, D.J., Whelan, D.S., 2003. Ultrastructural evidence of autoinfection in the gills of Atlantic cod *Gadus morhua* infected with *Loma* sp. (phylum microsporidia). Dis. Aquat. Organ. 57, 227–230.

Rosenlund, S.F., 2017. Effects of potential functional ingredients on gill pathology and gene expression in Atlantic salmon challenged with *Neoparamoeba perurans*. MSc thesis, University of Bergen, 86 p.

Rowley, H.M., Doherty, C., McLoughlin, M.F., Welsh, M.D., 1998. Isolation of salmon pancreas disease virus (SPDV) from Scottish farmed salmon. J. Fish Dis. 21, 469–471.

Sanchez, J.G., Speare, D.J., Markham, R.J.F., 2000. Normal and abarent tissue distribution of *Loma salmonae* (Microspora) within rainbow trout, *Oncorhynchus mykiss* (Walbaum), following experimental infection at water temperatures within and outside the xenoma-expression temperature boundaries. J. Fish Dis. 23, 235–242.

Sanchez, J.G., Speare, D.J., Markham, R.J.F., Wright, G.M., Kidenge, F.S.B., 2001. Localization of the initial developmental stages of *Loma salmonae* in rainbow trout (*Oncorhynchus mykiss*). Vet. Pathol. 23, 540–546.

Scarano, G., Sargolia, M.G., Gray, R.H., Thibaldi, E., 1984. Haematological responses of seabass (*Dicentrarchus labrax*) to sub lethal nitrite exposures. Trans. Am. Fish. Soc. 113, 360–364.

Scholz, F., Ruane, N.M., Morrissey, T., Marcos-López, M., Mitchell, S., O'Connor, I., Mirimin, L., MacCarthy, E., Rodger, H.D., 2017. Piscine myocarditis virus detected in corkwing wrasse (*Symphodus melops*) and ballan wrasse (*Labrus bergylta*). J. Fish Dis. https://doi.org/10.1111/jfd.12661.

Seierstad, S.L., Poppe, T.T., Koppang, E.O., Svindlund, A., Rosenlund, G., Frøyland, L., Larsen, S., 2005. Influence of dietary lipid composition on cardiac pathology in farmed Atlantic salmon, *Salmo salar* L. J. Fish Dis. 28, 677–690.

Seierstad, S.L., Svindland, A., Larsen, S., Rosenlund, G., Torstenen, B.E., Evensen, Ø., 2008. Development of intimal thickening of coronary arteries over the lifetime of Atlantic salmon, *Salmo salar* L., fed different lipid sources. J. Fish Dis. 31, 401–413.

Shetler, M.D., Hill, A.O., 1985. Reactions of haemoglobin with phenylhydrazine: a review of selected aspects. Environ. Health Perspect. 64, 265–281.

Silva, M.T., do Vale, A., dos Santos, N.M.S., 2008. Fish and apoptosis: studies in disease and pharmaceutical design. Curr. Pharma Des. 14, 170–183.

Simonot, D.L., Farrell, A.P., 2007. Cardiac remodeling in rainbow trout *Oncorhynchus mykiss* Walbaum in response to phenylhydrazine-induced anaemia. J. Exp. Biol. 210, 2574–2584.

Simonot, D.L., Farrell, A.P., 2009. Coronary vascular volume remodeling in rainbow trout, *Oncorhynchus mykiss*. J. Fish Biol. 75, 1762–1772.

Smith, C.E., McLain, L.R., Zaugg, W.S., 1971. Phenylhydrazine induced anaemia in Chinook salmon. Toxicol. Appl. Pharmacol. 20, 73–81.

Smith, J.L., Wootten, R., Sommerville, C., 2007. The pathology of the early stages of the crustacean parasite, *Lernaeocera branchialis* (L.), on Atlantic cod, *Gadus morhua* L. J. Fish Dis. 30, 1–11.

Snow, M., Black, J., Matejusova, I., McIntosh, R., Baretto, E., Wallace, I.S., Bruno, D.W., 2010. Detection of salmonid alphavirus RNA in wild marine fish: implications for the origins of salmon pancreas disease in aquaculture. Dis. Aquat. Organ. 91, 177–188.

Steffensen, J.F., Farrell, A.P., 1998. Swimming performance, venous oxygen tension and cardiac performance of coronary-ligated rainbow trout, *Oncorhynchus mykiss* exposed to progressive hypoxia. Comp. Biochem. Physiol. 119A, 585–592.

Stensløkken, K.-O., Sundin, L., Nilsson, G.E., 2002. Cardiovascular effects of prostaglandin $F_{2\alpha}$ and prostaglandin E_2 in Atlantic cod (*Gadus morhua*). J. Comp. Physiol. B 172, 363–369.

Takizawa, F., Dijkstra, J.M., Kotterba, P., Korytar, T., Kock, H., Köllner, B., Jaureguiberry, B., Nakanishi, T., Fischer, U., 2011. The expression of CD8α discriminates distinct T cell subsets in teleost fish. Dev. Comp. Immunol. 35, 752–763.

Taksdal, T., Olsen, A.B., Bjerkås, I., Hjortaas, M.J., Dannevig, B.H., Graham, D.A., McLoughlin, M.F., 2007. Pancreas disease in farmed Atlantic salmon, *Salmo salar* L., and rainbow trout, *Oncorhynchus mykiss* (Walbaum), in Norway. J. Fish Dis. 30, 545–558.

Tengs, T., Bockerman, I., 2012. A strain of piscine myocarditis virus infecting Atlantic argentine, *Argentina silus* (Ascanius). J. Fish Dis. 35, 545–547.

Terova, G., Rimoldi, S., Cora, S., Bernadini, G., Gornati, R., Saroglia, M., 2008. Acute and chronic hypoxia affects HIF-1α mRNA levels in sea bass (*Dicentrarchus labrax*). Aquaculture 279, 150–159.

Thorud, K., Djupvik, H.O., 1988. Infectious salmon anaemia in Atlantic salmon (*Salmo salar* L.). Bull. Eur. Assoc. Fish. Pathol. 8, 109–111.

Tierney, K.B., Farrell, A.P., 2004. The relationship between fish health, metabolic rate, swimming performance and recovery in return-run sockeye salmon, *Oncorhynchus nerka* (Walbaum). J. Fish Dis. 27, 663–671.

Timmerhaus, G., Krasnov, A., Nilsen, P., Alarcon, M., Afanasyev, S., Rode, M., Takle, H., abd Jorgensen, S.M., 2011. Transcriptome profiling of immune responses to cardiomyopathy syndrome (CMS) in Atlantic salmon. BMC Gen. 12, 459.

Villoing, S., Bearzotti, M., Chilmonczyk, S., Castric, J., Bremont, M., 2000. Rainbow trout sleeping disease virus is an atypical alphavirus. J. Virol. 74, 173–183.

Weston, J., Walsh, M.D., McLoughlin, M.F., Todd, D., 1999. Salmon pancreas disease virus, an alphavirus infecting farmed Atlantic salmon. J. Virol. 256, 188–195.

Weston, J., Villoing, S., Brémont, M., Castric, J., Pfeffer, M., Jewhurst, V., et al., 2002. Comparison of two aquatic alphaviruses, salmon pancreas disease virus and sleeping disease virus, by using genome sequence analysis, monoclonal reactivity, and cross-infection. J. Virol. 76, 6155–6163.

Wickramasinghe, S.N., 1993. Erythropoietin and the human kidney: evidence for an evolutionary link from studies of *Salmo gairdneri*. Comp. Biochem. Physiol. Comp. Physiol. 104, 63–65.

Wiik-Nielsen, C.R., Ski, P.M., Aunsmo, A., Lovoll, M., 2012. Prevalence of viral RNA from piscine reovirus and piscine myocarditis virus in Atlantic salmon, *Salmo salar* L., broodfish and progeny. J. Fish Dis. 35, 169–171.

Woo, P.T.K., Bruno, D.W., Lim, L.H.S., 2003. Diseases and Disorders of Finfish in Cage Culture. p. 354, CAB international publishers, Oxon, UK.

Wood, C.M., McMahon, B.R., McDonald, D.G., 1979. Respiratory ventilatory and cardiovascular responses to experimental anaemia in the starry flounder, *Platichthys stellatus*. J. Exp. Biol. 82, 139–162.

Wood, C.M., McDonald, D.G., McMahon, B.R., 1982. The influence of experimental anaemia on acid–base regulation in vivo and in vitro in the starry flounder (*Platichthys stellatus*) and the rainbow trout (*Salmo gairdneri*). J. Exp. Biol. 96, 221–237.

Yousaf, M.N., 2012. Cardiac morphology and pathology of disease in Atlantic salmon. PhD thesis, University of Nordland.

Yousaf, M.N., Poppe, T.T., 2017. Cartilage in the bulbus arteriosus of farmed Atlantic salmon (*Salmo salar* L.). J. Fish Dis. 40, 1249–1252.

Yousaf, M.N., Koppang, E.O., Hordvik, I., Jirillo, E., Köllner, B., Skjødt, K., Zou, J., Secombes, C., Powell, M.D., 2012. Cardiac pathological changes of Atlantic salmon (*Salmo salar* L.) affected with heart and skeletal muscle inflammation (HSMI). Fish Shellfish Immunol. 33, 305–315.

Yousaf, M.N., Koppang, E.O., Hordvik, I., Jirillo, E., Köllner, B., Skjødt, K., Zou, J., Secombes, C., Powell, M.D., 2013. Comparative pathological changes of cardiac diseases in Atlantic salmon (*Salmo salar* L.). Vet. Immunol. Immunopathol. 151, 49–62.

Zhang, L., Tarleton, R.L., 1999. Parasite persistence correlates with disease severity and localization in chronic Chargas' disease. J. Inf. Dis. 180, 480–486.

Zou, J., Secombes, C.J., Long, S., Miller, N., Clem, L.W., Chinhar, V.G., 2003. Molecular identification and expression analysis of tumor necrosis factor in channel catfish (*Ictalurus punctatus*). Devel. Comp. Immunol. 27, 845–858.

INDEX

Note: Page numbers followed by "*f*" indicate figures, "*t*" indicate tables, and "*np*" indicate footnotes.

B

C

M

N

Q

R

OTHER VOLUMES IN THE FISH PHYSIOLOGY SERIES

VOLUME 1	Excretion, Ionic Regulation, and Metabolism *Edited by W. S. Hoar and D. J. Randall*
VOLUME 2	The Endocrine System *Edited by W. S. Hoar and D. J. Randall*
VOLUME 3	Reproduction and Growth: Bioluminescence, Pigments, and Poisons *Edited by W. S. Hoar and D. J. Randall*
VOLUME 4	The Nervous System, Circulation, and Respiration *Edited by W. S. Hoar and D. J. Randall*
VOLUME 5	Sensory Systems and Electric Organs *Edited by W. S. Hoar and D. J. Randall*
VOLUME 6	Environmental Relations and Behavior *Edited by W. S. Hoar and D. J. Randall*
VOLUME 7	Locomotion *Edited by W. S. Hoar and D. J. Randall*
VOLUME 8	Bioenergetics and Growth *Edited by W. S. Hoar, D. J. Randall, and J. R. Brett*
VOLUME 9A	Reproduction: Endocrine Tissues and Hormones *Edited by W. S. Hoar, D. J. Randall, and E. M. Donaldson*
VOLUME 9B	Reproduction: Behavior and Fertility Control *Edited by W. S. Hoar, D. J. Randall, and E. M. Donaldson*

Printed in the United States
By Bookmasters